REMOTE COMPOSITIONAL ANALYSIS

How do planetary scientists analyze and interpret data from telescopic and spacecraft observations of planetary surfaces? What elements, minerals, and volatiles are found on the surfaces of our Solar System's planets, moons, asteroids, and comets? This comprehensive volume answers these topical questions by providing an overview of the theory and techniques of remote compositional analysis of planetary surfaces. Bringing together eminent researchers in Solar System exploration, it describes state-of-the-art results from spectroscopic, mineralogical, and geochemical techniques used to analyze the surfaces of planets, moons, and small bodies. The book introduces the methodology and theoretical background of each technique, and presents the latest advances in space exploration, telescopic observation, and laboratory instrumentation, and major new work in theoretical studies. This engaging volume provides a comprehensive reference on planetary surface composition and mineralogy for advanced students, researchers, and professional scientists.

JANICE L. BISHOP is a senior research scientist at the SETI Institute, where she is Chair of Astrobiology and on the Science Council. She investigates Mars surface composition, mineral spectroscopy, and volcanic alteration, and has worked with data from many martian missions, including MRO/CRISM for which she is a Co-I. She has served as editor of *Icarus* and special issues of *American Mineralogist* and *Clay Minerals*. She has received awards from the Clay Minerals Society, the Humboldt Foundation, and the Helmholtz Foundation, and is a Fellow of GSA and MSA.

JAMES F. BELL III is a professor at Arizona State University, where he specializes in astronomy and planetary science. He studies the geology, geochemistry, and mineralogy of Solar System objects using telescopes and spacecraft. He received the Carl Sagan Medal from the American Astronomical Society for excellence in public communication in the planetary sciences, and asteroid 8146 Jimbell was named after him by the International Astronomical Union. He edited *The Martian Surface: Composition, Mineralogy, and Physical Properties* (Cambridge, 2008) and coedited *Asteroid Rendezvous* (Cambridge, 2002).

JEFFREY E. MOERSCH is a professor of Earth and Planetary Sciences and Director of the Planetary Geosciences Institute at the University of Tennessee. His research focuses on the geology of planetary surfaces, remote sensing, terrestrial analog field work, and planetary instrument development. He has extensive spacecraft mission experience, including work as a member of the science teams of the Mars Exploration Rover mission, the Mars Odyssey mission, and the Mars Science Laboratory mission. He has conducted astrobiology-related research in many terrestrial analog field sites, including Death Valley and the Mojave Desert, the Atacama Desert, the Andes, and the Arctic. From 2010 to 2015 he was the Mars Editor for *Icarus*.

Cambridge Planetary Science

Series Editors: Fran Bagenal, David Jewitt, Carl Murray, Jim Bell, Ralph Lorenz, Francis Nimmo, Sara Russell

Books in the Series:

1. *Jupiter: The Planet, Satellites and Magnetosphere*[†]
   Edited by Bagenal, Dowling and McKinnon
   978-0-521-03545-3

2. *Meteorites: A Petrologic, Chemical and Isotopic Synthesis*[†]
   Hutchison
   978-0-521-03539-2

3. *The Origin of Chondrules and Chondrites*[†]
   Sears
   978-1-107-40285-0

4. *Planetary Rings*[†]
   Esposito
   978-1-107-40247-8

5. *The Geology of Mars: Evidence from Earth-Based Analogs*[†]
   Edited by Chapman
   978-0-521-20659-4

6. *The Surface of Mars*[†]
   Carr
   978-0-521-87201-0

7. *Volcanism on Io: A Comparison with Earth*[†]
   Davies
   978-0-521-85003-2

8. *Mars: An Introduction to Its Interior, Surface and Atmosphere*[†]
   Barlow
   978-0-521-85226-5

9. *The Martian Surface: Composition, Mineralogy and Physical Properties*
   Edited by Bell
   978-0-521-86698-9

10. *Planetary Crusts: Their Composition, Origin and Evolution*[†]
    Taylor and McLennan
    978-0-521-14201-4

11. *Planetary Tectonics*[†]
    Edited by Watters and Schultz
    978-0-521-74992-3

12. *Protoplanetary Dust: Astrophysical and Cosmochemical Perspectives*[†]
    Edited by Apai and Lauretta
    978-0-521-51772-0

13. *Planetary Surface Processes*
    Melosh
    978-0-521-51418-7

[†] Reissued as a paperback

14. *Titan: Interior, Surface, Atmosphere and Space Environment*
    Edited by Müller-Wodarg, Griffith, Lellouch and Cravens
    978-0-521-19992-6

15. *Planetary Rings: A Post-Equinox View (Second edition)*
    Esposito
    978-1-107-02882-1

16. *Planetesimals: Early Differentiation and Consequences for Planets*
    Edited by Elkins-Tanton and Weiss
    978-1-107-11848-5

17. *Asteroids: Astronomical and Geological Bodies*
    Burbine
    978-1-107-09684-4

18. *The Atmosphere and Climate of Mars*
    Edited by Haberle, Clancy, Forget, Smith and Zurek
    978-1-107-01618-7

19. *Planetary Ring Systems*
    Edited by Tiscareno and Murray
    978-1-107-11382-4

20. *Saturn in the 21st Century*
    Edited by Baines, Flasar, Krupp and Stallard
    978-1-107-10677-2

21. *Mercury: The View after MESSENGER*
    Edited by Solomon, Nittler and Anderson
    978-1-107-15445-2

22. *Chondrules: Records of Protoplanetary Disk Processes*
    Edited by Russell, Connolly Jr. and Krot
    978-1-108-41801-0

23. *Spectroscopy and Photochemistry of Planetary Atmospheres and Ionospheres*
    Krasnopolsky
    978-1-107-14526-9

24. *Remote Compositional Analysis: Techniques for Understanding Spectroscopy, Mineralogy, and Geochemistry of Planetary Surfaces*
    Edited by Bishop, Bell III and Moersch
    978-1-107-18620-0

REMOTE COMPOSITIONAL ANALYSIS

Techniques for Understanding Spectroscopy, Mineralogy, and Geochemistry of Planetary Surfaces

Edited by

JANICE L. BISHOP
SETI Institute

JAMES F. BELL III
Arizona State University

JEFFREY E. MOERSCH
University of Tennessee, Knoxville

CAMBRIDGE
UNIVERSITY PRESS

University Printing House, Cambridge CB2 8BS, United Kingdom

One Liberty Plaza, 20th Floor, New York, NY 10006, USA

477 Williamstown Road, Port Melbourne, VIC 3207, Australia

314–321, 3rd Floor, Plot 3, Splendor Forum, Jasola District Centre,
New Delhi – 110025, India

79 Anson Road, #06–04/06, Singapore 079906

Cambridge University Press is part of the University of Cambridge.

It furthers the University's mission by disseminating knowledge in the pursuit of
education, learning, and research at the highest international levels of excellence.

www.cambridge.org
Information on this title: www.cambridge.org/9781107186200
DOI: 10.1017/9781316888872

© Cambridge University Press 2020

This publication is in copyright. Subject to statutory exception
and to the provisions of relevant collective licensing agreements,
no reproduction of any part may take place without the written
permission of Cambridge University Press.

First published 2020

Printed in the United Kingdom by TJ International Ltd. Padstow Cornwall

A catalogue record for this publication is available from the British Library.

Library of Congress Cataloging-in-Publication Data
Names: Bishop, Janice L., editor. | Bell, Jim, 1965– editor. | Moersch, Jeffrey E., 1966– editor.
Title: Remote compositional analysis : techniques for understanding spectroscopy, mineralogy, and
geochemistry of planetary surfaces / edited by Janice L. Bishop (SETI Institute), James F. Bell, III (Arizona
State University), Jeffrey E. Moersch (University of Tennessee, Knoxville).
Description: Cambridge ; New York, NY : Cambridge University Press, [2020] | Includes bibliographical
references and index.
Identifiers: LCCN 2019002564 | ISBN 9781107186200 (alk. paper)
Subjects: LCSH: Planets – Spectra. | Astronomical spectroscopy. | Planetary science.
Classification: LCC QB603.S6 R46 2019 | DDC 523.4028/7–dc23
LC record available at https://lccn.loc.gov/2019002564

ISBN 978-1-107-18620-0 Hardback

Cambridge University Press has no responsibility for the persistence or accuracy of
URLs for external or third-party internet websites referred to in this publication
and does not guarantee that any content on such websites is, or will remain,
accurate or appropriate.

Contents

List of Contributors	*page* xi
Foreword	xix
CARLÉ M. PIETERS AND PETER A. J. ENGLERT	
Preface	xxi
Acknowledgments	xxii

Part I Theory of Remote Compositional Analysis Techniques and Laboratory Measurements — 1

1. Electronic Spectra of Minerals in the Visible and Near-Infrared Regions — 3
   GEORGE R. ROSSMAN AND BETHANY L. EHLMANN

2. Theory of Reflectance and Emittance Spectroscopy of Geologic Materials in the Visible and Infrared Regions — 21
   JOHN F. MUSTARD AND TIMOTHY D. GLOTCH

3. Mid-infrared (Thermal) Emission and Reflectance Spectroscopy: Laboratory Spectra of Geologic Materials — 42
   MELISSA D. LANE AND JANICE L. BISHOP

4. Visible and Near-Infrared Reflectance Spectroscopy: Laboratory Spectra of Geologic Materials — 68
   JANICE L. BISHOP

5. Spectroscopy of Ices, Volatiles, and Organics in the Visible and Infrared Regions — 102
   DALE P. CRUIKSHANK, LYUBA V. MOROZ, AND ROGER N. CLARK

6. Raman Spectroscopy: Theory and Laboratory Spectra of Geologic Materials — 120
   SHIV K. SHARMA AND MILES J. EGAN

7. Mössbauer Spectroscopy: Theory and Laboratory Spectra of Geologic Materials — 147
   M. DARBY DYAR AND ELIZABETH C. SKLUTE

8	Laser-Induced Breakdown Spectroscopy: Theory and Laboratory Spectra of Geologic Materials SAMUEL M. CLEGG, RYAN B. ANDERSON, AND NOUREDDINE MELIKECHI	168
9	Neutron, Gamma-Ray, and X-Ray Spectroscopy: Theory and Applications THOMAS H. PRETTYMAN, PETER A. J. ENGLERT, AND NAOYUKI YAMASHITA	191
10	Radar Remote Sensing: Theory and Applications JAKOB VAN ZYL, CHARLES ELACHI, AND YUNJIN KIM	239

Part II Terrestrial Field and Airborne Applications — 259

11	Visible and Near-Infrared Reflectance Spectroscopy: Field and Airborne Measurements ROGER N. CLARK	261
12	Raman Spectroscopy: Field Measurements PABLO SOBRON, ANUPAM MISRA, FERNANDO RULL, AND ANTONIO SANSANO	274

Part III Analysis Methods — 287

13	Effects of Environmental Conditions on Spectral Measurements EDWARD CLOUTIS, PIERRE BECK, JEFFREY J. GILLIS-DAVIS, JÖRN HELBERT, AND MARK J. LOEFFLER	289
14	Hyper- and Multispectral Visible and Near-Infrared Imaging Analysis WILLIAM H. FARRAND, ERZSÉBET MERÉNYI, AND MARIO C. PARENTE	307
15	Thermal Infrared Spectral Modeling JOSHUA L. BANDFIELD AND A. DEANNE ROGERS	324
16	Geochemical Interpretations Using Multiple Remote Datasets SUNITI KARUNATILLAKE, LYNN M. CARTER, HEATHER B. FRANZ, LYDIA J. HALLIS, AND JOEL A. HUROWITZ	337

Part IV Applications to Planetary Surfaces — 349

17	Spectral Analyses of Mercury SCOTT L. MURCHIE, NOAM R. IZENBERG, AND RACHEL L. KLIMA	351
18	Compositional Analysis of the Moon in the Visible and Near-Infrared Regions CARLÉ M. PIETERS, RACHEL L. KLIMA, AND ROBERT O. GREEN	368
19	Spectral Analyses of Asteroids JOSHUA P. EMERY, CRISTINA A. THOMAS, VISHNU REDDY, AND NICHOLAS A. MOSKOVITZ	393

20 Visible and Near-Infrared Spectral Analyses of Asteroids and Comets from
    Dawn and Rosetta   413
    M. CRISTINA DE SANCTIS, FABRIZIO CAPACCIONI,
    ELEONORA AMMANNITO, AND GIANRICO FILACCHIONE

21 Spectral Analyses of Saturn's Moons Using the *Cassini* Visual Infrared
    Mapping Spectrometer   428
    BONNIE J. BURATTI, ROBERT H. BROWN, ROGER N. CLARK,
    DALE P. CRUIKSHANK, AND GIANRICO FILACCHIONE

22 Spectroscopy of Pluto and Its Satellites   442
    DALE P. CRUIKSHANK, WILLIAM M. GRUNDY, DONALD E. JENNINGS,
    CATHERINE B. OLKIN, SILVIA PROTOPAPA, DENNIS C. REUTER, BERNARD
    SCHMITT, AND S. ALAN STERN

23 Visible to Short-Wave Infrared Spectral Analyses of Mars from Orbit
    Using CRISM and OMEGA   453
    SCOTT L. MURCHIE, JEAN-PIERRE BIBRING, RAYMOND E. ARVIDSON,
    JANICE L. BISHOP, JOHN CARTER, BETHANY L. EHLMANN,
    YVES LANGEVIN, JOHN F. MUSTARD, FRANCOIS POULET, LUCIE RIU,
    KIMBERLY D. SEELOS, AND CHRISTINA E. VIVIANO

24 Thermal Infrared Spectral Analyses of Mars from Orbit Using the Thermal
    Emission Spectrometer and Thermal Emission Imaging System   484
    VICTORIA E. HAMILTON, PHILIP R. CHRISTENSEN, JOSHUA L. BANDFIELD,
    A. DEANNE ROGERS, CHRISTOPHER S. EDWARDS, AND STEVEN W. RUFF

25 Thermal Infrared Remote Sensing of Mars from Rovers Using the
    Miniature Thermal Emission Spectrometer   499
    STEVEN W. RUFF, JOSHUA L. BANDFIELD, PHILIP R. CHRISTENSEN,
    TIMOTHY D. GLOTCH, VICTORIA E. HAMILTON, AND A. DEANNE ROGERS

26 Compositional and Mineralogic Analyses of Mars Using Multispectral
    Imaging on the Mars Exploration Rover, Phoenix, and Mars Science
    Laboratory Missions   513
    JAMES F. BELL III, WILLIAM H. FARRAND, JEFFREY R. JOHNSON,
    KJARTAN M. KINCH, MARK LEMMON, MARIO C. PARENTE,
    MELISSA S. RICE, AND DANIKA WELLINGTON

27 Mössbauer Spectroscopy at Gusev Crater and Meridiani Planum:
    Iron Mineralogy, Oxidation State, and Alteration on Mars   538
    RICHARD V. MORRIS, CHRISTIAN SCHRÖDER, GÖSTAR KLINGELHÖFER,
    AND DAVID G. AGRESTI

28 Elemental Analyses of Mars from Rovers Using the Alpha-Particle X-Ray
    Spectrometer   555
    RALF GELLERT AND ALBERT S. YEN

29 Elemental Analyses of Mars from Rovers with Laser-Induced Breakdown
    Spectroscopy by ChemCam and SuperCam   573
    NINA L. LANZA, ROGER C. WIENS, SYLVESTRE MAURICE,
    AND JEFFREY R. JOHNSON

30 Neutron, Gamma-Ray, and X-Ray Spectroscopy of Planetary Bodies   588
    THOMAS H. PRETTYMAN, PETER A. J. ENGLERT, NAOYUKI YAMASHITA,
    AND MARGARET E. LANDIS

31 Radar Remote Sensing of Planetary Bodies   604
    JEFFREY J. PLAUT

Index   624

Contributors

Editors

JANICE L. BISHOP
Carl Sagan Center, SETI Institute, Mountain View, CA, USA

JAMES F. BELL III
School of Earth and Space Exploration, Arizona State University, Tempe, AZ, USA

JEFFREY E. MOERSCH
Earth and Planetary Science Department, University of Tennessee, Knoxville, TN, USA

Contributing Authors

DAVID G. AGRESTI
Department of Physics, University of Alabama at Birmingham, Birmingham, AL, USA

ELEONORA AMMANNITO
Agenzia Spaziale Italiana, Via del Politecnico snc, Rome, Italy

RYAN B. ANDERSON
Astrogeology Science Center, United States Geological Survey, Flagstaff, AZ, USA

RAYMOND E. ARVIDSON
Department of Earth and Planetary Sciences, Washington University in St. Louis, St. Louis, MO, USA

JOSHUA L. BANDFIELD
formerly, Space Science Institute, Boulder, CO, USA

PIERRE BECK
Institut de Planétologie et d'Astrophysique de Grenoble, Université Grenoble Alpes, Saint-Martin-d'Hères, Grenoble Cedex, France

JAMES F. BELL III
School of Earth and Space Exploration, Arizona State University, Tempe, AZ, USA

JEAN-PIERRE BIBRING
Institut d'Astrophysique Spatiale, Orsay Cedex, France

JANICE L. BISHOP
Carl Sagan Center, SETI Institute, Mountain View, CA, USA

ROBERT H. BROWN
University of Arizona, Tucson, AZ, USA

BONNIE J. BURATTI
Jet Propulsion Laboratory, California Institute of Technology, Pasadena, CA, USA

FABRIZIO CAPACCIONI
Istituto di Astrofisica e Planetologia Spaziali, INAF, Rome, Italy

List of Contributors

JOHN CARTER
Institut d'Astrophysique Spatiale, Orsay Cedex, France

LYNN M. CARTER
Department of Planetary Sciences, University of Arizona, Tucson, AZ, USA

PHILIP R. CHRISTENSEN
School of Earth and Space Exploration, Arizona State University, Tempe, AZ, USA

ROGER N. CLARK
Planetary Science Institute, Tucson, AZ, USA

SAMUEL M. CLEGG
Chemistry Division, Los Alamos National Laboratory, Los Alamos, NM, USA

EDWARD CLOUTIS
Department of Geography, University of Winnipeg, Winnipeg, MB, Canada

DALE P. CRUIKSHANK
NASA Ames Research Center, Moffett Field, CA, USA

M. CRISTINA DE SANCTIS
Istituto di Astrofisica e Planetologia Spaziali, INAF, Rome, Italy

M. DARBY DYAR
Department of Astronomy, Mount Holyoke College, South Hadley, MA, USA, and Planetary Science Institute, Tucson, AZ, USA

CHRISTOPHER S. EDWARDS
Department of Physics and Astronomy, Northern Arizona University, Flagstaff, AZ, USA

MILES J. EGAN
Hawai'i Institute of Geophysics and Planetology, University of Hawai'i at Mānoa, Honolulu, HI, USA

BETHANY L. EHLMANN
Division of Geological and Planetary Sciences, California Institute of Technology, Pasadena, CA, USA

CHARLES ELACHI
Jet Propulsion Laboratory, California Institute of Technology, Pasadena, CA, USA

JOSHUA P. EMERY
Earth and Planetary Science Department, University of Tennessee, Knoxville, TN, and Astronomy and Planetary Science Department, Northern Arizona University, Flagstaff, AZ, USA

PETER A. J. ENGLERT
Hawai'i Institute of Geophysics and Planetology, University of Hawai'i at Mānoa, Honolulu, HI, USA

WILLIAM H. FARRAND
Space Science Institute, Boulder, CO, USA

GIANRICO FILACCHIONE
Istituto di Astrofisica e Planetologia Spaziali, INAF, Rome, Italy

HEATHER B. FRANZ
NASA Goddard Space Flight Center, Greenbelt, MD, USA

RALF GELLERT
Department of Physics, University of Guelph, Guelph, ON, Canada

List of Contributors

JEFFREY J. GILLIS-DAVIS
Hawai'i Institute of Geophysics and Planetology, University of Hawai'i at Mānoa, Honolulu, HI, USA

TIMOTHY D. GLOTCH
Department of Geosciences, Stony Brook University, Stony Brook, NY, USA

ROBERT O. GREEN
Jet Propulsion Laboratory, California Institute of Technology, Pasadena, CA, USA

WILLIAM M. GRUNDY
Lowell Observatory, Flagstaff, AZ, USA

LYDIA J. HALLIS
School of Geographical and Earth Sciences, University of Glasgow, Glasgow, Scotland

VICTORIA E. HAMILTON
Department of Space Studies, Southwest Research Institute, Boulder, CO, USA

JÖRN HELBERT
Deutsches Zentrum für Luft und Raumfahrt e.V. (DLR), Berlin, Germany

JOEL A. HUROWITZ
Department of Geosciences, Stony Brook University, Stony Brook, NY, USA

NOAM R. IZENBERG
Johns Hopkins University Applied Physics Laboratory, Laurel, MD, USA

DONALD E. JENNINGS
NASA Goddard Space Flight Center, Greenbelt, MD, USA

JEFFREY R. JOHNSON
Johns Hopkins University Applied Physics Laboratory, Laurel, MD, USA

SUNITI KARUNATILLAKE
Planetary Science Lab, Geology and Geophysics, Louisiana State University, Baton Rouge, LA, USA

YUNJIN KIM
Jet Propulsion Laboratory, California Institute of Technology, Pasadena, CA, USA

KJARTAN M. KINCH
Astrophysics and Planetary Science, Niels Bohr Institute, University of Copenhagen, Denmark

RACHEL L. KLIMA
Johns Hopkins University Applied Physics Laboratory, Laurel, MD, USA

GÖSTAR KLINGELHÖFER
formerly, Institut für Anorganische und Analytische Chemie, Johannes Gutenberg-Universität, Mainz, Germany

MARGARET E. LANDIS
Planetary Science Institute, Tucson, AZ, USA

MELISSA D. LANE
Fibernetics LLC, Lititz, PA, USA

YVES LANGEVIN
Institut d'Astrophysique Spatiale, Orsay Cedex, France

NINA L. LANZA
Space and Remote Sensing, Los Alamos National Laboratory, Los Alamos, NM, USA

MARK LEMMON
Texas A&M University, College Station, TX, USA

MARK J. LOEFFLER
Department of Physics and Astronomy, Northern Arizona University, Flagstaff, AZ, USA

SYLVESTRE MAURICE
Institut de Recherche en Astrophysique et Planétologie, Toulouse, France

NOUREDDINE MELIKECHI
Kennedy College of Sciences, University of Massachusetts Lowell, Lowell, MA, USA

ERZSÉBET MERÉNYI
Department of Statistics, and Department of Electrical and Computer Engineering, Rice University, Houston, TX, USA

ANUPAM MISRA
Hawai'i Institute of Geophysics and Planetology, University of Hawai'i at Mānoa, Honolulu, HI, USA

LYUBA V. MOROZ
University of Potsdam, Potsdam, Germany

RICHARD V. MORRIS
Exploration and Integration Science Division, NASA Johnson Space Center, Houston, TX, USA

NICHOLAS A. MOSKOVITZ
Lunar and Planetary Laboratory, University of Arizona, Tucson, AZ, USA, and Lowell Observatory, Flagstaff, AZ, USA

SCOTT L. MURCHIE
Johns Hopkins University Applied Physics Laboratory, Laurel, MD, USA

JOHN F. MUSTARD
Department of Earth, Environmental and Planetary Sciences, Brown University, Providence, RI, USA

CATHERINE B. OLKIN
Southwest Research Institute, Boulder, CO, USA

MARIO C. PARENTE
Department of Electrical and Computer Engineering, University of Massachusetts at Amherst, Amherst, MA, USA

CARLÉ M. PIETERS
Department of Earth, Environmental and Planetary Sciences, Brown University, Providence, RI, USA

JEFFREY J. PLAUT
Jet Propulsion Laboratory, California Institute of Technology, Pasadena, CA, USA

FRANCOIS POULET
Institut d'Astrophysique Spatiale, Orsay Cedex, France

THOMAS H. PRETTYMAN
Planetary Science Institute, Tucson, AZ, USA

SILVIA PROTOPAPA
Southwest Research Institute, Boulder, CO, USA

VISHNU REDDY
Lunar and Planetary Laboratory, University of Arizona, Tucson, AZ, USA

DENNIS C. REUTER
NASA Goddard Space Flight Center, Greenbelt, MD, USA

MELISSA S. RICE
Western Washington University, Bellingham, WA, USA

LUCIE RIU
Institut d'Astrophysique Spatiale, Orsay Cedex, France

A. DEANNE ROGERS
Department of Geosciences, Stony Brook University, Stony Brook, NY, USA

GEORGE R. ROSSMAN
Division of Geological and Planetary Sciences, California Institute of Technology, Pasadena, CA, USA

STEVEN W. RUFF
School of Earth and Space Exploration, Arizona State University, Tempe, AZ, USA

FERNANDO RULL
University of Valladolid/Unidad Asociada UVa-CSIC Centro de Astrobiología, Valladolid, Spain

ANTONIO SANSANO
University of Valladolid/Unidad Asociada UVa-CSIC Centro de Astrobiología, Valladolid, Spain

BERNARD SCHMITT
Université Grenoble Alpes, Saint-Martin-d'Hères, France

CHRISTIAN SCHRÖDER
Biological and Environmental Sciences, University of Stirling, Stirling, Scotland, UK

KIMBERLY D. SEELOS
Johns Hopkins University Applied Physics Laboratory, Laurel, MD, USA

SHIV K. SHARMA
Hawai'i Institute of Geophysics and Planetology, University of Hawai'i at Mānoa, Honolulu, HI, USA

ELIZABETH C. SKLUTE
Department of Astronomy, Mount Holyoke College, South Hadley, MA, USA, and Planetary Science Institute, Tucson, AZ, USA

PABLO SOBRON
Impossible Sensing, St. Louis, MO, USA, and Carl Sagan Center, the SETI Institute, Mountain View, CA, USA

S. ALAN STERN
Department of Space Studies, Southwest Research Institute, Boulder, CO, USA

CRISTINA A. THOMAS
Planetary Science Institute, Tucson, AZ, USA

CHRISTINA E. VIVIANO
Johns Hopkins University Applied Physics Laboratory, Laurel, MD, USA

DANIKA WELLINGTON
School of Earth and Space Exploration, Arizona State University, Tempe, AZ, USA

ROGER C. WIENS
Space and Remote Sensing, Los Alamos National Laboratory, Los Alamos, NM, USA

NAOYUKI YAMASHITA
Planetary Science Institute, Tucson, AZ, USA

ALBERT S. YEN
Jet Propulsion Laboratory, California Institute of Technology, Pasadena, CA, USA

JAKOB VAN ZYL
Jet Propulsion Laboratory, California Institute of Technology, Pasadena, CA, USA

Chapter Reviewers

S. MICHAEL ANGEL (University of South Carolina)

GABRIELA ARNOLD (German Aerospace Center, DLR-Berlin)

JAMES W. ASHLEY (Jet Propulsion Laboratory)

JOSHUA L. BANDFIELD formerly, (Space Science Institute)

ADRIAN BROWN (Plancius Research)

BONNIE J. BURATTI (Jet Propulsion Laboratory)

JAMES BYRNE (Universität Tübingen)

BRUCE CAMPBELL (Smithsonian National Air and Space Museum)

DONALD B. CAMPBELL (Cornell University)

LYNN M. CARTER (University of Arizona)

SAMUEL M. CLEGG (Los Alamos National Laboratory)

EDWARD CLOUTIS (University of Winnipeg)

EDDY DE GRAVE (University of Ghent)

M. CRISTINA DE SANCTIS (Istituto di Astrofisica e Planetologia Spaziali)

M. DARBY DYAR (Mount Holyoke College and Planetary Science Institute)

CHRISTOPHER S. EDWARDS (Northern Arizona University)

BETHANY L. EHLMANN (California Institute of Technology)

WILLIAM H. FARRAND (Space Science Institute)

ABIGAIL A. FRAEMAN (Jet Propulsion Laboratory)

PATRICK J. GASDA (Los Alamos National Laboratory)

MARTHA S. GILMORE (Wesleyan University)

TIMOTHY D. GLOTCH (Stony Brook University)

WILLIAM M. GRUNDY (Lowell Observatory)

VICTORIA E. HAMILTON (Southwest Research Institute)

CRAIG HARDGROVE (Arizona State University)

AMANDA HENDRIX (Planetary Science Institute)

TAKAHIRO HIROI (Brown University)

BRIONY HORGAN (Purdue University)

MELISSA D. LANE (Fibernetics LLC)

LUCY F. LIM (NASA Goddard Space Flight Center)

PAUL G. LUCEY (University of Hawai'i at Mānoa)

ROBERT L. MARCIALIS (University of Arizona)

THOMAS MCCORD (Bear Fight Institute)

FRANCIS M. MCCUBBIN (NASA Johnson Space Center)

LUCY ANN A. MCFADDEN (NASA Goddard Space Flight Center)

HARRY Y. MCSWEEN (University of Tennessee)

NOUREDDINE MELIKECHI (University of Massachusetts Lowell)

ALBERT E. METZGER (Jet Propulsion Laboratory)

JOSEPH R. MICHALSKI (University of Hong Kong)

DOUGLAS W. MING (NASA Johnson Space Center)

ANDRZEJ W. MIZIOLEK (US Army Research Laboratory, retired)

GARETH A. MORGAN (Planetary Science Institute)

RICHARD V. MORRIS (NASA Johnson Space Center)

JOHN F. MUSTARD (Brown University)

ION PRISECARU (Bruker, WMOSS)

VISHNU REDDY (Lunar and Planetary Laboratory)

ROBERT C. REEDY (Planetary Science Institute)

A. DEANNE ROGERS (Stony Brook University)

TED L. ROUSH (NASA Ames Research Center)

MARK SALVATORE (Northern Arizona University)

CHRISTIAN SCHRÖDER (University of Stirling)

SHIV K. SHARMA (University of Hawai'i at Mānoa)

GREGG A. SWAYZE (US Geological Survey Boulder)

STEFANIE TOMPKINS (Colorado School of Mines)

TOON VAN ALBOOM (University of Ghent)

ANNE VERBISCER (University of Virginia)

ALIAN WANG (Washington University in St. Louis)

SHOSHANA Z. WEIDER (Imperial College London)

JAMES J. WRAY (Georgia Institute of Technology)

Foreword

CARLÉ M. PIETERS AND PETER A. J. ENGLERT

So much has happened in the 25 years since the last collection of expert papers documenting principles and products of remote compositional analyses were brought together as a book! Of course, it is not that the physics and chemistry governing properties of planetary materials has changed much in a generation. However, great advances have been made during the intervening decades as new instruments were built and new spacecraft were flown across the Solar System. This is coupled with advances in information extraction techniques and instrument technology that enable the measurement of these properties with increasing detail both in the laboratory and remotely. Consequently, understanding nuances of the diagnostic properties forming the basis for compositional analyses has grown in leaps and bounds along with remarkable and expanding new data from the inner to outer Solar System, including both rocky and icy bodies with and without an atmosphere.

The 1993 book *Remote Geochemical Analyses: Elemental and Mineralogical Composition* was compiled near the end of the last millennium following a symposium bringing together planetary scientists across many disciplines. At that time, remote planetary exploration techniques were just beginning to grow in importance and impact. The impetus for bringing together information and technical background in one book was to make the scientific basis for this relatively new field readily available across a growing community. The initial discussions in *Remote Geochemical Analysis* laid the foundation for years of basic exploration of Solar System bodies in all their diversity and mystery. Subsequent expansion and maturation of remote sensing data obtained using telescopes and increasingly sophisticated exploratory spacecraft opened a wide range of data types and approaches with which to obtain information and understand the diverse and complex bodies of our Solar System.

Today, several decades later, the initial reconnaissance of the Solar System is complete. We have now looked at everything from Mercury to the Kuiper belt at least once. That has taught us there is a LOT more to learn. The path that exploration has taken has provided profound insight, awesome discoveries, and continuous inspiration. Nevertheless, it necessarily has not been a linear or complete process. We are now embarking on an era of detailed and serious exploration, that cries for in-depth knowledge of, and comparisons between, the diverse rocky, hydrated, icy, and gaseous bodies of our Solar System (including Earth) – and even planets of other star systems. Although the exploration focus and resulting data acquired have been uneven, a plethora of fundamental questions are posed and remain unanswered regarding the composition of each planetary body we have come to know. In parallel with the quest for deeper scientific understanding, modern technology provides increasingly sophisticated instruments to measure compositional properties remotely, and such exploration tools promise many exciting decades ahead.

The chapters in this completely new *Remote Compositional Analysis: Techniques for Understanding Spectroscopy, Mineralogy, and Geochemistry of Planetary Surfaces* provide a diverse and enormously updated taste of what we know and what we have been able to learn about the composition of planetary bodies using remote sensing techniques over the last few decades. Remote compositional analysis has become a mature interdisciplinary field of science and has evolved into an indispensable component of Earth and planetary exploration.

Preface

The field of remote sensing is integral to exploration of our Solar System and encompasses results acquired from telescopic, spacecraft, and landed missions. The remote sensing techniques described in this book span a range of processes, compositions, and planetary bodies. It is a dramatically updated version of the original first edition from 1993. Since then, significant advances have occurred in space exploration, including dozens of new missions from NASA, ESA, and other space agencies around the world; substantial advances in telescopic and laboratory instrumentation, and major new work in theoretical studies have also occurred. As a result, every topic has been updated from the original book edited by C. M. Pieters and P. A. J. Englert, and new content has been added to reflect major advancements since 1993. This work sprung out of a 1988 symposium on mineral spectroscopy organized by L. M. Coyne with assistance from J. L. Bishop that was sponsored by the Division of Geochemistry of the American Chemical Society. The purpose of that symposium was to bring together an interdisciplinary, international community to foster spectral identification of minerals.

We have attempted to provide an introduction to the field of planetary surface composition and mineralogy for upper-level undergraduates, graduate students, or professional researchers just moving into this topic. This book is organized into four sections including (I) theory and laboratory measurements, (II) terrestrial field and airborne applications, (III) analysis methods, and (IV) applications to planetary surfaces. Among the types of remote sensing techniques covered are visible to infrared reflectance spectroscopy, infrared emission spectroscopy (also called thermal infrared spectroscopy), Raman spectroscopy, Mössbauer spectroscopy, Laser-Induced Breakdown Spectroscopy (LIBS), neutron spectroscopy, X-ray spectroscopy, gamma-ray spectroscopy, and radar. The basic premise of each technique, information on how to perform measurements, and example spectra of rocks, regolith, minerals, and volatiles are provided.

This book covers the minerals, elements, and molecules found on airless rocky bodies including Mercury, the Moon, and asteroids. It describes the kinds of volatiles (ices, organics, hydrated minerals) found on the surfaces of our Solar System's planets, moons, asteroids, and comets, and how they are related to volatiles on Earth. Finally, several chapters specifically focus on the composition and processes taking place on Mars, the planet most studied besides Earth.

Acknowledgments

We are grateful to the countless planetary scientists and engineers around the world who have contributed toward developing and operating the phenomenal instruments, telescopes, and spacecraft that enabled the compositional remote sensing results presented here. Obtaining these data from planets, moons, asteroids, and comets in our Solar System has required laborious efforts and diligence from large teams of people from space centers, research institutes, and universities. We appreciate the hard work of the authors who contributed state-of-the-art summaries of current topics in planetary remote sensing to this book and made it possible.

Unfortunately, two of our authors, Joshua Bandfield and Göstar Klingelhöfer, passed unexpectedly this year and we will miss them. They contributed to Chapters 15, 24, 25, and 27, where memorial statements are included.

Many others contributed to this book as well. The generous time volunteered by numerous reviewers is much appreciated. We also thank C. Gross for assistance with the cover art, L. Gründler for assistance with the index and editing, and S. Perrin for assistance with the references and editing. Finally, assistance from E. Kiddle, Z. Pruce, S. Ramamoorthy, and T. Kornak from the Cambridge University Press office and copy-editing teams is much appreciated.

Part I

Theory of Remote Compositional Analysis Techniques and Laboratory Measurements

1

Electronic Spectra of Minerals in the Visible and Near-Infrared Regions

GEORGE R. ROSSMAN AND BETHANY L. EHLMANN

1.1 Origin of Electronic Spectra of Minerals

Many of the spectral features of minerals in the visible to near-infrared region (VNIR; defined here as ~0.4–2.5 μm) arise from electronic transitions within and between transition elements and the anions chemically bound to them. Thousands of minerals have color or wavelength-variable properties in this portion of the spectrum. Metal ions including vanadium, chromium, manganese, iron, cobalt, nickel, and copper, usually in either the 2+ or 3+ oxidation state, are responsible for the color of many minerals. However, only a few of these elements, typically iron, titanium, and oxygen, are important in most remote sensing applications of rocky bodies. Many features arise from electronic transitions of electrons between the d orbitals of a metal ion, while some spectroscopic features arise from interactions between atoms.

1.2 Units

Wavelengths are commonly expressed in nanometers (nm) or micrometers (μm) and, in older literature, Ångström units. Literature on mineral spectroscopy and mineral chemistry often uses nm, while the remote sensing literature typically uses μm. The conversion among them is:

$$1000 \text{ nm} = 1 \text{ μm} = 10{,}000 \text{ Å}. \tag{1.1}$$

The spectrum can also be presented in energy units, usually wavenumbers, which are the reciprocal of the wavelength, and are usually expressed in reciprocal cm. The advantage of wavenumbers is that absorptions are symmetrical in energy coordinates but not in wavelength coordinates. Spectroscopic energies can also be expressed in electron volts, but this is more commonly encountered in the physics literature.

$$\begin{aligned} &\text{Wavenumbers } (\text{cm}^{-1}) = 10{,}000{,}000/\text{nm} = 10{,}000/\text{μm}) \\ &1000 \text{ nm} = 1 \text{ μm} = 10{,}000 \text{ cm}^{-1}; \quad 400 \text{ nm} = 0.4 \text{ μm} = 25{,}000 \text{ cm}^{-1} \\ &1 \text{ cm}^{-1} = 1.23984 \times 10^{-4} \text{ eV}; \quad 8065.54 \text{ cm}^{-1} = 1 \text{ eV} = 1239.8 \text{ nm}. \end{aligned} \tag{1.2}$$

Spectra are usually displayed in either reflectance units or absorbance units. Reflectance spectra must be taken in comparison to a standard. In a laboratory setting, the standard can be a colorless polytetrafluoroethylene-based plastic such as Spectralon®, aluminum, or, in the near-IR (NIR) region, gold. Spectra are presented as the percent (0–100%) or fraction (0–1) of sample reflectance relative to the standard versus wavelength. In spacecraft applications, the comparison standard is typically the solar flux. Reflected light data collected by spacecraft are typically expressed as I/F [radiance/(solar irradiance/π)]. For a given viewing geometry, these data can be further corrected for angular dependencies in scattering properties (see Hapke, 1981, for definitions of different types of reflectance).

Even though remotely obtained spectra are the composite response of many components in a field of view, many fundamental studies of mineral spectra are conducted with single crystals. In chemistry, such studies are usually presented in absorbance units, where

$$\text{Absorbance} = -\log_{10}(\text{Transmission}). \tag{1.3}$$

An absorbance of 1 means that 10% of the incident light is passing through the crystal; an absorbance of 2 means that 1% of the light passes through. Beer's law formulations ($I = I_0 e^{-\alpha d}$) are also sometimes used to derive an absorption coefficient (α) where the intensity of initial light (I_0) is compared to the intensity after transmission (I) through a given thickness of material (d).

Because most mineral crystals are anisotropic, fundamental studies of single crystals usually measure the spectrum with polarized light vibrating along the fundamental optical directions of the crystal. The refractive indices of a crystal for light traveling in different directions relative to the crystal axes form an optical indicatrix, mathematically, an ellipsoidal surface. Crystals that belong to the orthorhombic, monoclinic, and triclinic crystal systems will have three independent spectra that can display very different absorption properties (biaxial indicatrix). Crystals in the tetragonal and hexagonal systems will have two different spectra (uniaxial indicatrix), while isotropic, cubic crystals will have only one spectrum (spherical indicatrix). In general, spectra can be named either according to the crystal axes in which the vibration occurs (e.g., E\\$c$ or the c-spectrum) or by the symbol for the index of refraction that would be measured in the vibration direction. For biaxial crystals with three independent optical directions we have the α, β, and γ spectra (also called the X, Y, and Z spectra). For uniaxial crystals with two different spectra there are two independent optical orientations: the E$\perp c$ direction, also called the ω-spectrum, and the E\\$c$ direction, which is also called the ε-spectrum.

1.3 Crystal Field Transitions

The spectra of metal ions, particularly those of first-row transition elements, Ti through Cu, are often interpreted with the use of Crystal Field Theory. The d-orbital electrons are the valence (outermost) electrons in the case of these metals. For an isolated transition metal ion, electrons occupy any d orbital with equal probability. However, in a mineral, electrostatic fields produced by the anions (usually oxygen) surrounding the central metal ion separate the metal ion's d orbitals into different energy levels. This allows the d-orbital electrons to undergo transitions between orbital energy states. Their transitions to different energy levels under the influence of VNIR light give rise to much of the color we see and the spectra we measure of minerals.

This can be understood for the case of a metal ion surrounded by six oxide ions (ligands) arranged in perfect octahedral symmetry. If the metal ion were floating in free space with no oxide or other anions near it, all five orbitals in the $3d$ level would have the same energy (Figure 1.1a). But when the metal ion is in an octahedral arrangement of oxide ions, the d orbitals split into groups of two different energies (Figure 1.1b) reflecting the different interactions the d-orbital electron clouds have with the surrounding ligands.

Iron in the 2+ oxidation state has six electrons, the valence electrons, in the $3d$ orbital. In an octahedral coordination environment, these go into the $3d$ orbitals as pictured in Figure 1.2a because the electrons are energetically more stable when pairing of electrons is minimized. An electronic transition will occur when light of an appropriate energy interacts with the Fe^{2+} ion and promotes an electron from a lower energy orbital to a higher energy

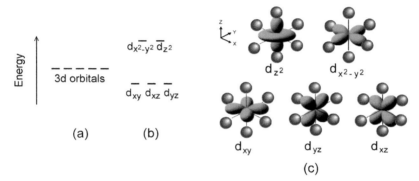

Figure 1.1 Energy diagram for $3d$ orbitals and their electron probability clouds. (a) Orbitals in free space. (b) Orbitals in an octahedron of oxide ions. (c) The electron clouds of the d orbitals in relationship to the oxide ions in an octahedral arrangement.

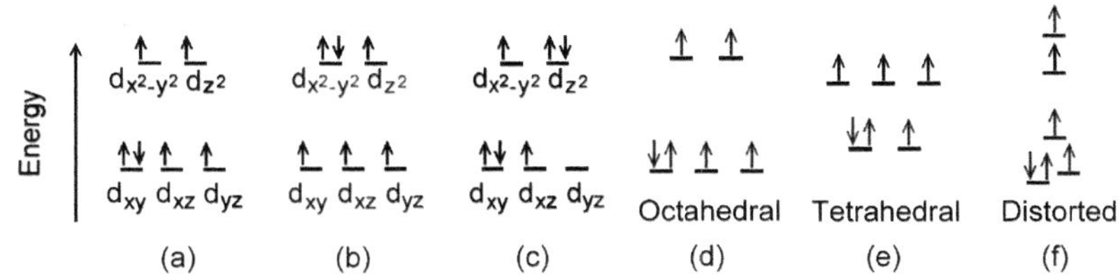

Figure 1.2 Electron configurations for Fe^{2+}. (a) The ground state in an octahedral coordination environment. (b) The spin-allowed excited state that gives rise to the primary NIR absorption bands. Here, the total number of unpaired electrons has not changed in the excited state. (c) A spin-forbidden state in which the total number of unpaired electrons has changed in the electronic excitation. A comparison of the relative splitting of ground state d orbitals for Fe^{2+} ion in (d) regular octahedral, (e) regular tetrahedral, and (f) a representative distorted coordination.

orbital (Figure 1.2b). Each configuration of the electrons is an electronic state of the system. If the total number of unpaired electrons is not changed during the transition from the electronic ground state to a higher energy state, this is called a spin-allowed transition. If the total number of unpaired electrons changes, this transition is called a spin-forbidden transition because such a transition is about 100 times less likely to occur than a spin-allowed one.

The intensities of electronic bands relate to their spin-allowed or spin-forbidden properties. Spin-forbidden transitions produce absorption bands that are commonly much weaker than the spin-allowed bands. However, interactions between cations, as explained in a following section, can dramatically increase the intensity of formally spin-forbidden bands and produce other features of high intensity when cations in different oxidation states interact. In addition, electronic transitions that involve transfer of charge from anions to cations (Sections 1.4 and 1.5) can also be of much higher intensity, but are usually centered in the ultraviolet portion of the spectrum.

Qualitative predictions of the spectrum of metal ion complexes can be obtained from Tanabe–Sugano diagrams. These diagrams usually present energy states for complexes in ideal, octahedral coordination that can be used to interpret the number of spin-allowed and spin-forbidden absorption bands and their widths, and, with suitable experimental parameters, can provide predictions of where bands will occur. Most ions in minerals are not in ideal octahedral coordination, so these diagrams often do not accurately interpret mineral spectra, but they do indicate which absorption bands will split into multiple components for metal ions in crystal sites of low symmetry. These diagrams, along with other concepts previously discussed, are reviewed in more detail in several books and articles about mineral spectroscopy (Karr, 1975; Burns, 1993; Rossman, 2014).

Another important factor in determining the number and wavelengths of absorption bands from a metal ion is the symmetry and distances of the ions surrounding the central metal ion. The number and energies of absorption bands strongly depend on the symmetry (Figure 1.2d–f). In a perfectly regular octahedron, Fe^{2+} will have one possible transition from the lower to the higher set of orbitals (Figure 1.2b). In a perfectly regular tetrahedral coordination environment, the energy difference between the orbitals will be smaller; consequently, the absorption will occur at longer wavelengths, but still with only one absorption band. However, coordination environments of ideal symmetry are almost never encountered. In nearly all minerals, the metal ion is in a coordination environment distorted from ideal symmetry. In such cases, the energies of the orbitals will split and multiple absorption bands will be possible. This fact is crucial for understanding the relatively broad nature of absorption bands in spectra of common rock-forming minerals. For example, in olivine the broad Fe-related electronic absorption observed is, in reality, a set of overlapping absorptions, caused by the existence of numerous 6-coordinated sites of different dimensions and symmetries that occur as the atoms around the iron vibrate due to thermal energy. In pyroxene, the wavelength of the Fe-related electronic absorption in the distorted 6-coordinated M(2) site shifts systematically with Ca, Fe, and Mg substitution that changes the dimensions of the octahedral site.

Group theory provides symbolic names for each of the electronic states of the system. These names convey the spin state, degeneracy, and symmetry of the electron cloud (for further reading see Harris & Bertolucci, 1989; Cotton, 1990). For example, in a perfectly octahedral coordination environment, the Fe^{2+} ground state has a $^5T_{2g}$ symmetry designation, where the T indicates that the state is triply degenerate, the 5 is the number of unpaired electrons +1, and the $2g$ relates to the symmetry of the electron cloud. The excited state has a designation ^5E_g, where the E symbolizes a doubly degenerate state. A single absorption band occurs when the electron is promoted from the $^5T_{2g}$ state to the ^5E_g state. In coordination environments of lower symmetry, the T state can split into three different electronic states and the E state can split into two, each with a different energy. In orthopyroxene, the electronic ground state of Fe^{2+} in the low symmetry M(2) site splits into three different states labeled $^5A_1$, $^5A_2$, and $^5B_2$, and the excited state splits into a $^5B_1$ and a $^5A_1$ state, each of which is no longer degenerate (Goldman & Rossman, 1977).

Electronic absorption bands can be temperature sensitive. They typically broaden at higher temperatures and sharpen at lower temperatures. Fundamental studies of minerals and chemicals are often conducted at liquid nitrogen or even liquid helium temperatures to sharpen absorptions and allow determination of band centers at high spectroscopic resolution. Particularly for targets below ~150 K, consideration of shifts may be relevant in interpretation of remotely collected spectra.

Absorptions can also shift position or change intensity as mineral sites are distorted and metal–oxygen bond distances change at elevated temperatures (e.g., Aronson et al., 1970; Sung et al., 1977). High-temperature spectra are important in planetary science for interpreting the composition of bodies that are several hundreds of degrees warmer than Earth such as Mercury, Venus, and lavas on Jupiter's moon, Io.

1.4 Oxygen-to-Metal Charge Transfer

Another common feature in the spectra of many minerals is the oxygen-to-metal charge transfer transition. This feature arises from absorption of photons with enough energy to transfer charge density from an oxygen ligand to the central metal ion. Oxygen-to-iron charge transfer is most commonly encountered in common rock-forming minerals where the band is usually centered in the ultraviolet region. The higher the charge state of the central metal ion, the lower the energy of the absorption band will be. Oxygen-to-Fe^{3+} charge transfer bands sometimes tail into the visible portion of the spectrum, where they absorb in violet and blue and often produce a rusty orange-red color. Oxygen-to-metal charge transfer absorptions are normally much more intense than those arising from transitions within the *d* orbitals of metal ions.

1.5 Intervalence Charge Transfer

Intervalence Charge Transfer (IVCT) refers to a process in which two metal ions in close proximity to each other in a structure transfer an electron between them, thereby temporarily changing the oxidation state of both cations. Absorption bands in the optical spectrum from IVCT can be comparatively intense, and only a little IVCT produces spectroscopic features and color in the visible spectral region. In the geological world, only two such interactions are commonly encountered: Fe^{2+}–Fe^{3+} and Fe^{2+}–Ti^{4+}. A third, Ti^{3+}–Ti^{4+}, is occasionally found in meteorites.

For these interactions to occur, cations need to be adjacent to each other in the mineral structure, often sharing a common edge or face of the coordination polyhedron. Both Fe^{2+}–Fe^{3+} and Fe^{2+}–Ti^{4+} IVCT are particularly common in terrestrial minerals such as micas, pyroxenes, amphiboles, and tourmalines and are the origin of the dark color of many minerals including magnetite and ilmenite. Fe^{2+}–Fe^{3+} IVCT in sites of near-octahedral coordination is found in the 630–820 nm region. Fe^{2+}–Ti^{4+} IVCT (Figure 1.3a) is typically found in the 425–460 nm region for 6-coordinated near-octahedral cations such as in pyroxenes (Mao et al., 1977; Mattson & Rossman, 1988). The Ti^{3+}–Ti^{4+} IVCT, observed in pyroxenes and hibonite from meteorites (Dowty & Clark, 1973; Burns & Vaughn, 1975), occurs near 690 nm in meteoritic hibonite from Murchison (Rossman, 2019).

In a number of terrestrial minerals adjacent sites may have different coordination polyhedra including edge-shared octahedra and tetrahedra in cordierite or edge-shared octahedra and distorted cubes in garnets. In these cases, the wavelengths of the IVCT bands will differ from those of the edge-shared octahedra. A number of different mineral examples are reviewed in Burns (1981).

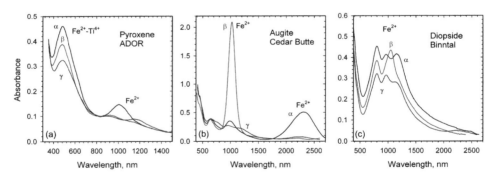

Figure 1.3 Transmission spectra. (a) Clinopyroxene from the Angra dos Reis meteorite showing Fe^{2+}–Ti^{4+} IVCT near 480 nm and the Fe^{2+} features near 1000 and 1200 nm discussed in Section 1.6. (Modified from Mao et al., 1977.) (b) A 200 μm thick augite crystal showing the absorption bands from Fe^{2+} in the geometrically distorted M(2) site near 1000 and 2400 nm, and the weaker bands from Fe^{2+} in the nearly octahedral M(1) site near 970 and 1200 nm. Weak absorption from Cr^{3+} appears near 450 and 650 nm. (c) A 200 μm thick diopside crystal showing comparatively weak absorption bands from Fe^{2+} in the geometrically distorted M(2) site near 1000 and 2400 nm, and the stronger bands from Fe^{2+} in the nearly octahedral M(1) site near 1000 and 1200 nm. Absorption near 800 nm arises from Fe^{2+}–Fe^{3+} IVCT.

1.6 Spectra of Key Minerals

There are currently more than 5400 known mineral phases, but only a small number of them contribute electronic absorptions routinely associated with remotely sensed spectra in the VNIR region. These phases include pyroxenes, olivines, feldspars, iron-bearing layered silicate minerals, and iron oxides. A number of other phases such as iron carbonates, iron sulfates, and other sulfur species are occasionally encountered. While they are only components contributing to the whole spectroscopic signature of an object, these minerals and their electronic absorptions carry important information for revealing the geological history of an object. In this section, we review the spectra of select, important phases. Many examples of the spectra of mineral single crystals with other cations are presented in Rossman (2014).

Iron is the element most commonly causing absorptions in the VNIR spectral region and is responsible for the color of common rock-forming minerals. In the primary igneous minerals, iron is usually found in the 2+ oxidation state, often either in sites that are somewhat distorted from ideal octahedral 6-coordination or in irregular sites of higher coordination number. Frequently, iron occurs in more than one distinct site in the crystal structure of the host mineral. Sulfur species of mixed oxidation state are important on some outer Solar System bodies (e.g., Io: Nash et al., 1980; Carlson et al., 1997). Other metal cations such as V, Cr, Mn, Ni, and Cu are important contributors to the spectra of terrestrial minerals and are responsible for the spectacular colors of many museum-quality minerals. To date, they have not played a significant role in remotely sensed spectra of other planetary bodies.

1.6.1 Pyroxenes

Pyroxenes, $(Ca, Mg, Fe)_2(Si, Al)_2O_6$, are important minerals in many planetary bodies and are an excellent example of how structural distortion affects spectral properties. The two components of the pyroxene absorption bands of Fe^{2+} become increasingly separated as the sites become more distorted from octahedral geometry due to cation substitutions. In the case of the pyroxene M(2) site, the two components can be separated by about 1000 nm (Figure 1.3b).

The spectrum of augite in Figure 1.3b, a terrestrial clinopyroxene, shows prominent absorptions at about 1000 nm in the beta polarization and near 2300 nm in the alpha direction. These two bands arise from Fe^{2+} in the M(2) site of pyroxene, which is highly distorted from an octahedral geometry. Two weaker bands near 970 and 1200 nm are due to Fe^{2+} in the less distorted M(1) site. Small, sharp spin-forbidden transitions are observed at wavelengths less than 1.0 μm. In contrast, the spectrum of diopside in Figure 1.3c has comparatively little contribution from the M(2) site and is primarily dominated by Fe^{2+} absorption from the M(1) site.

Pyroxenes are among the most widespread rock-forming minerals in the Solar System. The absorptions caused by Fe^{2+} in distorted M(1) and M(2) sites can be detected in remote sensing reflectance spectra and related to pyroxene crystal chemistry (Figure 1.4a–c),

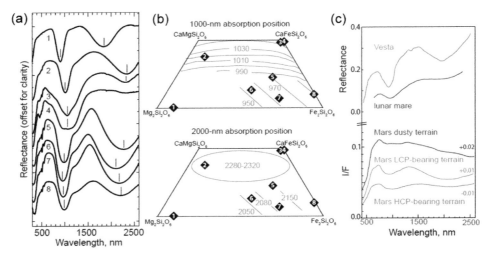

Figure 1.4 Reflectance spectra of pyroxenes. (a) Spectra of a variety of pyroxenes from Klima et al. (2011). (b) Pyroxene absorption band positions systematically shift with the crystal chemistry, here represented on a pyroxene quadrilateral. Apices are diopside ($CaMgSi_2O_6$), hedenbergite ($CaFeSi_2O_6$), enstatite ($Mg_2Si_2O_6$), and ferrosilite ($Fe_2Si_2O_6$). (c) These changes have been essential for identifying distinct geologic units with low-Ca and high-Ca pyroxene (LCP and HCP) on Mars (spectra from Mustard et al., 2005), pyroxenes on Vesta (spectra from DeSanctis et al., 2012), and high-Ca pyroxenes in lunar lavas (spectra from Pieters, 1986).

which in turn can be related to magmatic processes occurring on Solar System bodies. For example, 1 μm and 2 μm absorptions in dark lunar mare terrains were used to establish their volcanic origin and map distinct lava flows (e.g., Pieters, 1978; Staid et al., 2011; Whitten & Head, 2015). Strong pyroxene absorption bands observed for the asteroid Vesta and its family were used to identify it as the parent body for the HED meteorite suite and later mapped with spacecraft data (e.g., McCord et al., 1970; DeSanctis et al., 2012). On Mars, an observed transition from older lavas with low-Ca to younger lavas with high-Ca pyroxenes is inferred to result from thermal evolution of the martian mantle (Mustard et al., 2005; Baratoux et al., 2013). For more reading on pyroxene spectroscopy, see Klima et al. (2011).

1.6.2 The Olivine Series

The spectrum of forsterite provides another example of the role of Fe^{2+} in two distinct sites in the crystal of $(Mg, Fe)_2SiO_4$. Each of the 6-coordinated sites for the metal cations, known as the M(1) and M(2) sites, is significantly distorted from purely octahedral symmetry. Consequently, each site produces a pair of Fe^{2+} NIR absorption bands, which correspond to the crystal field splitting between the lower energy orbitals and the excited states (Figure 1.5). Because olivine is orthorhombic, the 3 spectra in Figure 1.5a represent polarizations along the *a*-, *b*-, and *c*-axes of the crystal which correspond to the γ, α, and

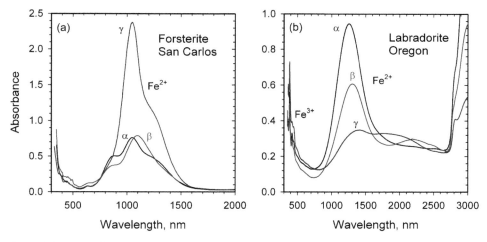

Figure 1.5 Absorption spectra. (a) Olivine (forsterite) from San Carlos, Arizona. (b) Fe-bearing plagioclase feldspar from Lake County, Oregon.

β spectra, respectively. The absorption in the 700–1600-nm region of the forsterite spectrum represents spin-allowed bands of Fe^{2+}. There are absorption bands centered near 830 nm, 1060 nm, 1100 nm, and 1310 nm. The two most intense bands displayed in the γ-spectrum are from iron in the M(2) site. Weak features at less than 800 nm are either spin-forbidden bands of Fe^{2+} or features from other minor components. The diffuse reflectance spectrum convolves the three spectra, as shown in Figure 1.6. The band positions shift with increasing Fe/(Mg + Fe) ratios. Figure 1.6a compares the spectra of an Mg-poor and an Mg-rich olivine. Olivine spectroscopy is further discussed in Sunshine et al. (1998), Isaacson et al. (2014), and Chapters 4 and 18.

1.6.3 Feldspars

Plagioclase feldspars (e.g., $CaAl_2Si_2O_8$) can have iron substitution and thus an Fe^{2+} absorption in the NIR region (Figure 1.5b). The dominant absorption centered near 1300 nm arises from Fe^{2+} in the Ca site, which is significantly distorted from any standard coordination geometry. Features in the 300–500 nm region are from Fe^{3+} in the Al sites, and features near 3000 nm are from the OH content of the feldspar. The Fe^{3+} bands are absent in the spectrum of lunar plagioclase returned by the Luna 20 mission (Bell & Mao, 1973; Chapter 18). Plagioclase spectroscopy is further discussed in Cheek (2014).

1.6.4 Spinels

The spinel group minerals of the general formula (XY_2O_4) are phases that commonly contain Fe^{2+} in a tetrahedral environment, substituting for Mg. Because there is less

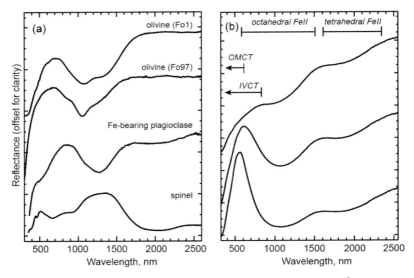

Figure 1.6 Reflectance spectra of iron-containing phases. (a) Spectra of Fe^{2+} in olivine of two compositions (from Sunshine & Pieters, 1998), plagioclase feldspar (from the NASA/Keck RELAB database at Brown University, spectra by C.M. Pieters), and a spinel (from Cloutis et al., 2004). OMCT = oxygen to metal charge transfer; IVCT = intervalence charge transfer. See text for absorption attributions. (b) Fe(II) and Fe(III) create prominent absorptions in Fe-bearing glasses (from Cannon et al., 2017).

electrostatic repulsion from the four oxygen atoms surrounding the iron compared to six oxygen atoms in octahedrally coordinated iron, the lowest energy (longest wavelength) Fe^{2+} absorption bands occur at lower energies (longer wavelengths) than those from 6-coordinated Fe^{2+} (Figure 1.7a).

1.6.5 Ferric Oxides

Ferric oxides contain VNIR absorption features due to both crystal field splitting and charge transfer. Hematite (Fe_2O_3) has a prominent absorption near 860 nm and a shoulder at 630 nm due to Fe^{3+} in a site of near-octahedral symmetry (Figure 1.7b). Starting at 530 nm, the visible light wing of a strong UV-visible charge transfer of oxygen–Fe^{3+} dominates the spectrum, making the reflectance very low and obscuring remaining crystal field absorptions. These properties change for nanocrystalline hematite, which has particle sizes <10 nm and is superparamagnetic. When hosted within a neutral matrix, the spectrum of nanocrystalline hematite has a small positive slope in the visible, within the charge transfer band, and lacks a deep 860-nm absorption, presumably because of the broadening of crystal field transitions (Morris et al., 1985). Goethite (Figure 1.7c) and lepidocrocite (FeOOH) are other commonly occurring iron oxides with octahedrally coordinated iron, generating absorptions in different positions. Magnetite (Fe_3O_4) has a VNIR absorption coefficient

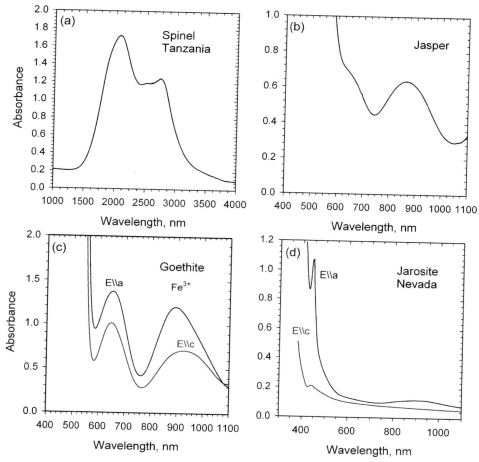

Figure 1.7 Absorption spectra of iron-containing phases. (a) The spectrum of Fe^{2+} in slightly distorted tetrahedral geometry in a spinel from Tanzania occurs in the 2000–3000 nm NIR region. (b) Spectrum of microparticles of hematite in red jasper. (c) Goethite single crystal spectrum showing two of the three polarization directions. (d) Jarosite spectrum showing strong anisotropy in the Fe^{3+} absorption bands.

more than an order of magnitude higher than other iron oxides due to intervalence charge transfer between ferric and ferrous iron, making it "opaque" in reflectance spectroscopy. It is a very efficient light absorber with low reflectance at all VNIR wavelengths.

Hematite and some other iron oxides/hydrous oxides have anomalously high absorption intensity compared to many other minerals and chemical compounds. Fe^{3+} has five d-orbital electrons that populate each of the five d orbitals with one electron each. As such, any electronic transition involves a flip in the spin direction and would require two electrons to pair, an energetically unfavorable process, making Fe^{3+} electronic transitions formally spin forbidden. However, when two adjacent Fe^{3+} cations share common oxygen ions, antiferromagnetic–magnetic interactions occur between and among the Fe^{3+} cations that can cause a large-intensity increase of the iron's electronic transition (Rossman, 1996). Effectively, one cation undergoes an electronic transition with a spin flip while an adjacent ion undergoes just a spin flip. The combined system of two iron ions would not undergo a total spin change in this process resulting, effectively, in a spin-allowed transition. See also reviews in Morris et al. (1985, 2000) and Chapter 4.

1.6.6 Iron Phyllosilicates

Iron-bearing phyllosilicates are a key indicator of water–rock interaction throughout the Solar System and often form via dissolution and reprecipitation reactions or transformation reactions from the iron-bearing silicates described earlier. Examples include iron–magnesium smectites, the most prevalent phyllosilicate on Mars, and smectites, serpentines, and chlorites found on Mars, asteroids, and meteorites. The spectra of these layered silicates have vibrational absorptions in the infrared longward of 1000 nm (see Chapter 4), as well as electronic absorptions (Figure 1.8a). Octahedrally coordinated Fe^{2+} and Fe^{3+} generate absorptions at 900 nm and 650 nm, as well as absorptions centered shortward of 500 nm. Mixed valence samples have a broad Fe^{2+}–Fe^{3+} charge transfer absorption centered near 700 nm. Also, Fe^{3+} in tetrahedral coordination can generate an absorption near 430 nm (see Sherman & Vergo, 1988). Collectively, these electronic absorptions allow phyllosilicates to be distinguished. Because of the complications of overlapping broad electronic absorptions with those of mafic silicates in samples, in remote sensing data, the sharp vibrational absorptions are often more diagnostic. Nevertheless, the electronic absorptions can be important, e.g., the 700 nm absorption in asteroid spectra inferred to result from iron phyllosilicates (Vilas et al., 1994) and crystalline hematite in select martian sedimentary deposits (e.g., Fraeman et al., 2013).

1.6.7 Carbonates

Fe and Mn substitution in the cation site in carbonates creates broad absorptions centered between 0.9 and 1.5 μm (as described in Gaffey, 1985). The absorption positions of Mn^{2+} and Fe^{2+} can be distinguished from each other, differ even for the same cation depending on the dimensions of the site (e.g., Fe within an Mg carbonate vs. Fe within a Ca carbonate), and, in the case of iron, exhibit a deep absorption with less than 1% of the transition metal (e.g., Bishop et al., 2013).

1.6.8 Iron Sulfates

To date, ferric and ferrous iron sulfates are found on Mars, Earth, and (rarely) meteorites. Many ferric iron sulfate crystals display large differences in the intensity of light absorption, with light polarized in different directions in the crystal. In particular, when the

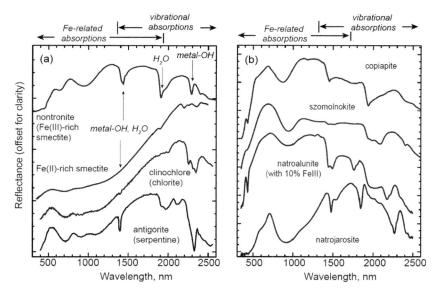

Figure 1.8 Reflectance spectra of phyllosilicates and sulfates. (a) In serpentine and chlorite, even small amounts of Fe(II) and Fe(III) produce VNIR absorptions (Spectra compiled from the authors' libraries; Clark et al., 2007; Chemtob et al., 2015.) (b) Fe(II) and Fe(III) create prominent absorptions with centers shortward of 1000 nm in sulfate minerals. Spectra of natrojarosite, $NaFe_3(SO_4)_2(OH)_6$, and natroalunite, $NaAl_3(SO_4)_2(OH)_6$, with 10% substitution of Fe for Al) are from McCollom et al. (2014). Szomolnokite and copiapite spectra are from Clark et al. (2007).

polarization direction is aligned in the direction of the Fe–OH–Fe bonds, the intensity of the lowest energy band ($^6A_{1g} \rightarrow {}^4T_{1g}$) and sharper ($^6A_{1g} \rightarrow {}^4A_{1g}, {}^4E_g$) band is increased. This is another example of intensity enhancement in antiferromagnetic systems, as discussed earlier for hematite. In the case of most ferric iron sulfates, and unlike hematite, the magnetic interaction occurs only in specific directions.

A practical consequence of this interaction is that because of the intensity enhancement, certain iron sulfates will be more readily detected than phases that do not have such interactions. Examples relevant to Mars include jarosite and magnesiocopiapite. For example, jarosite family minerals, $(K, Na, H_3O^+)Fe_3(SO_4)_2(OH)_6$, have four measured absorptions related to Fe electronic absorptions: a broad absorption centered at 930 nm ($^6A_{1g} \rightarrow {}^4T_{1g}$), a shoulder near 600 nm ($^6A_{1g} \rightarrow {}^4T_{2g}$), a strong, broad absorption near 500 nm that is the wing of the charge transfer band that is centered in the UV, and a sharp, narrow feature near 433 nm ($^6A_{1g} \rightarrow {}^4A_{1g}, {}^4E_g$) (Figure 1.7d) (Rossman, 1976; Sherman & Waite, 1985). As is commonly the case for iron in lower-symmetry sites, these absorptions are strongly anisotropic.

Ferrous and ferric sulfates can be distinguished based on a combination of their electronic absorptions in the VNIR as well as vibrational absorptions in the infrared related to OH, H_2O, and SO_4 (Figure 1.8b; Chapter 4). The spectra of iron sulfate minerals are reviewed in

Bishop and Murad (1996), Bishop et al. (2004), Cloutis et al. (2006), Pitman et al. (2014), Lane et al. (2015), Sklute et al. (2015), and Ling et al. (2016).

1.6.9 Other Sulfur Species

Metal sulfides occur in small amounts in silicate rocks and often act as darkening agents. Their presence is difficult to detect in VNIR remote sensing. SO_2 frost, native S, and complex S-bearing organics are constituents that occur in sufficient abundances on outer Solar System bodies to be observed in telescopic and spacecraft data. The prominent charge transfer absorptions extend into the UV and have been used to trace volcanic processes on Io and the products of radiolysis and space weathering on icy bodies. Spectra are reviewed in Nash et al. (1980). Unlike the transition metals discussed earlier that involve d orbitals, molecular sulfur spectra involve s and p orbitals (Meyer et al., 1972; Eclert & Steudel, 2003).

1.6.10 Iron-Containing Glasses

The effects of electronic processes on spectra are observed even in non- or poorly crystalline materials because the metal cations are still within an environment in which the electron orbitals are split energetically due to the presence of other ions. For example, iron-bearing glasses exhibit tetrahedrally and octahedrally coordinated Fe(II), intervalence charge transfer, and oxygen-to-iron charge transfer (Figure 1.6b; Burns, 1993; Cannon et al., 2017). The relative strengths of these absorptions depend on the amount of iron, its coordination (affected by other cations), and the Fe(II)/Fe(III) ratio (Figure 1.6b). The Fe(II)/Fe(III) ratio is indicative of the oxygen fugacity under which the glass formed. Other trace metals, such as Ti, also can generate absorptions in glasses.

1.7 Techniques for Mapping on Planetary Surfaces

The location of band centers in the VNIR is a crucial indicator that a particular mineral species is present. Thus, in the analyses of calibrated spectra, a first step is the identification of band minima. Their locations, shapes, and associations with other minima can be used to uniquely identify mineral species based on the results of the foregoing laboratory data and then spatially mapped. For example, broad absorptions centered at 1300 nm can be uniquely attributed to Fe–plagioclase feldspar if no ~2000 nm feature is apparent that might indicate glass. Olivine's broad absorption centered near 1000 nm (really a set of superimposed absorptions, as we know from Section 1.6.2) is unique in breadth and thus diagnostic (Figure 1.6a).

For olivines and pyroxenes, which occur in solid solution and whose crystal chemistry is uniquely diagnostic of magmatic processes, additional techniques have been devised to discriminate their solid solution chemistries. The most common implementation is the Modified Gaussian Model (e.g., Sunshine & Pieters, 1993, 1998) whereby the systematic shifts in position, width, breadth, and shape of absorptions in pyroxenes or olivines have been quantified as a function of solid solution chemistry. Information about pyroxene and olivine chemistry is obtained by fitting these parameters to remote sensing data. For more information, see Chapters 4, 11, and 14.

1.8 Geological Significance of Electronic Processes on Solar System Bodies

Because of a confluence of detector technology (abundant and relatively inexpensive charge coupled device (CCD) and complementary metal oxide semiconductor (CMOS) detectors sensitive over the ~400–1000 nm range and requiring no cooling), the relative transparency of the terrestrial atmosphere in the VNIR (for telescopic observation), and the presence of the fingerprints of electronic processes, VNIR spectroscopy of electronic processes has been fruitful in expanding our understanding of the Solar System. Within the asteroid belt, discrete families of pyroxene-rich, olivine-rich, and Fe–phyllosilicate-rich asteroids have been identified and linked to the early evolution and differentiation of planetary bodies (e.g., DeMeo et al., 2009; Chapters 19 and 20). On Mars, changes in the composition of lavas with time and location give insight into temporal and spatial variation in magmatic processes, thus providing key data on the interior evolution of our smaller neighboring planet (e.g., Mustard et al., 2005; Baratoux et al., 2013). Also on Mars, a rich array of secondary Fe-bearing minerals has been identified, which point to ubiquitous oxidation to form ferric oxides and discrete, geologic formations where water–rock interactions formed Fe-rich phyllosilicates, Fe carbonates, or Fe sulfates (e.g., Ehlmann & Edwards, 2014; Chapter 23). On the Moon, differences in pyroxene composition are found in the ancient highlands and later lavas, massive deposits of ferroan anorthosite support models of an early lunar magma ocean, and olivine-enriched deposits and spinel-anorthosite deposits point to excavation of lower crust materials (Chapter 18). The composition of the Venus surface is largely obscured because of atmospheric opacity at key wavelengths, though landed observations hint at the presence of oxidized Fe oxides, specifically hematite (Pieters et al., 1986). Mercury, intriguingly, has no evidence for electronic absorptions, indicating a uniquely iron-poor crust (e.g., Izenberg et al., 2014; Chapter 17). In the outer Solar System, vibrational absorptions of ices and electronic absorptions of sulfur species, rather than transition metal–bearing materials, dominate the surfaces and have been used to trace volcanism, upwelling of materials from subsurface oceans, and radiation-induced weathering of surfaces (e.g., Nash et al., 1980; Carlson et al., 1997).

References

Amthauer G. & Rossman G.R. (1984) Mixed valence of iron in minerals with cation clusters. *Physics and Chemistry of Minerals*, **11**, 37–51.

Aronson J.R., Bellotti L.H., Eckroad S.W., Emslie A.G., McConnell R.K., & Thüna P.C. (1970) Infrared spectra and radiative thermal conductivity of minerals at high temperature. *Journal of Geophysical Research*, **75**(17), 3443–3456.

Baratoux D., Toplis M.J., Monnereau M., & Sautter V. (2013) The petrological expression of early Mars volcanism. *Journal of Geophysical Research*, **118**, 59–64.

Bell P.M. & Mao H.K. (1973) Optical and chemical analysis of iron in Luna 20 plagioclase. *Geochimica et Cosmochimica Acta*, **37**, 755–759.

Berg B.L., Cloutis E.A., Beck P., et al. (2016) Reflectance spectroscopy (0.35–25 μm) of ammonium-bearing minerals and comparison to Ceres family asteroids. *Icarus*, **265**, 218–237.

Bishop J.L. & Murad E. (1996) Schwertmannite on Mars? Spectroscopic analyses of schwertmannite, its relationship to other ferric minerals, and its possible presence in the surface material on Mars. In: *Mineral spectroscopy: A tribute to Roger G. Burns*. Special publication (Geochemical Society). No. 5. Geochemical Society, Houston, TX, 337–358.

Bishop J.L., Dyar M.D., Lane M.D., & Banfield J.F. (2004) Spectral identification of hydrated sulfates on Mars and comparison with acidic environments on Earth. *International Journal of Astrobiology*, **3**, 275–285.

Bishop J.L., Perry K.A., Dyar M.D., et al. (2013) Coordinated spectral and XRD analyses of magnesite-nontronite-forsterite mixtures and implications for carbonates on Mars. *Journal of Geophysical Research*, **118**, 635–650.

Burns R.G. (1981) Intervalence transitions in mixed valence minerals of iron and titanium. *Annual Review of Earth and Planetary Sciences*, **9**, 345–383.

Burns R.G. (1993) *Mineralogical applications of crystal field theory*. Cambridge University Press, Cambridge.

Burns R.G. & Vaughan D.J. (1975) 2 – Polarized Electronic Spectra. In: *Infrared and Raman spectroscopy of lunar and terrestrial minerals* (C. Karr, ed.). Academic Press, New York, 39–72.

Cannon K.M., Mustard J.F., Parman S.W., Sklute E.C., Dyar M.D., & Cooper R.F. (2017) Spectral properties of martian and other planetary glasses and their detection in remotely sensed data. *Journal of Geophysical Research*, **122**, 249–268.

Carlson R.W., Smythe W.D., Lopes-Gautier R.M.C., et al. (1997) The distribution of sulfur dioxide and other infrared absorbers on the surface of Io. *Geophysical Research Letters*, **24**, 2479–2482.

Cheek L.C. (2014) *Foundations of lunar highland crustal mineralogy derived from remote sensing and laboratory spectroscopy of plagioclase-dominated Materials*. Brown University Earth, Environmental and Planetary Sciences Theses and Dissertations.

Chemtob S.M., Nickerson R.D., Morris R.V., Agresti D.G., & Catalano J.G. (2015) Synthesis and structural characterization of ferrous trioctahedral smectites: Implications for clay mineral genesis and detectability on Mars. *Journal of Geophysical Research*, **120**, 1119–1140.

Cloutis E.A., Sunshine J.M., & Morris R.V. (2004) Spectral reflectance-compositional properties of spinels and chromites: Implications for planetary remote sensing and geothermometry. *Meteoritics and Planetary Science*, **39**, 545–565.

Cloutis E.A., Hawthorne F.C., Mertzman S.A., et al. (2006) Detection and discrimination of sulfate minerals using reflectance spectroscopy. *Icarus*, **184**, 121–157.

Cotton F.A. (1990) *Chemical applications of group theory*, 3rd edn. Wiley-Interscience, New York.

DeMeo F.E., Binzel R.P., Slivan S.M., & Bus S.J. (2009) An extension of the Bus asteroid taxonomy into the near-infrared. *Icarus*, **202**, 160–180.

De Sanctis M.C., Ammannito E., Capria M.T., et al. (2012) Spectroscopic characterization of mineralogy and its diversity across Vesta. *Science*, **336**, 697.

Dowty E.C. & Clark J.R. (1973) Crystal structure refinement and visible-region absorption spectra of a Ti^{3+} fassaite from the Allende meteorite. *American Mineralogist*, **58**, 230–242.

Eckert B. & Steudel R. (2003) Molecular spectra of sulfur molecules and solid sulfur allotropes. In: *Elemental sulfur and sulfur-rich compounds II* (R. Steudel, ed.). Springer, Berlin, Heidelberg, 31–98.

Ehlmann B.L. & Edwards C.S. (2014) Mineralogy of the martian surface. *Annual Review of Earth and Planetary Sciences*, **42**, 291–315.

Fraeman A.A., Arvidson R.E., Catalano J.G., et al. (2013) A hematite-bearing layer in Gale crater, Mars: Mapping and implications for past aqueous conditions. *Geology*, **41**, 1103–1106.

Gaffey S.J. (1985) Reflectance spectroscopy in the visible and near-infrared (0.35–2.55 μm): Applications in carbonate petrology. *Geology*, **13**, 270–273.

Goldman D.S. & Rossman G.R. (1977) The spectra of iron in orthopyroxene revisited: The splitting of the ground state. *American Mineralogist*, **62**, 151–157.

Hapke B. (1981) Bidirectional reflectance spectroscopy, 1. Theory. *Journal of Geophysical Research*, **86**, 3039–3054.

Harris D.C. & Bertolucci M.D. (1989) *Symmetry and spectroscopy: An introduction to vibrational and electronic spectroscopy*. Dover Publications, Mineola, NY.

Horgan B.H.N., Cloutis E.A., Mann P., & Bell J.F. (2014) Near-infrared spectra of ferrous mineral mixtures and methods for their identification in planetary surface spectra. *Icarus*, **234**, 132–154.

Isaacson P.J., Klima R.L., Sunshine J.M., et al. (2014) Visible to near-infrared optical properties of pure synthetic olivine across the olivine solid solution. *American Mineralogist*, **99**, 467–478.

Izenberg N.R., Klima R.L., Murchie S.L., et al. (2014) The low-iron, reduced surface of Mercury as seen in spectral reflectance by MESSENGER. *Icarus*, **228**, 364–374.

Karr C. (1975) *Infrared and Raman spectroscopy of lunar and terrestrial materials.* Academic Press, New York.

Klima R.L., Dyar M.D., & Pieters C.M. (2011) Near-infrared spectra of clinopyroxenes: Effects of calcium content and crystal structure. *Meteoritics and Planetary Science,* **46**, 379–395.

Lane M.D., Bishop J.L., Dyar M.D., et al. (2015) Mid-infrared emission spectroscopy and visible/near-infrared reflectance spectroscopy of Fe-sulfate minerals. *American Mineralogist,* **100**, 66–82.

Ling Z., Cao F., Ni Y., Wu Z., Zhang J., & Li B. (2016) Correlated analysis of chemical variations with spectroscopic features of the K–Na jarosite solid solutions relevant to Mars. *Icarus,* **271**, 19–29.

Mao H.K., Bell P.M., & Virgo D. (1977) Crystal-field spectra of fassaite from the Angra dos Reis meteorite. *Earth and Planetary Science Letters,* **35**, 352–356.

Mattson S.M. & Rossman G.R. (1988) Fe^{2+}-Ti^{4+} charge transfer in stoichiometric Fe^{2+},Ti^{4+}-minerals. *Physics and Chemistry of Minerals,* **16**, 78–82.

McCollom T.M., Ehlmann B.L., Wang A., Hynek B., Moskowitz B., & Berquó T.S. (2014) Detection of iron substitution in natroalunite-natrojarosite solid solutions and potential implications for Mars. *American Mineralogist,* **99**, 948–964.

McCord T.B., Adams J.B., & Johnson T.V. (1970) Asteroid vesta – Spectral reflectivity and compositional implications. *Science,* **168**, 1445–1447.

Meyer B., Gouterman M., Jensen D., Oommen T.V., Spitzer K., & Stroyer-Hansen T. (1972) The spectrum of sulfur and its allotropes. *Advances in Chemistry,* **110**, 53–72.

Morris R.V., Lauer H.V. Jr., Lawson C.A., Gibson E.K. Jr., Nace G.A., & Stewart C. (1985) Spectral and other physicochemical properties of submicron powders of hematite (a-Fe_2O_3), maghemite (g-Fe_2O_3), magnetite (Fe_3O_4), goethite (a-FeOOH), and lepidocrocite (g-FeOOH). *Journal of Geophysical Research,* **90**, 3126–3144.

Morris R.V., Golden D.C., Bell J.F. III, et al. (2000) Mineralogy, composition, and alteration of Mars Pathfinder rocks and soils: Evidence from multispectral, elemental, and magnetic data on terrestrial analogue, SNC meteorite, and Pathfinder samples. *Journal of Geophysical Research,* **105**, 1757–1817.

Mustard J.F., Poulet F., Gendrin A., et al. (2005) Olivine and pyroxene diversity in the crust of Mars. *Science,* **307**, 1594–1597.

Nash D.B., Fanale F.P., & Nelson R.M. (1980) SO_2 Frost: UV-visible reflectivity and Io surface coverage. *Geophysical Research Letters,* **7**, 665–668.

Pieters C.M. (1978) Mare basalt types on the front side of the moon – A summary of spectral reflectance data. *Proc. 9th Lunar Planet. Sci. Conf.,* **3**, 2825–2849.

Pieters C.M. (1986) Composition of the lunar highland crust from near-infrared spectroscopy. *Reviews of Geophysics,* **24**, 557–578.

Pieters C.M., Head J.W. III, Patterson W., et al. (1986) The color of Venus. *Science,* **234**, 1379–1383.

Pitman K.M., Dobrea E.Z.N., Jamieson C.S., Dalton J.B., Abbey W.J., & Joseph E.C.S. (2014) Reflectance spectroscopy and optical functions for hydrated Fe-sulfates. *American Mineralogist,* **99**, 1593–1603.

Rossman G.R. (1975) Spectroscopic and magnetic studies of ferric iron hydroxy sulfates: Intensification of color in ferric iron clusters bridged by a single hydroxide ion. *American Mineralogist,* **60**, 698–704.

Rossman G.R. (1976) Spectroscopic and magnetic studies of ferric iron hydroxy sulfates: The series $Fe(OH)SO_4$ •nH_2O and the jarosites. *American Mineralogist,* **61**, 398–404.

Rossman G.R. (1988) Optical spectroscopy. In: *Spectroscopic methods in mineralogy and geology* (F. C. Hawthorne, ed.). Mineralogical Society of America, Washington, DC, 207–254.

Rossman G.R. (1996) Why hematite is red: Correlation of optical absorption intensities and magnetic moments of Fe^{3+} minerals. In: *Mineral spectroscopy: A tribute to Roger G. Burns.* Special publication (Geochemical Society). No. 5. Geochemical Society, Houston, TX, 23–27.

Rossman G.R. (2014) Optical spectroscopy. *Reviews in Mineralogy and Geochemistry,* **78**, 371–398.

Rossman G.R. (2019) minerals.gps.caltech.edu/FILES/Visible/Hibonite/Index.html. DOI:https://doi.org/10.7907/jywr-qq57.

Sherman D.M. & Waite T.D. (1985) Electronic spectra of Fe^{3+} oxides and oxide hydroxides in the near IR to near UV. *American Mineralogist,* **70**, 1262–1269.

Sherman D.M. & Vergo N. (1988) Optical (diffuse reflectance) and Mössbauer spectroscopic study of nontronite and related Fe-bearing smectites. *American Mineralogist,* **73**, 1346–1354.

Sklute E.C., Jensen H.B., Rogers A.D., & Reeder R.J. (2015) Morphological, structural, and spectral characteristics of amorphous iron sulfates. *Journal of Geophysical Research,* **120**, 809–830.

Staid M.I., Pieters C.M., Besse S., et al. (2011) The mineralogy of late stage lunar volcanism as observed by the Moon Mineralogy Mapper on Chandrayaan-1. *Journal of Geophysical Research,* 116, E00G10, DOI:10.1029/2010JE003735.

Sung C.-M., Singer R.B., Parkin K.M., & Burns R.G. (1977) Temperature dependence of Fe^{2+} crystal field spectra: Implications to mineralogical mapping of planetary surfaces. *Proc. 8th Lunar Sci. Conf,* 1063–1079.

Sunshine J.M. & Pieters C.M. (1993) Estimating modal abundances from the spectra of natural and laboratory pyroxene mixtures using the Modified Gaussian Model. *Journal of Geophysical Research*, **98**, 9075–9087.

Sunshine J.M. & Pieters C.M. (1998) Determining the composition of olivine from reflectance spectroscopy. *Journal of Geophysical Research*, **103**, 13,675–13,688.

Vilas F., Jarvis K.S., & Gaffey M.J. (1994) Iron alteration minerals in the visible and near-infrared spectra of low-albedo asteroids. *Icarus*, **109**, 274–283.

Whitten J. & Head J.W. (2015) Lunar cryptomaria: Mineralogy and composition of ancient volcanic deposits. *Planetary and Space Science*, **106**, 67–81.

Wildner M., Andrut M., & Rudowicz C.Z. (2004) Optical absorption spectroscopy in geosciences: Part I: Basic concepts of crystal field theory; Part 2: Quantitative aspects of crystal fields. In: *Spectroscopic methods in mineralogy* (A. Beran & E. Libowitzky, eds.). Mineralogical Society of Great Britain and Ireland.

2

Theory of Reflectance and Emittance Spectroscopy of Geologic Materials in the Visible and Infrared Regions

JOHN F. MUSTARD AND TIMOTHY D. GLOTCH

2.1 Introduction

Reflectance and emittance spectroscopy from visible (~0.35 μm) to middle infrared (~50 μm) wavelengths are among the core measurements made in support of planetary exploration. Data from instruments used for spectroscopic measurements are essential in determining surface mineralogy. Quantitative analysis of the observations is fundamental to understanding planetary formation and evolution. Our goal in this chapter is to present the theoretical foundation of these spectroscopic measurements, on which the determination of mineral presence and abundance is based. The details of absorption processes, central to the identification of mineral presence, are presented in Chapters 1, 3, and 4. Here we focus on the theories of radiative transfer that are used to predict and model reflectance and emittance spectroscopy. These models begin with the definition and determination of the optical constants $n(\lambda)$ and $k(\lambda)$, the real and imaginary indices of refraction. Starting from these optical constants, observation geometry and physical property constraints are used to arrive at models that can be used to quantify the physical state and composition of planetary surfaces.

2.2 Geometry and Optical Constants in Reflectance and Emission Spectroscopy

2.2.1 Measurement Geometry and Terminology

The terminology of reflectance spectroscopy is diverse because the measurement of reflected electromagnetic radiation is widely used in many fields. The instrumentation employed is often quite specific to the application, and the characteristics of the measurements must be factored into the calibration and utility of the measurements. In this chapter the larger goal is a description of reflectance and emission spectroscopy in planetary exploration. Thus, we focus on geometries and terminologies used in the acquisition and analysis of spacecraft data and supporting laboratory measurements.

The specific definition of reflectance is a function of the collimation of the source and the detector (Hapke, 1981). The light from the Sun arriving at a planetary surface can be treated as well collimated (parallel rays focused at infinity). The arriving incident light encounters the surface with a specific angle of incidence (*i*) relative to the surface normal. A portion of the scattered light is measured with an emergence angle (e) relative to the surface normal. In this geometry the reflectance is defined as the ratio of the radiance (watts/meter squared steradian, $W \cdot m^{-2} \cdot sr^{-1}$) diffusely reflected from the surface to the irradiance ($W \cdot m^{-2}$) measured on a plane perpendicular to the direction of propagation. The angle between the incident beam and emergence direction is defined as the phase angle, g. For these geometries, the scattered light and reflectance vary as a function of wavelength as well.

Thermal infrared (TIR) spectroscopic measurements in a remote sensing context are generally confined to passive measurements of emitted radiation. The total power per unit area thermally emitted by a surface is termed emittance (Hapke, 2012). Ideal isothermal surfaces emit radiation according to Planck's function modified by the spectral features of the material comprising the surface:

$$B_{\text{surf}}(\lambda, T) = \frac{2hc^2}{\lambda^5} \frac{1}{e^{\frac{hc}{\lambda k_B T}} - 1} \varepsilon(\lambda), \tag{2.1}$$

where h is Planck's constant, 6.626×10^{-34} J·s, c is the speed of light, 2.998×10^8 m/s, k_B is Boltzmann's constant, 1.38×10^{-23} m^2 · kg · s^{-2} · K^{-1}, $\varepsilon(\lambda)$ is the wavelength-dependent emissivity of the surface, and T is the temperature of the surface. Therefore, for an ideal isothermal surface, we can define emissivity as the ratio of emitted spectral radiance (W/m^2 · µm · sr) of the surface divided by the Planck blackbody (B_{bb}) radiance at the surface temperature (T):

$$\varepsilon(\lambda) = B_{\text{surf}}(\lambda, T) / B_{bb}(\lambda, T). \tag{2.2}$$

For isothermal surfaces in environments in which all of the reflected and emitted energy can be measured, reflectance and emissivity are related through Kirchhoff's law: $\varepsilon(\lambda) = 1 - R(\lambda)$, where $\varepsilon(\lambda)$ and $R(\lambda)$ are the wavelength-dependent emissivity and reflectance of a material, respectively (Christensen & Harrison, 1993; Hapke, 1993; Salisbury et al., 1994). This relationship also holds for directional emissivity measurements if surface emission/reflection is isotropic.

Anisothermal surfaces are common on airless bodies, where µm- to cm-scale surface roughness leads to shadowing, illumination differences, and variability of side-welling radiance, and on surfaces that have substantial fractions of both rocks and regolith. In these cases, emitted radiance cannot be divided by a single temperature blackbody curve. To calculate a true emissivity for an anisothermal surface, the temperature distribution (with depth and laterally) must be known. If the temperature distribution is not known, it is more proper to call the retrieved quantity emission or apparent emissivity (see Chapter 15 as well).

2.2.2 Calculation of n and k for Use in Radiative Transfer Models

The fundamental inputs into radiative transfer models are n and k, the real and imaginary components of the complex index of refraction, $\tilde{n} = n + ik$. At visible to near-IR (VNIR) wavelengths, where k is small in many geologic materials, optical constants can be derived using several methods. Traditionally, transmission measurements of thin films or crystal thin sections were used to calculate the wavelength-dependent imaginary index of refraction (e.g., Swanepoel et al., 1983). Reflectance measurements of single crystals at near-normal incidence, followed by Kramers–Kronig analysis, can yield both components of the complex refractive index (e.g., Huffman & Stapp, 1971). More recently, inverse models based on the Hapke (1981) or Shkuratov et al. (1999) radiative transfer models (described in detail in Sections 2.3.2 and 2.3.3) have been used to derive refractive indices from reflectance spectra of mineral particulates (e.g., Roush et al., 1991, 2007; Lucey, 1998; Denevi et al., 2007; Trang et al., 2013; Sklute et al., 2015; Robertson et al., 2016).

At TIR wavelengths, where k is large, the complex refractive index is typically derived from oriented, polarized reflectance spectra of highly polished mineral surfaces using the Lorentz–Lorenz coupled harmonic oscillator model (Spitzer & Kleinmann, 1961), also referred to as dispersion theory (see Chapter 3, Section 3.1.2 in this volume). The number of principal complex refractive indices that are required to fully model the spectral anisotropy of a mineral varies with crystal symmetry. For isometric minerals, a single set of n and k values is required. For minerals in the hexagonal, tetragonal, and trigonal crystal systems, two sets are required – one parallel to the extraordinary ray (E-ray) and one parallel to the ordinary ray (O-ray). For orthorhombic and lower symmetry minerals, three sets of optical constants are required. For example, in the orthorhombic case, polarized measurements parallel to the a, b, and c crystallographic axes are required to model the three principal indices of refraction. For each of these axial directions, the principal indices of refraction can be described as

$$n^2 - k^2 = \varepsilon_0 + \sum_j 4\pi \rho_j \nu_j^2 \frac{\nu_j^2 - \nu^2}{\left(\nu_j^2 - \nu^2\right)^2 + \gamma_j^2 \nu_j^2 \nu^2}, \quad (2.3a)$$

$$nk = \sum_j 2\pi \rho_j \nu_j^2 \frac{\gamma_j \nu \nu_j}{\left(\nu_j^2 - \nu^2\right)^2 + \gamma_j^2 \nu_j^2 \nu^2}, \quad (2.3b)$$

where the summation is over j oscillators; ρ_j, γ_j, and ν_j are the oscillator strength, damping coefficient, and frequency, respectively; and ε_0 is the short-wavelength dielectric constant (Spitzer & Kleinman, 1961).

Reflectance for normal incidence and large (compared to the wavelength) mineral grains can be described by the Fresnel equation as

$$R = \frac{(n-1)^2 + k^2}{(n+1)^2 + k^2}. \quad (2.4)$$

Laboratory specular reflectance spectra with incidence and emergence angles of less than ~15° can be well modeled using this simple relationship. For larger incidence and

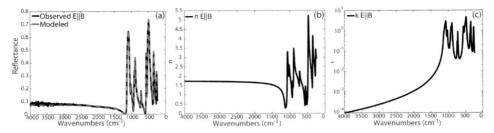

Figure 2.1 Measured and modeled mid-IR reflectance spectra (a), modeled real index of refraction (b), and modeled imaginary index of refraction (c), of the E ∥ b ray of enstatite.

emergence angles, reflectance is described by a much more complicated set of equations (Born & Wolf, 1980). An example of measured and modeled reflectance spectra and derived n and k values for the E∥b (electric vector parallel to the b-axis) orientation of the orthopyroxene mineral enstatite is shown in Figure 2.1.

For the lower symmetry cases of monoclinic and triclinic minerals, both the measurements and the models are more complicated. For monoclinic minerals, the oscillators are not parallel to the crystallographic axes, with the exception of the b-axis (Belousov & Pavinich, 1978; Arnold et al., 2014). In this case, reflectance measurements can be made parallel to the b-axis and modeled using the method of Spitzer and Kleinmann (1961). In the monoclinic a–c plane, measurements must be conducted at a minimum of three different angles (Ω) with respect to either the a- or c-axes, and additional oscillator parameters, φ and θ, which are the angle with respect to the a-axis and the angle between the a-axis and the measurement x-axis, respectively, are required to model the refractive indices. Mayerhöfer and Popp (2007) and Arnold et al. (2014) describe a general theory using the Berreman 4 × 4 matrix (Berreman, 1972) that can be applied to spectra acquired at any incidence angle. A similar formulation of dispersion theory has been reported for the triclinic mineral system (Emslie & Aronson, 1983; Mayerhöfer & Popp, 2007), although only one mineral, chalcanthite ($CuSO_4 \cdot 5(H_2O)$), has been modeled using that method (Aronson et al., 1985; Höfer et al., 2013).

Details on the measurements required for calculating optical constants as well as applications of these models can be found in Spitzer and Kleinman (1961), Aronson et al. (1983, 1985), Aronson (1986), Roush et al. (1991), Long et al. (1993), Wenrich and Christensen (1996), Lane et al. (1999), Mayerhöfer and Popp (2007), Glotch et al. (2007), Glotch and Rossman (2009), and Arnold et al. (2014), among many others.

2.3 Reflectance Theory

Two processes are at work as photons engage with a particulate or solid planetary surface: scattering and absorption. The interplay between scattering and absorption, governed by the optical characteristics of the materials, gives rise to the measured electromagnetic radiation

(EMR), and the variance in this reflected EMR as a function of wavelength gives rise to the characteristic absorptions and reflections that are frequently diagnostic of mineralogy. Photons that are reflected off mineral grains without passage through materials are not diagnostic of composition, but are affected by the texture, grain size, and porosity of the surface. Photons that are reflected from grain surfaces or refracted through a particle are said to be scattered.

Several processes lead to the absorption of photons transmitted into minerals. These processes are a foundation of the chemistry and structure of minerals. The specific processes of absorption and their utility as diagnostic indicators of mineralogy are covered in Chapters 1, 3, 4, and 5 in this volume.

2.3.1 Absorption, Reflection, and Scattering

Application of the theory of the radiative transfer models (RTMs) of Hapke (Hapke, 1981, 2012) and Shkuratov et al. (1999) to laboratory measurements and planetary surfaces has been examined in detail (e.g., Mustard & Pieters, 1987; Lucey, 1998; Poulet & Erard, 2004; Li & Li, 2011; Li & Milliken, 2015; Liu et al., 2016; Robertson et al., 2016). The reader is referred to these works for detailed descriptions of how the RTMs are implemented, their parameterization, and how they are solved using inverse methods. Briefly stated, RTMs are typically used to describe how light intensity changes as it enters, interacts with, and exits a specific medium. Hapke's (1993) radiative transfer (RT) theory is the most widely used foundation for nonlinear mixture models. Another geometric optics-based RTM described by Shkuratov et al. (1999) has been shown to be a viable alternative (Poulet & Erard, 2004).

A fundamental difference between the models is that Hapke's model is designed for intimate mixtures of particulates while the Shkuratov model treats particles as one-dimensional layers. The Shkuratov model is computationally faster, which is favorable for spectral unmixing of large datasets or image cubes acquired by instruments such as the Moon Mineralogy Mapper (M^3), the Compact Reconnaissance Imaging Spectrometer for Mars (CRISM), and the Observatoire pour la Minéralogie, l'Eau, les Glaces et l'Activité (OMEGA), but it does not explicitly account for the viewing geometry dependence of reflectance that is present in such image cubes (Poulet et al., 2009). In contrast, although the Hapke model may yield better estimates of modal mineralogy in some cases (e.g., Li & Milliken, 2015), it is at the expense of increased computation time. Both models are discussed in more detail in Sections 2.3.2 and 2.3.4.

2.3.2 Hapke Model

In Hapke's formulation of RT for particulate surfaces, we will be using the formulations for bidirectional reflectance, which is modeled as a function of single-scattering albedo (ω), observation geometry (i, e, g), back scattering function ($B(g, \phi)$, where ϕ is the filling factor), phase function ($p(g)$), and multiple scattering function (H, for both down-welling and up-welling radiance).

Single-scattering albedo (ω) is defined as the scattering efficiency divided by the extinction efficiency. Note that ω and extinction efficiency are properties of a single particle. When the particles are larger than the wavelength of light, the extinction efficiency is 1 and ω = scattering efficiency. The scattering efficiency of an equant particle is given by

$$\omega = S_e + (1 - S_e)\frac{(1 - S_i)\theta}{1 - S_i\theta}, \tag{2.5}$$

where S_e and S_i are the average Fresnel reflection coefficients for externally and internally incident light and θ is the internal-transmission factor. These terms are defined by

$$S_e = 0.0587 + \frac{(n-1)^2 + k^2}{(n+1)^2 + k^2} + \frac{(n-1)^2 + k^2}{(n+1)^2 + k^2}, \tag{2.6}$$

$$S_i = 1 - \frac{4}{n(n+1)^2}, \tag{2.7}$$

$$\theta = \frac{r_i + e^{-\sqrt{\alpha(\alpha+s)}\langle D \rangle}}{1 + r_i e^{-\sqrt{\alpha(\alpha+s)}\langle D \rangle}}, \tag{2.8}$$

where r_i is internal diffuse reflectance inside the particle, s is the near-surface internal scattering coefficient, $\langle D \rangle$ is the effective path length traveled by light through the particle, and α is the absorption coefficient defined as $\alpha = 4\pi k/\lambda$.

Spectral mixture analysis is based on the principle that the spectrum of a surface mixture composed of particulate materials is a systematic combination of the spectra of the individual components. This topic is covered for TIR measurements in Chapter 15. This combination will be linear if the components are arranged in physically distinct patches and photons interact with only one component (checkerboard mixing). However, when the materials are intimately mixed the photons interact with more than one constituent, resulting in a nonlinear convolution of the optical properties (see Mustard & Sunshine, 1999, and Keshava & Mustard, 2002, for a review). Unlike the TIR wavelength region in which the absorption coefficient is large, at VNIR wavelengths small absorption coefficients commonly result in the light interacting with multiple particles of multiple compositions along its path length before scattering back to the detector. Because assumptions of linear mixing within intimate mixtures can result in errors in abundance of up to 30% (Mustard et al., 1998), we will focus our attention on the Hapke and Shkuratov nonlinear mixture models based on RT theory.

The Hapke model (1993) can be used to compute the reflectance of a surface at any geometry. Hapke defines a range of equations for different types of reflectance, depending on the exact form (e.g., geometry) of the measurements. The radiance coefficient (brightness relative to the brightness of a Lambertian surface identically illuminated) is commonly measured in laboratory and field settings. This is the form of reflectance that we will use, and when the term reflectance is used it is actually referring to radiance coefficient. This is given in Eq. (2.9), where r is the reflectance at a given wavelength, μ and μ_o are the cosine of the emergence (e) and incidence (i) angles of the radiation, g is the phase angle, ω is the single-scattering albedo at the given wavelength, $B(g)$ is the backscattering function, $p(g)$ is the single-particle phase function, and $H(x)$ are the multiple-scattering functions of Chandrasekhar (1960).

$$r(i,e,g) = \frac{\omega}{4\pi}\frac{\mu_o}{\mu_o + \mu}\{[1 + B(g)]p(g) + H(\mu_o)H(\mu) - 1\}. \tag{2.9}$$

The H-function can be approximated using Eq. (2.10). This approximation is adequate for simplified mixing approaches but should be avoided for detailed photometric calculations (Hiroi, 1994). It is predicted that mixing systematics for intimate mixtures will be linear if reflectance is converted to single-scattering albedo; therefore this variable is the most important one when considering Eq. (2.9). The average ω of a mixture is a linear combination of the ω of each of the endmembers in the mixture weighted by their relative geometric cross section (RGXS; Eq. (2.11), where M_j, ρ_i, and d_i are the mass fraction, density, and particle diameter of the ith component in the mixture respectively). Thus, the endmember spectra combine linearly according to the projected area cross sections of the components in the mixture. The RGXS in Eq. (2.11) is computed for each mineral endmember in the mixture.

$$H(\mu) = \frac{(1 + 2\mu)}{(1 + 2\mu\sqrt{1-\omega})}, \tag{2.10}$$

$$\omega = \frac{\sum \frac{M_i}{\rho_i d_i}\omega_i}{\sum \frac{M_i}{\rho_i d_i}}. \tag{2.11}$$

Equation (2.9) can be used to calculate ω from reflectance spectra and thus linearize the mixing systematics. Equation (2.9) can be simplified to make the calculation more tractable.

Reflectance data acquired with a nadir viewing geometry and an incidence angle between 10° and 40° from normal follow a Lambertian approximation of surface scattering (Mustard & Pieters, 1989). In this case, the backscatter function $B(g)$ becomes 0, the single particle phase function, $p(g)$ equals 1, Eq. (2.9) is greatly simplified, and the conversion of reflectance to ω or ω to reflectance becomes trivial.

The conceptual approach to deriving mineral abundance information from spectra using these equations is straightforward. First, a plausible set of endmembers is defined for the problem under investigation. Second, all data and endmember spectra are converted to single-scattering albedo spectra using Eq. (2.9). Third, the endmembers are fit to the observation using least squares or other minimization techniques to solve Eq. (2.11), resulting in abundance estimates in terms of RGXS for each component. Knowledge of all endmembers and their particle sizes can yield abundance estimates with an accuracy of ± 5% (e.g., Mustard & Pieters, 1989). This approach has been applied with some success to remotely sensed data (e.g., Mustard & Pieters, 1987; Mustard et al., 1998; Clark et al., 2004;

Liu et al., 2016). While the validity of the model itself has been tested in a few carefully designed experimental studies (e.g., Li & Milliken, 2015; Robertson et al., 2016), it has not been rigorously tested for the range of mineral suites and conditions expected for planetary surfaces. Under these circumstances significant yet variable uncertainties may be introduced by unknown particle size and/or the presence of spectrally featureless minerals.

This approach also requires that the endmember reflectance spectra be known. In cases in which the mineral constituents may be known but their particle size or other properties affecting reflectance are not, it is advantageous to compute ω from the optical constants n and k. This approach is fully described by Lawrence and Lucey (2007). Briefly stated, Hapke (1993) derived equations to calculate ω from optical constants given by Eqs. 2.5 to 2.9. With these equations plus others as presented by Hapke (1993, 2001), and demonstrated by Lucey (1998) and Lawrence and Lucey (2007) to be reasonable, it is possible to compute reflectance from optical constants and vice versa with the Hapke RT formulation.

2.3.3 Shkuratov Model

The Shkuratov model (Eq. 2.12) for estimating modal mineralogy involves probabilistic treatment of forward and backward scattering through particles, modeled as a one-dimensional stacking of slabs (Shkuratov et al., 1999). Like the full calculations in the Hapke model, it relies on having wavelength-dependent endmember optical constants, n and k. The Shkuratov model employed here, adapted from Poulet and Erard (2004), uses downhill simplex methods to select endmember parameters, minimizing errors between measured and modeled spectra. This has been used recently for the computation of mineral abundance from OMEGA spectra of Mars (Poulet et al., 2009). However, as for the Hapke model, the performance of the model with mixtures involving hydrated materials or natural samples has not been validated.

The Shkuratov model is presented as

$$R(n,k,S,q,\lambda) = \frac{1 + \rho_b^2 - \rho_f^2}{2\rho_b} - \sqrt{(\frac{1 + \rho_b^2 - \rho_f^2}{2\rho_b})^2 - 1}, \qquad (2.12)$$

where R is reflectance, n and k are the real and imagery parts of the refractive index, S is the average optical path length, q denotes the volume filled by particles in an intersecting plane or a line, λ is the wavelength, and ρ_b and ρ_f are the one-dimensional indicatrixes for a layer of medium, with ρ_b describing the backscattering light and ρ_f the forward scattering light. All angle dependences of reflectance are ignored in the Shkuratov model, implying that the effect of surface roughness on reflectance cannot be taken into account (Shkuratov et al., 1999; Poulet et al., 2009; Poulet & Erard, 2004).

2.3.4 Similarities and Differences between the Hapke and Shkuratov Theories

Reflectance spectra as simulated by Hapke's model are dependent on the incidence (i), emergence (e), and phase (g). Conversely, the Shkuratov model has fixed normal incidence and emergence angles, and thus g is equal to 0. The scattering functions for this model are split into forward and backward scattering hemispheres. Quantitative inversion of the Hapke and Shkuratov models has been used with some success in laboratory investigations in the derivation of optical constants (e.g., Li & Milliken, 2015; Sklute et al., 2015). These investigations require that the particle size of the materials be well known; for the Shkuratov model, porosity must also be estimated.

Optical constants not available through previously published efforts can either be obtained through transmission spectroscopy or derived using spectra from multiple particle size separates of minerals (Roush et al., 2007). Given a suite of spectra of known, narrow particle size separates of an endmember and an estimate of the real index of refraction, it is possible to invert the Shkuratov model iteratively to determine the imaginary index of refraction. This approach is similar to that employed by Lucey (1998) using the Hapke (1993) model. However, as noted by Sklute et al. (2015), the refractive index values calculated from Shkuratov's model are dependent on particle size. In reality, these values should be independent of particle size.

Li and Milliken (2015) performed a detailed comparison of the Hapke and Shkuratov models for the derivation of optical constants of pyroxenes for their analysis of eucrite and diogenite meteorites. The model performances were almost identical in their ability to model laboratory reflectances from the derived values of n and k, differing in average absolute residual reflectance on the order of 0.003%. Li and Milliken (2015) hypothesize the differences may be due to how k is determined in the models and/or how scattering is accommodated.

For abundance determination, Li and Milliken (2015) showed that the Hapke and Shkuratov models also performed with a high degree of accuracy in estimating mineral abundances of laboratory measurements of controlled mixtures. Their calculated abundances were with within 5% of the known weight percentages for mixtures of olivine, low- and high-Ca pyroxene, and plagioclase. When the models were applied to spectra of eucrite and diogenite meteorites to estimate modal abundance and grain size, the models had a similar level of performance, with spectroscopic estimates within 5–10% of the laboratory-determined modal abundances and grain size. The models have different performance characteristics depending on the situation and circumstances, but in general are adequate when detailed sample information is available. The advantage of the Shkuratov model is that it allows simultaneous calculation of endmember abundance and particle size.

2.4 Emittance Theory

2.4.1 Introduction

Thermal infrared (2.5–100 μm; 100–4000 cm^{-1}, often used interchangeably with mid-infrared [mid-IR]) spectroscopy has been an important analytical method in Earth and planetary science since the early 1970s. While transmission and absorption measurements have been primary analytical tools for mineral analysis in the laboratory, the development of emission spectroscopy has enabled a revolution in remote mineralogical analysis. Our study of the composition and physical properties of solar system objects has benefitted tremendously from the use of thermal emission measurements. Both broadband and spectral measurements in the TIR region have led to an improved understanding of the Moon, Mars, asteroids, and even Saturn's atmosphere and icy satellites. Because TIR measurements are sensitive to many mineral, mineraloid, and amorphous phases, including non-Fe–bearing species, they provide unique information that is highly complementary to VNIR reflectance measurements.

Regardless of the nature of the observed surface, the major diagnostic features observable in emissivity spectra are the Christiansen feature, transparency features, and Reststrahlen bands. Reststrahlen bands are surface scattering phenomena that occur in regions where the imaginary index of refraction, k, is large, and light cannot propagate through a sample. For silicates, these bands occur primarily in the 8–25-μm (1250–400 cm^{-1}) range and appear as emissivity minima (Salisbury & Wald, 1992). The Christiansen feature (CF), on the other hand, is expressed as an emissivity maximum. It occurs in a region where the imaginary index, k, is at a minimum, and the real index, n, equals that of the surrounding medium. It occurs at wavelengths just shortward (higher frequency) of the fundamental Reststrahlen bands, and has been recognized as an indicator of silicate composition for some time (Conel, 1969; Salisbury & Walter, 1989; Greenhagen et al., 2010), with longer wavelength CF positions indicative of more mafic compositions and shorter wavelength CF positions indicative of more felsic compositions. Like Reststrahlen bands, transparency features (see Chapter 3, Section 3.2.6 and Figure 3.2b) are manifested as emissivity minima in TIR spectra. However, they typically occur in spectra of finely particulate silicate materials in regions where k is small. Between ~8 and 50 μm, transparency features will occur between Reststrahlen bands while at wavelengths shortward of the CF, transparency is manifested as a steep, steady decline in emissivity with decreasing wavelength.

2.4.2 Effects of Particle Size, Packing, and Texture

In the laboratory, sample preparation can have a strong effect on measured TIR emission spectra, with particle size and packing (porosity) leading to large changes in the shapes and strengths of Reststrahlen bands, transparency features, and the Christiansen feature. Salisbury and Wald (1992), Lane (1999), and Cooper et al. (2002) provide detailed overviews of the changes in TIR spectra that occur as a result of particle size and porosity. In the context of remote sensing measurements, spectral interpretations must take into account the highly nonuniform nature of planetary regoliths and how variables other than composition can affect spectra (see also Chapter 15).

The primary noncompositional variations seen in TIR emission spectra occur as a function of particle size. In strongly absorbing regions (the Reststrahlen bands), the spectral contrast is reduced with decreasing particle size due to multiple surface scattering (Aronson et al., 1967; Mustard & Hays, 1997). As the particles become very fine,

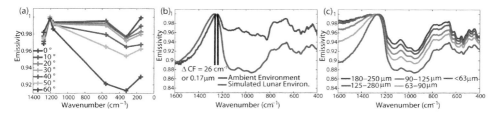

Figure 2.2 Effects of viewing angle, environment, and particle size on thermal emission spectra. (a) Increased incidence angle leads to pronounced changes to Diviner Lunar Radiometer spectra at long wavelengths. (Adapted from Shirley et al., 2015.) (b) The emissivity spectrum of anorthite in a simulated lunar environment and under ambient conditions. Under SLE conditions, spectral contrast is enhanced and the CF shifts by up to 0.2 μm to shorter wavelengths. (c) Under SLE conditions, particle size can cause substantial shifts in the CF. In general, spectral contrast increases with decreasing particle size. The opposite trend is observed under ambient conditions. (Adapted from Shirley & Glotch, 2019.)

volume scattering has an increased role in the portions of the Reststrahlen bands with slightly lower k, as the grains become optically thin at those wavelengths (Salisbury & Wald, 1992). The competing effects of multiple surface and volume scattering across the Reststrahlen band result in a nonuniform reduction of the Reststrahlen features, thereby affecting the spectral shape of the Reststrahlen features at very fine particle size, and not just the contrast. In weakly absorbing regions (where k is uniformly low), spectral contrast increases with decreasing particle size as a result of increased volume scattering. These "transparency" features are present in spectra for which the particle size is about equal to or smaller than the wavelength of light, as well as in weakly absorbing materials (e.g., halides) at any particle size.

An additional factor affecting the overall shape of TIR emission spectra is surface roughness. Roughness at the mm to cm scale results in local slopes – some of which are heated directly by sunlight, and some of which are shadowed – causing strong anisothermality in the detector field of view (Bandfield, 2009; Bandfield et al., 2015; Davidsson et al. 2015; Chapter 15, this volume). Figure 2.2a shows multispectral TIR data from the Diviner Lunar Radiometer Experiment on board the Lunar Reconnaissance Orbiter. These data demonstrate the effects of surface anisothermality on measured TIR spectra. As the solar incidence angle increases (resulting in increased exposure of both sunlit and shaded slopes compared to near-nadir measurements), a strong graybody slope is imposed on the spectrum. This effect, in conjunction with a Gaussian or other roughness model, can be used to determine the mm- to cm-scale surface roughness of planetary regoliths (Bandfield, 2009; Bandfield et al., 2015; Davidsson et al., 2015; Glotch et al., 2015).

2.4.3 Emission on Airless Bodies

TIR measurements acquired by the Lunar Reconnaissance Orbiter Diviner Lunar Radiometer, the OSIRIS-Rex Thermal Emission Spectrometer (OTES), and the future arrival of the Bepi-Colombo Mercury Radiometer and Thermal Infrared Spectrometer (MERTIS) instrument to Mercury have motivated renewed interest in understanding the effects of the unique thermal environments on airless planetary surfaces on disk-resolved TIR datasets. Early remote sensing (Murcray et al., 1970) and laboratory work (Logan & Hunt, 1970; Logan et al., 1973) showed that the Christiansen feature is the dominant spectral feature under vacuum conditions, with enhanced spectral contrast and its center position shifted to shorter wavelengths (often by up to 0.2 μm) compared to spectral measurements made under ambient terrestrial conditions. More recent laboratory measurements (Salisbury & Walter, 1989; Henderson et al., 1996; Thomson et al., 2012; Donaldson Hanna et al., 2012a,b, 2014, 2016; Shirley & Glotch, 2019) have confirmed these trends while also substantially increasing the range of compositions that have been measured under simulated airless body conditions.

Figure 2.2b shows spectra of forsteritic olivine acquired under both terrestrial ambient conditions and simulated lunar environment (SLE) conditions. The major differences between SLE and ambient spectra include (1) enhancement and shifting of the CF to shorter wavelengths, (2) an overall increase in spectral contrast, and (3) a reduction in the strength of the transparency feature (reduced emissivity feature at ~750 cm^{-1}). When several size fractions of the same mineral are measured under SLE conditions, additional trends emerge. In general, the CF shifts to longer wavelengths with decreasing particle size. In addition, overall spectral contrast, including the Reststrahlen bands, increases with decreasing particle size – the opposite of what occurs under terrestrial ambient conditions (Figure 2.2c).

The primary cause of the foregoing spectral changes is the unique thermal environment present on airless bodies. On Earth and Mars, the primary mechanism for heat transfer through the regolith is conduction through interstitial gas. In the absence of an atmosphere, heat is lost to cold space through radiation and grain-to-grain contact – both of which are inefficient processes. As a result, a strong thermal gradient (~100 K in the upper 100s of μm) occurs in the regolith, with a cool surface radiating directly to 3 K space, and a warmer interior heated by direct insolation. This thermal gradient has been modeled by several researchers (Henderson and Jakosky, 1994, 1997; Millán et al., 2011) who coupled heat transfer and radiative transfer theories and were able to reproduce most of the features seen in SLE spectra.

2.4.4 Emission Theories

A major goal of the analysis of laboratory and remotely acquired TIR spectra is the quantitative determination of bulk mineralogy. For samples or planetary surfaces composed of particulates that are large compared to the wavelength of light, simple linear mixing models can be used. When particle sizes approach the wavelength of light, linear mixing can no longer be assumed, and complex light scattering models must be used to determine mineralogy. In the following, we discuss the theories relevant to each of these scenarios and provide comparisons of several light scattering models.

2.4.4.1 Linear Mixing in the Geometric Optics Case

In the geometric optics case ($D \gg \lambda$), spectral mixing in the TIR is linear, and unmixing models have been shown to accurately reproduce sample mineralogy of igneous and metamorphic rocks (Feely & Christensen, 1999). This is thoroughly covered in Chapter 15.

2.4.4.2 Hapke's Models of Thermal Emission

Although it is not as heavily used as the Hapke scattering model for reflectance, Hapke (1993) devised a model for thermal emission that couples the equations of radiative transfer for visible and thermal light with the heat transfer equation. The model ignores reflected radiation and assumes isotropic scattering and emission. Therefore, emissivity depends only on the single-scattering albedo, ω, and the cosine of the emission angle, μ. The formulation is fundamentally the same as the scattering model for ω, S_e, S_i, and θ, which are defined in Eqs. (2.5) through (2.8).

The scattering model for reflectance, however, is in the regime where k is typically $\ll 1$. This is problematic for the TIR, where k can often be greater than unity. Caution must be exercised when utilizing this model to calculate TIR single-scattering albedos. In some cases, other methods, including Mie theory, or T-matrix calculations, discussed in more detail in the text that follows, may be better to relate optical constants to single-scattering albedo.

Hapke (1996) solves the coupled set of radiative transfer and heat transfer equations to calculate emissivity for two cases: hemispherical emissivity of an infinitely thick regolith heated from above by visible radiation and a finite, particulate surface that scatters anisotropically, heated from below at a constant temperature and radiating to space. In the first case, the absorbed visible radiation is in equilibrium with the emitted power, and hemispherical emissivity is estimated as

$$\varepsilon_h(\omega) = 2\gamma H_1(\omega), \qquad (2.13)$$

where $\gamma = \sqrt{1-\omega}$, and $H_1(\omega)$ is the first moment of Chandrasekhar's (1960) H-function. H_1 is approximated by

$$H_1(\omega) = \int_0^1 H(\omega, x) x \, dx \cong \frac{1}{1+\gamma}\left(1 + \frac{1}{6}\frac{1-\gamma}{1+\gamma}\right). \qquad (2.14)$$

For the second case, in which a medium is heated from below, emissivity is defined as

$$\varepsilon_h = \frac{2\gamma_T}{(\zeta_T + \gamma_T)}, \qquad (2.15)$$

where γ is the same as previously, $\zeta = \sqrt{1-\beta\omega}$ and β is the hemispherical asymmetry parameter typically used in the two-stream approximation (Hapke, 1981, 1993). For a particle subjected to uniform radiance incident upon one hemisphere, β is defined such that the fraction of radiation scattered back into the original hemisphere is $\frac{1-\beta}{2}$ and the fraction scattered in the forward direction is $\frac{1+\beta}{2}$.

Hapke's approximations of the radiative transfer equation contain an assumption that scattering is occurring in the geometric optics regime – that is, the size of the scattering particle is much larger than the wavelength of light, and diffraction does not occur. For many planetary surfaces, this assumption is not strictly (or even remotely) true. However, Mie theory or other methods, such as the T-matrix approach, which explicitly account for scattering by small particles, can be paired with Hapke's and other formulations of radiative transfer theory to achieve more realistic solutions. These hybrid models are discussed in more detail in the following text.

2.4.4.3 Hybrid Models Utilizing Mie/T-Matrix and Radiative Transfer Theory

The fundamental goal of several "hybrid" models that have been developed in the past is to combine the advantages of microphysical models that include diffraction, such as Mie theory, with RT models, such as Hakpe, or discrete ordinates radiative transfer approaches (e.g., Stamnes et al., 1988), that specifically include the effects of multiple scattering. In this way, the emissivity of compact regoliths, within any of the geometric optics, Mie, or Rayleigh scattering regimes, may be calculated. Among the first efforts to model the emissivity of compact regoliths was the cloudy atmosphere model of Conel (1969). This model uses Mie theory (Mie, 1908; van de Hulst, 1957) to calculate scattering properties of particulates that are input into a radiative transfer scattering model. Bohren and Huffman (2007) note that an individual who "has painstakingly followed the derivation (of Mie Theory) ... and thereby acquired virtue through suffering, may derive some comfort from the knowledge that it is relatively clear sailing from here on." That may be the case, but the use of Mie theory in a cloudy atmosphere model is a simplifying approximation that does not hold true for a close-packed regolith. For Mie theory to strictly hold, particles must be separated by at least three particle radii (van de Hulst, 1957). Nevertheless, this approach has paid dividends in several studies used to calculate reflectance of closely packed particulates. The outputs of the Mie calculation are the scattering (Q_{sca}) and total extinction (Q_{ext}) efficiencies, and

$$g = \langle \cos \theta \rangle = \frac{1}{2} \int_{-1}^{1} d(\cos \theta) p(\theta) \cos(\theta) \tag{2.16}$$

is the asymmetry parameter weighted by the phase function. The single-scattering albedo is related to the scattering and total extinction efficiencies as

$$\omega = \frac{Q_{sca}}{Q_{ext}}.$$

The cloudy atmosphere model starts with the equation of radiative transfer for a plane-parallel layer (Chandrasekhar, 1960):

$$\cos\theta \frac{dI}{d\tau} = I - \int pI \frac{d\Omega}{4\pi}, \quad (2.17)$$

where I is the intensity at an arbitrary wavelength, τ is the optical depth, p is the scattering phase function, θ is the angle of incident radiation with the surface normal, and the integration occurs over all solid angles Ω. Conel (1969), like Hapke after him, makes two simplifying assumptions to solve the radiative transfer equation: the phase function (p) is described as a Legendre polynomial, and the propagation of light is described by the two-stream approximation, in which the integration of the radiative transfer equation occurs over forward- and back-scattering hemispheres. After solving the radiative transfer equation in this way, Conel (1969) describes emissivity as

$$\varepsilon = \frac{2}{u+1}, \text{ where } u^2 = \frac{\left(1 - \omega \frac{\langle \cos\theta \rangle}{3}\right)}{(1-\omega)}. \quad (2.18)$$

Moersch and Christensen (1995) compared the results of the Mie/Conel model with the Hapke (1981) reflectance model converted to emissivity via Kirchhoff's law, the Hapke (1993) emissivity model, and a Mie/Hapke hybrid model. For the Mie/Hapke hybrid model, Moersch and Christensen (1995) calculate the single-scattering albedo in the same manner as Conel (1969) using Mie theory, and then apply a diffraction subtraction correction (Wald, 1994; Wald & Salisbury, 1995) to account for close packing:

$$\omega = 2\omega_{\text{Mie}} - 1. \quad (2.19)$$

This correction, or the alternative static structure factor correction (Mishchenko, 1994), is needed because in the geometric optics regime, forward-scattered light cannot be distinguished from unscattered light, and the total extinction efficiency approaches a value of 2. For closely packed particles, Q_{ext} cannot be greater than unity. Moersch and Christensen (1995) note that the assumptions inherent in the diffraction subtraction correction break down at small particle sizes. For particle sizes smaller than ~7% of the wavelength of light, they found that the correction produced negative single-scattering albedos.

Mustard and Hays (1997) also tested the utility of a Mie–Hapke hybrid model. They used Mie theory to calculate the single-scattering albedo and asymmetry parameter for several size fractions of quartz and olivine and then used the Hapke (1981) bidirectional reflectance model to calculate reflectance. They found that the model worked reasonably well for olivine, reproducing changes to the Christiansen feature, Reststrahlen bands, and transparency features that were observed in laboratory spectra. Their modeled spectra for quartz were generally similar to those of Moersch and Christensen (1995), and did not accurately reproduce the band strengths and changes in Reststrahlen bands seen with changes in particle size. Mustard and Hays (1997) note several possible reasons for this discrepancy, but the most likely is that either Mie theory or Hapke's reflectance model breaks down for large n and k values. In their work, the maximum absolute values of the optical constants for quartz were about two times those of olivine.

Pitman et al. (2005) utilized a hybrid model that, like previous hybrid models, used Mie theory to calculate the single-scattering properties of quartz grains. The scattering parameters derived from Mie theory were then input into a numerically exact discrete ordinates multiple scattering radiative transfer (DISORT) model (Stamnes et al., 1988). A major result of this work is a clear demonstration that both the diffraction subtraction correction (Wald, 1994; Wald & Salisbury, 1995) and the static structure factor correction (Mishchenko, 1994) reduce the asymmetry parameter and the single-scattering albedo more than is required to match laboratory spectra. These results suggest that a better method to address close packing of grains is required to accurately model emissivity spectra.

The T-matrix approach is a method to model close packing of spheres that is more representative of planetary regolith than Mie theory. Like Mie theory, the T-matrix method can calculate the scattering properties, including Q_{sca}, Q_{ext}, and g, that are required inputs into radiative transfer models. Essentially an extension of Mie theory developed for multi-sphere systems, the multiple sphere T-matrix model (MSTM), was formulated and described by Mackowski (1994), Mackowski and Mishchenko (1996), and Mackowski and Mishchenko (2011). A major strength of this model is the ability to control the size, position, composition (via assigned optical constants), and packing density of clusters of particles. However, as with Mie theory, particles are represented as perfect spheres, which may represent a major source of error in computed spectra (Pitmann et al., 2005).

Glotch et al. (2016) and Hardgrove et al. (2016) used a hybrid MSTM–Hapke model to investigate the TIR scattering and spectral properties of halite–silicate mixtures, and solid microcrystalline quartz, respectively. The models are set up analogously to the Mie–Hapke hybrid models described by Moersch and Christensen (1995) and Mustard and Hays (1997), but the scattering properties were derived from the MSTM model and emissivity was calculated from the Hapke (1996) thermal emission model. In both cases, the model yielded spectra that were comparable to laboratory or remote sensing emissivity spectra. In the case of halite–silicate mixtures (Glotch et al., 2016), the model accurately reproduced the shift in Christiansen feature position observed with increasing salt content as well as the overall changes in spectral shape. The results were used, in combination with measured laboratory spectra and infrared spectra from the Thermal Emission Imaging System (THEMIS), to constrain the salt content of chloride-rich deposits on Mars (Osterloo et al., 2008, 2010; Glotch et al., 2010) to <20 wt.%. Hardgrove et al. (2016) demonstrated that the model, through a carefully constructed cluster of particles, can reproduce the observed effects of surface roughness on the order of the wavelength of light. They showed that modeled spectra of a rough quartz surface display the same changes in spectral shape and contrast in the Reststrahlen band region that are seen in natural microcrystalline quartz samples.

2.4.5 Comparison of Theories

Moersch and Christensen (1995) provided a direct comparison of four models used to calculate the emissivity spectra of quartz: Hapke's (1981) theory of reflectance converted

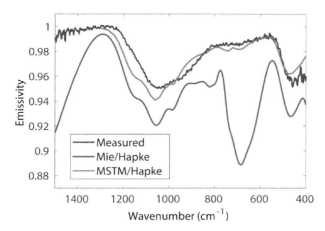

Figure 2.3 Comparisons of Mie–Hapke96 and MSTM–Hapke96 models for 19-μm separates of amorphous silica glass spheres. (Adapted from Arnold, 2014.)

to emissivity using Kirchhoff's law, Hapke's (1993) emission theory, Conel's cloudy atmosphere model, and a Mie–Hapke hybrid model. They found that, in general, the Mie–Hapke approach did the best job of replicating changes to spectra as a function of grain size.

Neither Glotch et al. (2016) nor Hardgrove et al. (2016) provided a direct comparison of the MSTM–Hapke hybrid model with the Mie–Hapke or Mie–Conel hybrid models. Arnold (2014) and Ito et al. (2017) conducted direct comparisons of Mie–Conel, Mie–Hapke, MSTM–Conel, and MSTM–Hapke hybrid models. Arnold (2014) collected micro-FTIR reflectance spectra of isotropic, monodisperse borosilicate glass spheres of four sizes (2, 19, 49, and 179 μm). This arrangement represents a best case scenario for testing the ability of various scattering models to accurately reproduce laboratory reflectance or emissivity spectra. After calculating the TIR optical constants using the model of Spitzer and Kleinman (1961), Arnold (2014) used either a Mie or MSTM model to determine Q_{sca} and Q_{ext} and the asymmetry parameter, calculated ω, and from that value, calculated emissivity spectra using the Hapke radiative transfer model. A comparison of the results is shown in Figure 2.3. In general, as would be expected, the MSTM–Hapke model provides a better match to measured emissivity spectra than the Mie–Hapke hybrid model. This is especially true for finer grain sizes, including the 2-μm size fraction, where the Mie–Hapke model greatly overestimates the spectral contrast, especially in regions of low k (near ~750 cm^{-1}). By comparison, for coarser particulates (the 179-μm, and to a lesser extent, the 49-μm size fractions), where multiple scattering is substantially reduced, both the Mie–Hapke and MSTM–Hapke models provide relatively good fits to the measured emissivity spectrum.

2.5 Summary

The theory of reflectance and emittance spectroscopy is very well established across the visible to mid-IR wavelength range. The VNIR wavelength range encompasses the regime where n is close to 1 and shows limited variation, and k is on the order of 10^{-3} where scattering dominates (0.4–≈5.0 μm). In the mid-IR region, where thermal IR processes are important, $k(\lambda)$ exhibits a wide range from large (>5.0) to small (<0.001) and $n(\lambda)$ can vary from <1 to as high as 8. Theory, nevertheless, has been shown through experimental and planetary observations to be very valuable for the modeling of reflectance and emission spectroscopy, and the application of models to invert observed spectroscopic observations to quantitative analyses of modal mineral abundance, grain size, and particle size.

Important challenges remain in the full application of quantitative models to the interpretation of laboratory, field, and spacecraft data. One of the key challenges is the lack of a uniform and well-documented library of optical constants. These values are fundamental to the application of theory to observations from first principles. This is in part due to the sheer diversity of geologic materials that need to be characterized and also in the verification, validation, and curation of the database. Another challenge is the performance and verification of models that encompass the transition from scattering processes to emission. This typically occurs where the optical constants are varying across many orders of magnitude, and while there has been success in the use of hybrid models, there is a need to have more well-designed laboratory experiments using a diverse suite of materials for model testing and validation. This problem is particularly acute when the particle size approaches the wavelength of light where the fundamental assumption of most common radiative transfer models break down.

References

Arnold J.A. (2014) *Refining mid-infrared emission spectroscopy as a tool for understanding planetary surface mineralogy through laboratory studies, computational models, and lunar remote sensing data.* PhD thesis, State University of New York at Stony Brook.

Arnold J.A., Glotch T.D., & Plonka A.M. (2014) Mid-infrared optical constants of clinopyroxene and orthoclase derived from oriented single-crystal reflectance spectra. *American Mineralogist*, **99**, 1942–1955.

Aronson J.R. (1986) Optical constants of monoclinic anisotropic crystals: Orthoclase. *Spectrochimica Acta A: Molecular Spectroscopy*, **42**, 187–190.

Aronson J.R., Emslie A.G., Allen R.V., & McLinden H.G. (1967) Studies of the middle- and far-infrared spectra of mineral surfaces for application in remote compositional mapping of the Moon and planets. *Journal of Geophysical Research*, **72**, 687–703.

Aronson J.R., Emslie A.G., Miseo E.V., Smith E.M., & Strong P.F. (1983) Optical constants of monoclinic anisotropic crystals: Gypsum. *Applied Optics*, **22**, 4093–4098.

Aronson J.R., Emslie A.G., & Strong P.F. (1985) Optical constants of triclinic anisotropic crystals: Blue vitriol. *Applied Optics*, **24**, 1200–1203.

Bandfield J.L. (2009) Effects of surface roughness and graybody emissivity on martian thermal infrared spectra. *Icarus*, **202**, 414–428.

Bandfield J.L., Hayne P.O., Williams J.-P., Greenhagen B.T., & Paige D.A. (2015) Lunar surface roughness derived from LRO Diviner Radiometer observations. *Icarus*, **248**, 357–372.

Belousov M.V. & Pavinich V.F. (1978) Infrared reflection spectra of monoclinic crystals. *Optics and Spectroscopy*, **45**, 771–774.

Berreman D.W. (1972) Optics in stratified and anisotropic media: 4×4-matrix formulation. *Journal of the Optical Society of America*, **62**, 502–510.

Bohren C.F. & Huffman D.R. (2007) *Absorption and scattering of light by small particles*. John Wiley & Sons, Hoboken, NJ.
Born M. & Wolf E. (1980) *Principles of optics*. Pergamon Press, Oxford.
Chandrasekhar S. (1960) *Radiative transfer*. Dover Publications, Mineola, NY.
Christensen P.R. & Harrison S.T. (1993) Thermal infrared emission spectroscopy of natural surfaces: Application to desert varnish coatings on rocks. *Journal of Geophysical Research*, **98**, 19,819–19,834.
Clark B.E., Bus S.J., Rivkin A.S., et al. (2004) E-type asteroid spectroscopy and compositional modeling. *Journal of Geophysical Research*, **109**, E02001, DOI:10.1029/2003JE002200.
Conel J.E. (1969) Infrared emissivities of silicates: Experimental results and a cloudy atmosphere model of spectral emission from condensed particulate mediums. *Journal of Geophysical Research*, **74**, 1614–1634.
Cooper C.D. & Mustard J.F. (2002) Spectroscopy of loose and cemented sulfate-bearing soils: Implications for duricrust on Mars. *Icarus*, **158**, 42–55.
Davidsson B.J.R., Rickman H., Bandfield J.L., et al. (2015) Interpretation of thermal emission. I. The effect of roughness for spatially resolved atmosphereless bodies. *Icarus*, **252**, 1–21.
Denevi B.W., Lucey P.G., Hochberg E.J., & Steutel D. (2007) Near-infrared optical constants of pyroxene as a function of iron and calcium content. *Journal of Geophysical Research*, **112**, E05009, DOI:10.1029/2006JE002802.
Donaldson Hanna K.L., Thomas I.R., Bowles N.E., et al. (2012a) Laboratory emissivity measurements of the plagioclase solid solution series under varying environmental conditions. *Journal of Geophysical Research*, **117**, E11004, DOI:10.1029/2012JE004184.
Donaldson Hanna K.L., Wyatt M.B., Thomas I.R., et al. (2012b) Thermal infrared emissivity measurements under a simulated lunar environment: Application to the Diviner Lunar Radiometer experiment. *Journal of Geophysical Research*, **117**, E00H05, DOI:10.1029/2011JE003862.
Donaldson Hanna K.L., Cheek L.C., Pieters C.M., et al. (2014) Global assessment of pure crystalline plagioclase across the Moon and implications for the evolution of the primary crust. *Journal of Geophysical Research*, **119**, 1516–1545.
Donaldson Hanna K.L., Greenhagen B.T., Patterson W.R., et al. (2017) Effects of varying environmental conditions on emissivity spectra of bulk lunar soils: Application to Diviner thermal infrared observations of the Moon. *Icarus*, **283**, 326–342.
Emslie A.G. & Aronson J.R. (1983) Determination of the complex dielectric tensor of triclinic crystals: Theory. *Journal of the Optical Society of America*, **73**, 916–919.
Feely K.C. & Christensen P.R. (1999) Quantitative compositional analysis using thermal emission spectroscopy: Application to igneous and metamorphic rocks. *Journal of Geophysical Research*, **104**, 24,195–24,210.
Glotch T.D. & Rossman G.R. (2009) Mid-infrared reflectance spectra and optical constants of six iron oxide/oxyhydroxide phases. *Icarus*, **204**, 663–671.
Glotch T., Rossman G.R., & Aharonson O. (2007) Mid-infrared (5–100 μm) reflectance spectra and optical constants of ten phyllosilicate minerals. *Icarus*, **192**, 605–622.
Glotch T.D., Bandfield J.L., Tornabene L.L., Jensen H.B., & Seelos F.P. (2010) Distribution and formation of chlorides and phyllosilicates in Terra Sirenum, Mars. *Geophysical Research Letters*, **37**, 1–5.
Glotch T.D., Bandfield J.L., Lucey P.G., et al. (2015) Formation of lunar swirls by magnetic field standoff of the solar wind. *Nature Communications*, **6**, 6189.
Glotch T.D., Bandfield J.L., Wolff M.J., Arnold J.A., & Che C. (2016) Constraints on the composition and particle size of chloride salt-bearing deposits on Mars. *Journal of Geophysical Research*, **121**, 454–471.
Greenhagen B.T., Lucey P.G., Wyatt M.B., et al. (2010) Global silicate mineralogy of the Moon from the Diviner Lunar Radiometer. *Science*, **329**, 1507.
Hapke B. (1981) Bidirectional reflectance spectroscopy, 1. Theory. *Journal of Geophysical Research*, **86**, 3039–3054.
Hapke B. (1993/2012) *Theory of reflectance and emittance spectroscopy*. Cambridge University Press, Cambridge.
Hapke B. (1996) A model of radiative and conductive energy transfer in planetary regoliths. *Journal of Geophysical Research*, **101**, 16817–16831.
Hardgrove C.J., Rogers A.D., Glotch T.D., & Arnold J.A. (2016) Thermal emission spectroscopy of microcrystalline sedimentary phases: Effects of natural surface roughness on spectral feature shape. *Journal of Geophysical Research*, **121**, 542–555.
Henderson B.G. & Jakosky B.M. (1994) Near-surface thermal gradients and their effects on mid-infrared emission spectra of planetary surfaces. *Journal of Geophysical Research*, **99**, 19063–19073.
Henderson B.G. & Jakosky B.M. (1997) Near-surface thermal gradients and mid-IR emission spectra: A new model including scattering and application to real data. *Journal of Geophysical Research*, **102**, 6567–6580.
Henderson B.G., Lucey P.G., & Jakosky B.M. (1996) New laboratory measurements of mid-IR emission spectra of simulated planetary surfaces. *Journal of Geophysical Research*, **101**, 14969–14975.

Hiroi T. (1994) Recalculation of the isotropic H-functions. *Icarus*, **109**(2), 313–317.
Höfer S., Werling S., & Beyerer J. (2013) Thermal pattern generation for infrared deflectometry. *AMA Conferences 2013 – Nürnberg Exhibition Centre, May 14–16, 2013 – SENSOR, OPTO and IRS²*, 785–790.
Huffman D.R. & Stapp J.L. (1971) Interstellar silicate extinction related to the 2200 Å band. *Nature Physical Science*, **229**, 45.
Ito G., Arnold J.A., & Glotch T.D. (2017) T-matrix and radiative transfer hybrid models for densely packed particulates at mid-infrared wavelengths. *Journal of Geophysical Research*, **122**, 822–838.
Keshava N. & Mustard J.F. (2002) Spectral unmixing. *IEEE Signal Processing Magazine*, **19**(1), 44–57, DOI:10.1109/79.974727.
Lane M.D. (1999) Midinfrared optical constants of calcite and their relationship to particle size effects in thermal emission spectra of granular calcite. *Journal of Geophysical Research*, **104**, 14099–14108.
Lawrence S.J. & Lucey P.G. (2007) Radiative transfer mixing models of meteoritic assemblages. *Journal of Geophysical Research*, **112**, E07005, DOI:10.1029/2006JE002765.
Li S. & Li L. (2011) Radiative transfer modeling for quantifying lunar surface minerals, particle size, and submicroscopic metallic Fe. *Journal of Geophysical Research*, **116**, E09001, DOI:10.1029/2011JE003837.
Li S. & Milliken R.E. (2015) Estimating the modal mineralogy of eucrite and diogenite meteorites using visible–near infrared reflectance spectroscopy. *Meteoritics and Planetary Science*, **50**, 1821–1850.
Liu Y., Glotch Timothy D., Scudder Noel A., et al. (2016) End-member identification and spectral mixture analysis of CRISM hyperspectral data: A case study on southwest Melas Chasma, Mars. *Journal of Geophysical Research*, **121**, 2004–2036.
Logan L.M. & Hunt G.R. (1970) Emission spectra of particulate silicates under simulated lunar conditions. *Journal of Geophysical Research*, **75**, 6539–6548.
Logan L.M., Hunt G.R., Salisbury J.W., & Balsamo S.R. (1973) Compositional implications of Christiansen frequency maximums for infrared remote sensing applications. *Journal of Geophysical Research*, **78**, 4983–5003.
Long L.L., Querry M.R., Bell R.J., & Alexander R.W. (1993) Optical properties of calcite and gypsum in crystalline and powdered form in the infrared and far-infrared. *Infrared Physics*, **34**, 191–201.
Lucey P.G. (1998) Model near-infrared optical constants of olivine and pyroxene as a function of iron content. *Journal of Geophysical Research*, **103**, 1703–1713.
Mackowski D.W. (1994) Calculation of total cross sections of multiple-sphere clusters. *Journal of the Optical Society of America A*, **11**, 2851–2861.
Mackowski D.W. & Mishchenko M.I. (1996) Calculation of the T matrix and the scattering matrix for ensembles of spheres. *Journal of the Optical Society of America A*, **13**, 2266–2278.
Mackowski D.W. & Mishchenko M.I. (2011) A multiple sphere T-matrix Fortran code for use on parallel computer clusters. *Journal of Quantitative Spectroscopy and Radiative Transfer*, **112**, 2182–2192.
Mayerhöfer T. & Popp J. (2007) Employing spectra of polycrystalline materials for the verification of optical constants obtained from corresponding low-symmetry single crystals. *Applied Optics*, **46**, 327–334.
Mie G. (1908) Beiträge zur Optik trüber Medien, speziell kolloidaler Metallösungen. *Annalen der Physik*, **330**, 377–445.
Millán L., Thomas I., & Bowles N. (2011) Lunar regolith thermal gradients and emission spectra: Modeling and validation. *Journal of Geophysical Research*, **116**, DOI: 10.1029/2011JE003874.
Mishchenko M.I. (1994) Asymmetry parameters of the phase function for densely packed scattering grains. *Journal of Quantitative Spectroscopy and Radiative Transfer*, **52**, 95–110.
Moersch J.E. & Christensen P.R. (1995) Thermal emission from particulate surfaces: A comparison of scattering models with measured spectra. *Journal of Geophysical Research*, **100**, 7465–7477.
Murcray F.H., Murcray D.G., & Williams W.J. (1970) Infrared emissivity of lunar surface features: 1. Balloon-borne observations. *Journal of Geophysical Research*, **75**, 2662–2669.
Mustard J.F. & Pieters C.M. (1987) Quantitative abundance estimates from bidirectional reflectance measurements. *Journal of Geophysical Research*, **92**, E617–E626.
Mustard J.F. & Pieters C.M. (1989) Photometric phase functions of common geologic minerals and applications to quantitative analysis of mineral mixture reflectance spectra. *Journal of Geophysical Research*, **94**, 13619–13634.
Mustard J.F. & Hays J.E. (1997) Effects of hyperfine particles on reflectance spectra from 0.3 to 25 μm. *Icarus*, **125**, 145–163.
Mustard J.F. & Sunshine J.M. (1999) Spectral analysis for Earth science: Investigations using remote sensing data. *Remote sensing for the Earth sciences: Manual of remote sensing*, 3 (A. Rencz, ed.). John Wiley & Sons, New York, 251–307.
Mustard J.F., Li L., & He G.Q. (1998) Nonlinear spectral mixture modeling of lunar multispectral data: Implications for lateral transport. *Journal of Geophysical Research*, **103**, 19419–19425.
Osterloo M.M., Hamilton V.E., Bandfield J.L., et al. (2008) Chloride-bearing materials in the southern highlands of Mars. *Science*, **319**, 1651–1654.

Osterloo M.M., Anderson F.S., Hamilton V.E., & Hynek B.M. (2010) Geologic context of proposed chloride-bearing materials on Mars. *Journal of Geophysical Research*, **115**, E10012, DOI:10.1029/2010JE003613.

Pitman K.M., Wolff M.J., & Clayton G.C. (2005) Application of modern radiative transfer tools to model laboratory quartz emissivity. *Journal of Geophysical Research*, **110**, E08003, DOI:10.1029/2005JE002428.

Poulet F. & Erard S. (2004) Nonlinear spectral mixing: Quantitative analysis of laboratory mineral mixtures. *Journal of Geophysical Research*, **109**, DOI:10.1029/2003JE002179.

Poulet F., Bibring J.P., Langevin Y., et al. (2009) Quantitative compositional analysis of martian mafic regions using the MEx/OMEGA reflectance data. *Icarus*, **201**(1), 69–83, DOI:10.1016/J.Icarus.2008.12.025.

Ramsey M.S. & Christensen P.R. (1998) Mineral abundance determination: Quantitative deconvolution of thermal emission spectra. *Journal of Geophysical Research*, **103**, 577–596.

Robertson K.M., Milliken R.E., & Li S. (2016) Estimating mineral abundances of clay and gypsum mixtures using radiative transfer models applied to visible-near infrared reflectance spectra. *Icarus*, **277**, 171–186.

Rogers A.D. & Aharonson O. (2008) Mineralogical composition of sands in Meridiani Planum determined from Mars Exploration Rover data and comparison to orbital measurements. *Journal of Geophysical Research*, **113**, E06S14, DOI:10.1029/2007JE002995.

Roush T.L., Pollack J.B., & Orenberg J. (1991) Derivation of midinfrared (5–25 µm) optical constants of some silicates and palagonite. *Icarus*, **94**, 191–208.

Roush T., Esposito F., Rossman G.R., & Colangeli L. (2007) Estimated optical constants of gypsum in the regions of weak absorptions: Application of scattering theories and comparisons to independent measurements. *Journal of Geophysical Research*, **112**, DOI:10.1029/2007JE002920.

Salisbury J.W. & Wald A. (1992) The role of volume scattering in reducing spectral contrast of reststrahlen bands in spectra of powdered minerals. *Icarus*, **96**, 121–128.

Salisbury J.W. & Walter L.S. (1989) Thermal infrared (2.5–13.5 µm) spectroscopic remote sensing of igneous rock types on particulate planetary surfaces. *Journal of Geophysical Research*, **94**, 9192–9202.

Salisbury F.B., Wald A., & D'Aria D.M. (1994) Thermal-infrared remote sensing and Kirchhoff's law 1. Laboratory measurements. *Journal of Geophysical Research*, **99**, 11897–11911.

Shirley K.A. & Glotch T.D. (2019) Particle size effects on mid-IR spectra of lunar analog materials in a simulated lunar environment. *Journal of Geophysical Research*, **124**, 970–988.

Shirley K.A., Glotch T.D., Greenhagen B.T., & White M. (2015) A multiplicative approach to correcting the thermal channels for the Diviner Lunar Radiometer Experiment. *46th Lunar Planet. Sci. Conf.*, Abstract #1992.

Shkuratov Y., Starukhina L., Hoffmann H., & Arnold G. (1999) A model of spectral albedo of particulate surfaces: Implications for optical properties of the Moon. *Icarus*, **137**, 235–246.

Sklute E.C., Glotch T.D., Piatek J., Woerner W., Martone A., & Kraner M. (2015) Optical constants of synthetic potassium, sodium, and hydronium jarosite. *American Mineralogist*, **100**, 1110–1122.

Spitzer W.G. & Kleinman D.A. (1961) Infrared lattice bands of quartz. *Physical Review*, **121**, 1324–1335.

Stamnes K., Tsay S.-C., Wiscombe W., & Jayaweera K. (1988) Numerically stable algorithm for discrete-ordinate-method radiative transfer in multiple scattering and emitting layered media. *Applied Optics*, **27**, 2502–2509.

Swanepoel R. (1983) Determination of the thickness and optical constants of amorphous silicon. *Journal of Physics E: Scientific Instruments*, **16**, 1214.

Thomas I.R., Greenhagen B.T., Bowles N.E., Donaldson Hanna K.L., Temple J., & Calcutt S.B. (2012) A new experimental setup for making thermal emission measurements in a simulated lunar environment. *Review of Scientific Instruments*, **83**, 124502.

Trang D., Lucey Paul G., Gillis-Davis Jeffrey J., Cahill Joshua T.S., Klima Rachel L., & Isaacson Peter J. (2013) Near-infrared optical constants of naturally occurring olivine and synthetic pyroxene as a function of mineral composition. *Journal of Geophysical Research*, **118**, 708–732.

Van de Hulst H.C. (1957) *Light scattering by small particles*. Dover Publications, Mineola, NY.

Wald A.E. (1994) Modeling thermal infrared (2–14 µm) reflectance spectra of frost and snow. *Journal of Geophysical Research*, **99**, 24,241–24,250.

Wald A.E. & Salisbury J.W. (1995) Thermal infrared directional emissivity of powdered quartz. *Journal of Geophysical Research*, **100**, 24665–24675.

Wenrich M.L. & Christensen P.R. (1996) Optical constants of minerals derived from emission spectroscopy: Application to quartz. *Journal of Geophysical Research*, **101**, 15921–15931.

3

Mid-infrared (Thermal) Emission and Reflectance Spectroscopy
Laboratory Spectra of Geologic Materials

MELISSA D. LANE AND JANICE L. BISHOP

3.1 Introduction/Background: Fundamental Vibrations of the Crystal Lattice

Electromagnetic (EM) radiation studies have continued to develop since the work of Newton. In the late 1800s, scientists recognized the utility of measuring EM energy after it had interacted with matter (gas, liquid, or solid) as a means of identifying materials based on the resulting spectral features. The earliest laboratory spectroscopic studies of minerals conducted in the short-infrared EM range were pioneered by Coblentz (e.g., 1905, 1906, 1908) over the ~10,000–1250 cm^{-1} (~1–8 µm) region using transmission (absorption) and reflectance spectroscopy. It wasn't until the 1960s, however, that infrared (IR) spectroscopy as a laboratory tool for geologists and planetary scientists emerged from the works of Lyon (e.g., 1964, 1965), Lyon and Burns (1963), and Conel (1965, 1969), demonstrating promising results from laboratory analyses of geologic materials, i.e., rocks and minerals. IR spectroscopy developed because of the broader availability of commercial spectrometers and technological advances allowing measurement of energy at longer wavelengths (Estep-Barnes, 1977). Expanding the measurable wavelength range was crucial because the fundamental spectral features for minerals and inorganic molecules occur within the mid-infrared (mid-IR, ~2000–200 cm^{-1} or ~5–50 µm) range. When energy propagates within a mineral, the energy is rarely blackbody in character because components of the wave will be altered by interaction with the repetitive molecular structure of the mineral. The absorption features result from vibrations of the chemical bonds within the minerals due to stretching (changing the interatomic distance between two atoms), bending (changing the angle between two bonds), or rotating, in the case of gases. These vibrations include a molecular symmetric stretch (v_1), symmetric bend (v_2), asymmetric stretch (v_3), and asymmetric bend (v_4). The bending vibrations include molecular scissoring, rocking, wagging, and twisting. Longer-wavelength features are due to lattice modes. Hence, early geologic mid-IR studies included the use of spectra to investigate the structural chemistry of minerals. Mid-IR (aka vibrational or thermal) spectroscopic tools are still in use by geologists, planetary scientists, and astronomers and commonly include the measurement of energy transmitted through, reflected off, or emitted from geologic materials.

3.1.1 Molecular Vibration and Simple Harmonic Oscillators

Each chemical bond in a crystal (mineral) vibrates at a quantized mid-IR frequency related to the properties of the bond (e.g., the strength of the bond and the masses and distances of the end atoms). These vibrations reposition the atoms, which offsets the associated electrons, thus initiating an oscillating electric dipole. The strength of the dipole moment correlates to the degree of offset of the dipole from its equilibrium position. The energy emitted from a mineral is rarely blackbody in nature, but rather exhibits wavelength-dependent reduced radiance resulting from the vibrations. Every mineral has a unique chemical composition and structure; thus the frequencies and strengths of the associated molecular vibrations of the functional groups and the molecules as a whole, and the electron displacements, are also unique. The oscillating dipole moments are manifested as absorption features in emissivity/reflectivity spectra. As a first approximation to estimate the position of an absorption feature, isolated molecules are modeled as "masses on springs" that are allowed to vibrate freely as simple harmonic oscillators using the equation

$$\nu_0 = (1/2\pi)(\kappa/m)^{1/2}, \tag{3.1}$$

where ν_0 is the resonant frequency (in Hz) of the dipolar molecular vibration, m is reduced mass which is equal to $m_1 m_2 / (m_1 + m_2)$ (where m_1 and m_2 are the masses of the atoms on the spring ends), and κ is the force constant. Generally, the vibrational frequency is inversely proportional to the masses at the ends of the bond; however, hybridization (type of bonding) affects the bond strength. Here $\kappa = -F/x$, where F is the Coulomb force of attraction between masses m_1 and m_2 and x is the displacement between the masses. The angular frequency ω_0 is given by

$$\omega_0 = 2\pi\nu_0 = \kappa/m^{1/2}. \tag{3.2}$$

However, for a dense, dielectric medium with an integrated framework, such as a mineral, a simple elastic oscillator model is not a completely accurate means of describing the bond resonances because coupling of the oscillators occurs, which introduces damping forces that act on the vibrations. Thus, dispersion theory was developed by Lorentz (1880) and Lorenz (1881) to accommodate the frictional (damping) forces and additive contributions from coupled oscillators in a solid-state, dense medium.

3.1.2 Dispersion Theory

Classical Lorentz–Lorenz dispersion analysis mathematically describes the vibrational behavior of a mineral as the summation of the crystal's individual molecular oscillations and how the atomically bound electrons interact with a propagating EM wave (e.g., Spitzer & Kleinman, 1961; Wenrich & Christensen, 1996; Lane, 1999; Glotch et al., 2007; Marino et al., 2007; Glotch & Rossman, 2009; Pitman et al., 2010; Arnold et al., 2014). The physical dipole oscillations are related to the optical properties, n and k per Eq. (2.3a) and (2.3b) in Chapter 2.

The frequencies and strengths of the associated molecular vibrations and electron displacements are unique to each chemical composition and related structure (i.e., mineral). For simple mineral structures (such as alkali halides that have repeated ionic bonds associated with a well-defined strong oscillation), spectral resonances may easily be correlated to vibrational modes. Spectral complexities arise, however, when considering minerals that contain molecules consisting of more than two atoms because the IR features do not necessarily correspond to single oscillations, making correlation of the spectral features with vibrational modes much more difficult. For additional details, including expansion of this model to low-symmetry minerals, see Chapter 2. Furthermore, when a fundamental vibration couples with an overtone or a combination band, a Fermi resonance will occur, complicating the spectrum. *Nonetheless, it is not a requirement to understand the physical reason for each spectral feature; the shape and position of the spectral features themselves are useful for identifying minerals and provide a unique spectral fingerprint of each mineral.*

This chapter describes and utilizes the typically nondestructive technique of mid-IR emission and reflectance spectroscopy for analyzing various geologic materials in the laboratory, whose spectra then may be used in spectral libraries (e.g., Christensen et al., 2000a) for determining the mineralogy and chemistry of unknowns (see Chapter 15). These laboratory-derived mid-IR spectra may be used as spectral end members in analyses of remotely acquired mid-IR data such as those from field and airborne, space-borne (Chapter 24), and rover-borne (Chapter 25) platforms.

The acquisition and calibration of raw data acquired in a laboratory is the focus of Section 3.2, including special considerations of crystal-axis and volume scattering effects on the spectral character, and the spectra of many minerals from various mineral classes are presented in Section 3.3.

3.2 Methods: Spectroscopy in the Laboratory – Ambient Measurements of Minerals and Other Geologic Materials of Planetary Interest

Laboratory studies using mid-IR spectroscopy of geologic materials include emission, reflectance, transmittance, attenuated total reflection (ATR), and IR microscopy. Emission spectroscopy is described in the most detail in this chapter, as this technique is most commonly used in remote sensing. King et al. (2004) describe laboratory methods for emission, reflectance, and transmittance spectra of minerals. Salisbury et al. (1991b) present IR reflectance spectra from 2.1 to 25 µm of most of the common minerals prepared as <75-µm particles, 75–250-µm particles, and a cleavage/fracture surface. They also include transmittance spectra for comparison of these minerals prepared as KBr pellets. Detailed analyses of IR transmittance spectra of minerals are available in Farmer (1974), which also provides explanations of the vibrational bands and how they relate to the mineral structure. Gadsden (1975), Kodama (1985), and Moenke (1962, 1966) also contain collections of IR transmittance spectra of minerals. Milosevic (2012) summarizes the use of ATR spectroscopy, and Che et al. (2011) discuss ATR as applied to minerals. Chen et al. (2015) provide a summary of IR transmission and reflectance microscopy.

The sample environment at the time of measurement plays an important role in the spectral properties of hydrous minerals. This effect is described in detail in Chapter 13.

3.2.1 Laboratory Equipment for Ambient Emission Measurements

Two types of spectrometers commonly have been used for mid-IR studies: (1) dispersive (or grating) spectrometers and (2) Fourier Transform Infra-Red (FTIR) spectrometers that house a Michelson interferometer. Both of these instruments are designed to produce a spectrum that shows the energy (reflected or emitted from, or transmitted through a sample) over a range of wavelengths. A dispersive spectrometer is a fixed system that works as a prism to spread the energy from the sample according to wavelength. Dispersive spectrometers will not be discussed further here because their performance is inferior to that of FTIR spectrometers due to lower throughput and a necessarily longer integration time for a similar signal-to-noise ratio (SNR) compared to the FTIR technique. FTIR spectrometers utilize a Michelson interferometer to collect all wavelengths of energy simultaneously over some range of the electromagnetic spectrum (another advantage over grating spectrometers). The spectral range is determined by the throughput of a beam splitter, which is commonly composed of germanium-coated KBr (7400–350 cm^{-1}) or CsI (6400–200 cm^{-1}), although other materials are used as well. The raw data from the detector (output energy) is called an interferogram (an oscillation over time) and is converted mathematically to a function of frequency (a spectrum) using a mathematical treatment called a Fourier transform. For emission measurements, the difference in temperature (ΔT) between the sample and the detector should be large for improved SNR, so these measurements require either a room-temperature sample and a cooled detector, or a room-temperature detector and a heated sample (traditional) or actively cooled sample (Baldridge & Christensen, 2009).

There are two main types of detectors: deuterated L-alanine-doped triglycene sulfate (DLaTGS, also known as DTGS or TGS) and mercury–cadmium–telluride (MCT). DTGS detectors are *thermal* (heat) detectors that run at room temperature and are less expensive and less sensitive; MCT detectors are *quantum* detectors that sense electrons and require cryogenic cooling, but are fast-sampling and more sensitive. Conventional reflectance/transmission spectrometers may be converted to obtain emission measurements, such that the radiation source (traditionally a heated rod of silicon carbide, such as a Globar^{TM}) is removed (or turned off) and a heated sample itself becomes the radiation source. The emitted energy coming from a sample is folded into the ray path of the spectrometer using a series of mirrors. The sample energy reaches a half-silvered beam splitter set at a 45° angle from the direction of the incoming energy, where half of the energy passes through and impinges on a stationary mirror (at a fixed path length, l) and the other half of the energy reflects off the beam splitter surface into the ray path of a moving mirror. The moving mirror slides over a distance Δl. The point at which the path lengths of the rays traveling to the stationary and moving mirrors are equal (where the optical path difference, δ, equals 0) is called the zero path difference (ZPD) and the beams are in constructive interference (in phase). Both the stationary and moving mirrors reflect the energy back to the beam splitter,

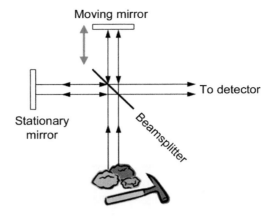

Figure 3.1 Schematic drawing of a conventional Michelson interferometer. This shows a moving (sliding) mirror as used in traditional FTIR spectrometers, where the "light" source is a heated geologic sample (in emission spectroscopy) whose emitted energy from the geologic material is being measured at the detector.

where it again either passes through into the ray path of the detector or is reflected to the detector (Figure 3.1). When the beams following these different paths are recombined, the energy is either in-phase, out-of-phase, or somewhere in between. When this combined beam energy finally impinges on the detector, it is measured as voltage (see Section 3.2.1.1). Although Michelson interferometers with a stationary and a moving mirror are the standard for FTIR spectroscopy and can be designed in numerous configurations, other innovative, compact Michelson interferometers are being made using rotating, rather than sliding, mirrors to change the relative path lengths of the energy and produce constructive and destructive interference.

3.2.1.1 Calibration of Emission Data

An object's temperature and the distribution of its energy are related, as defined by Planck's law, which relates the radiative properties of a blackbody to wavelength (see Chapter 2). Thermal IR radiation is measured in the spectrometer at the detector and converted to voltage that is both wavelength and temperature dependent. There are several ways to calibrate acquired data to produce emissivity spectra using a two-temperature, one-temperature, or direct-temperature strategy. Historically, a two-temperature method was used (described and dismissed in Ruff et al., 1997), for which a blackbody and a sample are both measured at two different temperatures. More commonly used is the one-temperature method (Christensen & Harrison, 1993; Ruff et al., 1997; Edwards & Christensen, 2013), which increases spectral measurement accuracy by stabilizing system temperatures and minimizing instrument- and environment-induced calibration errors. For these first two methods, it is assumed that at some wavelength the sample emissivity is equal to 1.0; hence, at some wavelength the sample reflectivity is zero (although this is not always true for minerals or rocks). This assumption is required for the fit of a Planck blackbody curve to the calibrated sample radiance curve for determination of sample temperature. The final calibration strategy is known as a direct-temperature method (Maturilli et al., 2016). Using an ideal experimental configuration, emissivity can be directly calibrated by dividing the radiance of the sample at an accurately known temperature by a blackbody curve at the same temperature. This approach does not require the assumption of unit emissivity at some point in the sample curve.

3.2.2 Laboratory Equipment for Ambient Reflectance Measurements

Kirchhoff's law describes the relationship between reflectivity (R) and emissivity (ε). Kirchhoff's law, $1 - R = \varepsilon$, is exactly valid only for the case of measured hemispherical reflectance (Hapke, 1993), as opposed to bidirectional or biconical reflectance measurements, or for surfaces that are Lambertian (e.g., Salisbury et al., 1994). Kirchhoff's law often holds for rough surfaces or larger particle sizes (>125 μm) of geologic materials, but does not hold for smooth surfaces or smaller particle sizes (<125 μm).

Unlike using a heated sample in emission spectroscopy, in reflectance spectroscopy a light source (e.g., an incandescent light such as a tungsten halogen lamp, or a SiC glowing rod) is shone upon the sample and the reflected light that reaches the detector is measured and quantified. Similar to emission spectroscopy, mid-IR reflectance spectroscopy requires a beam splitter (e.g., DTGS, CsI, ZnSe, KBr) and a detector (e.g., DTGS, MCT). FTIR spectrometers are most common for reflectance measurements, as in the case of emission spectroscopy.

3.2.2.1 Measurement Geometry and Specular versus Diffuse Reflectance

There are a few different measurement geometries possible for reflectance spectroscopy. Most common are hemispherical, biconical, and bidirectional. The type of measurement geometry can have implications for the reflectance spectra depending on the sample characteristics (e.g., if the sample reflects light specularly or diffusely). A smooth surface (i.e., a surface that is planar with respect to the wavelength of the light) will reflect light like a mirror, such that the reflected rays have the same angle as the incident rays. This behavior is called specular reflectance and is most commonly found for smoothly cut and polished rock surfaces, mineral cleavage faces, euhedral crystal faces, or glasses. In contrast, most rock surfaces are rougher and scatter reflected light in a diffuse manner, which means that the emergent rays exit the surface in many directions. A Lambertian surface is one that is perfectly diffuse. For this case, it can be assumed that the reflected light is equally dispersed in all directions. Most surfaces are largely diffuse with a small specular component, such that more light is reflected in the specular direction. Diffuse measurements of rough surfaces typically are preferred for geological applications.

Hemispherical systems collect the reflected light in nearly all directions and reflect it further to the detector. This system is also called an integrating (hemi-) sphere. The surface is coated with a Lambertian substance such as a rough gold surface or a white standard (e.g., barium sulfate, Halon, or Spectralon®), and the light reflected off the sample is then reflected diffusely inside the integrating sphere to the detector. Typically, the incident light arrives at the sample from one specified angle for hemispherical systems.

Biconical systems have cones of incident, i, and emergent, e, angles. In this case the light hitting the surface will have a range of angles and the light collected from the surface will be measured over a range of angles. For example, if the instrument is configured for i and e angles of 30° with 20° of width, the cones of i and e angles would expand this angle to cover $i = 20$–$40°$ and $e = 20$–$40°$. The emerging reflected light that hits the field of view of the collection cone would be reflected to a detector for measurement. Biconical systems are configured for on-axis or off-axis collection of the reflected light. The on-axis biconical systems are configured with the incident and emergent cones in one plane and will over-sample the specular component of reflected light, while the off-axis biconical systems will under-sample the specular component of reflected light. Many manufacturers prefer to design on-axis systems because the signal is stronger. They then use software to attempt to remove the specular component. Off-axis biconical configurations produce reflectance spectra more representative of hemispherical reflectance; i.e., they predominantly collect the diffusely scattered light from the surface. Hemispherical reflectance measurements produce spectra that are more similar to (inverse) emissivity spectra (according to Kirchhoff's law) than other types of reflectance measurements.

Bidirectional systems are less common for mid-IR spectroscopy of geologic materials. Bidirectional spectrometers are configured to shine light on the surface from one specific angle and collect the reflected light off the surface at another specific angle. Goniometers are bidirectional systems in which the incidence and emergence angles can be varied. This is important for remote sensing studies where the angle of the Sun and spacecraft can be variable and may affect the apparent spectral properties of the surface.

3.2.2.2 Sample Chamber for Reflectance Measurements

Samples are typically placed in horizontal dishes in a sample chamber for reflectance measurements. The number of samples that can be placed in the chamber is highly variable (one or multiple), depending on the system. The advantage of having multiple sample positions is that the standard and several samples in one project can be analyzed under the same conditions. This is especially important for hydrous or finely particulate samples that tend to adsorb or desorb H_2O molecules from the surface depending on the relative humidity of the sample chamber, which can change rapidly under ambient conditions if the lab air changes. Most current FTIR reflectance spectrometers allow for control of the environment in the sample chamber. This can involve flowing nitrogen gas through the sample chamber and FTIR housing or pumping out the H_2O and CO_2 from the room air. Either of these scenarios can provide a dry, controlled sample environment, similar to what is expected on most planets other than Earth. Measurements under specific environments are described in detail in Chapters 2 and 13.

3.2.2.3 Collection and Calibration of Reflectance Spectra

IR reflectance spectra can be measured for particulate samples or rock chips in sample dishes 5–10 mm in diameter. Spectra are typically collected of two or more spots on each sample and the data are compared to check for heterogeneities in the sample. The beam size is typically 1–2 mm in diameter for lab FTIR systems. Thus, heterogeneous rock surfaces or large mineral or rock grains may produce variable spectra of different spots on the same sample. The spectra of multiple runs on a sample can be averaged in order to achieve a spectrum representing multiple components of the sample. A rough gold surface is the typical standard material for calibrating diffuse IR spectra and mitigating any unscrubbed atmospheric components, while a first-surface gold-coated mirror is the common standard for specular reflectance studies of polished rock/mineral slabs. Spectra collected of the sample are ratioed to a spectrum of the gold standard taken about the same time as the sample spectrum. Whenever samples are measured with a new system, spectra should be acquired for a few geologic standards to ensure accuracy of the instrument.

3.2.3 Laboratory Equipment for Attenuated Total Reflection Spectroscopy

Attenuated total reflection (ATR) spectroscopy is also used in the geological sciences. Recently, ATR accessories have become available that can be used in most FTIR spectrometers, so this technique may be used by more researchers in the future. The sample is positioned in the ATR attachment in contact with a crystal of high refractive index such as ZnSe, Ge, or diamond, and an IR beam is directed onto the high-refractive-index crystal. That light is subsequently internally reflected toward the sample. This internally reflected IR beam creates an evanescent wave that propagates through the sample, and the attenuated beam is reflected back through the crystal, where it is collected by a detector. The evanescent wave attenuates at frequencies where the sample absorbs energy (e.g., fundamental vibrational frequencies). The ATR technique works well for highly absorbing materials, coatings on rocks, homogeneous rocks, or particles (finely ground or naturally fine-grained). Sample grains <10 μm have the advantage of close contact between the sample and the high-refractive-index crystal, thus minimizing the complexity of air interfering with the beam reflecting from the crystal to the sample. ATR spectra can be measured of a variety of minerals including phyllosilicates, sulfates, and oxides (e.g., Glotch et al., 2007). ATR is particularly useful for phyllosilicates (e.g., Che et al., 2011; Friedlander et al., 2015) and Fe hydroxides (e.g., Bishop et al., 2015) because strong spectral features are observed without the need for pressing the samples into pellets.

3.2.4 Laboratory Equipment for FTIR Microscopes

FTIR microscopy is becoming more widespread in geologic analyses of heterogeneous samples (e.g., Chen et al., 2015). Transmittance FTIR microscopy requires thin sections

(~5–15 μm thick) as with other microscopy techniques. Sample preparation for reflectance FTIR microscopy is less complex and thicker rock slices are typically prepared by polishing one side to create a smooth surface. ATR attachments can also be employed in FTIR microscopes. Some manufacturers offer micro-FTIR systems with polarized light and objective lenses configured with high incidence angles (source angle close to the sample surface, called grazing angle objective) for analysis of ultrathin coatings or monolayers on a surface.

Conventional micro-FTIR systems in the lab have a beam size ~20–100 μm, while micro-FTIR instruments using synchrotron radiation can have a beam size as small as 3–5 μm. Recent applications using an atomic force microscope and IR beam have even enabled submicron beam sizes (Dominguez et al., 2014). FTIR reflectance microscopes require a light source, beam splitter, and detector as with traditional reflectance FTIR measurements. A high-resolution IR camera or video camera is also required. FTIR microscopy has been applied successfully to analyses of mineral grains in soils (e.g., Weinger et al., 2009), inclusions in rocks (e.g., Klima and Pieters, 2006), meteorites (e.g., Morlok et al., 2006; Palomba et al., 2006; Kereszturi et al., 2015; Yesiltas et al., 2017), and interplanetary dust particles (e.g., Keller et al., 2006).

3.2.5 Crystal Axis Effects (Polarization Effects)

If a mineral is isotropic (i.e., symmetric in physical properties), then energy propagating through the crystal will behave similarly along each mineral axis. If the crystal is anisotropic, energy will behave differently in each axis direction. Regardless of which axis the energy is propagating along, an electromagnetic wave passing through a crystal will reach the interface with the air surrounding the sample, and it will be reflected back into the mineral or refracted into the surrounding medium (e.g., the air). The reflected electromagnetic wave in any plane can be represented as a combination of waves polarized either perpendicular ($R_\perp$) or parallel ($R_\parallel$) to the plane of incidence. The way the wave behaves is governed by the optical properties of the medium. Polished, single-crystal samples may be analyzed to minimize the surface scatter of emitted energy inherent to a particulate sample. When considering a smooth interface, the total Fresnel reflectivity (R_T) is predicted following the Fresnel reflection equations, in accordance with Born and Wolf (1980). These equations can be found in Lane (1999).

Materials are commonly anisotropic. This is the case for minerals that have unit cell crystal axes that are not the same length because of their molecular structure, which also governs the character of the energy propagating along each axis. Hence, the energy measured from one side of a crystal will appear to be different than energy measured from another crystal face. For example, emission measured along the c axis of a hexagonal mineral (i.e., emission from the **c**-face) differs from that measured along the a axis of the mineral (Figure 3.2a). Although not discussed in detail here, spectra may also be acquired using polarizers to isolate certain propagation directions of the emitted energy (e.g., Spitzer & Kleinman, 1961; Hellwege et al., 1970; Onomichi et al., 1971; Long et al., 1993; Glotch

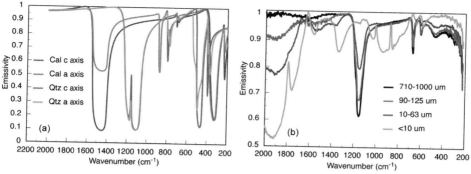

Figure 3.2 Dependence of emissivity spectra on crystal axis and particle size. (a) Emissivity spectra of calcite (Cal) and quartz (Qtz) measured along the c and a axes, showing axis dependency of the spectra. (Quartz spectra are from Wenrich and Christensen, 1996; calcite c-axis spectrum and derived a-axis spectrum are from Lane, 1999 and data therein). (b) Emissivity spectra acquired at ~1 atm (ambient conditions) of gypsum for various particle-size fractions. Note that fundamental (reststrahlen) bands shrink with decreasing particle size, and volume-scattering (transparency between the fundamental bands) features appear and deepen with decreasing particle size (as clearly seen in the <10-μm spectrum). (Gypsum spectra are adapted from Lane and Christensen, 1998, with the addition here of the <10-μm data.)

et al., 2006). The deepest spectral fundamental bands are associated with samples that are limited to one lossy air/sample interface (i.e., a first-surface/specular reflection at a cleavage plane or a polished face) because multiple reflections are minimized (see Section 3.2.6) (e.g., Lane, 1999).

Crystal-axis effects are important for interpreting data acquired of single grains (micro-FTIR) that may be dominated by a single axis. However, for a laboratory sample composed of many grains of a given mineral, the crystal axis issue is diminished because generally the particles are oriented randomly and the resulting spectrum contains energy emitted from all mineral axes; however, the spectral bands will not be as deep as those observed for a single solid sample. For remote sensing purposes, a spectral library composed of coarse, randomly oriented mineral particles may be compared to planetary surfaces because the surface spectra are averages over the spatial footprint of the instrument and these surfaces include randomly oriented mineral particles as well (i.e., rocks, dunes, soils). Most rocks in the lab or field, too, may be analyzed using laboratory coarse particulate spectra because the rocks are comprised of crystallites of minerals that generally are oriented randomly (Chapter 15), and typically, rock spectra are a linear mixture of the constituent mineral spectra (e.g., Ramsey & Christensen, 1998; Feely & Christensen, 1999). Furthermore, rock spectral bands will be deeper than the equivalent particles because the close proximity of grains in a solid rock (essentially, a rock may be viewed as one enormous particle) minimizes energy loss at each grain interface. In this chapter, we focus on minerals and their mid-IR spectral properties (Figures 3.2–3.6), but (reflectivity and emissivity) spectra are also available for many types of rock lithologies (e.g., Salisbury et al., 1988; Salisbury & Walter, 1989;

Salisbury & D'Aria, 1992; Feely & Christensen, 1999; Hamilton & Christensen, 2000; Wyatt et al., 2001). Spectra of selected meteorites (Figure 3.7) represent examples of rock spectra in this chapter.

3.2.6 Spectral Effects Due to Particle Size and the Resulting Volume Scattering

Samples composed of particles that are not large relative to the wavelength of the emitted/reflected radiation (e.g., particulate samples, hand samples with a particulate coating, friable samples) will show spectral character that results from volume scattering within the sample. As the grain size of any particulate sample is decreased so that it approaches the wavelength of the emitted/reflected energy, the resulting spectra change shape. Previous studies (e.g., Lyon, 1964; Aronson et al., 1966; Hunt & Vincent, 1968; Conel, 1969; Aronson & Emslie, 1973; Salisbury & Estes, 1985; Salisbury & Wald, 1992; Moersch & Christensen, 1995; Wald & Salisbury, 1995; Mustard & Hays, 1997; Cooper & Mustard, 1999; Lane, 1999; Bishop et al., 2014a) have addressed the issue of energy scattering and the effect of particle size on mid-IR spectra. The differences between the particulate emissivity/reflectivity spectra and the consolidated hand sample spectra are due primarily to the increasing number of grain/air interfaces per unit volume with decreasing particle size, and the associated increase in sample porosity that facilitates scattering due to reflections and refractions from both external (multiple scattering) and internal (volume scattering) sides of the grain. Scattering depends on the difference in optical properties (n and k – the real and imaginary indices of refraction; see Chapter 2) between the grains and the pore-filling atmosphere and by the properties of the grains themselves, such as shape, size, and transmissivity.

As an example, emissivity spectra of several particle sizes of gypsum ($CaSO_4 \cdot 2H_2O$) are shown in Figure 3.2b to emphasize the spectral behavior due to particle size alone. These particle-size samples include 710–1000, 90–125, 10–63, and <10-µm fractions. Spectral bands, including those for gypsum, display various behaviors with decreasing particle size, i.e., various "Type" behaviors as classified by Hunt and Vincent (1968). These behavioral types regarding the spectral behavior of gypsum are discussed in detail in Lane and Christensen (1998). Other in-depth discussions regarding spectral emissivity behaviors for fine particles may be found in Salisbury and Wald (1992) (calcite and quartz), Lane (1999) (calcite), or Mustard and Hays (1997) (olivine and quartz).

When more than a single reflection (at the sample/air interface) occurs, the spectrum will be affected by the atmospheric properties; the smaller the particles, the more impact the air properties surrounding the mineral sample will have. Under low atmospheric pressure conditions (minimal interstitial gas), fine particles are poor heat conductors and have an insulating effect, causing steep thermal gradients that change the shape of mid-IR spectra (e.g., Logan & Hunt, 1970; Logan et al., 1973; Henderson and Jakosky, 1997; Thomas et al., 2012; Donaldson Hanna et al., 2012, 2017) resulting in an emissivity/reflectivity spectrum that is different from that of the sample being studied in a 1-atm environment (a common ambient condition). Laboratory studies of fine-grained minerals in vacuum conditions are becoming more common because a suite of mineral spectra acquired under low to no atmospheric pressure (over a wide range of temperatures) are required for more accurate interpretation of planetary remote sensing data sets (Moon, asteroids, Mercury, Mars, etc.). For further discussion of spectra acquired under nonambient conditions, see Chapters 2 and 13.

The spectra presented in the text that follows were acquired at ambient pressure.

3.3 Spectra of Minerals and Meteorites of Planetary and Broader Interest

In this section mid-IR laboratory emissivity spectra acquired at 1 atm of various mineral groups will be presented through a sampling of example minerals of each type (mineral groups included are silicates, carbonates, sulfates, oxides/hydroxides, phosphates, sulfides, and chlorides/perchlorates). In addition, a variety of meteorite emissivity spectra will be presented to represent a range of parent bodies and rocks. These 93 spectra represent many never-published emissivity spectra from our collection, but also include some of our previously published spectra and 3 phyllosilicate spectra from the Arizona State University (ASU) spectral library (Christensen et al., 2000a) where noted. For additional mineral spectra beyond those presented here, the reader is referred to the ASU and other mid-IR laboratory databases that are available online, e.g., the Jet Propulsion Laboratory ASTER library (Baldridge et al., 2009), which contains data from the Johns Hopkins collection (Salisbury et al., 1991b) and the United States Geological Survey (Clark et al., 2007; Kokaly et al., 2017), the Stony Brook University Vibrational Spectroscopy Laboratory, Brown University's RELAB (Pieters & Hiroi, 2004), the University of Winnipeg (Cloutis, 2015), and the Berlin Emissivity Database (Maturilli et al., 2008). As often as possible, we present emissivity spectra here that are dominated by the fundamental spectral features (reststrahlen bands) and show minimal volume scattering effects. Operationally, fine particulate materials that would show volume scattering effects may be pressed into pellets (without the addition of a salt carrier such as is needed for transmission measurements) using a hydraulic press and die in order to increase their spectral contrast in mid-IR emissivity spectra (e.g., Michalski et al., 2006; Lane et al., 2011a, 2015).

The acquired data range is from ~2000 to 200 cm^{-1} (5–50 μm). However, the presented spectra are truncated in the figures at the high-frequency end, to minimize the spectrally flat region and emphasize the diagnostic spectral features. This approach is especially relevant to the sulfide spectra (Figure 3.6b) that exhibit spectral features only at less than ~450 cm^{-1}.

3.3.1 Silicates

Mid-IR spectra of silicates are dominated by vibrational modes of the SiO_4^{4-} tetrahedra and metal–oxygen polyhedra, and their linkages in the crystal structure. Mid-IR spectra of many silicate minerals have been published previously and are available in public spectral libraries (e.g., Salisbury et al., 1991b; Crowley & Hook, 1996; Ramsey & Christensen, 1998; Christensen et al., 2000a; Chihara et al., 2002; Johnson et al., 2003; Hamilton, 2000, 2010; Michalski et al., 2003, 2006; Ruff, 2004; Edwards & Christensen, 2013). Emissivity spectra of select silicates are shown in Figure 3.3. In addition to a few more common mineral spectra, we present mostly silicate spectra that are less available in the literature to further the number of mineral emissivity spectra in print.

Nesosilicates: This group is structured by independent SiO_4 tetrahedra. We have included forsterite here; for other mid-IR olivine spectra in the Mg–Fe solid solution and discussion, see Koike et al. (2003), Hamilton (2010), and Lane et al. (2011a). Andradite and almandine represent two different types of garnet. Titanite is the mineral formerly known as sphene. Andalusite and topaz spectra are also presented.

Cyclosilicates: Beryl and dioptase represent this group of close-ringed chains of SiO_4 tetrahedra.

Inosilicates: This group of silicates is structured by continuous single- or double-chain units of tetrahedra that share oxygen atoms. Pyroxenes including diopside, enstatite, ferrosilite, and spodumene are represented. Amphibole is represented by the hornblende spectrum. See Hamilton (2000) for details regarding pyroxene emission spectroscopy.

Tectosilicates: This group is structured by a continuous framework of silica tetrahedra. Feldspar is represented by the andesine spectrum, and sodalite represents the feldspathoids. Milam et al. (2007) present more details regarding feldspar spectroscopy. The final tectosilicate spectrum presented is that of opal CT.

We have separated the secondary silicates, phyllosilicates (Figure 3.3b) and zeolites (Figure 3.3c) into their own categories and are presenting them in Sections 3.3.2 and 3.3.3.

3.3.2 Phyllosilicates

Phyllosilicates are structured by a continuous sheet of tetrahedra, each sharing three oxygen atoms, and are very important rock-forming minerals. Here, this group is represented in Figure 3.3b by the minerals apophyllite, biotite, phlogopite, glauconite, Fe-smectite (SWa-1), montmorillonite (SWy-2), chlorite, clinochlore, kaolinite, and serpentine. The primary features are related to the tetrahedral (Si, Al)–O bonds and their vibrational modes.

Mid-IR spectra of fine grain sizes of some of these phyllosilicate minerals are available in Bishop et al. (2008) and show the expected decrease in fundamental band intensity and increased volume-scattering features. For further discussion about phyllosilicate mid-IR spectroscopy and the derivation of their optical constants, see Michalski et al. (2006), Glotch et al. (2007), and Ruff and Christensen (2007).

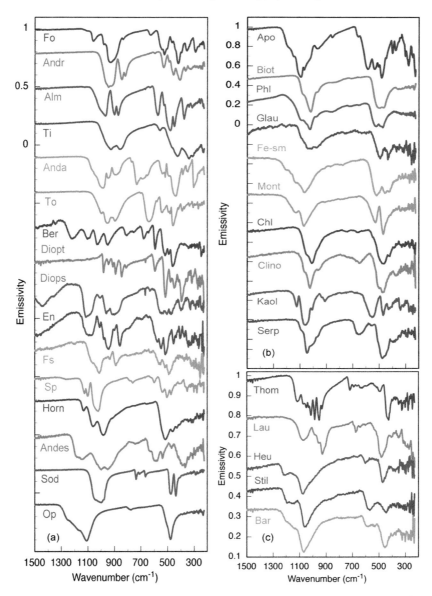

Figure 3.3 Variability in mid-IR silicate emissivity spectra. (a) Traditional silicate spectra of forsterite (Fo) (x0.5), andradite (Andr) (x0.6), almandine (Alm) (x0.7), titanite (Ti) (x0.6), andalusite (Anda) (x0.4), topaz (To) (x0.4), beryl (Ber), dioptase (Diopt) (x3), diopside (Diops) (x3), enstatite (En) (x3), ferrosilite (Fs) (x3), spodumene (Sp), hornblende (Horn), andesine (Andes) (x1.8), sodalite (Sod) (x0.4), opal CT (Op) (x0.6). (b) Phyllosilicate emissivity spectra of apophyllite (Apo), biotite (Biot) (x0.5), phlogopite (Phl) (x0.6), glauconite (Glau) (x5), Fe-smectite* SWa-1 (Fe-sm) (x2.2), montmorillonite* SWy-2 (Mont) (x2), chlorite (Chl) (x1.5), clinochlore (Clino) (x1.3), kaolinite (Kaol) (x2), serpentine* (Serp) (x2). Asterisks (*) denote the three spectra from the ASU spectral library. (c) Zeolite emissivity spectra of thomsonite (Thom), laumontite (Lau) (x1.3), heulandite (Heu) (x1.5), stilbite (Stil) (x2), barrerite (Bar) (x1.3). The multiplicative value was used to modify the original spectrum so all spectra share a similar depth of features for comparison. Spectra are offset for clarity.

3.3.3 Zeolites

Zeolites are tectosilicates (see earlier) that are hydrous. They are structured by a continuous framework of silica tetrahedra, much like feldspar, and so their spectra (Figure 3.3c) tend to resemble those of the feldspars. They commonly are associated with the alteration of volcanic rocks and hydrothermal environments. This group is represented here by thomsonite, laumontite, heulandite, stilbite, and barrerite.

Further discussions regarding the mid-IR spectroscopy of zeolites may be found in Cloutis et al. (2002), Ruff (2004), Mozgawa et al. (2011), and Che and Glotch (2012).

3.3.4 Carbonates

Carbonate spectra are dominated by the vibrational modes of the carbonate CO_3^{2-} anion. Carbonates are divided into two groups on the basis of their cation size and packing density (e.g., Weir & Lippincott, 1961). The 6-fold cation site in rhombohedral carbonates requires smaller cations. Larger cations will require a 9-fold site. Note that Ca can be found in both groups as calcite and aragonite.

The 6-fold carbonate spectra presented in Figure 3.4a are of calcite, magnesite, siderite, rhodochrosite, smithsonite, dolomite, kutnahorite, and minrecordite. Further discussion of these and other carbonate spectra can be found in Adler and Kerr (1963), Lane and Christensen (1997), Lane (1999), and Bishop et al. (2017).

The 9-fold carbonate spectra presented in Figure 3.4b are aragonite, strontianite, witherite, and cerussite.

Additional mid-IR (transmittance) spectra of synthetic 9-fold carbonates may be found in Böttcher et al. (1997), in which the witherite–strontianite solid solution is discussed.

3.3.5 Sulfates

A comprehensive collection of 37 sulfate mid-IR emissivity spectra were presented in Lane (2007) and a focused study of a more extensive collection of Fe sulfates by Lane et al. (2015) offered 21 emissivity spectra of iron-bearing mineral samples and detailed discussion. Figure 3.4c presents just a few sulfate spectra from those works as examples of this class of minerals. These include thenardite, glauberite, alunite, jarosite, sulfohalite, and coquimbite. Sulfate spectra are dominated by the vibrational modes of the sulfate SO_4^{2-} anion (e.g., Adler & Kerr, 1965; Ross, 1974a) and are discussed in detail in Lane (2007) and Lane et al. (2015). For further discussion about the derivation of select Fe–sulfate optical constants, see Pitman et al. (2014).

3.3.6 Oxides/Hydroxides

The oxide emissivity spectra of magnetite, franklinite, chromite, hematite, ilmenite, maghemite, rutile, pyrolusite, and goethite were first published in Christensen et al. (2000b) for

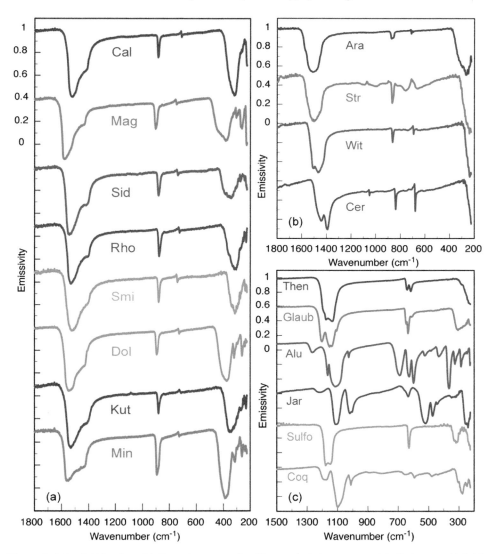

Figure 3.4 Variability in mid-IR carbonate and sulfate emissivity spectra. (a) Carbonate (6-fold) emissivity spectra of the minerals calcite (Cal), magnesite (Mag), siderite (Sid) (x1.5), rhodochrosite (Rho) (x1.2), smithsonite (Smi) (x1.5), dolomite (Dol) (x0.8), kutnahorite (Kut) (x1.3), and minrecordite (Min) (x1.5). (6-fold carbonate spectra are adapted from Lane and Christensen, 1997.) (b) Carbonate (9-fold) emissivity spectra of the minerals aragonite (Ara), strontianite (Str) (x1.3), witherite (Wit), and cerussite (Cer) (x1.2). (c) Sulfate emissivity spectra of the minerals thenardite (Then), glauberite (Glaub), alunite (Alu) (x1.5), jarosite (Jar) (x2), sulfohalite (Sulfo) (x2), and paracoquimbite/coquimbite (Coq) (x1.5). The multiplicative value was used to modify the original spectrum so all spectra share a similar depth of features for comparison. Spectra are offset for clarity.

comparison to the Mars Global Surveyor Thermal Emission Spectrometer data of Mars (see Chapter 24). Those laboratory emissivity spectra are presented in Figure 3.5 with an additional corundum spectrum. Oxides have various crystal structures, from simple (metal–oxygen, M–O) to more complex structures (M_xO_y) to double-oxide $A_xB_xO_u$ compounds (such as AB_2O_4 as rutile), where A and B are cations in tetrahedral and octahedral coordination, respectively (Farmer, 1974).

Generally, oxide spectra are mostly featureless (i.e., spectrally flat) and exhibit high emissivity at wavenumbers larger than ~900 cm^{-1}. The oxide/hydroxide spectra are dominated by features at lower frequencies (smaller wavenumbers) that are diagnostic of composition. Spectral features associated with the simple highly ionic metal–oxygen materials generally are due to crystal lattice modes. In the more complex minerals, the M–O tetrahedra vibrate causing additional spectral bands. For hydroxide minerals, the OH$^-$ group introduces observed bands as well. An in-depth discussion about the bonding and vibrations related to these various oxides may be found in Farmer (1974).

For a variety of hematite mid-IR emissivity spectra, and a discussion of hematite crystal-axis orientation effects, see Lane et al. (2002). For the derivation of oxide/hydroxide mid-IR optical constants, see Glotch and Rossman (2009).

3.3.7 Phosphates

Primary phosphates that crystallize from a fluid include apatite and triphylite. Secondary phosphates that form at low temperatures in the presence of water include strengite and vivianite. Whitlockite can be found in chondrites within meteorites. Mid-IR spectra of those phosphates plus other phosphate minerals including amblygonite, triplite, lazulite, kulanite, pyromorphite, metavariscite, baricite, strunzite, gormanite, wavellite, childrenite–eosphorite solid solution, and turquoise are shown in Figure 3.6a.

Phosphate mid-IR spectra are dominated by the vibrational modes of the PO_4^{3-} tetrahedra (e.g., Stutman et al., 1965; Ross, 1974b; Frost et al., 2002; Lane et al., 2011b). Additional information regarding the wide variety of phosphate minerals may be found in Huminicki and Hawthorne (2002).

3.3.8 Sulfides

Sulfide minerals are common on Earth in volcanic terrains as well as "black smokers" at ocean depths. Oxidation of sulfides may lead to the formation of sulfates. Sulfide mineral spectra are represented in Figure 3.6b (presented earlier in Section 3.3.7) by the minerals sphalerite, chalcopyrite, galena, stibnite, and pyrite. The mid-IR sulfide spectra generally are spectrally flat (graybody, at some value of ε <1.0) at the higher frequencies (higher wavenumbers), hence our spectra are shown for 800–200 cm^{-1} only. This spectral behavior is similar to that of the oxides (Figure 3.5); however, instead of showing features at

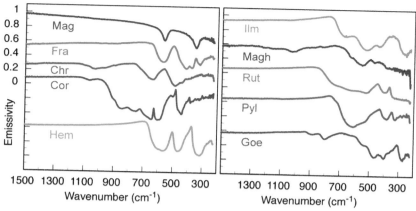

Figure 3.5 Oxide/hydroxide emissivity spectra. Minerals shown include magnetite (Mag), franklinite (Fra), chromite (Chr), Corundum (Cor), hematite (Hem) (x0.5), ilmenite (Ilm), maghemite (Magh) (x2), rutile (Rut) (x0.6), pyrolusite (Pyl), and goethite (Goe). The multiplicative value was used to modify the original spectrum so all spectra share a similar depth of features for comparison. Spectra are offset for clarity. (Spectra [except for corundum] are adapted from Christensen et al., 2000b, 2001.)

wavenumbers less than ~900 cm^{-1} as for oxides, the sulfide features dominate at wavenumbers less than ~450 cm^{-1} due to the crystal lattice modes.

3.3.9 Chlorides/Perchlorates

Similar to the oxides and more so to the sulfides, the chlorides are spectrally flat (graybody) throughout most of the mid-IR region and exhibit spectral features at longer wavelengths (i.e., <400 cm^{-1}; >25 μm) (Figure 3.6c). Chlorides are part of the mineral class of halides, which also includes fluorides, bromides, and iodides. The chlorides presented here are halite (NaCl), sinjarite (CaCl$_2$·2H$_2$O), and sylvite (KCl). The CsI beam splitter causes the cutoff of the usable data at 200 cm^{-1}, but it is clear this is near where the fundamental bands of the chlorides begin. The halite band initiates first, followed by the sinjarite band. The sylvite spectrum is just beginning to show this band as the data are cut off at 200 cm^{-1}. Chlorides are highly ionic, and the strong ionic bonding prevents the chemical bond between individual diatomic pairs (e.g., Na and Cl) from vibrating independently, hence the long-wavelength features are due to crystal lattice vibrations.

The perchlorates included in Figure 3.6c were initially measured in reflectance, then the reflectivity data were converted to emissivity through the application of Kirchhoff's law (see Section 3.3.2). The perchlorates exhibit more spectral features in the mid-IR than the chlorides, due to the vibrations of the ClO$_4^-$ anion. Mid-IR (and visible/near-IR) spectra and vibrational-mode discussion of these and other perchlorates may be found in Bishop et al. (2014b).

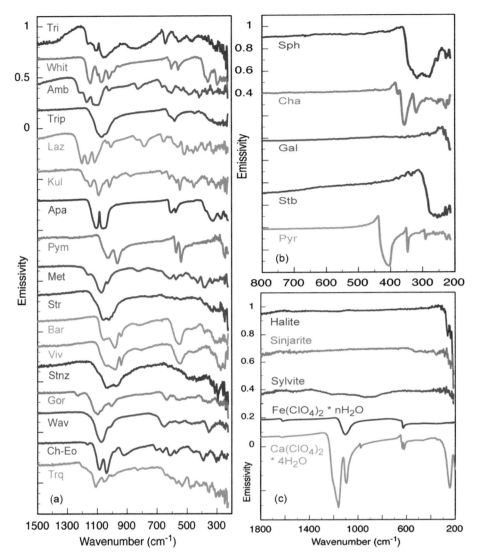

Figure 3.6 Variability in mid-IR phosphate, sulfide, and chloride/perchlorate emissivity spectra. (a) Phosphate emissivity spectra of the minerals triphylite (Tri) (x3), whitlockite (Whit), amblygonite (Amb) (x1.5), triplite (Trip) (x0.7), lazulite (Laz) (x1.5), kulanite (Kul) (x1.5), apatite (Apa) (x0.5), pyromorphite (Pym) (x1.5), metavariscite (Met), strengite (Str) (x2), baricite (Bar), vivianite (Viv) (x2), strunzite (Stnz) (x4), gormanite (Gor), wavellite (Wav), childrenite–eosphorite solid solution (Ch-Eo) (x1.2), and turquoise (Tur) (x2). (b) Sulfide emissivity spectra of the minerals sphalerite (Sph), chalcopyrite (Cha), galena (Gal) (x0.5), stibnite (Stb), and pyrite (Pyr) (x0.6). (c) Chloride and perchlorate emissivity spectra. The halite, sinjarite, and sylvite samples were measured in emission. The perchlorate samples were measured in reflectivity and converted to emissivity. The multiplicative value was used to modify the original spectrum so all spectra share a similar depth of features for comparison. Spectra are offset for clarity.

3.3.10 Meteorites

The meteorites presented here represent a wide range of meteorite (rock) types that can be related to a wide array of parent bodies (planets and asteroids). Included in Figure 3.7 are previously published spectra of Northwest Africa (NWA) 2737 (a Chassignite, Mars) (Pieters et al., 2008), Yamato 984028 (a shergottite, Mars) (Dyar et al., 2011), Northwest Africa (NWA) 7325 (an ungrouped achondrite) (Goodrich et al., 2017), and the following four meteorites published in Ashley (2011) and Ashley and Christensen (2012): Dar el Gani 862 (an H3 ordinary chondrite), Allende (a CV3 carbonaceous chondrite), Dhofar 007 (a eucrite), and Abee (an enstatite chondrite). Compositions of the meteorites and detailed discussions may be found in those references. A few examples of other works containing mid-IR laboratory spectra of meteorites include Salisbury et al. (1991a), Hamilton et al. (2003), Palomba et al. (2006), Maturilli et al. (2006), Helbert et al. (2007), Donaldson Hanna and Sprague (2009), and Vernazza et al. (2012, 2017). For mid-IR spectra of other rocks not derived from space (i.e., terrestrial rocks) see Feely and Christensen (1999), Hamilton and Christensen (2000), Hamilton et al. (2001), Cooper et al. (2002), and Rogers and Nekvasil (2015).

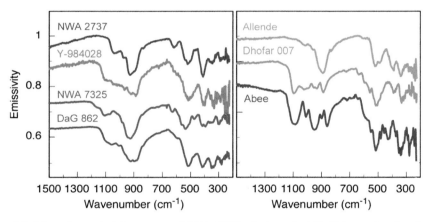

Figure 3.7 Meteorite emissivity spectra. NWA 2737 whole rock (45–125-μm size fraction, the interior of Y-984028 (x4), NWA 7325 (x0.4), DaG 862 (x0.6), Allende, Dhofar 007 (x0.8), and Abee. The spectra of DaG 862, Allende, Dhofar 007, and Abee were supplied by Dr. James Ashley (Jet Propulsion Laboratory). The multiplicative value was used to modify the original spectrum so all spectra share a similar depth of features for comparison. Spectra are offset for clarity. (Spectra are adapted from Pieters et al., 2008; Ashley, 2011; Dyar et al., 2011; Ashley and Christensen, 2012; and Goodrich et al., 2017).

3.4 Summary

Mid-IR thermal emission and reflectance spectroscopies produce similar (inverse) spectra according to Kirchhoff's law ($1 - R = \varepsilon$; see Section 3.2.2) and both are powerful techniques for the study and identification of geologic materials in the lab and in the field for studying in situ land surfaces. Laboratory emission and reflectance spectra are particularly valuable for application to remote sensing data for the observation of Solar System bodies (planets, asteroids, moons, etc.). In the mid-IR region, the spectral character of minerals is determined primarily by the vibrations of key functional chemical groups (e.g., SiO_4^{4-}, CO_3^{2-}, SO_4^{2-}, PO_4^{3-}, ClO_4^{-}) and the molecular structure, and secondarily by the presence of water or other subordinate chemistry. Minor impurities rarely affect the spectrum of the dominant mineralogy in the mid-IR (but often do affect the color of the sample as is manifested by variations at a higher frequency spectral range – i.e., in the visible spectrum). Rocks represent combinations of individual constituent minerals; hence their spectra can be unmixed spectrally using key single-mineral laboratory spectra such as those presented here to discern the rock (or meteorite) components. Meteorite spectra are useful for comparison to asteroid spectra in order to identify a parent body type. Mid-IR laboratory spectra are useful for comparing to remotely acquired data not only to help identify composition, but also to determine other traits such as particle size, water content, and crystal-axis orientation effects.

Acknowledgments

In addition to the samples measured for this chapter from the authors' own collections, the authors thank D. Dyar, R. Klima, R. Morris, E. Cloutis, and F. McCubbin for loaning samples from their collections for spectra included here. The authors are grateful to P. Christensen for the use of his emission lab at Arizona State University, to T. Glotch for the use of his emission lab at Stony Brook University, and to C. Pieters and R. Milliken for the use of RELAB at Brown University for reflectance spectroscopy.

References

Adler H.H. & Kerr P.F. (1963) Infrared spectra, symmetry and structure relations of some carbonate minerals. *American Mineralogist*, **48**, 839–853.

Adler H.H. & Kerr P.F. (1965) Variations in infrared spectra, molecular symmetry and site symmetry of sulfate minerals. *American Mineralogist*, **50**, 132–147.

Arnold J.A., Glotch T.D., & Plonka A.M. (2014) Mid-infrared optical constants of clinopyroxene and orthoclase derived from oriented single-crystal reflectance spectra. *American Mineralogist*, **99**, 1942–1955.

Aronson J.R. & Elmslie A.G. (1973) Spectral reflectance and emittance of particulate materials. 2: Application and results. *Applied Optics*, **12**, 2573–2585.

Aronson J.R., Emslie A.G., & McLinden H.G. (1966) Infrared spectra from particulate surfaces. *Science*, **152**, 345–346.

Ashley J.W. (2011) *Meteorites on Mars as Planetary Research Tools with Special Considerations for Martian Weathering Processes*. PhD dissertation, Arizona State University.

Ashley J.W. & Christensen P.R. (2012) Thermal emission spectroscopy of unpowdered meteorites. *43rd Lunar Planet. Sci. Conf.*, Abstract #2519.

Baldridge A.M. & Christensen P.R. (2009) A laboratory technique for thermal emission measurement of hydrated minerals. *Applied Spectroscopy*, **63**, 678–688.

Baldridge A.M., Hook S.J., Grove C.I., & Rivera G. (2009) The ASTER spectral library version 2.0. *Remote Sensing of the Environment*, **13**, 711–715.

Bishop J.L., Lane M.D., Dyar M.D., & Brown A.J. (2008) Reflectance and emission spectroscopy study of four groups of phyllosilicates: Smectites, kaolinite-serpentines, chlorites and micas. *Clay Minerals*, **43**, 35–54.

Bishop J.L., Lane M.D., Dyar M.D., King S.J., Brown A.J., & Swayze G. (2014a) Spectral properties of Ca-sulfates: Gypsum, bassanite, and anhydrite. *American Mineralogist*, **99**, 2105–2115.

Bishop J.L., Quinn R., & Dyar M.D. (2014b) Spectral and thermal properties of perchlorate salts and implications for Mars. *American Mineralogist*, **99**, 1580–1592.

Bishop J.L., Murad E., & Dyar M.D. (2015) Akaganéite and schwertmannite: Spectral properties, structural models and geochemical implications of their possible presence on Mars. *American Mineralogist*, **100**, 738–746.

Bishop J.L., King S.J., Lane M.D., et al. (2017) Spectral properties of anhydrous carbonates and nitrates. *48th Lunar Planet. Sci. Conf.*, Abstract #2362.

Born M. & Wolf E. (1980) *Principles of Optics*, 6th edn. Pergamon, Tarrytown, NY, 627–633.

Böttcher M.E., Gehlken P.-L., Fernandez-Gonzalez A., & Prieto M. (1997) Characterization of synthetic $BaCO_3$–$SrCO_3$ (witherite-strontianite) solid-solutions by Fourier transform infrared spectroscopy. *European Journal of Mineralogy*, **9**, 519–528.

Che C. & Glotch T.D. (2012) The effect of high temperatures on the mid-to-far-infrared and near-infrared reflectance spectra of phyllosilicates and natural zeolites: Implications for martian exploration. *Icarus*, **218**, 585–601.

Che C., Glotch T.D., Bish D.L., Michalski J.R., & Xu W. (2011) Spectroscopic study of the dehydration and/or dehydroxylation of phyllosilicate and zeolite minerals. *Journal of Geophysical Research*, **116**, DOI:10.1029/2010JE003740.

Chen Y., Zou C., Mastalerz M., Hu S., Gasaway C., & Tao X. (2015) Applications of micro-Fourier Transform Infrared Spectroscopy (FTIR) in the geological sciences: A review. *International Journal of Molecular Sciences*, **16**, 26227.

Chihara H., Koike C., Tsuchiyama A., Tachibana S., & Sakamoto D. (2002) Compositional dependence of infrared absorption spectra of crystalline silicates. I. Mg-Fe pyroxenes. *Astronomy & Astrophysics*, **391**, 267–273.

Christensen P.R. & Harrison S.T. (1993) Thermal infrared emission spectroscopy of natural surfaces: Application to desert varnish coatings on rocks. *Journal of Geophysical Research*, **98**, 19,819–19,834.

Christensen P.R., Bandfield J.L., Hamilton V.E., et al. (2000a) A thermal emission spectral library of rock-forming minerals. *Journal of Geophysical Research*, **105**, 9735–9739.

Christensen P.R., Bandfield J.L., Clark R.N., et al. (2000b) Detection of crystalline hematite mineralization on Mars by the Thermal Emission Spectrometer: Evidence for near-surface water. *Journal of Geophysical Research*, **105**, 9623–9642.

Christensen P.R., Morris R.V., Lane M.D., Bandfield J.L., & Malin M.C. (2001) Global mapping of martian hematite mineral deposits: Remnants of water-driven processes on early Mars. *Journal of Geophysical Research*, **106**, 23873–23885.

Clark R.N., Swayze G.A., Wise R., et al. (2007) USGS Digital Spectral Library splib06a: U.S. Geological Survey, Digital Data Series 231.

Cloutis E.A. (2015) The University of Winnepeg's Planetary Spectrophotometer Facility (aka HOSERLab): What's new. *46th Lunar Planet. Sci. Conf.*, Abstract #1187.

Cloutis E.A., Pranoti M.A., & Mertzman S.A. (2002) Spectral reflectance properties of zeolites and remote sensing implications. *Journal of Geophysical Research*, **107**, E9, DOI:1029/2000JE001467.

Coblentz W.W. (1905) *Investigations of infra-red spectra*, **35**. Carnegie Institution Publications, Washington, DC.

Coblentz W.W. (1906) *Investigations of infra-red spectra*, **65**. Carnegie Institution Publications, Washington, DC.

Coblentz W.W. (1908) *Investigations of infra-red spectra*, **97**. Carnegie Institution Publications, Washington, DC.

Conel J.E. (1965) Infrared thermal emission from silicates. Jet Propulsion Laboratory Technical Memorandum 33–243.

Conel J.E. (1969) Infrared emissivities of silicates: Experimental results and a cloudy atmosphere model of spectral emission from condensed particulate mediums. *Journal of Geophysical Research*, **74**, 1614–1634.

Cooper C.D. & Mustard J.F. (1999) Effects of very fine particle size on reflectance spectra of smectite and palagonitic soil. *Icarus*, **142**, 557–570.

Cooper B.L., Salisbury J.W., Killen R.M., & Potter A.E. (2002), Midinfrared spectral features of rocks and their powders. *Journal of Geophysical Research*, **107**, 5017, 10.1029/2001JE001462.

Crowley J.K. & Hook S.J. (1996) Mapping playa evaporate minerals and associated sediments in Death Valley, California, with multispectral thermal infrared images. *Journal of Geophysical Research*, **101**, 643–660.

Dominguez G., McLeod A.S., Gainsforth Z., et al. (2014) Nanoscale infrared spectroscopy as a non-destructive probe of extraterrestrial samples. *Nature Communications*, **5**, 5445.

Donaldson Hanna K. & Sprague A.L. (2009) Vesta and the HED meteorites: Mid-infrared modeling of minerals and their abundances. *Meteoritics and Planetary Science*, **44**(11), 1755–1770.

Donaldson Hanna K.L., Thomas I.R., Bowles N.E., et al. (2012) Laboratory emissivity measurements of the plagioclase solid solution series under varying environmental conditions. *Journal of Geophysical Research*, 117, E11004, DOI:10.1029/2012JE004184.

Donaldson Hanna K.L., Greenhagen B.T., Patterson W.R. III, et al. (2017) Effects of varying environmental conditions on emissivity spectra of bulk lunar soils: Application to Diviner thermal infrared observations of the Moon. *Icarus*, **283**, 326–342.

Dyar M.D., Glotch T.D., Lane M.D., et al. (2011) Spectroscopy of Yamato 984028. *Polar Science*, **4**, 530–549.

Edwards C.S. & Christensen P.R. (2013) Microscopic emission and reflectance thermal infrared spectroscopy: Instrumentation for quantitative in situ mineralogy of complex planetary surfaces, *Applied Optics*, **52**, 2200–2217.

Estep-Barnes P.A. (1977) Infrared spectroscopy. In J. Zussman (ed.), *Physical methods in determinative mineralogy*, 2nd edn. Academic Press, New York, 529–603.

Farmer V.C. (1974) *The infrared spectra of minerals*. The Mineralogical Society, London.

Feely K.C. & Christensen P.R. (1999) Quantitative compositional analysis using thermal emission spectroscopy: Application to igneous and metamorphic rocks. *Journal of Geophysical Research*, **104**, 24,195–24,210.

Friedlander L.R., Glotch T.D., Bish D.L., et al. (2015) Structural and spectroscopic changes to natural nontronite induced by experimental impacts between 10 and 40 GPa. *Journal of Geophysical Research*, **120**, 888–912.

Frost R.L., Kloprogge T., Martens W.N., & Williams P. (2002) Vibrational spectroscopy of the basic manganese, ferric and ferrous phosphate minerals: Strunzite, ferristrunzite, and ferrostrunzite. *Neues Jahrbuch für Mineralogie, Monatshefte*, **11**, 481–496.

Gadsden J.A. (1975) *Infrared spectra of minerals and related inorganic compounds*. Butterworth & Co, London.

Glotch T.D. & Rossman G.R. (2009) Mid-infrared reflectance spectra and optical constants of six oxide/oxyhydroxide phases. *Icarus*, **204**, 663–671.

Glotch T.D., Christensen P.R., & Sharp T.G. (2006) Fresnel modeling of hematite crystal surfaces and application to martian hematite spherules. *Icarus*, **181**, 408–418.

Glotch T.D., Rossman G.R., & Aharonson O. (2007) Mid-infrared (5–100 μm) reflectance spectra and optical constants of ten phyllosilicate minerals. *Icarus*, **192**, 605–622.

Goodrich C.A., Kita N.T., Yin Q., et al. (2017) Petrogenesis and provenance of ungrouped achondrite Northwest Africa 7325 from petrology, trace elements, oxygen, chromium and titanium isotopes, and mid-IR spectroscopy. *Geochimica et Cosmochimica Acta*, **203**, 381–403.

Hamilton V.E. (2000) Thermal infrared emission spectroscopy of the pyroxene mineral series. *Journal of Geophysical Research*, **105**, 9701–9716.

Hamilton V.E. (2010) Thermal infrared (vibrational) spectroscopy of Mg-Fe olivines: A review and applications to determining the composition of planetary surfaces. *Chemie der Erde*, **70**, 7–33.

Hamilton V.E. & Christensen P.R. (2000) Determining the modal mineralogy of mafic and ultramafic igneous rocks using thermal emission spectroscopy. *Journal of Geophysical Research*, **105**, 9717–9733.

Hamilton V.E., Wyatt M.B., McSween H.Y. Jr., & Christensen P.R. (2001) Analysis of terrestrial and martian volcanic compositions using thermal emission spectroscopy. 2. Application to martian surface spectra from the Mars Global Surveyor Thermal Emission Spectrometer. *Journal of Geophysical Research*, **106**(7), 14722–14746.

Hamilton V.E., Christensen P.R., McSween H.Y. Jr., & Bandfield J.L. (2003) Searching for the source regions of Martian meteorites using MGS TES: Integrating martian meteorites into the global distribution of igneous materials on Mars. *Meteoritics and Planetary Science*, **38**(6), 871–885.

Hapke B. (1993) *Theory of reflectance and emittance spectroscopy*. Cambridge University Press, Cambridge.

Helbert J., Moroz L.V., Maturilli A., et al. (2007) A set of laboratory analogue materials for the MERTIS instrument on the ESA BepiColombo mission to Mercury. *Advanced Space Research*, **40**, 272–279.

Hellwege K.H., Lesch W., Plihal M., & Schaack G. (1970) Zwei-Phononen-Absorptionsspektren und Dispersion der Schwingungszweige in Kristallen der Kalkspatstruktur. *Zeitschrift für Physik*, **232**, 61–86.

Henderson B.G. & Jakosky B.M. (1997) Near-surface thermal gradients and mid-IR emission spectra: A new model including scattering and application to real data. *Journal of Geophysical Research*, **102**, 6567–6580.

Huminicki D.M.C. & Hawthorne F.C. (2002) The crystal chemistry of the phosphate minerals. In: *Phosphates: Geochemical, geobiological, and materials importance* (M.J. Kohn, J. Rakovan, & J.M. Hughes, eds.). Reviews in Mineralogy and Geochemistry. Mineralogical Society of America, Washington, DC, **48**, 123–253.

Hunt G.R. & Vincent R.K. (1968) The behavior of spectral features in the infrared emission from particulate surfaces of various grain sizes. *Journal of Geophysical Research*, **73**, 6039–6046.

Johnson J.R., Hörz F. & Staid M.I. (2003) Thermal infrared spectroscopy and modeling of experimentally shocked plagioclase feldspars. *American Mineralogist*, **88**, 1575–1582.

Keller L.P., Bajt S., Baratta G.A., et al. (2006) Infrared spectroscopy of comet 81P/Wild 2 samples returned by Stardust. *Science*, **314**, 1728–1731.

Kereszturi A., Gyollai I., & Szabó M. (2015) Case study of chondrule alteration with IR spectroscopy in NWA 2086 CV3 meteorite. *Planetary and Space Science*, **106**, 122–131.

King P.L., Ramsey M.S., McMillan P.F., & Swayze G.A. (2004) Laboratory Fourier Transform Infrared Spectroscopy methods for geologic samples. In: *Infrared Spectroscopy in Geochemistry, Exploration Geochemistry and Remote Sensing* (P.L. King, M.S. Ramsey, & G.A. Swayze, eds.). Mineralogical Association of Canada, Short Course 33, 57–91.

Klima R.L. & Pieters C.M. (2006) Near- and mid-infrared microspectroscopy of the Ronda peridotite. *Journal of Geophysical Research*, 111, DOI:10.1029/2005JE002537.

Kodama H. (1985) *Infrared Spectra of Minerals: Reference Guide to Identification and Characterization of Minerals for The Study of Soils*. Agriculture Canada, Ottawa.

Koike C., Chihara H., Tsuchiyama A., Suto H., Sogawa H., & Okuda H. (2003) Compositional dependence of infrared absorption spectra of crystalline silicate. II. Natural and synthetic olivines. *Astronomy & Astrophysics*, **399**, 1101–1107.

Kokaly R.F., Clark R.N., Swayze G.A., et al. (2017) USGS Spectral Library Version 7, USGS Data Series, Reston, VA.

Lane M.D. (1999) Midinfrared optical constants of calcite and their relationship to particle size effects in thermal emission spectra of granular calcite. *Journal of Geophysical Research*, **104**, 14099–14108.

Lane M.D. (2007) Midinfrared emission spectroscopy of sulfate and sulfate-bearing minerals. *American Mineralogist*, **92**, 1–18.

Lane M.D. & Christensen P.R. (1997) Thermal infrared emission spectroscopy of anhydrous carbonates. *Journal of Geophysical Research*, **102**, 25581–25592.

Lane M.D. & Christensen P.R. (1998) Thermal infrared emission spectroscopy of salt minerals predicted for Mars. *Icarus*, **135**, 528–536.

Lane M.D., Morris R.V., Mertzman S.A., & Christensen P.R. (2002) Evidence for platy hematite grains in Sinus Meridiani, Mars. *Journal of Geophysical Research*, **107**(E12), 5126, DOI:10.1029/2001JE001832.

Lane M.D., Glotch T.D., Dyar M.D., et al. (2011a) Midinfrared spectroscopy of synthetic olivines: Thermal emission, specular and diffuse reflectance, and attenuated total reflectance studies of forsterite to fayalite. *Journal of Geophysical Research*, **116**, E08010, DOI:10.1029/2010JE003588.

Lane M.D., Mertzman S.A., Dyar M.D., & Bishop J.L. (2011b) Phosphate minerals measured in the visible-near infrared and thermal infrared: Spectra and XRD analyses. *42nd Lunar Planet. Sci. Conf.*, Abstract #1013.

Lane M.D., Bishop J.L., Dyar M.D., et al. (2015) Mid-infrared emission spectroscopy and visible/near-infrared reflectance spectroscopy of Fe-sulfate minerals. *American Mineralogist*, **100**, 66–82, DOI:10.2138/am-2015-4762.

Logan L.M. & Hunt G.R. (1970) Emission spectra of particulate silicates under simulated lunar conditions. *Journal of Geophysical Research*, **75**, 6539–6548.

Logan L.M., Hunt G.R., Salisbury J.W., & Balsamo S.R. (1973) Compositional implications of Christiansen frequency maximums for infrared remote sensing applications. *Journal of Geophysical Research*, **78**, 4983–5003.

Long L.L., Querry M.R., Bell R.J., & Alexander R.W. (1993) Optical properties of calcite and gypsum in crystalline and powdered form in the infrared and far-infrared. *Infrared Physics*, **34**, 191–201.

Lorentz H.A. (1880) Über die Beziehung zwischen der Fortpflanzungsgeschwindigkeit des Lichtes und der Körperdichte. *Annalen der Physik*, **245**, 641–665.

Lorenz L. (1881) Über die Refractionsconstante. *Annalen der Physik*, **247**, 70–103.

Lyon R.J.P. (1964) Evaluation of infrared spectrophotometry for compositional analysis of lunar and planetary soils. II. Rough and powdered surfaces. NASA Contract Report, CR-100.

Lyon R.J.P. (1965) Analysis of rocks by spectral infrared emission (8–25 μm). *Economic Geology*, **60**, 715–736.

Lyon R.J.P. & Burns E.A. (1963) Analysis of rocks and minerals by reflected infrared radiation. *Economic Geology*, **58**, 274–284.

Marino M., Carati A., & Galgani L. (2007) Classical light dispersion theory in a regular lattice. *Annals of Physics*, **322**, 799–823.

Maturilli A., Helbert J., Witzke A., & Moroz L. (2006) Emissivity measurements of analogue materials for the interpretation of data from PFS on Mars Express and MERTIS on Bepi-Colombo. *Planetary and Space Science*, **54**(11), 1057–1064.

Maturilli A., Helbert J., & Moroz L. (2008) The Berlin emissivity database (BED). *Planetary and Space Science*, **56**(3–4), 420–425. Spectral library now available at figshare.com/articles/BED_Emissivity_Spectral_Library/1536469.

Maturilli A., Helbert J., Ferrari S., Davidsson B., & D'Amore M. (2016) Characterization of asteroid analogues by means of emission and reflectance spectroscopy in the 1- to 100-m spectral range. *Earth Planets and Space*, **68(1)**, article ID 113, 1–11.

Michalski J.R., Kraft M.D., Diedrich T., Sharp T.G., & Christensen P.R. (2003) Thermal emission spectroscopy of the silica polymorphs and considerations for remote sensing of Mars. *Geophysical Research Letters*, **30**, DOI:10.1029/2003GL018354.

Michalski J.R., Kraft M.D., Sharp T.G., Williams L.B., & Christensen P.R. (2006) Emission spectroscopy of clay minerals and evidence for poorly crystalline aluminosilicates on Mars from Thermal Emission Spectrometer data. *Journal of Geophysical Research*, **111** (E3), DOI:10.1029/2005JE002438.

Milam K.A., McSween H.Y. Jr., & Christensen P.R. (2007) Plagioclase compositions derived from thermal emission spectra of compositionally complex mixtures: Implications for martian feldspar mineralogy. *Journal of Geophysical Research*, **112**, DOI:10.1029/2006JE002880.

Milosevic M. (2012) Internal Reflection and ATR Spectroscopy. In: *Chemical analysis: A series of monographs on analytical chemistry and its applications* (Mark F. Vitha, Series Editor). John Wiley & Sons, New York.

Moenke H. (1962) *Mineralspektren I: Die Ultrarotabsorption der Häufigsten und Wirtschaftlich Wichtigsten Halogenid-, Oxyd-, Hydroxyd-, Carbonat-, Nitrat-, Borat-, Sulfat-, Chromat-, Wolframat-, Molybdat-, Phosphat-, Arsenat-, Vanadat- und Silikatmineralien im Spektralbereich 400–4000 cm^{-1}*. Akademie Verlag, Berlin.

Moenke H. (1966) *Mineralspektren II: Die Ultrarotabsorption Häufiger und Paragenetisch oder Wirtschaftlich Wichtiger Carbonate-, Borat-, Sulfat-, Chromat-, Phosphat-, Arsenat-, und Vanadat- und Silikatmineralien im Spektralbereich 400–4000 cm^{-1} (25–2.5 microns)*. Akademie Verlag, Berlin.

Moersch J.E. & Christensen P.R. (1995) Thermal emission from particulate surfaces: A comparison of scattering models with measured spectra. *Journal of Geophysical Research*, **100**, 7465–7477.

Morlok A., Bowey J., Köhler M., & Grady M.M. (2006) FTIR 2–16 micron spectroscopy of micron-sized olivines from primitive meteorites. *Meteoritics and Planetary Science*, **41**, 773–784.

Mozgawa W., Krol M., & Barczyk K. (2011) FT-IR studies of zeolites from different structural groups. *Chemik*, **65**, 667–674.

Mustard J.F. & Hays J.E. (1997) Effects of hyperfine particles on reflectance spectra from 0.3 to 25 μm. *Icarus*, **125**, 145–163.

Onomichi M., Kudo K., & Arai T. (1971) Reflection spectra of calcite in far-infrared region. *Journal of the Physical Society of Japan*, **31**, 1837.

Palomba E., Rotundi A., & Colangeli L. (2006) Infrared micro-spectroscopy of the martian meteorite Zagami: Extraction of individual mineral phase spectra. *Icarus*, **182**, 68–79.

Pieters C.M. & Hiroi T. (2004) RELAB (Reflectance Experiment Laboratory): A NASA multiuser spectroscopy facility. *35th Lunar Planet. Sci. Conf.*, Abstract #1720.

Pieters C.M., Klima R.L., Hiroi T., et al. (2008) Martian dunite NWA 2737: Integrated spectroscopic analyses of brown olivine. *Journal of Geophysical Research*, **113**, E06004, DOI:10.1029/2007JE002939.

Pitman K.M., Dijkstra C., Hofmeister A.M., & Speck A.K. (2010) Infrared laboratory absorbance spectra of olivine: Using classical dispersion analysis to extract peak parameters. *Mon. Royal Astronomical Society*, **406**, 460–481.

Pitman K.M., Noe Dobrea E.Z., Jamieson C.S., Dalton III J.B., Abbey W.J., & Joseph E.C.S. (2014) Reflectance spectroscopy and optical functions for hydrated Fe-sulfates. *American Mineralogist*, **99**, 1593–1603.

Ramsey M.S. & Christensen P.R. (1998) Mineral abundance determination: Quantitative deconvolution of thermal emission spectra. *Journal of Geophysical Research*, **103**, 577–596.

Rogers A.D. & Nekvasil H. (2015) Feldspathic rocks on Mars: Compositional constraints from infrared spectroscopy and possible formation mechanisms. *Geophysical Research Letters*, **42**, 2619–2626.

Ross S.D. (1974a) Sulphates and other oxy-anions of Group VI. In: *The Infrared Spectra of Minerals* (V. C. Farmer, ed.). The Mineralogical Society, London, 423–444.

Ross S.D. (1974b) Phosphates and other oxyanions of Group V. In: *The Infrared Spectra of Minerals* (V. C. Farmer, ed.). The Mineralogical Society, London, 383–422.

Ruff S.W. (2004) Spectral evidence for zeolite in the dust on Mars. *Icarus*, **168**, 131–143.

Ruff S.W. & Christensen P.R. (2007) Basaltic andesite, altered basalt, and a TES-based search for smectite clay minerals on Mars. *Geophysical Research Letters*, 34, DOI:10.1029/2007GL029602.

Ruff S.W., Christensen P.R., Barbera P.W., & Anderson D.L. (1997) Quantitative thermal emission spectroscopy of minerals: A technique for measurement and calibration. *Journal of Geophysical Research*, **102**, 14899–14913.

Salisbury F.B. & D'Aria D.M. (1992) Emissivity of terrestrial materials in the 8–14 μm atmospheric window. *Remote Sensing Environment*, **42**, 83–106.

Salisbury J.W. & Eastes J.W. (1985) The effect of particle size and porosity on spectral contrast in the mid-infrared. *Icarus*, **64**, 586–588.

Salisbury J.W. & Wald A. (1992) The role of volume scattering in reducing spectral contrast of reststrahlen bands in spectra of powdered minerals. *Icarus*, **96**, 121–128.

Salisbury J.W. & Walter L.S. (1989) Thermal infrared (2.5–13.5 µm) spectroscopic remote sensing of igneous rock types on particulate planetary surfaces. *Journal of Geophysical Research*, **94**, 9192–9202.

Salisbury J.W., Walter L.S., & D'Aria D. (1988) Mid-infrared (2.5 to 13.5 µm) spectra of igneous rocks. USGS Open File Report 88–686.

Salisbury J.W., D'Aria D.M., & Jarosewich E. (1991a) Midinfrared (2.5–13.5 µm) reflectance spectra of powdered stony meteorites. *Icarus*, **92**, 280–297.

Salisbury J.W., Walter L.S., Vergo N., & D'Aria D.M. (1991b) *Infrared (2.1–25 µm) spectra of minerals*. Johns Hopkins University Press, Baltimore, MD.

Salisbury J.W., Wald A., & D'Aria D.M. (1994) Thermal-infrared remote sensing and Kirchhoff's law 1. Laboratory measurements. *Journal of Geophysical Research*, **99**, DOI:10.1029/93JB03600.

Spitzer W.G. & Kleinman D.A. (1961) Infrared lattice bands of quartz. *Physical Review*, **121**, 1324–1335.

Stutman J.M., Termine J.D., & Posner A.S. (1965) Vibrational spectra and the structure of the phosphate ion in some calcium phosphates. *Transactions of the New York Academy of Sciences*, **27**, 669–675, DOI:10.1111/j.2164-0947.

Thomas I.R., Greenhagen B.T., Bowles N.E., Donaldson Hanna K.L., Temple J., & Calcutt S.B. (2012) A new experimental setup for making thermal emission measurements in a simulated lunar environment. *Review of Scientific Instruments*, **83**, 124502.

Vernazza P., Delbo M., King P.L., et al. (2012) High surface porosity as the origin of emissivity features in asteroid spectra. *Icarus*, **221**, 1162–1172.

Vernazza P., Castillo-Rogez J., Beck P., et al. (2017) Different origins or different evolutions? Decoding the spectral diversity among C-type asteroids. *The Astronomical Journal*, **153**, 72.

Wald A.E. & Salisbury J.W. (1995) Thermal infrared directional emissivity of powdered quartz. *Journal of Geophysical Research*, **100**, 24665–24675.

Weinger B.A., Reffner J.A., & DeForest P.R. (2009) A novel approach to the examination of soil evidence: Mineral identification using infrared microprobe analysis. *Journal of Forensic Sciences*, **54**, 851–856.

Weir C.E. & Lippincott E.R. (1961) Infrared studies of aragonite, calcite, and vaterite type structures in the borates, carbonates, and nitrates. *Journal of Research of the National Bureau of Standards A: Physics and Chemistry*, **65A**, 173–183.

Wenrich M.L. & Christensen P.R. (1996) Optical constants of minerals derived from emission spectroscopy: Application to quartz. *Journal of Geophysical Research*, **101**, 15921–15931.

Wyatt M.B., Hamilton V.E., McSween H.Y., Jr., Christensen P.R., & Taylor L.A. (2001) Analysis of terrestrial and martian volcanic compositions using thermal emission spectroscopy, 1. Determination of mineralogy, chemistry, and classification strategies. *Journal of Geophysical Research*, **106**, 14,711–14,732.

Yesiltas M., Sedlmair J., & Peale R.E. (2017) Synchrotron-based three-dimensional Fourier-transform infrared spectro-microtomography of Murchison meteorite grain. *Applied Spectroscopy*, **71**(6), 1198–1208.

4

Visible and Near-Infrared Reflectance Spectroscopy
Laboratory Spectra of Geologic Materials

JANICE L. BISHOP

4.1 Introduction

Visible to near-infrared (VNIR) reflectance spectra of solid geologic materials are presented here for the purpose of interpreting VNIR remote sensing data from planetary surfaces. VNIR spectra are defined in this chapter as extending from ~0.3 to 5 µm, which is the spectral region dominated by reflected radiation and volume scattering (Chapter 3). The term visible to short-wave infrared (VSWIR) is also used for this spectral range in some remote sensing contexts. Spectral features are related to the structure of minerals and other materials, and the spectral bands in the VNIR region are largely due to overtones and combinations of the fundamental vibrations (e.g., Hunt & Salisbury, 1970; Gaffey et al., 1993), described in Chapter 3, and electronic transitions (Hunt & Salisbury, 1970; Burns, 1993), described in Chapter 1. The spectral properties in this region are affected by grain size (e.g., Adams & Filice, 1967; Sunshine & Pieters, 1993; Mustard & Hays, 1997), environmental conditions (e.g., Bishop et al., 1994; Bishop & Pieters, 1995; Milliken & Mustard, 2005; Cloutis et al., 2008), described in Chapter 14, and mixture components (e.g., Nash & Conel, 1974; Singer, 1981; Clark, 1983; Cloutis et al., 1986; Mustard & Pieters, 1989). Spectra of volatiles and organics are described in Chapter 5. The best results for VNIR spectra of minerals and rocks are achieved for those containing Fe (and other transition metals), H_2O, and anions such as OH^-, CO_3^{2-}, NO_3^-, SO_4^{2-}, PO_4^{3-}, and ClO_4^- (Hunt & Salisbury, 1970; Burns, 1993; Gaffey et al., 1993). These features are sometimes easier to isolate and characterize in VNIR spectroscopy than in mid-IR spectroscopy (see Chapter 3) because the fundamental region at longer wavelengths includes multiple, overlapping absorptions from several species including SiO_4, Al_2O_3, FeO, and MgO bonds that are largely not detected in the VNIR region. Raman spectroscopy tends to provide multiple narrow vibrational absorptions that are characteristic of minerals (see Chapter 6), but VNIR spectroscopy typically provides more distinctive bands for Fe, H_2O, and anions, and better signatures for fine-grained and poorly crystalline or amorphous phases. Thus, studies coordinating VNIR, mid-IR, and Raman spectroscopy often provide complementary information for geologic samples.

Lab spectra of several classes of minerals, mineral mixtures, natural samples, and rocks are presented in this chapter to illustrate the characteristic spectral fingerprints of these mineral groups and also to demonstrate differences in the spectral features as a function of mineral chemistry. This chapter builds on previous studies (e.g., Clark et al., 1990; Gaffey et al., 1993; Clark, 1999; Bishop, 2005) to provide the VNIR spectral characteristics of a variety of minerals and geologic samples currently available. VNIR lab studies enable characterization of minerals, identification of minerals in natural samples or mixtures, and detection of rocks and minerals on planetary bodies. This is an important technique for remote sensing on Earth (Chapter 11) and planetary bodies, e.g., Mercury (Chapter 17), our Moon (Chapter 18), Mars (Chapter 23), asteroids (Chapters 19 and 20), comets (Chapter 20), Saturn's satellites (Chapter 21), and the Pluto system (Chapter 22).

4.2 Background: The Properties of VNIR Reflectance Spectroscopy

Reflectance spectroscopy measures the properties of solid materials when radiation (light) is directed at a surface and the photons reflected off the surface are collected by a detector. The spectrum depends on the optical properties of the material and the energy that is absorbed by excitations of atoms or molecules in the structure (Chapters 1–3). Minerals are identified using reflectance spectroscopy through the "bands," or dips in reflectance that occur at wavelengths where energy is absorbed, which are often unique to a given mineral structure. These bands are typically narrower for highly crystalline materials and broaden for less crystalline materials or molecular groups where there is a distribution in the excitational energies involved. The intensity of these spectral bands is in general related to the abundance of the excited phase.

4.2.1 Electronic Absorptions

As described in Burns (1993) and Chapter 1, electronic absorptions due to Fe or other metals are produced by electronic transitions from crystal field theory (CFT), metal-to-metal intervalence charge transfers, and ligand-to-metal charge transfers. The latter are generally O to Fe charge transfer bands for geologic materials. The frequency of an electronic transition is affected by the electron density at the Fe–O bond, which is determined by the charge on the Fe, the bonding configuration (octahedral or tetrahedral), and the mineral structure. The strongest CFT band for $Fe^{3+}O_6$ octahedra in ferric oxides/oxyhydroxides typically occurs in the 0.85–0.93 μm range, while the dominant CFT band for $Fe^{2+}O_6$ octahedra in ferrous minerals such as olivine or carbonate is typically in the 0.95–1.2 μm range. Further, if the FeO_6 octahedra are edge-sharing, then some O atoms will be bound to two separate Fe atoms and both will attract the electrons of the O atoms. When OH groups are bound to Fe, the H^+ in the OH bond pulls some electron density away from the O^{2-}, and the bond between Fe and OH is typically weaker than FeO bonds.

4.2.2 Vibrational Absorptions

Fundamental stretching and bending vibrations are described in Chapter 3. These vibrations were defined in terms of symmetric stretching (v_1), symmetric bending (v_2), asymmetric stretching (v_3), and asymmetric bending (v_4) for most minerals in early studies (e.g., Herzberg, 1945). For comparison with VNIR bands these vibrations are often termed v for stretching and δ for bending when it is clear which bending and stretching modes are intended. Molecules or groups of atoms in a mineral structure vibrate corresponding to the energy of the bond connecting the atoms. Vibrations of groups of atoms in minerals depend on the mass and charge of the atoms, physical constraints of the mineral, and the electron density of nearby atoms or molecules in the structure. Vibrations can be envisioned as motion along a spring of two atoms or along two or three springs for connected atoms. Shorter bonds exist for atoms bound more tightly; these have higher frequencies or shorter wavelengths. Heavy atoms will vibrate at a lower frequency.

The ability of an atom to pull electron density from O in an ionic bond depends on the polarizing power of the cation and influences the vibrational energy of the bond, and thus the wavelength of the vibrational absorption. Polarizing power is determined from the charge per radius and is affected by the size of the cation and the octahedral or tetrahedral configuration (Table 4.1). Cations such as K, Na, Ca, Mg, Fe, Al, and Si are often connected to O in mineral structures. For smectite clays Na, Ca, Mg, and Fe can be present as interlayer cations, and the polarizing power of these cations (Na < Ca < Mg < Fe^{3+}) is related to how strongly they attract water (Bishop et al., 1994). Similarly, the relative polarizing power of cations bound to sulfate, carbonate, perchlorate, etc. can be evaluated. This can enable qualitative estimates of bond strength, bond length, and vibrational energy by varying the cation.

Calculations using group theory (e.g., Cotton, 1990) determine which of these fundamental vibrational modes are active (induce an absorption) in IR or Raman spectroscopy for each specific crystal structure. A vibration is IR active when the electric dipole moment of the bond changes on excitation (Chapter 3), while a vibration is Raman active when the polarizability of the molecule changes on excitation (Chapter 6). Because of this difference, IR and Raman spectra may identify different species in a sample and are often complementary techniques in geologic analyses. IR active groups in minerals include nonsymmetric groups (e.g., OH) and molecules with a permanent dipole moment such as H_2O. NIR vibrational absorptions are readily observed for geologic materials containing H_2O, OH^-, NO_3^-, ClO_4^-, CO_3^{2-}, SO_4^{2-}, PO_4^{3-}, and similar complexes.

When the wavelength of incident radiation on a sample corresponds to the vibrational energy of the bond, a photon with that energy is absorbed and the bond jumps from the ground state to an excited state. The amount of energy equal to the vibrational frequency is absorbed from the reflectance spectrum, producing a dip in reflectance or a "band" at that energy level. A fundamental IR vibration occurs when the energy provided corresponds to the first excitational level, while overtones occur when the energy provided matches the energy of the second, third, etc. excitational level. VNIR reflectance spectroscopy is usually measured in terms of wavelength in micrometers or microns (μm); however, frequencies or wavenumbers in inverse centimeters (cm^{-1}) must be used for calculating vibrational bands. These units are inversely proportional and can be converted using $\lambda = 10\,000/k$, where λ is

Table 4.1 *Polarizing power of selected cations in minerals*

	Tetrahedral coordination				Octahedral coordination		
Cation	R (Å)	Charge	PP	Cation	R (Å)	Charge	PP
Si	0.40	4	10.00	Al	0.675	3	4.44
Al	0.53	3	5.66	Fe	0.69	3	4.35
Fe	0.63	3	4.76	Fe	0.75	2	2.67
Fe	0.77	2	2.60	Mg	0.86	2	2.33
Mg	0.71	2	2.82	Cr	0.87	2	2.30
Ti	0.56	4	7.14	Zn	0.88	2	2.27
Cr	0.55	4	7.27	Li	0.90	1	1.11
Co	0.54	4	7.41	Ca	1.14	2	1.75
Mn	0.53	4	7.55	Na	1.16	1	0.86
Ga	0.61	3	4.92	K	1.52	1	0.66

Notes: R is the effective ionic radius from Huheey et al. (1993); PP is the polarizing power calculated as charge per radius.

wavelength in μm and k is wavenumber in cm^{-1}. The frequency of overtone vibrations is slightly less than double that of the fundamental vibration due to anharmonic oscillations as described by quantum mechanics. The frequency of an overtone vibration (2ν) can be calculated from the fundamental vibration using an anharmonicity constant of 85.6 cm^{-1} (Petit et al., 2004a) as shown in Eq. (4.1) and Table 4.2.

$$2\nu(AB) = 2 \times [\nu(AB) - 85.6 \text{ cm}^{-1}]. \quad (4.1)$$

Combination bands occur when two or more vibrations are excited simultaneously. Common NIR combination bands occur for coordinated stretch plus bend vibrations of H_2O and OH. Overtones and combination bands typically have weaker absorptions than fundamental vibrations; however, in many cases the overtones and combination bands are more diagnostic of minerals. The combination ($\nu+\delta$) bands can be calculated simply by adding up the frequencies of the fundamental ν and δ frequencies (in wavenumber units, cm^{-1}) as in Bishop et al. (2002a). Examples are shown for nontronite in Table 4.2. The measured OH overtone value OH2ν (1) was converted to a calculated OHν value (2) using Eq. (4.1), and the measured OHν (3) and OHδ (4) values were combined to produce a calculated OH($\nu+\delta$) value (5) that is close to the measured OH($\nu+\delta$) value (6). Similarly, the measured H_2O overtone value $H_2O2\nu$ (7) was converted to a calculated $H_2O\nu$ value (8) using Eq. 4.1, and the measured values of $H_2O(\nu+\delta)$ (9) and $H_2O\delta$ (10) were used to calculate a value for $H_2O\nu$ (11). The calculated values for $H_2O\nu$ using these two methods produce similar results: 3613 cm^{-1} (8) and 3610 cm^{-1} (11).

Isotopes also affect vibrational energies due to their different masses. Thus, vibrational bonds for OH and OD, where D is deuterium, will have different vibrational energies.

Table 4.2 *Calculations of overtones and combination bands for nontronite*

(1) Meas.	(2) Calc.	(3) Meas.	(4) Meas.	(5) Calc.	(6) Meas.
OH$2v$ cm^{-1} (or μm) 6980 (or 1.43)	OHv cm^{-1} 3576	OHv cm^{-1} 3567	OHδ cm^{-1} 819	OH($v+\delta$) cm^{-1} 4386	OH($v+\delta$) cm^{-1} (or μm) 4380 (or 2.28)
(7) Meas.	(8) Calc.	(9) Meas.	(10) Meas.	(11) Calc.	
H$_2$O$2v$ cm^{-1} (or μm) 7055 (or 1.42)	H$_2$Ov cm^{-1} 3613	H$_2$O($v+\delta$) cm^{-1} (or μm) 5235 (or 1.91)	H$_2$Oδ cm^{-1} 1625	H$_2$Ov cm^{-1} 3610	

Notes: Data from Bishop et al. (2002a); Meas. refers to band centers measured in the spectra; Calc. refers to calculated band centers; the fundamental vibrations are typically measured in transmittance spectra, while the combinations and overtones are measured in reflectance spectra; v refers to stretching vibration, δ to bending vibration, ($v+\delta$) to combination band, and $2v$ to stretching overtone band.

Bonds with heavier atoms vibrate more slowly, resulting in lower frequencies (longer wavelengths) for the OD vibration compared to the OH vibration and similarly for the D$_2$O vibration compared to the H$_2$O vibration. Exchanging D for H in mineral structures has been effective in making band assignments in minerals containing multiple groups or complex bonding (e.g., Powers et al., 1975).

4.2.3 Effects of Grain Size and Surface Type on VNIR Spectral Features

Mineral characterization by VNIR spectroscopy generally involves measurement of multiple size fractions of the mineral in order to evaluate any changes in properties of the mineral or composition with grain size. For fine-grained materials, such as phyllosilicates or nanophase minerals/mineraloids, there will likely be a difference between grain size and particle size. In this case the sample grains may be on the order of tens or 100s of nm or a few μm, but the particle size range could be <25, 45–75, 90–125 μm, etc. For these types of materials the particles are aggregates of many grains, and if the grains are compacted in these particles, then the photons will interact with the sample as if it were a single larger particle rather than multiple tiny grains. VNIR spectra of particle size fractions of the same mineral illustrate that spectra of finer grains are typically brighter than spectra of coarser grains and that larger grains tend to have stronger spectral bands (e.g., Pieters, 1983; Mustard and Hays, 1997; Bishop et al., 2014b; Roush et al., 2015).

Light reflected off a smooth rock surface typically produces a strong specular component, where all or most of the light is reflected at the same angle. Conversely, a rough rock surface or a particulate sample scatters light diffusely, which means that light is reflected evenly in all directions. Diffuse reflectance is desired in most cases for VNIR spectroscopy of geologic materials. A Lambertian surface scatters light diffusely (Chapter 3). If a sample scatters diffusely, then the incident (i) angle (angle between the incident beam and nadir) and emergent (e) angle (angle between the detector and nadir) can be set to convenient values, e.g., $i = 30°$ and $e = 0°$. Some spectrometers are configured with rotating sample dishes in order to minimize orientation effects of the grains that could produce a specular component.

4.3 Methods

Laboratory VNIR reflectance measurements of well-characterized minerals, rocks, and mixtures provide a database for interpreting reflectance spectra from planetary bodies. To characterize the spectral properties of a given mineral well, it is good practice to measure several samples of that mineral type because minerals contain lattice defects and disordering, especially on the surfaces of grains. VNIR reflectance spectrometers are configured with an incident radiation beam, a sample holder, and a detector to collect the reflected radiation. Samples are normally placed in a dish made of nonreflective or highly absorbing material (e.g., Teflon™), so that stray photons are not reflected off the sample container during measurement. The reference standard should be highly reflecting and Lambertian; commonly used white references are Spectralon® (composed of polytetrafluoroethylene), Halon, barium sulfate, or a rough gold surface.

A bidirectional spectrometer has i and e angles that can be set to specific values and it typically measures spectra from 0.3 to 2.5 μm using a goniometer or grating type instrument. Other spectrometers are configured to have fixed i and are designed to collect all or most of the reflected photons in a hemispherical or biconical geometry using Fourier Transform Infrared (FTIR) spectra. The viewing geometry of the spectrometer describes the configuration of i and e, and can be designed for on-axis or off-axis measurements. In an off-axis configuration, i and e may be the same (e.g., 30°), but they are tilted off-axis such that the specular component would not be collected by the detector. On-axis systems rely on software to remove the specular component.

Detectors commonly used in VNIR spectroscopy include photodiodes made from silicon (~0.35–1.1 μm), a photoelectric cell or phototube (~0.3–1.1 μm), Indium Antimonide (InSb) (~1–5 μm), Mercury Cadmium Telluride (MerCad Telluride or MCT) (~2–25 μm), or deuterium triglyceride (DTGS) (~1.8–50 μm). A limitation of MCT detectors is that they must be cooled to reduce noise from thermally excited current.

Many spectrometers are enclosed in a chamber in order to control the sample environment during measurement (see Chapter 13). Water adsorption on particulate samples is common under ambient laboratory conditions and increases for fine-grained materials and samples with high surface area or disorder. In many cases, spectral absorptions due to this adsorbed water can mask the spectral features of the mineral or material under investigation. For spectral analyses of planetary surfaces, it is generally desired to acquire spectra in the laboratory of samples under controlled conditions, where moisture in the air (or other sample environment) is minimized. Temperature also affects VNIR spectra (see Chapter 13) and can produce shifts or changes in the spectral features (e.g., Singer & Roush, 1985; Pieters et al., 1986; Morris et al., 1997; Moroz et al., 2000).

The example spectra of minerals and other geologic samples presented in this chapter were measured either at the Reflectance Experiment Laboratory (RELAB) at Brown University or at the USGS spectroscopy laboratory (SpecLab) at Denver. The RELAB instruments are described in several studies (e.g., Pieters, 1983; Mustard, 1992; Bishop et al., 1994; Pieters & Hiroi, 2004; Bishop et al., 2008b) and online at www.planetary.brown.edu/relab/. Briefly stated, biconical reflectance spectra were measured relative to a rough gold surface using a Nicolet FTIR spectrometer in an H_2O- and CO_2-scrubbed environment. These data were spliced and scaled at ~1.2 μm to spectra acquired relative to Halon under ambient conditions from 0.3 to 2.5 μm using a bidirectional instrument. The spectral resolution is 5 nm for the bidirectional data and 4 cm^{-1} for the FTIR data. Details on the SpecLab data can be found online at https://speclab.cr.usgs.gov/ or in Clark et al. (1990) and Swayze et al. (2018). Briefly stated, spectra were measured relative to Spectralon® using high-resolution (ASD) spectrometers from 0.35 to 2.5 μm and using either a dual-grating, dual-slit Beckman monochromater with an integrating sphere or a Nicolet 760 FTIR spectrometer using an InSb detector at 4 cm^{-1} spectral resolution.

4.4 Spectral Features of Minerals

4.4.1 Silicate Minerals

Silicate minerals with spectral features in the VNIR region due to Fe and other transition metal cations include pyroxene, olivine, feldspar, garnet, and spinel (e.g., Hunt & Salisbury, 1970; Adams, 1974; Cloutis & Gaffey, 1991; Gaffey et al., 1993; Pieters et al., 1996). The electronic absorptions in these minerals are described in detail in Chapter 1. VNIR spectra include these metal absorptions as well as H_2O and OH absorptions due to inclusions in mineral grains in some cases (Figure 4.1). Narrow bands near 3.5 μm are due to CH absorptions (Chapter 5) in organic inclusions or admixtures and are similarly not part of the mineral structure.

VNIR spectra of pyroxenes contain two strong bands near 1 and 2 μm that are sensitive to the mineral structure and chemical composition (Adams, 1974, 1975; Cloutis & Gaffey, 1991; Sunshine & Pieters, 1993; Schade et al., 2004; Klima et al., 2007, 2008, 2011). Adams (1975) first mapped out the band centers of electronic absorptions in pyroxenes and other minerals in order to identify mineral type. Cloutis and Gaffey (1991) refined this technique to a comparison of the band centers near 1 and 2 μm for pyroxene analysis. Sunshine et al. (1990) identified multiple bands for pyroxene due to individual sites in the crystal structure. Since then, detailed analyses of the Fe absorption bands in pyroxene have enabled determination of pyroxene type, composition, and abundance (Sunshine & Pieters, 1993; Klima et al., 2011).

Early VNIR spectral studies of olivine found that it could be discriminated from many other Fe-bearing minerals by its much broader Fe band, extending from ~0.8 to 1.3 μm

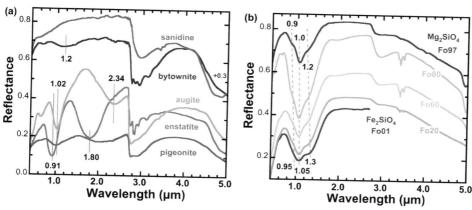

Figure 4.1 Reflectance spectra of selected feldspars, pyroxenes, and olivines from 0.3 to 5 μm. (a) Feldspars (from USGS library): sanidine (GDS19.19876) and bytownite (HS105.4014), and pyroxenes (from USGS library; Roush et al., 2015): augite (WS592.3259), enstatite (JB1183) and pigeonite (HS199.17856). (b) Olivines (from Sunshine and Pieters, 1998; Dyar et al., 2009; Bishop et al., 2013): Mg-rich forsterite (PO52, JB945), Fo80 (MDD87), Fo60 (MDD81), Fo20 (MDD96), to Fe-rich fayalite (PO58).

(Burns, 1970; Hunt & Salisbury, 1970; Burns & Huggins, 1972; Adams, 1975). Continued analyses of VNIR spectra of olivines with different compositions have enabled identification of multiple bands due to specific sites in the crystal structure (Sunshine & Pieters, 1998; Dyar et al., 2009; Isaacson et al., 2014). Modified Gaussian Modeling (MGM) of overlapping olivine bands in these studies enabled characterization of the three dominant bands for olivine. Modeling of synthetic olivine samples with variable Mg–Fe abundances found band centers at ~0.92 μm for site M1-1, ~1.07 μm for site M2, and ~1.29 μm for site M1-2 for fayalite (Fe_2SiO_4). For forsterite (Mg_2SiO_4), band centers were found at ~0.85 μm for site M1-1, ~1.04 μm for site M2, and ~1.23 μm for site M1-2 (Isaacson et al., 2014).

Additional silicate minerals such as feldspar, garnet, and spinel also exhibit VNIR bands when Fe is present in the structure, but these tend to be weaker and less diagnostic (Adams & Goullaud, 1978; Pieters, 1996; Johnson & Hörz, 2003; Cloutis et al., 2004; Cheek et al., 2009). The spectra of feldspars such as anorthite or bytownite include a weak, broad band near 1.1–1.3 μm that is related to the abundance of Fe (Adams & Goullaud, 1978).

4.4.2 Iron Oxide–Bearing Minerals

Iron oxide, hydroxide, and oxyhydroxide (FeOx) minerals are common on Earth and are an important component of the surface of Mars and likely Venus. Characterizing iron oxide–bearing minerals in geologic samples is important for assessing formation conditions of rocks and alteration of ferrous minerals. The properties of this mineral group have been described in detail by Cornell and Schwertmann (2003) and spectra have been presented in several studies (Hunt et al., 1971a; Sherman et al., 1982; Morris et al., 1985; Sherman &

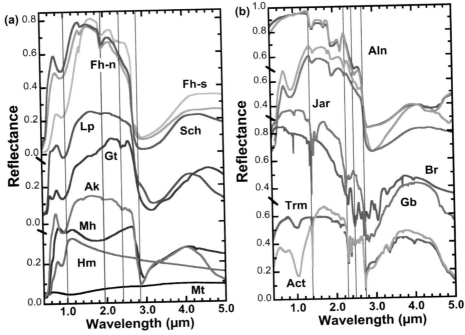

Figure 4.2 VNIR reflectance spectra of oxides and hydroxides from 0.3 to 5 μm (<125 μm size fraction for most samples). (a) Iron oxides/hydroxides (from Bishop et al., 1993; Bishop and Murad, 1996, 2002): hematite (Hm JB129), goethite (Gt JB047), ferrihydrite (Fh-s synthetic JB046, Fh-n natural, from Iceland JB499), magnetite (Mt JB307), maghemite (Mh JB301), lepidocrocite (Lp JB050), akaganéite (Ak JB048), schwertmannite (Sch JB130). (b) Selected OH-bearing minerals (from Bishop and Murad, 2005; Bishop library; USGS library): alunite (Aln: Na-Aln JB446 light green, K-Aln JB444 dark green), jarosite (Jar: Na-Jar JB440 orange, K-Jar JB441 red), gibbsite (Gb JB1644), brucite (Br JB944), tremolite (Trm HS18), actinolite (Act HS315). Lines mark spectral features in some of the spectra.

Waite, 1985; Bishop et al., 1993; Bishop & Murad, 1996; Morris et al., 1998; Crowley et al., 2003). The VNIR spectral region is dominated by Fe electronic and charge transfer bands from ~0.4 to 1.2 μm, plus OH vibrations near 1.41, 2.3–2.5, and 2.7 μm, and H_2O vibrations near 1.45, 1.95, and 3 μm. Reflectance spectra of selected fine-grained iron oxide–bearing minerals are shown in Figure 4.2a. Many FeOx minerals exhibit characteristic extended visible region spectra that can be used for identification of these minerals (e.g., Morris et al., 1985; Bishop et al., 1993), although this becomes more difficult for mixtures of FeOx–bearing minerals (Scheinost et al., 1998). Nanophase FeOx tend to have broader and weaker spectral features than crystalline structures, thus increasing the challenge of uniquely identifying these minerals in mixtures. Temperature also affects at least the Fe excitational bands in FeOx spectra (Pieters et al., 1986; Morris et al., 1997). Small amounts of Si, Al, or Ti are frequently observed in FeOx minerals and can alter their structures and properties

(Cornell & Schwertmann, 2003). The presence of Al in hematite was observed to reduce the crystallinity and broaden the spectral features (Morris et al., 1992).

Spectra of the oxides hematite (α-Fe$^{3+}_2$O$_3$), maghemite (γ-Fe$^{3+}_2$O$_3$), and magnetite (Fe^{2+}Fe$^{3+}_2$O$_4$) exhibit differences in the Fe bands due to the Fe–O bonding configuration and electronic environment, as described in Chapter 1. Spectra of fine-grained (~10–125 µm grain size) hematite have a band at 0.53 µm, a shoulder near 0.6 µm, and a band near 0.85 µm that shifts toward 0.88 µm in spectra of hematite with grains less than ~10 µm (Morris et al., 1989). In contrast, coarse-grained, gray hematite is spectrally dark and nearly featureless in the VNIR region. Maghemite has a characteristic ferric band near 0.92–0.93 µm and can exhibit a broad band near 1.85 µm when some Fe^{2+} is present. Magnetite is spectrally dark in this region and contains a broad, weak band centered near 1.1 µm that varies with grain size.

Goethite (α-Fe^{3+}OOH), lepidocrocite (γ-Fe^{3+}OOH), and akaganéite (β-Fe^{3+}OOH/Cl/H$_2$O) are ferric oxyhydroxides with structures built from arrangements of Fe octahedra with O or OH at the apices. The spectrum of goethite includes Fe electronic bands near 0.64–0.65 and 0.92–0.93 µm, an Fe-OH doublet near 2.4 and 2.5 µm, and a broad OH/H$_2$O band centered near 3.1–3.2 µm. The lepidocrocite spectrum has a shoulder near 0.65–0.7 µm, a band near 0.96–0.98 µm, a weak Fe-OH band near 2.5 µm, and a strong OH/H$_2$O band centered near 3.2 µm. Both goethite and lepidocrocite have weak bands near 1.9 µm due to H$_2$O as well. Although H$_2$O is not nominally in the structure of these minerals, there can be some H-bonding of the OH or adsorbed H$_2$O on grain surfaces. Akaganéite has a tunnel structure composed of Fe^{3+}-(O/OH) octahedra with Cl$^-$ and H$_2$O molecules in the tunnels (Song & Boily, 2013), and the spectra of akaganéite vary depending on the amount of Cl (Song & Boily, 2012). VNIR spectra of akaganéite include an Fe band near 0.92 µm and an OH combination band near 2.46 µm (Bishop et al., 2015). The H$_2$O combination band is shifted for akaganéite toward 2.01 µm due to the constrained environment in the tunnels. Both the OH and H$_2$O band positions occur at longer wavelength for akaganéite than for other minerals and are thus good indicators of this mineral.

Ferrihydrite and schwertmannite are nanophase hydrated FeOx, whose structures are not yet uniformly agreed upon (e.g., Bigham et al., 1996; Cornell & Schwertmann, 2003). Ferrihydrite has an approximate formula of Fe$^{3+}_5$O$_7 \cdot$n(OH/H$_2$O) and schwertmannite of Fe$_8$O$_8$(OH)$_{4-6}$(SO$_4$)$_{1-2} \cdot$nH$_2$O (Fernandez-Martinez et al., 2010). Small differences are observed in the spectral properties of these minerals reflecting variations in the structure due to changes in the synthesis procedures or formation conditions (e.g., Bigham et al., 1996; Bishop & Murad, 2002). Ferrihydrite typically has a broad Fe band near 0.92 µm, an H$_2$O overtone near 1.42 µm, an H$_2$O combination band near 1.93 µm, sometimes a shoulder or weak band near 2.3 µm, and a strong H$_2$O stretching vibration near 2.9–3 µm (Bishop & Murad, 2002). Schwertmannite is characterized by a broad Fe band near 0.92 µm, an H$_2$O overtone near 1.46 µm, an H$_2$O combination band near 1.98 µm, sometimes a shoulder or weak band near 2.3 µm, and a strong H$_2$O stretching vibration near 2.9–3 µm (Bishop et al., 2015). Understanding the spectra of such nanophase FeOx is important for planetary surfaces such as Mars, where poorly crystalline Fe phases are abundant and not yet well understood.

4.4.3 Hydroxide-Bearing Minerals

There are several important hydroxides and hydroxide-bearing minerals in addition to the Fe-bearing minerals presented in Section 4.4.1. The spectral features of brucite, $Mg(OH)_2$, and gibbsite, $Al(OH)_3$, in the NIR region are due only to OH vibrations (Hunt et al., 1971a). The brucite spectrum exhibits an OH stretching vibration at 2.71 μm and an overtone of this band at 1.40 μm, while the spectrum of gibbsite includes a triplet of OH stretching vibrations at 2.76, 2.83, and 2.94 μm, a set of OH overtone bands from 1.40 to 1.46 μm, and a doublet at 1.52 and 1.55 μm. Additional features occur at 2.33, 2.48, and 3.06 μm in the brucite spectrum. The OH combination stretching plus bending band at 2.33 μm for brucite is similar to the band observed for the Mg–OH combination vibration in some phyllosilicates. The OH combination band occurs at 2.27 μm for gibbsite, which is distinct from Al–OH vibrations in phyllosilicates.

Amphiboles are OH-bearing silicates, similar in structure to pyroxene, and early amphibole spectra are reported in Hunt et al. (1970). The most common amphiboles are the monoclinic actinolite group minerals that typically have the formula $Ca_2(Mg,Fe^{2+})_5Si_8O_{22}(OH)_2$, where the combined number of Mg or Fe^{2+} (or other) cations per structural unit is 5. The actinolite (Fe-rich amphibole) spectrum has a stronger Fe band centered near 1.04 μm (Figure 4.2) compared to that of tremolite (Mg-rich). The OH vibrations are similar for actinolite and tremolite, with an OH stretching band near 2.73 μm and an overtone of this at 1.40 μm and two bands due to combination vibrations near 2.29–2.32 and 2.39 μm. VNIR spectra of multiple actinolites and tremolites were analyzed and the OH band positions near 1.4 and 2.3 μm were used to estimate Mg abundance in the samples (Mustard, 1992).

OH-bearing sulfates include jarosite, alunite, amarantite, beaverite, butlerite, and others, but jarosite and alunite exhibit the strongest features due to structural OH (e.g., Bishop & Murad, 2005; Lane et al., 2015) and are thus included in Figure 4.2b. Alunite spectra have features near 1.76, 2.17, and 2.53 μm due to Al–OH vibrations, while jarosite spectra have bands near 1.85, 2.27, and 2.63 μm due to Fe–OH vibrations. Jarosite spectra also include Fe electronic bands at 0.43 and 0.91 μm. Additionally, multiple, overlapping sulfate bands are present in these spectra near 4.3–4.7 μm for alunite and near 4.6–4.9 μm for jarosite (e.g., Bishop & Murad, 2005; McCollom et al., 2014).

4.4.4 Phyllosilicates

Phyllosilicates, or clay minerals, include mica, talc, chlorite, smectite, the kaolin–serpentine group, the chain-structure group (e.g., prehnite, sepiolite, attapulgite), and vermiculite (Brindley & Brown, 1980; Deer et al., 1992; Swayze et al., 2018). Identifying phyllosilicates on planetary bodies provides constraints on the aqueous geochemical environments they formed in. Reflectance spectra of selected clay minerals are shown in Figure 4.3. These spectra include bands due to OH near 1.4, 2.2–2.3, and 2.7 μm, and due to water near 1.4, 1.9, and 3 μm; however, the exact position of these bands depends on the mineral structure and the type of cations in the structure. Fe-bearing phyllosilicates such as nontronite and chlorite also exhibit Fe electronic transitions from 0.6 to 1 µm. A number of phyllosilicates were collected and characterized (e.g., SAz-1, SWy-1, SapCa-1, IMt-1, KGa-1) by the Clay Minerals Society Source Clays Repository (Van Olphen & Fripiat, 1979) and spectra of many of these are included in Figure 4.3a. VNIR reflectance spectra of many types of phyllosilicates are described in Hunt et al. (1970), Clark et al. (1990), and Bishop et al. (2008b).

Smectites have received the most study of clay minerals in this spectral region (Cariati et al., 1981; Post, 1984; Clark et al., 1990; Post & Noble, 1993; Bishop et al., 1994, 1999, 2002b,c, 2008b, 2011b; Post et al., 1997; Gates, 2005; Decarreau et al., 2008). All smectite spectra exhibit an H_2O stretching overtone at 1.41 µm and an asymmetric H_2O combination band centered at 1.91 µm (Bishop et al., 1994). The M–OH features depend on the presence of Al, Fe, or Mg cations in the octahedral sites, as well as Al and Fe substitution for Si in the tetrahedral sites. Montmorillonites with high Al abundance in the octahedral sites exhibit an OH combination band near 2.20 µm, but this can be extended toward 2.21 µm for montmorillonites with some Fe or Mg in the octahedral sites. The Al–OH stretching overtone lies at 1.41 µm, at the same wavelength as the H_2O stretching overtone, which was a source of confusion for decades in making band assignments. Beidellite spectra exhibit vibrations at 1.40 and 2.18 µm (Bishop et al., 2011b), which are shorter wavelengths than the bands observed in montmorillonite spectra due to increased tetrahedral Al substitution for Si. Nontronite spectra include an Fe^{3+}–OH stretching overtone band at 1.43 µm and an Fe^{3+}–OH combination band at 2.29 µm. Spectra of nontronites containing some Al in octahedral sites (AlFeOH) have an additional shoulder or weak band near 2.23 µm. Saponite spectra include an Mg–OH stretching overtone band at 1.39 µm and an Mg–OH combination band at 2.31 µm. Smectites also include an additional OH band at longer wavelengths (~2.4–2.5 µm) that depends on the octahedral cations. This occurs at 2.38 µm for saponite, 2.39 µm for hectorite, 2.41 µm for nontronite, 2.44 µm for beidellite, and 2.45 µm for montmorillonite.

Several studies have investigated the VNIR spectral properties of kaolin–serpentine group minerals (King & Clark, 1989; Clark et al., 1990; Petit et al., 1999; Bishop et al., 2002b, 2008a,b). VNIR spectra of kaolinite and serpentine are readily distinguished from smectite spectra because the kaolin–serpentine group clays do not have H_2O bands. Kaolinite spectra exhibit a sharp pair of bands due to the Al–OH stretching overtone at 1.396 and 1.416 µm and an Al–OH combination band doublet at 2.17 and 2.21 µm. The halloysite spectrum is similar to that of kaolinite, but includes a water band near 1.9 µm as well due to H_2O in the structure (e.g., Clark et al., 1990). Because of this 1.9 µm band in halloysite spectra, it can be difficult to determine if kaolinite or halloysite is present in mixtures of kaolin–montmorillonite samples. Serpentine spectra include an Mg–OH stretching overtone at 1.39 µm and an Mg–OH combination band near 2.33 µm. Many serpentine spectra also include bands near 2.52–2.57 µm. Some serpentines have minor amounts of Fe^{2+} present and these exhibit weak bands near 1 µm, but no shifts were observed in the bands near 2.3 and 2.5 µm with minor variations in Fe content (Bishop et al., 2002b).

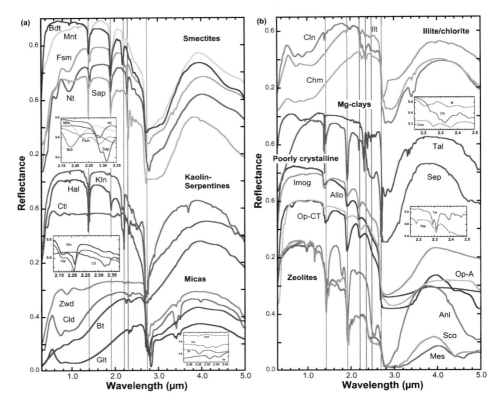

Figure 4.3 VNIR reflectance spectra of phyllosilicates and related materials from 0.3 to 5 μm (<45 or <125 μm size fraction). (a) Selected phyllosilicates (from Bishop et al., 2008): montmorillonite (Mnt, SAz-1 JB171), beidellite (Bdt, JB919), ferruginous smectite (Fsm, SWa-1 JB170), nontronite (Nt, JB175), saponite (Sap, SapCa-1 JB1184); kaolinite (Kln, KGa-1 JB225), halloysite (Hal JB1049), crysotile (Ctl JB732), zinnwaldite (Zwd JB729), celadonite (Cld JB727), biotite (Bt JB730), glauconite (Glt JB731). (b) Selected phyllosilicates, opal and zeolites (from Bishop et al., 2008b; Bishop library, USGS library): illite (Ilt, IMt-1 JB782), clinochlore (Cln JB738), chamosite (Chm JB739), talc (Tal), sepiolite (Sep, SepSP-1 JB290), allophane (Allo JB870), imogolite (Imog JB872), opal-A (Op-A JB632), opal-CT (Op-CT JB874), analcime (Anl GDS1.1919), scolecite (Sco GDS7.20124), mesolite (Mes GDS6.25925). Gray lines mark spectral features and close-up views are included for selected spectra near 2.2–2.3 μm.

VNIR spectra of chlorites include the expected OH bands near 1.39 and 2.33–2.35 μm. These band positions vary depending on the relative abundance of Fe and Mg (King & Clark, 1989; Calvin & King, 1997; Bishop et al., 2008b); however, they are distinguished from serpentines by the presence of an additional AlFeMg–OH combination band near 2.25 μm. The Mg-rich chlorites such as clinochlore have spectral bands near 1.39, 2.25, and 2.33 μm, while Fe^{2+}-rich chlorites such as chamosite have spectral bands near 1.42, 2.26, and 2.36 µm. Also, the Fe–OH stretching overtone band for chamosite is weaker in intensity, likely because of the strongly absorbing Fe^{2+} absorptions from ~0.7–1.2 µm. Sometimes H_2O bands are observed near 2.0 µm in chlorite spectra due to constrained H_2O sites, rather than near 1.9 µm as in most hydrated phyllosilicates. Spectra of some chlorites also contain a band near 2.48–2.57 µm.

Micas tend to exhibit much weaker NIR bands from 1.4 to 2.5 µm (e.g., Bishop et al., 2008b) than smectites, kaolin–serpentines, and chlorites, and the bands near 1.4 and 1.9 µm are generally too weak to be useful for remote detection. The OH combination vibrations typically occur as multiple bands in micas and vary with the octahedral cation composition. Al-rich micas such as zinnwaldite have spectral bands near 2.20 and 2.25 µm due to Al and Fe^{3+}, while celadonite has bands near 2.25, 2.30, and 2.35 µm due to combinations of Al, Fe, and Mg. Biotite with Al and Fe^{2+} cations has bands near 2.25 and 2.35 µm, while glauconite with Fe^{3+} and Fe^{2+} cations includes spectral bands near 2.30 and 2.36 µm.

Additional clays such as illite, palygorskite, sepiolite and attapulgite all exhibit spectral features due to both OH and H_2O in their structures (e.g., Clark et al., 1990; Bishop et al., 1998a). Illite spectra have bands near 1.41, 1.91, 2.22, 2.35, and 2.44 µm (Figure 4.3b). Palygorskite spectra contain bands near 1.42, 1.91, and 2.22 µm, and a weak band near 2.25 µm. Attapulgite spectra exhibit bands near 1.41 and 1.91 µm, and a doublet feature at 2.18 and 2.22 µm. The VNIR spectra of sepiolite include features near 1.38, 1.42, 1.91, 2.18, and 2.31 µm due to both Al and Mg in its structure.

The VNIR spectra of talc (e.g., Clark et al., 1990; Petit et al., 2004b) include a band at 1.39 µm, a doublet at 2.29 and 2.31 µm, and additional bands near 2.39 and 2.46 µm. Talc is often mixed with smectites or chlorites and can be identified in these mixtures by the strong, sharp band at 1.39 µm and the shape of the doublet near 2.3 µm.

4.4.5 Poorly Crystalline Silicates and Aluminosilicates

Remote detection of poorly crystalline aluminosilicates versus phyllosilicates can provide information about the water/rock ratio and temperature during formation (e.g., Parfitt, 2009; Bishop & Rampe, 2016). The VNIR reflectance spectra of poorly crystalline silicates and aluminosilicates are somewhat related to the spectral properties of smectites in that they include features due to structural OH and H_2O. The spectral properties of opal and Si–OH phases have been studied by several groups (Anderson & Wickersheim, 1964; Milliken et al., 2008; McKeown et al., 2011; Rice et al., 2013; Bishop et al., 2016b; Sun et al., 2016). The VNIR spectra of opal-A are characterized by features near 1.39, 1.91, 2.21, and 3 µm (Figure 4.3b), where each of these bands can be broadened toward longer wavelengths due to H-bonding when the sample is hydrated. Spectra of opal-CT typically have a doublet or triplet feature near 1.4 µm and a narrower band near 2.21 µm. The spectra of Fe-treated opal have a shoulder near 2.26 µm causing a very broad band from 2.21 to 2.27 µm (Baker et al., 2011).

The spectral properties of nanophase aluminosilicates such as allophane and imogolite have been investigated by Bishop et al. (2013b) and Jeute et al. (2017). Allophane and imogolite have broadened spectral features in the VNIR region compared to smectites due to a distribution of Al–OH and Si–OH sites in the structure. Another difference is an H_2O combination band centered at 1.92 μm instead of 1.91 μm. The OH bands are also shifted. The OH stretching overtones occur at 1.38 and 1.40 μm for allophane and at 1.37 and 1.39 μm for imogolite, while the OH combination band occurs near 2.19 μm for both of these materials (Figure 4.3b).

4.4.6 Zeolites

Zeolites form in a variety of conditions, but many (e.g., chabazite, mesolite, scolecite, laumontite) are consistent with hydrothermal conditions or burial diagenesis (e.g., Cloutis et al., 2002), and can thus be indicators of such environments. Zeolite spectra in the VNIR region contain bands due to bound and adsorbed H_2O that are influenced by the zeolite structure and cations present (Cloutis et al., 2002). Analyses of water in zeolites (Bish & Carey, 2001; Bish et al., 2003) indicate that if zeolites are present on the surface of Mars they will be partially hydrated. In addition to H_2O bands near 0.97–0.98, 1.16–1.17, 1.42–1.47, 1.90–1.94, and 2.75–3 μm, many zeolites exhibit spectral features near 1.7–1.8, ~2.15 and 2.4–2.55 μm (Figure 4.3b). These latter bands are highly variable and depend on the structure of each particular zeolite. The spectrum of analcime, for example, contains a band near 1.8 μm that is typically observed for sulfates and a band near 2.54 μm similar to that of calcite; however, analcime (and other zeolites) can be distinguished from these other minerals by using multiple bands for identification.

4.4.7 Carbonates

Carbonates form in a variety of anhydrous and hydrated forms. Anhydrous carbonate spectra follow trends in band positions with cation type and structure in the NIR region (Hunt & Salisbury, 1971; Gaffey, 1987; Bishop et al., 2017, 2019). The spectral properties of hydrous carbonates are largely dominated by water bands (Calvin et al., 1994; Harner & Gilmore, 2015). The spectral properties of anhydrous carbonates are dominated by strong, sharp bands due to CO_3^{2-} vibrations; these follow different trends for the calcite-type with trigonal structure and the aragonite-type with orthorhombic structure (Weir & Lippincott, 1961; Bishop et al., 2017, 2019).

The most common calcite group minerals include calcite ($CaCO_3$), siderite ($FeCO_3$), and magnesite ($MgCO_3$); their spectra are shown in Figure 4.4a. Other carbonates with less common cations or mixed cations also fall in this calcite group: dolomite ($Ca/MgCO_3$), ankerite ($Ca/FeCO_3$), rhodocrosite ($MnCO_3$), smithsonite ($ZnCO_3$), and otavite ($CdCO_3$), among others. Only three of the four possible vibrational modes of the CO_3^{2-} anion are allowed in IR spectra for trigonal structure carbonates, giving rise

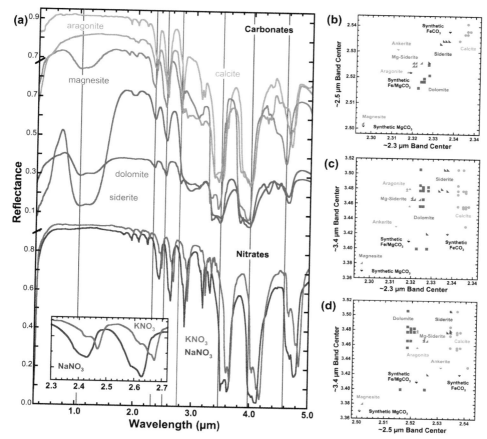

Figure 4.4 VNIR reflectance spectra of anhydrous carbonates and nitrates. (a) Spectra of selected anhydrous carbonates and nitrates from 0.3 to 5 μm (90–125 μm or <125 μm size fraction, from Bishop et al., 2017): magnesite (JB946), dolomite (JB1461), siderite (JB1462), calcite (JB1458), aragonite (JB1459), niter (JB998), and nitratine (JB997). Gray lines mark spectral features. Carbonate band center comparisons for (b) 2.3 vs. 2.5 μm, (c) 2.3 vs. 3.4 μm, and (d) 2.5 vs. 3.4 μm. (Band centers were determined from continuum-removed spectra.)

to mid-IR bands for the v_2, v_3, and v_4 modes (e.g., Lane & Christensen, 1997). Bands are observed near 2.30–2.34, 2.50–2.55, 3.5, 4.0, and 4.6 μm in the NIR region due to combinations and overtones of the mid-IR CO_3^{2-} vibrations. These NIR bands typically follow a pattern of having a shorter wavelength for magnesite, a longer wavelength for calcite, and a still longer wavelength for alkali carbonates. This trend can be observed by comparing the band centers near 2.3 and 2.5 μm (Figure 4.4b). Spectra were evaluated of multiple grain sizes of carbonates (Bishop et al., 2017) and the ~2.3 and 2.5 μm band centers appear extremely consistent in position across grain sizes for each sample. In contrast, variations were observed in shape and position for the bands at ~3.4 and 4 μm that are attributed to grain size effects. Siderite contains additional bands near 1.05 and 1.2 μm due to Fe electronic modes that result in a broad, strong feature. These Fe bands are nonlinear in carbonate spectra (Bishop et al., 2013a) and can be seen for the magnesite and dolomite samples in Figure 4.3a as well, although only small amounts of Fe are present (<0.3 wt.%).

Aragonite ($CaCO_3$) is the most common carbonate with orthorhombic structure; others include cerussite ($Pb\ CO_3$), strontianite ($Sr\ CO_3$) and witherite ($Ba\ CO_3$). All four of the fundamental CO_3^{2-} vibrational modes are IR active for carbonates with orthorhombic structure, resulting in mid-IR features for all of the v_1, v_2, v_3, and v_4 modes (Weir & Lippincott, 1961). Thus, NIR combinations and overtones of these mid-IR CO_3^{2-} vibrations are slightly different for aragonite group carbonates than for calcite group minerals and are observed near 2.32, 2.52, 3.48, 4.0, and 4.65 μm. Differences in carbonate spectral properties are demonstrated through comparison of band centers near 2.3, 2.5, and 3.4 μm (Figure 4.4b–d). For the 2.3/2.5 μm band comparison (Figure 4.4b), aragonite plots near dolomite and other mixed cation carbonates of the calcite group. However, the position of the bands near 3.4 and 4 μm in aragonite spectra are more similar to those in calcite spectra. The shape of the aragonite bands near 4.0 and 4.65 μm differ the most from the calcite group spectral features.

4.4.8 Nitrates

The nitrate minerals niter (KNO_3) and nitratine ($NaNO_3$) follow the aragonite and calcite structures, respectively, and hence exhibit related spectral features (Weir & Lippincott, 1961; Ross, 1974; Cloutis et al., 2016; Wang et al., 2018). NIR nitrate spectra exhibit NO_3^- bands similar to those of carbonates, but shifted toward longer wavelengths. These are observed near 2.43, 2.63, 2.87, ~3.55, ~4.1, and ~4.8 μm for $NaNO_3$ (Figure 4.4a). Similar bands are observed for KNO_3 at slightly longer wavelengths. As in the case of the carbonates, the spectra of nitrates following the aragonite and calcite structures differ the most in the shape of the bands near 4.1 and 4.8 μm.

4.4.9 Sulfates

Sulfate minerals form under a variety of pH conditions from very low to neutral, but are typically markers of acidic conditions and salty environments (e.g., Crowley, 1991). Sulfate minerals exhibit bands due to SO_4^{2-} in this region near 1.7–1.85 and 4–5 μm. Many also contain OH bands near 2.2–2.3 μm, H_2O bands near 1.45, 1.95, 2.4–2.55, and 3 μm, and Fe bands near 0.6–1.2 μm (e.g., Hunt et al., 1971b; Crowley, 1991; Crowley et al., 2003; Dalton, 2003; Bishop et al., 2005; Cloutis et al., 2006; Lane et al., 2015). Reflectance spectra of selected sulfates are shown in Figure 4.5 to illustrate these features. As an example, the spectral bands are marked at 1.44, 1.94, 2.42, 2.53, 4.2, 4.4, 4.6, and 4.8 μm for starkeyite ($MgSO_4 \cdot 4H_2O$) in Figure 4.5a. Related features are marked at 1.45,

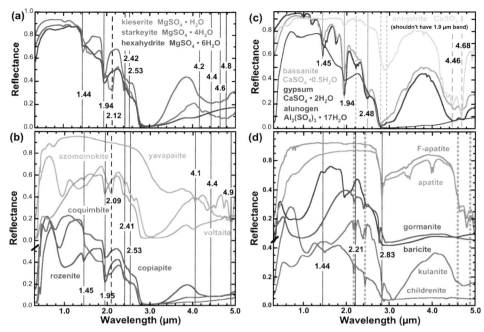

Figure 4.5 Reflectance spectra of selected sulfates and phosphates from 0.3 to 5 μm. (a) Mg sulfates (from Cloutis et al., 2006; Bishop et al., 2009): kieserite (JB736), starkeyite (JB711), and hexahydrite (LASF56). (b) Fe sulfates (from Lane et al., 2015): yavapaiite (JB634), szomolnokite (JB1167), voltaite (JB1165), coquimbite (JB621), copiapite (JB1172), and rozenite (JB626). (c) Ca and Al sulfates (from Bishop et al., 2014; USGS library and Bishop collection): anhydrite (GDS42), bassanite (GDS145), gypsum (JB557), and alunogen (JB1050). (d) Phosphates (from Lane et al., 2007; Bishop library): baricite (JB749), gormanite (JB750), childrenite (JB745), kulanite (JB746) F-apatite (JB275, offset +0.2), and apatite (JB320, offset +0.4). Gray lines mark spectral features.

1.95, 2.41, and 2.53 μm plus a broad band near 4.2–4.8 μm for rozenite (FeSO$_4$•4H$_2$O) in Figure 4.5b.

The VNIR sulfate bands near 1.7–1.85 and 4–5 μm fall where few other minerals exhibit spectral features, and thus may be useful for detection of sulfates on planetary surfaces. The bands in the 4–5 μm region are attributed to overtones and combinations of fundamental SO$_4^{2-}$ vibrations, while the band at 1.7–1.85 μm may be due to combinations of SO$_4^{2-}$ and OH$^-$ or H$_2$O vibrations (Bishop and Murad, 2005; Cloutis et al., 2006). Al- and Ca-bearing sulfates (e.g., alunite, gypsum) exhibit a band near 1.75 μm (Figure 4.5c), while this band varies for Fe-bearing sulfates. This band occurs at 1.85 μm for jarosite (Figure 4.2b), as a weak band near 1.72 μm for rozenite (Fe^{2+}SO$_4$•4H$_2$O) and voltaite (~K$_2$Fe$^{2+}_5$Fe$^{3+}_3$Al(SO$_4$)$_{12}$•18H$_2$O) (Figure 4.5b), and as a shoulder near 1.78 μm for coquimbite (~Fe$^{3+}_2$(SO$_4$)$_3$•9H$_2$O) and copiapite (~Fe^{2+}Fe$^{3+}_4$(SO$_4$)$_6$(OH)$_2$•20H$_2$O) (Figure 4.5b). The 4–5 μm region typically contains a cluster of multiple strong bands or a broad feature with overlapping absorptions for the more hydrated sulfates.

Hydrated sulfates can often be distinguished from hydrated phyllosilicates by the position of the H_2O combination band near 1.9 μm. Because of the polarizing nature of the sulfate anion, this water band is typically shifted toward longer wavelengths for sulfates and occurs near 1.93–1.98 μm for most polyhydrated sulfates. These minerals also frequently exhibit a broad band (sometimes doublet) near 2.4–2.5 μm. In contrast, monohydrated sulfates (one H_2O molecule per formula unit) have a distorted structure and the water bands are different; here the H_2O combination band occurs near 2.1 μm and varies with composition. It is observed at 2.12 μm for kieserite ($MgSO_4\cdot H_2O$) and at 2.09 μm for szomolnokite ($FeSO_4\cdot H_2O$). An additional feature is observed as a narrow band at 2.4 μm for both kieserite and szomolnokite.

Gypsum ($CaSO_4\cdot 2H_2O$) and bassanite ($CaSO_4\cdot 0.5H_2O$) exhibit characteristic bands near 1.4–1.5, 1.93–1.94, and 2.1–2.3 μm due to H_2O vibrations and additional broad bands near 4.5 and 4.7 μm due to SO_4 vibrations. For gypsum a triplet occurs at 1.446, 1.490, and 1.538 μm, again at 2.178, 2.217, and 2.268 μm, followed by a single band at 2.486 μm. For bassanite, a triplet is observed at 1.428, 1.476, and 1.540 μm, followed by multiple weaker features at 2.10, 2.164, and 2.219 μm, a doublet at 2.262 and 2.268 μm, and an additional band at 2.484 μm (Bishop et al., 2014b).

Anhydrous sulfates such as yavapaiite ($KFe(SO_4)_2$) and anhydrite ($CaSO_4$) should not have any bands due to H_2O or OH from 1.4 to 3.2 μm, but frequently have weak absorptions due to adsorbed water or mineral defects. Anhydrous sulfate spectra frequently have much sharper bands in the 4–5 μm region, similar to the features observed in spectra of yavapaiite at 4.07, 4.42, 4.53, 4.74, and 4.91 μm or those in spectra of anhydrite at 4.20, 4.29, 4.35, 4.47, 4.67, 4.69, and 4.92 μm.

4.4.10 Phosphates

Phosphate minerals are observed as minor components of rocks, including meteorites. They exhibit a variety of spectral features in the VNIR region due to Fe, H_2O, OH, and PO_4^{3-} groups in the structure (e.g., Bishop, 2005; Lane et al., 2007). Spectra of several phosphate examples are shown in Figure 4.5. Fe bands typically occur near 0.6–1.2 μm, OH bands near 1.45, 2.2, and 2.8 μm (gray lines in Figure 4.5d) and phosphate bands near 2.38–2.52 and 4.5–5 μm (light gray dashed lines in Figure 4.5d). Kulanite ($BaFe_2^{2+}Al_2(PO_4)_3(OH)_3$), childrenite ($Fe^{2+}AlPO_4(OH)_2\cdot H_2O$), and gormanite ($Fe_3^{2+}Al_4(PO_4)_4(OH)_6\cdot 2H_2O$) are all phosphates that have strong OH bands near 1.45, 2.2, and 2.8 μm. An OH stretching overtone is observed at 1.50 μm, an OH stretch plus bend combination band at 2.16, and an OH fundamental stretching vibration at 2.88 μm is observed for childrenite. (XRD of the childrenite sample shows some admixture of the Mn endmember eosphorite in the sample.) An additional band at 2.43 μm is assigned to a PO_4^{3-} overtone. Both kulanite and gormanite exhibit multiple bands near 2.2 μm due to different OH sites in the structure. The gormanite spectrum has an OH triplet at 2.12, 2.17, and 2.20 μm with a PO_4^{3-} band at 2.39 μm and an OH fundamental band at 2.76 μm, while the kulanite spectrum has an OH triplet at 2.15, 2.20, and 2.24 μm; a PO_4^{3-} doublet at 2.37 and 2.45 μm, and an OH fundamental band at

2.83 μm. These OH overtone and fundamental bands can be verified using Eq. (4.1). For kulanite the OH stretching vibration occurs at ~3530 cm^{-1} (2.83 μm). This gives 3530 − 85.6 = 3444.4 cm^{-1} and then 2 × 3444.4 = 6888.8 cm^{-1}, which compares well with the measured value of ~6890 cm^{-1} (1.45 μm) for the OH stretching overtone. The spectrum of baricite ((Mg,Fe^{2+})$_3$(PO$_4$)$_2$•8H$_2$O) is dominated by water bands near 1.45, 1.95, and 3 μm and also contains a band near 2.47 μm attributed to PO$_4^{3-}$. Spectra of apatite (Ca$_5$(PO$_4$)$_3$OH) and F-apatite (Ca$_5$(PO$_4$)$_3$F) both exhibit PO$_4^{3-}$ bands near 4.64 and 4.80 μm. Both spectra also have a sharp, narrow band at 2.83 μm due to an OH stretching vibration. Apatite can include a mixture of OH, F, or Cl in the OH site; the F-apatite sample must contain some OH because of the 2.83 μm band. Additional bands observed as a triplet near 3.37–3.48 μm and a doublet at 3.98 and 4.02 μm are attributed to OH as well because they are found only in OH-apatite spectra.

4.4.11 Cl-Bearing Salts and Minerals

Cl-bearing minerals include chlorides (Cl$^-$), chlorates (ClO$_3^-$), and perchlorates (ClO$_4^-$) and they frequently form in evaporative desert environments where salts are concentrated or near oceans. Their VNIR spectra are dominated by H$_2$O bands (e.g., Bishop et al., 2014a, 2016a; Hanley et al., 2014, 2015). These occur near 1.44 and 1.94 μm for hydrous perchlorates, along with a drop in reflectance near 2.48 μm. Features are also observed for the perchlorate ion near 3.34, 4.5, and 4.9 μm (Figure 4.6). Some Cl-bearing salts also include features near 1.8 and 2.1 μm that appear to vary with the degree of hydration and are

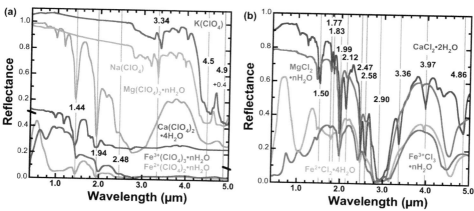

Figure 4.6 Reflectance spectra of selected chlorides and perchlorates from 0.3 to 5 μm. (a) Anhydrous perchlorates K(ClO$_4$) (JB983) and Na(ClO$_4$) (JB982) and hydrous perchlorates Mg(ClO$_4$)$_2$•nH$_2$O (JB981), Ca(ClO$_4$)$_2$•4H$_2$O (JB984), Fe^{2+}(ClO$_4$)$_2$•nH$_2$O (JB985), and Fe^{3+}(ClO$_4$)$_3$•nH$_2$O (JB986) (from Bishop et al., 2014a), and (b) MgCl$_2$ (JB1632) and Fe^{3+}Cl$_3$ (JB1637) were purchased as anhydrous chlorides, but they contain H$_2$O bands, and hydrous chlorides sinjarite or Ca^{2+}Cl$_2$•2H$_2$O (JB1629) and Fe^{2+}Cl$_2$•4H$_2$O (JB1634) (from Bishop et al., 2016a). Gray lines mark spectral features.

likely due to H_2O or OH in constrained sites. NaCl is bright and largely featureless in this spectral region, other than weak bands due to adsorbed water. All Cl-bearing salts are hygroscopic and adsorb water quickly in ambient environments, and many even deliquesce.

4.5 Spectral Features of Mineral Mixtures

The spectral properties of mineral mixtures are frequently not simply linear mixtures of the endmember components due to differences in the optical properties or particle sizes. Numerous studies have evaluated VNIR spectra of various combinations of silicate minerals in order to understand the spectra of planetary surfaces and estimate the abundances of individual mineral components (e.g., Nash & Conel, 1974; Singer, 1981; Clark, 1983; Cloutis et al., 1986; Mustard & Pieters, 1989; Mustard et al., 1993; Pieters et al., 1993; Sunshine & Pieters, 1993; Hiroi & Pieters, 1994; Lapotre et al., 2017). These studies have illustrated that smaller particles typically coat the surfaces of larger particles in mixtures and that the smaller particles tend to dominate the spectral properties of the mixture. For example, mixtures containing only a couple volume percent of ultra-fine-grained carbon or other dark materials can greatly darken the overall reflectance and dampen the absorption bands (e.g., Clark, 1983; Bishop et al., 1993; Fischer & Pieters, 1993). Studies involving mixtures of phyllosilicates and mafic components have found that only a small amount of Fe-rich silicates disproportionally influences the band strength in the extended visible region (~0.4–1.2 µm), while a small amount of phyllosilicate can produce OH absorptions stronger than expected based on their abundance (Orenberg & Handy, 1992; Honma et al., 2008; Amador et al., 2009; Ehlmann et al., 2009; McKeown et al., 2011; Saper & Bishop, 2011; Bishop et al., 2013a; Roush et al., 2015; De Angelis et al., 2016). This likely occurs because Fe is highly absorbing in the extended visible region and because phyllosilicates are typically in the form of small grains. It has been observed that while gypsum often dominates mixtures with other bright components, it can be masked in mixtures with dark components (Clark et al., 2008; King et al., 2013).

Examples of nonlinear mixing are illustrated in Figure 4.7. For mixtures of quartz and gypsum, only 10 wt.% gypsum produced an ~3 µm water band about half as strong as that of pure gypsum. For mixtures of magnesite (Mg-carbonate) and forsterite (Mg-olivine), 10 wt.% magnesite produced absorptions near 3.4 and 4 µm that are about one third as strong as those in spectra of pure magnesite; the 50 wt.% magnesite sample exhibited spectral features here that are nearly as strong as those of pure magnesite. However, mixtures of saponite (phyllosilicate) and enstatite (pyroxene) excluding the finest particles (Roush et al., 2015) resulted in mixture spectra where the phyllosilicate bands were underestimated compared to their abundance (Figure 4.7b).

As described in Chapter 2, several modeling techniques have been evaluated for determining mineral abundances in mixtures. These include radiative transfer modeling using methods developed by Hapke (e.g., Hapke, 1993) and Shkuratov (e.g., Shkuratov & Grynko, 2005) and Modified Gaussian Modeling (MGM) (e.g., Sunshine & Pieters, 1993). Analysis techniques have also been developed for VNIR reflectance spectra in

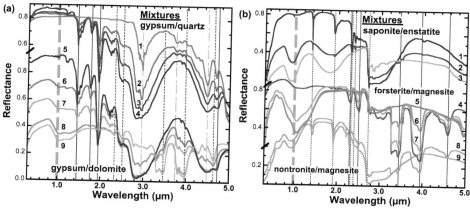

Figure 4.7 Reflectance spectra of selected mineral mixtures from 0.3 to 5 μm. (a) Mixtures of gypsum with dolomite and quartz (data from King et al., 2013). Size separates 90–150 (or 125) μm are shown. 1: quartz (JB1468), 2: 90 wt.% quartz and 10 wt.% gypsum (JB1466), 3: 80 wt.% quartz and 20 wt.% gypsum (JB1467), 4: 50 wt.% quartz and 50 wt.% gypsum (JB 1468), 5: gypsum (JB1464), 6: 80 wt.% gypsum and 20 wt.% dolomite (JB1475), 7: 50 wt.% gypsum and 50 wt.% dolomite (JB1474), 8: 20 wt.% gypsum and 80 wt.% dolomite (JB1473), 9: dolomite (JB1461). Gray solid lines mark gypsum features, gray dotted lines mark dolomite bands, light gray broken lines mark quartz bands, and a thick gray dashed line at 1 μm marks the region of Fe bands. (b) Mixtures of phyllosilicates, carbonates and mafic minerals (from Roush et al., 2015; Bishop et al., 2013a). 1: saponite (JB1184), 2: 50 wt.% saponite and 50 wt.% enstatite (JB1191), 3: enstatite (JB1183), 4: magnesite (JB946), 5: forsterite (JB945), 6: 90 wt.% forsterite and 10 wt.% magnesite (JB957), 7: 50 wt.% forsterite and 50 wt.% magnesite (JB955), 8: 50 wt.% nontronite and 50 wt.% magnesite (JB953), and 9: nontronite NAu-1 (JB790). Gray solid lines mark magnesite features, gray dotted lines mark saponite bands (similar to nontronite), and a thick gray dashed line at 1 μm marks the region of Fe bands.

order to remove a continuum and determine the band center (Clark & Roush, 1984; Clark et al., 2003). Quantitative analysis methods for VNIR reflectance spectra have also been developed through evaluating individual Gaussian bands for each Fe site in Fe-bearing minerals (Sunshine et al., 1990; Pieters et al., 1996). The MGM technique uses modified Gaussians to model the continuum together with Fe absorption bands in order to assess abundances of mineral components in mixtures (e.g., Sunshine et al., 1993; Sunshine & Pieters, 1993) and the amount of Fe in pyroxene or olivine samples (e.g., Sunshine & Pieters, 1993, 1998; Klima et al., 2007; Isaacson et al., 2014). The MGM approach applies modified Gaussians to model Fe absorptions and fits them to the reflectance spectrum using the Total Inversion algorithm (Tarantola & Valette, 1982). Parente et al. (2011) introduced an automated parameter initialization step to Gaussian modeling so that the user is not required to provide initial band estimates as with the MGM technique. This is based on the Levenberg–Marquardt algorithm and applies a nonconvex continuum removal and can be applied to vibrational bands as well as Fe bands (Bishop et al., 2011b). When the continuum is described well, both of these approaches model mineral absorption features successfully. MGM is desirable when band parameters are known or can be estimated prior to the modeling, while the Parente et al. (2011) approach is advantageous for situations in which the types of minerals in the sample are not already known. More recently, a Markov Chain Monte Carlo approach to spectral unmixing was implemented on laboratory mixture spectra to highlight the usefulness of probabilistic approaches to spectral unmixing of remote sensing data (Lapotre et al., 2017). Additional unmixing techniques are described in Chapter 14.

4.6 Spectral Features of Natural Geologic Materials

Rocks and soils represent variable textures and complex mixtures of mineral grains, glass, amorphous material, and altered components. VNIR spectra of these natural samples can be much more challenging to interpret than the spectra of pure mineral mixtures (e.g., Adams & Filice, 1967; Hunt & Ashley, 1979; Mustard & Pieters, 1987; Pieters & Mustard, 1988; Morris et al., 1995; Clark, 1999; Bishop et al., 2004; Davis et al., 2014). VNIR spectra of synthetic glasses have been characterized in an effort to better understand how glass composition affects the spectral properties of rocks (Minitti & Rutherford, 2000; Minitti et al., 2002; Cannon et al., 2017). Rock coatings can also mask the spectral features of rocks (Singer & Roush, 1983; Fischer & Pieters, 1993; Morris et al., 2003; Minitti et al., 2007; Bishop et al., 2011c). Studying the spectral properties of natural, altered volcanic rocks and soils (e.g., Allen et al., 1981; Morris et al., 1990; Farrand & Singer, 1992; Bell et al., 1993; Bishop et al., 1998a, 2002a; Ehlmann et al., 2012), sediments (e.g., Bishop et al., 2001; Cuadros et al., 2013; Nwaodua et al., 2014), evaporites (Crowley, 1991), and products from hot springs and hydrothermal sources (e.g., Bishop & Murad, 2002; Goryniuk et al., 2004; Lin et al., 2016) also contributes toward our understanding of the geologic signatures of planetary bodies.

Selected spectra of altered volcanic ash and rocks, volcanic glass, and hot spring precipitates are shown in Figure 4.8. The VNIR spectra of altered volcanic material in Figure 4.8a illustrate how the fines are frequently spectrally dominated by bands due to phyllosilicates and poorly crystalline aluminosilicates (~1.4, 1.9, 2.2–2.3 µm) and ferric oxide–bearing phases (~0.9 µm), while the coarser fractions can represent less altered material and often contain features consistent with pyroxene, olivine, feldspar, or basaltic glass. Thin coatings (few µm) on volcanic glass can greatly change the VNIR spectral character (spectrum 7, Figure 4.8a). The smooth surface of volcanic glass can appear dark and featureless (spectrum 10) in VNIR spectra, while particulate versions of such samples exhibit typical basalt features near 1 and 2 µm (spectra 8–9). The VNIR spectra of hot springs precipitates can include mixtures of silica, ferrihydrite, carbonate, gypsum, and other sulfates, as shown in Figure 4.8b.

Meteorites offer an opportunity to study rocks from other planetary bodies. They are classified by petrologic type, metamorphic grade, and origin (e.g., Wasson, 1985; Papike, 1989; Lauretta & McSween, 2006), and their spectral properties have been measured and analyzed in an effort to connect meteorite classes with types of asteroids (e.g., Chapman & Salisbury, 1973; Salisbury & Hunt, 1974; Gaffey, 1976; McFadden et al., 1980; Salisbury

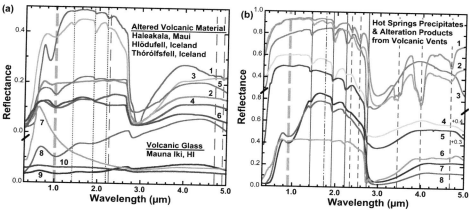

Figure 4.8 Reflectance spectra of selected basalt, tephra, ash, and sinter samples from 0.3 to 5 μm. (a) Altered volcanic material from Haleakala, Maui (Bishop et al., 2007): 1) JB394 45–125 μm, 2) JB395 45–125 μm, from Hlödufell, Iceland (Bishop et al., 2002a): 3) JB501b <45 μm, 4) JB501c 45–125 μm, from Thórólfsfell, Iceland (Bishop et al., 2002a): 5) JB500b <125 μm, 6) JB602f tephra surface, and volcanic glass from Mauna Iki, Hawaii (Bishop et al., 2003): 7) JB388y yellow coating, 8) JB388a ground <125 μm, 9) JB388b ground 150–250 μm, 10) JB388f fresh surface. (b) Precipitates from hot springs at Mammoth Formation, Yellowstone National Park (YNP) (Bishop et al., 2004): 1) JB264 <125 μm, 2) JB330 <125 μm, 3) JB331 <125 μm, from Octopus Springs, YNP (Bishop et al., 2004): 4) JB270 <125 μm, from Chocolate Pots, YNP: 7) JB644, and from the Landmannalaugar region of Iceland (Bishop & Murad, 2002): 8) JB499 <45 μm, and alteration products formed near volcanic vents at Kilauea, Hawaii (Bishop et al., 2016b): 5) JB632 <125 μm, and Nea Kameni, Santorini, Greece (Bishop et al., 1998a): 6) JB182 <125 μm. Gray lines mark spectral features.

et al., 1991; Moroz et al., 2000; Johnson & Hörz, 2003; Sunshine et al., 2004; Sunshine et al., 2007; Cloutis et al., 2010a, 2011a,b, 2012a,b,c; Ruesch et al., 2015). This is complicated by space weathering on asteroid surfaces (Chapter 13) and terrestrial weathering of meteorites. More than 100 meteorites from Mars are currently available and several VNIR spectroscopic studies have characterized meteorites from this collection (e.g., Sunshine et al., 1993; Bishop et al., 1998b,c; Schade & Wäsch, 1999; Hiroi et al., 2005; McFadden & Cline, 2005; Pieters et al., 2008; Bishop et al., 2011a; Parente et al., 2011) and these data have been used to identify possible source regions on Mars (Mustard et al., 1997; Ody et al., 2015). VNIR spectroscopy studies of lunar meteorites have also been performed (e.g., Pieters et al., 1983; Sugihara et al., 2004; Isaacson et al., 2009, 2010). These have been compared with the spectra of lunar rocks and soils as summarized by Isaacson et al. (2011) and VNIR spectra of the surface of the Moon described in detail in Chapter 18. Ureilite meteorites received some additional attention because of the fall in Sudan in 2008 and VNIR spectra of several fragments of Almahata Sitta (Hiroi et al., 2010) and other ureilites (Cloutis et al., 2010b) are available for comparison with the spectral properties of the incoming asteroid TC3 (Jenniskens et al., 2009). VNIR spectra of selected meteorites are examples of natural mixtures of mafic minerals and glasses (Figure 4.9). VNIR spectra documenting the

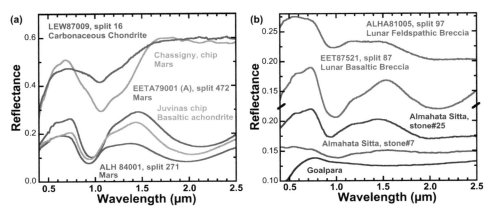

Figure 4.9 Reflectance spectra of selected meteorites chips and breccias from 0.4 to 2.5 µm (bidirectional spectra only from RELAB). (a) Carbonaceous chondrite LEW 87009 (C1LM11, McFadden collection), basaltic achondrite Juvinas (C2MB70, Hiroi collection), martian meteorites Chassigny (C1DD23, Dyar collection), EET A79001 (C1MT04, Bishop et al., 2011a), and ALH 84001 (C1MT03, Bishop et al., 1998c). (b) Lunar breccias ALH A81005 (C1LM34, Isaacson et al., 2009) and EET 87521 (C1LM37, Isaacson et al., 2009), and the ureilites Almahata Sitta stone #25 (C1MT94, Hiroi et al., 2010) and Almahata Sitta stone #7 (C1MT95, Hiroi et al., 2010), and Goalpara (C1PH29, Cloutis et al., 2010b).

properties of carbonaceous chondrites have enabled mapping these features to specific chondrite classes (Cloutis et al., 2011a,b, 2012a,b,c). Microimaging spectroscopy has also been used to survey the spectral diversity of howardite, eucrite, and diogenite (HED) meteorites (Fraeman et al., 2016) for comparison with Vesta and other bodies.

4.7 Summary of VNIR Spectral Features of Geologic Materials

VNIR reflectance spectroscopy is an important technique for characterizing the surface of planetary bodies using spectra of pure minerals, mineral mixtures, and analog materials. This chapter covers the region 0.3–5 µm and describes the species responsible for the absorption of radiation at specific wavelengths, creating spectral features used to identify minerals, rocks, and other geologic materials. Fe contributes greatly to the VNIR spectral signatures. Features near 1 and 2 µm characteristic of Fe^{2+} are present in spectra of pyroxene and glass, while a broad, strong band from ~0.9 to 1.3 µm is characteristic of Fe^{2+} in olivine, carbonate, and many sulfates. A weak band near 1.2 µm is due to Fe^{2+} in feldspar, and bands near 0.6 and 0.9 µm arise from Fe^{3+} in ferric oxides/hydroxides. Water bands occur near 1.4, 1.9, and 2.9 µm, while structural OH bands occur near 1.4, 2.1–2.5, and 2.7 µm. These H_2O and OH features are characteristic of phyllosilicates. Additional features are observed for carbonates near 2.3, 2.5, 3.4, 4.0, and 4.6 µm, for nitrates near 2.4, 2.6, 3.5, 4.1, and 4.8 µm, for sulfates near 1.75–1.85,

2.4–2.5, and 4–5 µm, for phosphates near 4.5–5 µm, for chlorides near 1.8, 2.1 and 2.5 µm, and for perchlorates near 2.1, 3.3, 4.5, and 4.9 µm. The spectral signatures of geologic samples are also affected by how photons interact with particles in the sample. Factors such as grain size, coatings, and mixtures influence the reflectance, transmittance, and absorption of photons at grain boundaries and contribute to the VNIR spectral properties of geologic materials.

Acknowledgments

Most spectra described here are from the author's collection and were measured at RELAB, Brown University. I am grateful to C. Pieters, R. Milliken, and support from NASA for the use of the RELAB facility and to T. Hiroi for running countless spectra. I also appreciate the use of spectra from the USGS library to fill in gaps, as well as meteorite spectra from the collections of L. McFadden, T. Hiroi, and D. Dyar.

References

Adams J.B. (1974) Visible and near-infrared diffuse reflectance spectra of pyroxenes as applied to remote sensing of solid objects in the Solar System. *Journal of Geophysical Research*, **79**, 4829–4836.

Adams J.B. (1975) Interpretation of visible and near-infrared diffuse reflectance spectra of pyroxenes and other rock-forming minerals. In: *Infrared and Raman spectroscopy of lunar and terrestrial minerals* (C. Karr, ed.). Academic Press, New York, 91–116.

Adams J.B. & Filice A.L. (1967) Spectral reflectance 0.4 to 2.0 microns of silicate rock powders. *Journal of Geophysical Research*, **72**, 5705–5715.

Adams J.B. & Goullaud L.H. (1978) Plagioclase feldspars: Visible and near infrared diffuse reflectance spectra as applied to remote sensing. *Proceedings of the 9th Lunar and Planetary Science Conference*, 2901–2909.

Allen C.C., Gooding J.L., Jercinovic M., & Keil K. (1981) Altered basaltic glass: A terrestrial analog to the soil of Mars. *Icarus*, **45**, 347–369.

Amador E.S., Bishop J.L., McKeown N.K., Parente M., & Clark J.T. (2009) Detection of Kaolinite at Mawrth Vallis, Mars: Analysis of laboratory mixtures and development of remote sensing parameters. *40th Lunar Planet. Sci. Conf.*, Abstract #2188.

Anderson J.H. & Wickersheim K.A. (1964) Near infrared characterization of water and hydroxyl groups on silica surfaces. *Surface Science*, **2**, 252–260.

Baker L.L., Strawn D.G., McDaniel P.A., et al. (2011) Poorly crystalline, iron-bearing aluminosilicates and their importance on Mars. *42nd Lunar Planet. Sci. Conf.*, Abstract #1939.

Bell J.F., III, Morris R.V. & Adams J.B. (1993) Thermally altered palagonitic tephra: A spectral and process analog to the soil and dust of Mars. *Journal of Geophysical Research*, **98**, 3373–3385.

Bigham J.M., Schwertmann U., Traina S.J., Winland R.L., & Wolf M. (1996) Schwertmannite and the chemical modeling of iron in acid sulfate waters. *Geochimica Cosmochimica Acta*, **60**, 2111–2121.

Bish D. & Carey J.W. (2001) Thermal behavior of natural zeolites. In: *Natural zeolites: Occurrence, properties, and applications*. Mineralogical Society of America Reviews in Mineralogy and Geochemistry (D.L. Bish & D. W. Ming, eds.). Mineralogical Society of America, Washington, DC, 403–452.

Bish D.L., Carey J.W., Vaniman D.T., & Chipera S.J. (2003) Stability of hydrous minerals on the martian surface. *Icarus*, **164**, 96–103.

Bishop J.L. (2005) Hydrated minerals on Mars. In: *Water on Mars and life*. Advances in Astrobiology and Biogeophysics. (T. Tokano, ed.). Springer, Berlin, 65–96.

Bishop J.L. & Murad E. (1996) Schwertmannite on Mars? Spectroscopic analyses of schwertmannite, its relationship to other ferric minerals, and its possible presence in the surface material on Mars. In: *Mineral spectroscopy: A tribute to Roger G. Burns* (M.D. Dyar, C. McCammon, & M.W. Schaefer, eds.). The Geochemical Society, Houston, TX, 337–358.

Bishop J.L. & Murad E. (2002) Spectroscopic and geochemical analyses of ferrihydrite from hydrothermal springs in Iceland and applications to Mars. In: *Volcano–ice interactions on Earth and Mars* (J.L. Smellie & M.G. Chapman, eds.). Special Publication No.202. Geological Society, London, 357–370.

Bishop J.L. & Murad E. (2005) The visible and infrared spectral properties of jarosite and alunite. *American Mineralogist*, **90**, 1100–1107.

Bishop J.L. & Pieters C.M. (1995) Low-temperature and low atmospheric pressure infrared reflectance spectroscopy of Mars soil analog materials. *Journal of Geophysical Research*, **100**, 5369–5379.

Bishop J.L. & Rampe E.B. (2016) Evidence for a changing martian climate from the mineralogy at Mawrth Vallis. *Earth and Planetary Science Letters*, **448**, 42–48.

Bishop J.L., Pieters C.M., & Burns R.G. (1993) Reflectance and Mössbauer spectroscopy of ferrihydrite-montmorillonite assemblages as Mars soil analog materials. *Geochimica Cosmochimica Acta*, **57**, 4583–4595.

Bishop J.L., Pieters C.M., & Edwards J.O. (1994) Infrared spectroscopic analyses on the nature of water in montmorillonite. *Clays and Clay Minerals*, **42**, 702–716.

Bishop J.L., Fröschl H., & Mancinelli R.L. (1998a) Alteration processes in volcanic soils and identification of exobiologically important weathering products on Mars using remote sensing. *Journal of Geophysical Research*, **103**, 31,457–31,476.

Bishop J.L., Pieters C.M., Hiroi T., & Mustard J.F. (1998b) Spectroscopic analysis of martian meteorite Allan Hills 84001 powder and applications for spectral identification of minerals and other soil components on Mars. *Meteoritics and Planetary Science*, **33**, 699–708.

Bishop J.L., Mustard J.F., Pieters C.M., & Hiroi T. (1998c) Recognition of minor constituents in reflectance spectra of Allan Hills 84001 chips and the importance for remote sensing on Mars. *Meteoritics and Planetary Science*, **33**, 693–698.

Bishop J.L., Murad E., Madejová J., Komadel P., Wagner U., & Scheinost A. (1999) Visible, Mössbauer and infrared spectroscopy of dioctahedral smectites: Structural analyses of the Fe-bearing smectites Sampor, SWy-1 and SWa-1. *11th International Clay Conference, June, 1997* (H. Kodama, A.R. Mermut, & J.K. Torrance, eds.). Ottawa, 413–419.

Bishop J.L., Lougear A., Newton J., et al. (2001) Mineralogical and geochemical analyses of Antarctic sediments: A reflectance and Mössbauer spectroscopy study with applications for remote sensing on Mars. *Geochimica Cosmochimica Acta*, **65**, 2875–2897.

Bishop J.L., Schiffman P., & Southard R.J. (2002a) Geochemical and mineralogical analyses of palagonitic tuffs and altered rinds of pillow lavas on Iceland and applications to Mars. In: *Volcano–ice interactions on Earth and Mars* (J.L. Smellie & M.G. Chapman, eds.). Special Publication No. 202. Geological Society, London, 371–392.

Bishop J.L., Murad E., & Dyar M.D. (2002b) The influence of octahedral and tetrahedral cation substitution on the structure of smectites and serpentines as observed through infrared spectroscopy. *Clay Minerals*, **37**, 617–628.

Bishop J.L., Madeová J., Komadel P., & Fröschl H. (2002c) The influence of structural Fe, Al and Mg on the infrared OH bands in spectra of dioctahedral smectites. *Clay Minerals*, **37**, 607–616.

Bishop J.L., Minitti M.E., Lane M.D., & Weitz C.M. (2003) The influence of glassy coatings on volcanic rocks from Mauna Iki, Hawaii and applications to rocks on Mars. *34th Lunar Planet. Sci. Conf.*, Abstract #1516.

Bishop J.L., Murad E., Lane M.D., & Mancinelli R.L. (2004) Multiple techniques for mineral identification on Mars: A study of hydrothermal rocks as potential analogues for astrobiology sites on Mars. *Icarus*, **169**, 331–323.

Bishop J.L., Dyar M.D., Lane M.D., & Banfield J.F. (2005) Spectral identification of hydrated sulfates on Mars and comparison with acidic environments on Earth. *International Journal of Astrobiology*, **3**, 275–285.

Bishop J.L., Schiffman P., Murad E., Dyar M.D., Drief A., & Lane M.D. (2007) Characterization of alteration products in tephra from Haleakala, Maui: A visible-infrared spectroscopy, Mössbauer spectroscopy, XRD, EPMA and TEM study. *Clays and Clay Minerals*, **55**, 1–17.

Bishop J.L., Dyar M.D., Sklute E.C., & Drief A. (2008a) Physical alteration of antigorite: A Mössbauer spectroscopy, reflectance spectroscopy and TEM study with applications to Mars. *Clay Minerals*, **43**, 55–67.

Bishop J.L., Lane M.D., Dyar M.D., & Brown A.J. (2008b) Reflectance and emission spectroscopy study of four groups of phyllosilicates: Smectites, kaolinite-serpentines, chlorites and micas. *Clay Minerals*, **43**, 35–54.

Bishop J.L., Parente M., Weitz C.M., et al. (2009) Mineralogy of Juventae Chasma: Sulfates in the light-toned mounds, mafic minerals in the bedrock, and hydrated silica and hydroxylated ferric sulfate on the plateau. *Journal of Geophysical Research*, **114**, E00D09, DOI:10.1029/2009JE003352.

Bishop J.L., Parente M., & Hamilton V.E. (2011a) Spectral signatures of martian meteorites and what they can tell us about rocks on Mars. *Meteoritical Society 74th Annual Meeting*, Abstract #5393.

Bishop J.L., Gates W.P., Makarewicz H.D., McKeown N.K., & Hiroi T. (2011b) Reflectance spectroscopy of beidellites and their importance for Mars. *Clays and Clay Minerals*, **59**, 376–397.

Bishop J.L., Schelble R.T., McKay C.P., Brown A.J., & Perry K.A. (2011c) Carbonate rocks in the Mojave Desert as an analog for martian carbonates. *International Journal of Astrobiology*, **10**, 349–358, DOI:10.1017/S1473550411000206.

Bishop J.L., Perry K.A., Dyar M.D., et al. (2013a) Coordinated spectral and XRD analyses of magnesite-nontronite-forsterite mixtures and implications for carbonates on Mars. *Journal of Geophysical Research*, **118**, 635–650.

Bishop J.L., Rampe E.B., Bish D.L., et al. (2013b) Spectral and hydration properties of allophane and imogolite. *Clays and Clay Minerals*, **61**, 57–74.

Bishop J.L., Quinn R.C., & Dyar M.D. (2014a) Spectral and thermal properties of perchlorate salts and implications for Mars. *American Mineralogist*, **99**, 1580–1592.

Bishop J.L., Lane M.D., Dyar M.D., King S.J., Brown A.J., & Swayze G. (2014b) Spectral properties of Ca-sulfates: Gypsum, bassanite and anhydrite. *American Mineralogist*, **99**, 2105–2115.

Bishop J.L., Murad E., & Dyar M.D. (2015) Akaganéite and schwertmannite: Spectral properties, structural models and geochemical implications of their possible presence on Mars. *American Mineralogist*, **100**, 738–746.

Bishop J.L., Davila A., Hanley J., & Roush T.L. (2016a) Dehydration-rehydration experiments with Cl salts mixed into Mars analog materials and the effects on their VNIR spectral properties. *47th Lunar Planet. Sci. Conf.*, Abstract #1645.

Bishop J.L., Schiffman P., Gruendler L., et al. (2016b) Formation of opal, clays and sulfates from volcanic ash at Kilauea Caldera as an analog for surface alteration on Mars. *Clay Minerals Society 53rd Annual Meeting*.

Bishop J.L., King S.J., Lane M.D., et al. (2017) Spectral properties of anhydrous carbonates and nitrates. *48th Lunar Planet. Sci. Conf.*, Abstract #2362.

Bishop J.L., King S.J., Lane M.D., et al. (2019) Spectral properties of anhydrous carbonates and nitrates. *Journal of Geophysical Research*, submitted.

Brindley G.W. & Brown G. (1980) *Crystal structures of clay minerals and their X-ray identification*. Mineralogical Society, London.

Burns R.G. (1970) Crystal field spectra and evidence of cation ordering in olivine minerals. *American Mineralogist*, **55**, 1608–1632.

Burns R.G. (1993) *Mineralogical applications of crystal field theory*. Cambridge University Press, Cambridge.

Burns R.G. & Huggins F.E. (1972) Cation determinative curves for Mg-Fe-Mn olivines from vibrational spectra. *American Mineralogist*, **57**, 967–985.

Calvin W.M. & King T.V.V. (1997) Spectral characteristics of Fe-bearing phyllosilicates: Comparison to Orgueil (C11), Murchison and Murray (CM2). *Meteoritics and Planetary Science*, **32**, 693–701.

Calvin W.M., King T.V.V., & Clark R.N. (1994) Hydrous carbonates on Mars? Evidence from Mariner 6/7 infrared spectrometer and groundbased telescopic spectra. *Journal of Geophysical Research*, **99**, 14659–14675.

Cannon K.M., Mustard J.F., Parman S.W., Sklute E.C., Dyar M.D., & Cooper R.F. (2017) Spectral properties of martian and other planetary glasses and their detection in remotely sensed data. *Journal of Geophysical Research*, **122**, 249–268.

Cariati F., Erre L., Gessa C., Micera G., & Piu P. (1981) Water molecules and hydroxyl groups in montmorillonites as studied by near infrared spectroscopy. *Clays and Clay Minerals*, **29**, 157–159.

Chapman C.R. & Salisbury J.W. (1973) Comparisons of meteorite and asteroid spectral reflectivities. *Icarus*, **19**, 507–522.

Cheek L.C., Pieters C.M., Dyar M.D., & Milam K.A. (2009) Revisiting plagioclase optical properties for lunar exploration. *40th Lunar Planet. Sci. Conf.*, Abstract #1928.

Clark J.T., Bishop J.L., Parente M., Brown A.J., & McKeown N.K. (2008) Constraining sulfate abundances on Mars using CRISM spectra and laboratory mixtures. *39th Lunar Planet. Sci. Conf.*, Abstract #1540.

Clark R.N. (1983) Spectral properties of mixtures of montmorillonite and dark carbon grains: Implications for remote sensing minerals containing chemically and physically adsorbed water. *Journal of Geophysical Research*, **88**, 10635–10644.

Clark R.N. (1999) Spectroscopy of rocks and minerals, and principles of spectroscopy. In: *Manual of remote sensing, 3: Remote sensing for the Earth sciences* (A.N. Rencz, ed.). John Wiley & Sons, New York, 3–58.

Clark R.N. & Roush T.L. (1984) Reflectance spectroscopy: Quantitative analysis techniques for remote sensing applications. *Journal of Geophysical Research*, **89**, 6329–6340.

Clark R.N., King T.V.V., Klejwa M., & Swayze G.A. (1990) High spectral resolution reflectance spectroscopy of minerals. *Journal of Geophysical Research*, **95**, 12653–12680.

Clark R.N., Swayze G.A., Livo K.E., et al. (2003) Imaging spectroscopy: Earth and planetary remote sensing with the USGS Tetracorder and expert systems. *Journal of Geophysical Research*, **108**, 5131, DOI:10.1029/2002JE001847.

Cloutis E.A. & Gaffey M.J. (1991) Pyroxene spectroscopy revisited: Spectral-compositional correlations and relationships to geothermometry. *Journal of Geophysical Research*, **96**, 22809–22826.

Cloutis E.A., Gaffey M.J., Jackowski T., & Reed K. (1986) Calibration of phase abundance, composition, and particle size distribution for olivine-orthopyroxene mixtures from reflectance spectra. *Journal of Geophysical Research*, **91**, 11641–11653.

Cloutis E.A., Asher P.M., & Mertzman S.A. (2002) Spectral reflectance properties of zeolites and remote sensing implications. *Journal of Geophysical Research*, **107**, 5067, DOI:10.1029/2000JE001467.

Cloutis E.A., Sunshine J.M., & Morris R.V. (2004) Spectral reflectance-compositional properties of spinels and chromites: Implications for planetary remote sensing and geothermometry. *Meteoritics and Planetary Science*, **39**, 545–565.

Cloutis E.A., Hawthorne F.C., Mertzman S.A., et al. (2006) Detection and discrimination of sulfate minerals using reflectance spectroscopy. *Icarus*, **184**, 121–157.

Cloutis E.A., Craig M.A., Kruzelecky R.V., et al. (2008) Spectral reflectance properties of minerals exposed to simulated Mars surface conditions. *Icarus*, **195**, 140–168.

Cloutis E.A., Hardersen P.S., Bish D.L., Bailey D.T., Gaffey M.J., & Craig M.A. (2010a) Reflectance spectra of iron meteorites: Implications for spectral identification of their parent bodies. *Meteoritics and Planetary Science*, **45**, 304–332.

Cloutis E.A., Hudon P., Romanek C.S., et al. (2010b) Spectral reflectance properties of ureilites. *Meteoritics and Planetary Science*, **45**, 1668–1694.

Cloutis E.A., Hiroi T., Gaffey M.J., Alexander C.M.O.D., & Mann P. (2011a) Spectral reflectance properties of carbonaceous chondrites: 1. CI chondrites. *Icarus*, **212**, 180–209.

Cloutis E.A., Hudon P., Hiroi T., Gaffey M.J., & Mann P. (2011b) Spectral reflectance properties of carbonaceous chondrites: 2. CM chondrites. *Icarus*, **216**, 309–346.

Cloutis E.A., Hudon P., Hiroi T., & Gaffey M.J. (2012a) Spectral reflectance properties of carbonaceous chondrites: 3. CR chondrites. *Icarus*, **217**, 389–407.

Cloutis E.A., Hudon P., Hiroi T., & Gaffey M.J. (2012b) Spectral reflectance properties of carbonaceous chondrites: 7. CK chondrites. *Icarus*, **221**, 911–924.

Cloutis E.A., Hudon P., Hiroi T., Gaffey M.J., & Mann P. (2012c) Spectral reflectance properties of carbonaceous chondrites: 8. "Other" carbonaceous chondrites: CH, ungrouped, polymict, xenolithic inclusions, and R chondrites. *Icarus*, **221**, 984–1001.

Cloutis E., Berg B., Mann P., & Applin D. (2016) Reflectance spectroscopy of low atomic weight and Na-rich minerals: Borates, hydroxides, nitrates, nitrites, and peroxides. *Icarus*, **264**, 20–36.

Cornell R.M. & Schwertmann U. (2003) *The iron oxides: Structure, properties, reactions, occurrences and uses*, 2nd edn. Wiley-VCH, Weinheim.

Cotton F.A. (1990) *Chemical applications of group theory*, 3rd edn. Wiley-Interscience, New York.

Crowley J.K. (1991) Visible and near-infrared (0.4–2.5 μm) reflectance spectra of Playa evaporite minerals. *Journal of Geophysical Research*, **96**, 16231–16240.

Crowley J.K., Williams D.E., Hammarstrom J.M., Piatak N., Chou I.-M. & Mars J.C. (2003) Spectral reflectance properties (0.4–2.5 μm) of secondary Fe-oxide, Fe-hydroxide, and Fe-sulphate-hydrate minerals associated with sulphide-bearing mine wastes. *Geochemistry: Exploration, Environment, Analysis*, **3**, 219–228.

Cuadros J., Michalski J.R., Dekov V., Bishop J., Fiore S., & Dyar M.D. (2013) Crystal-chemistry of interstratified Mg/Fe-clay minerals from seafloor hydrothermal sites. *Chemical Geology*, **360–361**, 142–158.

Dalton J.B. (2003) Spectral behavior of hydrated sulfate salts: Implications for Europa Mission spectrometer design. *Astrobiology*, **3**, 771–784.

Davis A.C., Bishop J.L., Veto M., et al. (2014) Comparing VNIR and TIR spectra of clay-bearing rocks. *45th Lunar Planet. Sci. Conf.*, Abstract #2699.

De Angelis S., Manzari P., De Sanctis M.C., Ammannito E., & Di Iorio T. (2016) VIS-IR study of brucite–clay–carbonate mixtures: Implications for Ceres surface composition. *Icarus*, **280**, 315–327.

Decarreau A., Petit S., Martin F., Vieillard P., & Joussein E. (2008) Hydrothermal synthesis, between 75 and 150C, of high-charge ferric nontronites. *Clays and Clay Mineral*, **56**, 322–337.

Deer W.A., Howie R.A., & Zussman J. (1992) *An introduction to the rock-forming minerals*. Longman, London.

Dyar M.D., Sklute E.C., Menzies O.N., et al. (2009) Spectroscopic characteristics of synthetic olivine: An integrated multi-wavelength and multi-technique approach. *American Mineralogist*, **94**, 883–898.

Ehlmann B.L., Mustard J.F., & Poulet F. (2009) Modeling modal mineralogy of laboratory mixtures of nontronite and mafic minerals from visible near-infrared spectra data. *40th Lunar Planet. Sci. Conf.*, Abstract #1771.

Ehlmann B.L., Bish D.L., Ruff S.W., & Mustard J.F. (2012) Mineralogy and chemistry of altered Icelandic basalts: Application to clay mineral detection and understanding aqueous environments on Mars. *Journal of Geophysical Research*, **117**, E00J16, DOI:10.1029/2012JE004156.

Farrand W.H. & Singer R.B. (1992) Alteration of hydrovolcanic basaltic ash: Observations with visible and near-infrared spectrometry. *Journal of Geophysical Research*, **97**, 17393–17408.

Fernandez-Martinez A., Timon V., Roman-Ross G., Cuello G.J., Daniels J.E., & Ayora C. (2010) The structure of schwertmannite, a nanocrystalline iron oxyhydroxysulfate. *American Mineralogist*, **95**, 1312–1322.

Fischer E. & Pieters C.M. (1993) The continuum slope of Mars: Bi-directional reflectance investigations and applications to Olympus Mons. *Icarus*, **102**, 185–202.

Fraeman A.A., Ehlmann B.L., Northwood-Smith G.W.D., Liu Y., Wadhwa M., & Greenberger R.N. (2016) Using VSWIR microimaging spectroscopy to explore the mineralogical diversity of HED meteorites. *8th Workshop on Hyperspectral Image and Signal Processing: Evolution in Remote Sensing (WHISPERS)*, 1–5.

Gaffey M.J. (1976) Spectral reflectance characteristics of the meteorite classes. *Journal of Geophysical Research*, **81**, 905–920.

Gaffey S.J. (1987) Spectral reflectance of carbonate minerals in the visible and near infared (0.35–2.55 µm): Anhydrous carbonate minerals. *Journal of Geophysical Research*, **92**, 1429–1440.

Gaffey S.J., McFadden L.A., Nash D. & Pieters C.M. (1993) Ultraviolet, visible, and near-infrared reflectance spectroscopy: Laboratory spectra of geologic materials. In: *Remote geochemical analysis: Elemental and mineralogical composition* (C.M. Pieters & P.A.J. Englert, eds.). Cambridge University Press, Cambridge, 43–77.

Gates W.P. (2005) Infrared spectroscopy and the chemistry of dioctahedral smectites. In: *The application of vibrational spectroscopy to clay minerals and layered double hydroxides* (J.T. Kloprogge, ed.). Clay Minerals Society, Aurora, CO, 125–168.

Goryniuk M.C., Rivard B.A., & Jones B. (2004) The reflectance spectra of opal-A (0.5–25 µm) from the Taupo Volcanic Zone: Spectra that may identify hydrothermal systems on planetary surfaces. *Geophysical Research Letters*, **31**, DOI:10.1029/2004GL021481.

Hanley J., Dalton J.B., Chevrier V.F., Jamieson C.S., & Barrows R.S. (2014) Reflectance spectra of hydrated chlorine salts: The effect of temperature with implications for Europa. *Journal of Geophysical Research*, **119**, 2370–2377.

Hanley J., Chevrier V.F., Barrows R.S., Swaffer C., & Altheide T.S. (2015) Near- and mid-infrared reflectance spectra of hydrated oxychlorine salts with implications for Mars. *Journal of Geophysical Research*, **120**, 1415–1426.

Hapke B. (1993) *Theory of reflectance and emittance spectroscopy*. Cambridge University Press, Cambridge.

Harner P.L. & Gilmore M.S. (2015) Visible–near infrared spectra of hydrous carbonates, with implications for the detection of carbonates in hyperspectral data of Mars. *Icarus*, **250**, 204–214.

Herzberg G. (1945) *Molecular spectra and molecular structure. II. Infrared and Raman spectra of polyatomic molecules*. D. Van Nostrand, New York.

Hiroi T. & Pieters C.M. (1994) Estimation of grain sizes and mixing ratios of fine powder mixtures of common geologic minerals. *Journal of Geophysical Research*, **99**, 10,867–10,879.

Hiroi T., Miyamoto M., Mikouchi T., & Ueda Y. (2005) Visible and near-infrared reflectance spectroscopy of the Yamato 980459 meteorite in comparison with some shergottites. *Antarctic Meteorite Research*, **18**, 83–95.

Hiroi T., Jenniskens P.M., Bishop J.L., Shatir T.S.M., Kudoda A.M., & Shaddad M.H. (2010) Bidirectional visible-NIR and biconical FT-IR reflectance spectra of Almahata Sitta meteorite samples. *Meteoritics and Planetary Science*, **45**, 1836–1845.

Honma A., Bishop J.L., McKeown N.K., Brown A.J., & Parente M. (2008) Constraining phyllosilicate abundances on Mars using CRISM spectra and laboratory mixtures. *39th Lunar Planet. Sci. Conf.*, Abstract #1457.

Huheey J.E., Keiter E.A., & Keiter R.I. (1993) *Inorganic chemistry: Principles of structure and reactivity*, 4th edn. HarperCollins, New York.

Hunt G.R. & Ashley R.P. (1979) Spectra of altered rocks in the visible and near infrared. *Economic Geology*, **74**, 1613–1629.

Hunt G.R. & Salisbury J.W. (1970) Visible and near-infrared spectra of minerals and rocks: 1. Silicate minerals. *Modern Geology*, **1**, 283–300.

Hunt G.R. & Salisbury J.W. (1971) Visible and near-infrared spectra of minerals and rocks: II. Carbonates. *Modern Geology*, **2**, 23–30.

Hunt G.R., Salisbury J.W., & Lenhoff C.J. (1971a) Visible and near-infrared spectra of minerals and rocks: III. Oxides and hydroxides. *Modern Geology*, **2**, 195–205.

Hunt G.R., Salisbury J.W., & Lenhoff C.J. (1971b) Visible and near-infrared spectra of minerals and rocks: IV. Sulphides and sulphates. *Modern Geology*, **3**, 1–14.

Isaacson P.J., Liu Y., Patchen A., Pieters C.M., & Taylor L.A. (2009) Integrated analyses of Lunar meteorites: Expanded data for lunar ground truth. *40th Lunar Planet. Sci. Conf.*, Abstract #2119.

Isaacson P.J., Liu Y., Patchen A.D., Pieters C.M., & Taylor L.A. (2010) Spectroscopy of Lunar meteorites as constraints for ground truth: Expanded sample collection diversity. *41st Lunar Planet. Sci. Conf.*, Abstract #1927.

Isaacson P.J., Basu Sarbadhikari A., Pieters C.M., et al. (2011) The lunar rock and mineral characterization consortium: Deconstruction and integrated mineralogical, petrologic, and spectroscopic analyses of mare basalts. *Meteoritics and Planetary Science*, **46**, 228–251.

Isaacson P.J., Klima R.L., Sunshine J.M., et al. (2014) Visible to near-infrared optical properties of pure synthetic olivine across the olivine solid solution. *American Mineralogist*, **99**, 467–478.

Jenniskens P., Shaddad M.H., Numan D., et al. (2009) The impact and recovery of asteroid 2008 TC$_3$. *Nature*, **458**, 485–488.

Jeute T.J., Baker L.L., Abidin Z., Bishop J.L., & Rampe E.B. (2017) Characterizing nanophase materials on Mars: Spectroscopic studies of allophane and imogolite. *48th Lunar Planet. Sci. Conf.*, Abstract #2738.

Johnson J.R. & Hörz F. (2003) Visible/near-infrared spectra of experimentally shocked plagioclase feldspars. *Journal of Geophysical Research*, **108**, 5120, DOI:10.1029/2003JE002127, E11.

King S.J., Bishop J.L., Fenton L.K., Lafuente B., Garcia G.C., & Horgan B.H. (2013) VNIR reflectance spectra of gypsum mixtures for comparison with White Sands National Monument, New Mexico (WSNM) dune samples as an analog study of the Olympia Undae region of Mars. *AGU Fall Meeting*, Abstract #P23C-1800.

King T.V.V. & Clark R.N. (1989) Spectral characteristics of chlorites and Mg-serpentines using high-resolution reflectance spectroscopy. *Journal of Geophysical Research*, **94**, 13,997–14,008.

Klima R.L., Pieters C.M., & Dyar M.D. (2007) Spectroscopy of synthetic Mg-Fe pyroxenes I: Spin-allowed and spin-forbidden crystal field bands in the visible and near-infrared. *Meteoritics and Planetary Science*, **42**, 235–253.

Klima R.L., Pieters C.M., & Dyar M.D. (2008) Characterization of the 1.2 micrometer M1 pyroxene band: Extracting cooling history from near-IR spectra of pyroxenes and pyroxene-dominated rocks. *Meteoritics and Planetary Science*, **43**, 1591–1604.

Klima R.L., Dyar M.D., & Pieters C.M. (2011) Near-infrared spectra of clinopyroxenes: effects of calcium content and crystal structure. *Meteoritics and Planetary Science*, **46**, 379–395.

Lane M.D. & Christensen P.R. (1997) Thermal infrared emission spectroscopy of anhydrous carbonates. *Journal of Geophysical Research*, **102**, 25581–25592.

Lane M.D., Dyar M.D., & Bishop J.L. (2007) Spectra of phosphate minerals as obtained by visible-near infrared reflectance, thermal infrared emission, and Mössbauer laboratory analyses. *38th Lunar Planet. Sci. Conf.*, Abstract #2210.

Lane M.D., Bishop J.L., Dyar M.D., et al. (2015) Mid-infrared emission spectroscopy and visible/near-infrared reflectance spectroscopy of Fe-sulfate minerals. *American Mineralogist*, **100**, 66–82.

Lapotre M.G.A., Ehlmann B.L., & Minson S.E. (2017) A probabilistic approach to remote compositional analysis of planetary surfaces. *Journal of Geophysical Research*, **122**, 983–1009.

Lauretta D.S. & McSween H.Y. Jr. (2006) *Meteorites and the early solar system II*. The University of Arizona Press, Tucson, AZ.

Lin T.J., Ver Eecke H.C., Breves E.A., et al. (2016) Linkages between mineralogy, fluid chemistry, and microbial communities within hydrothermal chimneys from the Endeavour Segment, Juan de Fuca Ridge. *Geochemistry, Geophysics, Geosystems*, **17**, 300–323.

McFadden L.A. & Cline T.P. (2005) Spectral reflectance of martian meteorites: Spectral signatures as a template for locating source region on Mars. *Meteoritics and Planetary Science*, **40**, 151–172.

McFadden L.A., Gaffey M.J., & Takeda H. (1980) Reflectance spectra of some newly found, unusual meteorites and their bearing on the surface mineralogy of asteroids. *Proceedings of the 13th Lunar and Planetary Symposium*, Tokyo, 273–280.

McKeown N.K., Bishop J.L., Cuadros J., et al. (2011) Interpretation of reflectance spectra of clay mineral-silica mixtures: Implications for martian clay mineralogy at Mawrth Vallis. *Clays and Clay Mineral*, **59**, 400–415.

Milliken R.E. & Mustard J.F. (2005) Quantifying absolute water content of minerals using near-infrared reflectance spectroscopy. *Journal of Geophysical Research*, **110**, E12001, DOI:10.1029/2005JE002534.

Milliken R.E., Swayze G.A., Arvidson R.E., et al. (2008) Opaline silica in young deposits on Mars. *Geology*, **36**, 847–850.

Minitti M.E. & Rutherford M.J. (2000) Genesis of the Mars Pathfinder "sulfur-free" rock from SNC parental liquids. *Geochimica Cosmochimica Acta*, **64**, 2535–2547.

Minitti M.E., Mustard J.F., & Rutherford M.J. (2002) The effects of glass content and oxidation on the spectra of SNC-like basalts: Application to Mars remote sensing. *Journal of Geophysical Research*, **107**(E5), DOI:10.1029/2001JE001518.

Minitti M.E., Weitz C.M., Lane M.D., & Bishop J.L. (2007) Morphology, chemistry, and spectral properties of Hawaiian rock coatings and implications for Mars. *Journal of Geophysical Research*, **112**, E05015, DOI: 10.1029/2006JE002839.

Moroz L., Schade U., & Wäsch R. (2000) Reflectance spectra of olivine-orthopyroxene-bearing assemblages at decreased temperatures: Implications for remote sensing of asteroids. *Icarus*, **147**, 79–93.

Morris R.V., Lauer H.V. Jr., Lawson C.A., Gibson E.K. Jr., Nace G.A., & Stewart C. (1985) Spectral and other physicochemical properties of submicron powders of hematite (a-Fe2O$_3$), maghemite (g-Fe2O$_3$), magnetite (Fe3O$_4$), goethite (a-FeOOH), and lepidocrocite (g-FeOOH). *Journal of Geophysical Research*, **90**, 3126–3144.

Morris R.V., Agresti D.G., Lauer H.V. Jr., Newcomb J.A., Shelfer T.D., & Murali A.V. (1989) Evidence for pigmentary hematite on Mars based on optical, magnetic and Mössbauer studies of superparamagnetic (nanocrystalline) hematite. *Journal of Geophysical Research*, **94**, 2760–2778.

Morris R.V., Gooding J.L., Lauer H.V. Jr., & Singer R.B. (1990) Origins of Marslike spectral and magnetic properties of a Hawaiian palagonitic soil. *Journal of Geophysical Research*, **95**, 14,427–14,434.

Morris R.V., Schulze D.G., Lauer Jr. H.V., Agresti D.G., & Shelfer T.D. (1992) Reflectivity (visible and near IR), Mössbauer, static magnetic, and X ray diffraction properties of aluminum-substituted hematites. *Journal of Geophysical Research*, **97**, 10257–10266.

Morris R.V., Golden D.C., Bell III J.F., & Lauer H.V. Jr. (1995) Hematite, pyroxene, and phyllosilicates on Mars: Implications from oxidized impact melt rocks from Manicouagan crater, Quebec, Canada. *Journal of Geophysical Research*, **100**, 5319–5328.

Morris R.V., Golden D.C., & Bell III J.F. (1997) Low-temperature reflectivity spectra of red hematite and the color of Mars. *Journal of Geophysical Research*, **102**, 9125–9133.

Morris R.V., Golden D.C., Shelfer T.D., & Lauer H.V. Jr. (1998) Lepidocrocite to maghemite to hematite: A pathway to magnetic and hematitic martian soil. *Meteoritics and Planetary Science*, **33**, 743–751.

Morris R.V., Graff T.G., Mertzman S.A., Lane M.D., & Christensen P.R. (2003) Palagonitic (not Andesitic) Mars: Evidence from thermal emission and VNIR spectra of Palagonitic alteration rinds on basaltic rock. *6th Int. Conf. on Mars*, Abstract #3211.

Mustard J.F. (1992) Chemical analysis of actinolite from reflectance spectra. *American Mineralogist*, **77**, 345–358.

Mustard J.F. & Hays J.E. (1997) Effects of hyperfine particles on reflectance spectra from 0.3 to 25 µm. *Icarus*, **125**, 145–163.

Mustard J.F. & Pieters C.M. (1987) Abundance and distribution of ultramafic microbreccia in moses rock dike: Quantitative application of mapping spectroscopy. *Journal of Geophysical Research*, **92**, 10376–10390.

Mustard J.F. & Pieters C.M. (1989) Photometric phase functions of common geologic minerals and applications to quantitative analysis of mineral mixture reflectance spectra. *Journal of Geophysical Research*, **94**, 13619–13634.

Mustard J.F., Sunshine J.M., Pieters C.M., Hoppin A., & Pratt S.F. (1993) From minerals to rocks: Toward modeling lithologies with remote sensing. *24th Lunar Planet. Sci. Conf.*, Abstract, 1041–1042.

Mustard J.F., Murchie S.L., Erard S., & Sunshine J.M. (1997) In situ compositions of martian volcanics: Implications for the mantle. *Journal of Geophysical Research*, **102**, 25,605–25,615.

Nash D.B. & Conel J.E. (1974) Spectral reflectance systematics for mixtures of powdered hypersthene, labradorite, and ilmenite. *Journal of Geophysical Research*, **79**, 1615–1621.

Nwaodua E.C., Ortiz J.D., & Griffith E.M. (2014) Diffuse spectral reflectance of surficial sediments indicates sedimentary environments on the shelves of the Bering Sea and western Arctic. *Marine Geology*, **355**, 218–233.

Ody A., Poulet F., Quantin C., Bibring J.P., Bishop J.L, & Dyar M.D. (2015) Candidates source regions of martian meteorites as identified by OMEGA/MEx. *Icarus*, **258**, 366–383.

Orenberg J. & Handy J. (1992) Reflectance spectroscopy of palagonite and iron-rich montmorillonite clay mixtures: Implications for the surface composition of Mars. *Icarus*, **96**, 219–225.

Papike J.J. (1989) Planetary materials. In: *Reviews in mineralogy*, 36. Mineralogical Society of America, Chantilly, VA.

Parente M., Makarewicz H.D., & Bishop J.L. (2011) Decomposition of mineral absorption bands using nonlinear least squares curve fitting: Application to martian meteorites and CRISM data. *Planetary and Space Science*, **59**, 423–442.

Parfitt R.L. (2009) Allophane and imogolite: Role in soil biogeochemical processes. *Clay Minerals*, **44**, 135–155.

Petit S., Madejova J., Decarreau A., & Martin F. (1999) Characterization of octahedral substitutions in kaolinites using near infrared spectroscopy. *Clays and Clay Minerals*, **47**, 103–108.

Petit S., Decarreau A., Martin F., & Buchet R. (2004a) Refined relationship between the position of the fundamental OH stretching and the first overtones for clays. *Physics and Chemistry of Minerals*, **31**, 585–592.

Petit S., Martin F., Wiewiora A., de Parseval P., & Decarreau A. (2004b) Crystal-chemistry of talc: A near infrared (NIR) spectroscopy study. *American Mineralogist*, **89**, 319–326.

Pieters C.M. (1983) Strength of mineral absorption features in the transmitted component of near-infrared reflected light: First results from RELAB. *Journal of Geophysical Research*, **88**, 9534–9544.

Pieters C.M. (1996) Plagioclase and maskelynite diagnostic features. *27th Lunar Planet. Sci. Conf.*, Abstract #1031.

Pieters C.M. & Hiroi T. (2004) RELAB (Reflectance Experiment Laboratory): A NASA multiuser spectroscopy facility. *35th Lunar Planet. Sci. Conf.*, Abstract #1720.

Pieters C.M. & Mustard J.F. (1988) Exploration of crustal/mantle material for the Earth and Moon using reflectance spectroscopy. *Remote Sensing Environment*, **24**, 151–178.

Pieters C.M., Hawke B.R., Gaffey M., & McFadden L.A. (1983) Possible lunar source areas of meteorite ALHA81005: Geochemical remote sensing information. *Geophysical Research Letters*, **10**, 813–816.

Pieters C.M., Mustard J.F., Pratt S.F., Sunshine J.M., & Hoppin A. (1993) Visible-infrared properties of controlled laboratory soils. *24th Lunar Planet. Sci. Conf.*, Abstract, 1147–1148.

Pieters C.M., Mustard J.F., & Sunshine J.M. (1996) Quantitative mineral analyses of planetary surfaces using reflectance spectroscopy. In: *Mineral spectroscopy: A tribute to Roger G. Burns* (M.D. Dyar, C. McCammon, & M.W. Schaefer, eds.). The Geochemical Society, Houston, TX, 307–325.

Pieters C.M., Klima R.L., Hiroi T., et al. (2008) The origin of brown olivine in martian dunite NWA 2737: Integrated spectroscopic analyses of brown olivine. *Journal of Geophysical Research*, **113**, E06004, DOI:10.1029/2007JE002939.

Post J.L. (1984) Saponite from near Ballarat, California. *Clays and Clay Minerals*, **32**, 147–152.

Post J.L. & Noble P.N. (1993) The near-infrared combination band frequencies of dioctahedral smectites, micas, and illites. *Clays and Clay Minerals*, **41**, 639–644.

Post J.L., Cupp B.L., & Madsen F.T. (1997) Beidellite and associated clays from the DeLamar mine and Florida mountain area, Idaho. *Clays and Clay Mineral*, **45**, 240–250.

Powers D.A., Rossman G.R., Schugar H.J., & Gray H.B. (1975) Magnetic behavior and infrared spectra of jarosite, basic iron sulfate, and their chromate analogs. *Journal of Solid State Chemistry*, **13**, 1–13.

Rice M.S., Cloutis E.A., Bell J.F. III, et al. (2013) Reflectance spectra diversity of silica-rich materials: Sensitivity to environment and implications for detections on Mars. *Icarus*, **223**, 499–533.

Ross S.D. (1974) Phosphates and Other Oxyanions of Group V. In: *The infrared spectra of minerals* (V.C. Farmer, ed.). The Mineralogical Society, London, 383–422.

Roush T.L., Bishop J.L., Brown A.J., Blake D.F., & Bristow T.F. (2015) Laboratory reflectance spectra of clay minerals mixed with Mars analog materials: Toward enabling quantitative clay abundances from Mars spectra. *Icarus*, **258**, 454–466.

Ruesch O., Hiesinger H., Cloutis E., et al. (2015) Near infrared spectroscopy of HED meteorites: Effects of viewing geometry and compositional variations. *Icarus*, **258**, 384–401.

Salisbury J.W. & Hunt G.R. (1974) Meteorite spectra and weathering. *Journal of Geophysical Research*, **79**, 4493–4441.

Salisbury J.W., D'Aria D.M., & Jarosewich E. (1991) Midinfrared (2.5–13.5 μm) reflectance spectra of powdered stony meteorites. *Icarus*, **92**, 280–297.

Saper L. & Bishop J.L. (2011) Reflectance spectroscopy of nontronite and ripidolite mineral mixtures in context of phyllosilicate unit composition at Mawrth Vallis. *42nd Lunar Planet. Sci. Conf.*, Abstract #2029.

Schade U. & Wäsch R. (1999) Near-infrared reflectance spectra from bulk samples of the two martian meteorites Zagami and Nakhla. *Meteoritics and Planetary Science*, **34**, 417–424.

Schade U., Wäsch R., & Moroz L. (2004) Near-infrared reflectance spectroscopy of Ca-rich clinopyroxenes and prospects for remote spectral characterization of planetary surfaces. *Icarus*, **168**, 80–92.

Scheinost A.C., Chavernas A., Barrón V., & Torrent J. (1998) Use and limitations of second-derivative diffuse reflectance spectroscopy in the visible to near-infrared range to identify and quantify Fe oxide minerals in soils. *Clays and Clay Minerals*, **46**, 528–536.

Sherman D.M. & Waite T.D. (1985) Electronic spectra of Fe^{3+} oxides and oxide hydroxides in the near IR to near UV. *American Mineralogist*, **70**, 1262–1269.

Sherman D.M., Burns R.G., & Burns V.M. (1982) Spectral characteristics of the iron oxides with application to the martian bright region mineralogy. *Journal of Geophysical Research*, **87**, 10169–10180.

Shkuratov Y.G. & Grynko Y.S. (2005) Light scattering by media composed of semitransparent particles of different shapes in ray optics approximation: Consequences for spectroscopy, photometry, and polarimetry of planetary regoliths. *Icarus*, **173**, 16–28.

Singer R.B. (1981) Near-infrared spectral reflectance of mineral mixtures: Systematic combinations of pyroxenes, olivine, and iron oxides. *Journal of Geophysical Research*, **86**, 7967–7982.

Singer R.B. & Roush T.L. (1983) Spectral reflectance properties of particulate weathered coatings on rocks: Laboratory modeling and applicability to Mars. *14th Lunar Planet. Sci. Conf.*, Abstract, 708–709.

Singer R.B. & Roush T.L. (1985) Effects of temperature on remotely sensed mineral absorption features. *Journal of Geophysical Research*, **90**, 12,434–12,444.

Song X. & Boily J.-F. (2012) Variable hydrogen bond strength in akaganéite. *The Journal of Physical Chemistry C*, **116**, 2303–2312.

Song X. & Boily J.-F. (2013) Water vapor diffusion into a nanostructured iron oxyhydroxide. *Inorganic Chemistry*, **52**, 7107–7113.

Sugihara T., Ohtake M., Owada A., Ishii T., Otsuki M., & Takeda H. (2004) Petrology and reflectance spectroscopy of lunar meteorite Yamato 981031: Implications for the source region of the meteorite and remote-sensing spectroscopy. *Antarctic Meteorite Research*, **17**, 209–230.

Sun V.Z., Milliken R.E., & Robertson K.M. (2016) Hydrated silica on Mars: Relating geologic setting to degree of hydration, crystallinity, and maturity through coupled orbital and laboratory studies. *47th Lunar Planet. Sci. Conf.*, Abstract #2416.

Sunshine J.M. & Pieters C.M. (1993) Estimating modal abundances from the spectra of natural and laboratory pyroxene mixtures using the Modified Gaussian Model. *Journal of Geophysical Research*, **98**, 9075–9087.

Sunshine J.M. & Pieters C.M. (1998) Determining the composition of olivine from reflectance spectroscopy. *Journal of Geophysical Research*, **103**, 13,675–13,688.

Sunshine J.M., Pieters C.M., & Pratt S.F. (1990) Deconvolution of mineral absorption bands: An improved approach. *Journal of Geophysical Research*, **95**, 6955–6966.

Sunshine J.M., McFadden L.A., & Pieters C.M. (1993) Reflectance spectra of the Elephant Moraine A79001 meteorite: Implications for remote sensing of planetary bodies. *Icarus*, **105**, 79–91.

Sunshine J.M., Bus S.J., McCoy T.J., Burbine T.H., Corrigan C.M., & Binzel P. (2004) High-calcium pyroxene as an indicator of igneous differentiation in asteroids and meteorites. *Meteoritics and Planetary Science*, **39**, 1343–1357.

Sunshine J.M., Bus S.J., Corrigan C.M., McCoy T.J., & Burbine T.H. (2007) Olivine-dominated asteroids and meteorites: Dinstinguishing nebular and igneous histories. *Meteoritics and Planetary Science*, **42**, 155–170.

Swayze G.A., Lowers H.A., Benzel W.M., et al. (2018) Characterizing the source of potentially asbestos-bearing commercial vermiculite insulation using in situ IR spectroscopy. *American Mineralogist*, **103**, 517–549.

Tarantola A. & Valette B. (1982) Generalized nonlinear inverse problems solved using the least squares criterion. *Reviews of Geophysics and Space Physics*, **20**, 219–232.

van Olphen H. & Fripiat J.J. (1979) *Data handbook for clay materials and other non-metallic minerals.* Pergamon Press, Oxford.

Wang F., Bowen B.B., Seo J.-H., & Michalski G. (2018) Laboratory and field characterization of visible to near-infrared spectral reflectance of nitrate minerals from the Atacama Desert, Chile, and implications for Mars. *American Mineralogist*, **103**, 197–206.

Wasson J.T. (1985) *Meteorites: Their record of early Solar System history.* W.H. Freeman, New York.

Weir C.E. & Lippincott E.R. (1961) Infrared studies of aragonite, calcite, and vaterite type structures in the borates, carbonates, and nitrates. *Journal of Research of the National Bureau of Standards A: Physics and Chemistry*, **65A**, 173–183.

5

Spectroscopy of Ices, Volatiles, and Organics in the Visible and Infrared Regions

DALE P. CRUIKSHANK, LYUBA V. MOROZ, AND ROGER N. CLARK

5.1 Spectral Properties of Ices in the Visible and Infrared Spectral Regions

5.1.1 Absorption Band Characteristics, Grain Size Effects

The spectral properties of ices observed on Earth and throughout the Solar System are seen in solar reflected light in wavelength regions where the Sun is brightest (~0.3–5 μm), or in thermal emission at longer wavelengths. The spectral properties are controlled by two primary effects: absorption and scattering. Absorption, in its simplest form, is given by Beer's law:

$$I = I_o e^{-\alpha x}, \tag{5.1}$$

where the intensity observed, I, is the incident intensity, I_o, times the exponential of the absorption coefficient, α, and the length of the optical path through the medium, x. The confounding issue for remote sensing is that scattering from grain boundaries in particulate media results in variable photon path length as a function of absorption coefficient and wavelength. Modeling the interactions of light with particulate media requires the use of a radiative transfer model (see Chapter 2). The model most often used in terrestrial and planetary remote sensing has been that by Hapke (1981, 1993, 2012) and has been extended to scattering by particles much smaller than the wavelength (Clark et al., 2012; Brown, 2014). The important role of ice particle size in quantitative spectroscopy is explored in detail for the case of H_2O on Mars by Brown et al. (2016).

The absorptions in ices in the visible and infrared are mainly due to overtones and combinations of vibrational modes in the lattice structure. For a molecule of N atoms, there are $2N - 1$ modes, but with overtones of stretching modes, combinations of different stretching modes, stretching plus bending modes, and librational modes, the spectra of even simple molecules can display very complex spectral structure. This is illustrated in Figure 5.1a. While the C–H stretch fundamental occurs in the 3.0–3.5-μm region, the spectra are rich in hundreds of overtones and combination bands.

Most ices observed in the laboratory exhibit rich spectra similar to those in Figure 5.1 with the exception of H_2O ice. The term ice most often refers to naturally occurring solid

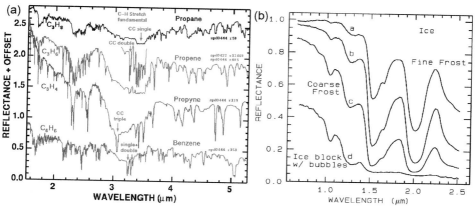

Figure 5.1 Reflectance spectra of organics and ices. (a) Simple organics including alkane (propane), alkene, alkyne, and an aromatic hydrocarbon (benzene) ice. As the CC bond strength increases from propane (single C–C bond) through propene (double C=C bond) to propyne (triple C≡C bond), the C–H stretching absorptions shift to shorter wavelengths, from 3.36+ μm, to 3.24+ μm, and to 3.05+ μm, demonstrating an increase in energy necessary to activate the bond vibration. For illustration purposes, the propane spectrum has been offset upward by 2.2; the propene spectrum offset by 1.4; the propene C–H stretch inset by 1.45; and the propyne spectrum by 0.4 (data from Clark et al., 2009). (b) Water ice at 112–140 K for different grain sizes (data from Clark et al., 1986) including: spectrum a of fine-grained (~50 μm) water frost, spectrum b of medium-grained (~200 μm) frost, spectrum c of coarse-grained (400–2000 μm) frost, and spectrum d of an ice block containing abundant tiny bubbles. The larger the effective grain size, the greater the mean photon path that light travels in the ice, and the deeper the absorptions become. Curve d is very low in reflectance because of the large path length in ice. The strong absorption near 1.5 μm is due to the first overtone of the O–H stretch near 3 μm. The ~2 μm feature is a combination of the H–O–H stretching and bending modes near 3 and 6 μm, respectively.

water, but in planetary science, ice can be any frozen volatile, so the type of ice must be specified, as in water ice or methane ice. Volatile solids observed on other planets that are not naturally occurring on Earth are not technically minerals, though they qualify in every other way, such as regular crystal structures. See Sill and Clark (1982), Cruikshank et al. (1985), Clark et al. (1986, 2013, 2014), Cull et al. (2010), and Kokaly et al. (2017) and references therein for detection of ices in the Solar System and representative laboratory spectra.

As on Earth, H_2O ice is the most widespread ice in the Solar System (Clark et al., 2013 and references therein). Although water ice is prevalent, particularly on planetary satellites and rings, at visible wavelengths (0.3–1.0 μm) its spectrum is very bland. At near-infrared (NIR) wavelengths (1–5 μm), however, it displays a very characteristic and easily recognized spectrum. Those characteristics result from the fact that although water ice has a regular crystal structure, it is orientationally disordered, meaning that while the oxygen atoms are in a well-defined crystal structure, the direction of the O–H bonds is more random. The disordering causes modes to shift in wavelength, resulting in overlap and loss of fine structure (Figure 5.1b).

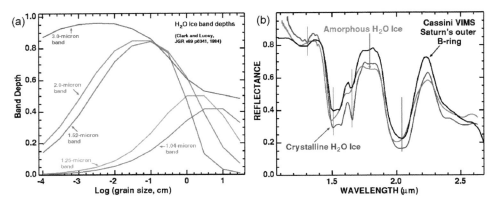

Figure 5.2 Spectral properties of water ice. (a) Water ice absorption band depths as a function of grain diameter, called curves of growth. The band depth increases with grain size up to a point and then decreases due to saturation. The shape of the absorption function curve shows that in the saturation regime the curves (to the right of the peak) have flatter bottoms (data from Clark & Lucey, 1984). (b) Water ice spectra, computed using Hapke theory and optical constants from Mastrapa et al. (2008), are compared to a spectrum of Saturn's rings. Note the ~2-μm absorption in the ring spectrum is asymmetric on the long-wavelength side, indicating the presence of submicron ice grains (data from Cuzzi et al., 2009).

The spectra in Figure 5.1b show another spectral property: as the grain size increases, path length increases, but not all light is absorbed. Reflections off grain surfaces return some light from the particulate surface, even when all light entering into a grain is absorbed. The path length in the ice is a local minimum at the absorption maxima and path length increases as the absorption decreases. These absorption/scattering effects provide diagnostic clues to the grain sizes in the surface. Note the shape of the ~2-μm absorption in Figure 5.1b; at small grain sizes the absorption has a relatively round bottom (spectrum *a*). At large grain sizes, the absorption width increases and the band bottom becomes flattened (spectrum *d*). Higher overtones and combination bands have low absorption coefficients and can probe deeper into the surface. These properties were quantified in Clark and Lucey (1984), Clark and Roush (1984), and Lucey and Clark (1984). The band depth curves of growth for water ice absorptions are shown in Figure 5.2a. All materials, both crystalline and amorphous, show curves of growth. For a pure ice surface, the band depths can be a direct measure of grain size and the different strength absorptions probe the grain size distribution and/or change in grain size with depth into the surface. As contaminants are added to the surface, the band depths decrease and a multiple-scattering radiative transfer model is needed to derive abundances and grain sizes.

5.1.2 Amorphous and Crystalline Phases of H_2O

The freezing of liquid water produces hexagonal ice, whereas amorphous, cubic, or hexagonal ice can be produced by condensation of the vapor at different temperatures. Water ice phases and spectral features as a function of temperature are discussed in detail by

Mastrapa et al. (2008, 2013). H_2O ice exists in multiple forms including cubic (Ic) and hexagonal (Ih) crystal structures and amorphous solids (e.g., Hobbs, 2010) that can be encountered on planetary surfaces in the Solar System. Mastrapa et al. (2008) measured the optical constants of crystalline and amorphous ice from 20 to 120 K and reviewed formation conditions for amorphous versus crystalline water ice. Below about 135 K, amorphous ice is expected to condense from the vapor phase if the rate of growth is slow. In view of the nominal temperatures in the outer Solar System, we might expect amorphous ice to be present in the Jupiter system and at greater heliocentric distances. However, as discussed in Clark et al. (2013), with the probable exception of the Jupiter system, where surfaces are being irradiated by charged particles caught in Jupiter's magnetic field, outer Solar System surfaces are dominated by crystalline H_2O. The predominance of crystalline ice on extremely cold Solar System bodies is a puzzle that is not completely solved. Annealing of originally amorphous ice by the heat of meteoroid impacts on a micro-scale may, over time, induce the transition to a crystalline phase.

In amorphous H_2O ice, the absorptions shift to shorter wavelengths (Figure 5.2b). The Fresnel peak near 3.1 µm also shifts to shorter wavelengths, and the temperature-sensitive 1.65-µm absorption becomes very weak (Figure 5.2b). Note, however, that a decrease in band depth can also be influenced by the presence of submicron H_2O ice grains (Clark et al., 2012). The search for amorphous H_2O ice in the outer Solar System needs to be a multispectral feature test and include the confounding effects of diffraction from submicron ice grains.

5.2 Special Circumstances

5.2.1 Clathrates, Hydrates, Adsorbates, and Solid Solutions

Ices occur in combinations of three basic kinds known to occur on planetary surfaces. *Clathrates* are molecular structures in which a host molecule, usually H_2O, encloses a guest molecule (e.g., CH_4, CO_2) in a cage having one of a few limited specific three-dimensional geometries. *Hydrates* are structures in which molecules are joined by weak forces in nonspecific geometrical configurations, such as NH_3, which can join with H_2O as a hemihydrate ($NH_3 \cdot 0.5H_2O$), a dihydrate ($NH_3 \cdot 2H_2O$), and other combinations. *Solid solutions* are molecular combinations, such as $N_2:CH_4$, in which the crystalline structure of the solvent is not changed by the addition of the solute. Allowed solid solution mixtures of two or more components are limited over a range of the fractions of each component at a given temperature, and are best described in a phase diagram.

Clathrates, hydrates, and solid solutions all have characteristic NIR spectra that can be detected in remotely sensed data for icy Solar System bodies, although the differences from spectra of the pure components in remotely sensed spectra are subtle in most cases. An additional configuration of mixed ices that may occur on planetary surfaces is the presence of an adsorbed species on a substrate of another ice, such as H_2O (Devlin & Buch, 1997).

This configuration has not yet been directly identified or clearly recognized. We consider each of these four molecular configurations as they apply to planetary surfaces.

5.2.1.1 Clathrates

Following the discovery of CH_4 clathrate with H_2O in natural gas pipelines, the possibility of similar structures on planetary surfaces was considered (e.g., Smythe, 1975). Subsequently, this clathrate has been found at moderate ocean depths in regions of continental shelves on Earth, but has not yet been found elsewhere in the Solar System, although calculations indicate generally favorable conditions for its formation on Titan and on some other planetary bodies (e.g., Choukroun et al., 2013). Dartois and Deboffle (2008) have identified unique spectral bands in CH_4 clathrate, with the potential that they might at some time be identified in low-temperature Solar System bodies.

CO_2 clathrate is thought to be present in the nucleus ices of comet 67P/ Churyumov–Gerasimenko to account for the release of CO_2 gas observed by the Rosetta spacecraft (Luspay-Kuti et al., 2016), and is implicated in the CO_2 reservoir on Mars (Chassefière et al., 2013). Dartois (2010) and Dartois and Schmitt (2009) have presented laboratory NIR spectra of CH_4 and CO_2 clathrate to demonstrate spectral features that might be found on icy planetary bodies, while Blake et al. (1991) conducted an experimental study of CO_2 formation in amorphous H_2O ice in the context of comet chemistry. Cruikshank et al. (2010) examined the 4.27-μm CO_2 stretching mode absorption band on three of Saturn's satellites, finding that the observed wavelength positions of the band centers are most consistent with CO_2 complexed with H_2O, but not in a clathrate structure. The complexing of CO_2 with H_2O (and CH_3OH) and the wavelength shift of the principal absorption bands have been demonstrated with ab initio computations by Chaban et al. (2007). In the laboratory, Oancea et al. (2012) found that the structure of the clathrate hydrate can be discerned in the splitting of the v_3 absorption band, producing a band at 4.26 μm (2347 cm^{-1}) for molecules trapped in small cages and a band at 4.28 μm (2334 cm^{-1}) for molecules trapped in large cages. In summary, there is currently no conclusive observational spectroscopic identification of either CH_4 or CO_2 clathrates on planetary bodies, although their presence is inferred on the basis of the known or suspected conditions and observed volatile behavior on some comets and icy planetary satellites.

5.2.1.2 Hydrates

The spectrum of Pluto's largest satellite, Charon, exhibits an absorption band at ~2.2 μm that is attributed to a hydrate of NH_3 (Brown & Calvin, 2000; Cook et al., 2007; Chapter 22). Spectral images of Charon obtained in 2015 with the New Horizons spacecraft confirm the presence of this band and map its distribution on the satellite's surface, showing local concentrations (Grundy et al., 2016; Dalle Ore et al., 2018). Similar spectral indications are found for Transneptunian Objects (TNOs) 50000 Quaoar (Barucci et al., 2015) and 90482 Orcus (Barucci et al., 2008). NH_3 does not form a clathrate structure with H_2O in the way that CH_4 does; its combination with H_2O molecules occurs in various other complex configurations, such as a hemihydrate (2NH$_3$:H$_2$O) and a dihydrate (NH$_3$:2H$_2$O). There is a paucity of laboratory spectroscopic data to allow the distinction among the possible hydrate states for application to Charon, but it seems clear that pure NH$_3$ cannot satisfy the observations currently available. The role of strong adsorbates discussed in the next section may be relevant.

5.2.1.3 Adsorbates

Nanocrystalline H$_2$O ice is a receptor to adsorbed species classed as weak adsorbates (e.g., H$_2$, N$_2$, and CO), moderate adsorbates (e.g., HCN, SO$_2$, H$_2$S), and strong adsorbates (e.g., NH$_3$) according to the degree to which H-bonds are formed. NH$_3$, for example, is a strong proton acceptor and tends to overtake the normal H-bonds in water ice, allowing the NH$_3$ to penetrate the ice to form hydrates at cryogenic temperatures (Devlin & Buch, 1997). This effect may be of special relevance to Charon as a means by which the observed NH$_3$ hydrate(s) forms and is sustained against rapid destruction at the surface by incident ultraviolet radiation (Hernandez et al., 1998).

The physical, chemical, and spectroscopic effects of NH$_3$ and other adsorbates such as CO$_2$ have not been explored in sufficient detail in the planetary science context, although their presence on a surface of H$_2$O ice alters some of the H$_2$O absorption bands in the NIR region where most planetary spectra are observed (e.g., Manca et al., 2003). This is a subject worthy of additional attention in planetary science to gain a deeper understanding of the compositions and dynamics of surfaces of icy Solar System bodies in the "optical layer" sampled by remote sensing spectroscopy (see also Devlin & Buch, 2003).

5.2.1.4 Solid Solutions

Minerals occurring as solid solutions of two or more components are well known in mineralogy; similar combinations of different volatile molecules also occur in the ice phase. The solid solution represented by N$_2$ and CH$_4$ is the best demonstrated and understood in the study of planetary ices. When solid N$_2$ and CH$_4$ were discovered spectroscopically on Triton (Cruikshank et al., 1993) and Pluto (Owen et al., 1993), a significant shift of the CH$_4$ bands toward shorter wavelengths compared to the wavelengths in pure CH$_4$ ice was found. N$_2$ and CH$_4$ are mutually soluble in the ice phase, and the observed band shift occurs when CH$_4$ is diluted in N$_2$ such that the nearest neighboring molecules of CH$_4$ are N$_2$ (see Quirico & Schmitt, 1997a). While the intrinsically weak, induced-dipole 2–0 band of solid N$_2$ is seen in the spectra of both Triton and Pluto, the wavelength shift of CH$_4$ bands seen in other objects for which the N$_2$ band itself is undetected (because of inferior data in the case of faint objects) is an indication of the presence of N$_2$.

Laboratory spectra of combinations of N$_2$ and CH$_4$ by Protopapa et al. (2015) explored the spectroscopic consequences of different degrees of dilution of one molecule in the other and at various temperatures. The binary phase diagram of N$_2$:CH$_4$ at temperatures relevant to outer Solar System bodies ($T < 50$ K) (Prokhvatilov & Yantsevich, 1983) demonstrates that pure CH$_4$ and pure N$_2$ do not exist in thermodynamic equilibrium, but occur as a saturated solid solution with distinct solubility limits at a given temperature (Trafton, 2015). At $T = 40$ K, for example, in the CH_4-rich mixture the saturation limit of N_2 is ~0.035, and at $T = 50$ K it is ~0.07. In the N_2-rich mixture at $T = 40$ K, the saturation limit of N_2 is 0.94, and at $T = 50$ K, the saturation limit of N_2 is 0.86.

Quirico and Schmitt (1997b) also explored solid solution mixtures of CO in N_2, finding variations with temperature of the width and intensity of the 2.35-μm CO band that is seen in the spectra of both Triton and Pluto. The characteristics of the CO band are also dependent on the phase of the N_2, which at $T <35.6$ K occurs in the α-phase, and at $T >35.6$ K is in the β-phase (Scott, 1976); on both Triton and Pluto only the lower density β-phase of N_2 has been detected to date.

5.2.2 Darkening and Coloring of Ices with Admixtures of Carbon, Minerals, Metal Grains, and Tholins

Planetary ices, which include H_2O, CO_2, N_2, CH_4, and other simple molecules, when present in the form of fine grains (less than a few hundred μm), are normally colorless and produce a surface of high reflectivity. Small amounts of mineral grains, metal particles, or tholins mixed with the ice have a large effect on reflectance as well as color. Clark (1981) measured reflectance spectra (0.33–2.5 μm) in the laboratory of a number of combinations of H_2O ice and minerals. He found that very small quantities (fractional areal coverage less than ~0.005) of some minerals could be detected. Conversely, to mask the presence of a mineral, a layer of H_2O frost on the order of 1 mm is needed. In the context of a planetary surface, the general conclusion drawn from this work is that a very small amount of a mineral (or carbon) mixed with an otherwise highly reflective ice can have a large effect on reducing the surface reflectance, depending on the characteristics of the mineral. At the same time, a relatively thick layer of frost is required to mask a mineral-rich surface.

Color effects of mineral– or metal–grain mixtures with ices have not been widely explored, but Clark et al. (2012) have modeled the spectral reflectance of Iapetus with H_2O ice and nm-size particles of iron metal and iron oxide (hematite). The reflectance of the leading hemisphere of Iapetus in the range of 0.4–2.5 μm rises steeply toward longer wavelengths, and both the reflectance level and the color can be matched with a model using the iron–H_2O mixture, with hematite comprising the main component. Model mixtures of the Khare et al. (1984) Titan tholin in ice (plus other components) also reproduce the reflectance spectrum of Iapetus (Clark et al., 2012), but less satisfactorily.

Model mixtures of tholins and H_2O ice have been used to calculate synthetic spectra of various outer Solar System bodies, such as the centaur 5145 Pholus (Cruikshank et al., 1998) and several TNOs (e.g., Dalle Ore et al., 2015). By virtue of their strong colors and generally low overall reflectances, small amounts of tholins mixed in ice have a pronounced effect on the shape and overall albedo level of the calculated spectra, consistent with Clark's (1981) conclusion that small quantities of admixed materials produce strong effects on color and albedo.

5.2.3 Volatile Ices Precipitating from Planetary Atmospheres Resulting from Photochemistry

Photochemical reactions in planetary atmospheres produce new chemical species that can appear as aerosols. The CH_4-rich atmospheres of Titan, Triton, Pluto, and the giant planets all have photochemical aerosols. In the cases of Titan, Triton, and Pluto, the aerosols precipitate to the surface and accumulate at rates dependent on the characteristics of the atmospheres and haze particle size. Other simpler chemical species such as C_2H_2, C_4H_2, HCN, and others can reach saturation abundances in the atmospheres of Triton and Pluto, for example (e.g., Krasnopolsky & Cruikshank, 1999), and precipitate to the surface. HCN, C_2H_2, and C_2H_4 have been found in Pluto's atmosphere from Earth-based observations and with the New Horizons spacecraft (Gladstone et al., 2016). Direct sampling of the Titan aerosol with the Huygens in situ probe (Raulin et al., 2007; Waite et al., 2007) and close-up remote sensing of Pluto's aerosol with the New Horizons spacecraft (Gladstone et al., 2016) have given detailed insight into the processes of tholin formation in two chemically similar, but otherwise somewhat different, planetary atmosphere environments.

5.3 Tholins: Origin and Relevance to Planetary Surfaces

Numerous laboratory studies of ices processed in conditions simulating the energetic photon and charged particle environments in space have been conducted both in the context of Solar System bodies and interstellar ices. Rather than review the detailed chemical transformations that occur in ices during the processes of photolysis and radiolysis, we focus here on the end state of those processes that result in relatively refractory organic solids. See Hudson et al. (2008) and Bennett et al. (2013) for comprehensive reviews of the irradiation and processing of ices in the Solar System.

5.3.1 Gas-Phase Tholins

Macromolecular organic solids produced in the laboratory under conditions approximately simulating the space environment acting on native materials on planetary surfaces and in their atmospheres are generally called tholins. Most of the early work on tholins was directed toward the atmospheric and aerosol chemistry of Titan and involved energy deposition by UV, cold-cathode emission, and electrons in gaseous mixtures of N_2 and CH_4. Khare et al. (1984) measured the complex refractive indices of their Titan tholin ($N_2:CH_4 = 10:1$) over a wide wavelength range, and these values continue to be widely used for models of both planetary atmospheres and planetary surfaces. More recent studies of Titan tholins and newly determined refractive indices are reported in Imanaka et al. (2004) and (2012), respectively. See also the comprehensive review by Cable et al. (2012).

The molecular structures of the early tholins were difficult to characterize, but modern analytical techniques have improved that situation. In $N_2:CH_4$ gas-phase tholins made by radiofrequency discharge and characterized as hydrogenated carbon nitrides, Quirico et al. (2008) found that large polyaromatic rings are absent, and that the colors of tholins are due instead primarily to conjugated carbon bonding in aliphatic structures. Structural and spectral similarities to an HCN polymer were also noted, although the composition of tholins is more complex because the range of possible polymerization reactions is very large.

Colors in macromolecular solids originate from electronic transitions when σ or π bonding and n nonbonding orbitals are promoted to antibonding orbitals σ^* or π^* by the absorption of UV photons. Chromophores thus produced are efficient in absorbing visible light, especially when conjugated molecules (unsaturated compounds with alternating single and double bonds) are present. In general, the longer the conjugated chain, the farther into the visible spectral region the absorption extends, thus producing a range of colors from yellow to red, and a significant decrease in the overall reflectance. The broadband absorption in solids arising in this manner does not occur at narrow or discrete wavelength intervals where astronomical observations can be made, and therefore is not directly diagnostic of specific molecules.

5.3.2 Ice-Phase Tholins

Colored tholins are also produced in the laboratory by the UV photolysis or charged-particle radiolysis of pure or mixed ices (e.g., McDonald et al., 1996). The analysis of these residues, for example from processing of a mixture of N_2, CH_4, and CO ices, reveals carboxylic acids, urea, HCN and other nitriles, alcohols, ketones, aldehydes, and amines, with other unidentified materials of high molecular weight (several hundred Da) (e.g., Materese et al., 2014, 2015). The atomic ratios in a tholin produced with N_2 as an initial component depend on the ability of the irradiating source to break the N≡N bond; ultraviolet sunlight is highly inefficient, while charged particles readily break the bond, allowing N atoms to participate in further chemical reactions. Materese et al. (2015) found that in UV tholin the N/C ratio is about 0.5, while the same ice mixture irradiated with kilovolt electrons produced tholin with N/C ~0.9.

The Pluto ice tholins made by Materese et al. (2014, 2015) were analyzed with X-ray absorption edge spectroscopy, laser-desorption mass spectroscopy, and other techniques, pointing to a structure characterized by individual aromatic rings and partial rings linked by aliphatic bridging units with varying degree of N-substitution, depending on the starting mixture (Figure 5.3a). The variety of structural components will increase with additional O- or S-bearing molecules included in the starting mix. Ice tholins of the kind made by Materese et al. (2015) are expected to be present and to provide the observed coloration to the surfaces of Neptune's satellite Triton, and Pluto, where N_2, CH_4, and CO ices are known to be present (Cruikshank et al., 1993; Owen et al., 1993). In addition to tholin generated directly in the surface ice, gas-phase tholins formed from atmospheric N_2, CH_4, and CO, create aerosols that precipitate to the surface (Gladstone et al., 2016).

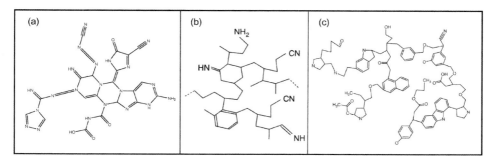

Figure 5.3 Notional structures of three amorphous mixed aromatic–aliphatic organic solids. (a) Structure derived for Pluto tholin ice (electron irradiation of an $N_2 + CH_4 + CO$ ice mixture, from Materese et al., 2015). (b) Structure of organic aerosols in the atmosphere of Titan (Raulin et al., 2007) adapted from Lebreton et al. (2009). (c) Organic solid proposed by Kwok and Zhang (2011) to account for the infrared emission spectral features observed in the interstellar medium and in circumstellar environments. This structure includes sulfur atoms, while the structures in (a) and (b) do not. Note: In these schematic diagrams, a carbon atom (not shown) occurs at each line intersection and all available bonds are saturated with H atoms (not shown).

5.4 Organic Solids Relevant to Meteorites, Interplanetary Dust Particles, Asteroids, Comets, and TNOs, and Their Reflectance Properties

Refractory organic/carbonaceous solids that are present on small Solar System bodies and potentially detected by remote sensing instruments mostly occur in a form of complex macromolecular material rich in polycyclic aromatic structures with aliphatic chains and a variety of organic functional groups (Cronin et al., 1988). Such materials could be formed/processed either in the solar nebula or have presolar origin (Kerridge, 1999). In addition, a part of organic refractories in comets (mostly Oort cloud comets) and some TNOs could be produced by irradiation of organic ices with cosmic rays and other energetic particles (Strazzulla et al., 1991; Cooper et al., 2003).

Organic-bearing samples available for laboratory studies of extraterrestrial refractory organic matter (OM) include carbonaceous chondrites (CCs), interplanetary dust particles (IDPs), some micrometeorites, and cometary dust samples delivered to Earth from comet 81P/Wild 2 by the Stardust mission. CI1, CM2, and CR CCs contain few wt.% organics of abiotic origin (Pizzarello et al., 2006) intimately mixed with matrix minerals. The largest portion (>90 wt.%) of meteoritic OM consists of a complex macromolecular material insoluble in organic solvents (Cronin et al., 1988; Kerridge, 1999; Pizzarello et al., 2006) and chemically/structurally resembling coals and type III kerogens from terrestrial sedimentary rocks (Kebukawa et al., 2011; Quirico et al., 2016). Meteoritic insoluble organic matter (IOM) is rich in highly substituted polycyclic aromatic hydrocarbons. CCs also contain a wide variety of well-characterized soluble organic matter (SOM), including aliphatic and polycyclic aromatic hydrocarbons (PAHs), carboxylic and amino acids, etc. (Sephton, 2002; Pizzarello et al., 2006). CCs of higher petrologic types than 2 contain much smaller amounts of organics that are significantly carbonized due to thermal metamorphism.

Some IDPs of presumably cometary origin ("anhydrous porous IDPs") are C-rich, with mean C content of 12 wt.% (Thomas et al., 1993). The OM from such IDPs and "cometary" Antarctic micrometeorites (AMMs) shows many similarities to that of primitive CCs – it is dominated by aromatic-rich structures (Allamandola et al., 1987; Quirico et al., 2005) and shows evidence of abundant aromatic and aliphatic ($-CH_2-$) and ($-CH_3$) groups in IR spectra (Flynn et al., 2004). However, anhydrous porous IDPs show a wider range of H/C ratios, contain a more diverse population of PAHs than meteorites, and have higher $-CH_2/-CH_3$, O/C, and N/C ratios (Clemett et al., 1993; Flynn et al., 2003, 2004). The nature of these organics indicates that the bulk of the organic material in "cometary" IDPs originated in protosolar nebula and interstellar medium environments rather than as a parent body (Flynn et al., 2003). A wide variety of complex organic compounds has been found in dust samples from comet 81P/Wild 2 (Sandford et al., 2006; Sandford, 2008) and in remote sensing data for comet 67P (Altwegg et al., 2017). These samples were severely damaged by heating during their capture (Velbel & Harvey, 2009); therefore one should treat the data with great caution. Cometary organic materials in Stardust samples resemble meteoritic organics in some respects but show lower aromaticity and have longer and less branched aliphatic chains than organics from CCs (Sandford et al., 2006; Sandford, 2008). High contents of O and N atoms replacing carbon, resulting in lower aromaticity compared to IOM from CCs suggests that organics from comet 81/P are more primitive in terms of thermal processing, and show similarities to organic materials of anhydrous porous IDPs (Sandford et al., 2006; Sandford, 2008). Finally, unusual UltraCarbonaceous Antarctic micrometeorites (UCAMMs) have very high contents of aromatic-rich OM significantly enriched in N-containing functional groups, suggesting that the OM was formed in the outermost regions of the Solar System, beyond the nitrogen snow line (Dartois et al., 2013).

Understanding the spectral reflectance properties of a pure "mineral-free" extraterrestrial OM from small Solar System bodies requires laboratory spectral reflectance studies of terrestrial analog materials of wide range of compositions and structures, such as solid oil bitumens (asphaltites, kerites, anthraxolites) and coals (Moroz et al., 1998; de Bergh et al., 2008; Quirico et al., 2016). The observed NIR spectral slopes depend on the degree of thermal processing of polyaromatic-rich OM, that leads to its carbonization accompanied by the decrease in (H + O + S + N)/C ratios, NIR spectral slopes, and the depths of fundamental absorptions between ~2.7 and 3.5 μm (Figure 5.4; Moroz et al., 1998; de Bergh et al., 2008). The bands due to aromatic C–H stretches become more pronounced compared to C–H stretches in aliphatic hydrocarbons (Figure 5.4). The composition and structure of meteoritic IOM suggest relatively "red" (positive) slopes in the NIR reflectance spectra and deep fundamental absorption bands (Moroz et al., 1998; Quirico et al., 2016). However, these features are not observed in reflectance spectra of IOM derived from primitive meteorites because of the damage produced by acid demineralization (Moroz et al., 1998) and the inability of acid treatment to remove fine-grained opaques (Fe-sulfides and/or magnetite) from the treated sample (Quirico et al., 2016). As mentioned earlier, meteoritic IOM is analogous to coals, while a wider range of H/C and carbon aromaticities in solid oil bitumens makes them comparable to refractory organics from comets, some TNOs, and "cometary" IDPs (Moroz et al., 1998; Quirico et al., 2016).

Identification and characterization of complex C-bearing materials in remote sensing reflectance spectra of small Solar System bodies is challenging because of a number of complicating factors. The related materials are visually dark, spectrally featureless in the visible/near-infrared (VNIR) spectral range, or show only weak NIR absorption features (Cloutis, 1989, 2003; Cloutis et al., 1994, 2011; Moroz et al., 1998; De Bergh et al., 2008). The presence of such a material on the surface of a small Solar System body is often inferred based only on low albedo and positive ("red") NIR spectral slope (Gradie & Veverka, 1980; Cruikshank & Khare, 2000; Jewitt, 2002; Lagerkvist et al., 2005) that is typical of aromatic-rich macromolecular materials with relatively high (H + O + S + N)/C ratios (Cloutis et al., 1994; Moroz et al., 1998; Figure 5.4). The reddish spectral slope and the weak NIR overtone/combination bands of organics can be easily suppressed/modified in reflectance spectra of remotely sensed surfaces because of the presence of intimately admixed/inter-grown fine-grained opaque phases (e.g., Capaccioni et al., 2015; Quirico et al., 2016) or surface alteration processes (Korochantsev et al., 1997; Moroz et al., 2004).

Fundamental vibrations of organic functional groups give rise to absorption features in the spectral region between 2.7 and 4 μm (Moroz et al., 1998; de Bergh et al., 2008; Quirico et al., 2016), providing better constraints on the composition of refractory OM, if detected.

Stretching C–H vibrations in aliphatic and aromatic hydrocarbons produce absorption bands between 3.2 and 3.6 μm (Figure 5.4). Four overlapping absorption features with

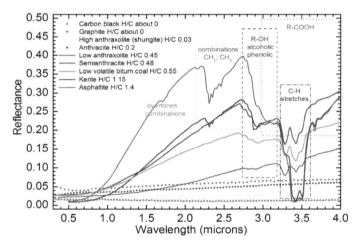

Figure 5.4 VNIR reflectance spectra of multiple forms of carbon and organics. The carbon particulates include carbon black (90–125-μm powder) and graphite (<45 μm). The high-rank coals (<100 μm) include low-volatile bituminous, semianthracite, and anthracite and the solid oil bitumens (<25 μm) include asphaltite, kerite, low and high anthraxolites. The viewing geometry was biconical for solid bitumens (see Moroz et al., 1998, for details), bidirectional (0.3–2.2 μm) and biconical (2.2–4 μm) for coals (spectra acquired at NASA's RELAB facility), hemispherical for graphite (reproduced from the ASTER spectral library), and hemispherical (0.3–2.7 μm) and biconical (2.1–4 μm) for carbon black (spectra acquired by R. Clark at the U.S. Geological Survey and J. Salisbury at Johns Hopkins University).

individual minima due to C–H symmetric and antisymmetric stretches in methyl (–CH$_3$) and methylene (–CH$_2$–) groups can be observed between ~3.38 and 3.5 µm. The C–H stretching band in aromatic structures can be detected near 3.3 µm. Relative depths of the bands mentioned earlier might provide some information regarding carbon aromaticity of OM and possibly CH$_2$/CH$_3$ ratios. However, such estimates should be treated with caution if the opaques are intimately mixed/intergrown with OM (Moroz & Arnold, 1999).

Absorption features due to O–H stretches in alcoholic and phenolic hydroxyl groups may be present in reflectance spectra between ~2.7 and 3.2 µm, but the diagnostic potential of these bands is rather limited, taking into account that this spectral region may include absorption bands of other phases (e.g., O–H stretches of water ice or hydroxyl in phyllosilicates). Hydrogen-bonded O–H stretching modes in carboxylic groups (e.g., COOH groups attached to aromatic structures) give rise to a broader absorption band extending from ~3 to 4 µm (Figure 5.4).

N-bearing functional groups in a macromolecular organic solid (such as NH and NH$_2$) can produce a broad feature that extends from ~2.7 to 3.2 µm with a substructure showing two individual minima centered at ~3 and 3.12 µm (Quirico et al., 2008). In general, N contents in macromolecular organic solids relevant to asteroids, comets, and TNOs appear too low to show signatures of amine groups in reflectance spectra (Quirico et al., 2016). The only exceptions are extremely N-rich UCAMMs, whose OM is very different from organic solids typically found/expected on small Solar System bodies (Dartois et al., 2013).

5.5 Summary and Conclusions

We have reviewed the basic characteristics of planetary ices in terms of their spectral characteristics when they are found as individual components, and have alluded to the properties of two or more species mixed in various configurations. Spectral bands of H$_2$O and other ice species detected on planetary surfaces reveal a number of complexities resulting from the sizes and shapes of grains comprising the surfaces. The phases (crystalline or disordered), temperature, and mixing of two or more species add greatly to the complexity of the spectral characteristics. Also, the incorporation of one species in another as a solid solution or a clathrate, or the adsorption of one species on another produces spectral characteristics observed in ground-based or spacecraft remote sensing data. The identification of individual species and modes of mixing are modeled with scattering theory, using various approximations of the properties of individual and aggregated particles on the scattering surface. A number of scattering theories are used in planetary science, each with its own strengths, limitations, and utility characteristics; no single theory in current use appears to be able to reliably model surfaces having the full range of composition and molecular configurations that occur in nature. As noted earlier, the Hapke theory is most often used in the analysis of planetary spectra. In addition to various simplifying assumptions, all the theories require knowledge of the complex refractive indices of the components. These data are incomplete for many of the ices known on planetary surfaces, often leaving ambiguities and open questions of the applicability of a computational model to a real surface. Fortunately, the most basic information – the identification of molecular species – is relatively straightforward, with available laboratory data on the most important and plausible components.

Non-ice components on icy planetary surfaces often include minerals and refractory complex organic materials (e.g., tholins). Spectral characteristics of minerals are covered in Chapter 3 for the mid-infrared (mid-IR) region and in Chapter 4 for the VNIR region. Here we provide an overview of some of the tholins that are currently being produced in the laboratory, analyzed, and applied to planetary problems. Tholins are a critical component that imparts color and influences the overall reflectance of otherwise colorless and high-albedo ices. Much more work is needed on these materials, particularly in the determination of reliable complex refractive indices of tholins for which chemical composition and molecular structure have been determined.

Complex organic materials other than tholins are found in many meteorites, interplanetary dust particles, and on comets; their presence is inferred on many other bodies having (usually) red coloration and relatively low albedos. While the analysis of extraterrestrial organic materials in the laboratory has made great progress, the application to remotely sensed planetary bodies presents many challenges, most of which are addressed through infrared spectroscopy. However, spectroscopy of these materials is subject to many of the limitations noted for ices, plus the fact that data in the mid-IR spectral region, where most organic materials have their strongest diagnostic features, cannot usually be obtained for most planetary bodies. The spectral properties of volatiles and ices are described in Chapter 20 for comets, in Chapter 21 for Saturn's moons, and in Chapter 22 for the Pluto system.

Abundant literature is available on the subjects covered in this chapter that can be found in comprehensive volumes on planetary ices by Schmitt et al. (1998) and Gudipati and Castillo-Rogez (2013), and in a volume containing detailed reviews of meteoritic materials edited by Lauretta and McSween (2006). Another useful review of laboratory data on ices and carbonaceous materials relevant to planetary surfaces is de Bergh et al. (2008).

Further progress in all the topics in this chapter will be made as the spectral range for remote sensing observations is expanded, with concurrent extensions of laboratory data for relevant species and their combinations. Improvements in scattering theories may be able to account for the properties of singular and combined materials in more physically realistic ways, yielding synthetic spectra that match more closely the remotely sensed data from telescopes on Earth and in space, and spacecraft in the proximity of planets and the full range of bodies in the Solar System.

Acknowledgments

D.P.C. acknowledges support from the *Cassini* and *New Horizons* missions, L.V.M. acknowledges the DFG grant MO 3007/1–1, and R.N.C. acknowledges support from the Cassini mission.

References

Allamandola L.J., Sandford S.A., & Wopenka B. (1987) Interstellar polycyclic aromatic hydrocarbons and carbon in interplanetary dust particles and meteorites. *Science*, **237**, 56–59.

Altwegg K., Balsiger H., Berthelier J.J., et al. (2017) Organics in comet 67P – a first comparative analysis of mass spectra from ROSINA-DFMS, COSAC, and Ptolemy. *Monthly Notices of the Royal Astronomical Society*, **469**, Issue Supplement 2, S130–S141.

Barucci M.A., Merlin F., Guilbert A., et al. (2008) Surface composition and temperature of the TNO Orcus. *Astronomy & Astrophysics*, **479**, L13–L16.

Barucci M.A., Dalle Ore C.M., Perna D., et al. (2015) (50000) Quaoar: Surface composition variability. *Astronomy & Astrophysics*, **584**, A107.

Bennett C.J., Pirim C., & Orlando T.M. (2013) Space-weathering of Solar System bodies: A laboratory perspective. *Chemical Reviews*, **113**, 9086–9150.

Blake D., Allamandola L., Sandford S., Hudgins D., & Freund F. (1991) Clathrate hydrate formation in amorphous cometary ice analogs in vacuo. *Science*, **254**, 548–551.

Brown A.J. (2014) Spectral bluing induced by small particles under the Mie and Rayleigh regimes. *Icarus*, **239**, 85–95.

Brown A.J., Calvin W.M., Becerra P., & Byrne S. (2016) Martian north polar cap summer water cycle. *Icarus*, **277**, 401–415.

Brown M.E. & Calvin W.M. (2000) Evidence for crystalline water and ammonia ices on Pluto's satellite Charon. *Science*, **287**, 107–109.

Cable M.L., Hörst S.M., Hodyss R., et al. (2012) Titan tholins: Simulating Titan organic chemistry in the Cassini-Huygens era. *Chemical Reviews*, **112**, 1882–1909.

Capaccioni F., Coradini A., Filacchione G., et al. (2015) The organic-rich surface of comet 67P/Churyumov-Gerasimenko as seen by VIRTIS/Rosetta. *Science*, **347**, aaa0628.

Chaban G.M., Bernstein M., & Cruikshank D.P. (2007) Carbon dioxide on planetary bodies: Theoretical and experimental studies of molecular complexes. *Icarus*, **187**, 592–599.

Chassefière E., Dartois E., Herri J.-M., et al. (2013) CO_2–SO_2 clathrate hydrate formation on early Mars. *Icarus*, **223**, 878–891.

Choukroun M., Kieffer S.W., Lu X., & Tobie G. (2013) Clathrate hydrates: Implications for exchange processes in the outer Solar System. In: *The science of Solar System ices* (M.S. Gudipati & J. Castillo-Rogez, eds.). Springer Science+Business Media, New York, 409–454.

Clark R.N. (1981) The spectral reflectance of water-mineral mixtures at low temperatures. *Journal of Geophysical Research*, **86**, 3074–3086.

Clark R.N. & Lucey P.G. (1984) Spectral properties of ice-particulate mixtures and implications for remote sensing: 1. Intimate mixtures. *Journal of Geophysical Research*, **89**, 6341–6348.

Clark R.N. & Roush T.L. (1984) Reflectance spectroscopy: Quantitative analysis techniques for remote sensing applications. *Journal of Geophysical Research*, **89**, 6329–6340.

Clark R.N., Fanale F.P., & Gaffey M.J. (1986) Surface composition of satellites. In: *Satellites* (J. Burns & M. S. Matthews, eds.), University of Arizona Press, Tucson, 437–491.

Clark R.N., Curchin J.M., Hoefen T.M., & Swayze G.A. (2009) Reflectance spectroscopy of organic compounds: 1. Alkanes. *Journal of Geophysical Research*, **114**, E03001, DOI:10.1029/2008JE003150.

Clark R.N., Cruikshank D.P., Jaumann R., et al. (2012) The surface composition of Iapetus: Mapping results from Cassini VIMS. *Icarus*, **218**, 831–860.

Clark R.N., Carlson R., Grundy W., & Noll K. (2013) Observed ices in the Solar System. In: *The science of Solar System ices* (M.S. Gudipati & J. Castillo-Rogez, eds.). Springer Science+Business Media, New York, 3–46.

Clark R.N., Swayze G.A., Carlson R., Grundy W., & Noll K. (2014) Spectroscopy from space. In: *Spectroscopic methods in mineralogy and material sciences* (G. Henderson, ed.). Reviews in Mineralogy & Geochemistry, **78**, 399–446.

Clemett S.J., Maechling C.R., Zare R.N., Swan P.D., & Walker R.M. (1993) Identification of complex aromatic molecules in individual interplanetary dust particles. *Science*, **262**, 721–725.

Cloutis E.A. (1989) Spectral reflectance properties of hydrocarbons: Remote-sensing implications. *Science*, **245**, 165–168.

Cloutis E.A. (2003) Quantitative characterization of coal properties using bidirectional diffuse reflectance spectroscopy. *Fuel*, **82**, 2239–2254.

Cloutis E.A., Gaffey M.J., & Moslow T.F. (1994) Spectral reflectance properties of carbon-bearing materials. *Icarus*, **107**, 276–287.

Cloutis E.A., Hiroi T., Gaffey M.J., Alexander C.M.O.D., & Mann P. (2011) Spectral reflectance properties of carbonaceous chondrites: 1. CI chondrites. *Icarus*, **212**, 180–209.

Cook J.C., Desch S.J., Roush T.L., Trujillo C.A., & Geballe T. (2007) Near-infrared spectroscopy of Charon: Possible evidence for cryovolcanism on Kuiper Belt objects. *The Astrophysical Journal*, **663**, 1406.

Cooper J.F., Christian E.R., Richardson J.D., & Wang C. (2003) Proton irradiation of Centaur, Kuiper Belt, and Oort Cloud objects at plasma to cosmic ray energy. *Earth, Moon, and Planets*, **92**, 961–277.

Cronin J.R., Pizzarello S., & Cruikshank D.P. (1988) Organic matter in carbonaceous chondrites, planetary satellites, asteroids and comets. In: *Meteorites and the early Solar System* (J.F. Kerridge & M.S. Matthews, eds.). University of Arizona Press, Tucson, 819–857.

Cruikshank D. & Khare B. (2000) Planetary surfaces of low albedo: Organic material throughout the Solar System. *A new era in bioastronomy* (G.A. Lemarchand & K.J. Meech, eds.) ASP Conference Series, **213**, 253–262.

Cruikshank D.P., Brown R., & Clark R. (1985) Methane ice on Triton and Pluto. In: *Ices in the Solar System* (J. Klinger, D. Benest, A. Dollfus, & R. Smoluchowski, eds.). Springer-Verlag, New York, 817–827.

Cruikshank D.P., Roush T.L., Owen T.C., et al. (1993) Ices on the surface of Triton. *Science*, **261**, 742–745.

Cruikshank D., Roush T., Bartholomew M., et al. (1998) The composition of centaur 5145 Pholus. *Icarus*, **135**, 389–407.

Cruikshank D.P., Meyer A.W., Brown R.H., et al. (2010) Carbon dioxide on the satellites of Saturn: Results from the *Cassini* VIMS investigation and revisions to the VIMS wavelength scale. *Icarus*, **206**, 561–572.

Cull S., Arvidson R.E., Mellon M., et al. (2010) Seasonal H_2O and CO_2 ice cycles at the Mars Phoenix landing site: 1. Prelanding CRISM and HiRISE observations. *Journal of Geophysical Research*, **115**, DOI:10.1029/2009JE003340.

Cuzzi J., Clark R., Filacchione G., et al. (2009) Ring particle composition and size distribution. In: *Saturn after Cassini/Huygens* (M.K. Dougherty, L.W. Esposito, & S.M. Krimigis, eds.). Springer Science+Business Media, New York, 459–509.

Dalle Ore C.M., Barucci M., Emery J., et al. (2015) The composition of "ultra-red" TNOs and Centaurs. *Icarus*, **252**, 311–326.

Dalle Ore C. M., Protopapa S., Cook J.C. et al. (2018) Ices on Charon: Distribution of H_2O and NH_3 from New Horizons LEISA observations. *Icarus*, **300**, 21–32.

Dartois E. (2010) Clathrates hydrates FTIR spectroscopy: Infrared signatures and their astrophysical significance. *Molecular Physics*, **108**, 2273–2278.

Dartois E. & Deboffle D. (2008) Methane clathrate hydrate FTIR spectrum: Implications for its cometary and planetary detection. *Astronomy & Astrophysics*, **490**, L19-L22.

Dartois E. & Schmitt B. (2009) Carbon dioxide clathrate hydrate FTIR spectrum-near infrared combination modes for astrophysical remote detection. *Astronomy & Astrophysics*, **504**, 869–873.

Dartois E., Engrand C., Brunetto R., et al. (2013) UltraCarbonaceous Antarctic micrometeorites, probing the Solar System beyond the nitrogen snow-line. *Icarus*, **224**, 243–252.

de Bergh C., Schmitt B., Moroz L., Quirico E., & Cruikshank D.P. (2008) Laboratory data on ices, refractory carbonaceous materials, and minerals relevant to transneptunian objects and Centaurs. In: *The Solar System beyond Neptune* (A. Barucci, H. Boehnhardt, D.P. Cruikshank, & A. Morbidelli, eds.). University of Arizona Press, Tucson, 483–506.

Devlin J.P. & Buch V. (1997) Vibrational spectroscopy and modeling of the surface and subsurface of ice and of ice-adsorbate interactions. *Journal of Physical Chemistry B*, **101**, 6095–6098.

Devlin J.P. & Buch V. (2003) Ice nanoparticles and ice adsorbate interactions: FTIR spectroscopy and computer simulations. In: *Water in confining geometries* (V. Buch & J.P. Devlin, eds.). Springer Science+Business Media, 425–462.

Flynn G., Keller L., Feser M., Wirick S., & Jacobsen C. (2003) The origin of organic matter in the Solar System: Evidence from the interplanetary dust particles. *Geochimica et Cosmochimica Acta*, **67**, 4791–4806.

Flynn G., Keller L., Jacobsen C., & Wirick S. (2004) An assessment of the amount and types of organic matter contributed to the Earth by interplanetary dust. *Advances in Space Research*, **33**, 57–66.

Gladstone G.R., Stern S.A., Ennico K., et al. (2016) The atmosphere of Pluto as observed by New Horizons. *Science*, **351**, aad8866.

Gradie J. & Veverka J. (1980) The composition of the Trojan asteroids. *Nature*, **283**, 840.

Grundy W.M., Binzel R.P., Buratti B.J., et al. (2016) Surface compositions across Pluto and Charon. *Science*, **351**, aad9189-8.

Gudipati M.S., Castillo-Rogez J., eds. (2013) *The science of Solar System ices*. Astrophysics and Space Science Library, **356**. Springer Science+Business Media, New York.

Hapke B. (1981) Bidirectional reflectance spectroscopy: 1. Theory. *Journal of Geophysical Research*, **86**, 3039–3054.

Hapke B. (1993) *Theory of reflectance and emittance spectroscopy*. Cambridge University Press, Cambridge.

Hapke B. (2012) *Theory of reflectance and emittance spectroscopy*, 2nd edn. Cambridge University Press, Cambridge.

Hernandez J., Uras N., & Devlin J.P. (1998) Coated ice nanocrystals from water–adsorbate vapor mixtures: Formation of ether–CO_2 clathrate hydrate nanocrystals at 120 K. *Journal of Physical Chemistry B*, **102**, 4526–4535.

Hobbs P.V. (2010) *Ice physics*. Oxford: Oxford University Press.
Hudson R., Palumbo M., Strazzulla G., Moore M., Cooper J., & Sturner S. (2008) Laboratory studies of the chemistry of Transneptunian Object surface materials. In: *The Solar System beyond Neptune* (A. Barucci, H. Boehnhardt, Cruikshank, & D.P. Morbidelli, eds.). University of Arizona Press, Tucson, 507–523.
Imanaka H., Khare B.N., Elsila J.E., et al. (2004) Laboratory experiments of Titan tholin formed in cold plasma at various pressures: Implications for nitrogen-containing polycyclic aromatic compounds in Titan haze. *Icarus*, **168**, 344–366.
Imanaka H., Cruikshank D.P., Khare B.N., & McKay C.P. (2012) Optical constants of laboratory synthesized complex organic materials: Part 1, Titan tholins at mid-infrared wavelengths (2.5–25 μm). *Icarus*, **218**, 247–261.
Jewitt D.C. (2002) From Kuiper Belt object to cometary nucleus: The missing ultrared matter. *The Astronomical Journal*, **123**, 1039.
Kebukawa Y., Alexander C.M.D., & Cody G.D. (2011) Compositional diversity in insoluble organic matter in type 1, 2 and 3 chondrites as detected by infrared spectroscopy. *Geochimica et Cosmochimica Acta*, **75**, 3530–3541.
Kerridge J.F. (1999) Formation and processing of organics in the early Solar System. *Space Science Review*, **90**, 275–288.
Khare B.N., Sagan C., Arakawa E., Suits F., Callcott T., & Williams M. (1984) Optical constants of organic tholins produced in a simulated Titanian atmosphere: From soft X-ray to microwave frequencies. *Icarus*, **60**, 127–137.
Kokaly R.F., Clark R.N., Swayze G.A., et al. (2017) USGS spectral library version 7. USGS Data Series.
Korochantsev A., Badjukov D., Moroz L., & Pershin S. (1997) Experiments on impact-induced transformations of asphaltite. *Experimental Geoscience*, **6**, 66–67.
Krasnopolsky V.A. & Cruikshank D.P. (1999) Photochemistry of Pluto's atmosphere and ionosphere near perihelion. *Journal of Geophysical Research*, **104**, 21979–21996.
Kwok S. & Zhang Y. (2011) Mixed aromatic–aliphatic organic nanoparticles as carriers of unidentified infrared emission features. *Nature*, **479**, 80.
Lagerkvist C.-I., Moroz L., Nathues A., et al. (2005) A study of Cybele asteroids-II. Spectral properties of Cybele asteroids. *Astronomy & Astrophysics*, **432**, 349–354.
Lauretta D. & McSween H.Y. Jr., eds. (2006) *Meteorites and the early Solar System II*. University of Arizona Press, Tucson.
Lebreton J.-P., Coustenis A., Lunine J., Raulin F., Owen T., & Strobel D. (2009) Results from the *Huygens* probe on Titan. *The Astronomy and Astrophysics Review*, **17**, 149–179.
Lucey P.G. & Clark R.N. (1985) Spectral properties of water ice and contaminants. In: *Ices in the Solar System* (J. Klinger, D. Benest, A. Dollfus, & R. Smoluchowski, eds.). Springer-Verlag, New York, 155–168.
Luspay-Kuti A., Mousis O., Hässig M., et al. (2016) The presence of clathrates in comet 67P/Churyumov-Gerasimenko. *Science Advances*, **2**, e1501781.
Manca C., Martin C., & Roubin P. (2003) Comparative study of gas adsorption on amorphous ice: Thermodynamic and spectroscopic features of the adlayer and the surface. *The Journal of Physical Chemistry B*, **107**, 8929–8934.
Mastrapa R., Bernstein M., Sandford S., Roush T., Cruikshank D., & Dalle Ore C. (2008) Optical constants of amorphous and crystalline H_2O-ice in the near infrared from 1.1 to 2.6 μm. *Icarus*, **197**, 307–320.
Mastrapa R., Grundy W., & Gudipati M.S. (2013) Amorphous and crystalline H_2O ice. In: *The science of Solar System ices* (M.S. Gudipati & J. Castillo-Rogez, eds.). Springer Science+Business Media, 371–408.
Materese C.K., Cruikshank D.P., Sandford S.A., Imanaka H., Nuevo M., & White D.W. (2014) Ice chemistry on outer Solar System bodies: Carboxylic acids, nitriles, and urea detected in refractory residues produced from the UV photolysis of N_2: CH4: CO-containing ices. *The Astrophysical Journal*, **788**, 111.
Materese C.K., Cruikshank D.P., Sandford S.A., Imanaka H., & Nuevo M. (2015) Ice chemistry on outer Solar System bodies: Electron radiolysis of N_2-, CH_4-, and CO-containing ices. *The Astrophysical Journal*, **812**, 150.
McDonald G.D., Whited L.J., DeRuiter C., et al. (1996) Production and chemical analysis of cometary ice tholins. *Icarus*, **122**, 107–117.
Moroz L. & Arnold G. (1999) Influence of neutral components on relative band contrasts in reflectance spectra of intimate mixtures: Implications for remote sensing: 1. Nonlinear mixing modeling. *Journal of Geophysical Research*, **104**, 14109–14121.
Moroz L.V., Arnold G., Korochantsev A.V., & Wäsch R. (1998) Natural solid bitumens as possible analogs for cometary and asteroid organics: 1. Reflectance spectroscopy of pure bitumens *Icarus*, **134**, 253–268.
Moroz L.V., Baratta G., Atrazzula G., et al. (2004) Optical alteration of complex organics induced by ion irradiation: 1. Laboratory experiments suggest unusual space weathering trend. *Icarus*, **170**, 214–228.
Oancea A., Grasset O., Le Menn E., et al. (2012) Laboratory infrared reflection spectrum of carbon dioxide clathrate hydrates for astrophysical remote sensing applications. *Icarus*, **221**, 900–910.

Owen T.C., Roush T.L., Cruikshank D.P., et al. (1993) Surface ices and the atmospheric composition of Pluto. *Science*, **261**, 745–748.

Pizzarello S., Cooper G., & Flynn G. (2006) The nature and distribution of the organic material in carbonaceous chondrites and interplanetary dust particles. In: *Meteorites and the early Solar System II* (D.S. Lauretta & H. Y. McSween, Jr., eds.). University of Arizona Press, Tucson, 625–651.

Prokhvatilov A. & Yantsevich L. (1983) X-ray investigation of the equilibrium phase diagram of CH_4–N_2 solid mixtures. *Soviet Journal of Low Temperature Physics*, **9**, 94–98.

Protopapa S., Grundy W., Tegler S., & Bergonio J. (2015) Absorption coefficients of the methane–nitrogen binary ice system: Implications for Pluto. *Icarus*, **253**, 179–188.

Quirico E. & Schmitt B. (1997a) Near-infrared spectroscopy of simple hydrocarbons and carbon oxides diluted in solid N_2 and as pure ices: Implications for Triton and Pluto. *Icarus*, **127**, 354–378.

Quirico E. & Schmitt B. (1997b) A spectroscopic study of CO diluted in N_2 ice: Applications for Triton and Pluto. *Icarus*, **128**, 181–188.

Quirico E., Schmitt B., Bini R., & Salvi P.R. (1996) Spectroscopy of some ices of astrophysical interest: SO_2, N_2 and N_2: CH_4 mixtures. *Planetary and Space Science*, **44**, 973–986.

Quirico E., Borg J., Raynal P.-I., Montagnac G., & d'Hendecourt L. (2005) A micro-Raman survey of 10 IDPs and 6 carbonaceous chondrites. *Planetary and Space Science*, **53**, 1443–1448.

Quirico E., Montagnac G., Lees V., et al. (2008) New experimental constraints on the composition and structure of tholins. *Icarus*, **198**, 218–231.

Quirico E., Moroz L., Schmitt B., et al. (2016) Refractory and semi-volatile organics at the surface of comet 67P/Churyumov-Gerasimenko: Insights from the VIRTIS/Rosetta imaging spectrometer. *Icarus*, **272**, 32–47.

Raulin F., Gazeau M.-C., & Lebreton J.-P. (2007) A new image of Titan: Titan as seen from Huygens. *Planetary and Space Science*, **55**, 1843–1844.

Sandford S.A. (2008) Terrestrial analysis of the organic component of comet dust. *Annual Review of Analytical Chemistry*, **1**, 549–578.

Sandford S.A., Aléon J., Alexander C.M.D., et al. (2006) Organics captured from comet 81P/Wild 2 by the Stardust spacecraft. *Science*, **314**, 1720–1724.

Schmitt B., de Bergh C., & Festou M., eds. (1998) *Solar System ices*. Kluwer Academic, Dordrecht.

Scott T.A. (1976) Solid and liquid nitrogen. *Physics Reports*, **27**, 89–157.

Sephton M.A. (2002) Organic compounds in carbonaceous meteorites. *Natural Product Reports*, **19**, 292–311.

Sill G.T. & Clark R.N. (1982) Composition of the surfaces of the Galilean satellites. In: *The satellites of Jupiter* (D. Morrison, ed.). University of Arizona Press, Tucson, 174–212.

Smythe W.D. (1975) Spectra of hydrate frosts: Their application to the outer Solar System. *Icarus*, **24**, 421–427.

Strazzulla G., Baratta G., Johnson R., & Donn B. (1991) Primordial comet mantle: Irradiation production of a stable organic crust. *Icarus*, **91**, 101–104.

Thomas K.L., Blanford G.E., Keller L.P., Klöck W., & McKay D.S. (1993) Carbon abundance and silicate mineralogy of anhydrous interplanetary dust particles. *Geochimica et Cosmochimica Acta*, **57**, 1551–1556.

Thomas P.J., Chyba C.F. & McKay C.P., eds. (2006) *Comets and the origin and evolution of life*. Springer Science +Business Media, New York.

Trafton L.M. (2015) On the state of methane and nitrogen ice on Pluto and Triton: Implications of the binary phase diagram. *Icarus*, **246**, 197–205.

Velbel M.A. & Harvey R.P. (2009) Along-track compositional and textural variation in extensively melted grains returned from comet 81P/Wild 2 by the Stardust mission: Implications for capture-melting process. *Meteoritics and Planetary Science*, **44**, 1519–1540.

Waite J., Young D., Cravens T., et al. (2007) The process of tholin formation in Titan's upper atmosphere. *Science*, **316**, 870–875.

6

Raman Spectroscopy
Theory and Laboratory Spectra of Geologic Materials

SHIV K. SHARMA AND MILES J. EGAN

6.1 Introduction

In 1928, C.V. Raman discovered inelastic scattering of light during the course of extended research on the molecular scattering of light (Raman, 1928; Raman & Krishnan, 1928). For this discovery, Raman was awarded the Nobel Prize in Physics in 1930 and the effect is known as the Raman Effect. Raman observed that when monochromatic light of frequency v_0 is incident on a transparent sample, the spectrum of the scattered light shows two features. One is an elastically scattered light of identical frequency, called the Rayleigh line, and the other is a pattern of spectral lines of altered frequency (Δv_i), called the Raman spectrum. The pattern on the low-frequency side of the exciting light ($-\Delta v_i$), which is at longer wavelength than the exciting light, resembles the Stokes shift found in luminescence spectra. For this reason these are referred to as Stokes Raman lines. The pattern is mirrored by an analogous pattern on the high-frequency side ($+\Delta v_i$) at shorter wavelengths, referred to as anti-Stokes Raman lines. The relationship of intensities between the Stokes Raman lines and the anti-Stokes Raman lines depends on the temperature of the sample, as described in Section 6.2.3. Raman spectroscopy provides complementary information to infrared (IR) spectroscopy because the selection rules are different.

Raman spectral analysis has a number of advantages over passive IR and optical spectroscopy, chief among them being the sharpness and selectivity of spectral features. This allows unambiguous detection of specific minerals, particularly mineral mixtures, rocks, and isochemical glasses. A number of authors (e.g., White & De Angelis, 1967; White, 1975; McMillan, 1985; McMillan & Hofmeister, 1988; Nasdala et al., 2004; Dubessy et al., 2012) have reviewed the applications of Raman techniques to geologic materials. In this chapter, the basic theory of normal (spontaneous) and resonance Raman scattering are discussed along with advanced Raman techniques applicable to geologic materials. Applications of Raman spectroscopy to field settings are described in Chapter 12.

6.2 Theory of Normal Raman Scattering

6.2.1 Classical Theory of Normal Raman Scattering

The phenomena of Rayleigh and Raman scattering can be explained in part by classic electrodynamics (e.g., Colthup et al., 1975). Light scattering may be considered as an interaction between electromagnetic (EM) waves with matter that perturbs the electron clouds of the sample atoms and molecules. The perturbation of the electron cloud results in the periodic separation of the center of positive and negative charges within the atoms and/or molecules, which produces an induced dipole (P). The oscillating induced dipole moment is a source of electromagnetic radiation that may produce scattered light. For example, a hypothetical atom with a spherical symmetric electron cloud has no permanent dipole moment.

$$P = \alpha E, \tag{6.1}$$

where α is a constant of proportionality (also known as polarizability), and E is the magnitude of the electric field as described by $E = E_0 \cos 2\pi \nu_0 t$.

When the atoms and molecules are placed in an oscillating field of an electromagnetic light wave of frequency ν_0, it will induce a dipole moment P given by the following equation:

$$P = \alpha E_0 \cos(2\pi \nu_0 t), \tag{6.2}$$

where ν_0 is the frequency in Hz (c/λ_0, where c is the velocity of light and λ_0 is the wavelength) of incident light, and t is time.

In the case of molecules, polarizability is not constant as certain vibrations and rotations of a molecule can cause α to vary. For example, during the vibration of a diatomic molecule, the molecular shape is alternately compressed and extended. If the electron cloud is not identical at the extremes of the vibration, a change in polarizability will result.

For small displacements the polarizability of a diatomic molecule can be given as

$$\alpha = \alpha_0 + \left(\frac{\partial \alpha}{\partial Q}\right) dQ, \tag{6.3}$$

where dQ is the difference between the internuclear distance at any time and the equilibrium internuclear distance, and $\partial \alpha / \partial Q$ is the change of α with respect to dQ. If the vibration is considered harmonic, dQ is given by

$$dQ = Q_0 \cos(2\pi \nu_{\text{vib}} t), \tag{6.4}$$

where Q_0 is the maximum displacement about the equilibrium position, and ν_{vib} is the vibrational frequency. Substituting the values of α and dQ from Eqs. (6.3) and (6.4) in Eq. (6.5) yields

$$P = \alpha_0 E_0 \cos(2\pi \nu_0 t) + \left(\frac{\partial \alpha}{\partial Q}\right) E_0 Q_0 \cos(2\pi \nu_0 t) \cos(2\pi \nu_{\text{vib}} t). \tag{6.5}$$

Using the product to sum trigonometric identity, Eq. (6.5) can be rewritten as

$$P = \alpha_0 E_0 \cos(2\pi\nu_0 t) + \left(\frac{E_0 Q_0}{2}\right)\left(\frac{\partial\alpha}{\partial Q}\right)\left[\cos\left(2\pi(\nu_0 + \nu_{\text{vib}})t\right) + \cos\left(2\pi(\nu_0 - \nu_{\text{vib}})t\right)\right]. \quad (6.6)$$

The spectrum of the scattered light will contain a line corresponding to incident light (Rayleigh scattering) as described in Eq. (6.6), and also two modified lines corresponding to frequencies ($\nu_0 \pm \nu_{\text{vib}}$) (Raman scattering). The intensity of the Rayleigh line depends on the square of the amplitude, i.e., $(E_0\alpha_0)^2$, and the intensities of the Raman lines depend on $(E_0 Q_0/2)^2 * (\partial\alpha/\partial Q)^2$, where $\partial\alpha/\partial Q$ is the rate at which α changes during a given normal mode of vibration. A necessary condition for Raman scattering is that $\partial\alpha/\partial Q$ must be nonzero. Thus we have a classical explanation of Stokes and anti-Stokes Raman lines. As demonstrated by Lord Rayleigh, the intensity of scattered light is proportional to ν_0^4, and is responsible for the blue color of the sky. According to the classical theory, the intensities of Stokes Raman and anti-Stokes Raman lines should be $(\nu_0 - \nu_{\text{vib}})^4/(\nu_0 + \nu_{\text{vib}})^4$; however, this prediction is not borne out by experiments, and is the main limitation of the classical Raman theory.

6.2.2 Polarization and Intensities of Raman Lines

Polarization of incident radiation and the orientation of single crystal samples affect the intensities of Raman scattered radiation. Studies of the intensities of Raman lines in varying polarization and crystal orientations provide a valuable guide in assigning the observed frequencies to specific modes of vibrations.

As discussed previously, a normal mode of vibration is allowed in the Raman spectrum if the polarizability of the molecule changes during the mode. By resolving P, α, and E in the x, y, and z directions, we anticipate the following relationships.

$$P_x = \alpha_x E_x; \quad P_y = \alpha_y E_y; \quad P_z = \alpha_z E_z. \quad (6.7)$$

Equation (6.7) is applicable only to a completely spherical molecule. Most molecules, however, have structures that are not totally spherical; therefore the equation does not hold in such cases because the direction of polarization does not coincide with the direction of the applied field. Instead of Eq. (6.7), we must use the following relationship:

$$\begin{bmatrix} P_x \\ P_y \\ P_z \end{bmatrix} = \begin{bmatrix} \alpha_{xx} & \alpha_{xy} & \alpha_{xz} \\ \alpha_{yx} & \alpha_{yy} & \alpha_{yz} \\ \alpha_{zx} & \alpha_{zy} & \alpha_{zz} \end{bmatrix} \begin{bmatrix} E_x \\ E_y \\ E_z \end{bmatrix}. \quad (6.8)$$

The first matrix on the right-hand side is called the second rank polarizability tensor. It is a symmetric tensor, i.e., $\alpha_{xy} = \alpha_{yx}$; $\alpha_{yz} = \alpha_{zy}$; and $\alpha_{xz} = \alpha_{zx}$. Equations (6.7) and (6.8) are for Rayleigh scattering. For Raman scattering, the polarizability components α_{xx}, α_{xy}, α_{xx}, etc. should be replaced by the change in polarizability with respect to coordinate $\partial\alpha_{xx}/\partial Q$, and so on. The polarizability tensor can be visualized by means of a polarizability ellipsoid.

To define the term depolarization ratio in the Raman spectra of fluids, which measures the degree of depolarization of Raman lines, let us assume that the x-axis is the direction of propagation of incident monochromatic light onto the sample and the direction of observation is perpendicular to the x-axis in the yz plane. If the plane polarized incident laser light is the excitation source, then the depolarization ratio ρ_p is defined as the ratio of intensity of the scattered light polarized perpendicular to the yz plane, $I_\perp$, to that polarized parallel to the yz plane, $I_\parallel$.

$$\rho_p = \frac{I_\perp}{I_\parallel}. \qquad (6.9)$$

In the case of fluid samples, including glasses, where all orientations of scattering molecules are possible, the degree of depolarization of a totally symmetric Raman line will be close to zero and the maximum degree of depolarization of an antisymmetric Raman line will be 3/4 (Colthup et al., 1975). A measurement of the depolarization ratio in gases, solutions, and glasses provides a means of distinguishing totally symmetric vibrations from the rest of the vibrational modes.

The effect of polarization of radiation and orientation of crystals on the Raman spectra was investigated using a 90° scattering geometry in the early days of Raman spectroscopic research on single crystals of sodium nitrate and calcite. It was found that the polarization of Raman lines varies with orientation of the crystals (e.g., Nadungadi, 1939; Bhagavantam, 1940). The polarized Raman spectral analysis of an oriented single crystal yields a detailed description of the symmetry of the crystal vibrations. In the 1960s, Porto's group at Bell Laboratories (Damen et al., 1966) proposed a useful notation to specify the relative orientation of the polarization vector. In the case of an oriented crystal excited by polarized laser light under a 90° scattering configuration, these authors proposed the notation $x(zx)y$, where x is the direction of propagation of incident light and y is the direction of propagation of scattered light. The terms within the parentheses refer to the polarization of the incident light and scattered light, respectively. These terms represent the α'_{zx} component of the change in the polarizability. By changing the polarization of the incident and observed scattered light, and the crystal orientation, all the components of the polarizability tensor can be observed. The polarized Raman spectra help in identifying symmetry species of various modes.

6.2.3 Energy Diagram and Phenomena of Luminescence, IR Absorption, and Scattering

The interaction of light with matter can be explained graphically with the aid of an energy diagram that depicts the quantized nature of the energy states, as shown in Figure 6.1.

In the case of luminescence, the molecules are first excited by absorption of a photon to the higher energy second excited electronic state (S_2, the singlet 2 state) (see Figure 6.1). The excited molecules relax to the first electronic excited state (S_1) by internal conversion (IC), a nonradiative process, and can decay to the ground state by a radiative process emitting photons. This phenomenon is called fluorescence and has a lifetime in the range

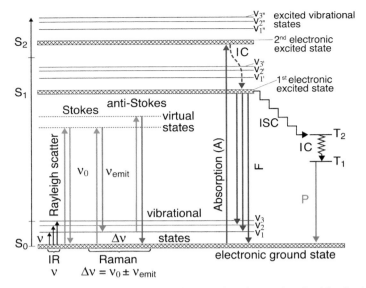

Figure 6.1 Jablonski energy-level diagram showing the virtual states involved in elastic (Rayleigh) and inelastic (Raman) scattering and the real electronic excited states involved in fluorescence and phosphorescence (see text for explanation). IC = internal conversion; ISC = intersystem crossing; F = fluorescence (lifetime 10×10^{-12} s (10 ps)–10^{-9} s (nanosecond), and P = phosphorescence (lifetime 10^{-7}–10^2 s) (modified from Panczer et al., 2012).

of 10 ps to a few ns for biological and organic molecules. If there is a triplet (T_1 and T_2) electronic state available, the excited molecule in the first (S_1, lowest energy) excited singlet state could decay by a nonradiative process to the triplet T_2 excited state by intersystem crossing (ISC), followed by relaxation to the T_1 excited state by a nonradiative IC process, and can decay to the ground state by phosphorescence. The lifetime for phosphorescence in minerals from transition metal ion and rare-earth ion impurities varies from submicroseconds to milliseconds (Gaft et al., 2005; Gaft & Nagli, 2009; Panczer et al., 2012).

During IR absorption, a molecule absorbs IR radiation and moves from the ground vibrational state to an excited rotational-vibrational state. A few molecules will be in higher vibrational energy states, which are known as excited vibrational states. The fraction of molecules occupying a given vibrational state at a given temperature can be calculated using the Boltzmann distribution.

Atoms or molecules may elastically scatter light by first absorbing light, exciting the atom or molecule to a virtual state, followed by relaxation of the atom or molecule and simultaneous reemission of a photon of identical frequency to that of the incident light in a process known as elastic scattering or Rayleigh scattering (Figure 6.1).

In Raman scattering, an absorbed photon is reemitted with lower energy, the difference between the incident and scattered photon corresponds to energy required to excite a molecule to a higher vibrational state (Figure 6.1). This process gives rise to Stokes Raman lines. Similarly, a photon absorbed by a molecule in the excited vibrational state is reemitted with a shorter wavelength by bringing the molecule to the ground state. This process gives rise to anti-Stokes Raman lines. This phenomenon of inelastic scattering is referred to as spontaneous or normal Raman scattering to distinguish it from stimulated Raman scattering, which is a nonlinear optical effect.

The intensity ratio of the Stokes Raman to anti-Stokes Raman lines of the sample depend on the population of molecules in the ground and excited states, according to following equation:

$$\frac{I_{\text{Stokes}}}{I_{\text{anti-Stokes}}} = \frac{(\nu_0 - \Delta\nu)^4}{(\nu_0 + \Delta\nu)^4} e^{-(hc\Delta\nu/kT)}, \quad (6.10)$$

where h is Planck's constant, k is Boltzmann's constant, T is the temperature in Kelvin, and $\Delta\nu$ is the Raman shift due to a normal mode of vibration of the molecule (ν_{vib}) in cm^{-1}. This intensity ratio predicted based on the quantized nature of vibrational modes is consistent with the observed ratio of Stokes Raman to anti-Stokes Raman lines. Based on Eq. (6.10), the Stokes to anti-Stokes ratio of Raman lines can be used to determine the temperature of the sample.

For a molecular bond, the individual atoms are confined to specific vibrational modes in which the vibrational modes are quantized (e.g., Kittel, 1976). For example, the vibrational energy of a particular vibrational mode in a diatomic molecule can be given by the solution of the Schrödinger equation:

$$\frac{d^2\Psi}{dq^2} + \frac{8\pi^2\mu}{h^2}\left(E - \frac{1}{2}Kq^2\right)\Psi = 0, \quad (6.11)$$

where Ψ is the wave function of the quantum system, q is the atomic displacement from the equilibrium position, and μ is the reduced mass of the system. For a diatomic molecule with molecular masses m_1 and m_2, the reduced mass is $1/\mu = 1/m_1 + 1/m_2$, h is Planck's constant, and K is the force constant of the molecular bond. Solving for the condition that the wave function must be single valued, the eigenvalues are

$$E_v = (v + 0.5)h\nu_{\text{vib}}, \quad (6.12)$$

where v is the vibrational quantum number, 0.5 is the zero point energy, ν_{vib} is the vibrational frequency, and the frequency of vibration is given by

$$\nu_{\text{vib}} = \frac{1}{2\pi}\sqrt{\frac{K}{\mu}}. \quad (6.13)$$

The vibrational frequencies depend on the reduced mass of atoms involved in a particular vibration. Isotopic substitutions have been used to assign the origin of the vibrational modes in silicate minerals and glasses (e.g., McMillan, 1985).

6.2.4 Resonance Raman Scattering

Resonance Raman scattering is a variant of the "normal" Raman scattering phenomenon. Normal Raman spectroscopy is typically performed using laser sources whose energy is above the vibrational or rotational energies of the molecule, but far below the first electronic excited state. Resonance Raman scattering takes place when the laser's energy nears that of an electronic excited state (e.g., first or second electronic excited state in Figure 6.1). In such a case, the Raman bands originating in the excited electronic transition may show very strong enhancements, with intensities 10^3–10^5 higher than predicted by the v_0^4 rule. When the excitation energy is in the vicinity of the electronic absorption, the term "pre-resonance Raman scattering" is commonly used. In minerals, Raman enhancement has been observed for the iron-containing oxides such as hematite (α-Fe_2O_3), maghemite (γ-Fe_2O_3), and magnetite (Fe_3O_4) (de Faria et al., 1997; Nieuwoudt et al., 2011). For this group of minerals, pre-resonance Raman occurs with 636-nm excitation due to a spectral absorption band located near 640 nm (Cornell & Schwertmann, 2003). The resonance Raman spectra are also observed in biominerals containing chromophore biomolecules, such as carotene in pink corals (e.g., Urmos et al., 1991).

6.2.5 Normal Modes of Vibrations and Selection Rules

A polyatomic molecule composed of N atoms has $3N - 6$ internal or vibrational degree of freedom ($3N - 5$ in the case of a linear molecule). This is obtained by excluding translational and rotational degrees of freedom from the total number of degrees of freedom. Corresponding to $3N - 6$ degrees of freedom, there are $3N - 6$ fundamental, or normal, modes of vibrations; every other vibrational mode can be expressed as a linear combination of these normal modes.

The physical nature of the vibrational modes depends mainly on the symmetry properties of the molecule. The activity of the normal modes in the Raman or IR spectrum also depends on the geometry of the molecule. The shape of the molecule can be specified by symmetry elements (e.g., axes of rotation, planes of symmetry, center of symmetry) that can be ascribed to it, which leads to a point-group notation. The point group implies that one point remains invariant under all of the operations of the group. The first step in identifying the symmetries, and subsequently the Raman and IR active modes, is to identify the point group symmetry of the molecule or the crystal. Following Colthup et al. (1975), various types of vibrational modes or species in Mulliken symbols are designated according to symmetry operations, which they represent as follows:

A   symmetric with respect to principal axis of symmetry
B   antisymmetric with respect to principal axis of symmetry
E   doubly degenerate vibrations, the irreducible representation is two dimensional
F   triply degenerate vibrations, i.e., a three-dimensional representation
  g and u (subscripts)   symmetric or antisymmetric with respect to a center of symmetry

1 and 2 (subscripts)   symmetric or antisymmetric with respect to rotational axis (Cp) or rotation–reflection axis (Sp) other than the principal axis or in those point groups with only one symmetry axis with respect to plane of symmetry

Prime and double prime (superscripts)   symmetric or antisymmetric with respect to plane of symmetry.

For the linear molecules belonging to the point group $C_{\infty v}$ and $D_{\infty h}$, capital Greek letter designations are used, the same as for electronic states for homonuclear diatomic molecules, as follows:

Σ^+ symmetric with respect to plane of symmetry through the molecular axis
Σ^- antisymmetric with respect to plane of symmetry through the molecular axis
Π Δ, Φ degenerate vibrations with degree of degeneration increasing in this order.

Group theory and its application to molecular vibrations have been described in several textbooks (e.g., Cotton, 1963; Bhagavantam & Venkatarayudu, 1969; Ferraro & Ziomek, 1969; Colthup et al., 1975; McMillan, 1985; Rull, 2012). Here, we focus on molecules of the type MX_2, MX_3, and MX_4, where M is a positive valence atom and X is a negative valence atom, which are of interest in geologic and planetary science.

6.2.5.1 Raman and IR Active Modes of the H_2O Molecule

The symmetry elements of the water molecule and the symmetric properties of the normal modes of vibration of the H_2O molecule in C_{2v} symmetry are shown in Figure 6.2 and classified in Table 6.1. For $N = 3$ atoms in the water molecule the normal modes of H_2O labeled T_x, T_y, and T_z correspond to translational modes of the molecule moving the molecule as a whole, and three labeled as R_x, R_y, and R_z correspond to the rotational modes of the molecule. These are referred to as the lattice or external vibrational modes of the molecule in solids.

In Table 6.1, the top row consists of the symmetry operations that form the point group. The first column lists the symmetry species that comprises the C_{2v} point group. The symmetry species' irreducible representations appear in the column immediately to the right of the Mulliken symbols. The individual characters indicate the result of the symmetry operation at the top of the column on a molecular basis for the symmetry. A comparison of the entries in Table 6.1 with the vibrations shown in Figure 6.2 shows that the vibrations do, in fact, fall into one or the other of these species as indicated in Figure 6.2. Table 6.1 also shows the symmetry species for T_x, T_y, and T_z, and R_x, R_y, and R_z. Only the vibrations of the symmetric species that include change in the electric dipole moment of the molecule (i.e., T_x, T_y, and T_z) are IR active, and those that include change in the polarizability components are Raman active.

The symmetric stretching mode $v_1(A_1)$, symmetric bending mode $v_2(A_1)$, and antisymmetric stretching mode $v_3(B_2)$ are the three normal modes of vibrations of water molecules that are active both in Raman and IR spectra. Accordingly, the three vibrational modes of water molecules can be classified according to the irreducible

Table 6.1 *Character table for point group C_{2v}*

C_{2v}	E	C_2 (z)	σ_v(xz)	σ_v'(yz)		
A_1	1	1	1	1	T_z	$\alpha_{xx}, \alpha_{yy}, \alpha_{zz}$
A_2	1	1	−1	−1	R_z	α_{xx}
B_1	1	−1	1	−1	$T_x; R_y$	α_{zx}
B_2	1	−1	−1	1	$T_y; R_x$	α_{yz}

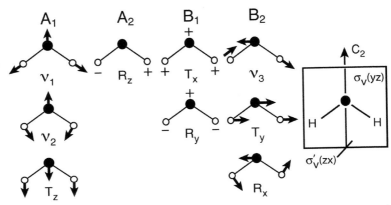

Figure 6.2 Insert on the right-hand side shows the symmetry elements of water molecule. Various normal modes of H_2O molecule are shown to the left-hand side. (The + and − signs denote vibration going upward and downward, respectively, perpendicular to the plane of the paper.)

representation of the C_{2v} point group as $\Gamma = 2A_1$ (R, IR) $+ B_2$ (R, IR) for Raman active (R) and IR active (IR) modes. The v_1, v_2, and v_3 modes of water vapors were observed in the first-order Raman and IR spectra at 3651, 1595, and 3755.8 cm^{-1}, respectively (Herzberg, 1945: 281).

6.2.5.2 Raman and IR Active Modes of the CO_2 Molecule

Figure 6.3 illustrates normal vibrational modes of linear CO_2 and tetrahedral MX_4 molecules. The linear CO_2 molecule consist of three atoms and belongs to point group $D_{\infty h}$ (e.g., Ferigle & Meister, 1952). We should expect $3N - 5$ or four normal modes of vibrations for this molecule (see Figure 6.3). Molecules that have a center of inversion, such as CO_2, adhere to the rule of mutual exclusion that states a vibration that is active in IR will be Raman inactive and vice versa. The $v_1(\Sigma_g^+)$ totally symmetric stretching of oxygen atoms does not involve any change in the electrical dipole moment of the molecule and will, therefore, be inactive in the IR but does involve change in the polarizability and therefore

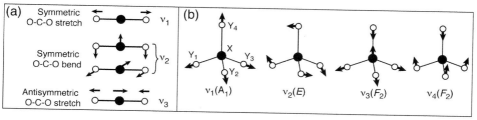

Figure 6.3 Normal modes of vibration of (a) CO_2 molecule and (b) MX_4 molecule.

will be active in the Raman spectrum. The doubly degenerate $v_2(\Pi_u)$ involves a motion of the carbon atom against the oxygen atoms in a line perpendicular to the symmetry axis of the molecule. The $v_3(\Sigma_u^+)$ involves the oscillation of a carbon atom with respect to oxygen atoms along the symmetry axis of the molecule. Both the v_2 and v_3 vibrations possess change in the dipole moment of the molecule and are active in IR but inactive in the Raman spectrum.

According to the group theoretical analysis, the four vibrational modes of CO_2 can be classified according to the irreducible representation of the $D_{\infty h}$ point group as

$$\Gamma = \Sigma_g^+(R) + \Sigma_u^+(IR) + \Pi_u(IR). \tag{6.14}$$

Strong IR bands observed for gaseous CO_2 are 667.3 and 2349.3 cm^{-1}, and are assigned to the v_2 and v_3 modes, respectively. Group theory predicted a single Raman line in the Raman spectrum of a CO_2 molecule; however, a doublet consisting of Raman lines at 1388.3 and 1285.5 cm^{-1} is observed. The doublet is assigned to an accidental degeneracy due to Fermi resonance between the v_1 and the overtone ($2v_2$) of the bending mode of CO_2 (e.g., Herzberg, 1945: 274).

6.2.5.3 Raman and IR Active Modes of the MX_3 Molecule

For MX_3 types of molecules in which all the X atoms are attached to central M atom, the simplest structures are planar and pyramidal. For planar structures, the three X atoms are at the corner of an equilateral triangle and the M atom is at its center. The symmetry in this case is that of point group D_{3h}. For a pyramidal structure, all the X atoms are at the corner of a pyramid and the M atom is at its center. The symmetry in this case is that of a point group C_{3v}. The total number of normal modes of vibration for the MX_3 molecules will be six ($=3N-6$). According to group theory, the numbers of normal modes of vibrations belonging to each symmetry species are different in these two cases (Bhagavantam & Vaenkatarayudu, 1969). These are summarized in the following respective irreducible representations:

$$\Gamma_{D_{3h}} = A_1'(R) + A_2''(IR) + 2E'(IR). \tag{6.15}$$

For the pyramidal molecule:

$$\Gamma_{C_{3v}} = 2A_1(R, IR) + 2E(R, IR). \tag{6.16}$$

For planar carbonate ions, the point group symmetry is D_{3h}. The Raman active totally symmetric stretching mode of carbonate $v_1(A_1')$ appears at 1088 cm^{-1}. The Raman and IR active doubly degenerate antisymmetric stretching mode $v_3(E')$ and the antisymmetric in-plane bending mode $v_4(E')$ appear at 1438 and 714 cm^{-1}, respectively. The IR active nondegenerate out-of-plane bending mode $v_2(A_2')$ appears at 880 cm^{-1}.

The pyramidal NH_3 molecules, which belong to the C_{3v} point group, should display four Raman lines, which are all present in the IR absorption spectra (e.g., Buback & Schultz, 1976). In the Raman spectrum of ammonia the totally symmetric $v_1(A_1)$ mode appears as a strong polarized line at 3336 cm^{-1} and as a weak line in the IR spectra. The symmetrical bending mode $v_2(A_1)$ appears as a polarized Raman line at 950 cm^{-1}. The Raman and IR active doubly degenerate antisymmetric stretching mode $v_3(E)$ and the antisymmetric bending mode $v_4(E)$ appear at 3444 and 1626 cm^{-1}, respectively (e.g., Buback & Schultz, 1976). This shows that it is possible to differentiate between the molecules in D_{3h} and C_{3v} point groups based on the polarized Raman spectra.

6.2.5.4 Raman and IR Active Modes of the MX_4 Tetrahedral Molecule

A number of minerals including perchlorate, silicates, and sulfates have basic building blocks that have tetrahedral symmetry. The symmetry of these MX_4 molecules or ions is that of point group T_d. The total number of normal modes of vibration for MX_4 is nine. According to group theory, these modes of vibrations can be classified according to the irreducible representation of the T_d point group as $\Gamma_{Td} = A_1 (R) + E (R) + 2F (R, IR)$. Figure 6.3b illustrates these four fundamental modes of vibration. For isolated silicate ions, the totally symmetric Raman active $v_1(A_1)$ line appears at 819 cm^{-1}. The doubly degenerate Raman active symmetric bending mode $v_2(E)$ appears at 340 cm^{-1}; and the triply degenerate antisymmetric stretching $v_3(F)$, and antisymmetric bending modes, $v_4(F)$, appear, respectively, at 956 and 527 cm^{-1} (e.g., White, 1975).

6.3 Classification of Crystal Vibrations

In crystalline solids, atoms are arranged in a periodic lattice, and strong forces exist between neighboring atoms. If one atom is displaced from its mean equilibrium position, the neighboring atoms also undergo displacements. Therefore, the atomic motions in solids are collective rather than individual. These motions give rise to normal modes of vibrations, which travel as a wave through the crystal. The energy of these waves is quantized. The pseudo-particles associated with these waves are referred to as phonons, an analogy to the term photon for a quantized unit of light energy.

A crystal containing N atoms in the primitive cell has $3N$ degrees of freedom, of which three are associated with the translation of the unit cell as a whole and thus become acoustic modes. The acoustic modes are responsible for the propagation of sound waves through the crystal lattice. Thus, the number of lattice vibrations that may be observed by Raman and IR spectroscopy is equal to $3N - 3$ and are referred to as "optical modes". The transverse and longitude modes are referred to as TO and LO modes, respectively.

A classification of the crystal vibrational modes into internal (molecular) and external (lattice) vibrations can be accomplished by taking into account the factor group of the crystal instead of the entire space group. A factor group is defined as the group formed by the symmetry elements present in the smallest unit cell, which is called a Bravais cell. The Bravais cell is defined in such a way that the pure lattice translations of the unit cell can be used to obtain the entire space group. A factor group is always isomorphous with one of the 32 crystallographic point groups (e.g., Bertie & Bell, 1971).

The analysis of the vibration at $k = 0$, the center of the Brillouin zone, for the entire crystal can be accomplished by two principal methods. The first is the factor group method of Bhagavantam and Venkatarayudu (1939) that considers the atoms in the Bravais unit cell as a large molecule. The crystallographic unit cell may be identical with the Bravais cell or it may be larger by some simple multiple. For example, for all X-ray crystal structures designated by a symbol P (for primitive) or R (rhombohedral primitive), the crystallographic unit cell and Bravais unit cell are identical. Crystal structures designated by capital letters I, A, and C have unit cells that contain two Bravais cells. Crystal structures designated by capital letter F (Face-centered) have unit cells that contain four Bravais cells. For factor group analysis the number of molecules per crystallographic unit cell should be divided by the cell multiplicity, e.g., by 4 for an F cell, by 2 for I, A, and C cells, and by 1 for P and R cells (e.g., Fateley et al., 1972; Salthouse & Ware, 1972). Adams and Newton (1970a,b) established tables for factor group analysis of vibrational modes for all space groups.

The other method is called the site group method of Halford (1946), which involves deriving the number of allowed modes of specific molecular entities based on site symmetry of the molecule in the unit cell. It is an approximation that can be used for molecular crystals such as carbonates, sulfates, and phosphates, where the forces between molecules are considerably weaker than those between the atoms inside the molecules. Hornig (1948) and Winston and Halford (1949) discuss the relationship of the factor group method to the site group method. These authors showed that it is possible to get equivalent results using either of the methods. Maraduddin and Vosko (1968), and Warren (1968) provide an in-depth discussion on factor group analysis for the full Brillouin zone. Based on factor group analysis, Ferraro (1975) listed selection rules for some common minerals. Fateley et al. (1972) developed the correlation of the site group to the factor group. DeAngeles et al. (1972) and Fateley et al. (1972) summarized the procedure to perform calculations for the zone center ($k = 0$) phonons. To assist with the analysis of vibrational modes in crystals, a set of online databases is available that includes data from international tables, space groups, and point groups, symmetry relations between space groups, and a k-vector database with Brillouin zone figures and classification tables of the wave vectors for space groups (Aroyo et al., 2006a, 2006b, 2011).

The essentials of factor group analysis are illustrated in the text that follows for calcite and aragonite polymorphs of calcium carbonate minerals. The rhombohedral primitive cell of calcite (space group #167, D_{3d}^{16} (R $\bar{3}$ C)) contains two $CaCO_3$ formula units, i.e., a total of 10 atoms; its 27 ($3N - 3$) vibrational modes can be classified according to the irreducible representations of the D_{3d} point group as follows:

$$\Gamma_{D_{3d}} = A_{1g}(R) + 4E_g(R) + 3A_{2g}(i.a.) + 2A_{1u}(i.a.) + 3A_{2u}(IR) + 5E_u(IR), \quad (6.17)$$

where R = Raman active, IR = IR active, and i.a. = inactive mode.

At room temperature, in the polarized Raman spectrum of an oriented single crystal of calcite the following fundamental internal modes of the CO_3 ion are detected (e.g., Porto et al., 1966): $v_1(A_{1g}) = 1088$ cm^{-1}, doubly degenerate $v_3(E_g) = 1434$ cm^{-1}, and $v_4(E_g) = 711$ cm^{-1}. The two doubly degenerate E_g translational and rotation lattice modes of calcite are observed at the 156 and 283 cm^{-1}, respectively. The Raman line observed at 1750 cm^{-1} is a combination ($v_1 + v_4$) mode.

The orthorhombic cell of aragonite (space group #62, D_{2h}^{16}-Pnma) contains four formula units, i.e., 20 atoms. According to Fateley et al. (1972), its 57 vibrational modes can be classified as follows.

The irreducible representations of the D_{2d} point group is

$$\Gamma_{D_{2d}} = 9A_g(R) + 6A_u(i.a.) + 6B_{1g}(R) + 8B_{1u}(IR) + 9B_{2g}(R) + 5B_{2u}(IR) \\ + 6B_{3g}(R) + 8B_{3u}(IR). \quad (6.18)$$

Of the 30 Raman active fundamental modes of aragonite, 28 have been identified in single crystal measurements at 80 K (De La Pierre et al., 2014). The following internal modes of vibration of CO_3 were detected: four A_g (705, 854, 1087.2, and 1466.2 cm^{-1}), two B_{1g} (705.9 and 1464.1 cm^{-1}), three B_{2g} (716.9, 911.1, and 1579.0 cm^{-1}), and one B_{3g} (700.1 cm^{-1}). On the basis of theoretical calculations, De La Pierre et al. (2014) concluded that the two internal modes B_{2g} (1091.6 cm^{-1}) and B_{3g} (1415.0 cm^{-1}) have zero intensity. Due to the low intensities of these bands, these could not be detected in the Raman spectra (De La Pierre et al., 2014).

6.4 Experimental Methods

Raman spectroscopy was initially considered a physics curiosity until the invention of lasers because the intensities of Raman lines are extremely weak without a high-intensity monochromatic light source. This was despite the fact that Raman studies of several minerals and pure compounds had been carried out at the macroscopic scale using 404.6, 435.8, and 253.65 nm light from mercury emission lamps as the excitation (λ_0) source. Lasers that became available during the 1960s were able to provide high irradiance of monochromatic light onto samples for recording the Raman spectra. The following subsections outline the advancements in (1) continuous wave (CW) and pulsed laser excitation sources; (2) Raman dispersive spectrometers and interferometers; (3) detectors used for Raman measurements with CW and pulsed lasers, respectively; and (4) advances in micro-Raman and remote Raman instrumentation.

Table 6.2 *CW laser radiation commonly used as Raman excitation*

Continuous-wave lasers	Wavelengths (nm)
Ar-ion	457.9, 488.0, 514.5
Kr-ion	532.9, 566.2, 647.1, 676.4
He–Ne	632.8
He–Cd	325, 441.6

6.4.1 Excitation Laser Sources

Various lasers used for exciting Raman spectra of samples can be classified into two basic categories: (1) CW laser sources and (2) pulsed laser sources.

6.4.1.1 CW Lasers

In modern Raman spectroscopy systems three types of CW ionized-gas lasers are used: Ar-ion, Kr-ion, and mixed Ar–Kr-ion lasers. The wavelengths of these lasers are stable and known within an accuracy of ~1 pm which corresponds to 0.04 cm^{-1} around 500 nm. In addition, CW He–Ne and He–Cd lasers are also used as Raman excitation sources. Various excitation wavelengths provided by these lasers are listed in Table 6.2.

With these lasers, laser line filters or pre-monochromators are used to eliminate all plasma line emissions. The plasma lines may, however, be used as a wavelength standard for calibrating the Raman spectrograph (e.g., Pandya et al., 1988).

The compact UV hollow cathode NeCu ion-lasers, which emit CW radiation at 248.6 nm with output power up to >50 mW (Storrie-Lombardi et al., 2011), have recently been selected as the excitation source for the SHERLOC instrument on NASA's Mars 2020 rover mission (e.g., Beegle et al., 2014).

Diode Pumped Solid State (DPSS) CW Nd:YAG lasers that emit fundamental radiation at 1064 nm with output power of several watts are commonly used as an excitation source with Fourier Transform-Raman (FT-Raman) spectrometers. The linewidth is <1 cm^{-1} in standard lasers. In addition to the 1064-nm laser wavelength, radiation at 532, 355, and 266 nm can be obtained with frequency doubling, tripling, and quadrupling crystals.

Single-mode, frequency stabilized near-infrared (NIR) 785 and 830 nm diode lasers with linewidth <1 cm^{-1} are increasingly finding applications in Raman spectroscopy as excitation sources (e.g., Angel et al., 1995; Cooney et al., 1995; Cooper et al., 1995; Wang et al., 1995, 2012; McCreery, 2000). When excited with NIR lasers, the Raman spectra of some minerals and related materials display strong laser-induced fluorescence lines from rare-earth elements, which may be mistakenly identified as Raman lines (e.g., Aminzadeh, 1997; Kaszowska et al., 2016).

6.4.1.2 Pulsed Laser Sources

In the past, flash lamp pumped or diode pumped solid state (DPSS) Nd:YAG pulsed 532-nm laser excitation sources were not usually used for exciting Raman spectra of geologic samples. This was because a focused ns-pulsed beam could damage the samples when used in 180° scattering geometry. Sharma (1989) demonstrated that a 532-nm pulsed laser could be used by focusing the expanded excitation beam in a 135° scattering geometry and collecting the scattered beam with a 20× objective. A spectrally narrowed (<0.1 nm), 532 nm, passively Q-switched microchip laser with 600 ps pulses (1.5 µJ, 40 kHz) has also been used to measure the micro-Raman spectra of minerals in 180° scattering geometry (e.g., Blacksberg et al., 2010, 2016).

6.4.2 Dispersive Spectrometers and Interferometers

Both grating-based dispersive and interferometer-based Raman spectrometers have been developed for measuring Raman spectra of minerals and inorganic and organic materials. The following gives a brief description of these two types of Raman spectrographs.

6.4.2.1 Dispersive Raman Spectrographs

In modern spectrographs, an important component is the dispersing element. Two types of gratings are available: transmission gratings and reflection gratings. Dubessy et al. (2012) reviewed the properties of various gratings. Holographic gratings are used to efficiently reject stray light and eliminate ghost lines in the spectrographs. Both plane and concave holographic gratings are used in commercial Raman spectrometers (e.g., Delhaye & Dhamelincourt, 1975; Dhamelincourt et al., 1979; Li & Deen, 2014). Most dispersive Raman spectrographs employ the Czerny–Turner configuration. The aperture of Czerny–Turner optics is limited to F/4 or slower in order to minimize chromatic and spherical aberrations.

In the past two decades, the technology of Volume Phase Holographic gratings (VPHGs) has been widely exploited in the field of Raman spectroscopy for instrumentation that works both in the visible and NIR (Arns, 1995; Arns et al., 1999). This is because they have very large diffraction efficiency even at very high dispersion and are easy to customize (each VPHG is a master grating). In VPHGs the diffraction of light occurs due to a periodic modulation of the refractive index in the volume of an active material such as dichromate gelatin (DCG) (e.g., Arns, 1995). These holographic gratings have found applications in Raman instrumentations (Battey et al., 1993; Arns, 1995; Owen, 2007). These gratings have been instrumental in developing compact and high-throughput Raman spectrographs for planetary applications (e.g., Wang et al., 2003; Sharma et al., 2005; Gasda et al., 2015).

Due to the low ($\sim 10^{-7}$) efficiency of Raman scattering and high efficiency of Rayleigh (10^{-3}) scattering combined with reflection of the laser excitation radiation (λ_0), rejection of λ_0 radiation before it reaches the detector must be achieved to avoid detection of unwanted light. In the past, double and triple Czerny–Turner spectrographs were developed to measure low-wavenumber Raman spectra (e.g., McCreery, 2000). The ability to efficiently reject the excitation λ_0 laser radiation has been possible with advanced holographic notch filters and dielectric edge filters, which allows the use of only a single stage Czerny spectrograph for Raman scattering measurements down to 50 cm^{-1}. Since 2011 ultra-narrow holographic notch filters have been developed that allow the measurements of Raman lines down to ~10 cm^{-1} (e.g., Lebedkin et al., 2011).

6.4.2.2 FT-Raman Spectrometers

The development of laboratory-based FT-Raman spectrometers based on Michelson interferometers was motivated with the possibility of (1) achieving higher resolution than is possible with the dispersive grating based instruments (Jennings et al., 1986), and (2) measuring NIR excited Raman spectra of materials that are highly fluorescent when excited with visible radiation (e.g., Chase, 1986; Hirschfeld & Chase, 1986). In FT spectrometers, the limiting resolution does not depend on the size of the aperture, so high spectral resolution can be achieved without sacrificing the optical throughput. Multichannel FT-Raman systems that combine a common-path interferometer (Sagnac) with NIR laser excitation (Zhao & McCreery, 1996, 1997), and spatial heterodyne interferometers with 532 nm and UV pulsed laser excitations have been developed (e.g., Gomer et al., 2011; Hu et al., 2015; Lamsal et al., 2016). The multichannel FT-Raman systems with pulsed laser and gated ICCD detectors do not have any moving parts, thus offering an advantage for standoff measurements under high ambient light environments.

6.4.3 Detectors for Raman Measurements with CW Lasers and Pulsed Lasers

Various detectors used in CW and time-resolved modes with pulsed lasers are briefly discussed in the following subsections.

6.4.3.1 Detectors for Raman Measurements with CW Lasers

Single-channel photomultiplier (PMT) tubes were used during 1960–1980 for Raman spectroscopy. These PMT detectors were used in the first Raman microprobes at ambient pressures (Delhaye & Dhamelincourt, 1975; Rosasco et al., 1975) as well as in high-pressure Raman with the diamond anvil cell (e.g., Adams et al., 1977; Sharma, 1979). In the mid-1980s, the first popular multichannel detectors were linear intensified photodiode arrays of 1024 pixels with dimensions of 3 mm × 24 μm (e.g., Denson et al., 2007). These diode array detectors significantly reduced the measurement time as a 1024-pixel spectral wavelength could be measured simultaneously in 1 s; however, the spectral resolution was degraded because of the size and number of pixels. Charge-coupled device (CCD) cameras appeared in mid-1990 and replaced all other detectors because of their high quality. The CCDs are made up of two-dimensional arrays of individual pixels. The common dimensions of CCDs used in laboratory Raman spectrographs are 1024 × 256 pixels. The pixel size is typically 26 μm × 26 μm. The dynamic range of most laboratory CCDs is 16 bit (e.g., maximum counts $65\,536 = 2^{16}$).

6.4.3.2 Detectors for Raman Measurements with Pulsed Lasers

Minerals and rocks containing transition metal and rare-earth ions can produce luminescence spectra that in some cases overlap with their Raman spectra. The lifetime of mineral photoluminescence is much longer (microseconds to milliseconds) compared to the lifetime of the Raman signal, which is $\sim 10^{-13}$ s. The difference in the lifetime of Raman scattering and fluorescence has been exploited to minimize interference by luminescence with the Raman spectra by using pulsed laser excitation and time-gated detection (e.g., Matousek et al., 1999; Carter et al., 2005a,b; Misra et al., 2005; Blacksberg et al., 2010). Raman spectrometers equipped with intensified CCD (ICCD) detectors or single-photon avalanche diode (SPAD) array detectors and pulsed laser excitation allow measurements of Raman spectra of minerals with minimal interference from photoluminescence and ambient light (e.g., Carter et al., 2005b; Sharma, 2007; Blacksberg et al., 2010). The ICCD detectors normally allow gating down to 2 ns and the SPAD arrays offer gating down to 0.5 ns, thus effectively minimizing interference from mineral photoluminescence and short-lived fluorescence (>0.5 ns) from organic and biological molecules, respectively.

6.4.4 Advances in Micro-Raman and Remote Raman Instruments

Since the late 1990s, significant advances have been made both in the micro-Raman and remote Raman instrumentation. These advances in Raman instrumentation are briefly outlined in the text that follows.

6.4.4.1 Advances in Micro-Raman Spectroscopy

Figure 6.4 shows a schematic of a micro-Raman system with CW Ar-ion laser, 488- and 514.5-nm radiation used in 180° excitation, and a pulsed frequency-doubled 532-nm Nd:YAG laser used in an oblique (typically 135°) angle excitation.

One of the principal advantages of micro-Raman spectroscopy is the ability to make measurements on a region of a sample the same size as the focused laser spot. The laser can be focused to a diffraction limited spot such that the system will accept light from other areas including above and below the nominal focus area in an on-axis configuration (e.g., backscattering geometry). Insertion of a pinhole in the optical path in the image plane, as shown in Figure 6.4, will make the Raman system confocal, allowing rejection of out of focus signals as well as spurious light from other sources. The 180° confocal microscope setup can map the sample in three dimensions. Use of 135° excitation geometry also avoids the generation of scattered light within the collection optics that may interfere with the weak light scattered from a sample.

Fully automated commercial confocal Raman instruments are available from many companies including HORIBA Instrument Inc. (Edison, NJ, USA), Renishaw inVia

Raman Spectroscopy: Theory and Lab Spectra of Geologic Materials

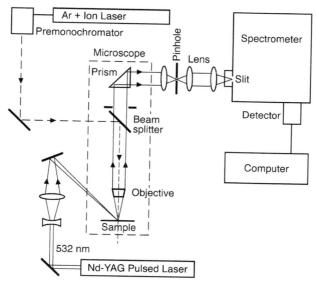

Figure 6.4 Schematics of a confocal micro-Raman system with 180° and 135° scattering geometry with a CW Ar-ion laser and a pulsed 532-nm laser, respectively.

Raman microscope (Renishaw, Gloucestershire UK), WITec alpha300 confocal Raman system (WITec GmbH, Ulm, Germany), and F/1.8 RamanRXN microprobe (described by Battey et al. [1993]; Kaiser Optical Systems, Inc., Arbor, MI, USA). The inVia Raman microscope has the option of focusing the laser beam on a spot or line on the sample. The spectral image of the sample can be obtained by rastering the sample under the laser focus and moving from point to point on the sample until the desired area is mapped. By focusing the laser on a line rather than a spot, the resulting image can be focused on the slit and subsequently imaged on the CCD detector. The x-axis on the CCD will be the Raman shift and y-axis will show variation in the spectra of the sample. The line focus allows the use of higher laser power on the sample as compared to the point focus, and allows a faster spectral image of the sample as multiple points along the line are measured simultaneously.

Advances in notch and sharp edge filters have significantly reduced the size of the dispersive Raman systems. SuperNotch filters allows the rejection of the Rayleigh scattered light by a factor of ~10^8, thereby eliminating the need for a filter stage in modern Raman spectrometers. These advances have been used to develop research-grade miniature Raman spectrometers for space applications (Wang et al., 1998; Gasda et al., 2015), as well as commercial miniature Raman spectrometers (for a review see Dubessy et al., 2012).

6.4.4.2 Advances in Remote Raman Instrumentation

Telescopic Raman systems equipped with an intensified CCD camera using a 532-nm pulsed laser with 10-ns pulses have been developed for measuring time-resolved remote

Raman spectra of minerals during daytime and nighttime (e.g., Lucey et al., 1998; Carter et al., 2005b; Misra et al., 2005; Sharma et al., 2005; Wiens et al., 2016). For a review, see Angel et al. (2012).

In recent years, remote Raman systems based on spatially heterodyne Raman spectrometers (SHRS) using both 532-nm and UV pulsed lasers are being developed for measuring time-resolved Raman spectra of both inorganic and organic compounds over a large area (e.g., Gomer et al., 2011; Lamsal et al., 2016). Using SHRS and 532-nm pulsed laser excitation, Egan et al. (2017) measured Raman spectra of feldspars and olivine minerals. These spectra show well-resolved Raman fingerprint modes of four-membered TO_4 rings in microcline at 475, 484, and 513 cm^{-1}, and in plagioclase at 481 and 509 cm^{-1} (e.g., McKeown, 2005; Freeman et al., 2008). The Raman fingerprint of olivine, which includes a doublet consisting of well-resolved lines at 824 and 855 cm^{-1}, was also observed with SHRS.

6.5 Raman Spectra of Planetary Ices and Geologic Materials

6.5.1 Raman Spectra of Planetary Ices

Raman spectroscopy can detect homopolar molecules such as H_2, N_2, and O_2 that do not absorb in IR as well as heteropolar molecules. Raman spectrometry has been successfully adopted by a number of scientists for in situ investigation of irradiated frozen ices containing organic molecules (e.g., Spinella et al., 1991; Ferini et al., 2004), and crystalline and amorphous water ice films on metal surfaces (Sonwalker et al., 1991). Micro-Raman techniques are used in the laboratory to study the effects of ion-induced lattice damage in carbonaceous solids and organic compounds in frozen planetary ice analogs (Elman et al., 1981; Strazzulla & Baratta, 1992; Strazzulla et al., 2001; Ferini et al., 2004). Bennett et al. (2013) developed a novel high-sensitivity Raman spectrometer to study pristine and irradiated interstellar ice analogs. These scientists measured in situ Raman spectra of thin films of CO_2 ices of 10–396 nm thickness at 4.5 K under ultrahigh vacuum. The Fermi resonance doublet of CO_2 ice was detected at 1385 and 1278 cm^{-1}, which is in good agreement with previous studies. To evaluate the effect of irradiation on CO_2 ice, a CO_2-ice film (450 nm thick) was deposited at 4.5 K and subjected to 1 h of irradiation from a 5 keV electron beam over an area of 1.6 cm^2. The Raman spectrum of the irradiated CO_2 film detected the presence of two new species in the film, namely CO at 2145 cm^{-1} and O_2 at 1545 cm^{-1} (Bennett et al., 2013).

6.5.2 Raman Spectra of Geologic Materials

Advances in the Raman instrumentation have enhanced the applications of Raman spectroscopy in nearly all geologic and planetary disciplines. Fluid inclusions in various geologic and extraterrestrial materials have been studied with Raman spectroscopy (e.g., Rosasco et al., 1975; Pasteris et al., 1986; Fries & Steele, 2011; Frezzotti et al., 2012). The major strength of Raman spectroscopy in studying fluid inclusions is that this technique is capable of nondestructively analyzing single small (>5 µm) inclusions (e.g., Roedder, 1984). Raman spectra of minerals, glasses, and melts of mineral compositions have also been investigated, as these provide information about short range and intermediate range structural orders (e.g., Sharma et al., 1979a,b, 1981, 1985, 1988, 1996, 1997; Sharma & Simons, 1981; Matson et al., 1983, 1986; Rai et al., 1983; Cooney & Sharma, 1990; Wang et al., 1993, 1995).

Raman spectroscopy can provide structural information about minerals, glasses, and melts (Figure 6.5). For example, a comparative study of the Raman spectra silicate minerals and corresponding glasses led to reassignment of the two weak, sharp, and polarized peaks at 490 and 606 cm^{-1} in the Raman spectrum of SiO_2 glass (see Figure 6.5a). The 490 and 606 cm^{-1} bands have been assigned, respectively, to symmetric stretching of oxygen atoms of four- and three-membered rings of SiO_4 tetrahedra in the disordered three-dimensional network (e.g., Sharma et al., 1981; Galeener, 1982a,b; McMillan & Wolf, 1995 and references therein).

These ring modes are vibrationally decoupled from the rest of the network, remaining highly localized within the rings, which explains their narrow width compared to the rest of the glass spectrum. A micro-Raman study of a new silica polymorph moganite, which contains four-membered rings of SiO_4, found a strong Raman band at 501 cm^{-1} (Kingma & Hemley, 1994; Götze et al., 1998; Nasdala et al., 2004). This observation further confirmed that the lines in the range of 500–512 cm^{-1} are the Raman fingerprints of the four-membered rings of tetrahedra in tectosilicate minerals, glasses, and melts of mineral compositions.

A number of scientists have reviewed extensive work on Raman spectroscopic investigations of the structure of silicate glasses and melts of interest in earth science (e.g., McMillan & Wolf, 1995; McMillan et al., 1996; Nasdala et al., 2004; Mysen & Richet, 2005; Rossano & Mysen, 2012). The results of these investigations have had a profound effect on our understanding of magmatic processes.

Raman spectra of selected minerals are shown in Figure 6.5 to illustrate the variance in peak positions that enables identification of minerals. Micro-Raman and confocal Raman instruments are being used for identification of mineral phases in rocks and meteorite samples (e.g., Haskin et al., 1997; Wang et al., 2001; Kuebler et al., 2006). For example, Acosta et al. (2013) used micro-Raman mapping of mineral phases in a thin section of a strongly shocked Taiban ordinary chondrite. The Raman spectra were excited with various laser excitation wavelengths (e.g., 244, 514.5, 785, and 830 nm), which allowed identification of mineral phases such as olivine, wadsleyite, ringwoodite, high-Ca clinopyroxene, majorite-pyrope, jadeite, maskelynite, and lingunite. Olivine was found to be Fe depleted (Fo_{88}) in contact with the ringwoodite as compared to olivine (Fo_{94}) separated from ringwoodite grains, which suggests chemical fractionation during a solid-state olivine-ringwoodite transformation. Raman imaging revealed a close correlation between the blue ringwoodite color and the peak observed at 877 cm^{-1}. This signal showed a strong near-resonance Raman enhancement when measured with NIR laser excitations (785 and 830 nm) close to the optical absorption bands of fourfold coordinated Fe^{2+} in the ringwoodite (Acosta et al., 2013). The NIR optical absorption is assigned to the transition $^5T_{2g} \rightarrow {}^5E_g$ of $^{VI}Fe^{2+}$ split by the trigonal distortion or Jahn–Teller effect (Taran et al., 2009).

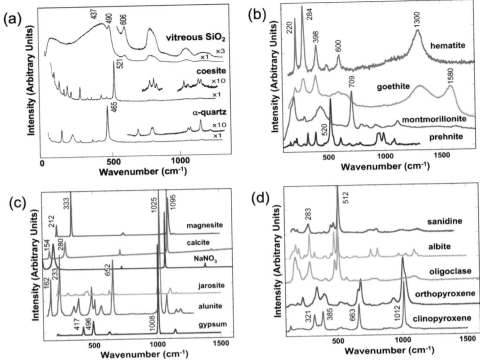

Figure 6.5 Raman spectra of selected minerals. (a) spectra of α-quartz, coesite and vitrous-SiO$_2$ (data from Sharma et al., 1981), (b) spectra of iron oxides/hydroxides and phyllosilicates (data from Edwards et al., 2004; Bishop and Murad, 2004), (c) spectra of carbonates, nitrates and sulfates (data from Bishop et al., 2017; RRUFF database: http://rruff.info), (d) spectra of feldspar and pyroxene minerals (data from Bishop et al., 2014; RRUFF database: http://rruff.info).

Micro-Raman spectroscopy has also been successfully used to investigate mineralogy and phase transitions under high pressure and temperature to gain insight into the Earth and planetary interiors (e.g., for reviews, see Hemley et al., 1987; Sharma, 1989; Gillet, 1993; McMillan et al., 1996; Gillet et al., 2000; Goncharov, 2012; Reynard et al., 2012).

6.6 Summary

In this chapter, we have introduced the theory of Raman scattering based both on classical electrodynamics and the quantized nature of the energy levels. Basic applications of group theory are discussed as they relate to the symmetry of the molecules and the crystal unit cells for prediction of the number of Raman and IR active modes and the degree of polarization. For crystalline solids, the factor group and site group approaches allow qualitative description of the association of normal modes with the symmetry species. Advances in instrumentation have facilitated the application of Raman spectroscopy in a number of geological and planetary disciplines. Laboratory measurements on a variety of geologic materials and planetary analogs with micro-Raman, confocal Raman, and small telescopic Raman systems with CW and pulsed-lasers are providing valuable reference datasets for nondestructively identifying minerals, disordered solids, fluids, glasses, and melts of mineral compositions. Miniaturization of Raman spectrometers with CCD, ICCD, and SPAD detectors and the use of CW and pulsed-laser excitation sources have made Raman spectroscopy much more practical for identifying Earth and planetary materials. Online databases, such as RRUFF (http://rruf.info), are useful and valuable resources for Earth and planetary scientists.

Acknowledgments

The authors would like to thank N. Hulbirt and M. Izumi for their valuable help with figures and editing, respectively.

References

Acosta T.E, Scott E.R.D., Sharma S.K., & Misra A.K. (2013) The pressures and temperatures of meteorite impact: Evidence from micro-Raman mapping of mineral phases in the strongly shocked Taiban ordinary chondrite. *American Mineralogist*, **98**, 859–869.

Adams D.M. & Newton D.M. (1970a) Tables for factor group analysis of the vibrational spectra of solids. *Journal of the Chemical Society* A, **1970**, 2822–2827.

Adams D.M. & Newton D.M. (1970b) *Tables for factor group analysis*. Beckman-RICC Ltd., Reading, UK.

Adams D.M., Sharma S.K., & Appleby R. (1977) Spectroscopy at very high pressures: Part 14. Laser Raman scattering in ultra-small samples in the diamond anvil cell. *Applied Optics*, **16**, 2572–2575.

Aminzadeh A. (1997) Fluorescence bands in the FT-Raman spectra of some calcium minerals. *Spectrochimica Acta*, **A53**, 693–797.

Angel S.M., Carrabba M., & Cooney T.F. (1995) The utilization of diode lasers for Raman spectroscopy. *Spectrochimica Acta*, **A51**, 1779–1799.

Angel S.M., Gomer N.R., Sharma S.K., & McKay C. (2012) Remote Raman spectroscopy for planetary exploration: A review. *Applied Spectroscopy*, **66**, 137–150.

Arns J.A. (1995) Holographic transmission gratings improve spectroscopy and ultrafast laser performances. *Proceedings of the Society of Photo-optical Instrumentation Engineers*, **2404**, 174–181.

Arns J.A., Colburn W.S., & Barden S.C. (1999) Volume phase gratings for spectroscopy, ultrafast laser compressors, and wavelength division multiplexing, *Proceedings of the Society of Photo-optical Instrumentation Engineers*, **3779**, 313–323.

Aroyo M.I., Perez-Mato J.M., Capillas C., et al. (2006a) Bilbao crystallographic server I: Databases and crystallographic computing programs. *Zeitschrift für Kristallographie*, **221**, 15–27.

Aroyo M.I., Kirov A., Capillas C., Perez-Mato J.M., & Wondratschek H. (2006b) Bilbao crystallographic server II: Representations of crystallographic point groups and space groups. *Acta Crystallographica*, **A62**, 115–128.

Aroyo M.I., Perez-Mato J.M., Orobengoa D., Tasci E., de la Flor G., & Kirov A. (2011) Crystallography online: Bilbao crystallographic server. *Bulgarian Chemistry Communications*, **43**, 183–197.

Battey D.E., Slater J.B., Wludyka R., Owen H., Pallister D.M., & Morris M.D. (1993) Axial transmissive f/1.8 imaging Raman spectrograph with volume-phase holographic filter and grating. *Applied Spectroscopy*, **47**, 1913–1919.

Beegle L.W., Bhartia R., DeFlores L., et al. (2014) SHERLOC: Scanning habitable environments with Raman and luminescence for organics and chemicals, an investigation for 2020. *45th Lunar Planet. Sci. Conf.*, Abstract #2835.

Bennett C.J., Brotton S.J., Jones B.M., Misra A.K., Sharma S.K., & Kaiser R.I. (2013) A novel high sensitivity Raman spectrometer to study pristine and irradiated interstellar ice analogs. *Analytical Chemistry*, **85**, 5659–5665.

Bertie J.E. & Bell J.W. (1971) Unit cell group and factor group in the theory of the electronic and vibrational spectra of crystals. *Journal of Chemical Physics*, **54**, 160–162.

Bhagvantam S. (1940) Effect of crystal orientation on the Raman spectrum of calcite. *Proceedings of the Indian Academy of Sciences A*, **11**, 62–71.

Bhagavantam S. & Venkatarayudu T. (1939) Raman effect in relation to crystal structure. *Proceedings of the Indian Academy of Sciences A*, **9**, 224–258.

Bhagavantam S. & Venkatarayudu T. (1969) *Theory of groups and its applications to physical problems*. Academic Press, New York.

Bishop J.L. & Murad E. (2004) Characterization of minerals and biogeochemical markers on Mars: A Raman and IR spectroscopy study of montmorillonite. *Journal of Raman Spectroscopy*, **35**, 480–486.

Bishop J.L., Englert P.A.J., Patel S., et al. (2014) Mineralogical analyses of surface sediments in the Antarctic Dry Valleys: Coordinated analyses of Raman spectra, reflectance spectra and elemental abundances. *Philosophical Transactions of the Royal Society of London A*, **372**, 20140198.

Bishop J.L., King S.J., Lane M.D., et al. (2017) Spectral properties of anhydrous carbonates and nitrates. *48th Lunar Planet. Sci. Conf.*, Abstract #2362.

Blacksberg J., Rossman G.R., & Gleckler A. (2010) Time-resolved Raman spectroscopy for in situ planetary mineralogy. *Applied Optics*, **49**, 4951–4962.

Blacksberg J., Alerstam E., Maruyama Y., Cochrane C.J., & Rossman G.R. (2016) Miniaturized time-resolved Raman spectrometer for planetary science based on a fast single photon avalanche diode detector array. *Applied Optics*, **55**, 739–748.

Buback M. & Schulz K.R. (1976) Raman scattering of pure ammonia at high pressures and temperatures. *Journal of Physical Chemistry*, **80**, 2478–2482.

Carter J.C., Scaffidi J., Burnett S., Vasser B., Sharma S.K., & Angel S.M. (2005) Stand-off Raman detection using dispersive and tunable filter based systems. *Spectrochimica Acta*, **A61**, 2288–2298.

Chase D.B (1986) Fourier transform Raman spectroscopy. *Journal of the American Chemical Society*, **108**, 7485–7488.

Colthup N.B., Daly L.H., & Wiberley S.E. (1975) *Introduction to infrared and Raman spectroscopy*, 2nd edn. Academic Press, New York.

Cooney T.F. & Sharma S.K. (1990) Structure of glasses in the system Mg_2SiO_4-Fe_2SiO_4, Mn_2SiO_4-Fe_2SiO_4, Mg_2SiO_4-$CaMgSiO_4$ and Mn_2SiO_4-$CaMnSiO_4$. *Journal of Non-Crystalline Solids*, **122**, 10–32.

Cooney T.F., Skinner H.T., & Angel S.M. (1995) Evaluation of external-cavity diode lasers for Raman spectroscopy. *Applied Spectroscopy*, **49**, 1846–1851.

Cooper J.B., Flecher P.E., Albin S., Vess T.M., & Welch W.T. (1995) Elimination of mode hopping and frequency hysteresis in diode laser Raman spectroscopy: The advantages of a distributed Bragg reflector diode laser for Raman excitation. *Applied Spectroscopy*, **49**, 1692–1698.

Cornell R.M. & Schwertmann U. (2003) *The iron oxides: Structure, reactions, occurrences and uses*, 2nd edn. Wiley-VCH Verlag GmbH & Co. KGaA, Weinheim, Germany.

Cotton F.A. (1963) *Chemical applications of group theory*. Wiley-Interscience, New York.

Damen T.C., Porto S.P.S., & Tell B. (1966) Raman effect in zinc oxide. *Physical Review*, **142**, 570–574.

DeAngeles B.A., Newnham R.E., & White W.B. (1972) Factor group analysis of the vibrational spectra of crystals: A review and consolidation. *American Mineralogist*, **57**, 255–268.

de Faria D.L.A., Venâncio Silva S., & de Oliveira M.T. (1997) Raman microspectroscopy of some iron oxides and oxyhydroxides. *Journal of Raman Spectroscopy*, **28**, 873–878.

De La Pierre M., Carteret C., Maschio L., André E., Orlando R., & Dovesi R. (2014) The Raman spectrum of $CaCO_3$ polymorphs calcite and aragonite: A combined experimental and computational study. *Journal of Chemical Physics*, **140**, 164509/1–12.

Delhaye M. & Dhamelincourt P. (1975), Raman microprobe and microscope with laser excitation. *Journal of Raman Spectroscopy*, **3**, 33–43.

Denson S.C., Pommier C.J.S., & Denton M.V. (2007) The impact of array detectors on Raman spectroscopy. *Journal of Chemical Education*, **84**, 67–74.

Dhamelincourt P., Wallart F., Leclercq M., N'Guyen A.T., & Landon D.O. (1979) Laser Raman molecular microprobe (MOLE). *Analytical Chemistry*, **51**, 414A–421A.

Dubessy J., Caumon M.-C., Rull F., & Sharma S. (2012) Instrumentation in Raman spectroscopy: Elementary theory and practice. In: *Applications of Raman spectroscopy to Earth sciences and cultural heritage* (J. Dubessy, F. Rull, & M.-C. Caumon, eds.). EMU Notes in Mineralogy, **12**. European Mineralogical Union and the Mineralogical Society of Great Britain & Ireland, 83–172.

Edwards H.G.M., Wynn-Williams D.D., & Jorge Villar S.E. (2004) Biological modification of haematite in Antarctic cryptoendolithic communities. *Journal of Raman Spectroscopy*, **35**, 470–474.

Egan M.J., Angel S.M., & Sharma S.K. (2017) Standoff spatial heterodyne Raman spectrometer for mineralogical analysis. *Journal of Raman Spectroscopy*, **48**, 1613–1617, DOI:10.1002/jrs.5121.

Elman B.S., Dresselhaus M.S., Dresselhaus G., Maby E.W., & Mazurek H. (1981) Raman scattering from ion-implanted graphite. *Physical Review B*, **24**, 1027–1034.

Fateley W.G., Dollish F.R., McDevitt N.T., & Bentley F.F. (1972) *Infrared and Raman selection rules for molecular and lattice modes*. Wiley-Interscience, New York.

Ferigle S.M. & Meister A.G. (1952) Selection rules for vibrational spectra of linear molecules. *American Journal of Physics*, **20**, 421–428.

Ferini G., Baratta G.A., & Palumbo M.E. (2004) A Raman study of ion irradiated icy mixtures. *Astronomy & Astrophysics*, **414**, 757–766.

Ferraro J.R. (1975) Factor group analysis for some common minerals. *Applied Spectroscopy*, **29**, 418–420.

Ferraro J.R. & Ziomek J.S. (1969) *Introductory group theory and its application to molecular structure*. Plenum Press, New York.

Freeman J.R., Wang A., Kuebler K.E., Jolliff B.L., & Haskin L.A. (2008) Characterization of natural feldspars by Raman spectroscopy for future planetary exploration. *Canadian Mineralogist*, **46**, 1477–1500.

Frezzotti M.L., Tecce F., & Casagli A. (2012) Raman spectroscopy for fluid inclusion analysis. *Journal of Geochemical Exploration*, **112**, 1–20.

Fries M. & Steele A. (2011) Raman spectroscopy and confocal Raman imaging in mineralogy and petrography. In: *Confocal Raman microscopy* (T. Dieing, O. Hollricher, & J. Toporski, eds.). Springer Series in Optical Sciences, **158**. Springer-Verlag, Berlin and Heidelberg, 111–133.

Gaft M. & Nagli L. (2009) Time-resolved laser based spectroscopies for mineralogical research and applications. In: *Micro-Raman spectroscopy and luminescence studies in the Earth and planetary sciences* (A. Gucsik, ed.). Mainz, Germany, April 2–4, 2009, *American Institute of Physics (AIP) Conference Proceedings*, **1163**, 3–14.

Gaft M., Reinsfeld R., & Panczer G. (2005) *Modern luminescence spectroscopy of minerals and materials*. Springer-Verlag, Berlin and Heidelberg.

Galeener F.L. (1982a) Planner rings in glasses. *Solid State Communication*, **44**, 1037–1040.

Galeener F.L. (1982b) Planner rings in vitreous silica. *Journal of Non-Crystalline Solids*, **49**, 53–62.

Gasda P.J., Acosta-Maeda T.E., Lucey P.G., Misra A.K., Sharma S.K., & Taylor G.J. (2015) Next generation laser-based standoff spectroscopy techniques for Mars exploration. *Applied Spectroscopy*, **69**, 173–192.

Gillet P. (1993) Stability of magnesite ($MgCO_3$) at mantle pressure and temperature: A Raman spectroscopic study. *American Mineralogist*, **78**, 1328–1331.

Gillet P., Daniel I., Guyot F., Matas J., & Chervin J.C. (2000) A thermodynamic model for $MgSiO_3$-perovskite derived from pressure and temperature and volume dependence of the Raman mode frequencies. *Physics of the Earth and Planetary Interiors*, **117**, 361–384.

Gomer N., Gordon C., Lucey P., Sharma S., Carter J., & Angel S. (2011) Raman spectroscopy using a spatial heterodyne spectrometer: Proof of concept. *Applied Spectroscopy*, **65**, 849–857.

Goncharov A.F. (2012) Raman spectroscopy at high pressures. *International Journal of Spectroscopy*, **2012**, 617528/1–16.

Götze J., Nasdala L. Kleeberg R., & Wenzel M. (1998) Occurrence and distribution of "moganite" in agate/chalcedony: A micro-Raman, Rietfeld, and cathodoluminescence study. *Contribution to Mineralogy and Petrology*, **133**, 96–105.

Halford R.S. (1946) Motions of molecules in condensed systems: I. Selection rules, relative intensities, and orientation effects for Raman and infrared spectra. *Journal of Chemical Physics*, **74**, 8–15.

Haskin L.A., Wang A., Rockow K.M., Jolliff B.L., Korotev R.L., & Viskupic K.M. (1997) Raman spectroscopy for mineral identification and quantification for in situ planetary surface analysis: A point count method. *Journal of Geophysical Research*, **102**, 19293–19306.

Hemley R.J., Bell P.M., & Mao H.K. (1987) Laser techniques in high-pressure geophysics. *Science*, **237**, 605–612.

Herzberg G. (1945) *Molecular spectra and molecular structure. II. Infrared and Raman spectra of polyatomic molecules*. Van Nostrand Reinhold, New York.

Hirschfeld T. & Chase B. (1986) FT-Raman spectroscopy: Development and justification. *Applied Spectroscopy*, **40**, 133–137.

Hornig D.F. (1948) The vibrational spectra of molecules and complex ions in crystals. I. General theory. *Journal of Chemical Physics*, **16**, 1063–1076.

Hu G., Xiong W., Shi H., Li Z., Shen J., & Fang X. (2015) Raman spectroscopic detection for liquid and solid targets using a spatial heterodyne spectrometer. *Journal of Raman Spectroscopy*, **47**, 289–298.

Jennings D.E., Weber A., & Brault J.W. (1986) Raman spectroscopy of gases with a Fourier transform spectrometer: The spectrum of D_2. *Applied Optics*, **25**, 284–290.

Kaszowska Z., Malek K., Staniszewska-Slezak E., & Niedzielska K. (2016) Raman scattering or fluorescence emission? Raman spectroscopy study on lime-based building and conservation materials. *Spectrochimica Acta*, **A169**, 7–15.

Kingma K.J. & Hemley R.J. (1994) Raman spectroscopic study of microcrystalline silica. *American Mineralogist*, **79**, 269–273.

Kittel C. (1976) *Introduction to solid state physics*, 5th edn. John Wiley & Sons, New York.

Kuebler K.E., Jolliff B.L., Wang A., & Haskin L.A. (2006) Extracting olivine (Fo-Fa) compositions from Raman spectral peak positions. *Geochimica et Cosmochimica Acta*, **70**, 6201–6222.

Lamsal N., Sharma S.K., Acosta T.E., & Angel S.M. (2016) UV standoff Raman measurements using a gated spatial heterodyne Raman spectrometer. *Applied Spectroscopy*, **70**, 666–675.

Lebedkin S., Blum C., Stürzl N., Hennrich F., & Kappes M.M. (2011) A low wavenumber extended confocal Raman microscope with very high laser excitation line discrimination. *Review of Scientific Instruments*, **82**, 013705/1–6.

Li Z. & Deen M.J. (2014) Towards a portable Raman spectrometer using a concave grating and a time-gated CMOS SPAD. *Optics Express*, **22**, 18736–18747.

Lopez-Reyes G., Rull F., Venegas G., et al. (2014) Analysis of the scientific capabilities of the ExoMars Raman laser spectrometer instrument. *European Journal of Mineralogy*, **25**, 721–733.

Lucey P.G., Cooney T.F., & Sharma S.K. (1998) A remote Raman analysis system for planetary landers. *29th Lunar Planet. Sci. Conf.*, Abstract #1354.

Maraduddin A.A. & Vosko S.H. (1968) Symmetry properties of the normal vibrations of a crystal. *Reviews of Modern Physics*, **40**, 1–37.

Matousek P., Towrie M., Stanley A., & Parker A.W. (1999) Efficient rejection of fluorescence from Raman spectra using picosecond Kerr gating. *Applied Spectroscopy*, **53**, 1485–1489.

Matson D.W., Sharma S.K., & Philpotts J.A. (1983) The structure of high-silica alkali-silicate glasses: A Raman spectroscopic investigation. *Journal of Non-Crystalline Solids*, **58**, 323–352.

Matson D.W., Sharma S.K., & Philpotts J.A. (1986) Raman spectra of some tectosilicates and of glasses along the orthoclase-anorthite and nepheline-anorthite joins. *American Mineralogist*, **71**, 694–704.

McCreery R.L. (2000) *Raman spectroscopy for chemical analysis*. John Wiley & Sons, New York.

McKeown D.A. (2005) Raman spectroscopy and vibrational analyses of albite: From 25°C through the melting temperature. *American Mineralogist*, **90**, 1506–1517.

McMillan P. (1985) Vibrational spectroscopy in the mineral sciences. In: *Microscopic to macroscopic: Atomic environments to thermodynamic properties* (S.W. Kieffer & A. Navrotsky, eds.). Reviews in Mineralogy, **14**. Mineralogical Society of America, Washington, DC, 9–63.

McMillan P.F. & Hofmeister A.M. (1988) Infrared and Raman spectroscopy. In: *Spectroscopic methods in mineralogy and geochemistry* (F.C. Hawthorne, ed.). Reviews in Mineralogy, **18**. Mineralogical Society of America, Washington, DC, 99–159.

McMillan P.F. & Wolf G.H. (1995) Vibrational spectroscopy of silicate liquids. In: *Structure, dynamics and properties of silicate melts* (J.F Stebbins, P.F. McMillan & D.B. Dingwell, eds.), Reviews in Mineralogy, **32**. Mineralogical Society of America, Washington, DC, 247–314.

McMillan P.F., Dubessy J., & Hemley R. (1996) Applications in Earth, planetary and environmental sciences. In: *Raman microscopy: Developments and applications* (G. Turrell & J. Corset, eds.). Academic Press, New York, 289–365.

Misra A.K., Sharma S.K., Chio C.H., Lucey P.G., & Lienert B. (2005) Pulsed remote Raman system for daytime measurements of mineral spectra. *Spectrochimica Acta*, **A61**, 2281–2287.

Mysen B.O. & Richet P. (2005) *Silicate glasses & melts: Properties and structure*. Elsevier, New York.

Nadungadi T.M.K. (1939) Effect of crystal orientation on the Raman spectrum of sodium nitrate. *Proceedings of the Indian Academy of Sciences*, **A10**, 197–212.

Nasdala L., Smith D.C., Kaindl R., & Ziemann M.A. (2004) Raman spectroscopy: Analytical perspectives in mineralogical research. In: *Spectroscopic Methods in mineralogy* (A. Beran & E. Libowitzky, eds.). EMU Notes in Mineralogy, **6**. European Mineralogical Union and Mineralogical Society of Great Britain and Ireland, 281–343.

Nieuwoudt M.K., Comins J.D., & Cukrowski I. (2011) The growth of the passive film on iron in 0.05 MNaOH studied *in situ* by Raman micro-spectroscopy and electrochemical polarisation. Part I: near-resonance enhancement of the Raman spectra of iron oxide and oxyhydroxide compounds. *Journal of Raman Spectroscopy*, **42**, 1335–1339.

Owen H. (2007) The impact of volume phase holographic filters and gratings on the development of Raman instrumentation. *Journal of Chemical Education*, **84**, 61–66.

Panczer G., De Ligny D., Mendoza C., Gaft M., Seydoux-Guillaume A.-M., & Wang X. (2012) Raman and fluorescence. In: *Applications of Raman spectroscopy to Earth sciences and cultural heritage* (J. Dubessy, F. Rull, & M.-C. Caumon, eds.). EMU Notes in Mineralogy, **12**. European Mineralogical Union and the Mineralogical Society of Great Britain & Ireland, 1–22.

Pandya N., Sharma S.K., & Muenow D.W. (1988) Calibration of a multichannel micro-Raman spectrograph with plasma lines of argon and krypton ion lasers. *Microbeam analysis – 1988: Proceedings of the 23rd Annual Conference of the Microbeam Analysis Society, Milwaukee, Wisconsin, August 8–12, 1988* (D. E. Newbury, ed.). San Francisco Press, San Francisco, 171–174.

Pasteris J.D., Kuehn C.A., & Bodnar R.J. (1986) Applications of the laser Raman microprobe Ramanor U-1000 to hydrothermal ore deposits: Carlin as an example. *Economic Geology*, **81**, 915–930.

Porto S.P.S., Giordmaine J.A., & Damen T.C. (1966) Depolarization of Raman scattering in calcite. *Physical Review*, **147**, 608–611.

Rai C.S., Sharma S.K., Muenow D.W., Matson D.W., & Byers C.D. (1983) Temperature dependence of CO_2 solubility in high-pressure quenched glasses of diopside composition. *Geochimica et Cosmochimica Acta*, **47**, 953–958.

Raman C.V. (1928) A change of wave-length in light scattering. *Nature*, **121**, 619–619.

Raman C.V. & Krishnan K.S. (1928) A new type of secondary radiation. *Nature*, **121**, 501–502.

Reynard B., Montagnac G., & Cardon H. (2012) Raman spectroscopy at high pressure and temperature for study of the Earth's mantle and planetary minerals. In: *Applications of Raman spectroscopy to Earth sciences and cultural heritage* (J. Dubessy, F. Rull, & M.-C. Caumon, eds.). EMU Notes in Mineralogy, **12**, European Mineralogical Union and the Mineralogical Society of Great Britain & Ireland, 367–390.

Roedder E. (1984) Nondestructive methods of determination of inclusion composition. In: *Fluid inclusions* (E. Roedder, ed.). Reviews in Mineralogy, **12**, Mineralogical Society of America, Washington, DC, 79–108.

Rosasco G.J., Etz E.S., & Cassatt W.A. (1975) The analysis of discrete fine particles by Raman spectroscopy. *Applied Spectroscopy*, **29**, 396–404.

Rossano S. & Mysen B.O. (2012) Raman spectroscopy of silicate glasses and melts in geological systems. *Applications of Raman spectroscopy to Earth sciences and cultural heritage* (J. Dubessy, F. Rull, & M.-C. Caumon, editors). EMU Notes in Mineralogy, **12**. European Mineralogical Union and the Mineralogical Society of Great Britain & Ireland, 321–366.

Rull F. (2012) The Raman effect and the vibrational dynamics of molecules and crystalline solids. In: *Applications of Raman spectroscopy to Earth sciences and cultural heritage* (J. Dubessy, F. Rull & M.-C. Caumon, eds.). EMU Notes in Mineralogy, **12**. European Mineralogical Union and the Mineralogical Society of Great Britain & Ireland, 1–60.

Salthouse J.A. & Ware M.J. (1972) *Point group character tables and related data*. Cambridge University Press, Cambridge.

Sharma S.K. (1979) Raman spectroscopy at very high pressure. *Carnegie Institution of Washington Year Book*, **78**, 660–665.

Sharma S.K. (1989) Applications of advanced Raman techniques in Earth sciences. *Vibrational Spectra and Structure*, **17B**, 513 568.

Sharma S.K. (2007) New trends in telescopic remote Raman spectroscopic instrumentation. *Spectrochimica Acta*, **A68**, 1008–1022.

Sharma S.K. & Simons B. (1981) Raman study of crystalline polymorphs and glasses of spodumene ($LiAlSi_2O_6$) composition quenched from various pressure. *American Mineralogist*, **66**, 118–126.

Sharma S.K., Hoering T.C., & Yoder H.S., Jr. (1979a) Quenched melts of akermanite compositions with and without CO_2-characterization by Raman spectroscopy and gas chromatography. *Carnegie Institution Washington Year Book*, **78**, 537–542.

Sharma S.K., Virgo D., & Mysen, B.O. (1979b) Raman study of the coordination of aluminum in jadeite melts as function of pressure. *American Mineralogist*, **64**, 779–787.

Sharma S.K., Mammone J.F., & Nicol M.F. (1981) Ring configurations in vitreous silica: A Raman spectroscopic investigation. *Nature*, **292**, 140–141.

Sharma S.K., Philpotts J.A., & Matson D.W. (1985) Ring distributions in alkali- and alkaline-earth alumino-silicate framework glasses: A Raman spectroscopic study. *Journal of Non-Crystalline Solids*, **71**, 403–410.

Sharma S.K., Yoder H.S., Jr., & Matson D.W. (1988) Raman study of some melilites in crystalline and glassy states. *Geochimica et Cosmochimica Acta*, **52**, 1961–1967.

Sharma S.K., Wang Z., & van der Laan S. (1996) Raman spectroscopy of oxide glasses at high pressure and high temperature. *Journal of Raman Spectroscopy*, **27**, 739–746.

Sharma S.K., Cooney T.F., Wang Z., & van der Laan S. (1997) Raman band assignments of silicate and germanate glasses in light of high pressure and high temperature spectral data. *Journal of Raman Spectroscopy*, **28**, 679–709.

Sharma S.K., Misra A.K., & Sharma B. (2005) Portable remote Raman system for monitoring hydrocarbon, gas hydrates and explosives in the environment. *Spectrochimica Acta*, **A61**, 2404–2412.

Sonwalker N., Sunder S.S., & Sharma S.K. (1991) Raman microprobe spectroscopy of icing on metal surfaces. *Journal of Raman Spectroscopy*, **22**, 551–557.

Spinella F., Barrata G.A., & Strazzulla G. (1991) An apparatus for in situ Raman spectroscopy of ion-irradiated frozen target. *Review of Scientific Instruments*, **62**, 1743–1745.

Storrie-Lombardi M.C., Hug W.F., McDonald G.D., Tsapin A.I., & Nealson K.H. (2011) Hollow cathode ion lasers for deep ultraviolet Raman spectroscopy and fluorescence imaging. *Review of Scientific Instruments*, **72**, 4452–4459.

Strazzulla G. & Baratta G.A. (1992) Carbonaceous material by ion irradiation in space. *Astronomy and Astrophysics*, **266**, 434–438.

Strazzulla G., Baratta G.A., & Palumbo M.E. (2001) Vibrational spectroscopy of ion-irradiated ices. *Spectrochimica Acta*, **A57**, 825–842.

Taran M., Koch-Müller M., Wirth R., Abs-Wurmbach I., Rhede D., & Greshake A. (2009) Spectroscopic studies of synthetic and natural ringwoodite, γ-(Mg, Fe)2SiO4. *Physics and Chemistry of Minerals*, **36**, 217–232.

Urmos J.P., Sharma S.K., & Mackenzie F.T. (1991) Characterization of some biogenic carbonates with Raman spectroscopy. *American Mineralogist*, **76**, 641–646.

Wang Z., Cooney T.F., & Sharma S.K. (1993) High-temperature structural investigation of iron-bearing glasses and melts. *Contributions to Mineralogy & Petrology*, **115**, 112–122.

Wang Z., Cooney T.F., & Sharma S.K. (1995) In situ structural investigation of iron-containing silicate melts and glasses. *Geochimica Cosmochimica Acta*, **59**, 1571–1577.

Wang A., Haskin L.A., & Cortez E. (1998) Prototype Raman spectroscopic sensor for in situ mineral characterization on planetary surfaces. *Applied Spectroscopy*, **52**, 477–487.

Wang A., Jolliff B.L., Haskin L.A., Kuebler K.E., & Viskupic K.M. (2001) Characterization and comparison of structural and compositional features of planetary quadrilateral pyroxenes by Raman spectroscopy. *American Mineralogist*, **86**, 790–806.

Wang A., Haskin L.A., Lane A.L., et al. (2003) Development of the Mars Microbeam Raman Spectrometer (MMRS). *Journal of Geophysical Research*, **108 E1**, 5005/1–18.

Wang W., Major A., & Paliwal J. (2012) Grating-stabilized external cavity diode lasers for Raman spectroscopy: A review. *Applied Spectroscopy Reviews*, **47**, 116–143.

Warren J.L. (1968) Further considerations on the symmetry properties of the normal vibrations of a crystal. *Reviews of Modern Physics*, **40**, 38–76.

White W.B. (1975) Structural interpretation of lunar and terrestrial minerals by Raman spectroscopy. In: *Infrared and Raman spectroscopy of lunar and terrestrial minerals* (C. Karr, Jr., ed.). Academic Press, New York, 325–358.

White W.B. & De Angelis B.A. (1967) Interpretation of the vibrational spectra of spinels. *Spectrochimica Acta*, **A23**, 985–995.

Wiens R.C., Maurice S., McCabe K., et al. (2016) The SUPERCAM remote sensing instrument suite for Mars 2020. *47th Lunar and Planetary Sci. Conf.*, Abstract #1332.

Winston H. & Halford R.S. (1949) Motions of molecules in condensed systems: V. Classification of motions and selection rules for spectra according to space symmetry. *Journal of Chemical Physics*, **17**, 607–616.

Zhao J. & McCreery R.L. (1996) Multichannel Fourier transform Raman spectroscopy: Combining the advantages of CCDs with interferometry. *Applied Spectroscopy*, **50**, 1209–1214.

Zhao J. & McCreery R.L. (1997) Multichannel FT-Raman spectroscopy: Noise analysis and performance assessment. *Applied Spectroscopy*, **51**, 1687–1697.

7

Mössbauer Spectroscopy
Theory and Laboratory Spectra of Geologic Materials

M. DARBY DYAR AND ELIZABETH C. SKLUTE

7.1 Introduction

Mössbauer spectroscopy is used for characterizing the redox state, coordination environment, and site occupancy of iron in geologic materials. It is widely recognized that the emergence of oxygen and the evolution of the interiors of terrestrial planets are recorded in the redox states of multivalent elements locked in rocks, glasses, and minerals on planetary surfaces. Central to this problem is the characterization of oxygen fugacity (f_{O2}) in minerals and glasses. This parameter is of paramount importance in constraining phase equilibria and crystallization processes of melts, as well as in understanding the partitioning of elements between the core and silicate portions of terrestrial planets. Studies of f_{O2} show extensive variations that range over nine orders of magnitude in terrestrial bodies (Carmichael, 1991; Parkinson & Arculus, 1999). Meteorite evidence suggests that rocky bodies in the Solar System also sample a wide range of f_{O2} conditions and potentially, multiple unique Solar System oxygen reservoirs (Wadhwa, 2001; Herd et al., 2001, 2002; McCanta et al., 2004, 2009). Currently one of the best proxies for assessing f_{O2} is the redox state of Fe, which records conditions at the time of crystallization or cooling from a melt.

However, the redox ratio of Fe^{3+}/Fe^{2+} is known to be susceptible to later modifications. While these modifications can complicate interpretation, they can sometimes also provide insights into processes that may have operated on the magma to change its f_{O2}, such as (1) metasomatism, degassing, and assimilation (McCanta et al., 2002), (2) subsolidus reequilibration (Lindsley et al., 1991), (3) dehydrogenation during ascent and cooling (Dyar et al., 1993), and (3) shock processes (Treiman et al., 2006).

Given the importance of f_{O2} for the interpretation of planetary processes and the utility of using Fe redox state to this end, the technique of Mössbauer spectroscopy has been used to measure Fe^{3+}/Fe^{2+} ratios since the 1960s when the first Apollo samples were returned. It remains the mainstay of iron redox analysis of geologic materials.

To facilitate understanding of this venerable technique, this chapter provides background on the fundamentals of resonant nuclear absorption as discovered by Rudolf Mössbauer (1958) for which he received a Nobel Prize in 1961. Isomer shift, quadrupole splitting, and hyperfine splitting, which are the three important Mössbauer parameters derived from a spectrum, are defined within this context, and their use in identifying the valence state and site occupancy of Fe in given sites and minerals is explained. In sum, this review aspires to enable understanding of how the Mössbauer effect works, and why its use is of paramount importance in studies of laboratory and extraterrestrial problems requiring an understanding of iron redox state.

7.2 Background

7.2.1 Resonant Absorption of Nuclear Gamma Rays

The remote sensing community is well acquainted with methods that exploit resonant absorption, which is the absorption of a photon whose energy is precisely matched to that of a transition in the absorbing medium. In visible region spectroscopy, such absorptions take the form of electronic transitions between $3d$ orbitals in transition metals (Chapter 1).

The Mössbauer effect also utilizes resonant absorption, but instead utilizes addition or loss of energy to energy levels within the nucleus itself. In this case, the energy source is the gamma ray given off by what will be referred to later as a "daughter" nucleus (such as ^{57}Fe) of a radioactive isotope with energy E_g in an excited nuclear state, E_e. The difference between those energy levels is E_0. If a departing gamma photon were to carry all the energy of the original nuclear transition, then it could be reabsorbed by another identical nucleus via resonant capture, because the absorber nucleus has exactly the right quantized transition energy. By virtue of the Heisenberg uncertainty principle, the energy of the emitted gamma ray coming from a state with a finite lifetime is not precisely defined. Instead it obeys a Breit–Wigner, or Lorentzian, probability distribution, centered on E_0 with full width at half maximum, Γ_0, given by $\Gamma_0 = \hbar / \tau$, where $\hbar$ is Planck's constant divided by 2π and τ is the finite lifetime of the excited nuclear level. Similarly, the required absorption energy is also a distribution. When these two probability distributions overlap, i.e., when the nuclear transition energies in the emitter and absorber are close, resonant absorption can occur.

The emitted photon, however, has momentum, $p_\gamma = E_\gamma/c$, where c is the speed of light, making it subject to recoil. If the emitting atom is isolated and initially at rest, then conservation of momentum dictates that it recoils with momentum, $p_{nucleus} = -p_\gamma$, and acquires a recoil energy, E_R, given by

$$E_R = \frac{(p_{nucleus})^2}{2M} = \frac{p_\gamma^2}{2M} = \frac{E_\gamma^2}{2Mc^2}, \quad (7.1)$$

where M is the mass of the emitting nucleus. Thus, the actual energy of the emitted photon is its initial energy minus a recoil term: $E_\gamma = E_0 - E_R$. Similarly, a photon that can be absorbed by an identical nucleus must possess energy $E_\gamma = E_0 + E_R$. So, in general, nuclear resonant emission–absorption processes cannot occur between stable free atoms, because the recoil energy is too much larger than the natural linewidth (10^{-7}–10^{-8} eV), and source and absorber transition energies will never overlap.

Mössbauer (1958), however, discovered that a certain fraction of nuclear transitions are recoil-less (given by the symbol f) (see also Visscher, 1960). The recoilless fraction arises because in a solid, recoil energy can be delivered to the solid by exciting vibrational or phonon modes; the quantization of vibrational energy leads to the probability that a zero-phonon transition can occur if E_R is less than the energy of the first vibrational mode, $\hbar\omega$. For these transitions, the recoil momentum is instead picked up by the entire solid. The recoil energy (Eq. 7.1) is thus negligible, leaving the emitted or absorbed gamma photon with the full transition energy. One important consequence of recoil-less emission is that the Mössbauer effect cannot occur appreciably in gases or nonviscous liquids because translational motion is allowed in those states.

7.2.2 Iron as an Ideal Mössbauer Absorber

The Mössbauer effect occurs for many unstable nuclides, but in practice, only a few combinations of unstable parent and stable daughter isotopes both produce a significant recoil-free fraction of events and have appropriate lifetimes. Because the lifetime of the excited nuclear state determines the linewidth, it should represent a compromise between being too long and vulnerable to vibration and being too short, with resultant loss of resolution. The unstable parent should also possess a long half-life to give off an abundance of gamma rays, and the value of f (recoil-free fraction) must be relatively high to achieve adequate signal-to-noise ratio. For planetary samples, a vast majority of Mössbauer studies use ^{57}Fe, although the effect has been observed in numerous other isotopes including ^{119}Sn, ^{61}Ni, and ^{197}Au (e.g., Kojima et al., 2012; Gee et al., 2016; Masai et al., 2016).

Although only 2.2% of natural Fe atoms are the ^{57}Fe isotope, its large value of f (~0.65–1) facilitates acquisition of ^{57}Fe Mössbauer spectra. The parent isotope is ^{57}Co, with a half-life of 270 days; it decays by capturing an electron. About 9% of the time (Figure 7.1),

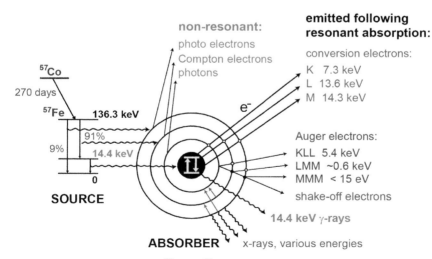

Figure 7.1 Nuclear decay scheme for ^{57}Co → ^{57}Fe, with related backscattering processes that follow resonant absorption of an incident gamma photon. (Modified from De Grave et al., 2005.)

deexcitation occurs directly to the ground state with emission of a 136.3-keV gamma photon. Otherwise, decay is to the 14.4-keV state and then to the ground state with a lifetime of 141 ns, which creates a natural linewidth of 4.67×10^{-9} eV (0.10 mm/s). The ^{57}Co decay process results in a wide variety of energetic particles that can be used in additional experiments (Figure 7.1). Dyar et al. (2006) gives a review of these processes.

7.2.3 Instrumentation

Conventionally, Mössbauer instruments consist of a ^{57}Co source, a sample spread over an area to allow the gamma rays to pass through a thin layer of material, a detector, and a velocity transducer to move either the source or the absorber. Oscillation causes the distance between source and sample to increase and decrease, such that velocity varies linearly with time; the emitted gamma ray's energy is thus Doppler-shifted. This motion, even at a velocity (v) of 1 mm/s, increases the energy of the emitted photons by 14.413 keV $\times$ (v/c), or 4.808×10^{-8} eV, approximately 10 natural linewidths. Thus, "mm/s" is the conventional "energy" unit used on the x-axis of plots of Mössbauer spectra.

The parent isotope of ^{57}Fe, ^{57}Co, is produced in an accelerator by bombarding ^{58}Ni with 20 MeV protons to produce the nuclear reaction ^{58}Ni(p, $2p$)^{57}Co, then the ^{57}Co is electroplated onto an Rh foil and finally diffused at high temperature throughout thickness of the Rh foil. The 2019 cost of a 100 mCi source is roughly US$11 000.

For synchrotron Mössbauer spectroscopy (SMS) experiments, radiation can be easily tuned to scan an energy range so no nuclear source is required. The basic concepts of this approach are described in Sturhahn et al. (1998) and many other publications (Alp et al., 1995; Sturhahn, 2004; Handke et al., 2005; Yan et al., 2012). This geometry can produce very high quality spectra because measurements can be performed on extremely small (≤ 5 μm) regions of a sample.

Instrumentation for typical laboratory Mössbauer experiments is well described by Bancroft (1973) and updated by Murad and Cashion (2004) and Gütlich et al. (2011). The vast majority of Mössbauer experiments in the geosciences are conducted in transmission mode, where the gamma path leads directly from source to detector and resonant gamma absorption is recorded. In this configuration, the velocity is well defined, and, for a thin absorber (sample) and a single line source (e.g., ^{57}Co in an Rh matrix), the resulting spectrum is a simple superposition of spectra of the individual mineral components.

Highly specialized spectrometers are also in use. The Mössbauer milliprobe developed by Catherine McCammon at Universität Bayreuth has the capability to measure spectra on absorbers with diameters smaller than 500 μm (McCammon, 1994), providing the potential for measurements on samples with extremely small grain sizes. At the University of Michigan, a wide-angle Mössbauer spectrometer design using 77 argon gas proportional counters has the capability for analyzing extremely small samples with very low iron contents at count rates 100 times those of conventional units (Moon et al., 1996). None of these advances can overcome the most formidable limitation of Mössbauer spectroscopy: long run times. Although high iron samples run using hot (100 mCi) sources can take as little as an hour, typical run times range from 12 hours to 2 weeks. A different configuration for Mössbauer technology was adapted for inclusion on martian landers (Klingelhofer et al., 1998, 2003), where it was selected as an instrument on the 2003 Mars Exploration Rovers (see Chapter 27 for more information).

7.2.4 Sample Preparation, Data Preprocessing, and Spectral Fitting

In the laboratory, samples are usually prepared as dilute powdered mixtures with some inert material with low atomic mass to minimize X-rays produced by scattering gamma rays. Common dilutants are sucrose or graphite for <300 K and >300 K experiments, respectively. For the temperature range from 4 to 300 K, powders are secured inside plastic or Plexiglas sample holders using Kapton® polyimide film tape with silicone adhesive, which adheres over a broad temperature range down to 4 K. If samples are air sensitive, they can be mounted in airtight Plexiglas discs under an inert atmosphere.

Absorber thickness, which is the amount of sample present in the gamma-ray flux (typically units are number of Mössbauer nuclei per square centimeter), must always be carefully considered in Mössbauer experiments. This subject has been dealt with in detail by Shimony (1965), Chandra and Lokanathan (1977), Sarma et al. (1980), Long et al. (1983), Ping and Rancourt (1992), and Rancourt et al. (1993). Two characteristic thicknesses can be calculated for any given material with known composition using expressions developed by Long et al. (1983). A sample mounted with a concentration to provide a thin absorber thickness, t_{thin}, produces spectra without the effects of thickness that can include incorrect spectral areas, peak heights, widths, and line shapes. A sample weighed to have an ideal absorber thickness, t_{thick}, provides the largest signal-to-noise ratio in a given time (Rancourt et al., 1993). Although it might seem obvious to always use the former, acquisition of such spectra can often require great lengths of time because so little sample is present, and the quality of the resultant spectra can be insufficient for measurements involving accurate determinations of site populations. Rancourt et al. (1993) and Dyar et al. (2006) provide detailed discussion of this issue.

Several other factors may influence Mössbauer measurements and contribute to their errors (Dyar, 1984, 1986, 1989; Waychunas, 1986, 1989). Of these, the most difficult to assess is the issue of sample heterogeneity at fine scales (atomic and molecular). Aside from fitting models (discussed later), other sources of error and their effects include external vibrations (line broadening), changes in temperature (changes Mössbauer parameters), counter saturation effects, and texture effects (caused by nonrandom orientation of grains in sample mounts that modify relative peak areas). Many of these can be corrected by adjustments to the experimental set-up. For example, Rancourt (1994) suggests that texture effects can be avoided by mounting micas in petroleum jelly; other workers suggest using a mixture of sugar and acetone or other transparent powder to coat sample grains before mounting (Clark, 1967; Annersten, 1975; Bancroft & Brown, 1975; Dyar, 1990). Changing the angle of gamma-ray flux has also been claimed to mitigate texture effects (Ericsson & Wäppling, 1976).

Once a Mössbauer spectrum has been collected, several steps are needed to produce a spectrum that can be interpreted. Data must be corrected for nonlinearity in the transducer using interpolation to a linear velocity scale. The reference used for this calibration is generally an Fe metal foil with a thickness of 25 µm. Its spectrum is fit to determine the locations (in channel numbers, the native x-axis unit for scintillation detectors) of each of the six sharp peaks that make up its sextet. Those locations are then compared to known peak positions for each of those six peaks. Interpolation is then accomplished by fitting a straight line to the points defined by the published values of the Fe metal peak positions (as y values) and the observed positions in channels (x values).

The next step is to fold each spectrum, which is generally collected as two mirror-image spectra that correspond to the source moving toward and away from the sample. Although different laboratories have varying protocols, the procedure ideally folds the spectrum at the channel value that produces the minimum least squares sum difference between the first half of the spectrum and the reflected second half of the spectrum.

Three different line shapes are commonly employed in modeling of Mössbauer spectra. Lorentzian (Cauchy) line shapes give good approximations for spectra of paramagnetic materials where all of the Fe nuclei are in identical electronic environments (e.g., in metals, simple compounds, and synthetic materials, but rarely natural materials). Voigt line shapes (Voigt, 1912) address variations in coordination polyhedra in natural samples by considering a Gaussian distribution of Lorentzian line shapes. Quadrupole splitting distributions (QSDs) are used to model Mössbauer spectra of minerals with poorly resolved quadrupole pairs, as seen in phyllosilicate spectra. QSDs reflect local distortions and atomic disorder surrounding the Fe atoms by allowing for asymmetric distributions of Lorentzian lines. A series of papers (Rancourt, 1994a,b; Rancourt et al., 1994), convincingly demonstrates that the QSD method performs better than Lorentzian fits.

Software written especially for processing Mössbauer data exists on many platforms (e.g., WMOSS is free software that performs well when compared to commercial forms: www.wmoss.org/Download1.html). Dyar et al. (2007b) performed an in-depth study comparing multiple programs and approaches (line shapes) to assess the associated errors. It is clear that interpretation of Mössbauer data is quite complex, and factors affecting its precision must be fully understood. The utility of Mössbauer spectroscopy is ultimately limited by cumulative errors. While the nature of the sample, sample preparation, experimental design, sample acquisition, and spectral calibration all contribute to the total error, software features, model assumptions, and the user's knowledge significantly contribute to the overall error. Mössbauer spectra of minerals frequently exhibit highly overlapping peaks and the final interpretation is greatly influenced by fitting techniques, model assumptions, and user knowledge of physically realistic parameters. A comprehensive understanding of cumulative errors is essential to successful spectral interpretation. As with any spectral fitting technique, it is possible to obtain high-quality fits with nonsensical parameters, but in Mössbauer spectra this may be more pronounced.

7.3 Fundamental Mössbauer Parameters

7.3.1 Parameters that Describe Peak Positions

The nucleus of an iron atom reacts to even small changes in the distribution of charge surrounding the atom. The properties that can be measured and quantified using Mössbauer spectroscopy are the result of hyperfine interactions of the nuclear charge distribution with (1) electrons that have a finite probability to penetrate inside the nuclear volume, (2) a nonhomogeneous electric field acting at the nucleus and generated by the extranuclear charges (creating an electric field gradient or EFG), and (3) the net magnetic field at the nucleus. In-depth treatments concerning the physics of hyperfine interactions are well covered in the works of Frauenfelder (1962), Greenwood and Gibb (1971), Gibb (1976), Herber (1984), Gütlich et al. (2011), and Cottenier (2016). A concise overview of these phenomena is presented here based on theoretical considerations laid out by these authors and with emphasis on the most practical Mössbauer isotope, ^{57}Fe. Full mathematical treatments are given in the aforementioned citations. The energy levels that are involved in ^{57}Fe Mössbauer spectroscopy are the ground state (nuclear spin $I_g = 1/2$) and the first excited state (nuclear spin $I_e = 3/2$) with energy 14.4 keV.

The *(intrinsic) isomer shift* (δ_i) is the result of the electric monopole, or Coulombic, interaction between the nuclear charge and the atomic *s*-electrons, which have a nonzero probability of entering the nucleus. Because the ground and first excited state of ^{57}Fe have different volumes, monopole interactions shift both levels but by different amounts. Furthermore, in a Mössbauer experiment where the source and absorber are different, substances containing the same Mössbauer nuclei (which is commonly the case in practice), the electric monopole interaction causes *different* shifts of the nuclear energy levels in source and absorber nuclei. This is because distinct chemical environments around the Mössbauer nuclei lead to differing *s*-electron density at the source and absorber nuclei. These combined effects results in a shift of the absorption lines (peaks) in a Mössbauer spectrum. This shift thus reflects the energy difference between absorber and source nuclear levels as well as the difference of the electric monopole interaction for the first-excited and ground-state energy levels in those nuclei. Consequently, the intrinsic isomer shift δ_i for a source–absorber combination can be expressed as follows:

$$\delta_i = \frac{2\pi}{3} Z e^2 \left\{ |\psi(0)|^2_{\text{absorber}} - |\psi(0)|^2_{\text{source}} \right\} \left\{ r^2_{\text{excited}} - r^2_{\text{ground}} \right\}, \qquad (7.2)$$

where e is the proton charge, Ze is the charge of the nucleus, r^2 is the mean square of nuclear radius (r), and $|\psi(0)|^2$ is the probability of electrons present at the center of the nucleus. Equation (7.2) shows that isomer shift is a relative parameter for the absorber nuclei with respect to the source nuclei and manifests itself as a uniform shift of the absorption lines in a Mössbauer spectrum (Figure 7.2).

In addition to the intrinsic isomer shift, absorption lines in a Mössbauer spectrum are also shifted by a second-order Doppler shift (δ_{SOD}) arising from a change in resonance energy of γ-quanta due to the nonzero value of the mean-squared velocity of the vibrating Mössbauer

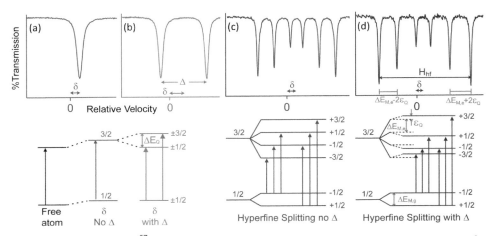

Figure 7.2 Positions of the ^{57}Fe nuclear energy levels and related transmission Mössbauer spectra for different combinations of electric and/or magnetic interactions. Energy levels of the absorber ^{57}Fe Mössbauer nucleus are shifted with respect to those of a free nucleus or nuclei in a source without electric quadrupole and magnetic dipole interactions. (a) The electric monopole interaction of absorber nuclei shifts the ground and first-excited states, but they remain degenerated. This manifests as a single absorption line in the spectrum, which is offset with respect to the source nuclei by the center shift, δ (or IS). (b) The combined electric monopole and quadrupole interactions of the absorber nuclei (without a magnetic dipole interaction) again shift the ground and first-excited levels by δ, with respect to the source nuclei but now the electric quadrupole interaction splits the first excited level ($I_e = 3/2$) into two twofold degenerated sublevels ($m_I = \pm 1/2$ and $m_I = \pm 3/2$, respectively). This gives rise to a doublet spectrum, where the distance between the two absorption lines is equal to the quadrupole splitting ΔE_Q (or QS), and the center of the two lines is offset by the center shift, δ, with respect to the source nuclei. (c) The combined influence of the electric monopole interaction and magnetic dipole interaction on the absorber nuclei in the absence of electric quadrupole interaction leads to Zeeman splitting of the nuclear levels, where the energetic difference between the outermost lines gives the magnitude of net hyperfine field at the nucleus, H_{hf}. Transitions for which $\Delta m_I = \pm 2$ are forbidden on the basis of the quantum mechanical selection rule for M1 radiation. Hence, a sextet pattern is obtained with peak area ratios equal to 3:2:1:1:2:3 (for ideal, random absorbers and in the absence of an external magnetic field). (d) The combined influence of the electric monopole interaction, electric quadrupole interaction, and magnetic dipole interaction on the absorber nuclei leads to Zeeman splitting of the nuclear levels where $I_e = 3/2$ sublevels are offset from the previous example by $\pm \varepsilon_Q$. H_{hf}, selection rules, and peak area ratios are as in the previous example, but the central four lines of the spectrum are offset with respect to the outer two lines by $2\varepsilon_Q$.

nuclei in the absorber. Because lattice vibrations are temperature dependent, this mean-squared velocity varies with temperature and, consequently, δ_{SOD} is temperature dependent. The total shift of the absorption lines is then given by the center shift (δ): $\delta = \delta_i + \delta_{\text{SOD}}$. The center shift (also termed chemical shift, CS) is generally expressed in velocity units, using the relation $v = c\delta/E$ (Gibb, 1976), and can be directly determined from the analysis of a Mössbauer spectrum.

To allow data from different studies to be compared, the velocity scale of a Mössbauer spectrum is generally calibrated relative to the center shift of a standard absorber at a given temperature (typically room temperature [RT]). Values are usually reported relative to α-Fe at RT, although sodium nitroprusside [$Na_2Fe(CN)_6NO \cdot 2H_2O$], which is shifted relative to α-Fe by −0.257 mm/s at RT, has been used as a reference in older papers.

Because the center or isomer shift, δ, results from the interaction of nuclear charge with electron density at the nucleus, it is straightforward to see the potential influence of the local structural environment and the chemical state of the Mössbauer probe on δ_i. Using a standard source to study a series of compounds containing the same Mössbauer nuclide, the quantity $|\psi(0)|^2_{absorber}$ will indeed be the only variable in Eq. (7.2), and δ_i will be a linear function of this quantity. Thus, the bonding environment (geometry, coordination, and covalency) and the valence state affect the isomer shift through indirect changes in s-electron density through $3d$ shielding as well as through direct participation of s-electrons in molecular orbital formation. Generally, isomer shifts are smaller for more covalently bonded Mössbauer atoms (Neese & Petrenko, 2011). For high-spin iron (the spin state of iron in most minerals), oxidation state (Fe^{2+} and Fe^{3+}) is inversely related to isomer shift.

Quadrupole splitting results from interactions between an electric field gradient (EFG) at the nucleus and the electric quadrupole moment (eQ) of the nucleus, the quadrupole interaction. The EFG at the nucleus arises from a nonspherical distribution of charges outside the nucleus. It includes contributions from valence electrons of the Mössbauer isotope and charges at lattice positions surrounding the Mössbauer atom. Q is a measure of the deviation of the nuclear charge distribution from spherical symmetry and may be different for different nuclear states. The sign of Q is positive for an elongated nucleus and negative for a flattened one. Q is zero for a spherical nuclear charge distribution, like the ^{57}Fe nuclear ground state. For the first-excited nuclear state in ^{57}Fe, Q is nonzero. The interaction with an EFG splits the first-excited energy level (nuclear spin $I_e = 3/2$) in two twofold degenerated sublevels ($m_I = \pm 1/2$ and $m_I = \pm 3/2$, respectively) while the ground-state level (nuclear spin $I_g = 1/2$) remains unsplit. As a consequence, in the absence of the magnetic dipole interaction (see later), the Mössbauer spectrum consists of two absorption lines ("doublet") with equal line areas for an ideal, random absorber (see Figure 7.2). The distance between those lines is the difference in energy between the two sublevels of the split first-excited nuclear energy level. This distance is the quadrupole splitting (ΔE_Q, sometimes denoted as QS or Δ), and is also typically expressed in velocity units (mm/s).

In the absence of the magnetic interaction (see later) and when the charge distribution outside the nucleus is totally spherically symmetric, the electric field gradients at the nucleus and the quadrupole interaction are zero. In those cases, the Mössbauer spectrum consists of only one absorption line (see Figure 7.2).

The type and number of bonds, the bonding environment, and the crystallographic point symmetry of the absorbing probe atom will all determine the EFG. Because the EFG components show a ($1/r^3$) dependence, the direct bonding environment has the most prominent effect on the magnitude of these components.

Finally, *magnetic hyperfine splitting* results from magnetic interactions between the nuclear magnetic dipole moment ($\vec{\mu}$) and a magnetic field acting at the nucleus. When a material containing Mössbauer nuclei exhibits spontaneous 3-D cooperative ordering, as in ferrimagnets, ferromagnets, and antiferromagnets, then the magnetic field is of internal origin and is called a magnetic hyperfine field (B_{hf}). However, the magnetic field at the nucleus can also be externally generated, or it can be a combination of both. The net magnetic field at the nucleus is the result of contributions from multiple phenomena, including, but not limited to, the following:

1. If there are an unequal number of spin-up versus spin-down electrons, polarization of s-shells can occur by exchange with d-shell electrons. This exchange leads to slight radial differences between spin-up and spin-down electrons in s-shells, resulting in a magnetic field induced at the nucleus. This is the so-called Fermi-contact term (B_{Fermi}) and is commonly the largest of all field contributions, when an externally applied magnetic field is absent (Munck et al., 1973).
2. Because electrons are spinning electrical charges, they each produce a magnetic field that can be felt by the nucleus.
3. Electrons moving in "orbits" around the nucleus give rise to current loops and thus to associated magnetic fields acting at the nucleus.
4. The magnetic fields induced by adjacent magnetic species in the lattice can also contribute to the net magnetic field experienced by the nucleus.
5. External magnetic fields such as Earth's own magnetic field or those experimentally applied also contribute to the net magnetic field experienced by the nucleus.

No matter the source of the magnetic field at the nucleus, the result is the same: due to the quantized nature of the nuclear energy states, the interaction between $\vec{\mu}$ and the net magnetic field leads to splitting of the energy levels (nuclear Zeeman splitting), which is dependent on the orientation of the nuclear magnetic moment with respect to the magnetic field. This interaction causes the ^{57}Fe nuclear ground-state level (with nuclear spin $I_g = 1/2$) to split into two sublevels, and the first-excited nuclear level (with nuclear spin $I_e = 3/2$) into four sublevels. In the absence of an external magnetic field, and when the quadrupole interaction is zero, the Mössbauer spectrum will consist of six equidistant absorption lines, forming a sextet spectrum. While eight transitions can be drawn between the six nuclear sublevels, quantum mechanical selection rules forbid transitions where $\Delta m_I \neq 0$ or $\Delta m_I \neq \pm 1$, so only six transitions will occur. For an ideal, random absorber, the line area ratios for these transitions are 3:2:1:1:2:3 (see Figure 7.2).

When the EFG is axially symmetric and the quadrupole interaction is weak relative to the magnetic dipole interaction, as in many mineral spectra, sublevels ($I_e = 3/2$, $m_I = +3/2$ and $m_I = -3/2$) and ($I_e = 3/2$, $m_I = +1/2$ and $m_I = -1/2$) of the first-excited state shift in opposite directions over an equal distance, ε_Q, respectively. The Mössbauer spectrum remains a sextet with line area ratios of 3:2:1:1:2:3 for an ideal, random absorber in the absence of an external magnetic field, but the line positions are no longer equidistant (Figure 7.2).

In both cases, the magnetic hyperfine field can be derived from the difference between the positions of the two outer lines. The quantity $2\varepsilon_Q$ is half of the difference between the sum of the positions of lines 1 and 6 and the sum of the positions of lines 2 and 5, respectively, and is called quadrupole shift. The center shift of the sextet is calculated as the average of the four outer peak positions.

When the quadrupole interaction is of the same order of magnitude as the magnetic interaction, the Mössbauer spectrum generally consists of eight absorption lines. This occurs for Fe^{2+} in the magnetic ordered region (below the magnetic ordering temperature) of many minerals with sufficient Fe^{2+} content (e.g., Van Alboom et al., 2009, 2011, 2015, 2016). Derivation of Mössbauer parameters from such spectra is quite complicated and based on the diagonalization of the total hyperfine-interaction Hamiltonian.

In summary, changes in the nuclear energy levels can be probed using Mössbauer spectroscopy just as shifts and splits in electronic orbitals are probed by visible spectroscopy. A typical ^{57}Fe Mössbauer spectrum consists of sets of peaks (doublets or sextets or combinations thereof), each arising from a particular Fe site in the sample. Different doublets or sextets can be associated with Mössbauer nuclei in different crystallographic and/or chemical environments, where Mössbauer parameters for each site depend on the number of electrons (Fe^0, Fe^{2+}, Fe^{3+}) of Mössbauer ions, the number and type of coordinating anions or ligands, bond covalency, site symmetry, and the presence/absence of long-range magnetic ordering (commonly temperature dependent).

In general, Mössbauer spectra of nearly all minerals are doublets at room temperature unless their spectra are acquired in an induced magnetic field (rapidly relaxing Curie and Curie–Weiss paramagnets). But many common rock-forming species also exhibit magnetic ordering with decreasing temperatures (slowly relaxing paramagnets), which can be observed and studied by Mössbauer spectroscopy. This can be simply understood by considering the timing of a Mössbauer measurement. While line width is determined by the lifetime of the ^{57}Fe nuclear excited state, the Larmor procession time of the nuclear spin sets the minimum time extent of a perturbation (such as a fluctuating magnetic field) that can be "detected" by the nucleus. If a fluctuation occurs in a sample on a timescale that is much faster than the Larmor procession time, the nucleus will "see" only an average of the states. Even though unpaired electrons of the Fe cation lead to hyperfine splitting of nuclear energy levels through polarization (Fermi contact), orbital, and spin effects (Greenwood and Gibb, 1971), fluctuation of this magnetic field at the nucleus is faster than it can "see" (Blume & Tjon, 1967). This is the case for paramagnetic samples showing only doublets.

Fluctuations in magnetic fields at the nucleus result from variations in electron spins that give rise to those magnetic fields (Oosterhuis & Spartalian, 1976). Above 0 K, a temperature-dependent population of states in the atom seeks its lowest energy configuration through relaxation processes (and is reexcited by thermal energy). For atoms in a lattice (minerals), the two most common relaxation processes are spin–lattice and spin–spin relaxation. In paramagnetic materials, these relaxations are often much faster than the Larmor procession time (see cited texts for a fully developed, mathematical explanation).

In spin–lattice relaxation, energy is transferred between lattice phonons and electron spin states. As the temperature decreases, lattice vibrations decrease, and this spin-flip (the change in spin quantum state resulting from this relaxation process and that dictates the configuration of the magnetic field) is slowed; sometimes the relaxation is slow enough that the nucleus "sees" a static magnetic field due to a specific spin configuration (Oosterhuis & Spartalian, 1976). High-spin Fe^{3+} ions have spherically symmetric (^6S) ground states that do not interact appreciably with the lattice (i.e., spin–lattice relaxation is not the dominant process). However, high-spin Fe^{2+} ions have cubic (^5D) symmetry when 6-coordinated, and couple strongly to the lattice, causing some minerals to display magnetically split Mössbauer spectra at low temperatures (Gibb, 1976). As the temperature is lowered and degeneracy between nuclear energy levels is removed, the Mössbauer spectrum displays doublet asymmetry, followed by broad, overlapped sextet and/or doublet distributions that finally resolves into sextets (or octets) of well-defined peaks.

In spin–spin relaxation, energy is transferred through magnetic dipole interactions between the spin states of adjacent ions, and is therefore dependent on the concentration of interacting ions and the distance between them. When concentrations are low enough or interatomic distances are large enough, relaxation slows and magnetic splitting of energy levels may be observed. Spin–spin relaxation is usually the dominant process for Fe^{3+}-bearing substances because orbital angular momentum (necessary for strong spin–lattice interactions and thus spin lattice relaxation) is zero for this ion (Herber, 1984). Although spin–spin relaxation is inherently temperature independent, the site populations of the spin states *become* temperature dependent at very low temperatures, leading to temperature-dependent relaxation. The resulting spectral changes with temperature are similar to spin-lattice coupling; as the temperature decreases, higher energy spin states are depopulated, relaxation time slows, and a split magnetic spectrum gradually develops (Reiff, 1984).

In ferrimagnets, ferromagnets, and antiferromagnets, magnetic splitting comes from coupling of spins between adjacent (direct) or nonmagnetically bridged (super-exchange) iron atoms. Coupling in these materials is strong enough to overcome thermal randomization effects. Thus, the bulk material develops a preferred spin orientation that leads to a static magnetic field experienced at the nucleus and a loss of degeneracy of the nuclear energy levels. Increasing available energy to the system by raising temperature will eventually result in collapse of sextet structure, but the collapse is usually much more rapid than in spin–spin or spin–lattice relaxation cases (Reiff, 1984). The temperature where the sextet appears is known as the magnetic ordering temperature (Gibb, 1976).

Even though magnetic coupling is strong in magnetically ordered solids, fluctuation of the magnetic field still occurs, with a relaxation time of

$$\tau = \tau_0 \exp(KV/k_B T), \qquad (7.3)$$

where K is the magnetic anisotropy constant, τ_0 is a weakly temperature-dependent value that approximates 10^{-13}–10^{-9} s, k_B is the Boltzmann constant, T is temperature, and V is the volume (Afanas'ev et al., 1999; Mørup, 2011). Normally the particle size is large enough that the barrier energy (effectively KV) is difficult to overcome. However, when particles

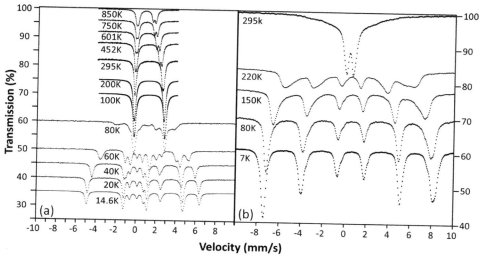

Figure 7.3 Comparison of spectra above and below the transition temperature for two different mineral species. (a) Olivine is shown from 850 to 14.6 K. This is an example of a paramagnetic mineral that orders at low temperatures (ca. 80 K). (b) The spectrum of nanophase synthetic akaganéite, which yields a superparamagnetic doublet at room temperature but orders magnetically ca. 80 K. Spectra are offset for clarity but the magnitude of the absorption is correct for each spectrum. Note the change in total percent transmission as temperature decreases.

are nanophase, the thermal energy ($\sim k_B T$) can easily approach KV. The sextet structure then collapses into a "superparamagnetic" doublet as the temperature is increased (Figure 7.3).

Temperature- and grain-size dependence of the Mössbauer parameters in these samples complicates interpretation of superparamagnetic spectra because of the substantial range of grain sizes present in most specimens (Figure 7.4).

Quadrupole splitting of the doublet, magnetic ordering temperature, quadrupole splitting of the sextet, and hyperfine field of the sextet all change with grain size. Crystallinity can affect these parameters, as well as isomer shift (although less so). Until the sample is completely ordered, these parameters also change with temperature. Finally, grain size distributions may result in multiple splitting regimes occurring in one sample, so the temperature range over which a sample magnetically orders may vary substantially (Sklute et al., 2016). This makes it incredibly difficult to differentiate nanophase (oxyhydr)oxides from one another based on Mössbauer data alone.

As the temperature decreases, doublets split into sextets or octets over a range of temperatures. At intermediate temperatures, doublets and sextets representing the same site may be observed. Spectra taken over such a transition temperature range are typically difficult to fit. Comparison of spectra acquired above and below the transition temperature are very valuable in determining the crystal environment of the iron in the sample.

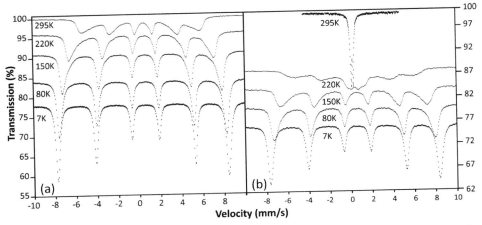

Figure 7.4 Comparison of multitemperature spectra of two different nanophase goethite samples. (a) The goethite is composed of poorly crystalline lathes that are ~ 3 ± 1 nm across. (b) The goethite is composed of well crystallized lathes that are ~ 6.5 ± 2 nm across.

7.3.2 Parameters that Affect Peak Intensities

Many workers erroneously assume that the area of the peaks in each Mössbauer doublet corresponds to the amount of Fe actually present in that site and/or valence state. However, it has long been known that the area of a Mössbauer doublet (pair of peaks) is actually a function of peak width Γ, sample saturation effects $G(x)$ (Bancroft, 1970), and the Mössbauer recoil-free fraction, f. The value of f depends greatly on site geometries and valence state, and thus is specific to different mineral species and even minor compositional variations (De Grave & Van Alboom, 1991; Dyar et al., 2007a, 2013b) (Figure 7.5). Whipple (1968) has suggested that variations in these parameters can influence the results of Mössbauer Fe^{3+}/Fe^{2+} ratios by up to 30%. Much of this error is due to variations in f, so it is important to understand and evaluate f for minerals of interest.

There are at least four ways to evaluate recoil free fraction as described in Dyar et al. (2006). Of these, the most common is to use the temperature dependence of the center shift using the formulation for the characteristic Mössbauer temperature (θ_M), which arises from the Debye approximation for the lattice vibrations of an ideal crystal. The center shift (δ) is modeled by fitting each Mössbauer spectrum at each temperature. Its temperature dependence ($\delta(T)$) arises from the second-order Doppler shift, which is related to the mean-square velocity of the emitting nucleus.

The intrinsic isomer shift (δ_I) is the center shift that would exist in the absence of temperature-dependent lattice vibrations in the absorber (LaFleur & Goodman, 1971) The parameters δ_I and θ_M may be determined from the change in center shift with temperature (De Grave et al., 1985):

$$\delta(T) = \delta_I - \frac{9\,k_B T}{2\,Mc}\left(\frac{T}{\theta_D}\right)^3 \int_0^{\theta_D/T} \frac{x^3}{e^x - 1}\,dx. \tag{7.4}$$

Here k_B is the Boltzmann constant, M is the mass of the ^{57}Fe nucleus, c is the speed of light, T is the absolute temperature, and $\theta_M \cong \theta_D$, the Debye temperature of the absorber, under the assumption of a Debye model for its vibrational spectrum.

To calculate f, Mössbauer spectra of the mineral of interest are acquired over a range of temperatures, usually from 4–12 K up to 600–800 K at 10–50 K increments. Next, the Mössbauer temperature (an approximation of the Debye temperature, θ_D) and the center shift (δ) are calculated based on a fit of Eq. (7.4) to the experimental data. Finally, the recoil-free fraction for each site is calculated using the relation

$$f(T) = \exp\left[-\frac{3}{2}\frac{E_R}{k_B \theta_D} + \left[1 + 4\left(\frac{T}{\theta_D}\right)^2 \int_0^{\theta_D/T} \frac{x\,dx}{e^x - 1}\right]\right], \tag{7.5}$$

where E_R is the recoil energy, related to the transition energy, E_γ, by $E_R = E\gamma^2/2Mc^2$. See Herberle (1971), De Grave et al. (1985), and Dyar et al. (2013b) for more information.

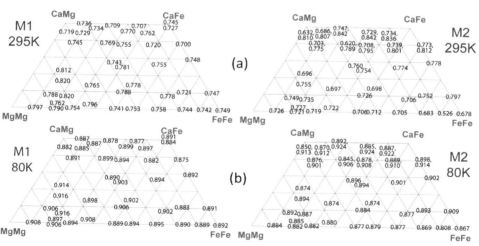

Figure 7.5 Recoil-free fractions as a function of pyroxene composition determined by Dyar et al. (2013b), with values at (a) 295 K and (b) 80 K. The value of f is specific to each site, so there are different values for the M1 and M2 sites in pyroxene.

7.3.3 Site Assignments Based on Fundamental Parameters

The primary application for Mössbauer spectroscopy of interest to planetary scientists is to determine the distribution of Fe cations between valence states (e.g., for oxybarometry) and/or between disparate sites in minerals (for geothermometry). For these purposes, δ and Δ parameters along with the hyperfine field are considered diagnostic, and when recoil-free fractions are known, these distributions can be accurately determined. Diagnostic parameter ranges for specific types of crystallographic sites have been determined empirically from Mössbauer spectra measured on minerals with known crystal structures from single crystal refinements. Ranges for Fe^0, Fe^{2+}, Fe^{3+}, Fe^{4+} and beyond (e.g., Perfiliev & Sharma, 2008; Popa et al., 2013) have been measured on samples for which there is independent or theoretical confirmation of oxidation state.

Exact values of Mössbauer parameters are difficult to predict from theory because long-range interactions in complicated mineral structures are difficult to anticipate but computational methods, such as density functional theory, are attempting to predict Mössbauer parameters of certain inorganic and bioinorganic compounds (Gütlich et al., 2011).

Room temperature isomer shift versus quadrupole splitting data for common rock-forming minerals, taken from our own laboratory data and Burns and Solberg (1988), are shown in Figure 7.6b. Some general trends are quickly apparent (Figure 7.6a). Fe^{3+} has lower isomer shift and quadrupole splitting than Fe^{2+}, and overall the range of Δ for Fe^{2+} is considerably larger (and thus likely more diagnostic) than for Fe^{3+}. Phosphate minerals tend to have slightly higher δ values due to the difference in size between the SiO_4 and PO_4 fundamental units. Nanophase iron oxides all fall in a small range of isomer shift. Not included on this plot are values for the higher valence states of Fe that have yet to be described in geologic materials. Those parameters extend into negative isomer shifts, with $\delta = -0.08$ to -0.16 mm/s for Fe^{4+} (Menil, 1985; Delattre et al., 2002), $\delta = -0.49$ to -0.55 mm/s for Fe^{5+} (Perfiliev & Sharma, 2008; Scepaniak et al., 2011), and $\delta = -0.87$ to -0.91 mm/s for Fe^{6+} (Shinjo et al., 1970; Ladriere et al., 1979; Herber & Johnson, 1979).

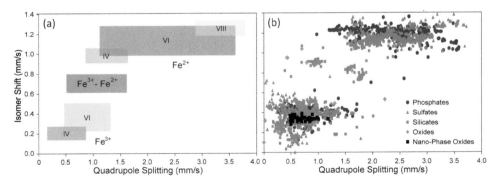

Figure 7.6 Parameter ranges for Mössbauer spectra of rock-forming minerals. (a) Ideal ranges for various coordination and valence of iron. Boundaries between ranges are indistinct, and there is considerable ambiguity in assigning sites when parameters lie near the boundaries on this plot. (b) Mössbauer parameters for various mineral groups. Note the substantial overlap.

There are some obvious drawbacks to using the data presented in Figure 7.6, particularly for the purpose of mineral identification. Most rock-forming silicate minerals contain Fe^{2+} in octahedral coordination, and thus have similar Mössbauer parameters. For example, pyroxene, amphibole, and mica spectra are nearly indistinguishable because of similarities in the octahedral site geometries (Dyar et al., 2006). Minerals containing Fe^{3+} all have parameters that are very close together on this plot, so minerals with Fe^{3+} in more than one site can rarely be distinguished. Furthermore, most minerals exhibit a range of Mössbauer parameters as a function of cation substitution, so clusters of parameters are loose for any particular mineral species. Finally, parameters vary with composition. For all these reasons, Mössbauer spectra in lab applications are typically used for characterization of minerals or samples, rather than for mineral identification in mixtures or natural samples. However, Mössbauer spectroscopy has been used for identification of Fe-bearing minerals on Mars where other options were limited (e.g., Chapter 27).

As a general rule, the larger the QS, the more distorted the coordination polyhedron surrounding the Fe ion. These characteristics allow QS to be used to distinguish between sites in a mineral where there is variable distortion (as well as covalency effects). Variations in Mössbauer parameters that are characteristic of each type of coordination polyhedron can be loosely related to polyhedral site distortion (Burns & Solberg, 1988; Dyar et al., 2013a, 2014). Note that there is a poorly defined region between the ranges for isomer shift of high-spin Fe^{3+} and Fe^{2+} is a where $0.5 > IS > 0.9$ mm/s (purple rectangle in Figure 7.6); doublets with those parameters are generally taken to represent delocalization of electrons between adjacent Fe^{3+} and Fe^{2+}, resulting in an averaged value of IS that can be assigned to $Fe^{2.5+}$.

For phases that show hyperfine splitting, the strength of the effective magnetic field can be useful in distinguishing among phases with similar IS and QS. Because the magnitude of the static magnetic field depends on the distribution of charges as well as the type and strength of magnetic coupling, this parameter is useful in identification of such phases as iron (oxyhydr)oxides. Even where superparamagnetic spectra exist at room temperature, combining room-temperature and low-temperature studies by comparing the hyperfine split low-temperature spectra to their simple quadrupole-split room-temperature counterparts can often lead to identifications. More robust interpretations can be gained by use of strong external magnetic fields (Prisecaru & Kent, 2012). Furthermore, in studies of a single mineral under variable synthesis conditions or with cation substitutions, the hyperfine splitting can be incredibly sensitive to minor changes in crystal structure.

Mössbauer parameters of minerals (and many other materials) can be found by searching the comprehensive online, subscription database maintained by the Mössbauer Effect Data Center (MEDC, at www.medc.dicp.ac.cn/). This resource, which includes all published Mössbauer papers back to and including Professor Rudolf L. Mössbauer's original paper in 1958, contains more than 100,000 records, is the result of a 40-year effort, and is updated monthly. Its strength lies in its incredible breadth of coverage, but its weakness is that parameters are reported directly from the literature without review. MEDC provides an excellent starting point for understanding Mössbauer parameters of any given material, but it is always necessary to consult the original papers to evaluate the conditions and constraints with which data were generated.

The MEDC compiles parameters only. Our research group maintains an online library of Mössbauer spectra (and ASCII data) of many rock-forming minerals at www.mtholyoke.edu/courses/mdyar/database. Although our site is limited to data collected in our own laboratory, it presents typical spectra from many mineral groups, and is intended as a teaching resource for the community.

One final and fundamental constraint on geological applications of Mössbauer spectroscopy is that results are frequently misunderstood. ^{57}Fe Mössbauer spectroscopy can determine only the *relative* amounts of iron in various types of sites and valence states. It cannot determine the total number of Fe ions that are present in a material (i.e., relative to the other ions present) because the other elements have no effect on the Mössbauer spectra.

7.4 Summary

This chapter summarizes underlying physical principles of the Mössbauer effect as they manifest in minerals found on planetary surfaces. The nuclear and electronic origins of the Mössbauer parameters are linked to common structural variations in minerals. However, the range over which each of those parameters varies in minerals is small. While individual iron sites and coordination environment can generally be differentiated by Mössbauer spectroscopy in a single phase, there is often substantial overlap attributed to continuum in bond lengths, bond types, and site distortions caused by structural and chemical distributions. The real power of Mössbauer spectroscopy lies in its application to the determination of Fe^{3+}/Fe^{2+}, and thus its ability to provide information about f_{O2}. Site areas in a Mössbauer spectrum can only be directly related back to relative amounts of iron in the modeled sites if the recoil free fraction for iron in each of those sites is known. Finally, the utility of Mössbauer spectroscopy, like other forms of spectroscopy, is dependent on the knowledge and control of cumulative errors associated with the sample, experiment, spectral quality, and fitting model (among others).

Acknowledgments

We dedicate this chapter to Eddy De Grave (1951–2018) for long-time mentorship and a lifetime of support for the authors. His kindness, dedication to Mössbauer spectroscopy, and generosity to all he knew made him a true asset to the field. We are indebted to him for all he has taught us and he will be greatly missed. We thank S. Cottenier for his useful online course in Mössbauer spectroscopy. This work was funded by NASA grants NNX14AK25G and NNA14AB04A (SSERVI RIS^{4}E).

References

Afanas'ev A.M., Chuev M.A., & Hesse J. (1999) Mössbauer spectra of Stoner-Wohlfarth particles in rf fields in a modified relaxation model. *Journal of Experimental and Theoretical Physics*, **89**, 533–546.

Alp E.E., Sturhahn W., & Toellner T. (1995) Synchrotron Mössbauer-spectroscopy of powder samples. *Nuclear Instruments and Methods in Physics Research B*, **97**, 526–529.

Annersten H. (1975) Mössbauer study of iron in natural and synthetic biotites. *Fortschritte der Mineralogie*, **52**, 583–590.

Bancroft G.M. (1970) Quantitative site populations in silicate minerals by the Mössbauer effect. *Chemical Geology*, **5**, 255–258.

Bancroft G.M. (1973) *Mössbauer spectroscopy: An introduction for inorganic chemists and geochemists*. McGraw Hill, New York.

Bancroft G.M. & Brown J.R. (1975) A Mössbauer study of coexisting hornblendes and biotites: Quantitative Fe^{3+}/Fe^{2+} ratios. *American Mineralogist*, **60**, 265–272.

Blume M. & Tjon J.A. (1968) Mössbauer spectra in a fluctuating environment. *Physics Reviews*, **165**, 446–456.

Burns R.G. & Solberg T.C. (1988) ^{57}Fe-bearing oxide, silicate, and aluminosilicate minerals. In: *Spectroscopic characterization of minerals and their surfaces* (L.M. Coyne, D.F. Blake, & S.W.S. McKeever, eds.). American Chemical Society, Symposium Series. Oxford University Press, Los Angeles, 263–282.

Carmichael I.S.E. (1991) The oxidation state of basic magmas: A reflection of their source region? *Contributions to Mineralogy and Petrology*, **106**, 129–142.

Chandra R. & Lokanathan S. (1977) Electric field gradient in biotite mica. *Physica Status Solidi*, **83**, 273–280.

Clark M.G., Bancroft G.M., & Stone A.J. (1967) Mössbauer spectrum of Fe^{2+} in a square planar environment. *Journal of Chemical Physics*, **47**, 4250–4261.

Cottenier S. (2016) www.hyperfinecourse.org : an open on-line course on hyperfine interaction methods by S. Cottenier (spring 2016 edition).

De Grave E. & Van Alboom A. (1991) Evaluation of ferrous and ferric Mössbauer fractions. *Physics and Chemistry of Minerals*, **18**, 337–342.

De Grave E., Verbeeck A.E., & Chambaere D.G. (1985) Influence of small aluminum substitutions on the hematite lattice. *Physics Letters*, **A107**, 181–184.

De Grave E., Vandenberghe R.E., & Dauwe C. (2005) ILEEMS: Methodology and applications to iron oxides. *Hyperfine Interactions*, **161**, 147–160.

Delattre J.L., Stacy A.M., Young V.G., Long G.J., Hermann R., & Grandjean F. (2002) Study of the structural, electronic, and magnetic properties of the barium-rich iron(IV) iron(IV) oxides, Ba_2FeO_4 and Ba_3FeO_5. *Inorganic Chemistry*, **41**, 2834–2838.

Dyar M.D. (1984) Precision and interlaboratory reproducibility of measurements of the Mössbauer effect in minerals. *American Mineralogist*, **69**, 1127–1144.

Dyar M.D. (1986) Practical application of Mössbauer goodness-of-fit parameters for evaluation of real experimental results: A reply. *American Mineralogist*, **71**, 1266–1267.

Dyar M.D. (1989) Applications of Mössbauer goodness-of-fit parameters to experimental spectra: Further discussion. *American Mineralogist*, **74**, 688–689.

Dyar M.D. (1990) Mössbauer spectra of biotites from metapelites. *American Mineralogist*, **75**, 656–666.

Dyar M.D., Mackwell S.J., McGuire A.V., Cross L.R., & Robertson J.D. (1993) Crystal chemistry of Fe^{3+} and H^+ in mantle kaersutite: Implications for mantle metasomatism. *American Mineralogist*, **78**, 968–979.

Dyar M.D., Agresti D.G., Schaefer M., Grant C.A., & Sklute E.C. (2006) Mössbauer spectroscopy of earth and planetary materials. *Annual Reviews in Earth and Planetary Science*, **34**, 83–125.

Dyar M.D., Klima R.L., & Pieters C.M. (2007a) Effects of differential recoil-free fraction on ordering and site occupancies in Mössbauer spectroscopy of orthopyroxenes. *American Mineralogist*, **92**, 424–428.

Dyar M.D., Schaefer M.W., Sklute E.C., & Bishop J.L. (2007b) Mössbauer spectroscopy of phyllosilicates: Effects of fitting models on recoil-free fractions and redox ratios. *Clay Minerals*, **43**, 1–31.

Dyar M.D., Breves E.A., Jawin E., et al. (2013a) Mössbauer parameters of iron in sulfate minerals. *American Mineralogist*, **98**, 1943–1965.

Dyar M.D., Klima R.L., Fleagle A., & Peel S.E. (2013b) Fundamental Mössbauer parameters of synthetic Ca-Mg-Fe pyroxenes. *American Mineralogist*, **98**, 1172–1186.

Dyar M.D., Jawin E., Breves E.A., et al. (2014) Mössbauer parameters of iron in phosphate minerals: Implications for interpretation of martian data. *American Mineralogist*, **99**, 914–942.

Ericsson T. & Wäppling R. (1976) Texture effects in 3/2–1/2 Mössbauer spectra. *Journal de Physique Colloques*, **37**, C6-719–C6-723.

Frauenfelder H. (1962) *The Mössbauer effect*. W.A. Benjamin, New York.

Gee L.B., Lin C.Y., Jenney F.E., et al. (2016) Synchrotron-based nickel Mössbauer spectroscopy. *Inorganic Chemistry*, **55**, 6866–6872.

Gibb T.C. (1976) *Principles of Mössbauer spectroscopy*. Springer-Verlag, Dordrecht.

Greenwood N.B. & Gibb T.C. (1971) *Mössbauer spectroscopy*. Chapman and Hall, London.

Gütlich P., Eckhard B., & Trautwein A.X. (2011) *Mössbauer spectroscopy and transition metal chemistry*. Springer-Verlag, Berlin and Heidelberg.

Handke B., Kozlowski A., Parlinski K., Przewoznik J., & Slezak T. (2005) Experimental and theoretical studies of vibrational density of states in Fe_3O_4 single-crystalline thin films. *Physical Review B: Condensed Matter and Materials Physics*, **71**, 144301.

Herber R.H. (1984) *Chemical Mössbauer spectroscopy*. Plenum, New York.

Herber R.H. & Johnson D. (1979) Lattice dynamics and hyperfine interactions in M_2FeO_4 (M = K^+, Rb^+, Cs^+) and M`$FeO_4$ (M`=Sr^{2+}, Ba^{2+}). *Inorganic Chemistry*, **18**, 2786–2790.

Herberle J. (1971) The Debye integrals, the thermal shift, and the Mössbauer fraction. In: *Mössbauer effect methodology* (I.J. Gruverman, ed.). Plenum, New York.

Herd C.D.K., Papike J.J., & Brearley A.J. (2001) Oxygen fugacity of martian basalts from electron microprobe oxygen and TEM-EELS analyses of Fe-Ti oxides. *American Mineralogist*, **86**, 1015–1024.

Herd C.D.K., Borg L.E., Jones J.H., & Papike J.J. (2002) Oxygen fugacity and geochemical variations in the martian basalts: Implications for Martian basalt petrogenesis and the oxidation state of the upper mantle of Mars. *Geochimica et Cosmochimica Acta*, **66**, 2025–2036.

Klingelhöfer G. (1998) In-situ analysis of planetary surfaces by Mössbauer spectroscopy. *Hyperfine Interactions*, **113**, 369–374.

Klingelhöfer G., Morris R.V., Bernhardt B., et al. (2003) Athena MIMOS II Mössbauer spectrometer investigation. *Journal of Geophysical Research*, **108**, 8067.

Kojima N., Ikeda K., Kobayashi Y., et al. (2012) Study of the structure and electronic state of thiolate-protected gold clusters by means of Au-197 Mössbauer spectroscopy. *Hyperfine Interactions*, **207**, 127–131.

Ladrière J., Meykens A., Coussement R., et al. (1979) Isomer shift calibration of ^{57}Fe by life-time variations in the electron capture decay of ^{57}Fe. *Journal de Physique Colloques*, **40**, C2-20–C2-22.

LaFleur L.D. & Goodman C. (1971) Characteristic temperatures of the Mössbauer fraction and thermal-shift measurements in iron and iron salts. *Physics Reviews B*, **4**, 2915–2920.

Lindsley D.H., Frost B.R, Ghiorso M.S., & Sack R.O. (1991) Oxides lie; the Bishop Tuff did not erupt from a thermally zoned magma body (abstr.). *Eos, Transactions AGU*, **72**, 312.

Long G.J., Cranshaw T.E., & Longworth G. (1983) The ideal Mössbauer effect absorber thicknesses. *Mössbauer Effect Reference Data Journal*, **6**, 42–49.

Masai H., Matsumoto S., Ueda Y., & Koreeda A. (2016) Correlation between valence state of tin and elastic modulus of Sn-doped Li_2O-B_2O_3-SiO_2 glasses. *Journal of Applied Physics*, **119**, 185104, DOI:10.1063/1.4948685.

McCammon C.A. (1994) A Mössbauer milliprobe: Practical considerations. *Hyperfine Interactions*, **92**, 1235–1239.

McCanta M.C., Rutherford M.J., & Muselwhite D.S. (2002) An experimental study of REE partitioning between a dry shergottite melt and pigeonite as a function of $f_{O(2)}$: Implications for the martian interior. *Meteoritics and Planetary Science*, **37**, A97–A97.

McCanta M.C., Rutherford M.J., & Jones J.H. (2004) An experimental study of rare earth element partitioning between a shergottite melt and pigeonite: Implications for the oxygen fugacity of the martian interior. *Geochimica et Cosmochimica Acta*, **68**, 1943–1952.

McCanta M.C., Elkins-Tanton L., & Rutherford M.J. (2009) Expanding the application of the Eu oxybarometer to the lherzolitic shergottites and nakhlites: Implications for the oxidation state heterogeneity of the martian interior. *Meteoritics and Planetary Science*, **44**, 725–745.

Menil F. (1985) Systematic trends of the ^{57}Fe Mossbauer isomer shifts in (FeOn) and (FeFn) polyhedra: Evidence of a new correlation between the isomer shift and the inductive effect of the competing bond T-X ($\rightarrow$Fe) (where X is O or F and T any element with a formal positive charge. *Journal of Physics and Chemistry of Solids*, **46**, 763–789.

Moon N., Coffin C.T., Steinke D.C., Sands R.H., & Dunham W.R. (1996) A high-sensitivity Mössbauer spectrometer facilitates the study of iron proteins at natural abundance. *Nuclear Instruments and Methods in Physics Research B*, **119**, 555–564.

Mørup S. (2011) Magnetic relaxation phenomena. In: *Mössbauer spectroscopy and transition metal chemistry* (P. E. Bill Gutlich & A.X. Trautwein, eds.). Springer-Verlag, Berlin, 201–234.

Mössbauer R.L. (1958) Kernresonanzfluoreszenz von Gammastrahlung in I^{191}. *Zeitschrift für Physik*, **151**, 124–143.

Munck E., Groves J.L., Tumolillo T.A., & Debrunner P.G. (1973) Computer simulations of Mössbauer-spectra for an effective spin S = 1/2 Hamiltonian. *Computer Physics Communications*, **5**, 225–238.

Murad E. & Cashion J. (2004) *Mössbauer spectroscopy of environmental materials and their industrial utilization*. Kluwer, Dordrecht.

Neese F. & Petrenko T. (2011) Quantum chemistry and Mössbauer spectroscopy. In: *Mössbauer spectroscopy and transition metal chemistry: Fundamentals and Applications* (P. Gütlich, E. Bill, & A.X. Trautwein, eds.). Springer, Berlin and Heidelberg, 137–199.

Oosterhuis W.T. & Spartalian K. (1976) Biological iron transport and storage compounds. In: *Applications of Mossbauer spectroscopy*, 1 (R.L. Cohen, ed.). Elsevier, New York, 142–170.

Parkinson I.J. & Arculus R.J. (1999) The redox state of subduction zones: Insights from arc-peridotites. *Chemical Geology*, **160**, 409–423.

Perfiliev Y.D. & Sharma V.K. (2008) Higher oxidation states of iron in solid state: Synthesis and their Mössbauer characterization. In: *Ferrates: Synthesis, properties, and applications in water and wastewater treatment* (V. K. Sharma, ed.). ACS Symposium Series. Oxford University Press, Los Angeles, 112–123.

Ping J.Y. & Rancourt D.G. (1992) Thickness effects with intrinsically broad absorption. *Hyperfine Interactions*, **71**, 1433–1436.

Popa T, Fan M. Argyle M.D., et al. (2013) H_2 and CO_x generation from coal gasification catalyzed by a cost-effective iron catalyst. *Applied Catalysis*, **464–465**, 207–217.

Prisecaru I. & Kent T.A. (2012) Manual for WMOSS4, www.wmoss.org/downloads/ WMOSS4F_Letter.pdf

Rancourt D.G. (1994a) Mössbauer spectroscopy of minerals I. Inadequacy of Lorentzian-line doublets in fitting spectra arising from quadrupole splitting distributions. *Physics and Chemistry of Minerals*, **21**, 244–249.

Rancourt D.G. (1994b) Mössbauer spectroscopy of minerals II. Problem of resolving cis and trans octahedral Fe^{2+} sites. *Physics and Chemistry of Minerals*, **21**, 250–257.

Rancourt D.G., McDonald A.M., Lalonde A.E., & Ping J.Y. (1993) Mössbauer absorber thickness for accurate site populations in Fe-bearing minerals. *American Mineralogist*, **78**, 1–7.

Rancourt D.G., Ping J.Y., & Berman R.G. (1994) Mössbauer spectroscopy of minerals III. Octahedral-site Fe^{2+} quadrupole splitting distributions in the phlogopite-annite series. *Physics and Chemistry of Minerals*, **21**, 258–267.

Reiff W.M. (1984) Zero and high field Mössbauer spectroscopy studies of the magnetic ordering behavior of one, two, and three dimensional systems. In: *Chemical Mössbauer spectroscopy* (R.H. Herber, ed.). Plenum Press, New York, 65–94.

Sarma P.R., Prakash V., & Tripathi K.C. (1980) Optimization of the absorber thickness for improving the quality of a Mössbauer spectrum. *Nuclear Instruments and Methods in Physics Research B*, **178**, 167–171.

Scepaniak J.J., Vogel C.S., Khusniyarov M.M., Heinemann F.W., Meyer K., & Smith J.M. (2011) Synthesis, structure, and reactivity of an iron(V) nitride. *Science*, **331**, 1049–1052.

Shimony U. (1965) Condition for maximum single-line Mössbauer absorption. *Nuclear Instruments and Methods in Physics Research B*, **37**, 348–350.

Shinjo T., Ichida T., & Takada T. (1970) Fe^{57} Mössbauer effect and magnetic susceptibility of hexavalent iron compounds – K_2FeO_4, $SrFeO_4$, and $BaFeO_4$. *Journal of the Physical Society of Japan*, **29**, 111–115.

Sklute E.C., Dyar M.D., Kashyap S., & Holden J. (2016) The challenge of dist8inguishing iron (hydr)oxides and what it means for Mars (abstr.). Geological Society of America National Meeting, Denver, CO, #197–10.

Sturhahn W. (2004) Nuclear resonant spectroscopy. *Journal of Physics – Condensed Matter*, **16**, S497–S530.

Sturhahn W., Alp E.E., Toellner T.S., Hession P., Hu M., & Sutter J. (1998) Introduction to nuclear resonant scattering with synchrotron radiation. *Hyperfine Interactions*, **113**, 47–58.

Treiman A.H., McCanta M., Dyar M.D., et al. (2006) Brown and clear olivine in Chassignite NWA 2737: water and deformation (abstr.). *37th Lunar Planet. Sci. Conf.*, Abstract #1314.

Van Alboom A. & De Grave E. (2016) Temperature dependences of the hyperfine parameters of Fe^{2+} in $FeTiO_3$ as determined by ^{57}Fe-Mössbauer spectroscopy. *American Mineralogist*, **101**, 735–743.

Van Alboom A., De Resende V.G., De Grave E., & Gomez J.M. (2009) Hyperfine interactions in szomolnokite ($FeSO_4·H_2O$). *Journal of Molecular Structure*, **924–926**, 448–456.

Van Alboom A., De Grave E., & Wohlfahrt-Mehrens M. (2011) Temperature dependence of the Fe^{2+} Mössbauer parameters in triphylite ($LiFePO_4$). *American Mineralogist*, **96**, 408–416.

Van Alboom A., De Resende V.G., da Costa G.M., & De Grave E. (2015) Mössbauer spectroscopic study of natural eosphorite, [(Mn, Fe)$AlPO_4(OH)_2H_2O$]. *American Mineralogist*, **100**, 580–587.

Visscher W.M. (1960) Study of lattice vibrations by resonance absorption of nuclear gamma rays. *Annals of Physics*, **9**, 194–210.

Voigt W. (1912) On the intensity distribution within lines of a gaseous spectrum, *Sitzungsberichte der Königlich Bayerischen Akademie der Wissenschaften zu München*, **1912**, 603–620.

Wadhwa M. (2001) Redox state of Mars' upper mantle and crust from Eu anomalies in shergottite pyroxenes. *Science*, **291**, 1527–1530.

Waychunas G.A. (1986) Performance and use of Mössbauer goodness of fit parameters: Response to spectra of various signal/noise ratios and possible misinterpretations. *American Mineralogist*, **71**, 1261–1265

Waychunas G.A. (1989) Applications of Mössbauer goodness-of-fit parameters to experimental spectra: A discussion of random noise versus systematic effects. *American Mineralogist*, **74**, 685–687.

Whipple E.R. (1968) *Quantitative Mössbauer spectra and chemistry of iron*. PhD thesis, Massachusetts Institute of Technology, Cambridge, MA.

Yan L., Zhao J., Toellner T.S., et al. (2012) Exploration of synchrotron Mössbauer microscopy with micrometer resolution: Forward and a new backscattering modality on natural samples. *Journal of Synchrotron Radiation*, **19**, 814–820.

8

Laser-Induced Breakdown Spectroscopy
Theory and Laboratory Spectra of Geologic Materials

SAMUEL M. CLEGG, RYAN B. ANDERSON, AND NOUREDDINE MELIKECHI

8.1 Introduction to Remote Laser-Induced Breakdown Spectroscopy

Laser-Induced Breakdown Spectroscopy (LIBS) has emerged as a highly versatile elemental analysis tool for planetary science because it can be used for both remote and in situ analysis without the need for sample preparation. The ChemCam instrument on the Mars Curiosity rover is the integration of a remote LIBS instrument capable of chemical measurements out to 7 m as well as a Remote Micro-Imager (RMI) designed to record context images of the samples probed by LIBS and is the focus of Chapter 29 (Maurice et al., 2012; Wiens et al., 2012). The SuperCam instrument selected for the Mars 2020 rover is based on the ChemCam instrument architecture and will record LIBS measurements out to 7 m as well as high-resolution color context images. The SuperCam instrument will also record mineralogic data from remote infrared and Raman spectroscopy as discussed in Chapters 1–7. This chapter focuses on the use of LIBS for remote planetary chemical analysis, where remote or stand-off refers to the ability to probe samples where contact with the sample is not necessary or possible.

Figure 8.1 contains a generic diagram of a remote LIBS measurement. LIBS involves focusing a relatively high power laser onto the sample surface to generate an expanding plasma containing electronically excited atoms, ions, and small molecules ablated from the sample. The excited species emit light at wavelengths that are diagnostic of the elements present in the sample. Although most LIBS measurements employ a 1064-nm neodymium-doped yttrium aluminum garnet (Nd:YAG) laser, it is worth noting that any laser that can generate the laser-induced plasma can be used. Some of the optical emission from the plasma is collected, directed into one or more dispersive spectrometers, and recorded with a detector. For the remote planetary applications discussed in this chapter, a telescope is typically used to focus the laser as well as collect the emission from meters away, such as is done with the ChemCam and SuperCam instruments. Figure 8.2 contains representative LIBS spectra of geologic standard samples collected with the ChemCam test bed at Los Alamos National Laboratory; major elements and some minor elements are identified.

The intensity of a LIBS spectrum is fundamentally sensitive to the atmospheric pressure and insensitive to ambient planetary temperatures in which the sample is found. This atmospheric pressure dependence is depicted in Figure 8.2a, where the relative intensities

Laser-Induced Breakdown Spectroscopy: Theory and Lab Spectra of Geologic Materials   169

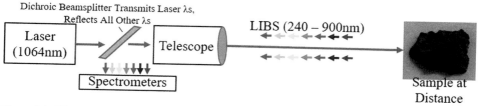

Figure 8.1 Diagram describing the components of a stand-off or remote LIBS instrument.

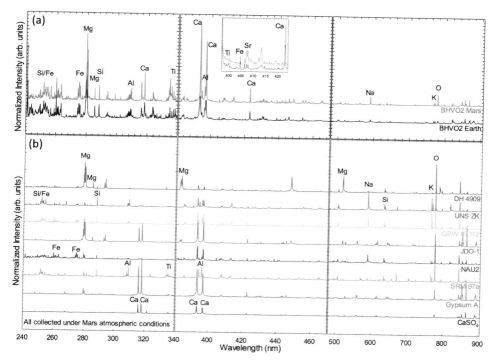

Figure 8.2 Example LIBS spectra. (a) The LIBS spectra produced on a BHVO2 geochemical standard using the ChemCam testbed at LANL under Mars (red) and Earth (black) atmospheric conditions. (b) The LIBS spectra of several disparate geological standards.

of the Mars elemental peaks (red) are higher than those observed under terrestrial atmospheric pressure (black). The inset in Figure 8.2a highlights the very high signal-to-noise quality of the ChemCam LIBS spectra as well as the enhanced spectral intensity observed under Mars atmospheric conditions relative to those of Earth. LIBS plasma is most intense when the atmospheric pressure is between 10 and 100 Torr, making the Mars surface atmosphere an ideal place for remote LIBS investigations. However, when the atmospheric pressure decreases from 10 Torr to atmosphere-less bodies, such as the Moon or asteroids, the plasma rapidly expands before the excited species emit light and produces a lower signal (Lasue et al., 2012). As the pressure increases to above 100 Torr, such as on Earth or Venus, the plasma is more constrained by the surrounding atmosphere and the relative intensity is again reduced. Regardless of the atmospheric pressure, instruments have been designed that can accommodate these various atmospheric conditions and can be exceedingly valuable exploratory analytical techniques. Finally, the LIBS plasma generated by ChemCam-like planetary instruments typically exceeds 5000 K. LIBS measurements are therefore insensitive to generating a plasma under most planetary conditions, including Venus. It is partly for this reason that LIBS has been selected for two Mars rover missions and has also been proposed for Venus and lunar surface investigations.

The LIBS spectra depicted in Figure 8.2b demonstrate the qualitative differences in the spectra collected from several different geologic materials. Starting from the top of the figure, DH4909 is an olivine sample and UNS ZK is a feldspar. The olivine spectrum demonstrates Mg enrichment while the feldspar spectrum recorded higher concentrations of Si, Ca, Al, Na, and K. GBW 07312 is a stream sediment that contains all of the major elements. In contrast, the JDO-1 dolomite spectrum depicts the expected Ca and Mg spectral signatures. NAu-2 (nontronite) is an Fe-rich smectite that contains many Fe lines in the UV spectrum. SRM 97A is an Al-rich clay that also contains Si, Ti and K and very little Ca. Finally, gypsum A is a geological standard and $CaSO_4$ is a chemical standard. The $CaSO_4$ sample appears to be pure, as expected, with only Ca and S emission lines (S is not observed on this scale) while the gypsum A geological standard is not pure gypsum and contains other major element impurities.

There are many synergies between the LIBS techniques discussed in this chapter and other analytical methods discussed in this volume. As for Alpha Proton X-ray Spectroscopy (APXS), which pioneered chemical investigations on the Pathfinder rover, Mars Exploration Rovers (MERs), and Mars Science Laboratory (MSL) Curiosity rover as described in Chapter 28, LIBS is fundamentally an elemental analysis technique. The LIBS methods demonstrated by ChemCam are capable of rapid remote analysis without sample preparation and the limitations associated with sample contact. LIBS is also sensitive to all elements, independent of the atomic mass, that are present above the instrumental detection limits. When properly calibrated, LIBS can produce accurate chemical compositions from which the sample mineralogy may be inferred. Consequently, LIBS analyses are also synergistic with direct mineralogic methods such as the MSL CheMin X-ray diffraction (XRD) instrument, passive infrared (IR) spectroscopy such as the MER Mini-TES (described in Chapter 25) and the Mars 2020 SuperCam instrument, and the SuperCam and SHERLOC Raman spectrometers on the Mars 2020 rover. Finally, the high-resolution context imagers on all of the Mars rovers have been critical to the interpretation of the samples probed and collected on Mars. The MSL ChemCam instrument has a black-and-white RMI integrated into a telescope that records the locations of the LIBS spots. These images have been so valuable to the ChemCam investigations that a color imager has been integrated into the Mars 2020 SuperCam architecture.

This chapter focuses on details of the LIBS technique for remote planetary chemical investigations. First, the fundamental physics associated with the generation of the LIBS plasma and the associated matrix effects are discussed. Second, we discuss the many analysis techniques that have been developed to extract chemical compositions and characterize the sample morphology and mineralogy. Finally, the use of LIBS under various planetary atmospheric conditions will be discussed.

8.2 Fundamental LIBS Technique

LIBS is an optical technique used for the qualitative and quantitative elemental analysis of solids, liquids, and gases (Miziolek et al., 2006; Singh et al., 2007; Cremers et al., 2013). One of the most significant advantages to planetary science is that LIBS provides multi-elemental spectroscopic information about rocks and soils with no sample preparation from stand-off distances. LIBS relies on focusing a laser of sufficient irradiance to generate a plasma with the same elemental composition as the irradiated material, a condition that is referred to as stoichiometric ablation, by vaporizing a few hundreds of nanograms of the material. As the plasma cools down, it emits light that can be analyzed to obtain information on its atomic and in some cases molecular constituents.

The initiation process that results in the formation of LIBS spectra depends on the type of material interrogated (either conductor or insulator, transparent or absorbing) and its state of aggregation (gas, liquid, or solid). However, the evolution of free electrons and ionized atoms once formed is very similar. In this section, we describe the optical, physical, and chemical processes that take place from excitation of rocks and soils by a nanosecond laser pulse, which represents the most widespread LIBS application to date and can effectively exemplify the main phenomena taking place in laser-induced breakdown with short pulses (Carroll et al., 1981; Hahn et al., 2010). Breakdown induced by ultrashort pulses involves a different set of phenomena, mainly related to the lack of thermal effects, and the interested reader may refer to Labutin et al. (2016) for details about this different regime of laser-induced breakdown.

Figure 8.3 illustrates the process by which laser ablation and excitation generate a LIBS spectrum. The LIBS process begins with optical breakdown of the material, a threshold process initiated by a laser pulse of sufficient energy to induce ablation and massive ionization of the vaporized material (Figure 8.3b–e). Following laser irradiation, resonant or nonresonant multiphoton absorption (MPA) can take place, which together with avalanche ionization (AI) occurring subsequently, is responsible for generating a very high density of free electrons (about 10^{18} electrons/cm^3) (Figure 8.3a). MPA occurs when a material absorbs more than one photon simultaneously to excite high-lying energy levels, which may lead to multiphoton ionization (MPI), in particular if low ionization potential impurities are present in the laser focal volume. The probability of MPA and MPI taking place depends critically on the laser intensity in the interaction volume, and both require a minimum – though different – threshold intensity before breakdown is initiated. In general, a material can be effectively ablated when the incoming laser irradiance exceeds the material breakdown threshold, which depends on the detailed atomic and molecular characteristics of the material, i.e., bandgap or ionization potential, as well as the laser characteristics, i.e., its wavelength and incident laser irradiance.

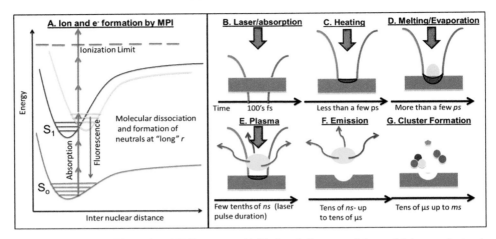

Figure 8.3 Diagram illustrating LIBS process. (a) Notional diagram of the multiphoton electronic excitation and ionization responsible for generating the LIBS plasma. (b–g) Depictions of the six steps from laser excitation to optical LIBS emission and the production of clusters of atoms.

Moreover, if the difference between the energy of a single photon of the exciting laser and the different accessible atomic or molecular energy gaps present in the material approaches zero (resonant case), the probability of MPA is substantially enhanced. Free electrons generated by MPI collide with heavier particles, i.e., neutrals, ions, and molecules. The laser-induced excitation is transferred via collisions from electrons to the lattice, which heats up the material and results in the vaporization of its interaction volume, provided that the energy density deposited on the surface of the material exceeds its latent heat of vaporization.

Following breakdown, the LIBS process is dominated by the formation and expansion of a plume of free electrons and ionized atoms (Figure 8.3e). The ablated material plume expands at an initial velocity of about 10^5–10^7 cm/s, much faster than the speed of sound, and forms a shock wave that displaces dust from the rock surface and enables direct access to the underlying surface for imaging and other analytical investigations. The expanding plume also displaces the surrounding atmosphere, with pressures at the surface of the material approaching 10^6 atm, far greater than the 92 atm Venus surface pressure. Within a few ps, the free electrons in the laser-induced plume absorb energy from the laser beam still available (its typical duration being from few ns to some tens of ns) through inverse Bremsstrahlung. This phenomenon causes the electron temperature and the frequency of electron collisions to increase, promoting avalanche ionization to create the highly ionized luminous LIBS plasma (Figure 8.3e–f).

As the energy provided by the laser pulse ends, so does the generation of free electrons and ions. Within hundreds of ns to tens of μs depending on the atmospheric pressure, the plasma expands, slows down via collisions with ambient gas species, and extinguishes. The shock wave detaches from the plasma front and continues propagating at a lower speed, eventually approaching the speed of sound (Figure 8.3e). Electron–ion recombination processes yielding to the decay of the plasma can also lead to the formation of high-density neutral species including molecules, mostly from recombination of the plasma species with the background gas. This usually occurs within hundreds of μs (up to ms) after the plasma has been ignited (Figure 8.3f).

LIBS plasmas are not necessarily uniform; the plasma plume is typically influenced by the laser characteristics (such as temporal pulse length, spatial pulse shape, wavelength, and irradiance), by rock or soil properties (such as atomic weight, density, surface reflectivity, conductivity, melting and boiling points) that are not always fully known, and the environment surrounding the sample of interest that cannot always be fully controlled and/or monitored. A quantitative analysis of the elemental composition of a sample based on the use of such plasma remains a challenging task. In particular, matrix effects may limit the domain of application of LIBS and our ability to provide quantitative analyses with a very high degree of accuracy. Many authors have addressed matrix effects in LIBS spectra. Wisburn et al. (1994) reported on grain size effects on the measured intensities of LIBS spectra of Cd. Such effects stem from the laser's sampling process. These authors observed that for samples contaminated with the same amount of weight of Cd relative to the weight of the solution but with grain sizes that vary from 0.38 mm to 1.1 mm, the LIBS intensities produced are proportional to the grain sizes. Eppler et al. (1996) investigated the effects of chemical speciation and matrix composition on Pb and Ba concentrations in sand and soil matrices using LIBS. With changes of the bulk matrix from pure sand to pure soil composition, the Ba (II)/C (I) signal was found to decrease. These matrix effects have challenged the development of quantitative analysis techniques and led to analytical methods that can exploit these matrix effects as discussed in Section 8.3.

Several approaches for matrix effect correction have been proposed (Chaléard et al., 1997). One of the most convenient approaches is to determine the element concentrations by the line intensity signals obtained from standard references having a similar matrix. Unfortunately, it is not always possible to obtain standard references and their corresponding calibration curves, especially for remote planetary investigations where the sample matrix is unknown. Ciucci et al. (1999) proposed an alternative approach of calibration. This method relies on the determination of plasma parameters (electron density and temperature) and on the assumption that the plasma is in local thermodynamic equilibrium (LTE), and optically thin. However, these assumptions are not always valid, particularly when experimental parameters vary widely (Colgan et al., 2015, 2016; Johns et al., 2015). Another approach is to employ statistical Multivariate Analysis (MVA) methods such as those discussed in the next section. They allow signals from other emission lines at other points in the spectrum to partially correct for chemical matrix effects that may be induced by the presence of these other elements. Studies that have investigated MVA include Sirven et al. (2006), Clegg et al. (2009), Tucker et al. (2010), Anderson et al. (2011a,b), and Dyar et al. (2011, 2012, 2016).

The properties and expansion dynamics of the plasmas produced by LIBS are sensitive to the conditions used to generate them. As mentioned earlier, laser pulse characteristics and

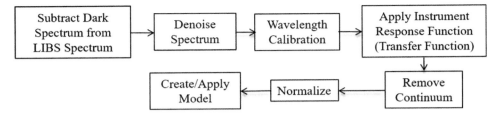

Figure 8.4 Flow chart illustrating the data processing and corrections required for remote LIBS spectra collected on other planets.

physical-chemical properties of the target material are key parameters that affect the spectra collected. Therefore, it is critical that care be taken to ensure that the atomic and molecular species formed in the plasma are representative of the rock and soil under investigation as much as possible.

8.3 LIBS Geochemical Analysis

The interpretation of LIBS spectra incorporates plasma physics, analytical chemistry, chemometrics, statistics, and machine learning, among other fields. An exhaustive discussion of the various methods that have been, or could be, applied to LIBS spectra is beyond the scope of this chapter. Figure 8.4 contains a summary of some of the most important data processing steps required for LIBS planetary remote sensing investigations in the field.

8.3.1 Data Processing

LIBS spectra contain a significant amount of information about the targets that have been analyzed, but to use this information effectively for geochemical analysis, a number of data processing steps are typically required. The preprocessing described here is based primarily on the steps used for ChemCam data (Wiens et al., 2013), but most will be relevant to any LIBS instrument though the order of the steps may vary.

The first step in processing LIBS spectra is to minimize the influence of ambient light and background instrumental noise. This is typically achieved by collecting "passive" spectra of the target without firing the laser and then subtracting these passive spectra from the active LIBS spectra. This results in spectra representing strictly the emission produced by the laser-induced plasma. Dark subtraction is especially important when the light source has emission and absorption lines of its own (e.g., the Sun), because those features can influence the apparent strength of emission lines in LIBS observations (Schröder et al., 2015).

Next, denoising can be applied to minimize the effects of high-frequency noise in the spectra. Denoising is desirable because it can improve the limit of detection for elements with faint emission lines and help to mitigate the effects of varying distance and signal intensity on the signal-to-noise ratio. The primary class of methods used to denoise LIBS spectra are wavelet-based methods (e.g., Schlenke et al., 2012; Zhang et al., 2013). These methods vary in their details but all fundamentally work by decomposing the spectra into wavelets of varying scales and eliminating those that represent very high frequencies likely to be the result of noise while preserving the larger-scale structure of the data that represents a meaningful signal. The ChemCam data processing pipeline uses a stationary undecimated wavelet transform that is described in Starck et al. (2002) and Wiens et al. (2013).

Rigorous wavelength calibration is a critical step in processing LIBS spectra to ensure that the signal recorded in each pixel can be accurately assigned to the element. This is especially true for fielded instruments such as ChemCam, where the spectrometer calibration changes with the spectrometer body temperature. LIBS spectra produce many relatively sharp spectra signatures for every element, and proper elemental identification requires an accurate spectral calibration. Wiens et al. (2013) demonstrated that a consistent spectral calibration is critical to the partial least squares calibration and the LIBS spectra should be resampled so that each pixel represents the same spectral range. This is accomplished by observing the spectrum of a target with numerous known emission lines and applying an appropriate shift. The details of wavelength calibration will be unique to each instrument, depending on the conditions under which measurements are made and the properties of the instrument itself. ChemCam uses a "matched filter" technique to align spectra of Ti taken on Mars to those taken in the laboratory. In addition, a wavelength-dependent temperature correction is applied to ensure that the spectra are properly wavelength calibrated under varying observational conditions (Wiens et al., 2013).

It is often desirable to convert spectra from the arbitrary DN (digital number in counts) produced by the instrument to physically meaningful units (photons), improving the compatibility of spectra taken on different instruments or with different optical components on the same system. This requires a light source that has been calibrated for the full spectral range of the instrument, as well as knowledge of the source area imaged, the solid viewing angle, the integration time, and the instrument gain. Details of the ChemCam instrument response correction are provided by Wiens et al. (2012).

Although the instrument response correction mitigates differences between instruments, it is not always possible to apply an instrument response correction and it does not always remove instrumental differences entirely. Thus, when comparing data or multivariate models between different instruments, calibration transfer methods may be useful (Wang et al., 1992). To improve the agreement between laboratory spectra and those collected by the ChemCam instrument on Mars, a simple calibration transfer method based on the ratio of calibration target spectra collected by both instruments is used (Clegg et al., 2017). Other, more advanced calibration transfer methods show promise for LIBS spectra (Boucher et al., 2015a,b), and methods that are commonly used in other spectroscopy applications such as direct standardization and piecewise direct standardization (Wang et al., 1992; Feudale et al., 2002), may be applicable to LIBS spectra as well.

The LIBS plasma produces a significant background or "continuum" signal due to Bremsstrahlung radiation and electron–ion recombination that does not contain much information about the target being analyzed (Wiens et al., 2013). Time-gated LIBS instruments can mitigate much of the continuum emission by limiting their integration time to when most of the atomic emission occurs, excluding the earlier period of plasma evolution during which the continuum emission occurs. In non-gated instruments (such as ChemCam), the entire evolution of the plasma is captured and the continuum emission is usually removed as a preprocessing step (Wiens et al., 2013). Continuum removal is particularly important for stand-off or remote analyses that observe targets at varying distances and is typically ignored for in situ analyses when observations are all made at the same distance (Clegg et al., 2009; Tucker et al., 2010; Melikechi et al., 2014; Mezzacappa et al., 2016).

Many different continuum removal algorithms are available in the spectroscopy literature that have been developed for different types of spectra (e.g., Giguere et al., 2015, 2017; Dyar et al., 2016). For the ChemCam data processing pipeline, a wavelet method very similar to that used for denoising is applied to the spectra, using low-frequency components of the decomposition to find local minima, which are then fit with a spline function representing the continuum (Wiens et al., 2013).

Normalization of the spectra to the total emission intensity is commonly applied to LIBS spectra. This normalization step can help to mitigate differences in intensity caused by fluctuations in laser energy and coupling (Wiens et al., 2013). Depending on the instrument and the element(s) of interest, normalization can be performed for each spectrometer, such that the sum of the full spectrum equals the number of spectrometers, or to the sum of the intensity across all spectrometers, such that the sum of the full spectrum equals 1.

Finally, extracting quantitative chemical information from a remote or stand-off LIBS analysis currently requires a distance correction. As discussed in this section, LIBS requires the development of a calibration model in which the calibration data are collected at one distance and applied to all other distances. One current method is to normalize the data to the total emission intensity and requires the removal of the continuum. Other methods exploit an internal signal to generate a distance correction. Using ChemCam data, Melikechi et al. (2014) and Mezzacappa et al. (2016) exploited the ubiquitous Mars dust signal from the first laser shot of each ChemCam analysis as a proxy for distance.

8.3.2 Dimensionality Reduction

LIBS spectra are high-dimensionality data products, typically with many thousands of spectral channels. These spectral channels are also highly correlated with each other: most elements have multiple emission lines within the LIBS spectrum, and element abundances are naturally correlated in geochemical samples. This high degree of correlation means that much of the variation in a set of LIBS spectra can be captured using a dramatically smaller number of variables. Two of the most common methods used to reduce the dimensionality of spectral data are principal component analysis (PCA) and independent component analysis (ICA) (Vance et al., 2010; Pokrajac et al., 2014). The process of PCA can be visualized by considering a set of LIBS spectra as a cloud of points in a high-dimensionality space, with each spectral channel corresponding to a dimension. PCA identifies the axis through that cloud corresponding to the largest variation. This axis represents the first principal component (PC). The algorithm finds the next PC by identifying the axis orthogonal to PC1 along which the data have the most variation. This process can be repeated until all of the variability in the data is explained (although in practice it is usually stopped after reaching a threshold such as 95% of the total variability).

PCA "scores" and "loadings" are equivalent to the eigenvalues and eigenvectors of the data. The scores can be thought of as the values of each spectrum when projected onto the PC axes, and the loadings are the weights (one per spectral channel) that convert the spectra to scores. Scores are often plotted on a scatter plot to visualize the variability in the LIBS spectra. Figure 8.5 illustrates PCA scores and loadings for the ChemCam laboratory database generated with the free open-source PySAT program (Anderson, 2016).

The advantage of PCA is that it finds an efficient representation of the data, often requiring only a few PCs to explain much of the variance in a LIBS dataset. The disadvantage is that this involves loadings that contain strong contributions from several different elements in the sample, making it less straightforward to interpret the variation along axes of the score plot.

ICA is quite similar to PCA, but instead of seeking the most "efficient" representation of the data, it seeks to separate out the signal into statistically independent sources. There are a number of ICA algorithms that give varying results, but the Joint Approximate Diagonalization of Eigenmatrices (JADE) method has been shown (Forni et al., 2013) to break down LIBS spectral data into sources (the ICA equivalent of PCA loadings) that closely correspond to the elements in geologic samples with the strongest emission lines (Figure 8.5). Thus, by judiciously choosing the number of ICA sources to minimize mixing between elements, the ICA scores can be interpreted as a proxy for the strength of that element's contribution to the spectrum.

Other methods of reducing the dimensionality of LIBS spectral data, such as locally linear embedding (LLE; Boucher et al., 2015a), also show promise and are an area of active research.

A complementary approach to reducing the dimensionality of LIBS spectra is to simply choose a small subset of the full spectrum to use for subsequent analysis. This process, called "feature selection" or "variable selection," can be accomplished manually by selecting the emission line or lines of interest, or algorithmically. Algorithmic feature selection is particularly common in conjunction with multivariate methods for classification and quantification, discussed in the text that follows. Multiple algorithms can be used for spectroscopic feature selection (Balabin & Smirnov, 2011), including genetic algorithms (Leardi & Gonzalez, 1998; Anderson et al., 2011b) and least absolute shrinkage and selection operator (LASSO) regression (Dyar et al., 2012b). Studies of feature selection methods as applied to LIBS data (e.g., Anderson et al., 2011b; Dyar et al., 2012b) have shown that they can yield comparable accuracy compared to using the full spectrum, and the significant reduction in the number of variables potentially makes interpretation more straightforward.

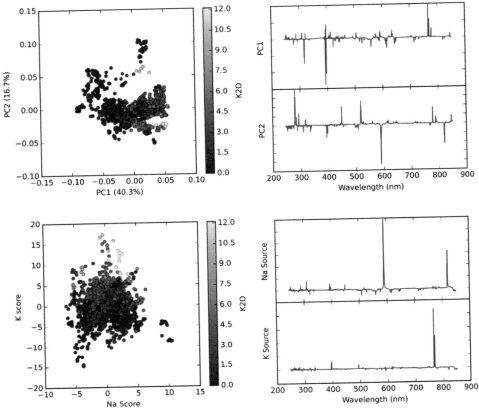

Figure 8.5 Qualitative classification of samples using LIBS. (a) Classification scores using PCA. (b) Loadings plots for PCA analysis. (c) Classification scores using ICA. (d) Loadings plots for ICA analysis. Each of the spectral features in (b) and (d) are associated with one or more atomic emissions.

8.3.3 Qualitative Methods

Once the LIBS spectra have been suitably preprocessed, there are many ways to proceed in analyzing and interpreting the chemical information in the rock or soil sample. We will first discuss qualitative methods that can be used to compare and contrast planetary samples based on their LIBS spectra without quantitative chemical details. These can be divided into two categories: clustering and classification.

8.3.3.1 Clustering Methods

Clustering methods can be implemented using the full spectra, but often the scores from PCA or ICA are used instead to speed calculation and improve performance by eliminating unnecessary or redundant information. There are two broad types of clustering: hierarchical and partitional (Jain et al., 1999). Hierarchical methods produce a tree or "dendrogram" of successive clusterings, while partitional methods result in a single division of the dataset into a user-specified number of clusters. Clustering is distinguished from classification by the fact that it is unsupervised: the data are grouped according to some inherent similarity without the use of training data.

One of the simplest and most common partitional clustering methods is K-means clustering, which divides the dataset into k clusters. The algorithm is initialized with random cluster centers and data points are assigned to the nearest cluster center. The cluster centers are then updated to be the centroid of the points that have been assigned to the cluster. This process is repeated until the cluster centers do not move. Because it is susceptible to local minima, K-means clustering typically is run many times and the result that minimizes some evaluation metric (often the sum of squared distances from the centroids) is chosen as the "optimum" solution (MacKay, 2003; Anderson et al., 2012).

Hierarchical clustering can be performed in an "agglomerative" or "divisive" manner. In agglomerative clustering, each observation starts out in its own cluster and the clusters are merged until a single cluster remains. Divisive clustering is the opposite: all observations start out in the same cluster and that cluster is divided until each observation is its own cluster. The clusters can be grouped or split based on a variety of metrics such as the minimum or maximum distance between cluster elements (single and complete respectively)(King, 1967; Sneath & Sokal, 1973), or minimization of within-cluster variance (Ward, 1963).

8.3.3.2 Classification Methods

Classification has a goal similar to clustering: to divide the observations into groups. However, classification methods differ from clustering in that they are supervised methods. Observations are assigned to classes based on a set of training data for which the classes are known. Similar to clustering, there are many different classification algorithms that could be used with LIBS spectra.

Partial Least Squares Discriminant Analysis (PLS-DA) is a classification method that makes use of the PLS algorithm that is commonly used for regression, as described in detail in the text that follows. The primary difference is that with PLS-DA, instead of using PLS to determine the value of a continuous variable such as chemical composition, PLS-DA is trained on binary classification data to determine whether an unknown observation should be classified as within a group or not. PLS-DA has been used to accurately determine the provenance of gemstones based on LIBS data (Kochelek et al., 2015).

Soft Independent Modeling of Class Analogy (SIMCA) is another supervised classification method that has been used with LIBS spectra (Clegg et al., 2009; Anderson et al., 2012). SIMCA works by performing PCA on each of the classes of observations in the training set, and then projecting new observations into the space defined for each class to determine which class is most similar (Wold & Sjöström, 1977). SIMCA has the advantage that if a sample does not belong to any of the classes defined in the training set, it is not "forced" into one of them. Likewise, an observation can be assigned to multiple classes when appropriate. While LIBS is fundamentally an elemental analysis technique, SIMCA has been used to accurately infer rock-type based on the spectral variations (Clegg et al., 2009).

8.3.4 Quantitative Methods

A wide variety of LIBS quantitative analysis methods have been developed over the decades since the advent of the laser and LIBS. The original methods relied on standard univariate analysis methods in which a peak height, or area, is plotted against concentration. However, matrix effects, especially those found in geochemical materials, quickly complicated the accuracy of these univariate methods (Dyar et al., 2016). Since the late 1990s, more complex multivariate methods have been used to qualitatively exploit these matrix effects as discussed in the previous section as well as compensate for these effects and produce more accurate calibration models.

8.3.4.1 Univariate Analysis

The simplest method of determining chemical composition based on LIBS spectra is to identify an emission line from the element of interest and create a calibration curve relating the strength of that line (typically based on integrated area of a peak fit) to concentration in training data. Univariate analysis is advantageous because it is relatively straightforward and easy to understand, and has been used for a number of minor and trace elements in ChemCam data (Fabre et al., 2014; Ollila et al., 2014; Lasue et al., 2016). The disadvantage of univariate analysis is that LIBS spectra suffer from "matrix effects" that can cause the strength of individual features to vary for reasons that are not related to the concentration of the element of interest, as discussed above (Clegg et al., 2009).

8.3.4.2 Multivariate Analysis

As their name implies, multivariate analysis methods use multiple variables (often the entire spectrum) to develop more complex regression models based on training data. Multiple studies have shown that multivariate methods generally produce more accurate quantitative results of complex samples such as rocks and soils than univariate methods (e.g., Clegg et al., 2009; Dyar et al., 2016). This is because, by incorporating information from the rest of the spectrum, multivariate methods can partially correct for matrix effects, resulting in more accurate solutions. The downside of multivariate methods is that they can be difficult to understand compared to univariate calibration curves and require more detailed investigations to interpret. There is also the risk that they can become "black boxes." Multivariate methods are also susceptible to overfitting, a situation in which the regression model is overly complex, resulting in excellent accuracy on training data, but very poor performance on novel data. This is analogous to using a polynomial of arbitrarily high degree to obtain a perfect fit to a scatter plot of noisy data, when a simple linear fit more accurately represents the underlying process that gave rise to the data. Overfitting can be mitigated by performing cross-validation to estimate the performance of a model on novel data and choosing model parameters accordingly.

Cross-validation is a procedure in which the training set is divided into several groups or "folds," and each fold is held out and predicted by a model trained only on the remaining training data. This process is repeated for each fold, and for each combination of the model parameters that are being tuned. The folds can be selected randomly, but in cases in which a dataset contains more than one observation of the same target, it is very important that those observations all be in the same fold. If data from the same target are spread among multiple folds, then the model will appear to be more accurate than it actually is. An alternative to random folds is to define "stratified" folds that are chosen to be as similar to each other as possible. A simple way to implement this for LIBS data is to sort the training set observations by their known composition, and sequentially assign samples to folds, resulting in a similar distribution of compositions in each fold.

The accuracy of MVA models can also be rigorously characterized. The accuracy of a regression model is often expressed as the root-mean-squared error of calibration (RMSEC), root-mean-squared error of cross-validation (RMSECV), and root-mean-squared error of prediction (RMSEP). RMSEC is the error of the regression model when predicting the composition of the samples that were used to train the model. Because the training process optimizes the model for these samples, RMSEC is typically very optimistic and should not be used to report the accuracy of the model on unknown data. RMSECV is a measurement of the accuracy of the predictions obtained during cross-validation. Because cross-validation holds out each fold and trains models using the remaining folds, RMSECV is more representative of the performance of the final model on unknown data. For this reason, RMSECV should be used to tune the parameters of the regression model. RMSEP is the accuracy of the model trained on the entire training set, at predicting a separate "test" set. RMSEP is the preferred metric for reporting a regression model's accuracy on novel data.

8.3.4.3 Multivariate Methods

There are a wide variety of multivariate regression algorithms that can be applied to LIBS data to quantify elemental abundances. Two of the most common methods (Anderson et al., 2011b; Sirven et al., 2006; Ferreira et al., 2008; Dyar et al., 2012) are artificial neural networks (ANN; Sarle, 1994), and partial least squares (PLS; Wold et al., 2001), though other methods such as support vector regression (SVR; Smola & Schölkopf, 2004) and principal component regression (PCR; Jolliffe, 1982) have also been used (e.g., Yaroshchyk et al., 2012; Shi et al., 2015).

PLS is widely used in chemometrics (Wold et al., 2001; Abdi, 2003; Rosipal et al., 2006) because it is well suited to spectroscopic data for which the number of observed variables (spectral channels) is typically significantly greater than the number of observations, and many of the variables are highly correlated. PLS is similar to PCA in that it represents the data as a small set of "latent vectors" or "components" but it differs in that PLS utilizes both the X variables (spectra) and Y variables (compositions). The number of components used in the PLS regression determines the model complexity, and is typically chosen by using cross-validation with a range of components and plotting RMSECV against number of components. The optimum number of components is generally the point at which the RMSECV curve flattens out, indicating that additional model complexity does not significantly improve the results. PLS is an appealing method because it produces regression vectors that can be viewed to gain insight into how the model is responding to different regions of the spectrum, thus making the method less of a "black box" than others.

ANNs are a popular class of nonlinear regression algorithms that are loosely based on the functioning of biological neurons. ANNs complete calculations by passing information through a large number of interconnected nodes, each of which performs a relatively simple task. The regression model itself is stored in the form of weights between these nodes. Multi-Layer Perceptrons (MLPs; Sarle, 1994) are the most commonly used ANNs for LIBS, and have been shown to outperform PLS in some cases, particularly when there is significant nonlinearity in the quantity being predicted (Sirven et al., 2006; Ferreira et al., 2008). Other studies have showed performance similar to that of PLS (Anderson et al., 2011).

One of the fundamental assumptions of any regression model (univariate or multivariate) is that all of the data being modeled represent the same population (i.e., they have similar geochemical correlations and therefore similar matrix effects). However, for LIBS spectra from very different target materials (e.g., a calcium sulfate sample, a basalt sample, and a silica sinter sample), this assumption does not fully hold. When targets are too different, matrix effects will also be quite different and it can be difficult to find a single regression method that accurately predicts their compositions. This can be partially overcome by using "submodel" regression, a method in which multiple regression models are optimized using restricted ranges of the dataset, and then their results are blended together to improve overall performance (Anderson et al., 2017). Other similar methods, using clustering or other methods to create custom local regression models for unknown spectra, have been used in other types of spectroscopy (Cleveland & Devlin, 1988; Shenk et al., 1997) and may prove beneficial for LIBS analysis as well. Note that submodels and other local regression methods cannot generally correct for differences caused by different instruments or experimental conditions, which should be handled prior to this stage of analysis using calibration transfer methods as discussed earlier.

8.3.4.4 Calibration-Free LIBS

As an alternative to multivariate methods, which are usually strictly empirical and rely on the statistical properties of the LIBS spectra, it is possible to take a more theoretical approach to quantifying the composition of a target by inferring the conditions in the plasma that produced the spectrum. The most commonly used theory-based approach is called calibration-free (CF) LIBS. CF-LIBS assumes that the plasma is in local thermal equilibrium (LTE) and uses a Boltzmann plot of spectral lines to determine the plasma temperature and species concentration, and has been the subject of considerable effort (Ciucci et al., 1999; Tognoni et al., 2010). CF-LIBS is generally more successful when analyzing metal alloys but does not yet perform well on organic and geologic materials (Tognoni et al., 2010). However, CF-LIBS and the physics of LIBS plasmas are active areas of research, and the accuracy of CF-LIBS and related methods is likely to continue to improve. Alternatively, purely theoretical methods of predicting LIBS spectra are also under development (Sivakumar et al., 2013, 2014). Colgan et al. (2015, 2016) developed purely theoretical spectra of pure metals and a geochemical standard powder that resulted in an accurate relative composition of the major elements. These purely theoretical methods are highly computationally expensive and are still under development.

8.3.4.5 Accuracy and Precision

Finally, when interpreting quantitative results, it is important to distinguish between accuracy and precision. As discussed earlier, whenever possible, accuracy should be determined using an independent test set that the model (whether univariate or multivariate) has not been exposed to while tuning the model parameters, and is typically reported as RMSEP. If the available dataset is not sufficiently large to hold out a separate test set and still train a robust model, RMSECV can be used as an estimate of model accuracy. In general, the precision of LIBS quantitative analyses (i.e., the repeatability of a measurement of the same target) is considerably better than the accuracy (Blaney et al., 2014).

8.4 Sensitivities to Planetary Atmospheric Pressure

As is discussed in Sections 8.1 and 8.2, the intensity of a LIBS spectrum depends on many experimental parameters including the pulsed laser power and the atmospheric pressure in which the sample is probed. Consequently, the remote sensing LIBS instrument must be designed to specifically meet the scientific requirements of the planetary investigation. In this section, we will highlight some of the specific experimental and instrumental characteristics that are currently under consideration for terrestrial, Mars, and Venus applications. However, despite the variety of approaches to planetary instrumentation, data processing, and soil and rock quantitative analysis, some guidelines are nearly universal and are worth discussing here.

First, observational conditions, including the instrument used, atmospheric pressure and composition, stand-off distance, focus, and the efficiency with which the laser energy is transferred to the target can all have significant effects on the LIBS spectrum that is recorded. Any LIBS analysis should begin by inspecting the spectra themselves to ensure adequate signal-to-noise ratio and to identify anomalous spectra (e.g., cases in which continuum removal failed, resulting in a distorted spectrum). Spectra should also be inspected alongside quantitative results to ensure that diagnostic emission lines for the elements being studied are present and consistent with the quantitative prediction (Clegg et al., 2009). The NIST Atomic Emission Database is universally used to assign the elements responsible for a given LIBS spectral feature. However, the NIST database also contains so much information that it is easy to misinterpret the elements responsible for a spectral feature if the spectrum is not carefully wavelength calibrated as discussed in Section 8.3.1. While the NIST database contains empirical and theoretical emissions from highly ionized atoms, most LIBS spectra designed for remote planetary exploration do not produce more than doubly ionized atoms.

8.4.1 LIBS Analysis under Terrestrial Surface Atmospheric Conditions

LIBS instruments capable of remote geochemical analysis under terrestrial atmospheric conditions vary widely. First, the pulsed power of the laser used to generate the plasma is typically much higher than under any other planetary investigation because there are no power constraints and this depends on the distance from the instrument to the sample. LIBS analyses can be collected up to many tens of meters with standard Nd: YAG lasers with a nanosecond pulse width. Longer distances can be achieved when the laser pulsed energy is increased to more than 1 J/pulse (compared to ChemCam <15 mJ/pulse) and a larger telescope is used to focus the laser on to the sample and collect the emission.

Terrestrial LIBS instruments also tend to use a gated intensified charge coupled device (ICCD) detector rather than the simpler CCD detector. Gated detectors enable collection of a small temporal slice of the LIBS emission where the plasma tends to be more thermally stable. In contrast, CCDs are limited to relatively long ms exposures where the entire plasma emission is recorded and is characteristic of the thermal average of the plasma. ICCDs also have the advantage of collecting time-resolved LIBS spectra where different temporal parts of the plasma are observed. For example, the Russo group has been developing Laser Ablation Molecular Isotope Spectroscopy (LAMIS) where the ICCD is timed such that one can record the emission from the excited molecules that recombine within the plasma (Russo et al., 2012; Gonzalez et al., 2013). Spectral shifts due to the isotopic composition of the sample are larger and more easily resolved when they are in molecular form rather than as atomic species. As gated detectors such as ICCDs become more widely used in planetary exploration, methods such as LAMIS can produce valuable isotopic analysis of rocks and soils that could be used for geochronology (Cohen et al., 2014).

8.4.2 LIBS Analysis under Martian Surface Atmospheric Conditions

As was discussed in Section 8.1, LIBS chemical analysis under the 7 Torr martian surface atmospheric pressure is ideal. Consequently, the ChemCam instrument discussed in Chapter 29 can record quantifiable elemental compositions from a single laser shot with a simpler instrument architecture than would be required under other planetary atmospheres. First, a very compact laser that focuses <15 mJ/pulse onto the sample is enough to generate a plasma out to 7 m. The ChemCam instruments use a relatively small 110 mm diameter telescope to focus the LIBS laser and collect the emission. Finally, relatively small and simple crossed Czerny–Turner spectrometers with CCD detectors are sufficient to collect geochemical spectra (Maurice et al., 2012; Wiens et al., 2012; Chapter 29).

The SuperCam instrument selected for the Mars 2020 rover includes many enhancements compared to ChemCam. The SuperCam instrument uses a gated ICCD detector primarily to enable the remote Raman investigation that will directly measure the Mars mineralogy. However, this ICCD will also enable time-resolved LIBS analysis of part of the LIBS spectrum and perhaps enable a LAMIS-type isotopic analysis of some of the major elements. If gated detectors are used to record the entire LIBS spectrum on future Mars missions, then perhaps more routine geochronology measurements could be realized.

8.4.3 LIBS Analysis under Venus Surface Atmospheric Conditions

The Venus surface consists of 92 atm of primarily supercritical CO_2 at ~740 K. This thick atmosphere and the clouds that surround the planet significantly limit orbital analysis of the surface. Consequently, investigations of the Venus surface have been limited to the Venera and Vega missions by the Soviet Union between 1961 and 1983 that provide the best and only compositional information using X-ray diffraction.

Under Venus surface conditions, chemical (or mineralogic) analyses must be accomplished rapidly. A LIBS instrument is capable of making more than 1000 chemical measurements within 1 hour on the surface and produce the first measurement of the local chemical heterogeneity. However, data from such a LIBS instrument would require careful consideration to achieve the accuracy and precision obtained under terrestrial and martian atmospheric conditions.

Knight et al. (2000) were the first to conduct exploratory LIBS analyses under Venus pressures. They also collected images of the plasma generated at both 1 and 92 atm that document the production of the plasma under Venus pressures. This study demonstrated that the Venus pressure constrains the plasma expansion and produces an attenuated intensity. More recently, Clegg et al. (2014) completed LIBS quantitative analysis on many geochemical standards under both Venus pressure and temperature. This involved the use of a 2 m long, 110 mm diameter Venus chamber that enabled the simulation of remote LIBS analysis from within the safety of the lander.

A LIBS instrument capable of rapid chemical analyses from within the safety of a Venus lander requires many enhancements compared to one operating on a Mars rover. First, the technique requires more pulse energy on the Venus surface to generate a reproducible LIBS plasma. Second, the plasma lifetime under Venus conditions is <1 μs compared to the 10–20 μs lifetime observed under terrestrial or martian conditions. Consequently, gated detectors on a ChemCam-like crossed Czerny–Turner spectrometer body have the advantage of amplifying the relatively weaker LIBS signal while also gating out the noise that would be recorded with the long integration time CCDs. These enhancements should result in elemental accuracy, precision, and detection limits comparable to Mars exploration.

8.4.4 General LIBS Detection Limits

LIBS elemental quantitative analysis has made significant improvements over the last decade and has become competitive with many standard laboratory techniques. The detection limits of a LIBS analysis depend on the physics discussed in Section 8.2, on the

Table 8.1 *General LIBS detection limits*

Major elements	Si, Ti, Al, Fe	100–1000 ppm
	Mg, Ca, Na, K, O	5–100 ppm
Minor elements	Transition Metals	100–1000 ppm
	H, B, Ga-Se, In-Te, Tl-Bi	100–1000 ppm
	C, N, P, S, Br, I	0.1–1 wt.%
	F, Cl, noble gases	Difficult to excite, detect, and quantify

calibration model methods discussed in Section 8.3 and the planetary science operational environment discussed in Section 8.4. Table 8.1 contains a general summary of the detection limits for elements that are of most interest for planetary science independent of the environment. Generally, the alkali and alkali earth metals are the easiest to detect with <100 ppm detection limits. The noble gases and halides are the most difficult to detect by atomic emission. However, Forni et al. (2015) have demonstrated much better F and Cl detection limits when these species recombine with Ca in the plasma. Detection of C, N, O, P, and S presents many unique challenges to quantitative analysis. For example, C and O signals are both produced by the CO_2 atmospheres of Mars and Venus and cannot be easily distinguished from these elements within the ablated material. Furthermore, P and S produce relatively weak emission lines that are complicated by stronger, overlapping emission lines from other elements. Finally, the transition metals generally produce the largest number of emission lines and are generally responsible for the complex spectral structures. Transition metals generally have detection limits <1000 ppm.

8.5 Summary

The Mars chemical analyses observed by ChemCam on the Curiosity rover as discussed in Chapter 29 have demonstrated that LIBS has a transformational impact on planetary science. The LIBS technique is an accurate elemental analysis technique capable of in situ and remote analyses without the need for sample preparation. LIBS analyses are very rapid, limited only by the repetition rate of the laser. Whether on a Mars rover or a Venus lander, rapid chemical analyses are enabled, making it possible to acquire more than a thousand chemical measurements within an hour. While the LIBS technique emerged in 1983, the fundamental physics and the methods to exploit the chemical analysis continue to advance (e.g., Radziemski et al., 1983; Clegg et al., 2009; Anderson et al., 2016; Giguere et al., 2017). Many analytical techniques have been developed to exploit the matrix effects observed in LIBS spectra that result in a more complete understanding of the rocks and soils probed. The LIBS technique has now been selected for a second Mars rover mission, the Mars 2020 rover, and is mature enough for other planetary missions to Venus and the Moon.

Acknowledgments

The authors gratefully acknowledge the NASA MSL and Mars 2020 programs for funding the demonstration of the LIBS technique on Mars. SMC gratefully acknowledges support from the NASA MIDP and the Los Alamos National Laboratory LDRD, DR, and ER Programs.

References

Abdi H. (2003) Partial least square regression (PLS regression). *Encyclopedia for Research Methods for the Social Sciences*, **6**, 792–795.

Abrahamsson C., Johansson J., Sparén A., & Lindgren F. (2003) Comparison of different variable selection methods conducted on NIR transmission measurements on intact tablets. *Chemometrics and Intelligent Laboratory Systems*, **69**, 3–12.

Anderson R.B., Morris R., Clegg S., Bell III J., Humphries S., & Wiens R. (2011a) A comparison of multivariate and pre-processing methods for quantitative Laser-Induced Breakdown Spectroscopy of geologic samples. *42nd Lunar Planet. Sci. Conf.*, Abstract #1308.

Anderson R.B., Morris R.V., Clegg S.M., et al. (2011b) The influence of multivariate analysis methods and target grain size on the accuracy of remote quantitative chemical analysis of rocks using laser induced breakdown spectroscopy. *Icarus*, **215**, 608–627.

Anderson R.B., Bell III J.F., Wiens R.C., Morris R.V., & Clegg S.M. (2012) Clustering and training set selection methods for improving the accuracy of quantitative laser induced breakdown spectroscopy. *Spectrochimica Acta B: Atomic Spectroscopy*, **70**, 24–32.

Anderson R.B., Clegg S.M., Frydenvang J., et al. (2016) Improved accuracy in quantitative Laser-Induced Breakdown Spectroscopy using sub-model partial least squares. *Spectrochimica Acta B: Atomic Spectroscopy*, **129**, 49–57.

Balabin R.M. & Smirnov S.V. (2011) Variable selection in near-infrared spectroscopy: Benchmarking of feature selection methods on biodiesel data. *Analytica Chimica Acta*, **692**, 63–72.

Blaney D.L., Wiens R.C., Maurice S., et al. (2014) Chemistry and texture of the rocks at Rocknest, Gale crater: Evidence for sedimentary origin and diagenetic alteration. *Journal of Geophysical Research*, **119**, 2109–2131.

Boucher T., Carey C.J., Dyar M.D., Mahadevan S., Clegg S., & Wiens R. (2015a) Manifold preprocessing for Laser-Induced Breakdown Spectroscopy under Mars conditions. *Journal of Chemometrics*, **29**, 484–491.

Boucher T., Dyar M.D., Carey C.J., et al. (2015b) Calibration transfer of LIBS spectra to correct for Mars-Earth lab differences. *46th Lunar Planet. Sci. Conf.*, Abstract #2773.

Carroll P. & Kennedy E. (1981) Laser-produced plasmas. *Contemporary Physics*, **22**, 61–96.

Chaleard C., Mauchien P., Andre N., Uebbing J., Lacour J., & Geertsen C. (1997) Correction of matrix effects in quantitative elemental analysis with laser ablation optical emission spectrometry. *Journal of Analytical Atomic Spectrometry*, **12**, 183–188.

Ciucci A., Corsi M., Palleschi V., Rastelli S., Salvetti A., & Tognoni E. (1999) New procedure for quantitative elemental analysis by laser-induced plasma spectroscopy. *Applied Spectroscopy*, **53**, 960–964.

Clegg S.M., Sklute E., Dyar M.D., Barefield J.E., & Wiens R.C. (2009) Multivariate analysis of remote Laser-Induced Breakdown Spectroscopy spectra using partial least squares, principal component analysis, and related techniques. *Spectrochimica Acta B: Atomic Spectroscopy*, **64**, 79–88.

Clegg S.M., Wiens R., Misra A.K., et al. (2014) Planetary geochemical investigations using Raman and Laser-Induced Breakdown Spectroscopy. *Applied Spectroscopy*, **68**, 925–936.

Clegg S.M., Wiens R.C., Anderson R., et al. (2017) Recalibration of the Mars Science Laboratory ChemCam instrument with an expanded geochemical database. *Spectrochimica Acta B: Atomic Spectroscopy*, **129**, 64–85.

Cleveland W.S. & Devlin S.J. (1988) Locally weighted regression: An approach to regression analysis by local fitting. *Journal of the American Statistical Association*, **83**, 596–610.

Cohen B.A., Miller J.S., Li Z.-H., Swindle T.D., & French R.A. (2014) The potassium-argon laser experiment (KArLE): In situ geochronology for planetary robotic missions. *Geostandards and Geoanalytical Research*, **38**, 421–439.

Colgan J., Judge E.J., Johns H.M., et al. (2015) Theoretical modeling and analysis of the emission spectra of a ChemCam standard: Basalt BIR-1A. *Spectrochimica Acta B: Atomic Spectroscopy*, **110**, 20–30.

Colgan J., Barefield J., Judge E.J., et al. (2016) Experimental and theoretical studies of Laser-Induced Breakdown Spectroscopy emission from iron oxide: Studies of atmospheric effects. *Spectrochimica Acta B: Atomic Spectroscopy*, **122**, 85–92.

Cremers D. & Radziemski L.J. (2013) *Handbook of Laser-Induced Breakdown Spectroscopy*. John Wiley & Sons, Oxford.

Dyar M.D., Tucker J., Humphries S., Clegg S.M., Wiens R.C., & Lane M.D. (2011) Strategies for Mars remote Laser-Induced Breakdown Spectroscopy analysis of sulfur in geological samples. *Spectrochimica Acta B: Atomic Spectroscopy*, **66**, 39–56.

Dyar M.D., Carmosino M.L., Breves E.A., Ozanne M.V., Clegg S.M., & Wiens R.C. (2012a) Comparison of partial least squares and lasso regression techniques as applied to Laser-Induced Breakdown Spectroscopy of geological samples. *Spectrochimica Acta B: Atomic Spectroscopy*, **70**, 51–67.

Dyar M.D., Carmosino M.L., Tucker J.M., et al. (2012b) Remote Laser-Induced Breakdown Spectroscopy analysis of East African Rift sedimentary samples under Mars conditions. *Chemical Geology*, **294–295**, 135–151.

Dyar M.D., Fassett C.I., Giguere S., et al. (2016) Comparison of univariate and multivariate models for prediction of major and minor elements from laser-induced breakdown spectra with and without masking. *Spectrochimica Acta B: Atomic Spectroscopy*, **123**, 93–104.

Eppler A.S., Cremers D.A., Hickmott D.D., Ferris M.J., & Koskelo A.C. (1996) Matrix effects in the detection of Pb and Ba in soils using Laser-Induced Breakdown Spectroscopy. *Applied Spectroscopy*, **50**, 1175–1181.

Fabre C., Cousin A., Wiens R., et al. (2014) In situ calibration using univariate analyses based on the onboard ChemCam targets: First prediction of martian rock and soil compositions. *Spectrochimica Acta B: Atomic Spectroscopy*, **99**, 34–51.

Ferreira E.C., Milori D.M., Ferreira E.J., Da Silva R.M., & Martin-Neto L. (2008) Artificial neural network for Cu quantitative determination in soil using a portable laser induced breakdown spectroscopy system. *Spectrochimica Acta B: Atomic Spectroscopy*, **63**, 1216–1220.

Feudale R.N., Woody N.A., Tan H., Myles A.J., Brown S.D., & Ferré J. (2002) Transfer of multivariate calibration models: A review. *Chemometrics and Intelligent Laboratory Systems*, **64**, 181–192.

Forni O., Maurice S., Gasnault O., et al. (2013) Independent component analysis classification of Laser-Induced Breakdown Spectroscopy spectra. *Spectrochimica Acta B: Atomic Spectroscopy*, **86**, 31–41.

Forni O., Gaft M., Toplis M.J., et al. (2015) First detection of fluorine on Mars: Implications for Gale crater's geochemistry. *Geophysical Research Letters*, **42**, 1020–1028.

Giguere S., Carey C.J., Boucher T., Mahadevan S., & Dyar M.D. (2015) An optimization perspective on baseline removal for spectroscopy. *Proceedings of the 5th IJCAI Workshop on Artificial Intelligence in Space*.

Giguere S., Boucher T., Carey C.J., Mahadevan S., & Dyar M.D. (2017) A fully customized baseline removal framework for spectroscopic applications. *Applied Spectroscopy*, **71**, 1457–1470.

Gonzalez J.J., Chirinos J.R., Dong M., et al. (2013) Simultaneous Laser Ablation Molecular Isotopic Spectrometry (LAMIS), Laser-Induced Breakdown Spectroscopy (LIBS) and Laser Ablation Inductively Coupled Plasma Spectrometry (LA-ICP-MS) for elemental analysis of geological samples. *Mineralogical Magazine*, **77**(5), Abstract #1193.

Hahn D.W. & Omenetto N. (2010) Laser-Induced Breakdown Spectroscopy (LIBS), Part I: Review of basic diagnostics and plasma–particle interactions: still-challenging issues within the analytical plasma community. *Applied Spectroscopy*, **64**, 335A–366A.

Jain A.K., Murty M.N., & Flynn P.J. (1999) Data clustering: A review. *ACM Computing Surveys (CSUR)*, **31**, 264–323.

Johns H., Kilcrease D., Colgan J., et al. (2015) Improved electron collisional line broadening for low-temperature ions and neutrals in plasma modeling. *Journal of Physics B: Atomic, Molecular and Optical Physics*, **48**, 224009.

Jolliffe I.T. (1982) A note on the use of principal components in regression. *Applied Statistics*, **31**, 300–303.

King B. (1967) Step-wise clustering procedures. *Journal of the American Statistical Association*, **62**, 86–101.

Knight A.K., Scherbarth N.L., Cremers D.A., & Ferris M.J. (2000) Characterization of Laser-Induced Breakdown Spectroscopy (LIBS) for application to space exploration. *Applied Spectroscopy*, **54**, 331–340.

Kochelek K.A., McMillan N.J., McManus C.E., & Daniel D.L. (2015) Provenance determination of sapphires and rubies using Laser-Induced Breakdown Spectroscopy and multivariate analysis. *American Mineralogist*, **100**, 1921–1931.

Labutin T.A., Lednev V.N., Ilyin A.A., & Popov A.M. (2016) Femtosecond Laser-Induced Breakdown Spectroscopy. *Journal of Analytical Atomic Spedctrometry*, **31**, 90–118.

Lasue J., Wiens R., Clegg S., et al. (2012) Remote Laser-Induced Breakdown Spectroscopy (LIBS) for lunar exploration. *Journal of Geophysical Research*, **117**, DOI:10.1029/2011JE003898.

Lasue J., Clegg S.M., Forni O., et al. (2016) Observation of >5 wt % zinc at the Kimberley outcrop, Gale crater, Mars. *Journal of Geophysical Research*, **121**, 338–352.

Leardi R. & Gonzalez A.L. (1998) Genetic algorithms applied to feature selection in PLS regression: How and when to use them. *Chemometrics and Intelligent Laboratory Systems*, **41**, 195–207.

MacKay D.J.C. (2003) *Information theory, inference, and learning algorithms*. Cambridge University Press, Cambridge.

Maurice S., Wiens R., Saccoccio M., et al. (2012) The ChemCam Instrument Suite on the Mars Science Laboratory (MSL) rover: Science objectives and mast unit description. *Space Science Reviews*, **170**, 95–166.

Melikechi N., Mezzacappa A., Cousin A., et al. (2014) Correcting for variable laser-target distances of Laser-Induced Breakdown Spectroscopy measurements with ChemCam using emission lines of martian dust spectra. *Spectrochimica Acta B: Atomic Spectroscopy*, **96**, 51–60.

Mezzacappa A., Melikechi N., Cousin A., et al. (2016) Application of distance correction to ChemCam Laser-Induced Breakdown Spectroscopy measurements. *Spectrochimica Acta B: Atomic Spectroscopy*, **120**, 19–29.

Miziolek A.W., Palleschi V., & Schechter I. (2006) *Laser-Induced Breakdown Spectroscopy (LIBS): Fundamentals and applications*. Cambridge University Press, Cambridge.

Ollila A.M., Newsom H.E., Clark B., et al. (2014) Trace element geochemistry (Li, Ba, Sr, and Rb) using Curiosity's ChemCam: Early results for Gale crater from Bradbury Landing Site to Rocknest. *Journal of Geophysical Research*, **119**, 255–285.

Pokrajac D., Lazarevic A., Kecman V., et al. (2014) Automatic classification of Laser-Induced Breakdown Spectroscopy (LIBS) data of protein biomarker solutions. *Applied Spectroscopy*, **68**, 1067–1075.

Radziemski L.J., Loree T.R., & Cremers D.A. (1983) Laser-Induced Breakdown Spectroscopy (LIBS): A new spectrochemical technique. In: *Optical and laser remote sensing* (D.K. Killinger & A. Mooradian, eds.). Springer-Verlag, Berlin and Heidelberg, 303–307.

Rosipal R. & Krämer N. (2006) Overview and recent advances in partial least squares. In: *Subspace, latent structure and feature selection* (C. Saunders, M. Grobelnik, S. Gunn, & J. Shawe-Taylor, eds.). SLSFS 2005. Lecture Notes in Computer Science, 3940. Springer-Verlag, Berlin and Heidelberg, 34–51.

Russo R.E., Mao X.L., Bol'shakov A.A., & Yoo J. (2012) Real-time elemental and isotopic analysis at atmospheric pressure in a laser ablation plasma. *Goldschmidt*, **76**, 2308.

Sarle W.S. (1994) Neural networks and statistical models. *Proceedings of the 19th Annual SAS Users Group International Conference*, 1538–1550.

Schlenke J., Hildebrand L., Moros J., & Laserna J.J. (2012) Adaptive approach for variable noise suppression on Laser-Induced Breakdown Spectroscopy responses using stationary wavelet transform. *Analytica Chimica Acta*, **754**, 8–19.

Schröder S., Meslin P.-Y., Gasnault O., et al. (2015) Hydrogen detection with ChemCam at Gale crater. *Icarus*, **249**, 43–61.

Shenk J.S., Westerhaus M.O., & Berzaghi P. (1997) Investigation of a LOCAL calibration procedure for near infrared instruments. *Journal of Near Infrared Spectroscopy*, **5**, 223–232.

Shi Q., Niu G., Lin Q., Xu T., Li F., & Duan Y. (2015) Quantitative analysis of sedimentary rocks using Laser-Induced Breakdown Spectroscopy: Comparison of support vector regression and partial least squares regression chemometric methods. *Journal of Analytical Atomic Spectrometry*, **30**, 2384–2393.

Singh J.P. & Thakur S. (2007) *Laser-Induced Breakdown Spectroscopy*. Elsevier, Philadelphia.

Sirven J.B., Bousquet B., Canioni L., & Sarger L. (2006) Laser-Induced Breakdown Spectroscopy of composite samples: Comparison of advanced chemometrics methods. *Analytical Chemistry*, **78**, 1462–1469.

Sivakumar P., Taleh L., Markushin Y., Melikechi N., & Lasue J. (2013) An experimental observation of the different behavior of ionic and neutral lines of iron as a function of number density in a binary carbon–iron mixture. *Spectrochimica Acta B: Atomic Spectroscopy*, **82**, 76–82.

Sivakumar P., Taleh L., Markushin Y., & Melikechi N. (2014) Packing density effects on the fluctuations of the emission lines in Laser-Induced Breakdown Spectroscopy. *Spectrochimica Acta B: Atomic Spectroscopy*, **92**, 84–89.

Smola A.J. & Schölkopf B. (2004) A tutorial on support vector regression. *Statistics and Computing*, **14**, 199–222.

Sneath P.H. & Sokal R.R. (1973) *Numerical taxonomy: The principles and practice of numerical classification*. W.H. Freeman, New York.

Starck J.L., Pantin E., & Murtagh F. (2002) Deconvolution in astronomy: A review. *Publications of the Astronomical Society of the Pacific*, **114**, 1051.

Tognoni E., Cristoforetti G., Legnaioli S., & Palleschi V. (2010) Calibration-Free Laser-Induced Breakdown Spectroscopy: State of the art. *Spectrochimica Acta B: Atomic Spectroscopy*, **65**, 1–14.

Tucker J., Dyar M., Schaefer M., Clegg S., & Wiens R. (2010) Optimization of Laser-Induced Breakdown Spectroscopy for rapid geochemical analysis. *Chemical Geology*, **277**, 137–148.

Vance T., Pokrajac D., Lazarevic A., et al. (2010) Classification of LIBS protein spectra using multilayer perceptrons. *Transactions on Mass-Data Analysis of Images and Signals*, **2**, 96–111.

Wang Y., Lysaght M.J., & Kowalski B.R. (1992) Improvement of multivariate calibration through instrument standardization. *Analytical Chemistry*, **64**, 562–564.

Ward J.H. Jr. (1963) Hierarchical grouping to optimize an objective function. *Journal of the American Statistical Association*, **58**, 236–244.

Wiens R., Maurice S., Barraclough B., et al. (2012) The ChemCam Instrument Suite on the Mars Science Laboratory (MSL) rover: Body unit and combined system tests. *Space Science Reviews*, **170**, 167–227.

Wiens R.C., Maurice S., Lasue J., et al. (2013) Pre-flight calibration and initial data processing for the ChemCam Laser-Induced Breakdown Spectroscopy instrument on the Mars Science Laboratory Rover. *Spectrochimica Acta B: Atomic Spectroscopy*, **82**, 1–27.

Wisbrun R., Schechter I., Niessner R., Schroeder H., & Kompa K.L. (1994) Detector for trace elemental analysis of solid environmental samples by laser plasma spectroscopy. *Analytical Chemistry*, **66**, 2964–2975.

Wold S. & Sjöström M. (1977) Method for analyzing chemical data in terms of similarity and analogy. *Chemometrics: Theory and application.* (B.R. Kowalski, ed.). ACS Symposium Series. American Chemical Society, Washington, DC, 243–282.

Wold S., Sjöström M., & Eriksson L. (2001) PLS-regression: A basic tool of chemometrics. In: *Chemometrics and Intelligent Laboratory Systems*, **58**, 109–130.

Yaroshchyk P., Death D., & Spencer S. (2012) Comparison of principal components regression, partial least squares regression, multi-block partial least squares regression, and serial partial least squares regression algorithms for the analysis of Fe in iron ore using LIBS. *Journal of Analytical Atomic Spectrometry*, **27**, 92–98.

Zhang B., Sun L., Yu H., Xin Y., & Cong Z. (2013) Wavelet denoising method for Laser-Induced Breakdown Spectroscopy. *Journal of Analytical Atomic Spectrometry*, **28**, 1884–1893.

Zhang P. (1993) Model selection via multifold cross validation. *The Annals of Statistics*, **21**(1), 299–313.

9

Neutron, Gamma-Ray, and X-Ray Spectroscopy
Theory and Applications

THOMAS H. PRETTYMAN, PETER A. J. ENGLERT, AND NAOYUKI YAMASHITA

9.1 Introduction

Neutron, gamma-ray, and X-ray spectroscopy are remote sensing methods used to determine the elemental composition of the solid surfaces of planetary bodies and their atmospheres. As we shall see, these methods have much in common; however, the depths probed and sensitivity to elemental composition depend on a number of factors, including the interrogation mode (active vs. passive), the type of spectrometer used, and how it is deployed and operated. All require acquisition of data in close proximity to the target, for example, in a low-altitude orbit or on the surface of a planetary body. Nuclear spectroscopy refers to the measurement of neutrons, gamma rays, and other subatomic particles that originate from the nucleus. X-rays are created by atomic transitions and by the deceleration of charged particles (Bremsstrahlung).

Neutrons and gamma rays are highly penetrating and are sensitive to subsurface composition to depths of about a meter (e.g., Prettyman, 2014), whereas X-rays are sensitive to submillimeter depths. Nuclear and X-ray spectroscopy techniques are insensitive to the chemical form of planetary materials. This contrasts with optical reflectance measurements (e.g., visible to near-infrared spectroscopy), which are sensitive to the rotational and vibrational modes of molecules and minerals within the uppermost few hundreds of μm of the surface. Nuclear and X-ray spectroscopy techniques can detect and quantify the concentration of specific elements, regardless of the molecular or mineralogical species to which the element is bound.

Gamma rays are generated within the regolith of Solar System bodies (planets, moons, asteroids, and comets) by the decay of natural radioelements K, Th, and U. For bodies with little or no intervening atmosphere (e.g., Mercury, Mars, the Moon, and small Solar System bodies), neutrons and gamma rays are made by the steady bombardment of the solid surface by galactic cosmic rays (GCRs). Secondary particles produced by GCR showers in the regolith, especially neutrons, undergo nuclear reactions to produce gamma rays. The radiation can be measured by a spectrometer deployed on an orbiter, lander, rover, or sonde. Nuclear spectroscopy has been successfully used on missions to the Moon, Venus, Mars, Mercury, 433 Eros, 4 Vesta, and the dwarf planet Ceres (e.g., Metzger et al., 1973; Surkov et al., 1987; Evans et al., 2001; Feldman et al., 2001; Boynton et al., 2007;

Peplowski et al., 2011; Prettyman et al., 2012, 2017; Lawrence et al., 2013b). Chapter 30 summarizes missions using nuclear and X-ray spectrometers.

The spectrum of gamma rays escaping planetary surfaces provides information about the composition of the regolith to depths of about a meter. Elements that make up igneous rocks, aqueous alteration products, and ices can be analyzed, including H, C, O, Mg, Al, Si, S, Cl, K, Ca, Ti, Fe, and Ni. For bodies with thick atmospheres (>1000 g/cm^2), such as Earth, Venus, and Titan, the flux of cosmic rays at the surface is small. Consequently, active interrogation with a neutron generator or radioisotope neutron source is needed to analyze elements other than K, Th, and U on planetary bodies with thick atmospheres. Neutron measurements provide constraints on elemental composition, including the average atomic mass and neutron absorption cross section of the regolith. Neutron spectroscopy is highly sensitive to the presence of hydrogen and can be used to characterize layering of hydrogen (e.g., in the form of subsurface water ice).

X-ray fluorescence experiments have been part of several orbital missions to the Moon, planets, and asteroids providing information on their chemical composition. The Moon was the first target for orbital XRF (Mandel'shtam et al., 1968) during the Apollo era (Adler et al., 1972a,b). This was followed many years later by 433 Eros, 25143 Itokawa, and Mercury (Nittler et al., 2001, 2011; Arai et al., 2008) and missions to the Moon by Europe and India (e.g., Grande et al., 2007; Narendranath et al., 2011). The most recent XRF remote sensing study of a planet was the MESSENGER mission to Mercury (e.g., Weider et al., 2012a).

The emission of fluorescence X-rays from planetary surfaces is dependent on the properties of the excitation source and surface composition. In the case of rover experiments, excitation sources can be deployed. For orbital experiments, solar X-rays and, sometimes, low-energy charged particles, induce the signal to be measured and analyzed. The solar X-ray and particle spectra are variable in intensity and shape on time scales of minutes to hours. X-ray fluorescence remote sensing experiments therefore require continuous monitoring of the solar spectrum, and their success depends on the Sun's cooperation. Major elements Mg, Al, Si, S, Ca, Ti, and Fe can be identified by their Kα lines with energies between 0.5 keV and 10 keV, and abundances of these elements can be determined from their respective line intensities.

With nuclear and X-ray spectroscopy, active and passive interrogation modes are feasible, with active interrogation requiring very close proximity to the target (e.g., from contact to meter-scale standoff distances). Examples of active interrogation include the use of a pulsed neutron generator to measure bulk regolith hydrogen content (e.g., Litvak et al., 2008, 2016; Hardgrove et al., 2011) and radioisotope sources, such as alpha-particle induced X-ray spectroscopy (APXS, see Chapter 28), to measure the elemental composition of rocks and soil (e.g., Clark et al., 1977; Surkov Yu et al., 1984; Rieder et al., 1997). In this chapter, we focus our attention on passive measurements, which require relatively simple instrumentation and afford greater flexibility in their deployment. Because the underlying theory is the same, our emphasis of passive measurements provides a point of entry for those unfamiliar with nuclear and X-ray spectroscopy.

This chapter provides a general introduction to nuclear and X-ray spectroscopy for readers with a background in science and technology. Basic radiation production and interaction mechanisms are summarized and the information conveyed by neutrons, gamma rays, and X-rays is described. Methods to calculate the leakage flux of GCR-induced neutrons and gamma rays and flux of particles at the detector are presented. Monitoring solar X-ray input spectra, and the effect of solar incidence and instrument viewing angles, are described. Advanced topics, including peak fitting, mapping, spatial deconvolution, shape and topography corrections, are discussed.

9.2 Particle Production and Interactions with Matter

9.2.1 Neutron Production by Galactic Cosmic Rays

Neutrons are slightly heavier than protons, but – as their name implies – have no charge (Evans, 1955). Neutrons and protons are tightly bound within the nucleus; however, the bombardment of nuclei by energetic, subatomic particles can liberate nucleons. Neutrons produced by GCR interactions with regolith materials enable passive measurements of surface composition.

The solid surfaces of bodies with little or no atmosphere, such as asteroids, the Moon, Mercury, and Mars, are directly exposed to GCRs. GCRs are high-energy ions that originate outside the Solar System, primarily from galactic sources, and arrive steadily from all directions (e.g., Reedy et al., 1983). GCRs include heavy ions, including an actinide component (Donnelly et al., 2012); however, most are protons and alpha particles (respectively, 88.6% and 10.2% of the ions) (Figure 9.1a, inset). The flux of GCR ions reaching the inner Solar System is modulated by the magnetic field of the Sun embedded in the solar wind. GCR fluxes are highest during periods of low solar activity, when there is minimal resistance to the inflow of particles from the interstellar medium (see Section 9.3.2 for a rigorous definition of flux). The convection and drift of GCR ions through the heliosphere is described by the Fokker–Planck equation for which approximate solutions are available (Gleeson & Axford, 1968). The shape of the cosmic ray spectrum can be described by a single parameter, the modulation potential, which is the energy loss experienced by particles that penetrate the heliosphere. For illustration, differential ion fluxes at 1 AU were calculated using the Badhwar–O'Neill model (O'Neill, 2010) (Figure 9.1a). Below kinetic energies of about 10 GeV/nucleon the GCR flux is modulated by the strength of the interplanetary magnetic field, which varies with the solar cycle (Figures 9.1 and 9.2). At very high energies, the GCR flux is unperturbed.

GCRs can penetrate deep into the regolith, interacting via electronic, radiative, and nuclear processes. Occasional direct collisions between GCRs and nuclei produce a spray of secondary particles, primarily neutrons and protons. Particles are produced by spallation of the target nucleus followed by evaporation of the collisional fragments. This process is repeated multiple times, producing a particle shower (e.g., Figure 9.1b). Depending on the kinetic energy and mass of the incident ion, the shower can extend to depths of several

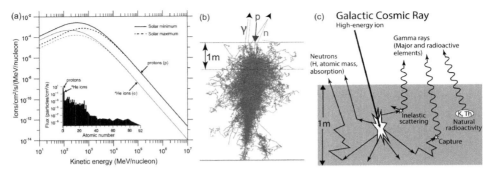

Figure 9.1 Flux of GCRs. (a) The differential flux of GCR protons and ^{4}He ions arriving at 1 AU is shown for solar minimum and maximum. Fluxes were estimated using the Badhwar–O'Neill model (O'Neill, 2010). The ion fluxes for solar minimum are representative of conditions on Jan. 20, 2009 whereas fluxes for solar maximum are representative of June 1, 2004 (Figure 9.2 shows how GCR fluxes vary with solar activity). The inset shows the relative abundance of ion species. Nearly 99% of GCR nuclei are protons and alpha particles (^{4}He ions). (b) Regolith shower illustrated: A high-energy (100 GeV) proton (p) smashes into a planetary surface producing a spray of secondary particles, including gamma rays (green) and neutrons (magenta). The shower extends to depths of several meters. (c) Particles escaping the surface convey information about the elemental composition. On average, the escaping particles sample depths up to a meter.

meters; however, the particles that escape into space originate from the uppermost meter of the planetary surface (Figure 9.1c).

The number of secondary particles produced generally increases with the atomic mass of the target nuclei. For lunar materials, Gasnault et al. (2001) found that the leakage flux of fast neutrons (kinetic energies greater than about 1 MeV) was strongly correlated with the average atomic mass of the regolith, given by

$$\langle A \rangle = \sum_{i=0}^{n-1} f_i A_i, \tag{9.1}$$

where n is the number of atomic constituents, and f_i and A_i are the atom fraction and atomic mass, respectively, of the ith constituent. Gasnault et al. (2001) used the correlation to map the average atomic mass of the lunar regolith, providing a constraint on elemental composition. The same approach was used to map the average atomic mass of Vesta (Lawrence et al., 2013a; Beck et al., 2015).

Energetic neutrons produced in cosmic ray showers undergo successive collisions with nuclei in planetary materials, gradually losing kinetic energy and approaching thermal equilibrium with their surroundings. Neutrons are ultimately absorbed by nuclei in the regolith; however, a portion of the neutrons escape into space. The energy distribution of escaping neutrons is sensitive to the composition of the regolith, including the average atomic mass, the concentration of light elements (such as H and C), and strong neutron absorbers, such as rare-earth elements, and the rock-forming elements Cl, Ti, and Fe. In

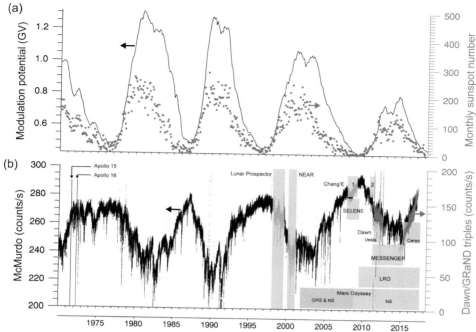

Figure 9.2 The solar cycle modulates the flux of GCRs, resulting in temporal variations in the production of secondary neutrons and gamma rays. (a) The intensity of GCRs below a few GeV/nucleon depends on the modulation potential, the energy loss experienced by particles as they penetrate the heliosphere. Modulation varies with solar activity, as measured by sunspot counts, during the roughly 11-year solar cycle. (b) During solar minimum when the Sun's magnetic fields are weakest, more low-energy GCRs reach 1 AU (Figure 9.1a), resulting in increased production of secondary particles, as measured by neutron monitors at McMurdo Station in Antarctica. The GCR flux varies weakly with heliocentric distance. The relative variation in the GCR flux as measured by Dawn's Gamma Ray and Neutron Detector (GRaND), between 2.7 and 3 AU, closely follows the McMurdo counting data. Although measurements during solar minimum or during a weak solar cycle are advantageous, nuclear spectroscopy has been proven for the full range of solar conditions. Successful missions are shown in (b).

GCR showers, gamma rays are made primarily by neutron collisions. The spectrum of gamma rays escaping planetary surfaces can be analyzed to determine the concentration of selected elements. Gamma-ray production is discussed in Section 9.2.3.

9.2.2 Neutron Moderation and Transport

Free neutrons collide with nuclei in the regolith, losing energy in successive collisions. They are ultimately absorbed in the regolith or escape into space. Neutrons travel in straight lines until they collide or escape. The probability that a neutron will collide is determined by the number density N and microscopic cross section of the target nuclei σ. The microscopic cross section, a property of the atomic nucleus, is the effective area projected by a nucleus and has units of barns (1 barn = 10^{-24} cm^2). The magnitude of σ depends on the target, reaction type, and kinetic energy of the neutron (Figure 9.3).

The probability of a collision per unit path length is given by the macroscopic cross section, $\Sigma = N\sigma$ (cm^{-1}), such that $e^{-\Sigma s}$ is the probability that a neutron will travel a distance

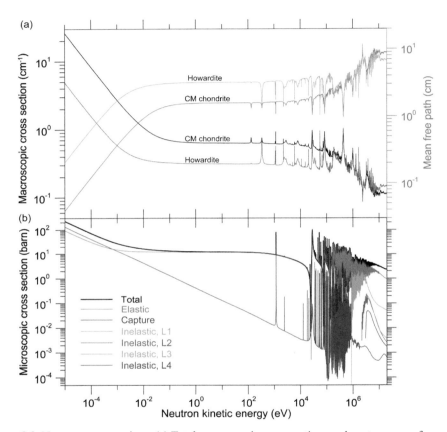

Figure 9.3 Neutron cross sections. (a) Total macroscopic cross sections and neutron mean free paths are plotted for regolith materials, howardite and CM chondrite, with compositions from Lodders and Fegley (1998). The howardite is hydrogen free, whereas the CM chondrite contains 1.4 wt.% hydrogen. This results in an increased cross section for elastic scattering, most noticeable for neutrons in the thermal and epithermal energy range. Compositional differences are also reflected in the pattern of resonances (sharp peaks) at high energy. (b) Microscopic cross sections for ^{56}Fe are shown. In the epithermal range (~0.5 eV to a few hundred keV), the cross section is dominated by elastic scattering. At very low neutron energies, radiative capture, which varies inversely with neutron speed, is important. High-energy neutrons probe the structure of the nucleus. Neutrons with kinetic energies greater than 1 MeV undergo resonance elastic scattering and inelastic scattering. Inelastic scattering is a threshold reaction, for which the kinetic energy of the neutron exceeds the energy of the first excited state of the residual nucleus (see Figure 9.5).

s (cm) without colliding. This is an example of the Beer-Lambert law, which describes the attenuation of neutral particles, including neutrons and photons. Macroscopic cross sections and attenuation lengths are shown in Figure 9.3. High-energy neutrons (>100 MeV) made directly by cosmic ray interactions can travel large distances (cm to m) between collisions.

Three types of neutron interactions are important for planetary applications: (1) inelastic scattering; (2) elastic scattering; and (3) radiative neutron capture. With inelastic scattering, the incident neutron is absorbed to form a compound nucleus. The compound nucleus decays by the emission of a neutron, leaving the residual nucleus in an excited state, which transitions to ground state by emitting characteristic gamma rays. The energies and relative intensities of the emitted gamma rays can be used to identify the target nucleus. For inelastic collisions, the kinetic energy of the incident neutron is not conserved. Inelastic scattering is significant for neutrons with kinetic energy greater than a few MeV (Figure 9.3). Inelastic scattering is an important energy-loss mechanism for fast neutrons.

Elastic scattering is the dominant energy loss mechanism for neutrons with kinetic energies below about 1 MeV. These are billiard-ball–like collisions, in which the kinetic energy of the incident neutron is conserved. A neutron striking a hydrogen nucleus (proton) head on results in the transfer of nearly all of the neutron's kinetic energy to the proton. For any other target, the mass of the nucleus is substantially larger than that of the neutron, and the neutron will retain a portion of its initial kinetic energy. Consequently, neutrons lose energy more rapidly in hydrogen-rich regolith materials than in anhydrous materials. As a result, neutrons are a sensitive probe of regolith hydrogen content (e.g., Feldman et al., 2001, 2004a; Prettyman et al., 2012, 2017; Lawrence et al., 2013b).

Neutron slowing down in the elastic regime has been the subject of decades of study, particularly in nuclear reactor physics (e.g., Duderstadt & Hamilton, 1976). Fermi's slowing down theory (Fermi, 1950) provides some insights useful for the analysis of planetary data (e.g., Feldman et al., 2000; Prettyman et al., 2011). The theory is valid for a monoenergetic source of fast neutrons with energy E_0 with a volumetric production rate of S_0 (neutrons/cm^3/s). The neutrons slow down via successive elastic collisions in an infinite medium. Neutron absorption is ignored. Under these conditions, the equilibrium flux of neutrons is given by

$$\varphi(E) = S_0/(\xi \Sigma_S E) \quad (\text{neutrons/cm}^2/\text{s/eV}), \tag{9.2}$$

where E is the kinetic energy of the neutron, Σ_S is the macroscopic cross section for elastic scattering, and ξ is the mean lethargy gain per collision (flux is defined in Section 9.3.2). Lethargy is defined as $\ln(E_0/E)$ and thus increases with decreasing neutron kinetic energy. For neutrons in the elastic regime, both the cross section for elastic scattering and ξ are approximately constant with neutron energy and their values are tabulated for all elements of interest in reactor physics texts (e.g., Duderstadt & Hamilton, 1976). For elemental mixtures, these parameters are calculated as follows:

$$\langle \xi\Sigma_s \rangle = \sum_{i=0}^{n-1} N_i \sigma_i \xi_i, \quad \Sigma_s = \sum_{i=0}^{n-1} N_i \sigma_i, \text{ and } \xi = \langle \xi\Sigma_s \rangle / \Sigma_s, \qquad (9.3)$$

where N_i, σ_i, and ξ_i are respectively, the atom density (atoms/cm^3), microscopic scattering cross section (cm^2), and mean lethargy gain per collision (dimensionless) for element i of n (Prettyman et al., 2011). The atom density is given by $N_i = N_A \rho w_i / A_i$, where N_A is Avogadro's number (atoms/mol), ρ is the bulk material density (g/cm^3), and w_i and A_i are, respectively, the concentration (g/g) and atomic mass (g/mol) of element i.

Fermi's slowing down theory predicts the number of collisions on average required to change the energy of a neutron from E_0 to E, given by $C = \ln(E/E_0)/\xi$. This expression is useful for visualizing the effectiveness of hydrogen at moderating neutrons, i.e., bringing fast neutrons into thermal equilibrium with surface materials. The mean lethargy gains per collision for H, C, and O are respectively 1, 0.16, and 0.12 and their microscopic scattering cross sections are 38, 4.8, and 4.2 barns, respectively (Duderstadt & Hamilton, 1976). Using Eq. (9.3), we find that $\xi = 0.95$ for water and $\xi = 0.13$ for carbon dioxide. These are found as ice on some planetary surfaces, including the high-latitude regions of Mars. A neutron slowing down in dry ice requires seven times the number of collisions to lose the same amount of energy as in water ice. For dry silicates, the mean lethargy gain per collision is lower than CO_2, due to the presence of heavier elements such as Fe. For example, $\xi = 0.1$ for anhydrous howardite; however, the addition of just 1000 μg/g of H to howardite increases ξ to 0.19 (Prettyman et al., 2011). The addition of a small amount of H has a large effect on the number of collisions required for thermalization.

The number of elastic collisions affects the magnitude of the spectrum in the epithermal energy range (for neutron kinetic energies from about 0.5 eV to 1 MeV). The equilibrium flux of epithermal neutrons varies in proportion to the product of the number of collisions and the mean free path for neutron elastic scattering (Eq. 9.2). Moderation of neutrons in hydrogenous materials generally requires fewer collisions with shorter mean-free paths than anhydrous materials, resulting in suppressed epithermal flux. Simulated leakage spectra for materials with different hydrogen content are shown in Figure 9.4. The relative variation in the epithermal flux is approximated by Fermi's model (Eq. 9.2).

Neutrons below a few tenths of an eV begin to thermally equilibrate with the regolith. The energy distribution of neutrons at these low energies can be described by the Maxwell–Boltzmann distribution (Figure 9.4); however, the mean kinetic energy is always higher than would be expected given the temperature of the subsurface medium in which they are slowing down. A portion of neutrons are removed by leakage into space and by absorption before reaching thermal equilibrium. The primary loss mechanism is radiative neutron capture, in which a neutron is absorbed to form a compound nucleus in an excited state, which deexcites by the emission of characteristic gamma rays. Temperature effects are apparent in thermal neutron counting data acquired by Lunar Prospector over the lunar highlands (Little et al., 2003).

The cross section for radiative neutron capture varies approximately as the inverse of neutron speed (Figure 9.3). Elements found in Solar System materials such as H, Cl, Fe, Gd,

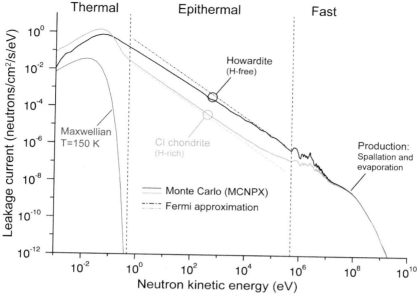

Figure 9.4 The leakage current of neutrons from a planetary surface consisting of howardite and CI chondrite, with compositions from Lodders and Fegley (1998). The leakage spectra were simulated by Monte Carlo (MCNPX). High-energy neutrons (a few hundred MeV) are produced by galactic cosmic ray interactions. They slow down via elastic and inelastic scattering. The population of neutrons with kinetic energies greater than about 1 MeV are sensitive the average atomic mass and hydrogen content of the regolith. Below 1 MeV, elastic scattering is dominant and the neutron flux varies inversely with kinetic energy (Fermi approximation, Eq. 9.2). The presence of hydrogen suppresses the flux of neutrons in the epithermal energy range. The CI chondrite composition contains 2 wt.% H, whereas howardite is H-free. At low energies, neutrons approach thermal equilibrium with the regolith; however, thermal equilibrium (represented by the Maxwell–Boltzmann distribution) is not achieved due to neutron loss by absorption and escape. The temperature of the surface in the Monte Carlo simulation was 150 K.

Eu, and Sm are strong neutron absorbers. Measurement of the flux of neutrons in the thermal and epithermal energy ranges can be combined to determine the macroscopic neutron absorption cross section of regolith materials Σ_A (Elphic et al., 1998, 2000; Feldman et al., 2000; Lawrence et al., 2010; Prettyman et al., 2013), providing an additional constraint on elemental composition:

$$\Sigma_A = \sum_{i=0}^{n-1} N_i \sigma_{ai}. \qquad (9.4)$$

The elemental microscopic absorption cross sections σ_{ai} are flux-weighted averages of the microscopic cross section over all neutron energies, and, consequently depend somewhat on overall surface composition. The general approach for calculating σ_{ai} is described by Lingenfelter et al. (1961, 1972) and a recent compilation is available for achondrites (Prettyman et al., 2013), which has been used in the analysis of igneous materials (e.g., Peplowski et al., 2015, 2016).

9.2.3 Gamma-Ray Production

Neutron-induced nuclear reactions are described using a notation similar to that used for chemical reactions. For example, radiative capture of neutrons by ^{28}Si is denoted

$$n + {}^{28}\text{Si} \rightarrow {}^{29}\text{Si} + Q,$$

where the "Q value" is the total amount of energy released (or absorbed) by the reaction. The energy liberated, given by the mass difference between the reactants and the products, is 9.0968×10^{-3} atomic mass units (amu), equivalent to 8473.6 keV. Here, Q is positive, indicating the reaction is exothermic. When Q is negative, the kinetic energy of the incident neutron must exceed a threshold in order for the reaction to occur. The foregoing reaction leaves ^{29}Si in an excited state, denoted by an asterisk, which transitions to ground state by the emission of electromagnetic radiation, ^{29}Si* $\rightarrow$ ^{29}Si + γ, with prominent gamma rays at 3.539 and 4.934 MeV. Gamma rays (γ) are photons emitted by the nucleus and typically have energies greater than a few hundred keV. They are characteristic of the energy level structure of the residual nuclei produced by nuclear reactions (Figure 9.5).

Gamma rays produced by neutron reactions with isotopes of elements commonly found in planetary materials are listed in Tables 9.1 and 9.2. The tables use a shorthand notation for nuclear reactions, T(n, x)R→G, where T is target isotope, x indicates the particles produced by the reaction, R is the residual nucleus, and G is the decay product, if any. The residual nucleus and decay product are usually implied. For example, neutron capture with Si is denoted ^{28}Si(n, γ). The tabulated data are from Reedy (1978); however, some gamma-ray energies and yields have been updated and for inelastic reactions, the identity of the excited level and level following gamma-ray emission have been added. Leakage fluxes for a nominal lunar material, calculated by Reedy (1978), are provided.

Gamma rays produced by the decay of natural radioelements (K, Th, and U) are listed in Table 9.3. The volumetric production rate of gamma rays by radioactive decay is given by

$$P = y\lambda \left(wf \frac{N_A}{A} \right) \quad (\text{gamma rays}/\text{cm}^3/\text{s}/\text{g-regolith}), \tag{9.5}$$

where y is the yield, the number of gamma rays produced per decay of the parent radionuclide, $\lambda = \ln(2)/T_{1/2}$ is the decay constant, which depends on the half-life $T_{1/2}$, and the term in parentheses is the number density of the radionuclide. The number density is determined by the weight fraction w of the radioelement (e.g., Th), the isotopic abundance f and atomic mass A of the parent radioisotope (e.g., ^{232}Th), and Avogadro's number N_A. For production of gamma rays in a decay chain, secular equilibrium is assumed. Namely, the decay rate of daughter products in the decay chain (e.g., ^{208}Tl) is the same as that of the

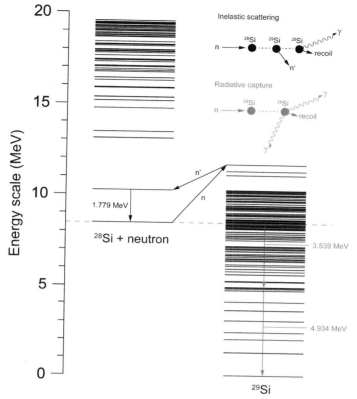

Figure 9.5 Level diagrams for ^{28}Si and ^{29}Si. Neutron inelastic scattering and radiative capture are illustrated. Radiative capture (red arrows): A low-energy neutron is absorbed by ^{28}Si, leaving ^{29}Si in an excited state, which decays promptly by the emission of two gamma rays (3.539 and 4.934 MeV, red). Inelastic scattering (black arrows): A fast neutron is absorbed by ^{28}Si, forming a compound nucleus (^{29}Si) that emits a neutron, leaving ^{28}Si in its first excited state (e.g., Hauser & Feshbach, 1952). ^{28}Si returns to ground state by the emission of a 1.779 MeV gamma ray. For inelastic scattering, the kinetic energy of the neutron and recoiling nucleus is not conserved. For inelastic scattering to occur, the kinetic energy of the incident neutron must exceed the first excited state of ^{28}Si.

parent radionuclide in the chain (e.g., ^{232}Th). Consequently, the gamma-ray yields in Table 9.3 for Th and U are gamma rays per decay of the parent.

For example, consider regolith materials containing Th. Thorium is almost entirely made of ^{232}Th, such that $f = 1$ g/g. The atomic mass of ^{232}Th, the parent of the decay chain, is 232.04 g/mol. The decay constant for ^{232}Th is 1.57×10^{-18} s^{-1} (with a half-life of 14 billion years). The 2.615 MeV gamma ray from the decay of ^{208}Tl is prominent in lunar gamma-ray spectra. From Table 9.3, the yield of 2.615 MeV gamma rays is 0.360 gamma rays per decay of ^{232}Th. Thus, for 1 μg/g of Th, which is comparable to concentrations found in the

Table 9.1 *Gamma rays produced by neutron inelastic scattering, the decay of GCR-produced radionuclides, and reactions other than radiative capture*[a]

Element	Reaction	Excited energy level	Transitioned energy level	Gamma-ray energy (MeV)	Yield	Flux ($\gamma/cm^2/min$)
O	$^{16}O(n, n\gamma)$	5	Ground	8.869	0.072	0.059
	$^{16}O(n, n\gamma)$	4	Ground	7.115	1	0.808
	$^{16}O(n, n\gamma)$	3	Ground	6.916	1	0.736
	$^{16}O(n, n\gamma)$	2	Ground	6.129	1	2.592
	$^{16}O(n, p)^{16}N \to {}^{16}O$	2	Ground	6.129	0.66	0.266
	$^{16}O(n, n\gamma)$	11	2	4.966	0.55	0.128
	$^{16}O(n, n\alpha\gamma)^{12}C$	1	Ground	4.438	1	1.214
	$^{16}O(n, n\gamma)$	11	3	4.179	0.45	0.096
	$^{16}O(n, \alpha\gamma)^{13}C$	3	Ground	3.853	0.62	0.336
	$^{16}O(n, n\gamma)$	9	4	3.840	1	0.058
	$^{16}O(n, \alpha\gamma)^{13}C$	2	Ground	3.684	0.99	0.687
	$^{16}O(n, \alpha\gamma)^{13}C$	1	Ground	3.089	1	0.283
	$^{16}O(n, n\gamma)$	5	2	2.742	0.78	0.367
	$^{16}O(n, n\gamma)$	5	4	1.755	0.11	0.042
Mg	$^{24}Mg(n, n\gamma)$	3	Ground	4.238	0.79	0.111
	$^{24}Mg(n, n\gamma)$	4	1	3.866	0.97	0.059
	$^{24}Mg(n, p)^{24}Na \to {}^{24}Mg$	2	1	2.754	0.9994	0.106
	$^{24}Mg(n, n\gamma)$	2	1	2.754	1	0.091
	$^{26}Mg(n, n\gamma)$	1	Ground	1.809	1	0.152
	$^{25}Mg(n, n\gamma)$	3	Ground	1.612	1	0.060
	$^{24}Mg(n, n\gamma)$	1	Ground	1.369	1	0.727
	$^{24}Mg(n, p)^{24}Na \to {}^{24}Mg$	1	Ground	1.369	1	0.069
	$Mg(n, x)^{22}Na \to {}^{22}Ne$	1	Ground	1.275	0.9996	0.074
	$^{26}Mg(n, n\gamma)$	2	1	1.130	0.91	0.039
Al	$^{27}Al(n, n\gamma)$	12	Ground	4.580	0.71	0.058
	$^{27}Al(n, n\gamma)$	10	Ground	4.410	0.57	0.048
	$^{27}Al(n, n\gamma)$	8	Ground	3.956	0.86	0.054
	$^{27}Al(n, n\gamma)$	9	1	3.211	0.86	0.060
	$^{27}Al(n, n\gamma)$	6	Ground	3.004	0.88	0.372
	$^{27}Al(n, n\gamma)$	5	Ground	2.982	0.97	0.117
	$^{27}Al(n, \alpha)^{24}Na \to {}^{24}Mg$	2	1	2.754	0.9994	0.209
	$^{27}Al(n, n\gamma)$	4	Ground	2.735	0.22	0.064
	$^{27}Al(n, n\gamma)$	11	Ground	2.299	0.77	0.058
	$^{27}Al(n, n\gamma)$	3	Ground	2.211	1	0.675
	$^{27}Al(n, 2n)^{26}Al \to {}^{26}Mg$	1	Ground	1.809	0.998	0.290
	$^{27}Al(n, d\gamma)^{26}Mg$	1	Ground	1.809	1	0.238
	$^{27}Al(n, n\gamma)$	4	2	1.720	0.76	0.166
	$^{27}Al(n, \alpha)^{24}Na \to {}^{24}Mg$	1	Ground	1.369	1	0.139
	$^{27}Al(n, x)^{22}Na \to {}^{22}Ne$	1	Ground	1.275	0.9996	0.093

Table 9.1 (cont.)

Element	Reaction	Excited energy level	Transitioned energy level	Gamma-ray energy (MeV)	Yield	Flux (γ/cm²/min)
Si	^{27}Al(n, nγ)	2	Ground	1.014	0.97	0.634
	^{27}Al(n, nγ)	1	Ground	0.844	1	0.305
	^{27}Al(n, p)^{27}Mg→^{27}Al	1	Ground	0.844	0.72	0.067
	^{28}Si(n, nγ)	9	Ground	7.414	0.94	0.116
	^{28}Si(n, nγ)	6	Ground	6.877	0.70	0.214
	^{28}Si(n, nγ)	8	1	5.600	0.63	0.052
	^{28}Si(n, nγ)	7	1	5.108	0.99	0.113
	^{28}Si(n, nγ)	6	1	5.099	0.27	0.073
	^{28}Si(n, nγ)	4	1	4.497	0.88	0.102
	^{28}Si(n, nγ)	3	1	3.201	1	0.088
	^{28}Si(n, nγ)	2	1	2.838	1	0.329
	^{28}Si(n, pα)^{24}Na→^{24}Mg	2	1	2.754	0.9994	0.309
	^{30}Si(n, nγ)	1	Ground	2.235	1	0.117
	^{28}Si(n, x)^{26}Al→^{26}Mg	1	Ground	1.809	0.972	0.298
	^{28}Si(n, nγ)	1	Ground	1.779	1	3.223
	^{28}Si(n, p)^{28}Al→^{28}Si	1	Ground	1.779	1	0.700
	^{28}Si(n, pα)^{24}Na→^{24}Mg	1	Ground	1.369	1	0.212
	^{28}Si(n, x)^{22}Na→^{22}Ne	1	Ground	1.275	0.9996	0.152
	^{29}Si(n, nγ)	1	Ground	1.273	1	0.067
Ca	^{40}Ca(n, nγ)	6	Ground	5.249	0.80	0.042
	^{40}Ca(n, nγ)	3	Ground	3.904	1	0.232
	^{40}Ca(n, nγ)	2	Ground	3.737	1	0.346
	^{40}Ca(n, αγ)^{37}Ar	2	Ground	1.611	1	0.106
	^{40}Ca(n, pγ)^{40}K	5	3	1.159	0.82	0.031
	^{44}Ca(n, nγ)	1	Ground	1.157	1	0.034
	^{40}Ca(n, pγ)^{40}K	3	Ground	0.891	0.99	0.031
	^{40}Ca(n, pγ)^{40}K	2	1	0.770	1	0.101
Ti	^{48}Ti(n, nγ)	2	1	1.312	1	0.037
	^{48}Ti(n, nγ)	1	Ground	0.984	1	0.163
	^{46}Ti(n, nγ)	1	Ground	0.889	1	0.016
	^{46}Ti(n, p)^{46}Sc→^{46}Ti	1	Ground	0.889	1	0.013
Fe	^{56}Fe(n, nγ)	14	Ground	3.602	0.69	0.040
	^{56}Fe(n, nγ)	11	1	2.599	0.78	0.048
	^{56}Fe(n, nγ)	9	1	2.523	0.85	0.052
	^{56}Fe(n, nγ)	5	1	2.113	0.98	0.076
	^{56}Fe(n, nγ)	3	1	1.811	0.97	0.119
	^{54}Fe(n, nγ)	1	Ground	1.408	1	0.061
	^{56}Fe(n, 2nγ)^{55}Fe	3	Ground	1.316	0.93	0.085
	^{56}Fe(n, nγ)	2	1	1.238	1	0.256

Table 9.1 (cont.)

Element	Reaction	Excited energy level	Transitioned energy level	Gamma-ray energy (MeV)	Yield	Flux (γ/cm^2/min)
	^{56}Fe(n, nγ)	8	2	1.038	0.99	0.041
	^{56}Fe(n, 2nγ)^{55}Fe	2	Ground	0.931	0.98	0.084
	^{56}Fe(n, nγ)	1	Ground	0.847	1	1.149
	^{54}Fe(n, p)^{54}Mn→^{54}Cr	1	Ground	0.835	1	0.090

[a] These reactions produce a residual nucleus in an excited state, which transitions to ground state by the production of one or more gamma rays. For each gamma ray listed, the excited energy level of the residual nucleus and transition level is indicated. The yields are gamma rays produced per transition. The gamma energies and yields are from the Evaluated Nuclear Structure Data File. Leakage fluxes for a representative lunar composition were calculated by Reedy (1978).

Table 9.2 *Gamma rays produced by radiative capture*[a]

Element	Reaction	Gamma-ray energy (MeV)	Yield	Flux (γ/cm^2/min)
Mg	^{24}Mg(n, γ)	3.917	0.48	0.0205
	^{24}Mg(n, γ)	2.828	0.36	0.0119
	^{24}Mg(n, γ)	0.585	0.47	NC[b]
Al	^{27}Al(n, γ)	7.724	0.21	0.123
	^{27}Al(n, γ)	7.693	0.035	0.020
	^{27}Al(n, γ)	4.734	0.055	0.024
	^{27}Al(n, γ)	4.260	0.066	0.028
	^{27}Al(n, γ)	4.133	0.065	0.026
	^{27}Al(n, γ)	3.034	0.077	0.024
	^{27}Al(n, γ)	2.960	0.09	0.028
	^{27}Al(n, γ)^{28}Al→^{28}Si	1.779	1	0.209
Si	^{28}Si(n, γ)	7.199	0.073	0.049
	^{28}Si(n, γ)	6.380	0.12	0.077
	^{28}Si(n, γ)	4.934	0.65	0.359
	^{28}Si(n, γ)	3.539	0.69	0.300
	^{28}Si(n, γ)	2.093	0.19	0.056
	^{28}Si(n, γ)	1.273	0.17	0.033
Ca	^{40}Ca(n, γ)	6.420	0.41	0.246
	^{40}Ca(n, γ)	5.900	0.06	0.034
	^{40}Ca(n, γ)	4.419	0.16	0.080
	^{40}Ca(n, γ)	2.010	0.09	0.025
	^{40}Ca(n, γ)	1.943	0.82	0.209

Table 9.2 (cont.)

Element	Reaction	Gamma-ray energy (MeV)	Yield	Flux (γ/cm²/min)
Ti	^{48}Ti(n, γ)	6.760	0.49	0.503
	^{48}Ti(n, γ)	6.556	0.05	0.055
	^{48}Ti(n, γ)	6.418	0.32	0.322
	^{48}Ti(n, γ)	4.967	0.03	0.028
	^{48}Ti(n, γ)	4.881	0.05	0.044
	^{48}Ti(n, γ)	1.762	0.05	0.020
	^{48}Ti(n, γ)	1.586	0.10	0.038
	^{48}Ti(n, γ)	1.499	0.05	0.017
	^{48}Ti(n, γ)	1.382	0.85	0.279
	^{48}Ti(n, γ)	0.342	0.30	0.039
Fe	^{54}Fe(n, γ)	9.298	0.03	0.080
	^{56}Fe(n, γ)	7.646	0.21	0.538
	^{56}Fe(n, γ)	7.631	0.26	0.640
	^{56}Fe(n, γ)	7.279	0.05	0.132
	^{56}Fe(n, γ)	6.019	0.09	0.197
	^{56}Fe(n, γ)	5.920	0.09	0.193
	^{56}Fe(n, γ)	4.810	0.02	0.032
	^{56}Fe(n, γ)	4.218	0.04	0.070
	^{56}Fe(n, γ)	1.725	0.07	0.064
	^{56}Fe(n, γ)	1.613	0.06	0.050
	^{56}Fe(n, γ)	0.692	0.05	0.024
	^{56}Fe(n, γ)	0.352	0.11	NC
H	^{1}H(n, γ)	2.223	1	0.0035
S	^{32}S(n, γ)	5.421	0.58	0.0034
	^{32}S(n, γ)	2.380	0.39	0.0013
	^{32}S(n, γ)	0.841	0.65	NC
Cl	^{35}Cl(n, γ)	7.790	0.08	0.0009
	^{35}Cl(n, γ)	7.414	0.10	0.0011
	^{35}Cl(n, γ)	6.628	0.04	0.0005
	^{35}Cl(n, γ)	6.620	0.08	NC
	^{35}Cl(n, γ)	6.111	0.20	0.0020
	^{35}Cl(n, γ)	1.959	0.12	NC
	^{35}Cl(n, γ)	1.951	0.19	0.0009
	^{35}Cl(n, γ)	1.165	0.27	NC
	^{35}Cl(n, γ)	0.788	0.16	NC
	^{35}Cl(n, γ)	0.786	0.10	NC
	^{35}Cl(n, γ)	0.517	0.23	NC
Cr	^{53}Cr(n, γ)	9.719	0.08	0.0033
	^{53}Cr(n, γ)	8.884	0.25	0.0094
	^{52}Cr(n, γ)	7.938	0.14	0.0160
	^{53}Cr(n, γ)	0.835	0.45	NC

Table 9.2 (cont.)

Element	Reaction	Gamma-ray energy (MeV)	Yield	Flux (γ/cm²/min)
Ni	^{58}Ni(n, γ)	8.998	0.34	0.0067
	^{58}Ni(n, γ)	8.534	0.16	0.0032
	^{58}Ni(n, γ)	0.465	0.19	NC

[a] The yields are gamma rays per capture of thermal neutrons from ENDSF. Leakage fluxes for a representative lunar composition were calculated by Reedy (1978).
[b] NC indicates gamma-ray fluxes that were not calculated.

Table 9.3 *Gamma rays produced by the decay of natural radioelements*[a]

Element	Nuclide	Gamma-ray energy (MeV)	Yield	Flux (γ/cm²/min)
K	^{40}K	1.461	0.107	2.406
Th	^{208}Tl	2.615	0.360	2.193
	^{228}Ac	1.588	0.037	0.177
	^{228}Ac	0.969	0.175	0.656
	^{228}Ac	0.965	0.054	0.202
	^{228}Ac	0.911	0.290	1.054
	^{228}Ac	0.795	0.048	0.163
	^{212}Bi	0.727	0.070	0.229
	^{208}Tl	0.583	0.307	0.916
	^{208}Tl	0.511	0.083	0.234
	^{228}Ac	0.338	0.120	0.285
	^{212}Pb	0.239	0.47	0.964
U	^{214}Bi	2.204	0.050	0.224
	^{214}Bi	1.765	0.159	0.637
	^{214}Bi	1.730	0.031	0.123
	^{214}Bi	1.378	0.040	0.142
	^{214}Bi	1.238	0.059	0.199
	^{214}Bi	1.120	0.150	0.481
	^{214}Bi	0.768	0.049	0.131
	^{214}Bi	0.609	0.461	1.118
	^{214}Pb	0.352	0.371	0.714
	^{214}Pb	0.295	0.192	0.343
	^{214}Pb	0.242	0.075	0.123

[a] For the K gamma ray, the yield is the number of gamma rays emitted per decay of ^{40}K. For Th and U, gamma rays are emitted by daughter products in the ^{232}Th and ^{238}U decay series. The gamma-ray yields were calculated by Reedy (1978), assuming decay products are in secular equilibrium. Thus, the yields are gamma rays emitted per decay of either ^{232}Th or ^{238}U. Tabulated fluxes were calculated by Reedy (1978) for a representative lunar composition.

lunar highlands (e.g., Lawrence et al., 2000; Prettyman et al., 2006; Yamashita et al., 2010), the production rate of 2.615 MeV gamma rays is about 5.3 gamma rays/cm^3/hour per gram of regolith (Eq. 9.5). Production rates for principle gamma rays for K and Th are reported by Prettyman et al. (2015).

9.2.4 X-Ray Production

Characteristic X-rays are obtained by exposure of a substrate to exciting radiation. X-ray production is limited to interactions within the electron shell and is therefore dependent on the electronic structure of atoms.

Atoms consist of a nucleus and an electron shell. The number of shell electrons of an atom corresponds to the number of protons in the nucleus, specified by the atomic number Z. Each shell electron is in a particular energy state that can be characterized by four quantum numbers. These are n, the main quantum number; l, the angular momentum quantum number; m, the magnetic quantum number; and s, the spin quantum number.

In classical nomenclature electrons are located in *K, L, M, N*, etc. shells of atoms, and placed in orbitals according to quantum mechanical principles. The K-shell has one orbital (1s) that can be filled with two electrons of opposing spin (+1/2,–1/2), followed by the L-shell with 8 electrons in 2s and 2p orbitals; the M-shell with 18 electrons in 3s, 3p, and 3d orbitals; and the N-shell with 32 electrons in 4s, 4p, 4d, and 4f shells. For major rock-forming elements on silicate bodies, Z is less than 29. In this range, the N-shell is the outermost shell for Z>18.

For the analysis of planetary surfaces, a key process is the removal of an electron from the K-shell, by an incoming X-ray, electron, proton or alpha particle, creating a vacancy in the bound shell from which it was emitted. Creating such a vacancy by any atomic mechanism ionizes the atom and is setting the condition for the emission of X-rays.

The vacancy can be filled by an electron from a higher electron shell of the atom, leading to the emission of characteristic X-ray photons. The energy emitted as an X-ray photon, after K-shell ionization of an atom (K+) and subsequent transformation into an L+ state by filling the vacancy with an L electron (2p), is called the K_α X-ray. The energy of that transition and the K_α X-ray can be calculated from the difference of the electron binding energy of the K-shell and the L-shell. In case of Cu, the K and L electron binding energies are 8.973 keV and 0.933 keV, respectively, leading to a K_α line energy of 8.040 keV (Jenkins, 1999).

When an $L \rightarrow K$ transition occurs, e.g., from a $2p \rightarrow 1s$ orbital, more than one transition energy is available. The 2p electron can transition from two different spin states (+1/2,-1/2), leading to a $2p^{3/2} \rightarrow 1s$ transition or a $2p^{1/2} \rightarrow 1s$ transition of different energies, with the nomenclature of $K_{\alpha 1}$ and $K_{\alpha 2}$, respectively.

In elements with sufficiently high atomic numbers, K shell vacancies can be filled from the M shell (3s, 3p, or 3d) or the N shell (4s, 4p, 4d, or 4f). Selection rules describe the allowed and forbidden transitions. Without going into details, M, N→K transitions have a K_β designation. For oxygen, where 3p electrons are not available, K_β lines cannot be observed. Elements with $3p$ (M-shell) electrons show K_β lines that vary in relative energy (to K_α) and absolute abundances as a function of atomic number. Vacancies in the L-shell, and more so in the M-shell, lead to complex arrays of electron transitions. As abundances of planetary surface elements measurable by X-ray fluorescence with occupied M- and L-shells are rare, gamma and delta lines are not discussed here.

An alternative process to X-ray producing electron transitions, Auger electron emission decreases the energy of an excited atom through self-ionization without emission of electromagnetic radiation, generally leading to multiple ionizations in upper shells. The distribution between the competing effects of fluorescence and self-ionization is described by the fluorescent yield. The Auger effect, for example, is the dominant process of producing double ionization. The second vacancy generally increases the binding energy of the remaining electron such that electron transitions will produce X-rays of slightly different energies, visible as satellites in an X-ray spectrum.

Major K lines from planetary surfaces typically include those of Mg, Al, Si, Ca, and Fe. Of these, only Mg, Al, and Si can be analyzed using ambient solar X-rays. Iron and Ca require excitation by intense solar flares (e.g., see Nittler et al., 2011). In addition, solar X-ray fluxes vary with distance from the Sun, limiting analyses to inner Solar System targets. K_α X-ray production yields as a function of atomic number are shown in Figure 9.6.

9.2.5 X-Ray and Gamma-Ray Transport

X-rays and gamma rays are electromagnetic radiation and their properties and interaction modes with matter are identical. Early in the history of radiation detection and measurement, energy range was used to distinguish gamma rays from X-rays, where the former were considered of high and the latter of low energy (much less than 1 MeV). A distinction by energy or wavelength is sometimes useful. In the electromagnetic spectrum X-rays cover energies between 0.5- and 12-keV, flanked by gamma rays at higher energies and by vacuum-UV at lower energies (Jenkins, 1999). A distinction by origin was established later and is maintained to date, whereby X-rays are generated by the interaction of the exciting radiation with orbital electrons (e.g., resulting in electron transitions), whereas gamma rays are produced primarily by nuclear transitions.

Interaction mechanisms of electromagnetic radiation with matter important for remote sensing spectroscopy include photoelectric absorption, Compton scattering (incoherent scattering), and pair production. For both the photoelectric effect and Compton scattering, electromagnetic energy is transferred to electrons, ionizing the atom. Liberated electrons can undergo subsequent interactions with matter, resulting in additional ionization. For example, high-energy gamma rays can produce a cascade of secondary electrons and photons. Electromagnetic radiation also undergoes coherent (Rayleigh) scattering.

The aforementioned electromagnetic interaction processes as well as macroscopic interaction properties, total absorption and total attenuation are all dependent on the energy of the photons and the properties of the materials they interact with (Figures 9.6 and 9.7). The

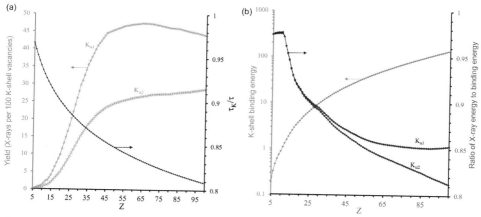

Figure 9.6 X-ray production by fluorescence. (a) The production yield of K_α X-rays is shown as a function of Z along with the ratio of the K-shell to total photoelectric absorption cross section. (b) The K-shell binding energy and ratios of K_α X-ray energies to the binding energy are shown. Production of Auger electrons competes with fluorescence, with Auger being more important at low Z. (Data source: Browne & Firestone, 1986.)

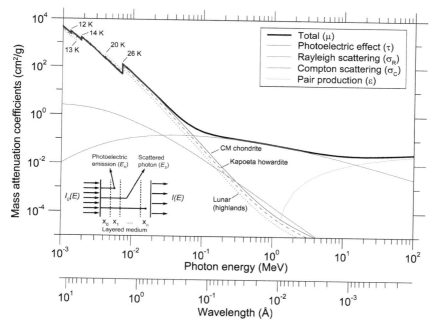

Figure 9.7 Interaction of electromagnetic radiation with selected planetary regolith materials. Mass attenuation coefficients, with units of cm^2/g, as a function of photon energy for the different photon interaction mechanisms are shown for a CM chondrite. The total mass attenuation coefficient (μ) is given by the sum of the contributions from the photoelectric effect, Compton scattering, and pair production. Variations in the photoelectric component are displayed for howardite and lunar highlands compositions. Prominent K-shell absorption edges are indicated (e.g., "26 K" for the Fe K-edge). The inset illustrates the interaction of a beam of photons with a multilayered material (see Section 9.2.5.4).

energy E of a photon depends only on its wavelength λ: $E = hc/\lambda$, where h is Planck's constant and c is the speed of light.

9.2.5.1 K-Absorption Edge and the Mass Attenuation Coefficient

For photon energies less than twice the electron rest mass, the mass attenuation coefficient (μ) with units of cm^2/g is a function of photoelectric absorption τ and scattering σ: $\mu = \tau + \sigma$. For energies less than a few keV, τ is much larger than σ, so $\mu \cong \tau_K + [\tau_{LI} + \tau_{LII} + \tau_{LIII}] + \ldots + \tau_n$. The mass attenuation coefficient decreases approximately with increasing photon energy (power of 3.5) and increases with the atomic number (Z) of the absorber (power of 4 to 5) (Knoll, 1989).

The photoelectric absorption process can occur for photons with energies sufficient to liberate an orbital electron. The photon completely disappears and the photoelectron ejected from the atom carries the energy of the original photon minus the binding energy of the electron, $E_{e-} = E - E_b$, with E_b representing the binding energy of the photoelectron.

There appear to be discontinuities in the absorption curve, so-called absorption edges, at photon energies that match the binding energies of electrons in the absorber material. These absorption edges are labeled according to the energy level of the corresponding electron: K, L_I, L_{II}, L_{III} etc.

If the absorbed photon energy is greater than the energy of the absorption edge, then an electron of the respective level can be removed. In general, when the photon energy is lower than that of the absorption edge electrons of that level can no longer undergo photoelectric absorption.

The attenuation coefficient of a mixture is given by the sum of all contributions from all elements: $\mu(E) = \sum \mu_j(E) w_j$, where μ_j is the mass attenuation coefficient for element j at energy E, and w_j is the weight fraction of element j. Figure 9.7 shows mass absorption coefficients and contributions for the main interactions mechanisms for selected regolith materials.

9.2.5.2 Scattering

Compton scattering (incoherent scattering) is important for photons of all energies. The incoming photon is deflected by an angle θ with respect to its original direction, with a part of the photon's energy being transferred to a recoil electron. The Compton energy–angle relationship gives the energy of the scattered photon E_s as a function of the initial energy E and cosine of the scattering angle, θ: $E_s = E/[1 + (1 - \cos\theta)E/E_R]$, where E_R is the electron rest mass (0.511 MeV). The distribution of scattering angles ($0 \leq \theta \leq 180°$) is described by the Klein–Nishina probability distribution, which assumes scattering from free electrons. For low-energy photons, electron binding effects must be considered. The maximum energy loss depends on photon energy. Photons with energies similar to the electron rest mass can lose up to two thirds of their energy in a head on collision ($\theta = 180°$). The cross section for Compton scattering depends on the electron density of the target.

Rayleigh scattering (coherent scattering) is most important for low-energy photons, which includes all characteristic X-rays of interest. In this process the photon interacts with atoms, e.g., of a detector or within the surface of a planet, by changing direction without loss of energy. The deflection angle decreases with incident photon energy and is insignificant above several hundred keV.

9.2.5.3 Pair Production

When the photon energy exceeds twice the rest mass of the electron (1.022 MeV) the process of pair production is energetically possible; however, pair production becomes significant only for very high photon energies (Figure 9.7). The interaction takes place in the Coulomb field of a nucleus, whereby the excess energy of the photon above 1.022 MeV consumed by pair production is maintained in the kinetic energy of the electron and positron thus created. A consequence of positron production in this process is the creation of two back-to-back 511-keV photons through annihilation of the slowed-down positron.

9.2.5.4 Interaction Mechanisms Sum

All interaction mechanisms influence what we can see with a detector of gamma rays and X-rays produced in a planetary surface and how a detector responds to incident electromagnetic radiation. In macroscopic terms, the linear attenuation coefficient for a given material describes removal of electromagnetic radiation (of known energy) between a source and a detector caused by any of the interaction mechanisms mentioned earlier, including the photoelectric effect, Compton scattering, and pair production. The sum total of these effects determines detector responses to a field of X- or gamma-radiation. They also determine, for X- and gamma-radiation, the depth (or volume) from which electromagnetic radiation–based information can be obtained.

For example, Figure 9.7 illustrates the interaction of a beam of electromagnetic radiation (intensity I_0 and energy E) with a layered medium. The i_{th} layer is x_i cm thick with density ρ_i g/cm^3. The intensity of the uncollided photons is given by the Beer-Lambert law, $I = I_0 \exp\left(-\sum_{i=0}^{n} \rho_i \mu_i x_i\right)$, where μ_i is the mass attenuation coefficient of the i_{th} layer.

9.3 Modeling Radiation Transport and Instrument Response

Models that merge geologic constraints with the physics of radiation transport and detection provide a useful and often necessary guide to the interpretation of remote sensing data acquired by nuclear and X-ray spectrometers. Geologic models provide constraints on the composition, distribution, and layering of subsurface materials, for example, based on studies of geochemical and geophysical processes for regolith emplacement, weathering, and contamination. The radiation transport model describes how the spectrometer responds to different material configurations in the subsurface of the target body. A deep dive into either of these topics is beyond the scope of this chapter. Here, we present fundamental concepts in radiation transport relevant to planetary science upon which the reader can build. The emphasis is on nuclear spectroscopy; however, the methods presented here can be extended to remote sensing with X-rays.

9.3.1 Definitions

For planetary applications, neutrons, gamma rays, and X-rays can be modeled as point-like particles, described by their position $\vec{r}$ and velocity $\vec{v}$ in three-dimensional Cartesian space. The particles interact with planet and spacecraft materials, but not with each other. Velocity and direction are related by $\vec{v} = v\hat{\Omega}$, where v is the speed of the particle and $\hat{\Omega}$ is the unit direction vector, containing the cosines of the angles between the three coordinate axes. In the following discussion, we will also specify directions in spherical coordinates with polar and azimuthal angles.

X-rays and gamma rays propagate at the speed of light. For particles with mass, velocity and kinetic energy E are related by $E/E_0 = \left(1/\sqrt{1-\beta^2}\right) - 1$, where E_0 is the rest mass (e.g., 939.565 MeV for a neutron) and $\beta = v/c$, where c is the speed of light in vacuum. For example, a relativistic neutron with $\beta = 0.9$ has a kinetic energy of 1.22 GeV. The vast majority of secondary neutrons travel at speeds much lower than the speed of light for which the classical approximation, $E/E_0 = \beta^2/2$, is accurate.

As cross sections for nuclear reactions are energy dependent, particle populations are typically described in terms of position, energy, and direction. Note also that for production of particles by natural sources, we will only consider the steady-state population, averaged over a long time in comparison to the time-scale for the evolution of a single cosmic ray shower.

9.3.2 Particle Populations and Computational Methods

Following Duderstadt and Hamilton (1976), let $n(\vec{r}, \hat{\Omega}, E)d^3\vec{r}\,d\hat{\Omega}\,dE$ be the number of particles within the differential volume $d^3\vec{r}$ about $\vec{r}$, with directions within $d\hat{\Omega}$ about $\hat{\Omega}$ and energies between E and $E + dE$. The particle density n is described by the transport equation, an integro-differential equation formulated as a balance that treats particle production, migration, and loss. The transport equation can be solved for particle flux, which is the product of speed and number density: $\phi(\vec{r}, \hat{\Omega}, E) = v\,n(\vec{r}, \hat{\Omega}, E)$, with units of particles per cm^2 per second per unit energy per steradian.

Deterministic and stochastic numerical methods have been developed to solve the transport equation (e.g., Metropolis & Ulam, 1949; Lewis & Miller, 1984). Monte Carlo provides the most detailed simulations of GCR interactions with planetary materials. With Monte Carlo, the random walk of individual particles and their progeny are simulated in a manner analogous to how they would behave in nature. Attributes of the particle population, such as fluxes and the energy–angle distribution of particles escaping into space, are determined by statistical analysis of many particle histories.

General-purpose Monte Carlo codes such as MCNP, GEANT, and FLUKA can provide detailed simulations of GCR air and regolith showers (Agostinelli et al., 2003; Battistoni et al., 2007; Goorley et al., 2013). They model the intranuclear cascade produced by high-energy collisions (kinetic energies in the GeV–TeV range), the production of secondary particles by spallation and evaporation, and the subsequent evolution of the hadronic and electromagnetic components of the shower as particles undergo many additional collisions (e.g., Figure 9.1b). These codes can also model the complex geometries of instrument-spacecraft systems, enabling detailed models of their response to energetic ions and neutral particles, such as neutrons, gamma rays, and X-rays. Note that analytic solutions to the transport equation are available in some cases. For example, the leakage current of uncollided gamma rays from radioactive sources distributed uniformly with depth has a simple analytic form (see Section 9.3.7).

9.3.3 Reaction Rate

Given the flux of neutrons, the rate at which neutron-induced nuclear reactions occur can be determined:

$$R(\vec{r}) = \int_{4\pi} d\hat{\Omega} \int_0^\infty dE\, \Sigma(\vec{r}, E) \phi(\vec{r}, \hat{\Omega}, E) \quad \text{(reactions/cm}^3\text{/s)}, \tag{9.6}$$

where $\Sigma(\vec{r}, E)$ (units of cm^{-1}) is the macroscopic cross section for the reaction of interest. This expression is useful for calculating the rate of gamma-ray production as a function of depth within the regolith when the flux is known, for example, from a radiation transport calculation, or the rate in which reactions occur within a detector.

9.3.4 Currents: Surface Crossings

Let's now consider a differential surface element, $d\vec{s} = \hat{n}\, dA$ located at $\vec{r}$, where $\hat{n}$ is the unit outward normal of the surface element, determining its orientation in space, and dA is its area. Particles crossing through the surface from the inside must satisfy $\hat{n}\cdot\vec{v} > 0$. Particles crossing from the outside have a negative dot product. The rate at which particles cross the surface is given by

$$j(\vec{r}, \hat{\Omega}, E) d\hat{\Omega}\, dE\, dA = d\vec{S}\cdot\vec{v}\, n(\vec{r}, \hat{\Omega}, E) d\hat{\Omega}\, dE \quad \text{(particles crossing per second)}, \tag{9.7}$$

or alternatively,

$$j(\vec{r}, \hat{\Omega}, E) d\hat{\Omega}\, dE = \hat{n}\cdot\hat{\Omega} \phi(\vec{r}, \hat{\Omega}, E) d\hat{\Omega}\, dE \quad \text{(particles crossing per second per unit area)}. \tag{9.8}$$

The current j has units of particles per second per unit area per unit energy per steradian.

9.3.5 Current of GCRs Entering a Planetary Surface

Consider a planetary surface parcel. How many GCRs cross into the surface per second? The flux of galactic cosmic rays in interplanetary space is nearly isotropic, namely $\phi(\vec{r}, \hat{\Omega}, E) = \phi_G(E)/4\pi$. Plots of $\phi_G(E)$ are shown in Figure 9.1a. The cosine of the angle between the incident cosmic ray and the surface outward normal is $\omega = \hat{n} \cdot \hat{\Omega}$ (Figure 9.8). Thus, $j(\vec{r}, \hat{\Omega}, E) = \omega \phi_G(E)/4\pi$. To answer the question, we simply integrate the current over angle and energy.

In spherical coordinates, $d\hat{\Omega} = d\omega d\varphi$, where φ is the azimuthal angle about the surface normal. Given the intervening planetary materials, particles crossing outward must be excluded such that $\omega < 0$. The integral becomes

$$J = \int_{-1}^{0} d\omega \int_{0}^{2\pi} d\varphi \int_{E_{min}}^{\infty} dE \, j(\vec{r}, \hat{\Omega}, E) = -\frac{1}{4} \int_{E_{min}}^{\infty} dE \phi_G(E) \quad \left(\text{particles/cm}^2/\text{s}\right). \quad (9.9)$$

The integral on the right is the total flux of galactic cosmic rays with kinetic energies greater than an arbitrary minimum (E_{min}). The total flux of galactic cosmic rays (all ion species, $Z = 0$–92) with kinetic energies greater than 10 MeV ranges from about 2.4 to 5.0 cm^{-2} s^{-1}. So, the number of GCRs crossing into the surface ranges from 0.6 to 1.3 cm^{-2} s^{-1} from solar max to solar min. Important results of the analysis include the ¼ scaling between inward surface crossings and GCR flux and the linear dependence of the direction of the ingoing particles on the cosine of the polar angle. This is the boundary condition used for simulations of GCR interactions with planetary surfaces (e.g., McKinney et al., 2006; Prettyman et al., 2006).

9.3.6 Flux of Particles at Orbital Altitudes, Solid Angle, and Spatial Resolution

Given the input cosmic ray spectrum, the current of neutrons or gamma rays escaping a planetary surface can be calculated by Monte Carlo (Section 9.3.2) for any surface composition. In some cases, analytical solutions of the transport equation are available, for example, for gamma rays made by the decay of radioelements in a homogeneous regolith (Section 9.3.7). Here, we consider a parcel of a planetary surface dS with unit outward normal $\hat{n}$ (Figure 9.8). We assume that the leakage current is symmetrical about the normal direction and varies only with the cosine of the polar emission angle, $\omega = \cos\theta = \hat{n} \cdot \hat{\Omega}$, where $\hat{\Omega}$ is the direction of the emitted particle. This assumption is generally valid for gamma rays and neutrons produced by GCR interactions and is a reasonable approximation for XRF. Note that for XRF, the magnitude of the leakage current depends on the angle of incidence of the interrogating solar X-rays (e.g., Weider et al., 2011). With this assumption, the angular leakage current is given by $j_s(\hat{n} \cdot \hat{\Omega}, E)$, with units of particles/cm^2/s/MeV/steradian. The contribution of the parcel to the flux at any point $\vec{r}$ outside the surface in direction $\hat{\Omega}$ is given by $j_s(\hat{n} \cdot \hat{\Omega}, E)dA/r^2$ (particles/cm^2/s/MeV), where r is the distance from the parcel to the

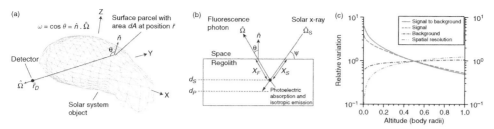

Figure 9.8 Geometry flux calculations. (a) Geometry and terms for orbital flux calculations (Eq. 9.10). The emission of particles from a surface parcel and their detection by an orbiting detector are illustrated for an irregular Solar System object, asteroid 25143 Itokawa (Gaskell et al., 2008). (b) Geometry for calculation of the production and emission of X-rays from a surface parcel. A solar X-ray traveling in direction $\hat{\Omega}_s$ travels distance x_s through the regolith before undergoing photoelectric effect. This occurs at depth d_s. A characteristic fluorescence X-ray is emitted in direction $\hat{\Omega}$ and must travel a distance x_F to escape the surface uncollided. Because the emission of fluorescence X-rays is isotropic, the emission of X-rays from the surface depends only on the polar emission angle θ, and the flux of particles arriving at the orbiting detector can be determined from Eq. (9.10). X-rays produced below depth contribute negligibly to the leakage flux. (c) Signal and background strength for an omnidirectional sensor is plotted as a function of the altitude of an orbiting spectrometer (see Section 9.3.6). The values have been arbitrarily normalized to 1 at 0.5 body radii. The signal-to-background ratio is higher by an order of magnitude at the surface than at 1 body radius. Near the surface, the background is suppressed because of shielding of cosmic rays by the planetary body. The ability to resolve spatial units improves with proximity.

point. This is an expression of the inverse square law. The total flux is given by the integral over all surface parcels visible from $\vec{r}$:

$$\phi(\vec{r}, E) = \int_S dA \frac{j_s(\hat{n} \cdot \hat{\Omega}, E)}{r^2} \quad \text{(particles/cm}^2\text{/s)}. \tag{9.10}$$

The rate at which particles interact with a detector is determined by multiplying the flux by the product of the projected area and the intrinsic efficiency of the detector, known as the efficiency–area product, denoted εA. Generally, the efficiency–area product depends on particle energy and the direction of incidence in the frame of the detector. The energy–angle response of detectors is typically characterized prior to flight, using a combination of experimental measurements and transport models (e.g., Prettyman et al., 2011). For the simple case of a spherical detector, εA depends only on particle energy, and the particle interaction rate within a detector at $\vec{r}$ is given by

$$C(\vec{r}) = \varepsilon A(E) \varphi(\vec{r}, E) \quad \text{(counts/s)}. \tag{9.11}$$

When the surface leakage current varies linearly with the cosine of the emission angle ($j_s \propto \hat{n} \cdot \hat{\Omega}$), the flux and counting rate vary in proportion to the solid angle subtended by the surface at $\vec{r}$. The solid angle is given by

$$\Omega = \int_S dA \frac{\hat{n} \cdot \hat{\Omega}}{r^2} \quad \text{(steradians)}. \tag{9.12}$$

The assumption of linearity is often a reasonable approximation and is strictly valid for gamma rays emitted by the decay of radioelements within a homogeneous regolith; however, we note that particles produced by cosmic ray interactions are typically forward peaked, such that $j_s \propto (\hat{n} \cdot \hat{\Omega})^k$, where $k > 1$. Solid angle corrections are nevertheless routinely applied to nuclear spectroscopy counting data to remove variations due to altitude and the geometry of the planetary body. The corrected counts are independent of geometry and sensitive to spatial variations in surface composition.

Equation (9.13) can equivalently be written in spherical coordinates as an integral over the limiting angles of the surface:

$$\Omega = \int_{\mu_{min}}^{\mu_{max}} d\mu \int_{\phi_{min}(\mu)}^{\phi_{max}(\mu)} d\phi \quad \text{(steradians)}, \quad (9.13)$$

where μ is the cosine of the polar angle and ϕ is the azimuthal angle. The equivalence of Eqs. (9.13) and (9.14) can be exploited to efficiently evaluate Eqs. (9.10) and (9.11) for irregularly shaped objects using Monte Carlo (Prettyman et al., 2011, 2017, 2019). For spherical objects, Eq. (9.13) has a simple analytical form:

$$\frac{\Omega}{2\pi} = 1 - \sqrt{1 - (R_0/R)^2}, \quad (9.14)$$

where R and R_0 are respectively the orbital radius and radius of the planet. For measurements made on the surface ($R = R_0$), $\Omega = 2\pi$ steradians. The solid angle decreases with increasing radial distance, with the measured counting rate. Close proximity ($R < 2R_0$) is typically required for nuclear spectroscopy. Background is produced by GCR interactions with the orbiting spacecraft. When in close proximity, the planetary body shields the spacecraft from a portion of the cosmic ray flux. Spacecraft contributions to the background are reduced by the fraction of the sky that is not obscured by the body is $(4\pi - \Omega)/4\pi$. Thus, measurements close to the surface have higher signal (Eq. 9.14) and lower background than those made further away. The relative variation of signal and background with spacecraft altitude is shown in Figure 9.8c.

For sensors with omnidirectional sensitivity, radiation emitted from all portions of the surface visible from the spacecraft contributes to the response. Nuclear spectrometers are omnidirectional, or nearly so depending on sensor geometry and shielding, and thus measure broad regions of the surface. Their ability to distinguish surface features is relatively poor compared to optical imagers. In addition, the size of surface regions sampled depends on altitude. The ability to distinguish surface regions with different composition is given by the spatial resolution. Spatial resolution is represented by the distance apart (arc length on the surface) two point sources of equal intensity must be before they can be distinguished. For altitudes less than a body radius, spatial resolution is approximately 1.5 × altitude (Figure 9.8c). Altitude determines the spatial scale on which orbital data can be mapped. For example, Lunar Prospector's low-altitude orbit was ~30 km above the lunar surface, enabling characterization of surface compositional units with scales greater than 45 km (about 1.5° of arc length).

9.3.7 Radioelements and the Isotropic Approximation for Orbital Fluxes

For gamma rays produced by radioelements, the total surface leakage current of uncollided gamma rays, integrated over all emission angles, has a simple analytical form for a homogeneous regolith (e.g., Prettyman et al., 2015):

$$J_s = \frac{1}{4} P/\mu \quad \left(\text{gamma rays/cm}^2/\text{s}\right), \tag{9.15}$$

where P is the volumetric production rate of decay gamma rays, given by Eq. (9.5), and μ is the mass attenuation coefficient of the regolith (Figure 9.7). The angular variation of the current is linear in the cosine of the emission angle. It follows from Eq. (9.8) that the flux of uncollided gamma rays at the surface is $\varphi_s = 2J_s$ (gamma rays/cm²/s). Recall that the production rate of 2.615 MeV gamma rays by the decay of ^{232}Th was 5.3 gamma rays/cm³/hour per gram of regolith when the concentration of Th was 1 µg/g, similar to concentrations found in the lunar highlands (Section 9.2.3). The mass attenuation coefficient of regolith materials is relatively insensitive to composition for gamma rays with energies greater than about 1 MeV. At 2.615 MeV, the regolith mass attenuation coefficient is ~0.04 cm²/g. In this case, the total number of uncollided gamma rays escaping the surface is ~0.55 gamma rays/cm²/min, which corresponds to a flux of 1.1 gamma rays/cm²/min. From Table 9.3, Reedy (1978) reports a flux of 2.19 gamma rays/cm²/min for 1.9 mg/g Th, equivalent to a flux of 1.15 gamma rays/cm²/min for 1 µg/g Th, which is consistent with our calculation.

When surface emission is linear in the cosine of the emission angle, the flux at orbit is given by (Prettyman et al., 2015)

$$\varphi_s = 2J_s \frac{\Omega}{2\pi} \quad \left(\text{gamma rays/cm}^2/\text{s}\right). \tag{9.16}$$

This simple expression enables accurate estimates of the flux of uncollided gamma rays produced by radioelements. It can also provide rough estimates of fluxes of particles produced by galactic cosmic rays and X-rays, given J_s from analytical models or Monte Carlo calculations.

9.3.8 X-Ray Production and Emission

The laboratory approach of production of X-rays in an infinitely thick and chemically homogeneous sample can be calculated analytically (Shiraiwa & Fujino, 1966; Jenkins, 1999) and is essentially applicable to orbital X-ray spectrometry. Figure 9.8b shows a planetary geometry where an incident flux $\varphi_0(E)$ (photons cm^{-2} s^{-1}) of solar X-rays is crossing into the planetary surface with an incidence angle ψ. E is the photon energy. This angle can vary with time and location. The figure depicts production of fluorescent X-rays from element i at depth d_s after the solar X-rays have traversed a distance x_s. Depth d_p represents the penetration depth of the incident X-rays from where no contributions to the emergent X-ray flux are to be expected.

In planetary X-ray florescence experiments K X-ray lines dominate analysis of surface element abundances and emission lines are not resolved for most of the elements of interest. The emission probability Q_{if} of fluorescent K lines,(see Figure 9.6), is then given by

$$Q_{if} = W_i \mu_i(E) \omega_{Ki}(\tau_K/\tau), \tag{9.17}$$

where W_i is the weight fraction of element i, $\mu_i(E)$ is the partial mass absorption coefficient of element i, ω_{Ki} is the K-shell fluorescence yield, and (τ_K/τ) is the K-edge jump ratio, shown in Figure 9.6 (Shiraiwa & Fujino, 1966).

The current of uncollided fluorescent X-ray photons from element leaving the surface with an angle θ relative to surface normal $\hat{n}$ is denoted J_i (photons cm^{-2} s^{-1}), and can be calculated as an integral over the energy range of exciting X-rays (Shiraiwa & Fujino, 1966):

$$J_i(E) = \int_{E_{Ki}}^{E_{max}} \frac{Q_{if}(E)\varphi_0(E)}{\mu(E_i) + \mu_i(E)\cos\theta/\sin\psi} dE, \tag{9.18}$$

where E_{Ki} is the K-shell binding energy for element i, and E_i is the energy of the fluorescence X-ray.

The observed fluorescent X-ray intensity of an element i that reaches an orbital detector is influenced by several factors, including the variability of the incident solar spectrum, the effects of the overall composition of the planetary surface and its physical constitution, as well as the overall observation geometry, therefore affecting the translation of current into elemental surface abundances.

The uncollided X-rays produce distinct peaks or lines in the spectrum measured by the spacecraft-based detector. These are superimposed on a broad continuum of radiation produced primarily by inelastic Compton scattering of solar and fluorescence photons within the regolith (ref. Section 9.5.3 and Figure 9.9). Determination of line intensities requires removal of the continuum using peak analysis methods similar to those used in nuclear spectroscopy (ref. Section 9.5.2).

The incident solar X-ray spectrum is highly variable in time, intensity, and energy, and ranges from so-called quiet Sun conditions to very strong X-class solar flares. Solar X-ray emission conditions are tied to the solar cycle (see Figure 9.2), where the number and intensity of solar flares follows the 11-year solar cycle. The quiet Sun X-ray spectrum can be modeled as emission from a hot, single-temperature, optically thin coronal plasma with coronal elemental abundances, resulting in a low-energy spectrum. Solar flares are dynamic, have high temperature multithermal and nonthermal components, and develop over time. In a flare during the precursor stage the evolution of soft (<10 keV) and hard X-rays (>20 keV) can be observed. During the impulsive phase a sudden fast increase of nonthermal hard X-rays is observed, lasting only for a few (~1–2) minutes, accompanied by a strong heating of the coronal plasma to tens of mega-Kelvin. After the peak of the hard X-ray flux, in the gradual phase the thermal soft X-ray flux begins to dominate. The overall time of each of the phases can vary. It is this high-temperature gradual phase that is useful

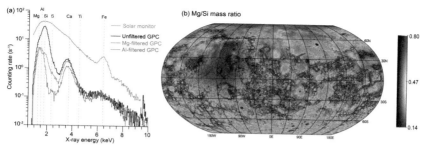

Figure 9.9 A solar flare solar monitor (SAX) spectrum, corresponding Gas Proportional Counter (GPC) X-ray spectra of filtered and unfiltered detectors, and a final data product Mg/Si map of Mercury as obtained by MESSENGER are presented side by side. (a) Example of an averaged solar monitor spectrum and corresponding XRS spectra obtained from a solar flare (Oct. 26, 2013, from 9:32 through 9:36 UTC) (MESSENGER X-Ray Spectrometer Calibrated Data Record archived at NASA's Planetary Data System) with a flare temperature of 14.9 MK (Weider et al., 2015). X-ray spectra of each of the three GPCs are shown in black for the unfiltered, in red for the Mg-filtered, and in blue for the Al-filtered detector; backgrounds have not been subtracted. The solar monitor spectrum shows a drop off at energies <2 keV due to attenuation by the detector's Be-window. The SAX spectrometer resolution at the time of measurement was in the order of 635 eV at 6.49 keV (Starr et al., 2016). The response of the SAX Si-PIN detector shows the 6.4 keV Fe emission line from highly ionized Fe atoms in the solar corona. From this spectrum best-fit solar coronal emission can be calculated using CHIANTI (Dere et al., 2009). During flares, Mercury surface fluorescence from elements up to Fe are observed. (b) Map of the Mg/Si elemental weight ratio derived from MESSENGER X-Ray spectrometer measurements (MESSENGER X-Ray Spectrometer Reduced Data Record archived at NASA's Planetary Data System). The map was determined from solar flare and quiet-Sun X-ray spectra.

for planetary X-ray spectroscopy, as it emits a high-flux and high-energy incident X-ray spectrum with several emission lines of coronal elements superimposed at plasma temperatures exceeding 8 MK.

Quiet Sun and solar flare thermal X-rays are described by an isothermal continuum and spectral lines model, CHIANTI 8.0 (Del Zanna et al., 2015). It provides spectra as a function of the plasma temperature in MK (which changes shape and spectral line intensities) and the emission measure $EM = n^2 V$ (cm^{-3}), where V is the volume (cm^3) of the emitting plasma, and n is density of the emitting plasma (cm^{-3}). CHIANTI-modeled photon fluxes from 5 to 30 MK are used in planetary X-ray data evaluation.

Matrix effects, including attenuation and secondary emission, can generally influence emission intensity of a fluorescence line. In addition to direct excitation as described in Eq. (9.18), if X-rays from an element i have energies greater than the binding energy of K-shell electrons of element j, then X-ray fluorescence from element j is excited. Thus a fraction of the photons of element i is consumed to generate XRF photons from j. Element j X-ray current increases while the current of element i is reduced. Thus the emergent intensity j_i is sensitive to the "matrix" of elements in the sample. A detailed analysis of these effects can be found in Shiraiwa and Fujino (1966).

In planetary surface regions where the mean particle size is larger than the penetration depth of X-rays, particle size can affect the X-ray fluorescence line intensity. Laboratory experiments have shown that the X-ray fluorescence line intensity for a given element decreases with increasing particle size (e.g., Näränen et al., 2009). More recent studies (e.g., Weider et al., 2011) provide evidence of a grain-size effect, where XRF line intensity decreases with increasing sample grain size. The studies also demonstrated an almost ubiquitous increase in XRF line intensity above incidence angles of $60°$.

Implicit in Eq. (9.18) is a notion that the emergent flux of a fluorescent X-ray depends on the geometry of the observation. Theoretical and experimental studies have shown that for observations made at a fixed emission angle θ, the XRF current increases with increasing incidence angle Ψ; for measurements made at a fixed incidence angle, the XRF current decreases with increasing emission angle (e.g., Parviainen et al., 2011; Weider et al., 2011). High phase angles ϕ ($90°$ or more), i.e., the angle between incident solar X-rays and detector normal, are also related to increased XRF current (Okada, 2004). Measurements made with phase angles $<70°$ are least likely to be subject to the phase-angle effects. The magnitude of these geometry-dependent effects varies with X-ray energy, grain size, and sample composition (Weider et al., 2011). These effects bear on elemental abundance determination discussed in Section 9.5.3.

9.4 Instrumentation

Detection of neutral particles (X-rays, gamma rays, and neutrons) requires their conversion to charged particles via interactions with the detector medium. The liberated charge produces a signal that can be counted or recorded as a pulse height spectrum. The most common types of detectors used for planetary science are gas proportional counters, scintillators, and semiconductors. A thorough description of radiation sensors and detection methods is provided by Knoll (1989).

Planetary instruments must be capable of separating the signal produced by particles originating from the planetary surface from the background. In the case of nuclear spectroscopy, sources of background include high-energy cosmic rays (solar and galactic) and secondary particles, including gamma rays and neutrons, produced by interactions with the spacecraft and instrument. Spectrometers must have ample pulse height resolution to enable the identification and measurement of radiation from specific nuclear reactions. Finally, the sensors must have high detection efficiency to maximize the precision of the measurements.

The aforementioned considerations must be balanced against the limited resources allocated to the payload. Instruments must meet payload requirements for size, mass and power. Data volume is also a consideration; however, X-ray and nuclear spectrometers have minimal requirements compared to optical imagers typically included on planetary payloads.

9.4.1 Detection Efficiency

As discussed in Section 9.3.6, the flux of gamma rays and neutrons arriving at orbital altitudes can be converted to counts given the product of the intrinsic efficiency ε and area of the sensor projected in the direction of the incident particle. The efficiency–area product can be determined experimentally in the laboratory and/or calculated by Monte Carlo (e.g., Prettyman et al., 2011). For computational purposes, it is useful to imagine a bounding sphere that contains the sensor of interest, the entire instrument, or possibly the entire spacecraft (e.g., see Figure 9.10). With a sphere, the projected area is independent of the direction of incidence, and the current of particles crossing through the projected surface is identical to the flux. If the detector was truly omnidirectional, then the efficiency–area product would also be independent of particle direction. In many situations, this is a reasonable approximation; however, if the detector is planar or is shielded by surrounding materials, active or passive, then the response may vary with direction.

In this work, we define the intrinsic efficiency as the fraction of particles crossing through the projected surface of the bounding sphere for which an event is registered by the counting system. We note that this deviates from the conventional definition of intrinsic efficiency, which is based on the projected area of the sensor's active volume (Knoll, 1989). Use of a bounding sphere is attractive for calculating the response of instruments with multiple sensors or complex sensor geometry.

The intrinsic efficiency can be determined experimentally, by placing the detector into a calibrated beam of particles and measuring the response as a function of beam incidence. In an analogous fashion, the response can also be calculated by Monte Carlo, by sending a beam of particles through the projected surface and tallying the number of event-producing reactions that occur per incident particle. If scattering from the instrument or spacecraft structure is important, then the bounding sphere could be quite large, in which case the intrinsic efficiency would be very low; however, the product of intrinsic efficiency and projected area is independent of the size of the bounding sphere. When only collisions with the sensor are important, then the sphere can closely bound the sensor, resulting in higher throughput for Monte Carlo calculations of efficiency. Measurements and calculations of the efficiency–area product for primary signatures measured by planetary nuclear spectrometers are described in the literature (e.g., Feldman et al., 2002a, 2004b; Prettyman et al., 2011; Peplowski et al., 2012).

9.4.2 Gamma-Ray Spectrometers

Gamma-ray spectrometers convert the charge liberated by gamma-ray interactions within the sensitive volume of the detector into electronic pulses. Ideally, the pulse amplitude is proportional to the energy deposited by the gamma ray in the detector. Fundamental to detector design is the goal of capturing all of the energy of the incident gamma ray within the sensitive volume, which drives the selection of materials with high density and atomic number and/or by making the volume of the sensor as large as possible. This is particularly

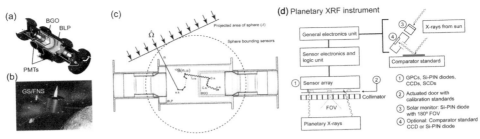

Figure 9.10 Nuclear spectrometer. (a) A cutaway view of the Lunar Prospector Gamma-Ray Spectrometer (GS) and Fast Neutron Sensor (FNS), which contains a bismuth germanate (BGO) crystal surrounded by boron-loaded plastic (BLP) (S. Storms, Los Alamos National Laboratory). The scintillators (BGO and BLP) are read out by separate photomultiplier tubes (PMT). (b) The sensor head is boom mounted to minimize spacecraft background. The diagram in (c) shows the geometry used to calculate the efficiency–area product (εA) for incident particles. The interaction rate is given by the flux of incident particles (Eq. 9.10) multiplied by the projected area A of the bounding sphere and the probability ε that a particle crossing the projected surface will interact with the sensor. The efficiency–area product depends on angle of incidence; however, a spherical approximation is sometimes used (Eq. 9.11). The efficiency–area product also depends on particle type and energy and the specific signature recorded by the spectrometer. For example, the interaction of an epithermal neutron is depicted. The neutron undergoes multiple elastic collisions, "e.s.," before being absorbed by ^{10}B, denoted "^{10}B(n,α)." This reaction produces light sensed by the photomultiplier and often leaves ^{6}Li in an excited state, which deexcites by the emission of a 478-keV gamma ray. For the event depicted, the gamma ray interacts with the BGO scintillator, depositing all of its energy by successive Compton scatters "C.s." and terminating with photoelectric effect "p.e." Thus, one of the signatures of a neutron is light output from the BLP in coincidence with detection of light in the BGO, with pulse amplitude near 478 keV. The εA product for the different event categories recorded by a spectrometer is usually calculated by Monte Carlo and verified via measurements of neutron and gamma-ray sources in the laboratory. (d) Components of a planetary XRF instrument include a general electronics unit, a sensor electronics and logic unit aligned with the sensor array and its functions, an actuated door often with radioisotope calibration sources, and a collimator to constrain the field of view. Sensor arrays consist of multiple X-ray detector units. The detector types used include gas proportional counters (GPCs) (Adler & Trombka, 1970; Adler et al., 1972a; Bielefeld et al., 1977), Si-PIN diodes (Peng et al., 2009; Ouyang et al., 2010a), charged coupled devices (CCDs) (Arai et al., 2008), and swept charge devices (SCDs) (Crawford et al., 2009; Grande et al., 2009; Howe et al., 2009). A major component of the XRF instrument is a solar monitor, most commonly an Si-PIN diode, with a large field of view (Huovelin et al., 2002). A small set of flight experiments carried a comparator standard exposed to the same solar X-ray fluxes as the surface under observation with fluorescence observed by a CCD or Si-PIN diode (Okada et al., 2009b).

important for the measurement of cosmogenic gamma rays, which span a wide range of energies from a few hundred keV to ~10 MeV. High-energy gamma rays have long mean free paths and interact primarily by pair production and Compton scattering (e.g., Figure 9.7). Some will pass through the sensor entirely or scatter out, depositing only a portion of their energy. These produce a continuum of pulses to accompany distinct peaks that result from full energy deposition (Figure 9.11). The pulse-height resolution of the spectrometer – the ability

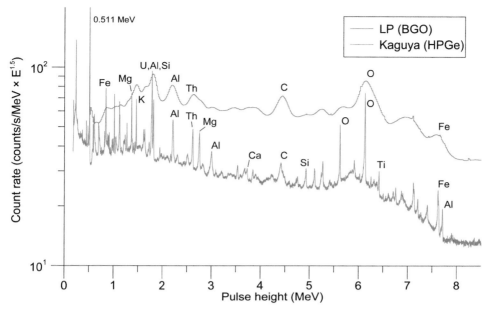

Figure 9.11 Comparison of globally averaged lunar spectra acquired by Lunar Prospector's BGO scintillator (blue, LP) and Kaguya's cryogenically cooled, high-purity germanium (HPGe) semiconductor. The x-axis is pulse amplitude, calibrated to gamma-ray energy. Distinct peaks in the spectra are reaction gamma rays, primarily made by neutron interactions in the lunar surface. These are superimposed on a continuum that is produced by various processes, including Compton scattering of gamma rays in the lunar regolith, instrument, and spacecraft.

to separate individual gamma rays in a spectrum – must also be sufficient to discern contributions from individual elements.

The requirements for resolution and efficiency have led to the implementation of two general types of gamma-ray spectrometers for planetary applications: semiconductors and inorganic scintillators. For semiconductors, an external bias is applied via electrical contacts to establish an electric field within a dielectric medium. The medium is ionized by the passage of energetic charged particles (e.g., electrons made by Compton scattering and the photoelectric effect). In this process, electrons are promoted to the conduction band, leaving behind a positive hole in the valance band. The electrons and holes are swept away by the electric field and collected by the contacts. The motion of carriers induces a current, which is proportional to the number of electron-hole pairs liberated, which is in turn proportional to the amount of energy deposited (e.g., Prettyman, 1999). Resolution is controlled by carrier statistics and trapping effects. For high-purity Ge (HPGe), trapping effects are small, and very high energy resolution can be achieved for large-volume detectors (Figure 9.11 and Table 9.4); however, cryogenic cooling is required to minimize thermal generation of carriers. The requirement for cryogenic cooling increases the

Table 9.4 *Comparison of selected sensor materials for planetary gamma-ray spectroscopy*

Material	Density (g/cm^3)	Effective atomic number[a]	Energy resolution (%)[b]	Perceived limitations	Flight heritage?
HPGe (intrinsic germanium)	5.3	32	0.2–0.6	Requires cryocooling, susceptible to radiation damage	Yes
$Cd_{1-x}Zn_xTe$, $x = 0.1$	5.8	50	1–3	Arrays required to achieve target volume, susceptible to radiation damage	Yes
$Bi_4Ge_3O_{12}$ "BGO"	7.13	71	9–10	Low resolution	Yes
NaI(Tl)	3.67	50	7–8	Low resolution, hygroscopic	Yes
CsI(Tl)	4.51	54	7–8	Low resolution	Yes
$LaBr_3$(Ce)	5.1	45	2.8–3.2	Self-activity, requires PMT readout,[b] hygroscopic	Yes
$CeBr_3$(Ce)	5.1	46	4–4.5	Requires PMT readout,[c] hygroscopic	No
Cs_2YLiCl_6(Ce) "CLYC"	3.3	43	4	Requires PMT readout,[c] crystal growth limits size – arrays needed to achieve target volume	No
SrI_2(Eu)	4.6	50	2.8–3.2	Hygroscopic	No

a For mixtures of elements, the single atomic number of that best represents the attenuation coefficient of the material, evaluated for intermediate gamma-ray energies using the expression suggested by Murty (1965).
b Full-width-at-half-maximum at 662 keV.
c To achieve best resolution for γ-ray spectroscopy.

complexity of thermal design of flight instruments as well as risk when actively cooled. Nevertheless, spectrometers using HPGe have successfully flown on missions to Mars, Mercury, and the Moon (Boynton et al., 2004; Goldsten et al., 2007a; Kobayashi et al., 2013).

Inorganic scintillators such as sodium iodide (NaI), bismuth germanate (BGO), lanthanum bromide ($LaBr_3$), and strontium iodide (SrI_2) make efficient gamma-ray detectors because they have high density and atomic number. These transparent, crystalline solids are insulators. Luminescence is activated by the addition of an impurity (Birks, 1964), such as Eu^{2+} in the case of SrI_2 (Cherepy et al., 2008). The passage of ionizing radiation through a scintillator produces electron-hole pairs and excitons (an excited electron bound to a hole), which migrate from their production sites via diffusion. Luminescence centers formed by the impurity ions are excited by the absorption of excitons and the recombination of free carriers. The transition to ground state results in the emission of visible light, which can be detected by a photosensor, such as a photomultiplier tube, avalanche photodiode (APD), or silicon photomuliplier (SiPM). However, radiation-less transitions (quenching) can compete with emission. In addition, the scintillator may not be entirely transparent to its emitted light. For some materials, absorption and subsequent reemission of photons must be considered for large crystals. The efficiency of luminescence depends on the density of excitons produced along the track of an ionizing particle. For this reason, light output is influenced by the linear energy transfer of ionizing particle, which is dependent on particle type and kinetic energy.

Scintillation light yield and the nonproportionality between the energy deposited by electrons and light output are among the factors that limit the resolution of scintillators (Knoll, 1989; Payne et al., 2009). In recent years, significant progress has been made in improving resolution. Ultrabright scintillators, including $LaBr_3$, $SrI_2(Eu)$, and the elpasolite CLYC, have improved resolution over previously flown scintillators because of increased light yield and low nonproportionality (Table 9.4) (e.g., Cherepy et al., 2008; Glodo et al., 2008). Spectrometers using large-volume crystals suitable for use in planetary science applications and with energy resolution of about 3–4% full-width-at-half-maximum at 662 keV have been manufactured (e.g., Cherepy et al., 2013). A $LaBr_3$ sensor was flown to the Moon on the Chang'E 2 mission (e.g., Zhu et al., 2013); however, the decay of radiolanthanum overwhelms peaks from natural radioelements and reaction gamma rays below ~3 MeV, making the technology undesirable for planetary applications. Strontium iodide and CLYC have negligible self-activity and are currently being evaluated by NASA for planetary applications.

Scintillators are routinely used for planetary gamma-ray spectroscopy, particularly for missions that require rugged sensors, with minimal complexity and consumption of payload resources. They do not require cryogenic cooling and are resilient to radiation damage effects. In contrast, the resolution of semiconductors, such as HPGe and CdZnTe, degrades with exposure to the radiation in the space environment. Periodic high-temperature annealing during flight is required to achieve optimal performance (e.g., Brückner et al., 1991; Goldsten et al., 2007; Prettyman et al., 2011; Kobayashi et al., 2013). Scintillator-based gamma-ray spectrometers were flown on Apollo (NaI) (Adler et al., 1973b), Venera/Vega (NaI) (e.g., Surkov et al., 1987), Lunar Prospector (BGO) (Feldman et al., 2004b), the Near Earth Asteroid Rendezvous (NEAR) (Goldsten et al., 1997) (BGO and CsI), Chang'E 1 and 2 (CsI and $LaBr_3$) (Zhu et al., 2013), and Dawn (BGO) (Prettyman et al., 2011).

9.4.3 Neutron Spectrometers

Neutron spectrometers convert neutrons to charge via nuclear reactions. A combination of shielding and coincidence counting can be used to separate neutrons into three broad energy ranges: thermal (<0.5 eV), epithermal (0.5 eV to a few hundred keV), and fast (greater than a few hundred keV to about 10 MeV) (Figure 9.4). Two types of neutron detectors have been used on planetary missions: ^{3}He proportional counters and loaded plastic scintillators.

The former are gas ionization chambers, cylindrical in geometry, consisting of a tube with a central wire anode (Knoll, 1989). The tube is filled with ^{3}He at high pressure, up to about 10 atm. The anode is biased relative to the housing to produce a radial electric field. Neutrons react with ^{3}He via n + ^{3}He → p + ^{3}H + Q (764 keV), which has a cross section of a few thousand barns for slow neutrons. The recoiling reaction products, the proton (p) and triton (^{3}H), ionize the gas and the electrons and ions are swept in opposite directions in the electric field, inducing charge on the electrodes. The pulse height spectrum contains a distinct peak corresponding to the deposition of all of the energy of the reaction products in the gas. The area under this peak gives the neutron interaction rate within the gas (e.g., Feldman et al., 2004b).

Helium-3 proportional counters were used on Lunar Prospector to separately measure the leakage flux of thermal and epithermal neutrons (e.g., Feldman et al., 2004b; Lawrence et al., 2006). This was accomplished using two separate counters, one covered with Cd and the other covered with an equivalent thickness of Sn. Neutrons in the thermal and epithermal energy range can penetrate the Sn, whereas only neutrons above the Cd cutoff (about 0.5 eV) can penetrate the Cd. Thus, the Sn-covered tube is sensitive to a combination of thermal and epithermal neutrons, whereas the Cd-covered tube is sensitive to epithermal neutrons. The thermal neutron interaction rate can be determined by subtracting the counts measured by the Cd-covered tube from those measured by the Sn-covered tube.

Scintillators loaded with a neutron absorber can also be used as neutron detectors. A boron-loaded plastic (BLP) scintillator, read out by a photomultiplier tube (PMT), was included in the Gamma-ray Sensor (GS) head on Lunar Prospector (Figure 9.10), serving both as an anticoincidence shield (ACS) and fast neutron spectrometer (Figure 9.10) (Feldman et al., 2004b). The BLP contained about 5 wt.% natural boron, which contains about 20 at.% ^{10}B, which is a strong neutron absorber. Neutrons are captured via n + ^{10}B → ^{4}He + ^{7}Li, or ^{10}B(n, α) in reaction shorthand notation (Figure 9.10). The reaction products produce a peak in the pulse height spectrum at about 93 keV$_{ee}$, where the subscript "ee" stands for "electron equivalent." In other words, the interaction of a 93-keV electron would produce the same amount of light as that made by the reaction products. The area of the peak is proportional to the rate at which neutrons are captured by ^{10}B in the scintillator, which is proportional to the incident flux of neutrons primarily in the thermal and epithermal energy range.

Boron-loaded plastic enables the detection of fast neutrons as illustrated in Figure 9.10 (Feldman et al., 1991). Fast neutrons undergo successive collisions with hydrogen in the plastic, producing recoil protons, which ionize the medium resulting in the production of a prompt flash of light. Once the neutron kinetic energy falls below about 0.5 MeV light is no longer produced by this mechanism. The neutron continues to undergo collisions until it is captured by ^{10}B, producing a second flash of light. The time between pulses is typically on the order of a few microseconds. For double-pulse events, the first pulse count rate gives the interaction rate of neutrons with energies >0.5 MeV. The spectrum of first-pulse amplitudes can be analyzed to determine the energy distribution of fast neutrons (Feldman et al., 2002b). The double-pulse signature was used for fast neutron spectroscopy on Lunar

Prospector, Mars Odyssey, MESSENGER, and Dawn (e.g., Feldman et al., 2002, 2004b; Prettyman et al., 2011; Lawrence et al., 2013a,b).

Arrangements of Li-loaded glass and B-loaded plastic can be used to separately measure thermal, epithermal, and fast neutrons (e.g., Goldsten et al., 2007; Prettyman et al., 2011). For example, the primary neutron detector for Dawn's GRaND consisted of a block of boron-loaded plastic optically coupled on one side to a thin layer of another scintillator, lithium-loaded glass. Both scintillators were viewed by the same PMT. The remaining five sides of the block were wrapped with gadolinium. The Li-loaded glass and Gd absorb all thermal neutrons. Consequently, only neutrons in the epithermal and fast energy range reach the boron-loaded plastic. Epithermal neutrons <0.5 MeV produce a single pulse when captured by boron; whereas, fast neutrons make the aforementioned double-pulse signature.

The lithium is enriched in ^{6}Li, which is a strong neutron absorber. Neutrons are absorbed by n + ^{6}Li → ^{3}H + ^{4}He, which makes about 260 keV$_{ee}$ of light in GRaND's composite scintillator. Neutrons that are captured by ^{6}Li appear as a distinct peak, separately resolved from the 93 keV$_{ee}$ peak from capture with ^{10}B. The ^{6}Li peak contains contributions from thermal and epithermal neutrons, from which the epithermal contributions can be subtracted. The approach for separating thermal and epithermal neutrons is similar to that taken with Sn- and Cd-covered ^{3}He tubes. An emerging scintillator for dual neutron and gamma-ray spectroscopy is CLYC, which contains Li and has higher intrinsic detection efficiency than ^{3}He with the same sensitive volume.

9.4.4 X-Ray Spectrometers

X-ray spectrometers in planetary remote sensing have to work under limitations of being on a spacecraft and the constraints of the space radiation environment. Gas Proportional Counters (GPCs) are advantageous because they are resistant to radiation damage and consume minimal power. A disadvantage of GPCs is their low energy resolution (>300 eV at 5.9 keV), which requires multiple detectors and filtering techniques to measure elements with low atomic number or low fluorescent X-ray energies (Yin et al., 1993).

Semiconductor detectors have higher energy resolution (<150 keV at 5.9 keV) and therefore can better resolve fluorescence X-rays from elements with low atomic number (Jenkins, 1999). A disadvantage of semiconductor detectors is their sensitivity to radiation damage and the requirement to operate at low temperatures (–40°C or lower).

Figure 9.10d shows a common concept of planetary XRF spectrometer experiments. The design includes general and specific detector electronics and logic units; an array of detectors of choice for the mission; an actuated door, where needed; the capacity for in-flight calibration, mostly with radioisotopes; and a collimator limiting the field of view, as well as solar monitors.

Detailed descriptions of the instrument assembly for Apollo, NEAR, and MESSENGER missions using gas proportional counters are provided by Yin et al. (1993), Trombka et al. (1997, 1999); Starr et al. (2000), Trombka et al. (2000), and Nittler et al. (2001).

Information on Chang'e 1 and 2 missions using Si-PIN detectors is given by Ouyang et al. (2008, 2010a,b) and by Peng et al. (2009). Hayabusa and SELENE mission CCD detector technology information is provided by Okada et al. (2002a,b, 2009a,b) and Shirai et al. (2008), and the successful test and use of swept charge devices (SCDs) as X-ray detectors is described in detail by Grande (2001), Grande et al. (2002, 2003, 2007), Alha et al. (2009), and Narendranath et al. (2010).

9.5 Data Reduction and Analysis

With numerous completed missions, data reduction and analysis methods for X-ray and nuclear spectroscopy are now quite mature (see Figure 9.2 and Chapter 30). Nevertheless, each mission and target provides unique challenges and opportunities to refine the state of the art. This section provides a summary of data reduction and analysis methods and how they have been applied to planetary datasets.

9.5.1 Data Reduction for Nuclear Spectroscopy

Nuclear spectrometers typically acquire a time series of counting data, including histograms of pulse heights recorded by the sensors. The accumulation time for each interval in the series depends on the objectives of the observation. Low-altitude orbital mapping requires fine time intervals in order to characterize specific surface regions (e.g., Maurice et al., 2004). The data for each accumulation interval are compressed, packetized, and stored on the instrument or sent to a buffer on the spacecraft to await downlink to the ground. The data volume is usually small compared to imagers and optical spectrometers that may be included on the payload.

Once data are on the ground, the raw data are decoded, decompressed, and processed to create a data set suitable for scientific analysis (e.g., Lawrence et al., 2004; Maurice et al., 2004, 2007). Typical processing steps include removal of corrupt data packages, identified via checksum comparisons and other metrics; elimination of data during conditions that are not valid for analyses, such as solar energetic particle events; and conversion of data numbers to physical units. Artifacts introduced by the differential nonlinearity of the analog to digital converter are removed from histograms. The histograms are corrected for variations in the instrument gain, which cause the channel number of reaction peaks to drift over time. Once corrected, the histograms are calibrated to produce pulse height spectra with energy units. Finally, the location of each measurement in the time series is determined using navigation data along with the measurement geometry, including the solid angle subtended by the target body at the spacecraft (e.g., Prettyman et al., 2019).

At NASA, raw science data records and instrument housekeeping data are subjected to reversible processes to form a complete set of Experimental Data Records (EDRs). These are archived in the Planetary Data System along with higher level data products. Reduced Data Records (RDRs) are derived from the EDR. The RDR typically contain a time series of counts and calibrated pulse height spectra, corrected for measurement geometry and variations in the flux of galactic cosmic rays. Specific signatures, such as the triple coincidence rate on GRaND (Prettyman et al., 2011), can serve as proxies for the GCR flux and are used for GCR corrections. The corrected time-series or mapped counting rates are sensitive to the surface composition of the planetary body (e.g., Prettyman et al., 2012). Higher level counting products are derived from the RDR. These include maps of corrected pulse height spectra and/or peak intensities.

For the purpose of mapping, the surface of the target body is divided into a grid. For example, simple cylindrical and quasi-equal-area grids were developed for Lunar Prospector and have been used on most planetary missions (e.g., Lawrence et al., 2004). The simplest mapping procedures involves averaging time series counts and spectra for which the subsatellite point falls within each map pixel. Contributions from multiple flyovers of a pixel are often needed to achieve adequate precision in the average pixel counts. More complex mapping procedures include the use of smoothing to remove noise in the resulting map (e.g., Yamashita et al., 2010 and Figure 9.12) and models of the instrument spatial response function to properly distribute counts for each measurement location within surface regions visible from the spacecraft (e.g., Lawrence et al., 2003).

9.5.2 Nuclear Spectroscopy Analysis Methods

The intensity of peaks for gamma-ray full-energy interactions and neutron reactions are related to elemental concentration following geometry and, in the case of cosmogenic reactions, variations in the flux of galactic cosmic rays. For isolated peaks, simple region-of-interest methods can be used to subtract the background to determine the net peak area (e.g., Lawrence et al., 2002; Prettyman et al., 2017). Overlapping peaks can be analyzed using peak fitting methods, as illustrated in the analysis of uranium on the Moon in Figure 9.12 (Yamashita et al., 2010). In this case, the intensity of the 1.765 MeV gamma ray, once corrected for measurement geometry, is proportional to the concentration of uranium within the lunar regolith (see Section 9.3.7). Peak analysis methods are usually applied to spectral regions. It is also possible to simultaneously determine the concentration of multiple elements using the entire spectrum. For example, elemental spectral unmixing was used to globally map the concentrations of O, Mg, Al, Si, K, Fe, Ti, and Th within the lunar regolith (Prettyman et al., 2006). The methods applied depend on the resolution of the spectrometer, signal-to-background ratio, and counting statistics.

To reliably extract peak areas, multiple accumulation intervals are typically combined to minimize the statistical uncertainty in the counting data. This can be accomplished, for example, with the time series (e.g., using a boxcar average), or by mapping the counts or spectra onto a surface grid. When properly corrected, counts measured by a nuclear spectrometer can be modeled as Poisson random variates (e.g., Lawrence et al., 2004). For a single, isolated peak for which background contributions are negligible, the relative uncertainty in measured peak intensity varies inversely with the square root of the product of the counting rate (peak area) and accumulation time. Many signals measured by nuclear spectroscopy are faint, requiring long accumulation times in close proximity to the target to

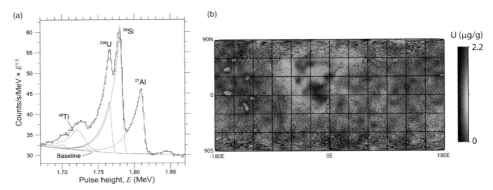

Figure 9.12 Analysis of uranium on the Moon. (a) The full-energy peak for the 1.765-MeV gamma ray produced by the decay of ^{214}Bi in the U decay series (Table 9.3) overlaps adjacent peaks from cosmogenic reactions in the instrument and regolith. The peak area of the 1.765 MeV peak, which is proportional to the concentration of U in the regolith (Eqs. 9.5, 9.15, 9.16), is determined by simultaneously fitting peaks in the energy window shown. The peak shapes are modeled as a Gaussian and exponential tail. The gamma-ray continuum is modeled as a polynomial. (b) The map of U was determined from peak areas extracted from many spectra acquired in a circular polar mapping orbit using methods described by Yamashita et al. (2010). For context, the map of uranium is superimposed on shaded relief. For regions with no data (gray), the peak area determined by the fitting procedure was negative.

maximize the counting rate (signal) and to reduce background contributions (Figure 9.8c). A thorough understanding of counting experiments and propagation of errors is needed to plan operations and to estimate uncertainties in derived compositional data (Knoll, 1989; Bevington & Robinson, 1992).

In contrast to the analysis of radioelements, peak areas for cosmogenic reactions do not vary in exact proportion to the concentration of the target isotope/element. For example, in the analysis of capture gamma rays, peak areas vary with the production rate of gamma rays within the subsurface (Eq. 9.6), which is proportional to the product of elemental weight fraction and neutron number density (Lawrence et al., 2002). For Lunar Prospector, the neutron number density was determined from thermal and epithermal neutron measurements, enabling the determination of the concentration of Fe and Ti (Feldman et al., 2000; Lawrence et al., 2002; Prettyman et al., 2006). A similar approach was taken by Yamashita et al. (2013) to determine the distribution of Fe on Vesta.

Fermi's theory of neutron slowing down (Section 9.2.2) motivates the determination of hydrogen from measurements of the leakage flux of epithermal neutrons, which is largely insensitive to variations in composition other than hydrogen. Monte Carlo modeling is required to accurately determine parameters used to determine hydrogen concentration from neutron measurements. Analyses of epithermal neutrons provided evidence for water in permanently shadowed regions near the lunar poles (e.g., Feldman et al., 2001; Elphic et al., 2005), subsurface ice at high latitudes on Mars (Feldman et al., 2004c), and delivery of hydrogen to the surface of Vesta by the infall of carbonaceous impactors (Prettyman et al., 2012). Measurements of epithermal and fast neutrons by MESSENGER's neutron spectrometer provided evidence for water ice at the poles of Mercury (Lawrence et al., 2012), and measurements of neutrons in the thermal and epithermal energy range by Dawn revealed a global subsurface ice table on Ceres.

Geochemical and geophysical constraints can provide added information to interpret data acquired by nuclear spectrometers. For example, two-layer models of hydrogen stratigraphy motivated by ice stability modeling were used to map the depth and concentration of subsurface ice on Mars using neutron and gamma-ray signatures with different depth sensitivities (e.g., Boynton et al., 2002; Feldman et al., 2011). Analyses of neutron and gamma-ray data guided by atmospheric modeling provided measurements of the thickness of Mars' seasonal caps and seasonal variations in the compositions of Mars' polar atmosphere.

The width of the spatial response function of the instrument is often broader than surface features of interest (e.g., Figure 9.8c). When geologic constraints are available, forward modeling of the instrument's response can be used to characterize sub-resolution surface units. Parameters such as the thickness of the south polar perennial CO_2 cap on Mars (e.g., Prettyman et al., 2004) and the concentration of Th for lunar red spots were determined using this approach (Hagerty et al., 2006). Forward modeling can also be used to smooth higher resolution data sets (e.g., mineral maps determined by reflectance spectroscopy) for direct comparison with elemental maps determined by nuclear spectroscopy (Prettyman et al., 2019). Spatial deconvolution can be applied to improve the spatial resolution of regional and global maps. The aim is to "sharpen" the map without introducing spurious artifacts in the reconstructed image. Spatial deconvolution has been successfully applied to various lunar and martian datasets (e.g., Elphic et al., 2007; Lawrence et al., 2007; Prettyman et al., 2009; Wilson et al., 2018).

9.5.3 X-Ray Analysis Procedures

The observed line strength $I_{obs}(i)$ of an element, as indicated in Section 9.3.8, is not a direct measure of surface abundance. There is a relationship between $I_{obs}(i)$ to surface current J_i that leads to the true abundance of element i:

$$I_{obs}(i) = f\big(\varphi_0(E), i\big) [J_i * \omega\big(\varphi_0(E), E\big) * S(E, \psi, \phi, i)], \tag{9.19}$$

where $f\big(\varphi_0(E), i\big)$ represents the incident solar spectrum $\varphi_0(E)$ for element i. $\omega\big(\varphi_0(E), E\big)$ represents chemistry matrix effects and $S(E, \psi, \phi, i)$ represents cumulative effects particle size, incidence, and phase angle, ψ and ϕ respectively, cross-correlated via the * convolution operator. The relation does not provide analytical solutions and has been used in a forward folding approach to derive the true elemental abundances (Narendranath et al., 2010, 2011). Based on Eqs. (9.17) and (9.18), this approach analytically predicts the expected X-ray line flux for a set of elemental composition matrices and compares the calculated X-ray line flux fraction (line flux/sum of the flux in all lines) with the observed line flux. Then the best set of elemental abundances is determined using χ^2 fitting. A larger parameter space is searched to establish a smooth convergence. Details of the XRF code and its validation through laboratory tests are given in Athiray et al. (2013).

The Chandrayaan-1 mission to the Moon also uses forward folding based on Eq. (9.19) to derive surface composition. The comparatively high energy-resolution of the SCD X-ray detectors and the properties of the XSM solar monitor (Grande, 2001) enabled accurate determination of elemental abundances. First results of lunar rock-forming elements of selected solar flares of the C1XS observations are provided by Narendranath et al. (2011) and Weider et al. (2012b). As in the fundamental-parameter approach of Clark and Rilee (2010), the intensity of a fluorescent X-ray emitted from a target can be calculated by first principles, as can coherent and incoherent scattering of the solar radiation. Factors that affect the XRF intensity are the primary solar X-ray flux, matrix effects, particle size effects, and the geometry of observation. Successful application of this analytical procedure is demonstrated through the discovery of enhanced abundances of sodium in the lunar surface derived from C1SX spectra (Athiray et al., 2014).

The previous application is the latest outcome based on the method originally developed for the Apollo 15 and 16 XRS experiments. For Apollo, theoretical X-ray line intensities were calculated using a fundamental parameter method for a set of anticipated lunar compositions, which were then compared with the observed line intensities, and the best match determined. Using this method, the Apollo experiments mapped about 10% of the equatorial region on the near side of the Moon (Adler et al., 1973a,b; Clark & Hawke, 1981, 1991). A forward modeling procedure was developed to fit lunar and planetary X-ray spectra (Clark & Adler, 1978; Clark & Trombka, 1997; Nittler et al., 2001, 2011; Weider et al., 2012a). The procedure uses 22 fundamental parameters which include distance to the Sun in au, the primary source flux of the incident solar spectrum, the viewing geometry, several atomic physics parameters, and the concentration of constituent elements in the target. It calculates the fluorescent line intensity in a spectrum and the scattered solar spectrum for each element (Clark & Trombka, 1997).

While quiet Sun X-ray spectra are common, the most comprehensive analyses are possible with X-ray spectra from solar flares (e.g., Nittler et al., 2011; Weider et al., 2012a). The primary solar flare spectra from the solar monitor are fitted to characterize the emitting solar plasma (e.g., its temperature and elemental composition), using CHIANTI (e.g., Dere et al., 2009; Del Zanna et al., 2015) to generate model spectra to be used in the forward modeling procedure. For solar flare MESSENGER X-ray spectra, theoretical X-ray spectra generated by the fundamental-parameter model are compared with the measured spectra to determine surface elemental abundances (e.g., Nittler et al., 2011). Nonlinear χ^2 minimization routines are used to fit the measured XRF spectra with theoretical spectra by varying the parameters. The fitted parameters include the energy calibration and resolution of the detectors, as well as the elemental abundances (wt.%) of Mg, Al, S, Ca, Ti, Cr, Mn, and Fe.

Weider et al. (2015) used data acquired by the MESSENGER X-ray Spectrometer under quiet-Sun conditions and during solar flares to measure elemental weight ratios. Global maps of Mg/Si and Al/Si and partial maps of S/Si, Ca/Si, and Fe/Si were determined (e.g., Figure 9.9). This was the first global survey of a Solar System body using X-ray fluorescence spectroscopy. The compositional data provided insights into magmatic processes that shaped Mercury's surface.

Acknowledgments

This work was supported by NASA's SSERVI Toolbox for Research and Exploration (TREX) project and by NASA's PICASSO program. The authors thank undergraduate research assistants D. Mozurkewich, who helped compile the references, and S. Borowski, who provided a student's perspective on the manuscript.

References

Adler I. & Trombka J. (1970) *Geochemical exploration of the Moon and planets*. Springer-Verlag, New York.

Adler I., Trombka J., Gerard J., et al. (1972a) Apollo 15 geochemical X-ray fluorescence experiment: Preliminary report. *Science*, **175**, 436–440.

Adler I., Trombka J., Gerard J., et al. (1972b) Apollo 16 geochemical X-ray fluorescence experiment: Preliminary report. *Science*, **177**, 256–259.

Adler I., Trombka J.I., Yin L.I., Gorenstein P., Bjorkholm P., & Gerard J. (1973a) Lunar composition from Apollo orbital measurements. *Naturwissenschaften*, **60**, 231–242.

Adler I., Trombka J.I., Lowman P., et al. (1973b) Apollo 15 and 16 results of the integrated geochemical experiment. *The Moon*, **7**, 487–504.

Agostinelli S., Allison J., Amako K., et al. (2003) Geant4—a simulation toolkit. *Nuclear Instruments and Methods in Physics Research A: Accelerators, Spectrometers, Detectors and Associated Equipment*, **506**, 250–303.

Alha L., Huovelin J., Nygård K., et al. (2009) Ground calibration of the Chandrayaan-1 X-ray Solar Monitor (XSM). *Nuclear Instruments and Methods in Physics Research A: Accelerators, Spectrometers, Detectors and Associated Equipment*, **607**, 544–553.

Arai T., Okada T., Yamamoto Y., Ogawa K., Shirai K., & Kato M. (2008) Sulfur abundance of asteroid 25143 Itokawa observed by X-ray fluorescence spectrometer onboard Hayabusa. *Earth, Planets and Space*, **60**, 21–31.

Athiray P.S., Sudhakar M., Tiwari M.K., et al. (2013) Experimental validation of XRF inversion code for Chandrayaan-1. *Planetary and Space Science*, **89**, 183–187.

Athiray P.S., Narendranath S., Sreekumar P., & Grande M. (2014) C1XS results—First measurement of enhanced sodium on the lunar surface. *Planetary and Space Science*, **104**, 279–287.

Battistoni G., Cerutti F., Fasso A., Ferrari A., & Muraro S. (2007) The FLUKA code: Description and benchmarking. *AIP Conference Proceedings*, **896**, 31–49.

Beck A.W., Lawrence D.J., Peplowski P.N., et al. (2015) Using HED meteorites to interpret neutron and gamma-ray data from asteroid 4 Vesta. *Meteoritics and Planetary Science*, **50**, 1311–1337.

Bevington P.R. & Robinson D.K. (1992) *Data reduction and error analysis for the physical sciences*, 2nd edn. McGraw-Hill, New York.

Bielefeld M.J., Andre C.G., Clark P.E., Adler I., Eliason E., & Trombka J. (1977) Imaging of lunar surface chemistry from orbital X-ray data. *8th Lunar Planet Sci. Conf.*

Birks J.B. (1964) *Theory and practice of scintillation counting*. Pergamon Press, Oxford.

Boynton W., Feldman W., Squyres S., et al. (2002) Distribution of hydrogen in the near surface of Mars: Evidence for subsurface ice deposits. *Science*, **297**, 81–85.

Boynton W., Feldman W., Mitrofanov I., et al. (2004) The Mars Odyssey gamma-ray spectrometer instrument suite. *Space Science Reviews*, **110**, 37–83.

Boynton W.V., Taylor G.J., Evans L.G., et al. (2007) Concentration of H, Si, Cl, K, Fe, and Th in the low- and mid-latitude regions of Mars. *Journal of Geophysical Research*, **112**, E12S99, DOI:10.1029/2007JE002887.

Brown E. & Firestone R.B. (1986) *Table of radioactive isotopes* (V.S. Shirley, ed.). John Wiley & Sons, New York.

Brückner J., Korfer M., Wanke H., et al. (1991) Proton-induced radiation damage in germanium detectors. *IEEE Transactions on Nuclear Science*, **38**, 209–217.

Cherepy N.J., Hull G., Drobshoff A.D., et al. (2008) Strontium and barium iodide high light yield scintillators. *Applied Physics Letters*, **92**, 083508.

Cherepy N.J., Payne S.A., Sturm B.W., et al. (2013) Instrument development and gamma spectroscopy with strontium iodide. *IEEE Transactions on Nuclear Science*, **60**, 955–958.

Clark B.C., Baird A.K., Rose H.J., et al. (1977) The Viking X Ray Fluorescence Experiment: Analytical methods and early results. *Journal of Geophysical Research*, **82**, 4577–4594.

Clark P.E. & Adler I. (1978) Utilization of independent solar flux measurements to eliminate nongeochemical variation in X-ray fluorescence data. *9th Lunar Planet. Sci. Conf.*, 3029–3036.

Clark P.E. & Hawke B.R. (1981) Compositional variation in the Hadley Apennine region. *12th Lunar Planet. Sci. Conf.*, 727–749.

Clark P.E. & Hawke B.R. (1991) The lunar farside: The nature of highlands east of Mare Smythii. *Earth, Moon, and Planets*, **53**, 93–107.

Clark P.E. & Rilee M.L. (2010) *Remote sensing tools for exploration: Observing and interpreting the electromagnetic spectrum.* Springer Science+Business Media, New York.

Clark P.E. & Trombka J.I. (1997) Remote X-ray spectrometry for NEAR and future missions: Modeling and analyzing X-ray production from source to surface. *Journal of Geophysical Research*, **102**, 16361–16384.

Crawford I.A., Joy K.H., Kellett B.J., et al., (2009) The scientific rationale for the C1XS X-ray spectrometer on India's Chandrayaan-1 mission to the moon. *Planetary and Space Science*, **57**, 725–734.

Del Zanna G., Dere K.P., Young P.R., Landi E., & Mason H.E. (2015) CHIANTI – An atomic database for emission lines. Version 8. *A&A*, **582**, A56.

Dere K.P., Landi E., Young P.R., Del Zanna G., Landini M., & Mason H.E. (2009) CHIANTI – an atomic database for emission lines. *A&A*, **498**, 915–929.

Donnelly J., Thompson A., O'Sullivan D., et al. (2012) Actinide and ultra-heavy abundances in the local galactic cosmic rays: an analysis of the results from the LDEF ultra-heavy cosmic-ray experiment. *The Astrophysical Journal*, 747(1), DOI:10.1088/0004-637X/747/1/40.

Duderstadt J.J. & Hamilton L.J. (1976) *Nuclear reactor analysis.* John Wiley & Sons, New York.

Elphic R.C., Lawrence D.J., Feldman W.C., et al. (1998) Lunar Fe and Ti abundances: Comparison of Lunar Prospector and Clementine data. *Science*, **281**, 1493–1496.

Elphic R.C., Lawrence D.J., Feldman W.C., et al. (2000) Lunar rare earth element distribution and ramifications for FeO and TiO2: Lunar Prospector neutron spectrometer observations. *Journal of Geophysical Research*, **105**, 20333–20345.

Elphic R., Lawrence D., Feldman W., et al. (2005) Using models of permanent shadow to constrain lunar polar water ice abundances. *36th Lunar Planet. Sci. Conf.*, Abstract #2297.

Elphic R.C., Eke V.R., Teodoro L.F.A., Lawrence D.J., & Bussey D.B.J. (2007) Models of the distribution and abundance of hydrogen at the lunar south pole. *Geophysical Research Letters*, **34**, L13204, DOI:10.1029/2007GL029954.

Evans L.G., Starr R.D., Brückner J., et al. (2001) Elemental composition from gamma-ray spectroscopy of the NEAR-Shoemaker landing site on 433 Eros. *Meteoritics and Planetary Science*, **36**, 1639–1660.

Evans R.D. (1955) *The atomic nucleus.* McGraw-Hill, New York.

Feldman W.C., Auchampaugh G.F., & Byrd R.C. (1991) A novel fast-neutron detector for space applications. *Nuclear Instruments and Methods in Physics Research A: Accelerators, Spectrometers, Detectors and Associated Equipment*, **306**, 350–365.

Feldman W.C., Lawrence D.J., Elphic R.C., Vaniman D.T., Thomsen D.R., & Barraclough B.L. (2000) Chemical information content of lunar thermal and epithermal neutrons. *Journal of Geophysical Research*, **105**, 20,347–20,363.

Feldman W.C., Maurice S., Lawrence D.J., et al. (2001) Evidence for water ice near the lunar poles. *Journal of Geophysical Research*, **106**, 23231–23251.

Feldman W., Prettyman T., Tokar R., et al. (2002) Fast neutron flux spectrum aboard Mars Odyssey during cruise. *Journal of Geophysical Research*, **107**, DOI:10.1029/2001JA000295.

Feldman W., Prettyman T., Maurice S., et al. (2004a) Global distribution of near-surface hydrogen on Mars. *Journal of Geophysical Research*, **109**, E09006, DOI:10.1029/2003JE002160.

Feldman W.C., Ahola K., Barraclough B.L., et al. (2004b) Gamma-ray, neutron, and alpha-particle spectrometers for the Lunar Prospector mission. *Journal of Geophysical Research*, **109**, E07S06, DOI:10.1029/2003JE002207.

Feldman W.C., Pathare A., Maurice S., et al. (2011) Mars Odyssey neutron data: 2. Search for buried excess water ice deposits at nonpolar latitudes on Mars. *Journal of Geophysical Research*, **116**, E11009.

Fermi E. (1950) Nuclear Physics: A course given by Enrico Fermi at the University of Chicago. Notes compiled by Jay Orear, A.H. Rosenfeld, and R.A. Schluter. University of Chicago Press, Chicago.

Floyd S.R., Trombka J.I., Leidecker H.W., et al. (1999) Radiation effects on the proportional counter X-ray detectors on board the NEAR spacecraft. *Nuclear Instruments and Methods in Physics Research A: Accelerators, Spectrometers, Detectors and Associated Equipment*, **422**, 577–581.

Gaskell R.W., Barnouin-Jha O.S., Scheeres D.J., et al. (2008) Characterizing and navigating small bodies with imaging data. *Meteoritics and Planetary Science*, **43**, 1049–1061.

Gasnault O., Feldman W.C., Maurice S., et al. (2001) Composition from fast neutrons: Application to the Moon. *Geophysical Research Letters*, **28**, 3797–3800.

Gleeson L.J. & Axford W.I. (1968) Solar modulation of galactic cosmic rays. *The Astrophysical Journal*, **154**, 1011–1026.

Glodo J., Higgins W.M., van Loef E.V.D., & Shah K.S. (2008) Scintillation properties of 1 Inch Cs2LiYCl6: CeCrystals. *Nuclear Science, IEEE Transactions on*, **55**, 1206–1209.

Goldsten J.O., Mcnutt R.L., Gold R.E., et al. (1997) The X-ray/gamma-ray spectrometer on the Near Earth Asteroid Rendezvous Mission. In: *The near Earth asteroid rendezvous mission* (C.T. Russell, ed.). Springer, Dordrecht, 169–216.

Goldsten J.O., Rhodes E.A., Boynton W.V., et al. (2007) The MESSENGER gamma-ray and neutron spectrometer. *Space Science Reviews*, **131**, 339–391.

Goorley J.T., James M.R., Booth T.E., et al. (2013) Initial MCNP6 release overview: MCNP6 version 1.0. Los Alamos National Laboratory document LA-UR-13–22934.

Grande M. (2001) The D-CIXS X-ray spectrometer on Esa's Smart-1 Mission to the Moon. In: *Earth–moon relationships* (C. Barbieri & F. Rampazzi, eds.). Springer, Dordrecht, 143–152.

Grande M., Dunkin S., Heather D., et al. (2002) The D-CIXS X-ray spectrometer, and its capabilities for lunar science. *Advances in Space Research*, **30**, 1901–1907.

Grande M., Browning R., Waltham N., et al. (2003) The D-CIXS X-ray mapping spectrometer on SMART-1. *Planetary and Space Science*, **51**, 427–433.

Grande M., Kellett B.J., Howe C., et al. (2007) The D-CIXS X-ray spectrometer on the SMART-1 mission to the Moon: First results. *Planetary and Space Science*, **55**, 494–502.

Grande M., Maddison B., Sreekumar P., et al. (2009) The Chandrayaan-1 X-ray spectrometer. *Current Science*, **96**, 517–519.

Hagerty J.J., Lawrence D.J., Hawke B.R., Vaniman D.T., Elphic R.C., & Feldman W.C. (2006) Refined thorium abundances for lunar red spots: Implications for evolved, nonmare volcanism on the Moon. *Journal of Geophysical Research*, **111**, E06002, DOI:10.1029/2005JE002592.

Hardgrove C., Moersch J., & Drake D. (2011) Effects of geochemical composition on neutron die-away measurements: Implications for Mars Science Laboratory's Dynamic Albedo of Neutrons experiment. *Nuclear Instruments and Methods in Physics Research A: Accelerators, Spectrometers, Detectors and Associated Equipment*, **659**, 442–455.

Hauser W. & Feshbach H. (1952) The inelastic scattering of neutrons. *Physical Review*, **87**, 366–373.

Howe C.J., Drummond D., Edeson R., et al. (2009) Chandrayaan-1 X-ray Spectrometer (C1XS)—Instrument design and technical details. *Planetary and Space Science*, **57**, 735–743.

Huovelin J., Alha L., Andersson H., et al. (2002) The SMART-1 X-ray solar monitor (XSM): Calibrations for D-CIXS and independent coronal science. *Planetary and Space Science*, **50**, 1345–1353.

Jenkins R. (1999) *X-ray fluorescence spectrometry*. Wiley-Interscience, New York.

Knoll G.F. (1989) *Radiation detection and measurement*. John Wiley & Sons, New York.

Kobayashi M., Hasebe N., Miyachi T., et al. (2013) The Kaguya gamma-ray spectrometer: Instrumentation and in-flight performances. *Journal of Instrumentation*, **8**, P04010–P04010.

Lawrence D., Feldman W., Barraclough B., et al. (2000) Thorium abundances on the lunar surface. *Journal of Geophysical Research*, **105**, 20307–20331.

Lawrence D., Feldman W., Elphic R., et al. (2002) Iron abundances on the lunar surface as measured by the Lunar Prospector gamma-ray and neutron spectrometers. *Journal of Geophysical Research*, **107**, 5130, DOI: 10.1029/2001JE001530.

Lawrence D.J., Elphic R.C., Feldman W.C., Prettyman T.H., Gasnault O., & Maurice S. (2003) Small-area thorium features on the lunar surface. *Journal of Geophysical Research*, **108**, 5102, DOI:10.1029/2003JE002050, E9.

Lawrence D.J., Maurice S., & Feldman W.C. (2004) Gamma-ray measurements from Lunar Prospector: Time series data reduction for the gamma-ray spectrometer. *Journal of Geophysical Research*, **109**, E07S05, DOI:10.1029/2003JE002206.

Lawrence D.J., Feldman W.C., Elphic R.C., et al. (2006) Improved modeling of Lunar Prospector neutron spectrometer data: Implications for hydrogen deposits at the lunar poles. *Journal of Geophysical Research*, **111**, E08001, DOI:10.1029/2005JE002637.

Lawrence D.J., Puetter R.C., Elphic R.C., et al. (2007) Global spatial deconvolution of Lunar Prospector Th abundances. *Geophysical Research Letters*, **34**, L03201, DOI:10.1029/2006GL028530.

Lawrence D.J., Feldman W.C., Goldsten J.O., et al. (2010) Identification and measurement of neutron-absorbing elements on Mercury's surface. *Icarus*, **209**, 195–209.

Lawrence D., Feldman W., Evans L., et al. (2012) Hydrogen at Mercury's north pole? Update on MESSENGER Neutron Measurements, 1802.

Lawrence D.J., Peplowski P.N., Prettyman T.H., et al. (2013a) Constraints on Vesta's elemental composition: Fast neutron measurements by Dawn's gamma ray and neutron detector. *Meteoritics and Planetary Science*, **48**, 2271–2288.

Lawrence D.J., Feldman W.C., Goldsten W.C., et al. (2013b) Evidence for water ice near Mercury's north pole from MESSENGER neutron spectrometer measurements. *Science*, **339**, 292–296.

Lewis E.E. & Miller W.F. (1984) *Computational methods of neutron transport*. John Wiley & Sons, New York.

Lingenfelter R.E., Canfield E.H., & Hess W.N. (1961) The lunar neutron flux. *Journal of Geophysical Research*, **66**, 2665–2671.

Lingenfelter R.E., Canfield E.H., & Hampel V.E. (1972) The lunar neutron flux revisited. *Earth and Planetary Science Letters*, **16**, 355–369.

Little R.C., Feldman W.C., Maurice S., et al. (2003) Latitude variation of the subsurface lunar temperature: Lunar Prospector thermal neutrons. *Journal of Geophysical Research*, **108**, 5046, DOI:10.1029/2001JE001497, E5.

Litvak M.L., Mitrofanov I.G., Barmakov Y.N., et al. (2008) The Dynamic Albedo of Neutrons (DAN) Experiment for NASA's 2009 Mars Science Laboratory. *Astrobiology*, **8**, 605–612.

Litvak M.L., Mitrofanov I.G., Hardgrove C., et al. (2016) Hydrogen and chlorine abundances in the Kimberley formation of Gale crater measured by the DAN instrument on board the Mars Science Laboratory Curiosity rover. *Journal of Geophysical Research*, **121**, 836–845.

Lodders K. & Fegley B., Jr. (1998) *The planetary scientist's companion*. Oxford University Press on Demand.

Mandel'shtam S.L., Tindo I.P., Cheremukhin G.S., Sorokin L.S., & Dmitriev A.B. (1968) X radiation of the Moon and X-ray cosmic background in the lunar Sputnik Luna-12. *Kosmicheskie Issledovaniia*, **6**, 119–127.

Maurice S., Lawrence D.J., Feldman W.C., Elphic R.C., & Gasnault O. (2004) Reduction of neutron data from Lunar Prospector. *Journal of Geophysical Research*, **109**, E07S04, DOI:10.1029/2003JE002208.

Maurice S., Feldman W., Prettyman T., Diez B., & Gasnault O. (2007) Reduction of Mars Odyssey neutron data, 38th Lunar Planet. Sci. Conf., Abstract #2036.

McKinney G.W., Lawrence D.J., Prettyman T.H., et al. (2006) MCNPX benchmark for cosmic ray interactions with the Moon.*Journal of Geophysical Research*, **111**, E06004, DOI:10.1029/2005je002551.

Metropolis N. & Ulam S. (1949) The Monte Carlo method. *Journal of the American Statistical Association*, **44**, 335–341.

Metzger A.E., Trombka J.I., Peterson L.E., Reedy R.C., & Arnold J.R. (1973) Lunar surface radioactivity: Preliminary results of the Apollo 15 and Apollo 16 gamma-ray spectrometer experiments. *Science*, **179**, 800–803.

Murty R.C. (1965) Effective atomic numbers of heterogeneous materials. *Nature*, **207**, 398.

Näränen J., Carpenter J., Parviainen H., et al. (2009) Regolith effects in planetary X-ray fluorescence spectroscopy: Laboratory studies at 1.7–6.4keV. *Advances in Space Research*, **44**, 313–322.

Narendranath S., Sreekumar P., Maddison B.J., et al. (2010) Calibration of the C1XS instrument on Chandrayaan-1. *Nuclear Instruments and Methods in Physics Research A: Accelerators, Spectrometers, Detectors and Associated Equipment*, **621**, 344–353.

Narendranath S., Athiray P.S., Sreekumar P., et al. (2011) Lunar X-ray fluorescence observations by the Chandrayaan-1 X-ray Spectrometer (C1XS): Results from the nearside southern highlands. *Icarus*, **214**, 53–66.

Nittler L.R., Starr R.D., Lev L., et al. (2001) X-ray fluorescence measurements of the surface elemental composition of asteroid 433 Eros. *Meteoritics and Planetary Science*, **36**, 1673–1695.

Nittler L.R., Starr R.D., Weider S.Z., et al. (2011) The major-element composition of Mercury's surface from MESSENGER X-ray spectrometry. *Science*, **333**, 1847–1850.

Okada T. (2004) Particle size effect in X-ray fluorescence at a large phase angle: Importance on elemental analysis of asteroid Eros (433). *35th Lunar Planet. Sci. Conf.*, Abstract #1927.

Okada T., Kato M., Shirai K., et al. (2002a) Elemental mapping of asteroid 1989ML from MUSES-C orbiter. *Advances in Space Research*, **29**, 1237–1242.

Okada T., Kato M., Yamashita Y., et al. (2002b) Lunar X-ray spectrometer experiment on the SELENE mission. *Advances in Space Research*, **30**, 1909–1914.

Okada T., Shiraishi H., Shirai K., et al. (2009a) X-Ray Fluorescence Spectrometer (XRS) on Kaguya: Current status and results. *40th Lunar Planet. Sci. Conf.*, Abstract #1897.

Okada T., Shirai K., Yamamoto Y., et al. (2009b) X-Ray fluorescence spectrometry of Lunar surface by XRS onboard SELENE (Kaguya). *Transactions of the Japan Society for Aeronautical and Space Sciences, Space Technology Japan*, 7, Tk_39–Tk_42.

O'Neill P.M. (2010) Badhwar–O'Neill 2010 galactic cosmic ray flux model—revised. *IEEE Transactions on Nuclear Science*, **6**, 3148–3153.

Ouyang Z., Jiang J., Li C., et al. (2008) Preliminary scientific results of Chang'E-1 Lunar Orbiter: Based on payloads detection data in the first phase. *Chinese Journal of Space Science*, **28**, 361–369.

Ouyang Z., Li C., Zou Y., et al. (2010a) Chang'E-1 lunar mission: An overview and primary science results. *Chinese Journal of Space Science*, **30**, 392.

Ouyang Z., Li C., Zou Y., et al. (2010b) Primary scientific results of Chang'E-1 Lunar mission. *Science China Earth Sciences*, **53**, 1565–1581.

Parviainen H., Näränen J., & Muinonen K. (2011) Soft X-ray fluorescence from particulate media: Numerical simulations. *Journal of Quantitative Spectroscopy and Radiative Transfer*, **112**, 1907–1918.

Payne S.A., Cherepy N.J., Hull G., Valentine J.D., Moses W.W., & Choong W.-S. (2009) Nonproportionality of scintillator detectors: Theory and experiment. *IEEE Transactions on Nuclear Science*, **56**, 2506–2512.

Peng W.-X., Wang H., Zhang C.-M., et al. (2009) Calibration of CE-1 X-ray spectrometer. *Nuclear Electronics and Detection Technology*, **29**, 235–239.

Peplowski P.N., Evans L.G., Hauck S.A., et al. (2011) Radioactive elements on Mercury's surface from MESSENGER: Implications for the planet's formation and evolution. *Science*, **333**, 1850–1852.

Peplowski P.N., Lawrence D.J., Rhodes E.A., et al. (2012) Variations in the abundances of potassium and thorium on the surface of Mercury: Results from the MESSENGER Gamma-Ray Spectrometer. *Journal of Geophysical Research*, **117**, E00L04, DOI:10.1029/2012JE004141.

Peplowski P.N., Lawrence D.J., Feldman W.C., et al. (2015) Geochemical terranes of Mercury's northern hemisphere as revealed by MESSENGER neutron measurements. *Icarus*, **253**, 346–363.

Peplowski P.N., Beck A.W., & Lawrence D.J. (2016) Geochemistry of the lunar highlands as revealed by measurements of thermal neutrons. *Journal of Geophysical Research*, **121**, 388–401.

Prettyman T. (1999) Method for mapping charge pulses in semiconductor radiation detectors. *Nuclear Instruments and Methods in Physics Research A: Accelerators, Spectrometers, Detectors and Associated Equipment*, **422**, 232–237.

Prettyman T.H. (2014) Remote sensing of chemical elements using nuclear spectroscopy. In: *Encyclopedia of the Solar System*, 3rd edn (T. Spohn, T. Johnson, & D. Breuer, eds.). Elsevier, Philadelphia, 1161–1183.

Prettyman T.H., Feldman W., Mellon M., et al. (2004) Composition and structure of the martian surface at high southern latitudes from neutron spectroscopy. *Journal of Geophysical Research*, **109**, E05001, DOI:10.1029/2003je002139.

Prettyman T.H., Hagerty J.J., Elphic R.C., et al. (2006) Elemental composition of the lunar surface: Analysis of gamma ray spectroscopy data from Lunar Prospector. *Journal of Geophysical Research*, **111**, E12007, DOI:10.1029/2005JE002656.

Prettyman T.H., Feldman W.C., & Titus T.N. (2009) Characterization of Mars' seasonal caps using neutron spectroscopy. *Journal of Geophysical Research*, **114**, E08005, DOI:10.1029/2008je003275.

Prettyman T.H., Feldman W.C., McSween H.Y., Jr., et al. (2011) Dawn's gamma ray and neutron detector. *Space Science Reviews*, **163**, 371–459.

Prettyman T.H., Mittlefehldt D.W., Yamashita N., et al. (2012) Elemental mapping by Dawn reveals exogenic H in Vesta's regolith. *Science*, **338**, 242–6.

Prettyman T.H., Mittlefehldt D.W., Yamashita N., et al. (2013) Neutron absorption constraints on the composition of 4 Vesta. *Meteoritics and Planetary Science*, **48**, 2211–2236.

Prettyman T.H., Yamashita N., Reedy R.C., et al. (2015) Concentrations of potassium and thorium within Vesta's regolith. *Icarus*, **259**, 39–52.

Prettyman T.H., Yamashita N., Toplis M.J., et al. (2017) Extensive water ice within Ceres' aqueously altered regolith: Evidence from nuclear spectroscopy. *Science*, **355**, 55–59.

Prettyman T.H., Yamashita N., Ammannito E., et al. (2019) Elemental composition and mineralogy of Vesta and Ceres: Distribution and origins of hydrogen-bearing species. *Icarus*, **318**, 42–55.

Reedy R.C. (1978) Planetary gamma-ray spectroscopy. *9th Lunar Planet. Sci. Conf.*, Abstract, 2961–2984.

Reedy R.C., Arnold J.R., & Lal D. (1983) Cosmic-ray record in Solar System matter. *Science*, **219**, 127–135.

Rieder R., Economou T., Wänke H., et al. (1997) The chemical composition of martian soil and rocks returned by the mobile alpha proton X-ray spectrometer: Preliminary results from the X-ray mode. *Science*, **278**, 1771–1774.

Shirai K., Okada T., Yamamoto Y., et al. (2008) Instrumentation and performance evaluation of the XRS on SELENE orbiter. *Earth, Planets and Space*, **60**, 277–281.

Shiraiwa T. & Fujino N. (1966) Theoretical calculation of fluorescent X-ray intensities in fluorescent X-ray spectrochemical analysis. *Japanese Journal of Applied Physics*, **5**, 886–899.

Starr R., Clark P.E., Murphy M.E., et al. (2000) Instrument calibrations and data analysis procedures for the NEAR X-ray spectrometer. *Icarus*, **147**, 498–519.

Starr R.D., Schlemm Ii C.E., Ho G.C., Nittler L.R., Gold R.E., & Solomon S.C. (2016) Calibration of the MESSENGER X-ray spectrometer. *Planetary and Space Science*, **122**, 13–25.

Surkov Yu A., Barsukov V.L., Moskalyeva L.P., Kharyukova V.P., & Kemurdzhian A.L. (1984) New data on the composition, structure, and properties of Venus rock obtained by Venera 13 and Venera 14. *Journal of Geophysical Research*, **89**, B393–B402.

Surkov Y.A., Kirnozov F.F., Glazov V.N., Dunchenko A.G., Tatsy L.P., & Sobornov O.P. (1987) Uranium, thorium, and potassium in the Venusian rocks at the landing sites of Vega 1 and 2. *Journal of Geophysical Research*, **92**, E537–E540.

Trombka J.I., Floyd S.R., Boynton W.V., et al. (1997) Compositional mapping with the NEAR X ray/gamma ray spectrometer. *Journal of Geophysical Research*, **102**, 23729–23750.

Trombka J.I., Squyres S.W., Brückner J., et al. (2000) The elemental composition of Asteroid 433 Eros: Results of the NEAR-Shoemaker X-ray spectrometer. *Science*, **289**, 2101–2105.

Weider S.Z., Swinyard B.M., Kellett B.J., et al. (2011) Planetary X-ray fluorescence analogue laboratory experiments and an elemental abundance algorithm for C1XS. *Planetary and Space Science*, **59**, 1393–1407.

Weider S.Z., Nittler L.R., Starr R.D., et al. (2012a) Chemical heterogeneity on Mercury's surface revealed by the MESSENGER X-Ray Spectrometer. *Journal of Geophysical Research*, **117**, E00L05, DOI:10.1029/2012je004153.

Weider S.Z., Kellett B.J., Swinyard B., et al. (2012b) The Chandrayaan-1 X-ray spectrometer: First results. *Planetary and Space Science*, **60**, 217–228.

Weider S.Z., Nittler L.R., Starr R.D., et al. (2015) Evidence for geochemical terranes on Mercury: Global mapping of major elements with MESSENGER's X-ray spectrometer. *Earth and Planetary Science Letters*, **416**, 109–120.

Wilson J.T., Lawrence D.J., Peplowski P.N., et al. (2018) Image reconstruction techniques in neutron and gamma-ray spectroscopy: Improving Lunar Prospector data. *Journal of Geophysical Research*, **123**, 1804–1822.

Yamashita N., Hasebe N., Reedy R.C., et al. (2010) Uranium on the Moon: Global distribution and U/Th ratio. *Geophysical Research Letters*, **37**, L10201, DOI:10.1029/2010gl043061.

Yamashita N., Prettyman T.H., Mittlefehldt D.W., et al. (2013) Distribution of iron on Vesta. *Meteoritics and Planetary Science*, **48**, 2237–2251.

Yin L.I., Trombka J.I., Adler I., & Bielefeld M. (1993) X-ray remote sensing techniques for geochemical analysis of planetary surfaces. In: *Remote geochemical analysis: Elemental and mineralogical composition* (C.M. Pieters & P.A.J. Englert, eds.). Cambridge University Press, Cambridge, 99–212.

Zhu M.H., Chang J., Ma T., et al. (2013) Potassium map from Chang'E-2 constraints the impact of Crisium and Orientale basin on the Moon. *Science Reports*, **3**, 1611, DOI:10.1038/srep01611.

10

Radar Remote Sensing
Theory and Applications

JAKOB VAN ZYL, CHARLES ELACHI, AND YUNJIN KIM

10.1 Introduction

The word RADAR is an acronym for Radio Detection and Ranging. A radar system measures the distance, or *range*, to an object by transmitting an electromagnetic signal and receiving an echo reflected from the object. Because electromagnetic waves propagate at the speed of light, one only has to measure the time it takes the radar signal to propagate to the object and back to calculate the range to the object.

Radar instruments provide their own signals to detect the presence of objects. Therefore, radar systems are known as *active* remote sensing instruments. Because the radar provides its own signal, it can operate during day or night. In addition, radar signals typically penetrate clouds and rain, which means that radar images can be acquired not only during day or night, but also under (almost) all weather conditions. For this reason, radar sensors are often referred to as *all-weather* instruments. Imaging remote sensing radar systems such as Synthetic Aperture Radars (SARs) produce high-resolution (from sub-meter to few tens of meters) images of surfaces. Geophysical information can be derived from these high-resolution images by using proper postprocessing techniques.

This article focuses on the specific class of implementation of radar known as SAR, with particular emphasis on advanced implementations of space-borne SAR, such as SAR polarimetry and SAR interferometry. We will start with a short introduction to SAR before describing these techniques. Even though radar remote sensing has been used for planetary studies for decades (e.g., Zisk et al., 1974; Pettengill et al., 1980; Ostro et al., 1992; Harmon et al., 1994; Plat et al., 2009), these techniques are still evolving and improving, and have seen limited application to planetary remote sensing, primarily because of the large data rates generated by these radar systems. For a comprehensive review of current and past planetary radar studies the reader is referred to Chapter 31.

10.2 Basic Principles of Radar Imaging

Imaging radar systems generate surface images that are at first glance very similar to the more familiar images produced by instruments that operate in the visible or infrared (IR) parts of the electromagnetic spectrum. However, the principle behind the image generation is fundamentally different in the two cases. Visible and IR sensors (e.g., Chapter 11) use a lens or mirror system to project the radiation from the scene onto a "two-dimensional array of detectors" that could be an electronic array or, in earlier remote sensing instruments, a film using chemical processes. The two-dimensionality can also be achieved by using scanning systems, or by moving a single line array of detectors. This imaging approach, similar to what we are all familiar with when taking photographs with a camera, preserves the relative angular relationships between objects in the scene and their images in the focal plane. Because of this conservation of angular relationships, the resolution of the images depends on how far away the camera is from the scene it is imaging. The closer the camera is, the higher the resolution and the smaller the details that can be recognized in the images. As the camera moves farther away from the scene, the resolution degrades, and only larger objects can be discerned in the image.

Imaging radar systems use a quite different mechanism to generate images, with the result that the image characteristics are also quite different from those of visible and IR images. There are two different ways radar can be used to produce images. These two types of radar are broadly classified as *real aperture* and *synthetic aperture* radar. To separate objects in radar images in the cross-track direction and the along-track direction, two different methods are implemented. The *cross-track* direction, also known as the *range* direction in radar imaging, is the direction perpendicular to the direction in which the imaging platform is moving. In this direction, radar echoes are separated using the *time delay* between the echoes that are back-scattered from the different surface elements. This is true for both real aperture and synthetic aperture radar imagers. The *along-track* direction, also known as the *azimuth* direction, is the direction parallel to the movement of the imaging platform. The angular size of the antenna beam (in the case of the real-aperture radar), or the Doppler history of the received signal (in the case of the synthetic-aperture radar) is used to separate surface pixels in the along-track dimension in the radar images. Using the time delay and Doppler history results, SAR images have resolutions that are independent of how far away the radar is from the scene it is imaging. This fundamental advantage enables high-resolution space-borne SAR without requiring an extremely large antenna.

Another difference between images acquired by cameras and radar images is the way in which they are acquired. Cameras typical look straight down; or at least have no fundamental limitation that prevents them from taking pictures looking straight down from the spacecraft or aircraft. Not so for imaging radar systems. To avoid so-called *ambiguities,* the imaging radar sensor has to use an antenna that illuminates the surface to the side of the flight track. Usually, the antenna has a fan beam that illuminates a highly elongated elliptically shaped area on the surface as shown in Figure 10.1. The illuminated area across-track defines the image *swath* in traditional single-beam imaging radar instruments. More modern implementations are capable of imaging much wider swaths than the single antenna beam using techniques such as scanSAR or sweepSAR (Freeman et al., 2009). These techniques are beyond the scope of this chapter, and will not be discussed here.

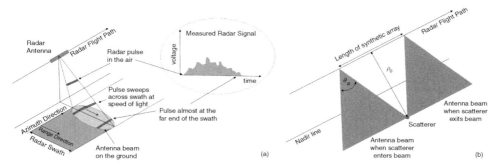

Figure 10.1 Geometry of imaging radar instruments. (a) Imaging radar systems typically use antennas that have elongated gain patterns that are pointed to the side of the radar flight track. The pulse sweeps across the antenna beam spot, creating an echo. (b) A synthetic aperture radar integrates the signal from the scatter for as long as the scatterer remains in the antenna beam.

Within the illumination beam, the radar sensor transmits a very short effective pulse of electromagnetic energy. Echoes from surface points farther away along the cross-track coordinate will be received at proportionally later times (see Figure 10.1a). Thus, by dividing the received time in increments of equal time bins, the surface can be subdivided into a series of *range bins*. The width in the along-track direction of each range bin is equal to the antenna footprint along the track x_a. As the platform moves, the sets of range bins are covered sequentially, thus allowing strip mapping of the surface line by line. This is comparable to strip mapping with a so-called pushbroom imaging system using a line array in the visible part of the electromagnetic spectrum. The brightness associated with each image pixel in the radar image is proportional to the echo power contained within the corresponding time bin.

10.2.1 Radar Cross-Track Resolution

The *resolution* of an image is defined as the separation between the two closest features that can still be resolved in the final image. First, consider two point targets that are separated in the slant range direction by x_r. Because the radar waves propagate at the speed of light, the corresponding echoes will be separated by a time difference equal to $\Delta t = 2x_r/c$, where c is the speed of light and the factor 2 is included to account for the signal round trip propagation as described previously. Radar waves are usually not transmitted continuously; instead, they usually transmit short bursts of energy known as radar *pulses*. The two features can be discriminated if the leading edge of the pulse returned from the second object is received later than the trailing edge of the pulse received from the first feature, as shown in Figure 10.1a. Therefore, the smallest separable time difference in the radar receiver is equal to the effective time length τ of the pulse. Thus, the slant range resolution of a radar signal is

$$x_r = \frac{c\tau}{2}. \qquad (10.1)$$

Now let us consider the case of two objects separated by a distance x_g on the ground. The corresponding echoes will be separated by a time difference equal to $\Delta t = 2x_g \sin\theta/c$. The angle θ is the local incidence angle, which is the angle between the incoming wave and the local normal to the surface. Therefore, the ground range resolution of the radar signal is given by

$$x_g = \frac{c\tau}{2 \sin\theta}. \tag{10.2}$$

In other words, the range resolution is equal to half the footprint of the radar pulse on the surface. The cross-track resolution in the radar case is independent of the distance between the scene and the radar instrument.

10.3 Along-Track Resolution for Real and Synthetic Aperture Radar

As mentioned previously, the antenna usually has a fan beam that illuminates a highly elongated elliptically shaped area on the surface as shown in Figure 10.1a. A real aperture radar relies on the resolution afforded by the antenna beam in the along-track direction for imaging. This means that the resolution of a real aperture radar signal in the along-track direction is driven by the size of the antenna as well as the range to the scene. Assuming an antenna length of L, the antenna beam width in the along-track direction is

$$\theta_a \approx \lambda/L. \tag{10.3}$$

At a distance ρ from the antenna, this means that the antenna beam width illuminates an area with the along-track dimension equal to

$$x_a \approx \rho\theta_a = \frac{\lambda\rho}{L} \approx \frac{\lambda h}{L \cos\theta}. \tag{10.4}$$

The practical along-track resolution for real aperture radars is quite coarse. As a consequence, the real-aperture technique is not often used for surface imaging, especially from space.

Synthetic aperture radar refers to a particular implementation of an imaging radar system that utilizes the movement of the radar platform and specialized signal processing to generate high-resolution images. The main difference between real and synthetic aperture radar techniques is in the way in which the azimuth resolution is achieved.

As the radar instrument moves along the flight path, it transmits pulses of energy and records the reflected signals, as shown in Figure 10.1a. When the radar data are processed, the position of the radar platform is taken into account when adding the signals to integrate the energy for the along-track direction. Consider the geometry shown in Figure 10.1b. As the radar instrument moves along the flight path, the distance between the radar sensor and the scatterer changes, with the minimum distance, ρ_0 occurring when the scatterer is directly broadside of the radar platform.

The range between the radar sensor and the scatterer as a function of position along the flight path is given by

$$\rho(s) = \sqrt{\rho_0^2 + v^2 s^2}, \quad (10.5)$$

where v is the velocity of the radar platform, and s is the time along the flight path, with zero time at the time of closest approach. To a good approximation for remote sensing radar systems, we can assume that $vs \ll \rho_0$. In this case, we can approximate the range as

$$\rho(s) = \rho_0 + \frac{v^2 s^2}{2\rho_0}. \quad (10.6)$$

The phase of the signal is given by

$$\varphi(s) = -\frac{4\pi \rho(s)}{\lambda} \approx -\frac{4\pi \rho_0}{\lambda} - \frac{2\pi v^2 s^2}{\lambda \rho_0}. \quad (10.7)$$

The instantaneous frequency of this signal is

$$f(s) = \frac{1}{2\pi} \frac{\partial \varphi(s)}{\partial s} \approx -\frac{2v^2}{\lambda \rho_0} s; \quad -\frac{\lambda \rho}{2Lv} \leq s \leq \frac{\lambda \rho}{2Lv}. \quad (10.8)$$

The frequency changes linearly with time. The maximum "integration time" is given by the amount of time that the scatterer will be in the antenna beam. If this signal is filtered using a matched filter, the resulting compressed signal will have a width

$$x_a = \frac{L}{2}. \quad (10.9)$$

This result shows that the azimuth (or along-track) surface resolution for a synthetic aperture radar system is equal to half the size of the physical antenna and is independent of the distance between the sensor and the surface. This comes at a price, however. To adequately sample the signal, at least one sample (i.e., one pulse) should be taken every time the sensor moves by half an antenna length. This requirement leads to large data rates for imaging radar platforms.

10.4 Geometric Distortion

Time delay is used to separate objects in the range dimension in radar images. This means that two neighboring pixels in the image plane correspond to two areas in the scene with slightly different range to the sensor. In this case, the scene is projected in a cylindrical geometry on the imaging plane. This leads to distortions, as illustrated in Figure 10.2a. Three segments of equal length, but with different slopes, are projected in the image plane as segments of different lengths. The side of a hill facing the sensor will be shortened, while the far side will appear stretched. This is called *foreshortening*. If the topography is known, this distortion can be corrected. Figure 10.2b shows a radar image acquired with the NASA/

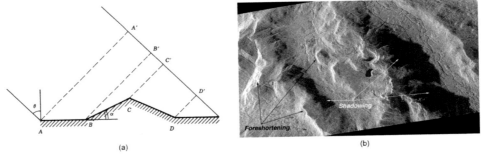

Figure 10.2 Distortions in radar images resulting from a combination of viewing geometry and surface topography. (a) The viewing geometry causes the front slope of a mountain to appear compressed; this is known as foreshortening. (b) Image showing both foreshortening and shadowing. The latter results when slopes facing away from the radar instrument are steep enough that the radar signal does not illuminate the terrain. (Image courtesy of the Jet Propulsion Laboratory.)

JPL AIRSAR system that shows examples of foreshortening and shadowing. Shadowing occurs in regions where the surface slope is such that a portion of the terrain is not illuminated by the radar signal.

In the extreme case where the surface slope α is larger than the incidence angle θ there is *layover* where the top of the hill will appear to be laid over the region in front of it. In this extreme case, it is not possible to correct for the distortion.

10.5 Signal Fading and Speckle

In addition to the normal thermal noise, SAR images contain a special type of noise, known as speckle, that makes interpretation of radar images more challenging, especially for new users. In general, the thermal noise level determines the darkest areas that the radar system can image. Speckle noise, on the other hand, is larger for brighter areas, and gives the radar image a grainy appearance. Speckle noise results from the fact that SAR systems make coherent measurements.

At every instant of time, the radar pulse illuminates a certain surface area that consists of many scattering points. Thus, the returned echo is the coherent addition of the echoes from a large number of points. The returns from these points add vectorially and result into a single vector that represents the amplitude V and phase ϕ of the total echo. The phase ϕ_i of each elementary vector is related to the distance between the sensor and the corresponding scattering point. If the sensor moves by a small amount, all the phases ϕ_i will change, leading to a change in the composite amplitude V. Thus, successive observations of the same surface region as the sensor moves over it will result in a different value V. This variation is called fading. Thus, an image of a homogeneous surface with a constant backscatter cross section will show brightness variations from one pixel to the next. This is called speckle.

To measure the backscatter cross section of the surface, the returns from many neighboring pixels will have to be averaged to reduce the effects of speckle. This is called the "number of looks." The larger the number of looks, the better is the quality of the image from the radiometric point of view. However, this degrades the spatial resolution of the image. It should be noted that for more than about 25 looks, a large increase in the number of looks leads to only a small decrease in the signal fluctuation. This small improvement in the radiometric resolution should be traded off against the large decrease in the spatial resolution. For example, if one were to average 10 resolution cells in a four-look image, the speckle noise will be reduced to about 0.5 dB. At the same time, however, the image resolution will be reduced by an order of magnitude. Whether this loss in resolution is worth the reduction in speckle noise depends on both the aim of the investigation, as well as the kind of scene imaged.

Figure 10.3 shows the effect of multilook averaging. The same image, acquired by the NASA/JPL AIRSAR system, is shown displayed at 1, 4, 16, and 32 looks, respectively. This figure clearly illustrates the smoothing effect, as well as the decrease in resolution resulting from the multilook process.

10.6 Radar Scattering

Visible or IR imagers provide detailed information about the chemistry and mineralogy of the surface (e.g., Chapters 11, 14, 17, 18, 20, 22, 23, 24). Radar scattering from a surface is

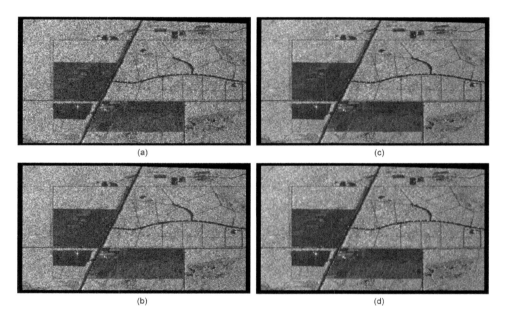

Figure 10.3 The effects of speckle can be reduced by incoherently averaging pixels in a radar image, a process known as multilooking. Shown in this image is the same image, processed at (a) 1 look, (b) 4 looks, (c) 16 looks, and (d) 32 looks. Note the reduction in the granular texture as the number of looks increase, while at the same time the resolution of the image decreases. The images cover an area of approximately 7 × 4 km. (Image courtesy of the Jet Propulsion Laboratory.)

strongly affected by the surface geometrical and bulk electrical properties. As such, radar images provide information about the surface that is complementary to that provided by the shorter wavelength imagers. In addition, visible and IR systems typically respond to the chemistry of the top few microns of the surface (e.g., Chapters 2, 11). Radar waves, especially from longer wavelength systems, penetrate deeper into the surface and typically provide information about the electrical properties of the top few cm of the surface. Under the right conditions, radar systems can penetrate several meters to several kilometers into a surface to provide information about the subsurface.

The surface small-scale geometric shape (also called roughness) can be statistically characterized by its standard deviation relative to a mean flat surface. The surface standard deviation (commonly referred to as the surface roughness) is the root mean square of the actual surface deviation from this average surface. However, knowing the surface standard deviation is not yet a complete description of the surface geometrical properties. It is also important to know how the local surface deviation from the mean surface is related to the deviation from the mean surface at other points on the surface. This is mathematically described by the surface height autocorrelation function. The surface correlation length is the separation after which the deviation from the mean surface for two points is statistically independent. Mathematically, it is the length after which the autocorrelation function is less than $1/e$.

Consider first the case of a perfectly smooth surface of infinite extent that is uniformly illuminated by a plane wave. This surface will reflect the incident wave into the specular direction with scattering amplitudes equal to the well-known Fresnel reflection coefficients. In this case, no scattered energy will be received in any other direction. If now the surface is made finite in extent, or the infinite surface is illuminated by a finite extent uniform plane wave, the situation changes. The maximum amount of reflected power still appears in the specular direction, but a lobe structure, similar to an "antenna pattern," appears around the specular direction. The exact shape of the lobe structure depends on the size and the shape of the finite illuminated area, and the pattern is adequately predicted using physical optics calculations. This component of the scattered power is often referred to as the *coherent component* of the scattered field. For angles far away from the specular direction, there will be very little scattered power in the coherent component.

The next step is to add some roughness to the finite surface such that the root-mean-square height of the surface is still much less than the wavelength of the illuminating source. The first effect is that some of the incident energy will now be scattered in directions other than the specular direction. The net effect of this scattered energy is to fill the nulls in the "antenna pattern" of the coherent field. The component of the scattered power that is the result of the presence of surface roughness is referred to as the *incoherent component* of the scattered field. At angles significantly away from the specular direction, such as the backscatter direction at larger incidence angles, the incoherent part of the scattered field usually dominates.

As the roughness of the surface increases, less power is contained in the coherent component, and more in the incoherent component. In the limit where the root-mean-square height becomes significantly larger than the wavelength, the coherent component is typically no longer distinguishable, and the incoherent power dominates in all directions. In this limit the strength of the scattering in any given direction is related to the number of surface facets that are oriented such that they reflect specularly in that direction. This is the same phenomenon that causes the shimmering of the Moon on a roughened water surface.

Depending on the angle of incidence, two different approaches are used to model radar scattering from rough natural surfaces. For small angles of incidence, scattering is dominated by reflections from appropriately oriented facets on the surface. In this regime, physical optics principles are used to derive the scattering equations. As a rule of thumb, facet scattering dominates for angles of incidence less than 20–30°. For the larger angles of incidence, scattering from the small-scale roughness dominates. The best known model for describing this type of scattering is the small perturbation model, often referred to as the Bragg model. This model, as its name suggests, treats the surface roughness as a small perturbation from a flat surface. More recently, Fung et al. (1992) proposed a model based on an integral equation solution to the scattering problem that seems to describe the scattering adequately in both limits. The reader is referred to chapter 10 of Ulaby and Long (2014) for a more complete discussion of many radar scattering models.

10.7 Radar Polarimetry

Electromagnetic wave propagation is a vector phenomenon, i.e., all electromagnetic waves can be expressed as complex vectors. If one observes a wave transmitted by a radar antenna when the wave is a large distance from the antenna (in the far-field of the antenna), the radiated electromagnetic wave can be adequately described by a two-dimensional (2-D) complex vector. If this radiated wave is now scattered by an object, and one observes this wave in the far-field of the scatterer, the scattered wave can again be adequately described by a 2-D vector. In this abstract way, one can consider the scatterer as a mathematical operator that takes one 2-D complex vector (the wave impinging upon the object) and changes that into another 2-D vector (the scattered wave). Mathematically, therefore, a scatterer can be characterized by a complex 2×2 scattering matrix $[\mathbf{S}]$. This scattering matrix is a function of the radar frequency and the viewing geometry.

The voltage measured by the radar system is the scalar product of the radar antenna polarization and the incident wave electric field, which can be written as (van Zyl & Kim, 2011)

$$V_r = \mathbf{p}^{\text{rec}} \cdot [\mathbf{S}] \mathbf{p}^{\text{tr}}. \tag{10.10}$$

Here, $\mathbf{p}^{\text{tr}}$ and $\mathbf{p}^{\text{rec}}$ are the normalized polarization vectors describing the transmitting and receiving radar antennas. The power received by the radar instrument is the magnitude of the voltage squared:

$$P = |V_r|^2 = |\mathbf{p}^{\text{rec}} \cdot [\mathbf{S}] \mathbf{p}^{\text{tr}}|^2. \tag{10.11}$$

Once the complete scattering matrix is known and calibrated, one can synthesize the radar cross section for any arbitrary combination of transmit and receive polarizations using Eq. (10.11). This expression forms the basis of radar polarimetry. Figure 10.4 shows a number of such synthesized images for the San Francisco Bay area in California. The data were acquired with the NASA/JPL AIRSAR system. The Golden Gate Bridge is the

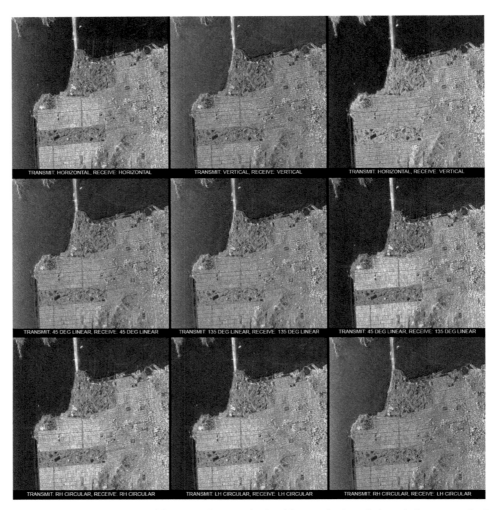

Figure 10.4 L-band images of San Francisco synthesized from a single polarimetric image acquired by the NASA/JPL AIRSAR system at L-band. The nine images show the co-polarized (transmit and receive polarizations are the same) and the cross-polarized (transmit and receive polarizations are orthogonal) images for the three axes of the Poincaré sphere. Note the relative change in brightness between the city of San Francisco, the ocean, and the Golden Gate Park, which is the large rectangle about one-third from the bottom of the images. The radar illumination is from the left.

linear feature at the top middle of the image. Golden Gate Park is the large rectangle about one-third from the bottom of the image. Note the strong variation in the relative return from the urban areas in the top left image. These are related to the orientation of the buildings and the streets relative to the radar look direction (Zebker et al., 1987). The contrast between the urban areas and the vegetated areas such as Golden Gate Park is maximized using the 45° linear polarization for transmit, and 135° linear polarization for receive (see chapter 7 of Ulaby & Elachi, 1990 for a discussion). These images show how scattering is a function of the polarization of the transmit and receive antennas in urban areas. While not quite as dramatic, many natural terrains also exhibit radar scattering that is different for different combination of transmit and receive polarizations. These differences form the basis of some geophysical algorithms to infer surface properties, as we will discuss in more detail later in this chapter.

10.8 Radar Interferometry

Interferometry refers to a class of techniques where additional information is extracted from SAR images that are acquired from different vantage points, or at different times. Various implementations allow different types of information to be extracted. For example, if two SAR images are acquired from slightly different viewing geometries, information about the topography of the surface can be inferred. On the other hand, if images are taken at slightly different times, a map of surface velocities can be produced. Finally, if sets of interferometric images are combined, subtle changes in the scene can be measured with extremely high accuracy.

10.8.1 Radar Interferometry for Measuring Topography

SAR interferometers for the measurement of topography can be implemented in one of two ways. In the case of single-pass or fixed baseline interferometry, the system is configured to measure the two images at the same time through two different antennas usually arranged one above the other. The physical separation of the antennas is referred to as the baseline of the interferometer. So far most single-pass interferometers have been implemented using airborne SARs. The Shuttle Radar Topography Mission (SRTM), a joint project between the United States National Imagery and Mapping Agency (NIMA) and the National Aeronautics and Space Administration (NASA), was the first space-borne implementation of a single-pass interferometer (Farr & Kobrick, 2000).

An alternative way to form the interferometric baseline is to use a single-channel radar pulse to image the same scene from slightly different viewing geometries at different times. This technique, known as *repeat-track* interferometry, has been mostly applied to space-borne data starting with data collected with the L-band SEASAT SAR. Other investigators used data from the L-band SIR-B, SIR-C, and JERS and the C-band ERS-1/2, Radarsat, and Envisat ASAR radars.

SAR interferometry was first demonstrated by Graham (1974), who demonstrated a pattern of nulls or interference fringes by vectorially adding the signals received from two SAR antennas, one physically situated above the other. Later, Zebker and Goldstein (1986) demonstrated that these interference fringes can be formed after SAR processing of the individual images if both the amplitude and the phase of the radar images are preserved during the processing.

The basic principles of interferometry can be explained using the geometry shown in Figure 10.5a. Using the law of cosines on the triangle formed by the two antennas and the point being imaged, it follows that

$$(R + \delta R)^2 = R^2 + B^2 - 2RB \cos\left(\frac{\pi}{2} - \theta + \alpha\right), \tag{10.12}$$

where R is the slant range to the point being imaged from the reference antenna, δR is the path length difference between the two antennas, B is the physical interferometric baseline length, θ is the look angle to the point being imaged, and α is the baseline tilt angle with respect to the horizontal.

From Eq. (10.12) it follows that we can solve for the path length difference δR. If we assume that $R \gg B$ (a very good assumption for most interferometers), one finds that

$$\delta R \approx -B \sin(\theta - \alpha). \tag{10.13}$$

The radar system does not measure the path length difference explicitly, however. Instead, what is measured is an interferometric phase difference that is related to the path length difference through

$$\phi = a2\pi \frac{\delta R}{\lambda} = -a2\pi \frac{B \sin(\theta - \alpha)}{\lambda}. \tag{10.14}$$

For single-pass interferometers, $a = 1$, while for repeat-pass interferometers, $a = 2$. From Figure 10.5a, it also follows that the elevation of the point being imaged is given by

$$z(y) = h - R \cos\theta, \tag{10.15}$$

with h denoting the height of the reference antenna above the reference plane with respect to which elevations are quoted.

To understand better how the information measured by an interferometer is used, consider again the expression for the measured phase difference given in Eq.(10.14). This expression shows that even if there is no relief on the surface being imaged, the measured phase will still vary across the radar swath, as shown in Figure 10.5b for a radar system with parameters similar to that of the Shuttle Radar Topography Mission (Farr & Kobrick, 2000). To show how this measurement is sensitive to topographical relief, consider now the case where there is indeed topographical relief in the scene. For illustration purposes, we used a digital elevation model of Mount Shasta in California, shown in Figure 10.5c, and calculated the expected interferometric phase of the SRTM system for this area. The result, shown in Figure 10.5d, demonstrates that the parallel lines of phase difference for a scene without relief are distorted by the presence of the relief. If we now subtract the expected "smooth" earth interference pattern shown in Figure 10.5b from the distorted pattern, the resulting interference pattern is known as the *flattened interferogram*. The expression for this flattened phase is

$$\phi_{\text{flat}} \approx -\frac{a2\pi}{\lambda} B \cos(\theta_0 - \alpha) \frac{z}{R \sin\theta_0}. \quad (10.16)$$

Here θ_0 is the look angle in the absence of any relief, z is the elevation of the point in the image, and R is the range to the point as shown in Figure 10.5a. For the Mount Shasta example, the result is shown in Figure 10.5e. This figure shows that the flattened interferogram resembles a contour map of the topography.

10.8.2 Differential Interferometry for Surface Deformation Studies

One of the most exciting applications of radar interferometry is implemented by subtracting two interferometric pairs separated in time from each other to form a so-called *differential interferogram*. In this way surface deformation can be measured with unprecedented accuracy. This technique was first demonstrated by Gabriel et al. (1989) using data from SEASAT to measure cm-scale ground motion in agricultural fields. Since then this technique has been applied to measure cm-scale coseismic displacements and to

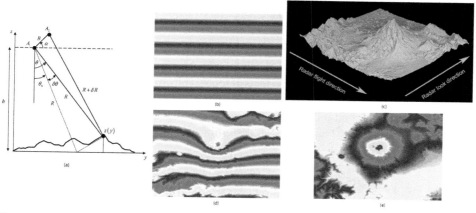

Figure 10.5 Basic interferometric radar geometry. The path length difference between the signals measured at each of the two antennas is a function of the elevation of the scatterer as shown in (a). The rest of the figure shows how the topography of a scene is expressed in the interferometric phase. If there is no topography, all interferometric fringes will be parallel to the radar flight path as shown in (b). Using the topography of Mount Shasta, California, shown in perspective view in (c), the expected fringes for the SRTM system are shown in (d). Once the contribution from a smooth earth as shown in (b) is subtracted from (d), the resulting flattened interferogram (e) resembles a contour map of the topography.

measure cm-scale volcanic deflation. The added information provided by high spatial resolution coseismic deformation maps was shown to provide insight into the slip mechanism that would not be attainable from the seismic record alone. A summary of the results can be found in Rosen et al. (2000).

Differential interferometry is implemented using repeat pass interferometric measurements made at different times. If the ground surface moved between observations by an amount Δr toward the radar sensor, the phase difference in the repeat pass interferogram will include this ground displacement in addition to the topography. In that case, the phase difference in the flattened interferogram becomes

$$\phi_{\text{flat}} \approx -\frac{a2\pi}{\lambda} B \cos(\theta_0 - \alpha) \frac{z}{R \sin\theta_0} + \frac{4\pi\Delta r}{\lambda}. \tag{10.17}$$

This expression shows that the phase difference is far more sensitive to changes in topography, i.e., surface displacement or deformation, than to the topography itself. Typically, the elevation has to change by hundreds of meters to cause one cycle in phase difference, whereas a surface displacement of half a wavelength would cause the same amount of phase change. This is why surface deformations of a few cm can be measured from orbital altitudes using SAR systems.

To extract the surface deformation signal from the interferometric phase, one has to separate the effects of the topography and the surface deformation. Two methods are commonly used to do this. If a good digital elevation model (DEM) is available, one can use that to form synthetic fringes and subtract the topography signal from the measured phase difference. The remaining signal then is due only to surface deformation. In this case, the relative lack of sensitivity to topography works to our benefit. Therefore, DEM errors on the order of several meters usually translate to deformation errors of only cm in the worst case.

The second way in which the deformation signal is isolated is to use images acquired during three overpasses. This method generates a DEM from one pair of images, and then uses that as a reference to subtract the effects of topography from the second pair to form the differential interferogram. If the motion occurred only between the first and second passes, or only between the second and third passes, one interferogram will include the deformation signal, while the other will not. The resulting differential interferogram will then contain only the deformation signal.

10.9 Examples of Radar Images

Imaging radar data are being used in a variety of applications, including geologic mapping, ocean surface observation, polar ice tracking, and vegetation monitoring. Qualitatively, radar images can be interpreted using the same photo interpretation techniques used with visible and near-IR images.

One of the most attractive aspects of radar imaging is the ability of the radar waves to penetrate dry soils to image subsurface features. This capability was first demonstrated with SIR-A imagery of southern Egypt and northern Sudan in the eastern Sahara desert.

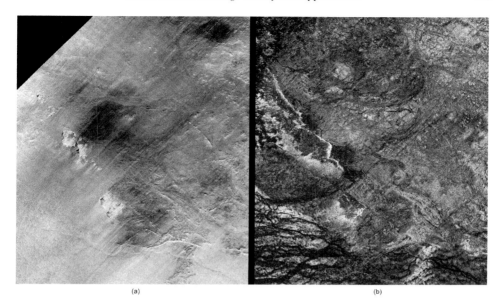

Figure 10.6 A comparison of images of the Safsaf Oasis area in south central Egypt. (a) A LandSat Thematic Mapper image showings bands 7, 4, and 1 displayed as red, green, and blue. (b) SIR-C/X-SAR image displaying L-band ($\lambda \sim 24$cm) HH, C-band ($\lambda \sim 6$ cm) HH, and X-band ($\lambda \sim 3$cm) VV as red, green, and blue, respectively. Each image represents an area of ~30 km × 25 km. (Image courtesy of NASA/JPL.)

Figure 10.6 shows a comparison of a LandSat image acquired in the visible and IR parts of the spectrum to a multifrequency Shuttle Imaging Radar (SIR-C/X-SAR) image acquired over south-central Egypt in 2004. This image was taken in the same area where SIR-A data previously showed buried ancient drainage channels. The Safsaf Oasis is located near the bright yellow feature in the lower left center of the Landsat image. While some features at the surface are visible in both images, in much of the rest of the image, however, the radar waves show a wealth of information about the subsurface geologic structure that is not otherwise visible in the LandSat image. For example, the dark drainage channels visible in the bottom of the radar image are filled with sand as much as 2 m thick. These features are dark in the radar image, because the sand is so thick that even the radar waves cannot penetrate all the way to the bottom of these channels. Only the most recently active drainage channels are visible in the LandSat image. Another example is the many rock fractures visible as dark lines in the radar image. Also visible in the radar image are several blue circular granite bodies. These show no expression in the LandSat image. Field studies conducted previously indicate that at L-band (24 cm wavelength) the radar waves can penetrate as much as 2 m of very dry sand to image subsurface structures.

The second unique contribution of radar imagers for planetary remote sensing is the ability of the radar waves to penetrate thick atmospheres to image surface features that would otherwise be obscured by the atmosphere. This was demonstrated in a spectacular way by NASA's *Magellan* mission to Venus (e.g., Saunders et al., 1992; Ford et al., 1993). The *Magellan* spacecraft was launched in 1989 and arrived at Venus in 1990. During a four-year mission, it used a 12.6-cm wavelength radar imager to produce the first global map of the surface of Venus. Figure 10.7a shows a 550 km × 500 km Magellan image of the Lavinia region on Venus centered at 27° south latitude and 339° east longitude. The image shows three large-impact craters ranging from 37 km to 50 km in diameter. Because of the thick Venusian atmosphere, only large meteorites actually make it to the surface, resulting in only relatively large impact craters being present. The craters in this image are surrounded by bright radar returns from the rough ejecta blanket. The crater floors have much less radar return, indicating that they are generally smoother compared to the radar wavelength. Also clearly visible are bright (rough) central peaks within the craters, characteristic of meteorite impacts. The relative symmetry of the ejecta indicates that the meteorites hit the surface at angles relatively close to vertical.

Another Solar System body with an opaque atmosphere is Titan, the largest moon of Saturn. The *Cassini* mission used a 2.2-cm wavelength radar to image the surface of Titan, returning images that show many different surface features such as large lakes, river systems, dune fields and rough and smooth plains. The image on the right in Figure 10.7b shows a *Cassini* radar image of the Ligeia Mare region near the north pole of Titan. It is the second largest body of liquid on the surface of Titan, and is larger than Lake Superior on Earth. The liquid is composed mostly of methane, but also contains dissolved nitrogen and ethane and

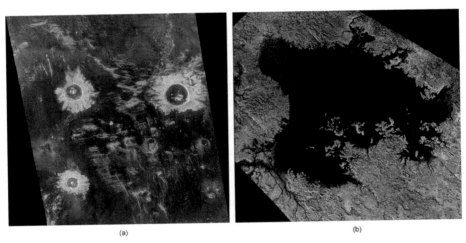

Figure 10.7 Radar images of Venus and Titan. (a) A *Magellan* radar image of the surface of Venus showing three large impact craters in the Lavinia region with diameters ranging from 37 km to 50 km. The image is about 420 km wide and 500 km from top to bottom. (b) A *Cassini* radar image of Ligeia Mare on Titan, the largest moon of Saturn is shown on the right. The image covers an area of ~530 km × 490 km. Besides the Earth, Titan is the only other body in the Solar System to have stable bodies of liquid. These images show the capability of radars to image the surface of planets with opaque atmospheres. (Images courtesy of NASA/JPL.)

other organic components (Le Gall et al., 2016). Also clearly visible in the image are river channels that drain into Ligeia Mare, confirming that Titan is the only other body in the Solar System where stable liquid bodies exist. The image shows that the liquid surface is smooth compared to the radar wavelength as evidenced by the lack of backscatter. The surrounding terrain shows little variation in roughness, and most of the brightness variations in the radar image are due to local slope variations resulting from the local topography.

As mentioned before, radar scattering is different for different combinations of transmit and received polarizations. Figures 10.8a and b show the predicted backscatter cross section as a function of incidence angle for different surface roughness values and different dielectric constants. Figure 10.8a shows that increasing the surface roughness generally causes an increase in the radar cross sections for all polarization combinations. Notice how the difference between the HH and VV cross sections becomes smaller as the surface gets rougher. Figure 10.8b shows that increasing the dielectric constant (soil moisture or soil density) also increases the radar cross sections for all polarizations. In this case, however, increasing the dielectric constant also increases the difference between the HH and VV cross sections. These observations form the basis for algorithms that use polarimetric data to quantitatively estimate surface roughness and dielectric constant for bare surfaces (van Zyl & Kim, 2011). Figure 10.8c shows an example of estimating surface soil moisture using two frequencies. The image shows a feature known as Cottonball Basin in Death Valley, California. Cottonball Basin is at the northern end of the larger Death Valley salt pan, and

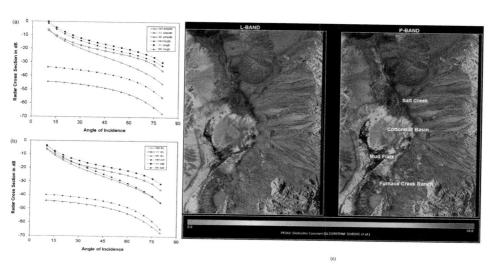

Figure 10.8 Graphs showing predicted radar cross sections for a slightly rough surface assuming an exponential correlation function. (a) The effect of changing surface roughness and constant dielectric constant, while (b) shows the effect of changing dielectric constant for constant roughness. The images on the right show dielectric constant maps inferred from L-band (left) and P-band (right) AIRSAR data of Cottonball Basin in Death Valley, California. Images cover an area about 12 km wide by 20 km long. (Images courtesy of NASA/JPL.)

receives most of its inflow from groundwater. The southern edge of Cottonball Basin are covered with mud flats resulting from seeping salty water. These areas stay wet the longest after an inflow event.

Figure 10.8c shows two dielectric constant maps, the one inferred from L-band data on the left, and from P-band data on the right. There are many similarities between the two maps. First, both show relatively large dielectric constants in the mudflats and lower dielectrics in the rest of Cottonball Basin. There are also some important differences, however. When looking closer at the mud flat areas, we note that the P-band dielectric constants show higher values over larger areas toward the edges of the mudflats. These are the shallower areas of the mud flats. As the mud flats dry out, the surface of the shallower areas dry first, while the subsurface in these areas can stay wetter longer. The longer wavelength P-band signals more than likely penetrate deeper into these surfaces and sense more of the wetter subsurface than the L-band signals. We also note the same behavior to the northern part of the Basin, next to the word "Salt Creek" in the image. There is an area, located in the Salt Creek, that shows a higher dielectric constant at P-Band. Another area is also visible further north following the Salt Creek further up in the image. Both these areas more than likely represent subsurface moisture. Unfortunately, no actual ground measurements were made during the data collection, so these explanations cannot be verified. But given that the algorithm consistently infers higher moistures at the longer wavelength in only some areas, and similar values in others, supports this conjecture.

Differential interferometry has been applied successfully to the study of a number of different geophysical processes, including tectonic movement, volcanic inflation, and surface subsidence due to oil and groundwater pumping. Earlier results focused on measuring the coseismic deformation signal of earthquakes, introducing the technique. Figure 10.9

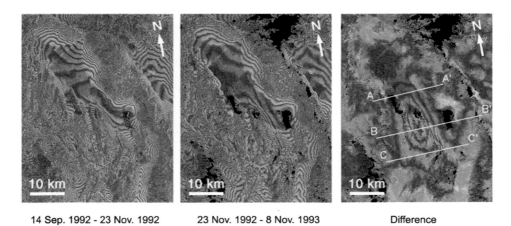

Figure 10.9 Deformation signals measured at C-band following the M 6.1 Eureka Valley earthquake in California. The left two images are the individual interferograms constructed from three acquisitions. The earthquake occurred between the second and third acquisitions. (From Peltzer et al., 1995.)

shows the deformation signal resulting from a magnitude 6.1 earthquake that was centered in the Eureka Valley area of California (Peltzer & Rosen, 1995). The earthquake occurred on May 17, 1993, and caused a few cm of subsidence over an area of about 35 km × 20 km. Profiles through the deformation signal shows that the noise on the measurement is ~3 mm. Using the measured deformation signal, Peltzer and Rosen (1995) argued that the rupture started at depth, and then propagated diagonally upward and southward on a north-east fault plane that dips to the west. They came to this conclusion based on the fact that the observed subsidence signature is elongated in the north–northwest direction.

10.10 Summary

Imaging radars are all-weather instruments that can image planetary surfaces regardless of local atmospheric or solar illumination conditions. Radar images provide information about surfaces that are complementary to the chemistry usually inferred from visible and IR images. Instead, radar images are strongly influenced by surface roughness and geomorphology, and to a lesser extent by the bulk electrical properties of the surface. Radar imagers have returned spectacular information about the surfaces of both Venus and Titan, bodies with dense, opaque atmospheres that are difficult to image using traditional camera systems. Newer radar techniques, such as polarimetry and interferometry, hold great promise to provide additional information about the physical properties of planetary surfaces in the future.

Acknowledgments

This work was performed at the Jet Propulsion Laboratory, California Institute of Technology, under contract with NASA.

References

Dobson M.C., Ulaby F.T., Hallikainen M.T., & El-Rayes M.A. (1985) Microwave dielectric behavior of wet soil. Part II: Dielectric mixing models. *IEEE Transactions on Geoscience and Remote Sensing*, 35–46.
Farr T.G. (1992) Microtopographic evolution of lava flows at Cima Volcanic Field, Mojave Desert, California. *Journal of Geophysical Research*, **97**, 15171–15179.
Farr T.G. & Kobrick M. (2000) Shuttle radar topography mission produces a wealth of data. *Eos, Transactions American Geophysical Union*, **81**, 583–585.
Ford J.P., Plaut J.J., Weitz C.M., et al. (1993) *Guide to Magellan image interpretation*. Jet Propulsion Laboratory Publications, Pasadena, CA.
Freeman A., Krieger G., Rosen P., et al. (2009) SweepSAR: Beam-forming on receive using a reflector-phased array feed combination for spaceborne SAR. *Proceedings of the 2009 IEEE Radar Conference*, 1–9.
Fung A.K., Li Z., & Chen K.S. (1992) Backscattering from a randomly rough dielectric surface. *IEEE Transactions on Geoscience and Remote Sensing*, **30**, 356–369.
Gabriel A.K., Goldstein R.M., & Zebker H.A. (1989) Mapping small elevation changes over large areas: Differential radar interferometry. *Journal of Geophysical Research*, **94**, 9183–9191.
Harmon J.K., Slade M.A., Vélez R.A., Crespo A., Dryer M.J., & Johnson J.M. (1994) Radar mapping of Mercury's polar anomalies. *Nature*, **369**, 213–215.
Le Gall A., Malaska M.J., Lorenz R.D., et al. (2016) Composition, seasonal change, and bathymetry of Ligeia Mare, Titan, derived from its microwave thermal emission. *Journal of Geophysical Research*, **121**, 233–251.

Ostro S.J., Campbell D.B., Simpson R.A., et al. (1992) Europa, Ganymede, and Callisto: New radar results from Arecibo and Goldstone. *Journal of Geophysical Research*, **97**, 18227–18244.

Peltzer G. & Rosen P. (1995) Surface displacement of the 17 May 1993 Eureka Valley, California, earthquake observed by SAR interferometry. *Science*, **268**, 1333–1336.

Peplinski N.R., Ulaby F.T., & Dobson M.C. (1995) Dielectric properties of soils in the 0.3–1.3-GHz range. *IEEE Transactions on Geoscience and Remote Sensing*, **33**, 803–807.

Pettengill G.H., Eliason E., Ford P.G., Loriot G.B., Masursky H., & McGill G.E. (1980) Pioneer Venus Radar results altimetry and surface properties. *Journal of Geophysical Research*, **85**, 8261–8270.

Plaut J.J., Safaeinili A., Holt J.W., et al. (2009) Radar evidence for ice in lobate debris aprons in the mid-northern latitudes of Mars. *Geophysical Research Letters*, **36**, L02203, DOI:10.1029/2008GL036379.

Rosen P.A., Hensley S., Joughin I.R., et al. (2000) Synthetic aperture radar interferometry. *Proceedings of the IEEE*, **88**, 333–382.

Saunders R.S., Spear A.J., Allin P.C., et al. (1992) Magellan mission summary. *Journal of Geophysical Research*, **97**, 13067–13090.

Shi J., Wang J., Hsu A.Y., O'Neill P.E., & Engman E.T. (1997) Estimation of bare surface soil moisture and surface roughness parameter using L-band SAR image data. *IEEE Transactions on Geoscience and Remote Sensing*, **35**, 1254–1266.

Ulaby F.T. & Elachi C.E. (1990) *Radar polarimetry for geoscience applications*. Artech House, London.

Ulaby F.T. & Long D. (2014) *Microwave radar and radiometric remote sensing*. University of Michigan Press, Ann Arbor, MI.

Valenzuela G. (1967) Depolarization of EM waves by slightly rough surfaces. *IEEE Transactions on Antennas and Propagation*, **15**, 552–557.

van Zyl J.J. & Kim Y. (2011) *Synthetic aperture radar polarimetry*. John Wiley & Sons, Hoboken, NJ.

Zebker H.A. & Goldstein R.M. (1986) Topographic mapping from interferometric synthetic aperture radar observations. *Journal of Geophysical Research*, **91**, 4993–4999.

Zebker H.A., van Zyl J.J., & Held D.N. (1987) Imaging radar polarimetry from wave synthesis. *Journal of Geophysical Research*, **92**, 683–701.

Zisk S.H., Pettengill G.H., & Catuna G.W. (1974) High-resolution radar maps of the lunar surface at 3.8-cm wavelength. *The Moon*, **10**, 17–50.

Part II
Terrestrial Field and Airborne Applications

11

Visible and Near-Infrared Reflectance Spectroscopy
Field and Airborne Measurements

ROGER N. CLARK

11.1 Introduction

Analyzing reflected solar radiation is the main strategy for remotely detecting minerals and compounds. By measuring the spectral response of sunlight reflected from a surface with sufficient spectral range and resolution, the signatures of numerous components can be unscrambled to determine surface composition. If spectra are measured in an array of points, the spatial distribution of individual compounds can be mapped. This method is called imaging spectroscopy and benefits from spatial context. The spectral–spatial combination can contain orders of magnitude more scientific information than a point spectral measurement, or a black-and-white single-channel (wavelength) image of a site.

Untangling the multiple competing processes that can influence the light measured by a sensor can be challenging. Sunlight incident on a planetary surface undergoes absorption and scattering from the top of the atmosphere to the surface. The surface reflects a fraction of the incident light and the brightness returned is a function of the reflectance of the surface, the angle of the incident sunlight, and the angle of the scattered light (see Chapters 2 and 4 for more details). Between the surface and the sensor, the light is then attenuated and scattered by the atmosphere. Although some solid bodies in the Solar System have no appreciable atmosphere (e.g., most moons of planets), this chapter considers only the more complicated case of Solar System bodies with atmospheres.

Adding to the complexity of atmospheric scattering and absorption is the fact that most surfaces are composed of mixtures. The type of mixture and the abundance and grain sizes of each component determine which compounds in a surface contribute most to the returned signal, and this can vary with wavelength.

An imaging detector receives a signal at each wavelength that is affected by several factors: (1) solar irradiance, (2) atmospheric transmission, (3) atmospheric scattering, (4) viewing geometry, (5) surface reflectance, (6) component abundances, (7) component grain sizes (and wavelength-dependent scattering), and (8) mixture type. For imagers with just one or a few wavelengths (e.g., Landsat), it is impossible to determine the factors influencing the measured intensity. As the number of spectral channels included in these measurements increases, some of these confounding effects can be isolated and understood.

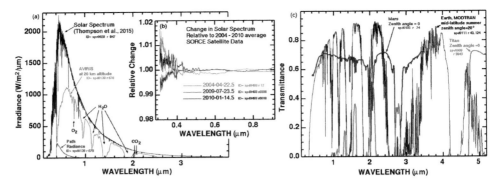

Figure 11.1 Solar spectra and atmospheric transmittance. (a) Solar irradiance at Earth's orbital distance (1 Astronomical Unit). (From Thompson et al., 2015a.) Observed signal measured by AVIRIS of sunlight reflected from Earth's surface from a height of 20 km is seen below the solar spectrum. The blue line is the derived path radiance from the signal, light scattered by the atmosphere in the direction of the detector. (b) Example solar relative variability in the UV and visible wavelengths (SORCE satellite data). (c) Atmospheric transmission, one-way, for Earth, Mars, and Titan (moon of Saturn) for the reflected solar region. Thermal emission becomes significant beyond 3 μm on Earth, about 3.5 μm on Mars, and beyond 6 μm on Titan. The main absorber in Earth's atmosphere is water vapor except in the UV, where it is ozone. The main absorber in the martian atmosphere is CO_2 and the UV absorber is dust. The main absorber in Titan's atmosphere is methane and the large decrease toward shorter wavelengths is due to absorption and scattering by organic aerosols. Earth spectrum from Clark (1999), Mars from P. Irwin (personal communication, 1997), and Titan from Clark et al. (2010a). The transmission of Venus' atmosphere would be too low to register on this plot; transmission at 1 μm is <~ 0.00002 (Baines et al., 2000).

11.2 The Problem of Remote Sensing Through the Atmosphere: Calibration to Surface *I/F* and Reflectance

Sunlight (Figure 11.1a) passing through a planetary atmosphere is subject to absorption and scattering by atmospheric components. At some wavelengths the atmosphere is nearly opaque, making remote sensing of the surface impossible. The one-way transmission of three planetary atmospheres is shown in Figure 11.1c. For space-based remote sensing, the path length would be doubled, squaring the transmission spectra in Figure 11.1c. For field-based methods, e.g., a rover or person measuring spectra at close range, and using the Sun as a source, the transmission is effectively the one-way path in Figure 11.1c. If a local light source is used, the atmospheric path is greatly reduced, opening up more of the spectrum because there is less atmospheric absorption as a result of the shorter path length.

Strictly speaking, the solar spectrum illustrated in Figure 11.1a applies only to a single point in time, as the Sun is continually varying at some wavelengths (Figure 11.1b), particularly in the ultraviolet, and to some extent, in the visible (Harder et al., 2009; Krivova et al., 2011). Tracking solar variability becomes increasingly important as spectral resolution increases, desired precision increases, and as shorter wavelengths are measured.

To derive surface reflectance the sensor needs to be calibrated to radiance, the solar spectrum removed, the path radiance subtracted, and the atmospheric absorption removed. In the simplest terms this calibration takes the form:

$$I/F = (R - P) / (F * T), \qquad (11.1)$$

where I/F is the apparent reflectance uncorrected for local incidence, R is the at-sensor radiance, P is the path radiance, πF is the solar irradiance (e.g., as shown in Figure 11.1a), and T is the atmospheric transmittance (e.g., from Figure 11.1c). Example at-sensor radiance from an aircraft at 20 km altitude is illustrated in Figure 11.1a along with the derived path radiance. Correction to surface reflectance is more difficult in the ultraviolet region than at visible wavelengths because at wavelengths shorter than about 0.4 μm, the path radiance is similar in magnitude to the direct reflected radiance.

Atmospheric transmission models require significant computation time and the detailed transmission and path radiance is dependent on the atmospheric temperature and pressure profile as a function of altitude, as well as gas and aerosol abundance. The total atmospheric path length depends on solar incidence, local topographic elevation, and path length from the surface to the sensor. At present it is not feasible to invert all of these parameters for each pixel in an imaging spectrometer scene. The strategy adopted by the remote sensing community to remove atmospheric effects is to precompute a table of possible solutions, examine the data for one pixel, and interpolate the data table to the conditions found in that pixel (Gao & Goetz, 1990; Thompson et al., 2015b and references therein). Because conditions are not likely to be completely covered by the data table, this strategy is an approximation, especially at wavelengths near strong atmospheric absorption bands. Similarly, McGuire et al. (2008) and Seelos et al. (2012) have developed a sophisticated radiative transfer model for correcting spectral image data from the Compact Reconnaissance Imaging Spectrometer for Mars (CRISM) instrument (Murchie et al., 2007) on the Mars Reconnaissance Orbiter.

To compensate for the residuals in the radiative transfer model inversions, investigators also have used ground calibration targets where spectra of those targets were used to correct residual artifacts in the atmospheric models, as well as in the radiometric calibration of the instrument, or in the solar irradiance. This dual calibration became known as Radiative Transfer, Ground Calibrated (RTGC or rtgc) (Clark et al., 1993, 1995, 2003b) and is now adopted by Thompson et al. (2015b) to deliver apparent reflectance data to investigators. The new calibration strategy (e.g., McGuire et al., 2009; Thompson et al., 2015b) enables investigators to use delivered data, generally without the need to apply additional calibration methods. This new calibration technique may sometimes need further residual reduction in cases of unusual atmospheric conditions, necessitating a local (second) ground calibration, potentially rtgc2. Atmospheric corrections could be improved with additional measurements such as Light Detection and Ranging (LiDAR) that would provide atmospheric path lengths and local slope as well as local topography information (e.g., see Asner et al., 2007).

In parallel to terrestrial atmospheric model development, radiative transfer inversions of imaging spectrometer data for Mars (Seelos et al., 2012) and Titan (Cornet et al., 2017) have been under development. Surface reflectance data for CRISM data at Mars are available at the Planetary Data System, Geological Sciences Node: http://pds-geosciences.wustl.edu/missions/mro/crism.htm. Inversion models for Titan, with the most difficult inversion problem due to both strong absorption and very high aerosol scattering, are still under development.

Aerosol scattering remains one of the hardest effects to quantify. On Earth, Mars, and Titan, aerosol scattering varies spatially with local conditions and topography. Imaging spectrometers need to measure shorter wavelengths to quantify path radiance effects. For example, on Earth and Mars, measuring spectra down to 0.3 µm would improve the lever arm to constrain the correction at longer wavelengths, even if those short wavelengths were not used to measure surface reflectance. For Titan, the shortest wavelength for which the surface can be observed from above the atmosphere is about 0.95 µm and the surface signal is a small fraction of the path radiance signal.

11.3 Mixtures and Complications

Surfaces in the real world are a complex mixture of materials at just about any scale from the laboratory to planetary bodies. In general, there are four types of mixtures:

1. *Linear mixture.* The materials in the field of view are optically separated so there is no multiple scattering between components. The combined signal is simply the sum of the fractional area times the spectrum of each component, and is called an areal mixture or checkerboard mixture.
2. *Intimate mixture.* An intimate mixture occurs when different materials are in intimate contact in a scattering surface, such as the mineral grains in a soil or rock. Depending on the optical properties of each component, the resulting signal is a highly nonlinear combination of the end-member spectra (see also Chapters 2 and 4).
3. *Coatings.* Coatings occur when one material coats another. Each coating is a scattering/transmitting layer where the optical thickness varies with material properties and wavelength (see also Chapter 4).
4. *Molecular mixtures.* Molecular mixtures occur on a molecular level, and include mixtures of two liquids, or a liquid and a solid (see also Chapter 5). Examples include water adsorbed onto a mineral and gasoline spilled onto soil. The close contact of the mixture components can cause band shifts in the adsorbate, such as the interlayer water in montmorillonite, or the water in plants.

An example mixture comparison is shown in Figure 11.2a for alunite and jarosite. Note in the intimate mixture how the jarosite spectrum dominates in the 0.4 to 1.3 µm region. Darker materials, e.g., jarosite, dominate in intimate mixtures because photons are absorbed when they encounter a dark grain. However, in areal mixtures, the brighter material dominates. In an intimate mixture of light and dark grains (e.g., quartz and magnetite) the photons have such a high probability of encountering a dark grain that only a few percent of dark grains can reduce the reflectance of the sample to a much greater

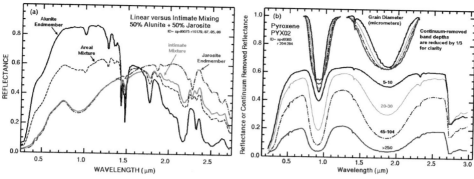

Figure 11.2 Effects of mixtures and grain size on band shape. (a) Reflectance spectra of alunite, jarosite, and mixtures of the two. Two mixture types are shown: intimate and areal. In the intimate mixture the darker of the two spectral components tends to dominate at any given wavelength. In an areal mixture, the brighter component dominates. The areal mixture is strictly a linear combination and was computed from the endmembers, whereas the intimate mixture is nonlinear, and the spectrum of the physical mixture was measured in the laboratory. Data from Clark et al. (2003a). (b) Pyroxene reflectance spectra as a function of grain size. As the grain size becomes larger, more light is absorbed, the reflectance decreases, and the absorption feature flattens. Note that trace tremolite contamination causes the narrow absorption features near 1.4 and 2.3 μm. The broader pyroxene absorptions are the continuum background to the narrow tremolite features. This example shows how the components in a mixture can be readily identified even though no unmixing analysis is performed. The component features are "spectrally separated" in wavelength. Continuum-removed feature fits (top) show the similarity in shape of features at different grain sizes. The small change in shape can be used to determine grain size from the spectra, independent of abundance. (Data from Clark et al., 2003a.)

extent than their weight fraction would imply (e.g., Clark, 1983, 1999 and references therein).

The scattering between particles in the surface enables us to see light returned from the surface, but adds to the complications of mixture type (see also Chapter 2). When light enters a particulate surface, scattering scrambles the direction of light. More absorbing grains inhibit further travel of the light (Figure 11.2a). Thus, as the wavelength varies, absorption changes, and depth of penetration changes. This variable path length into a particulate surface with varying absorption strength alters the shape of absorption bands. As a result, the mean optical path length in a particulate surface is roughly inversely proportional to the square root of the absorption coefficient (Clark & Roush, 1984), and absorption bands have a different width and shape in reflectance compared to transmittance.

The observed strength of absorption features changes with scattering and therefore also changes with grain size, as shown in Figures 11.2b and 11.3a. As grain size increases, reflectance levels drop and first-surface reflections contribute proportionally more to the returned light, flattening the band bottoms. A graph of absorption band depth versus grain size shows increasing depth at small grains, reaches a peak, and then decreases (Clark & Lucey, 1984).

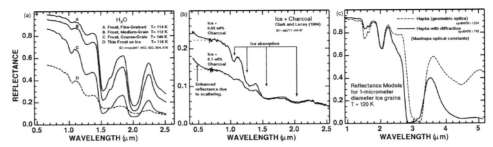

Figure 11.3 Water ice and frost spectral properties. (a) Illustration of changing ice absorption band shapes and strengths with grain size. The near-infrared spectral reflectance of *A*) a fine-grained (~200 μm diameter) water frost, *B*) medium-grained (~300 μm) frost, *C*) coarse grained (400–2000 μm) frost, and *D*) an ice block containing bubbles and frost on its surface. The larger the effective grain size, the greater the mean photon path travelled through the ice, and the deeper the absorptions become. Curve *D* is very low in reflectance because of the large path length in ice but scattering from fine frost at the surface raises the reflectance. (Adapted from data in Clark, 1981 and Clark & Lucey, 1984, with level corrections from the reflectance standard.) (b) Ice + charcoal dust mixtures showing enhanced scattering, a bluing, at shorter wavelengths. (Data from Clark & Lucey, 1984.) (c) Radiative transfer model showing the effects of diffraction from small particles. The dashed line indicates a traditional model with no diffraction. The solid line indicates a model with diffraction. In both models, the grain diameter was 1 μm. (Data from Clark et al., 2012; optical constants from Mastrapa et al., 2008).

The dominant radiative transfer theory used in planetary and terrestrial remote sensing for decades is that developed by Hapke (1981, 1993, 2012). See Chapter 2 of this volume for a more in-depth discussion of Hapke and other radiative transfer models. This chapter includes one additional aspect of scattering and modeling that has recently come to light. Hapke argued that particles in close contact scatter so that diffraction, a far-field effect, did not apply. However, observations by the *Cassini* Visual and Infrared Mapping Spectrometer (VIMS) in the Saturn system (see Chapter 21) found unusually shaped ice absorptions that were traced to diffraction effects (Clark et al., 2012). Diffraction by small particles increases scattering, thereby decreasing wavelength (Figure 11.3b) and altering band shape (Figure 11.3c). For cases in which the particles are much smaller than the wavelength, the particles Rayleigh scatter (Clark et al., 2012). This effect is now called bluing of the spectrum and the effects have been further quantified by Brown (2014).

11.4 Remote Detection of Mixtures and Grains Sizes

One challenge in remote sensing is identifying all components contributing to the surface spectrum and deriving their abundances. Ideally, this would be a straightforward radiative transfer inversion model; however, there are several challenges with achieving this goal. Of the four types of mixtures – (1) linear (areal), (2) intimate, (3) coatings, and (4) molecular – only linear mixtures have simple mathematical equations describing the physics behind this process. Modeling intimate mixtures depends on radiative transfer theory, which yields approximations and requires significant computational resources (Chapter 2). Hapke

(2012) extended his theory to include coatings, but this coating model needs verification. Molecular mixtures could be complicated if there is coupling between the molecules, causing absorption bands to change intensity and shift in wavelength, e.g., through hydrogen bonding. Laboratory experiments testing the properties of molecular mixtures are necessary to confirm which effects are important for a particular set of compounds.

A remotely sensed surface may contain multiple mixture types and it is usually not obvious which mixture types are indicated in a spectrum. Usually one mixture type is assumed by an analyst. Adding to the complexity of mixture type, the contribution of diffraction scattering by sub-wavelength particles is variable. For example, Clark et al. (2012) modeled spectra of Iapetus using simultaneous intimate, molecular, and areal mixtures with components of sub-μm particles exhibiting both Rayleigh scattering and Rayleigh absorption.

Rayleigh scattering occurs when the particle is much smaller than the wavelength of light and separated from other tiny particles, e.g., an nm-sized particle sitting on a much larger grain several μm across. Rayleigh absorption occurs when a high index of refraction material (e.g., metals) is much smaller than the wavelength of light and embedded in another material (e.g., a silicate) such that the index of refraction difference is small enough to reduce scattering. A good example is nanophase metallic iron embedded in a silicate matrix, commonly created in space weathering (Hapke, 2001). Rayleigh absorption increases with decreasing wavelength and creates the reddening observed in space-weathered silicates. Clark et al. (2012) showed Rayleigh absorption is a better explanation for the reddening on Saturn's satellites Iapetus and Phoebe than the traditional explanation of organic compounds.

The spectral continuum is another important component (e.g., Clark & Roush, 1984) and represents absorption by other processes. For example, Figure 11.2b shows spectra of pyroxene size fractions, where the reflectance level and slope changes with grain size. Small tremolite absorptions are superimposed on the absorptions from the pyroxene; thus, the pyroxene absorption is the continuum for the tremolite spectrum. The tremolite–pyroxene example in Figure 11.2b illustrates another concept: unmixing by isolating absorptions through continuum removal. The position, width, and shape of each feature can be isolated through visual examination. With that information, admixture can be determined using a library of known spectra. Analytically, this is performed through continuum removal. Another method employs fitting Gaussian profiles, e.g., Sunshine et al. (1990). Identification of components in the spectra and constraints on grain size through the shape of the absorptions can then be used in radiative transfer models to derive abundances of those components and other spectrally neutral components (e.g., Clark & Roush, 1984; Hapke, 2012; Clark et al., 2012).

11.5 Imaging Spectroscopy: The Spatial–Spectral Connection

An imaging spectrometer provides a spectrum for each pixel in an image, and it creates an image cube consisting of two spatial dimensions and one spectral dimension. A spectral image can show geologic context, raising confidence in the calibration and detection of a specific mineralogy or chemistry. Beyond basic calibration and identification of materials, the spatial extent of composition and its relationship to the land and other compositions can provide context, processes, and origin. One example of the utility of the spatial–spectral connection is its use in locating pollution sources undetectable in visible color images or broadband sensors.

Analysis of the surface reflectance data is generally more complex than for atmospheric data. The composition of the atmosphere is generally better known and varies much less from pixel to pixel. Absorption bands of solids include broader features that can overlap. Some absorption features are diagnostic of specific compositions (Chapters 1 and 4), while others are more general, e.g., the presence of ferrous iron (see Clark, 1999; Clark et al., 2003b and references therein). That complexity means it is challenging to perform a full radiative transfer model solution on imaging spectrometer datasets, and the potential variability within a scene means that precomputing a set of possible solutions is difficult. Prevailing methods vary from using band ratios and simple equations designed to isolate a spectral feature (e.g., Viviano-Beck et al., 2014), to continuum removal and least-squares spectral feature matching (e.g., Parente et al., 2011). With a derived map of different spectral features, researchers commonly extract an average spectrum of a particular composition for more detailed analyses (e.g., Murchie et al., 2009), including computing radiative transfer models, and/or constructing analogs and measuring spectra in a laboratory to derive abundances. Simple solutions of band ratios, 3-point band depths, and simple least squares comparisons of spectral features result in many false-positive or -negative detections (Clark et al., 2003a; Swayze et al., 2003). Heylen et al. (2014, 2017) presented and reviewed other methods. Chapter 15 discusses image analysis options including spectral matching with an expert system, spectral parameterization, spectral unmixturing techniques, and machine learning approaches that will not be discussed here.

This chapter focuses on a system called Tetracorder that has a proven strategy to detect hundreds of compounds simultaneously including all mixture types and for some materials, their grain sizes (Clark et al., 2003a, 2010b, 2015 and references therein). Heylen et al. (2017) used a new statistical method to identify the number of endmembers in a scene and reported up to 56 in an AVIRIS scene of Cuprite, Nevada. Tetracorder version 5.1, used for the mapping results in Figure 11.4a, employed 390 reference spectra on the same Cuprite AVIRIS scene and found with high confidence, 176 endmembers, including minerals; vegetation types; other compounds; coatings; and intimate, molecular, and areal mixtures.

11.6 The Tetracorder Technique

The Tetracorder technique has been used to analyze data from all over the Solar System, including mapping ice and other compounds on icy satellite surfaces in the Saturn (e.g., Clark et al., 2012 and references therein) and Jupiter (Carlson et al., 1996) systems, minerals on Mars (e.g., Hoefen et al., 2003; Ehlmann et al., 2016 and references therein), and was critical in making the discovery of widespread water on the Moon possible (Clark, 2009; Pieters et al., 2009). Tetracorder is used for mapping minerals on Earth (e.g., Swayze

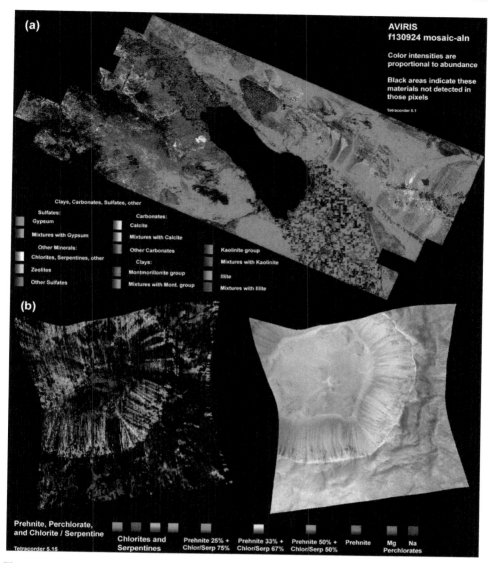

Figure 11.4 Tetracorder mineral map expert system standard products. (a) AVIRIS classic image cube mosaic processed with Tetracorder for a JPL-generated surface apparent reflectance calibration. Results here show minerals detected using absorption features in the near-infrared. Additional mapping of this scene identified hundreds of minerals, chemicals, and vegetation spectral signatures. (b) Tetracorder results for CRISM image FRT000148C1 for a crater at 7.7°S, 84.5°E in the Terra Tyrrhena region of Mars. Multiple types of phyllosilicates and other minerals were detected here using Tetracorder.

et al., 2000, 2002, 2009, 2014; Livo et al., 2007) and it was used in assessing the environmental damage from the World Trade Center disaster (Clark et al., 2001), the 2010 Deepwater Horizon oil spill in the Gulf of Mexico (Clark et al., 2010b), and mapping ecosystems (Kokaly et al., 2003, 2007). Derivative software using an early version of the Tetracorder algorithms and expert system coded in IDL has been used to map minerals in an entire country (Kokaly et al., 2013) based on a simpler, pre-1993, methodology of Tetracorder.

Tetracorder is a system for applying multiple algorithms to a single spectrum or a group of spectra (an imaging spectrometer dataset) to identify and map mineralogy, chemistry, and grain size. Tetracorder is different from other systems because it applies one or more algorithms to identify materials expressed in spectra, evaluating the results of those algorithms to make decisions, including identifications and applying additional algorithms to solve specific problems based on results of the previous analysis. Tetracorder employs data from other sources, e.g., temperature and pressure, if available, to further constrain identifications. In that sense, Tetracorder is an "algorithm to apply algorithms" through commanding by an expert system. Tetracorder benefits from ~30 years of development by a team of expert spectroscopists at the U.S. Geological Survey and additional collaborators who developed the expert system. This technique can now be applied anywhere in the Solar System (Clark et al., 2015, 2016, Figures 11.4a and b), and the same expert system can be used with any spectral range and resolution (Tetracorder version 5.1+). Tetracorder 5+ mapping results are automatically analyzed and color-coded maps are automatically produced (Figures 11.4a–b). Tetracorder produces maps of hundreds of materials, including chemical substitutions in some minerals. Tetracorder results could be fed to other systems, e.g., real-time robotic systems to guide a robot to resources (in situ resource utilization, ISRU). Indeed, a new NASA project, Toolbox for Research and Exploration, TREX, will use Tetracorder for real-time field assessment (https://trex.psi.edu/).

Tetracorder solves the mixture problem by implementing spectra of measured or computed mixtures, but only for overlapping absorption features that cannot be separated by continuum removal, thus greatly reducing the number of mixtures needed. For example, consider the narrow tremolite absorptions superimposed on the broad pyroxene bands in Figure 11.2b. No mixture reference spectra are needed by Tetracorder, because a short continuum adequately isolates the different absorptions. Only if similar width absorptions overlap is a mixture analysis needed. For example, if the pyroxene in Figure 11.2b contained ice (as in Figure 11.3a), the 2-μm ice band would overlap the 2-μm pyroxene band and a set of mixture spectra would be needed by Tetracorder. The advantage of this strategy is that fewer mixtures are required and computational time is reduced. Other solutions, e.g., linear equations of an areal mixture problem, involve a matrix inversion where the compute time for a solution increases as the square of the number of endmembers.

The rapid response results demonstrated with Tetracorder during the 2001 World Trade Center disaster and the 2010 Gulf of Mexico oil spill, where results were supplied to first responders within a few weeks, demonstrate that real-time results may soon be possible. Calibration and new laboratory measurements were required for these studies, but as spectral libraries mature to include more real-world cases and real planetary materials, accurate mapping of compounds will not need as much postevent laboratory work, if any. Combining updated spectral libraries with rapid production of calibrated surface reflectance (or apparent reflectance) with the atmospheric features removed enables faster mapping of minerals and other components. As computers become faster, we are advancing toward near real-time results for identifying and mapping many compounds.

11.7 Conclusions

There has been steady advancement in radiative transfer modeling and understanding of the absorption and scattering effects in planetary rocks, soils, and terrestrial ecosystems over the past few decades. Advancements have also occurred in processing large spectral datasets (imaging spectrometer data cubes). Of note is better calibration of sensors and improved atmospheric correction to surface reflectance. Derivation of surface reflectance, combined with more advanced spectral analyses, both in terms of identifying materials in imaging spectrometer datasets and in deriving abundances with radiative transfer models, means remote sensing spectroscopy and imaging spectroscopy are entering a new era in which more rapid turnaround is possible, perhaps with real-time results available soon. Such capability could mean that future spacecraft, including orbiters and landers/rovers, could be sensing mineralogy and chemistry in real time, facilitating the search for resources, in situ resource utilization (ISRU), or biosignatures.

Acknowledgments

The author was funded by the *Cassini* mission, VIMS team, and the Mars Reconnaissance Orbiter CRISM team.

References

Asner G.P., Knapp D.E., Kennedy-Bowdoin T., et al. (2007) Carnegie airborne observatory: In-flight fusion of hyperspectral imaging and waveform light detection and ranging for three-dimensional studies of ecosystems. *Journal of Applied Remote Sensing*, **1**, 013536.

Baines K.H., Yanamandra-Fisher P.A., Lebofsky L.A., et al. (1998) Near-infrared absolute photometric imaging of the uranian system. *Icarus*, **132**, 266–284.

Baines K.H., Bellucci G., Bebring J.P., et al. (2000) Detection of sub-micron radiation from the surface of Venus by the *Cassini*/VIMS. *Icarus*, **148**, 307–311.

Brown A.J. (2014) Spectral bluing induced by small particles under the Mie and Rayleigh regimes. *Icarus*, **239**, 85–95.

Carlson R., Smythe W., Baines K.H., et al. (1996) Near-infrared spectroscopy and spectral mapping of Jupiter and the Galilean satellites: First results from *Galileo*'s initial orbit. *Science*, **274**, 385–388.

Clark R.N. (1981) Water frost and ice: The near-infrared spectral reflectance 0.65–2.5 μm. *Journal of Geophysical Research*, **86**, 3087–3096.

Clark R.N. (1983) Spectral properties of mixtures of montmorillonite and dark carbon grains: Implications for remote sensing minerals containing chemically and physically adsorbed water. *Journal of Geophysical Research*, **88**, 10635–10644.

Clark R.N. (1999) Spectroscopy of rocks and minerals, and principles of spectroscopy. *Manual of Remote Sensing*, **3**, 2–2.

Clark R.N. (2009) Detection of adsorbed water and hydroxyl on the Moon. *Science*, **326**, 562–564.

Clark R.N. & Lucey P.G. (1984) Spectral properties of ice-particulate mixtures and implications for remote sensing: 1. Intimate mixtures. *Journal of Geophysical Research*, **89**, 6341–6348.

Clark R.N., Swayze G., Heidebrecht K., Goetz A.F., & Green R.O. (1993) Comparison of methods for calibrating AVIRIS data to ground reflectance. *Proceedings of the 5th Annual Airborne Geoscience Workshop*, 35–36.

Clark R.N., Swayze G.A., Heidebrecht K., Green R.O., & Goetz F. (1995) Calibration to surface reflectance of terrestrial imaging spectrometry data: Comparison of methods. *Proceedings of the 5th JPL Airborne Earth Science Workshop*, Abstract, 41–42.

Clark R.N., Green R.O., Swayze G.A., et al. (2001) Environmental studies of the World Trade Center area after the September 11, 2001 attack. U.S. Geological Survey, Open File Report OFR-01-0429.

Clark R.N., Swayze G.A., Livo K.E., et al. (2003a) Imaging spectroscopy: Earth and planetary remote sensing with the USGS Tetracorder and expert systems. *Journal of Geophysical Research*, **108**, E12, 5131, DOI:10.1029/2002JE001847.

Clark R.N., Swayze G., Livo K.E., et al. (2003b) Surface reflectance calibration of terrestrial imaging spectroscopy data: A tutorial using AVIRIS. *Proceedings of the 11th JPL Airborne Earth Science Workshop*, 43–63.

Clark R.N., Curchin J.M., Barnes J.W., et al. (2010a) Detection and mapping of hydrocarbon deposits on Titan. *Journal of Geophysical Research*, **115**, E10005, DOI:10.1029/2009JE003369.

Clark R.N., Swayze G.A., Leifer I., et al. (2010b) A method for quantitative mapping of thick oil spills using imaging spectroscopy. *US Geological Survey Open-File Report*, **20101167**, 1–51.

Clark R.N., Cruikshank D.P., Jaumann R., et al. (2012) The surface composition of Iapetus: Mapping results from Cassini VIMS. *Icarus*, **218**, 831–860.

Clark R.N., Swayze G.A., Murchie S.L., Seelos F.P., Seelos K., & Viviano-Beck C.E. (2015) Mineral and other materials mapping of CRISM data with Tetracorder 5. *46th Lunar Planet. Sci. Conf.*, Abstract #2410.

Clark R.N., Swayze G.A., Murchie S.L., Seelos F.P., Viviano-Beck C.E., & Bishop J. (2016) Mapping water and water-bearing minerals on Mars with CRISM. *47th Lunar Planet. Sci. Conf.*, Abstract #2900.

Cornet T., Rodriguez S., Maltagliati L., et al. (2017) Radiative transfer modelling in Titan's atmosphere: Application to *Cassini*/VIMS data. *48th Lunar Planet. Sci. Conf.*, Abstract #1847.

Ehlmann B.L., Swayze G.A., Milliken R.E., et al. (2016) Discovery of alunite in Cross crater, Terra Sirenum, Mars: Evidence for acidic, sulfurous waters. *American Mineralogist*, **101**, 1527–1542.

Gao B.C. & Goetz A.F. (1990) Column atmospheric water vapor and vegetation liquid water retrievals from airborne imaging spectrometer data. *Journal of Geophysical Research*, **95**, 3549–3564.

Hapke B. (1981) Bidirectional reflectance spectroscopy: 1. Theory. *Journal of Geophysical Research*, **86**, 3039–3054.

Hapke B. (1993) *Introduction to the theory of reflectance and emittance spectroscopy*. Cambridge University Press, New York.

Hapke B. (2001) Space weathering from Mercury to the asteroid belt. *Journal of Geophysical Research*, **106**, 10039–10073.

Hapke B. (2012) *Theory of reflectance and emittance spectroscopy*. Cambridge University Press, Cambridge.

Harder J.W., Fontenla J.M., Pilewskie P., Richard E.C., & Woods T.N. (2009) Trends in solar spectral irradiance variability in the visible and infrared. *Geophysical Research Letters*, **36**, L07801, DOI:10.1029/2008GL036797.

Heylen R., Parente M., & Gader P. (2014) A review of nonlinear hyperspectral unmixing methods. *IEEE Journal of Selected Topics in Applied Earth Observations and Remote Sensing*, **7**, 1844–1868.

Heylen R., Parente M., & Scheunders P. (2017) Estimation of the number of endmembers in a hyperspectral image via the hubness phenomenon. *IEEE Transactions on Geoscience and Remote Sensing*, **55**, 2191–2200.

Hoefen T.M., Clark R.N., Bandfield J.L., Smith M.D., Pearl J.C., & Christensen P.R. (2003) Discovery of olivine in the Nili Fossae region of Mars. *Science*, **302**, 627–630.

Kokaly R.F., Despain D.G., Clark R.N., & Livo K.E. (2003) Mapping vegetation in Yellowstone National Park using spectral feature analysis of AVIRIS data. *Remote Sensing of Environment*, **84**, 437–456.

Kokaly R., Despain D.G., Clark R., & Livo K.E. (2007) Spectral analysis of absorption features for mapping vegetation cover and microbial communities in Yellowstone National Park using AVIRIS data. In: *Integrated Geoscience Studies in the Greater Yellowstone Area: Volcanic, Tectonic, and Hydrothermal Processes in the Yellowstone Geoecosystem*. USGS Professional Paper 1717 (L.A. Morgan, ed.). U.S. Geological Survey.

Kokaly R.F., King T.V., & Hoefen T.M. (2013) Surface mineral maps of Afghanistan derived from HyMap imaging spectrometer data, version 2. US Department of the Interior, US Geological Survey Data Series, **787**, 29pp.

Krivova N., Solanki S., & Unruh Y. (2011) Towards a long-term record of solar total and spectral irradiance. *Journal of Atmospheric and Solar-Terrestrial Physics*, **73**, 223–234.

Livo K.E., Kruse F.A., Clark R.N., Kokaly R.F., & Shanks W.C.I. (2007) Hydrothermally altered rock and hot-spring deposits at Yellowstone National Park—Characterized using airborne visible- and infrared-spectroscopy data. Integrated geoscience studies in the Greater Yellowstone area: Volcanic, tectonic, and hydrothermal processes in the Yellowstone geoecosystem. USGS Professional Paper 1717 (L.A. Morgan, ed.). US Geological Survey, 463–489.

Mastrapa R., Bernstein M., Sandford S., Roush T., Cruikshank D., & Dalle Ore C. (2008) Optical constants of amorphous and crystalline H_2O-ice in the near infrared from 1.1 to 2.6 μm. *Icarus*, **197**, 307–320.

McGuire P.C., Wolff M.J., Smith M.D., et al. (2008) MRO/CRISM retrieval of surface Lambert albedos for multispectral mapping of Mars with DISORT-based radiative transfer modeling: Phase 1—Using historical climatology for temperatures, aerosol optical depths, and atmospheric pressures. *IEEE Transactions on Geoscience and Remote Sensing*, **46**, 4020–4040.

McGuire P.C., Bishop J.L., Brown A.J., et al. (2009) An improvement to the volcano-scan algorithm for atmospheric correction of CRISM and OMEGA spectral data. *Planetary and Space Science*, **57**, 809–815.

Murchie S., Arvidson R., Bedini P., et al. (2007) Compact reconnaissance imaging spectrometer for Mars (CRISM) on Mars reconnaissance orbiter (MRO). *Journal of Geophysical Research*, **112**, E05S03, DOI:10.1029/2006JE002682.

Murchie S.L., Seelos F.P., Hash C.D., et al. (2009) The Compact Reconnaissance Imaging Spectrometer for Mars investigation and data set from the Mars Reconnaissance Orbiter's primary science phase. *Journal of Geophysical Research*, **114**, E00D07, DOI:10.1029/2009JE003344.

Parente M., Makarewicz H.D., & Bishop J.L. (2011) Decomposition of mineral absorption bands using nonlinear least squares curve fitting: Application to martian meteorites and CRISM data. *Planetary and Space Science*, **59**, 423–442.

Pieters C.M., Goswami J., Clark R., et al. (2009) Character and spatial distribution of OH/H2O on the surface of the Moon seen by M3 on Chandrayaan-1. *Science*, **326**, 568–572.

Seelos F., Morgan M., Taylor H., et al. (2012) CRISM Map Projected Targeted Reduced Data Records (MTRDRs): High level analysis and visualization data products. *Planetary Data: A Workshop for Users and Software Developers*, 159–162.

Sunshine J.M., Pieters C.M., & Pratt S.F. (1990) Deconvolution of mineral absorption bands: An improved approach. *Journal of Geophysical Research*, **95**, 6955–6966.

Swayze G.A., Smith K.S., Clark R.N., et al. (2000) Using imaging spectroscopy to map acidic mine waste. *Environmental Science & Technology*, **34**, 47–54.

Swayze G., Clark R., Sutley S., et al. (2002) Mineral mapping Mauna Kea and Mauna Loa shield volcanos on Hawaii using AVIRIS data and the USGS Tetracorder spectral identification system: Lessons applicable to the search for relict martian hydrothermal systems. *Proceedings of the 11th JPL Airborne Earth Science Workshop*, 373–387.

Swayze G.A., Clark R.N., Goetz A.F., Chrien T.G., & Gorelick N.S. (2003) Effects of spectrometer band pass, sampling, and signal-to-noise ratio on spectral identification using the Tetracorder algorithm. *Journal of Geophysical Research*, **108**, 5105, DOI:10.1029/2002JE001975.

Swayze G.A., Kokaly R.F., Higgins C.T., et al. (2009) Mapping potentially asbestos-bearing rocks using imaging spectroscopy. *Geology*, **37**, 763–766.

Swayze G.A., Clark R.N., Goetz A.F., et al. (2014) Mapping advanced argillic alteration at Cuprite, Nevada, using imaging spectroscopy. *Economic Geology*, **109**, 1179–1221.

Thompson D.R., Seidel F.C., Gao B.C., et al. (2015a) Optimizing irradiance estimates for coastal and inland water imaging spectroscopy. *Geophysical Research Letters*, **42**, 4116–4123.

Thompson D.R., Gao B.-C., Green R.O., Roberts D.A., Dennison P.E., & Lundeen S.R. (2015b) Atmospheric correction for global mapping spectroscopy: ATREM advances for the HyspIRI preparatory campaign. *Remote Sensing of Environment*, **167**, 64–77.

Viviano-Beck C.E., Seelos F.P., Murchie S.L., et al. (2014) Revised CRISM spectral parameters and summary products based on the currently detected mineral diversity on Mars. *Journal of Geophysical Research*, **119**, 2014JE004627.

12

Raman Spectroscopy
Field Measurements

PABLO SOBRON, ANUPAM MISRA, FERNANDO RULL, AND ANTONIO SANSANO

12.1 Introduction

Raman spectroscopy is a well-established analytical technique for molecular and structural analyses across many disciplines. Advances in optical materials and miniaturization of electronics over the past decades have had a transformative impact on the way we perform Raman spectroscopy today. Raman systems that once occupied large volumes now fit in the palm of a hand. Indeed, portable Raman instruments are widely available commercially for scientific and industrial applications. Predictably, the National Aeronautics and Space Administration (NASA) and the European Space Agency (ESA) are developing Raman spectrometers to explore the surface and subsurface of Mars: two Raman instruments (SHERLOC and SuperCam) will fly on NASA's Mars 2020 mission, and ESA's 2020 ExoMars rover features the Raman Laser Spectrometer (RLS). SHERLOC will enable detection and characterization of organics and minerals (Beegle et al., 2014; Beegle et al., 2015); RLS will identify organic compounds and mineral products as indicators of biological activity (Hutchinson et al., 2014); SuperCam's Raman instrument will provide stand-off mineral and organic detection (Maurice et al., 2015). The selection of three Raman instruments for these upcoming missions testifies to the maturity of this technique and its perceived ability to perform mineralogical and astrobiologic investigations on Mars. In this chapter we provide a review of two stand-off Raman spectroscopy instrument prototypes, pioneered by groups at the University of Hawaii (United States) and Universidad de Valladolid (Spain), for the exploration of terrestrial and planetary environments, and discuss field applications of such instruments that have spurred new Raman spectroscopy technologies and instrumentation for planetary exploration. We describe technical aspects of the instruments and report results from tests with reference materials and natural samples in planetary analog terrains. We note that the use of Raman spectroscopy for planetary analog studies is not new. Close-up, compact, field-ready, Raman instruments have been developed and used extensively in such applications (e.g., Rull et al., 2010; Vítek et al., 2010; Sobron et al., 2014; Wei et al., 2015; Sobron et al., 2016).

12.2 Stand-off Raman Instrumentation and Planetary Analog Applications

The possibility of performing Raman spectroscopy at a distance rapidly and autonomously makes stand-off Raman a very powerful tool for planetary exploration. Stand-off Raman spectroscopy can be used for near real-time mineralogical surveys (fast analysis of major minerals) and detection and characterization of possible signs of life and potential habitats in support of current and future missions to Mars and other planets. Mounted on a rover, a stand-off Raman system can identify targets rapidly for further sampling and analysis and support decision-making processes on the fly, e.g., drive away or stay? More importantly, these tactical decisions can be made autonomously, without driving the rover and deploying arm-mounted sensors on targets – a time- and energy-taxing process – thus facilitating cost-effective measurements during Mars operations. ChemCam, a stand-off Laser-Induced Breakdown Spectroscopy (LIBS) instrument (see Chapter 29) on board NASA's Curiosity rover, has demonstrated that stand-off spectroscopy is extremely useful for analyzing samples at a distance and obtaining rapid, preliminary evaluations of targets without the need to drive to such targets and perform complex operations with the rover's arm and arm-mounted tools, operations that can span several sols. SuperCam, which will fly on board NASA's Mars 2020 mission, includes a stand-off Raman spectrometer in addition to a next-generation LIBS system similar to ChemCam (see Chapter 29).

Unfortunately, the benefits of stand-off Raman analysis come with two major drawbacks. First, stand-off Raman measurements are seriously influenced by the intrinsic weakness of the Raman effect. The Raman effect generates Raman-scattered photons which distribute isotropically (see Chapter 6). Therefore, the longer the distance between sample and light-collection optics, the fewer Raman photons reach the detection system; the amount of return signal photons over a predefined collection aperture decreases as the square of the distance between collection optics and sample. An instrument with large clear aperture collection optics (e.g., a telescope) is required to collect a significant amount of Raman photons in stand-off applications. A second challenge in planetary exploration with stand-off Raman systems is measuring under full sunlight – due to thermal and energy requirements, most landed missions require daytime science operations. This requirement significantly increases the number of non–Raman-scattered photons (background light) that reach the detector, thus reducing the sensitivity of the instrument. This problem is reduced to that of improving the ratio of Raman scattered photons/nonscattered photons, i.e., background.

The preferred approach to address both weak signal and low signal-to-noise ratio (SNR) in stand-off, daytime Raman applications is to use a pulsed excitation source and a gated detection system. This configuration enhances the Raman signal through high-energy laser pulses combined with very short integration times in the detector. This technique maximizes the Raman photons to background ratio and is commonly known as time-resolved pulsed Raman spectroscopy. A very short (pulse width is on the order of 10 ns) and high irradiance (~3 GW cm^{-2}) laser pulse is used to excite the sample under analysis. By synchronizing the laser pulse with the electronic-shuttering inside an intensified charge-coupled device (ICCD), the Raman spectrum of the excited state is recorded alone with little-to-no background interference. Using the pulsed laser plus ICCD approach, several research teams have developed stand-off time-resolved pulsed Raman spectroscopy systems with wide-ranging applicability (Sharma et al., 2006; Gaft & Nagli, 2008; Sharma et al., 2009; Pettersson et al., 2010; Scaffidi et al., 2010; Fulton, 2011; Loeffen et al., 2011; Moros & Laserna, 2011; Moros et al., 2011; Rull et al., 2011; Sharma et al., 2011; Zachhuber et al., 2011; Angel et al., 2012; Izake et al., 2012; Dogariu & Gauthier, 2013; Lin et al., 2013; Bremer & Dantus, 2014; Dantus, 2014; Hokr et al., 2014; Bykov et al., 2015; Gasda et al., 2015; Hopkins et al., 2016).

Profs. S. Sharma and F. Rull and their teams at the University of Hawaii (United States) and Universidad de Valladolid (Spain), respectively, pioneered the development of stand-off Raman spectroscopy systems for planetary applications. Both teams contributed to the development of SuperCam, a stand-off Raman system on board NASA's Mars 2020 mission; Dr. R. Wiens of Los Alamos National Laboratory (United States) leads the SuperCam team. Here we describe the latest prototypes developed by these groups and some applications in remote, autonomous detection and identification of materials relevant for planetary exploration.

Figure 12.1 shows a sketch of the latest generation stand-off Raman system developed by Sharma's team at the University of Hawaii (UH)(Acosta-Maeda et al., 2016). The entire system is mounted on a reinforced portable trolley. The optical path is coaxial – the expanded laser beam is aligned with the optical axis of the detector using two elliptical turning mirrors. The optical collector is an 8-inch (203.2 mm) diameter telescope (Meade LX-200 R Advanced Ritchey-Chretien, f/10). The excitation source is a frequency-doubled, actively Q-switched neodymium-doped yttrium aluminum garnet (Nd:YAG) pulsed laser source (Quantel Laser, Ultra model, 532 nm, 100 mJ/pulse, 15 Hz, pulse width 10 ns). The spectra are analyzed and recorded with a Kaiser Optical Systems f/1.8 HoloSpec spectrometer fitted with a 1024 × 256 pixels ICCD camera (PI-IMAX, Princeton

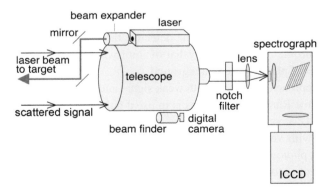

Figure 12.1 Schematic diagram of directly coupled pulsed standoff Raman+LIBS+Fluorescence system at the University of Hawaii.

Instrument, Trenton, NJ, USA). The telescope is directly coupled to the spectrometer through a camera lens (50 mm, f/1.8) that focuses the Raman signal into the 50-μm entrance slit of the spectrometer. A 532-nm SuperNotch filter (Kaiser Optical Systems) removes the reflected and Rayleigh-scattered laser light. Both the excitation and collection subsystems are mounted on a pan and tilt scanner (QuickSet, Northbrook, IL, USA) that enables precise aiming at large distances. This coaxial geometry ensures that the optical path remains aligned when moving the system or during operational and wind vibrations and oscillations.

Sharma's team has used this system to record spectra of materials relevant to planetary exploration at distances up to 430 m at UH (Figure 12.2). The UH team successfully measured dry ice, $CaSO_4 \cdot 2H_2O$, $KClO_3$, $KClO_4$, KNO_3, NH_4NO_3, sulfur, urea, water and water ice, acetonitrile, anhydrite, and Mg sulfate (Epsom salts, $MgSO_4 \cdot 7H_2O$) at a range of 430 m with integration times of 10 s or less. Figure 12.2 shows Raman spectra of these materials. Materials with strong Raman cross sections were detected with just 1 s integration time. These results are significant: the capability to analyze and detect

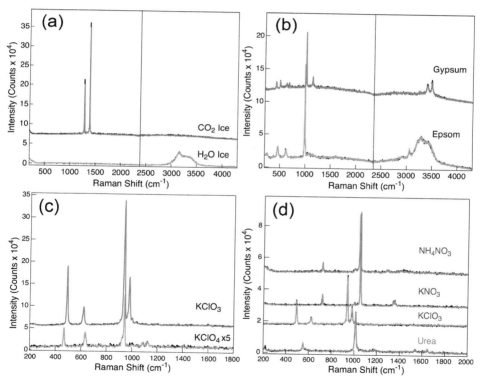

Figure 12.2 Remote Raman spectra of several materials of interest for planetary exploration recorded in 10 s at 430-m range. All measurements were recorded in broad daylight. A description of the vibrational modes and Raman peak assignment for these spectra was reported by Acosta-Maeda et al. (2016).

water, hydrous minerals, anhydrous minerals, and organic materials is critical in planetary science and facilitates new types of measurements on-the-fly, including the recognition of biosignatures and habitability features.

In these experiments, the laser beam diameter is adjusted on the sample through a 10× beam expander mounted in front of the laser, which generates a parallel beam and circular laser spot of ~10 cm in diameter at 430 m. The system was aligned by optimizing the measured biofluorescence signal from tree leaves >1600 behind the sample. Biofluorescence signals are several orders of magnitude stronger than Raman signals and also have fast life time, providing a suitable method for system alignment during daytime operation. This approach is described in detail elsewhere (Misra et al., 2011, 2012). For planetary exploration, this feature may augment the science output of a stand-off Raman system by generating column density measurements of atmospheric molecules, including ice, dust, and aerosols.

For their tests, the UH team used 100 mJ per pulse laser power at 15 Hz repetition rate with 2900 ns delay and 50 ns measurement gate width. The spectrometer uses a holographic volume-phase transmission grating that allows for measurement between 50 and 4200 cm^{-1} with a spectral resolution of 8 cm^{-1}. Raman spectra were generated and wavelengths calibrated using custom MATLAB® software. Wavelength calibration was performed using the Raman peaks of calcium carbonate, cyclohexane, acetonitrile, Ca sulfate dihydrate, and atmospheric nitrogen as references. Spectra in this section are shown as measured, with no baseline or cosmic ray correction, intensity calibration, or stitching procedures.

The stand-off Raman system developed by Rull's team at the University of Valladolid (UVa) is pictured in Figure 12.3. The main body of the instrument is composed of a frequency-doubled mini-Nd:YAG pulsed laser source (Ultra CFR Big Sky Laser at 532 nm) attached to a Meade DS-2114S telescope. The laser delivers up to 30 mJ per pulse at 532 nm with 2.7-mm near-field beam diameter, 6.8 ns pulse width, 0.5 mrad divergence, and 1.7% pulse-to-pulse stability. The Newtonian telescope has 1000-mm effective focal length (focus point outside the tube), 114 mm clear aperture, and f/8.8 focal ratio. The laser beam is configured coaxial with the telescope's optical axis through two high-energy BK7 beam-steering mirrors (Melles Griot), optimized for 45° incidence with reflectivity >99% at 532 nm. The laser/telescope structure is attached to a pan and tilt mount (BEWATOR model P25). A motorized Crayford focuser with 63.5 mm long drawtube (Feather Touch, Starlight Instruments) focuses the telescope's output onto either a 752 × 480 pixels imaging camera (µEye CCD model 1224) or a fiber optics coupler (Edmund Optics fiber focuser, 0.25 NA, 11 mm focal length) through a 45° flat flip-mirror (Meade UHTC 644 Flip-Mirror System). A Kaiser Super Notch Holographic Filter is placed before the flip-mirror system to reduce elastic dispersion (Rayleigh scattering) and to prevent damage to the imaging camera. The fiber optics coupler, mounted on a positioning stage (OWIS TRANS 40T-D25-XYZ) that enables optical alignment in the *XY* plane and two tilt angles, focuses light into a fused silica multimode optical fiber (Edmund Optics, 0.22 NA and 100-µm core diameter). The 100-µm core acts as the entrance slit for the spectrograph. The UVa team

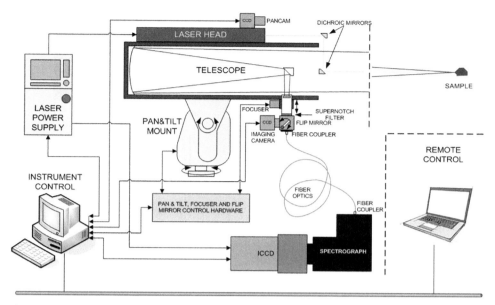

Figure 12.3 Sketch of University of Valladolid stand-off pulsed Raman spectroscopy instrument.

developed a customized compact and robust spectrograph that features high throughput (f/1.8, ideal for low-intensity applications such as Raman). The spectral range covered with the volume-phase holographic (VPH™, Holoplex) transmission grating, optimized for 532-nm Raman excitation, is 0–4500 cm^{-1}. An intensified gated detector (Andor iStar DH 720-25U-03) is attached to the spectrograph. It mounts a 1024 × 256 pixel matrix, each pixel being 26 × 26 μm^2. The detector can operate with 8-ns gating widths. It records the Raman spectra, which appear as two 10-pixel wide tracks (Figure 12.4a). The images are automatically converted to calibrated Raman spectra (Figure 12.4b).

A focusing device translates the telescope's output optics in order to project the sharpest image onto the imaging CCD. A Fourier-transform–based routine provides the instrument with auto-focus capabilities. A panoramic camera attached to the telescope–laser structure provides a full-color, wide-angle image of the targeted area. Panoramic images are also used to identify targets at long distances (from 30 m onwards). The pan and tilt mount positions the laser/collection system on the azimuthal and equatorial planes (±180° and ±90°, respectively), thus providing ultrawide field of view for target selection. Very fine target adjustment is achieved via stepper motors integrated in the pan and tilt unit. The UVa system is controlled via a graphical user interface that manages acquisition parameters (laser energy and frequency and detector gating, delay, gain level, and exposure), motorized position-feedback focusing hardware, imaging CCD, panoramic camera, and pan and tilt motors.

The UVa team has deployed the stand-off Raman system in several planetary analog terrains. Figure 12.5 shows a picture of the system in Rio Tinto (Spain) and several spectra

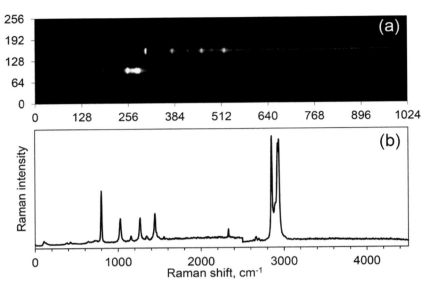

Figure 12.4 Raman spectrum of cyclohexane at 8 m distance shown in image mode (a), and in 2-D mode scaled to Raman shift units, cm^{-1} (b).

recorded in acidic sulfate systems. Sulfates on Mars could host information on the aqueous past of the planet and could play a role in defining potentially habitable past environments. Characterizing key sulfates in terrestrial Mars analog sites is therefore critical to design future Mars exploration missions. Stand-off Raman spectroscopy is uniquely suited to meet this goal. The spectra shown in the figure were recorded from materials mainly composed of hydrated metal–sulfate oxy/hydroxysulfates at a distance of 8 m from the instrument, in broad daylight. Acquisition parameters for each spectrum are specified in the plots. The system recorded unambiguous spectra of copiapite, halotrichite, gypsum, and quartz within the samples. Atmospheric oxygen (1555 cm^{-1}) and nitrogen (2332 cm^{-1}) are visible in the spectra.

In addition to detecting and identifying minerals, Raman spectroscopy can measure ice structure and crystallinity; there are strong, Raman-active intermolecular couplings between neighboring molecules as a consequence of hydrogen bonding networks, and the Raman spectral bands of in- and out-of-phase couplings and other intermolecular water interactions facilitate the accurate identification of icy materials (Furić & Volovšek, 2010). The UVa team used their stand-off Raman system to analyze icebergs in the Svalbard archipelago (Norway) during the Arctic Mars Analog Svalbard Expedition (AMASE; Steele et al., 2011). AMASE aims at testing and refining new technologies for planetary exploration in a subglacial volcanic complex that features hydrothermal, glacial, and geomorphological systems analogous to planetary environments where life may have existed. The specific aims of the stand-off Raman analyses were to demonstrate that the relative intensities of the OH stretching vibrations at 3138 cm^{-1} and 3360 cm^{-1} can be used

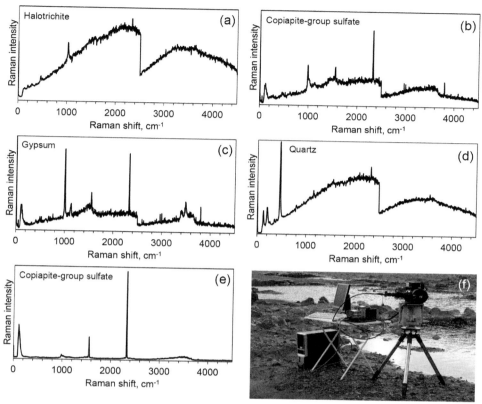

Figure 12.5 Picture of University of Valladolid's remote Raman system in Rio Tinto (Spain), and spectra of sulfates and quartz recorded in situ at 8 m in < 5 s.

as a proxy for crystallinity in remote Raman sensing, as they provide an implicit measure of the proportion of OH vibrators in water clusters of different sizes (Rull et al., 2011). Figure 12.6 shows Raman spectra recorded in four different regions of an iceberg at 120 m distance. The UVa team found a >15% decrease in the intensity of the 3138 cm^{-1} band relative to the 3360 cm^{-1} band in region (a) relative to (c). This indicates lower crystallinity and different ice hexagonal structure in region (c). This is not only an advantage of planetary stand-off Raman spectroscopy but also an important one, as measures of Europa's (or other Ocean Worlds with astrobiologic potential) ice crystallinity can be used to infer surface material formation and transformation pathways (Hansen & Mccord, 2004).

Figure 12.7a–d shows microimages of the four iceberg regions discussed in Figure 12.6; images were captured with the integrated CMOS camera and are coaligned with the Raman data. Images (a) and (c) in Figure 12.7 were processed using a MATLAB® edge detection algorithm that calculates total entropy (Canny, 1986; González et al., 2004). In this particular application, edges indicate changes in image intensity, i.e., crystal size, and

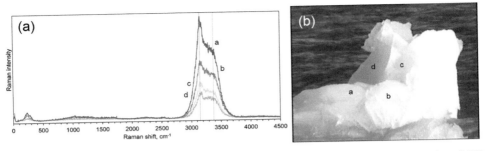

Figure 12.6 Raman spectra of four regions of an iceberg showing variable relative intensity of OH bands due to different ice crystallinity.

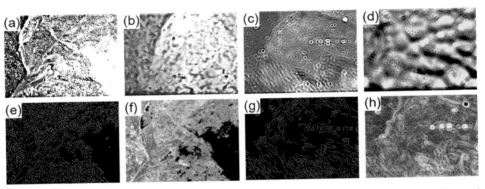

Figure 12.7 (Top) Micro images of the four regions in the iceberg. Image size is ~2 x 3 mm. (Bottom) Edge detection and local entropy for images (a) and (c), respectively.

local entropy is a measure of disorder in the image, i.e., texture. The longer edges and blurrier texture of image (c)(Figure 12.7g,h) relative to image (a)(Figure 12.7e,f) are indicative of more gradual intensity changes. The total entropies of images (a) and (c) are 5.12 and 7.28, respectively. Taken together, these edge and entropy results suggest that region (c) is more amorphous than region (a), consistent with the Raman-based measure of crystallinity discussed earlier: (c) is less crystalline than (a). The results of the UVa team highlight two critical advantages of stand-off Raman spectroscopy: (1) spectral changes observed from target to target can be correlated to modifications on Ice-Ih structure induced by ice formation processes, e.g., temperature and pressure effects, and (2) this remote sensing technique is a powerful tool for nondestructive, in situ structural analysis of ice.

12.3 Concluding Remarks

Stand-off Raman spectroscopy is a mature technology. NASA's Mars 2020 mission will, for the first time, deploy a stand-off Raman system on another planet, as part of the scientific payload that will look for signs of past or present life on Mars. SuperCam has high heritage from NASA- and ESA-funded instrument development efforts at the University of Hawaii (United States) and Universidad de Valladolid (Spain). The two systems we described in the preceding sections have reached a high level of technology and science readiness through intensive testing, optimization, and system-level demonstration of the instruments in the field.

The basic modular architectures of these systems feature variable focus optical assemblies that provide ranging, automated focusing, and range-gating parameters for the Raman sensors; compact Nd:YAG pulsed lasers emitting at 532 nm; spectrographs based on holographic transmission gratings; and ultrafast intensified gated detectors. The baseline concept of operation of these instruments is (1) point the instrument suite using pan and tilt actuators and context imagery; (2) select target(s) through supervised or unsupervised machine learning algorithms for visual target identification; (3) illuminate target(s) with the laser source to induce Raman scattering – laser power is manually or automatically controlled based on range and type of measurement; (4) collect and relay returned light onto the spectrometer, where its intensity and spectral distribution are measured and recorded; and (5) process the recorded spectral information.

In planetary exploration applications, stand-off Raman instruments will provide remote, rapid target assessment, which will in turn optimize selection of drilling and sampling locations and enable implementation of more advanced, unsupervised exploration strategies. High-fidelity laboratory and field tests of stand-off Raman instrument breadboards and prototypes pave the way for the design of flight instruments for Mars 2020 and future planetary exploration missions, e.g., NASA's Europa Lander mission. For instance, work led by Prof. Daly's team at York University (Canada) demonstrated the feasibility of a time-resolved stand-off UV-Raman prototype for detection and identification of pure organics, organics mixed in a quartz matrix, and minerals relevant to astrobiology and planetary exploration (Skulinova et al., 2014).

The combination of dedicated field and laboratory analyses with relevant instrumentation (benchtop and field portable prototypes, as we discuss in this Chapter) allows research teams to determine optimal working parameters for the next generation of stand-off Raman systems and evaluate instrument synergies to characterize relevant materials. Lessons learned from these research efforts enable the definition and validation of instrument and science requirements for spaceflight missions and planetary exploration technologies.

References

Acosta-Maeda T.E., Misra A.K., Muzangwa L.G., et al. (2016) Remote Raman measurements of minerals, organics, and inorganics at 430 m range. *Applied Optics*, **55**, 10283–10289.

Angel S.M., Gomer N.R., Sharma S.K., & McKay C. (2012) Remote Raman spectroscopy for planetary exploration: A review. *Applied Spectroscopy*, **66**, 137–150.

Beegle L.W., Bhartia R., DeFlores L., et al. (2014) SHERLOC: Scanning Habitable Environments with Raman and Luminescence for Organics and Chemicals, an investigation for 2020. *45th Lunar Planet. Sci. Conf.*, **178**, Abstract #2835.

Beegle L., Bhartia R., White M., et al. (2015) SHERLOC: Scanning Habitable Environments with Raman and Luminescence for Organics and Chemicals. *Proceedings of the 2015 IEEE Aerospace Conference*, 1–11.

Bremer M.T. & Dantus M. (2014) Detecting micro-particles of explosives at ten meters using selective stimulated Raman scattering. *CLEO: 2014*, JTh2A.5.

Bykov S.V., Mao M., Gares K.L., & Asher S.A. (2015) Compact solid-state 213 nm laser enables standoff deep ultraviolet Raman spectrometer: Measurements of nitrate photochemistry. *Applied Spectroscopy*, **69**, 895–901.

Canny J. (1986) A computational approach to edge detection. *IEEE Transactions on Pattern Analysis and Machine Intelligence*, PAMI-8, 679–698.

Dantus M. (2014) Single-beam stimulated Raman scattering for sub-microgram standoff detection of explosives. *Frontiers in Optics 2014*, LW5I.1.

Dogariu A.E.D.D.J.P. & Gauthier D. (2013) Standoff explosive detection and hyperspectral imaging using coherent anti-Stokes Raman spectroscopy. *Frontiers in Optics 2013*, LTh4G.4.

Fulton J. (2011) Remote detection of explosives using Raman spectroscopy. *Proceedings of SPIE 8018, Chemical, Biological, Radiological, Nuclear, and Explosives (CBRNE) Sensing*, **XII**, 80181A, DOI:10.1117/12.887101.

Furić K. & Volovšek V. (2010) Water ice at low temperatures and pressures: New Raman results. *Journal of Molecular Structure*, **976**, 174–180.

Gaft M. & Nagli L. (2008) UV gated Raman spectroscopy for standoff detection of explosives. *Optical Materials*, **30**, 1739–1746.

Gasda P.J., Acosta-Maeda T.E., Lucey P.G., Misra A.K., Sharma S.K., & Taylor G.J. (2015) Next generation laser-based standoff spectroscopy techniques for Mars exploration. *Applied Spectroscopy*, **69**, 173–192.

González R.C., Woods R.R.E., & Eddins S.L. (2004) *Digital image processing using Matlab*. Dorling Kindersley, London.

Hansen G.B. & McCord T.B. (2004) Amorphous and crystalline ice on the Galilean satellites: A balance between thermal and radiolytic processes. *Journal of Geophysical Research*, **109**, E01012.

Hokr B.H., Bixler J.N., Noojin G.D., et al. (2014) Single-shot stand-off chemical identification of powders using random Raman lasing. *Proceedings of the National Academy of Sciences of the USA*, **111**, 12320–12324.

Hopkins A.J., Cooper J.L., Profeta L.T., & Ford A.R. (2016) Portable Deep-Ultraviolet (DUV) Raman for standoff detection. *Applied Spectroscopy*, **70**, 861–873.

Hutchinson I.B., Ingley R., Edwards H.G.M., et al. (2014) Raman spectroscopy on Mars: Identification of geological and bio-geological signatures in martian analogues using miniaturized Raman spectrometers. *Philosophical Transactions of the Royal Society of London A: Mathematical, Physical and Engineering Sciences*, **372**, DOI:10.1098/rsta.2014.0204.

Izake E.L., Cletus B., Olds W., Sundarajoo S., Fredericks P.M., & Jaatinen E. (2012) Deep Raman spectroscopy for the non-invasive standoff detection of concealed chemical threat agents. *Talanta*, **94**, 342–347.

Lin Q., Niu G., Wang Q., Yu Q., & Duan Y. (2013) Combined laser-induced breakdown with Raman spectroscopy: Historical technology development and recent applications. *Applied Spectroscopy Reviews*, **48**, 487–508.

Loeffen P.W., Maskall G., Bonthron S., Bloomfield M., Tombling C., & Matousek P. (2011) Spatially offset Raman spectroscopy (SORS) for liquid screening. *Proceedings of SPIE 8189 Optics and Photonics for Counterterrorism and Crime Fighting VII*, 81890C.

Maurice S., Wiens R.C., Le Mouélic S., et al. (2015) The SuperCam instrument for the Mars 2020 rover. *European Planetary Science Congress Abstracts*, **10**, EPSC2015-185.

Misra A.K., Sharma S.K., Acosta T.E., & Bates D.E. (2011) Compact remote Raman and LIBS system for detection of minerals, water, ices, and atmospheric gases for planetary exploration. *Proceedings of the SPIE 8032, Next-Generation Spectroscopic Technologies IV*, 80320Q.

Misra A.K., Sharma S.K., Acosta T.E., Porter J.N., & Bates D.E. (2012) Single-pulse standoff Raman detection of chemicals from 120 m distance during daytime. *Applied Spectroscopy*, **66**, 1279–1285.

Moros J. & Laserna J.J. (2011) New Raman-Laser-Induced Breakdown Spectroscopy identity of explosives using parametric data fusion on an integrated sensing platform. *Analytical Chemistry*, **83**, 6275–6285.

Moros J., Lorenzo J.A., & Laserna J.J. (2011) Standoff detection of explosives: Critical comparison for ensuing options on Raman spectroscopy–LIBS sensor fusion. *Analytical and Bioanalytical Chemistry*, **400**, 3353–3365.

Pettersson A., Wallin S., Östmark H., et al. (2010) Explosives standoff detection using Raman spectroscopy: From bulk towards trace detection. *Proceedings of SPIE, Detection and Sensing of Mines, Explosive Objects, and Obscured Targets*, **XV**, 76641K, DOI:10.1117/12.852544.

Rull F., Sansano A., Sobron P., & Amase T. (2010) In-situ Raman-LIBS analysis of regolithes during AMASE 2008 and 2009 expeditions. *41st Lunar Planet. Sci. Conf.*, Abstract #2731.

Rull F., Vegas A., Sansano A., & Sobron P. (2011) Analysis of Arctic ices by remote Raman spectroscopy. *Spectrochimica Acta A: Molecular and Biomolecular Spectroscopy*, **80**, 148–155.

Scaffidi J.P., Gregas M.K., Lauly B., Carter J.C., Angel S.M., & Vo-Dinh T. (2010) Trace molecular detection via surface-enhanced Raman scattering and surface-enhanced resonance Raman scattering at a distance of 15 meters. *Applied Spectroscopy*, **64**, 485–492.

Sharma S.K., Misra A.K., Lucey P.G., Angel S.M., & McKay C.P. (2006) Remote pulsed Raman spectroscopy of inorganic and organic materials to a radial distance of 100 meters. *Applied Spectroscopy*, **60**, 871–876.

Sharma S.K., Misra A.K., Lucey P.G., & Lentz R.C.F. (2009) A combined remote Raman and LIBS instrument for characterizing minerals with 532 nm laser excitation. *Spectrochimica Acta A: Molecular and Biomolecular Spectroscopy*, **73**, 468–476.

Sharma S.K., Misra A.K., Clegg S.M., et al. (2011) Remote-Raman spectroscopic study of minerals under supercritical CO_2 relevant to Venus exploration. *Spectrochimica Acta A: Molecular and Biomolecular Spectroscopy*, **80**, 75–81.

Skulinova M., Lefebvre C., Sobron P., et al. (2014) Time-resolved stand-off UV-Raman spectroscopy for planetary exploration. *Planetary and Space Science*, **92**, 88–100.

Sobron P., Sanz A., Thompson C., Cabrol N., & Team P.L.L.P. (2014) In-situ lake bio-geochemistry using laser Raman spectroscopy and optrode sensing. *11th International GeoRaman Conference*, Abstract #5027.

Sobron P., Andersen D.T., & Pollard W.H. (2016) In-situ exploration of habitable environments and biosignatures in Arctic cold springs and Antarctic paleolakes. *Conference on Biosignature Preservation and Detection in Mars Analog Environments*, Abstract #1912.

Steele A., Amundsen H.E.F., Fogel M., et al. (2011) The Arctic Mars Analogue Svalbard Expedition (AMASE) 2010. *42nd Lunar Planet. Sci. Conf.*, Abstract #1588.

Vítek P., Edwards H.G.M., Jehlička J., et al. (2010) Microbial colonization of halite from the hyper-arid Atacama Desert studied by Raman spectroscopy. *Philosophical Transactions of the Royal Society A: Mathematical, Physical and Engineering Sciences*, **368**, 3205–3221.

Wei J., Wang A., Lambert J.L., et al. (2015) Autonomous soil analysis by the Mars Micro-beam Raman Spectrometer (MMRS) on-board a rover in the Atacama Desert: A terrestrial test for planetary exploration. *Journal of Raman Spectroscopy*, **46**, 810–821.

Zachhuber B., Gasser C., Chrysostom E.t.H., & Lendl B. (2011) Stand-off spatial offset Raman spectroscopy for the detection of concealed content in distant objects. *Analytical Chemistry*, **83**, 9438–9442.

Part III

Analysis Methods

13

Effects of Environmental Conditions on Spectral Measurements

EDWARD CLOUTIS, PIERRE BECK, JEFFREY J. GILLIS-DAVIS,
JÖRN HELBERT, AND MARK J. LOEFFLER

13.1 Introduction

The portion of a Solar System body's (SSB) surface accessible to optical remote sensing techniques (e.g., reflectance, thermal emission) is generally on the order of the uppermost few μm to few mm. As a result, it is in intimate contact with its surrounding environment, affecting it, and being affected by it. The importance of the local environment on surficial geology is readily apparent on Earth, where the styles and intensity of weathering that occur vary widely with location. As an example, surficial geology in hyperarid environments, such as the Atacama Desert, allows for the formation and preservation of mineral phases that are unstable in more humid environments (e.g., Michalski et al., 2004). Similarly, cold environments allow for the formation and preservation of mineral phases that are otherwise unstable at higher temperatures (e.g., Peterson et al., 2007).

The realization that the surface environment can have a significant effect on our ability to accurately determine an SSB's surficial geology remotely has been growing over the past few decades and is being increasingly demonstrated by results coming out of the ever-growing network of facilities that simulate various SSB surface environments.

Accurately reproducing the full range of conditions that prevail on an SSB's surface is practically impossible. However, this limitation does allow us to better isolate how individual processes may operate. By combining results from multiple facilities and experiments, a more complete picture of the relative importance of different surface environmental conditions can emerge. On the other hand, interactions between different surface modification processes and conditions can also exist, but are less tractable. In addition, laboratory experiments will often be of short duration or will accelerate these conditions so that results can be obtained in a reasonable time; therefore extrapolation of these results to geologic time scales can be uncertain.

Laboratory data are also enabling us to better model and understand the processes that can operate to modify the optical and geologic properties of SSB surfaces, as well as their relative importance. This chapter provides a brief overview of selected results from various SSB simulation facilities, organized along the lines of (1) relevance to specific SSBs and (2) "generic" surface modification processes (i.e., space weathering). In this context, this chapter is organized as follows: (1) bodies with "thin atmospheres" (i.e., Mars); (2) bodies with no atmospheres and Earth-like temperatures (e.g., main belt asteroids); (3) bodies with no atmospheres and elevated temperatures (i.e., Mercury, Moon); (4) bodies with no atmospheres and low temperatures (e.g., icy satellites, outer Solar System minor bodies); and (5) "space weathering" of atmosphereless bodies.

Absent from this review are data for bodies with thick atmospheres that are opaque at visible wavelengths but can have atmospheric transparency windows at other wavelengths (e.g., Venus, Titan). Optical remote sensing of their surfaces is possible but usually restricted to narrow wavelength intervals (Smith et al., 1996; Helbert et al., 2008) or from the surface or lower atmosphere, and laboratory facilities and laboratory-based optical spectroscopy studies relevant to these bodies are coming on stream (e.g., Wasiak et al., 2013; Singh et al., 2016). Also important but not dealt with here is a discussion of the effects of hypervelocity impacts, whose effects can be long lasting (e.g., shock disruption of crystal structures), and which are applicable to most planetary bodies. The material presented in this chapter is complemented by discussions of the spectral reflectance (Chapter 4) and emissivity (Chapter 3) properties of geologic materials.

13.2 Background

Environment chambers designed to reproduce some subset of surface conditions present on one to many SSBs, or some surface modification process, and that allow for in situ optical spectroscopy, are becoming increasingly widespread. As mentioned, such facilities enable exploration of surface modification processes under controlled conditions, so that factors such as their relative importance, time scales of operation, and kinetics can be explored. However, experimental runs are necessarily of limited duration (e.g., weeks to a few years); thus long-term effects can generally not be properly simulated. Investigation of long-term processes may be accessible under carefully considered "accelerated" conditions. As an example, Cloutis et al. (2008) examined how Mars surface conditions affect the stability and spectral reflectance properties of Mars analog minerals. "Shortcuts" used in these 40-day duration runs included high-intensity ultraviolet (UV) irradiation (where one day of UV irradiation in the chamber corresponded to roughly one decade of Mars UV exposure), and keeping the samples at ambient terrestrial temperatures (so as to not inhibit any kinetically slow temperature-related changes in compositional or structural properties).

As an early example of the importance of surface environmental conditions on optical remote sensing, Logan et al. (1973) showed that thermal emission spectra of geologic materials are affected by how samples are heated (from above or below), whether the sample spectra are measured in vacuum or air, as well as known factors such as composition and particle size. Some compounds can also undergo phase changes as a function of temperature (Wu et al., 2007), so that laboratory spectra of temperature-sensitive compounds should ideally be acquired under temperature conditions appropriate to a specific planetary body.

13.3 Environmental Condition Effects by Solar System Body Types

A plethora of modification processes can operate on planetary surfaces, each of variable importance and a function of factors such as surface environment, presence/absence of an atmosphere, surficial geology, and heliocentric distance. The Solar System hosts a diversity of objects ranging from small airless bodies such as asteroids and comets to gas giant planets such as Jupiter. The next sections discuss selected laboratory studies that highlight how planetary surface conditions can affect our ability to determine surficial geology from optical remote sensing.

13.4 Bodies with "Thin Atmospheres": Mars

Mars is one of the most actively explored bodies by visible to infrared remote sensing (e.g., Chapters 23–26). The importance of considering, and ability to investigate, surface environment effects on spectroscopic analysis of Mars surficial geology, particularly for water-bearing minerals, was first demonstrated by Bishop and Pieters (1995). The range of materials investigated has grown over time. Observed spectroscopic changes range from undetectable to significant, and the rate of change is also variable (e.g., Cloutis et al., 2007, 2008).

Results to date allow some generalizations to be made. Water is generally more susceptible to loss than structural hydroxyl, as measured by changes in absorption band depths and shapes of composite absorption features, while the degree and rate of water loss are a function of how tightly the water is bound into a mineral structure. UV irradiation does not appear to appreciably affect the type and rate of spectroscopic changes. The rate of reduction in absorption band depths is usually most rapid during the first day of exposure to Mars-like conditions.

When spectroscopic changes occur, they fall into a few broad categories, including (1) changes in overall spectral slope (likely due to changes in water content and/or iron oxidation state and its crystallographic site); (2) changes in the position and intensity of ferric iron associated absorption bands (likely due to change in crystallographic site properties, such as loss of bridging oxygen/hydroxyl molecules (Sherman, 1985), and possibly iron oxidation state); (3) reduction in water and/or hydroxyl absorption band depths, sometimes to the point of complete disappearance of a band; and (4) change in the position of an absorption band minimum, due to the loss or reduction in relative depth of an absorption band that contributes to a composite absorption feature – most often seen in water and hydroxyl-bearing minerals.

As an example of spectral changes in water-bearing minerals exposed to Mars surface conditions (Figure 13.1a), the sulfate fibroferrite shows a likely loss of adsorbed and perhaps structural water as evidenced by a decrease in the downturn in reflectance beyond 1.25 μm (toward the intense fundamental water absorption bands in the 3-μm region), and the depth of the H_2O absorption feature near 1.4 μm. A possible structural change is suggested by the change in the shape of the H_2O-associated absorption feature in the

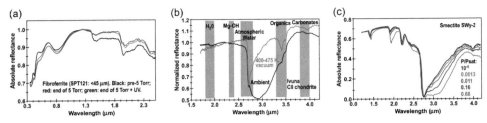

Figure 13.1 Changes in spectral properties with changes in relative humidity. (a) Reflectance spectra of fibroferrite measured under ambient terrestrial conditions, after 10 days of exposure to Mars-like pressures of CO_2, and an additional 12 days of exposure to the same conditions and with the addition of UV irradiation (after Cloutis et al., 2008). (b) Changes are observed in spectra of the CI chondrite Ivuna on heating under vacuum (after Takir et al., 2013). The broad H_2O band from ~2.8 to ~3.1 μm was greatly reduced, the weak H_2O combination band near 1.9 μm disappeared, and the CH organic features near 3.4 μm were enhanced due to dehydration of the sample. (c) Evolution of water bands in spectra of a clay mineral (SWy-2 smectite) acquired under increasing water vapor pressure conditions (after Pommerol et al., 2009). This illustrates growth of the features near 1.93 and 3.0 μm.

1.9-μm region. Ferric iron–associated absorption features are the sharp absorption band near 0.43 μm, the shoulder near 0.53 μm, and the absorption band near 0.80 μm; after exposure, the 0.43-μm band is weaker, the 0.53 μm shoulder is much less pronounced, and the 0.8-μm band has shifted to longer wavelengths. These changes are likely attributable to a loss of Fe^{3+}-bridging OH molecules, which is also consistent with the reduction in the depth of the 1.4-μm region OH/H_2O composite overtone absorption band. These results indicate that the use of reflectance spectra acquired under terrestrial ambient conditions for detection of specific minerals in Mars observational data, particularly for OH/H_2O-bearing species, must be approached cautiously.

13.5 Bodies with No Atmospheres and Earth-Like Temperatures: Main-Belt Asteroids

13.5.1 The Impact of Main-Belt Asteroid Conditions on Visible to Infrared Spectra

Main-Belt Asteroid (MBA) conditions are characterized by moderately low temperatures (typically around 200 K for the illuminated area) and by the exposure to ultrahigh vacuum at the surface. Many asteroid surfaces are fine grained and have a low thermal inertia (Delbo et al., 2015), implying the presence of strong temperature gradient through the first few mm of the surface (Gundlach & Blum, 2012). The presence of a strong thermal gradient in the near surface will have a marginal impact on the reflectance spectra but strong effects are expected for thermal emission spectroscopy (see Chapter 19 for the spectral properties of asteroids). In the following we describe the impact on visible to infrared spectra of low-pressure and moderately low-temperature conditions typical of MBA.

13.5.2 The Effect of High Vacuum

For a granular refractory material, the presence of high vacuum will have a very minor impact on the reflectance spectra when compared to ambient condition measurements. The difference in the refractive index contrast between air/mineral and vacuum/mineral is below 0.1% (Ciddor, 1996), and will produce a barely detectable effect on the visible to infrared reflectance spectra. However, a number of constituents of planetary materials can be sensitive to exposure to vacuum.

This is particularly true for H_2O-related absorption features. Many primitive meteorites contain hydrogen in the form of $-OH/H_2O$ within a mineral structure (structural hydroxyl/water), interlayer water within phyllosilicates, and finally adsorbed water (Beck et al., 2010; Garenne et al., 2014). Interlayer water and adsorbed water are unstable under the vacuum conditions of asteroidal surfaces (Beck et al., 2010). Different setups have been developed to measure the reflectance spectra of these materials under vacuum and at various temperatures to assess the effect of removing weakly bonded water molecules (Pommerol et al., 2009; Takir et al., 2013; Garenne et al., 2016).

This impact can be seen by observing a change in the shape of the so-called 3-μm band (Figure 13.1b). This spectral feature is a composite absorption of OH/H_2O stretching fundamentals and an H_2O bending overtone. When exposed to vacuum, the contribution from molecular water is removed while that from metal–OH within phyllosilicate remains (–OH groups are more tightly bonded and stable). In Figure 13.1c, the results of a reverse experiment are presented, where a phyllosilicate-bearing sample that was fully desiccated is subsequently exposed to progressively increasing water vapor pressure. This increase in humidity leads to an increase in the depth and change in shape of the absorption feature at 1.93 μm (combination of bending and stretching) and around 3 μm. This exercise emphasizes the fact that the feature around 1.93 μm is related to weakly bonded H_2O and hence is not expected to be visible on asteroidal surfaces. Such an absorption feature has not been observed on Ceres, in spite of the fact that its surface is rich in phyllosilicates (De Sanctis et al., 2015). The salt-rich white spots observed on Ceres do not appear to be hydrated (De Sanctis et al., 2016).

Other phases that are also sensitive to exposure to strong vacuum include sulfates, which can desiccate under low relative humidity conditions and transform to amorphous phases (Vaniman et al., 2004). More generally, hydrated salts are expected to be unstable at MBA conditions.

Many primitive meteorites contain organic matter in the form of macromolecular carbon compounds. Some is refractory and should be stable at the pressure and temperature conditions of asteroidal surfaces, but an important fraction of these organics can be considered as labile and might sublimate on exposure to strong vacuum. Changes in the form of the 3.4–3.5 μm C–H-absorption bands have been observed for primitive meteorites when exposed to secondary vacuum (Beck et al., 2010; Takir et al., 2013). Also, while there are abundant observations of a 3-μm region absorption feature on asteroids, C–H stretching modes (3.4–3.5 μm) are almost never observed but are often present in laboratory spectra of primitive meteorites. This could be due to multiple factors, such as space weathering and impact heating; or these absorption bands may be too shallow to be detected in lower signal-to-noise ratio spectra.

13.5.3 Thermal Effects

Spatial temperature variations are strong for MBAs (typically on the order of 150 K between day and night side), but only the warmest (illuminated) regions are probed with ground-based observations. The temperatures are generally lower than standard laboratory conditions, which can influence the spectra. This is particularly the case for the crystal field bands in Fe-bearing minerals that are modified under low temperature conditions (Roush & Singer, 1986, 1987; Hinrichs & Lucey, 2002) (discussed later), and this effect clearly needs to be accounted for when comparing observations to laboratory measurements (Moroz et al., 2000). Some other mineral phases can be sensitive to exposure to low temperatures, as is the case of sulfates, where a reorganization of the water molecule network occurs at low T, inducing similar changes to those observed for water ice (Dalton & Pitman, 2012; De Angelis et al., 2017). Some spectrum-altering effects can also be expected for absorption features related to a transition from an excited state, as was measured for brucite (Beck et al., 2015), a candidate to explain the peculiar 3-μm feature of Ceres.

In the case of near-Earth asteroids, surface temperatures are higher than MBAs, and the physical conditions are similar to those of the lunar surface. In some specific cases where the perihelion is unusually low (e.g., Phaethon), very high-T (1000 K) conditions can prevail, in addition to exposure to high vacuum. These conditions are more akin to those at the surface of Mercury.

13.6 Bodies with No Atmospheres and Elevated Temperatures: Mercury, Moon

Spectroscopy of planetary analog materials at elevated temperatures has typically been limited to the temperature range up to ~450 K (e.g., Singer & Roush, 1985; Burns, 1993; Lucey et al., 1998; Schade & Wasch, 1999). For most objects in the Solar System, including the dayside of the Moon, this is sufficient. With two missions to Mercury (Chapter 17) – the completed NASA MESSENGER mission and the just-launched ESA-JAXA Bepi-Colombo mission – this has changed. The temperature of the surface of Mercury can range between 70 and 725 K (Strom & Sprague, 2003), and vary by >400 K at the equator between sunrise and midday (Vasavada et al., 1999; Bauch et al., 2011). These temperature variations significantly affect the crystal structure and density of minerals, and both the reflected and thermal infrared (TIR) spectral signature. In addition, a large number of extrasolar planets close to their host star have increased the interest in the effects of high temperatures on characteristic spectral signatures.

Helbert et al. (2013) presented a new in situ multimethodologic laboratory approach complemented by a numerical analysis of the expected spectral change with temperature for olivine – a major mineral on inner SSBs. The laboratory approach consisted of high-temperature X-ray diffraction (XRD) and high-temperature TIR spectroscopy in the range of 1400–700 cm^{-1} (7–14 μm). This spectral range can be used very effectively to identify the fine-scale structural properties of silicates (e.g., stretching and bending modes in Si–O, metal–O, and lattice vibrations [Hamilton, 2010]). In addition, for some important mineral families, the TIR peak positions are a good indicator of composition (Christensen et al., 2000; Koike et al., 2003).

The four main bands between 1000 and 850 cm^{-1} are the most intense in olivine in this spectral range (Figure 13.2a) and their exact position is a strong function of Fe abundance (Hamilton, 2010). The Helbert et al. (2013) experiment showed a shift of 8 cm^{-1} in the position of the band near 990 cm^{-1} (Figure 2 in Helbert et al., 2013), between the 352 K spectrum and the 773 K spectrum. This is significant as the same wavenumber shift varies with olivine composition (from Fo$_{92}$ to Fo$_{77}$; Hamilton, 2010; Nestola et al., 2011a,b). The results of the TIR spectroscopy experiments are consistent with unit-cell volume calculations and the high-temperature XRD measurements. Increasing the iron content increases the unit-cell volume significantly due to the larger cation radius of Fe^{2+} versus Mg (Shannon, 1976). Hence, the increase in temperature and the increase in Fe have the same effect on olivine, increasing the unit-cell volume.

Ferrari et al. (2014), following a similar approach, focused on Mg-rich samples for interpreting spectral data for Mercury. They observed that an increase in unit-cell volume by 5.11 Å^3 could be obtained by an increase of 400 K or by a decrease in the molar proportion of Mg to (Mg + Fe) of 0.43. Their data suggest further that changes in the spectral signatures of Ca-rich pyroxenes as a function of temperature can be restricted to specific wavelength regions, and are closely related to the thermal expansion volume coefficients. Therefore they could discriminate between bands that are primarily sensitive to temperature effects and bands that are primarily sensitive to composition (Figure 4 in Ferrari et al., 2014). Maturilli et al. (2014a,b) extended these spectral investigations to a set of komatiites as a more natural analog for the surface of Mercury.

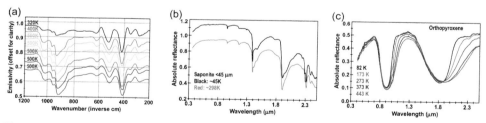

Figure 13.2 Changes in spectral properties with changes in temperature. (a) Emissivity spectra of Fo$_{89}$ olivine (125–250 μm grain size) measured under vacuum between 320 K and 900 K (data courtesy of C. Stangarone, Planetary Spectroscopy Laboratory). (b) Reflectance spectra of saponite at different temperatures (unpublished data: Cloutis). Changes are observed in the width and position of features at ~0.95, 1.40–1.47, 1.75, and 1.9 μm. (c) Reflectance spectra of an orthopyroxene (measured at different temperatures (after Roush and Singer (1987) – their sample Orthopyroxene 02).

Also of relevance to Mercury, Helbert et al. (2013) obtained spectra in the visible and near-infrared (VNIR) for thermally processed synthetic sulfides as a potential analog for hollow-forming materials (Blewett et al., 2013). From these measurements, they inferred that the spectral contrast of a diagnostic feature at or near 0.6 μm that is present in all sulfides (Izawa et al., 2013) can be strongly affected by heating. Both the spectral slope and the color observed before and after thermal processing showed significant changes attributed to the annealing of color centers in the structure. This led to the first tentative identification of MgS on the surface of Mercury by Vilas et al. (2016).

To illustrate the effects of temperature on TIR spectral signatures, Figure 13.2a shows TIR spectra of an Fo_{89} sample in a 125–250 μm grain size fraction obtained in vacuum between 320 K and 900 K. The shift of band positions in the 800–1000 cm^{-1} region is apparent. Reliance on measurements obtained under ambient conditions can lead to misinterpretation of compositional information. Therefore, it is important to obtain spectra of analog materials under realistic conditions for each target body.

13.7 Bodies with No Atmospheres and Low Temperatures: Icy Satellites, Outer Solar System Minor Bodies

Bodies with no atmospheres, or thin exospheres, include a diversity of bodies, such as the moons of the outer planets (Jupiter, Saturn (see Chapter 21), Uranus, Neptune), Pluto and its moons (see Chapter 22), comets (see Chapter 20), and outer Solar System minor planets, such as Kuiper Belt objects. With multiple missions to these bodies in the past, present, and future, and the diversity of compositions of these bodies, laboratory experiments conducted at conditions relevant to these bodies are important. In addition, because many of them possess ices, conducting laboratory spectroscopic studies at the conditions prevalent on them is essential (see Chapter 5 for the spectral properties of volatiles and ices). Finally, data from missions to these bodies suggest that the radiation environment of some small bodies, in particular the Galilean satellites, needs to be taken into account when interpreting observational spectroscopic data for them (e.g., Ip et al., 1998; Cooper et al., 2001; Mastrapa & Brown, 2006).

Laboratory spectroscopic studies conducted under realistic outer Solar System conditions are difficult to execute (Dalton et al., 2005; Mastrapa & Brown, 2006; De Angelis et al., 2017). However, successful studies of this sort date back to the 1960s (e.g., Kieffer, 1969; Clark, 1981; Hapke et al., 1981; Fink & Sill, 1982; Roush et al., 1990). These and other studies have shown that reflectance spectra of ices and hydrated phases vary as a function of temperature and grain size. Generally, with decreasing temperature, overlaps between adjacent absorption bands decrease as the individual absorption bands become narrower (Figure 13.2b). This improvement in band resolution can allow for better discrimination of different phases and provide constraints on temperature.

Many of the icy bodies in the Solar System show absorption features that can superficially be assigned to a particular mineral species. However, low-temperature and low-pressure laboratory studies are essential in elucidating the causes of differences between laboratory and observational spectra. Laboratory spectra can also be used to derive optical constants for subsequent quantitative abundance modeling of surface composition (Dalton & Pitman, 2012). The radiation environment of an icy body surface can also have major effects on observational spectra due to temperature-dependent phase changes, amorphization, and ion implantation (Mastrapa & Brown, 2006; Fama et al., 2010). The results arising from facilities able to measure optical spectra under low-pressure–low-temperature conditions have amply demonstrated that interpretation of observational data from icy satellites requires laboratory spectroscopic data acquired under similar conditions (De Angelis et al., 2017).

As discussed earlier, changes in temperature can cause asymmetric expansion or contraction of minerals along different axes. The behavior of absorption bands in reflectance spectra can also be complex in cases where an absorption feature may be composed of two or more individual absorption bands (e.g., Roush & Singer, 1986, 1987; Hinrichs & Lucey, 2002). With increasing temperature, absorption bands generally move to longer wavelengths (but sometimes to shorter wavelengths) and band widths increase, while depths are generally unchanged (Figure 13.2c).

13.8 "Space Weathering" of Atmosphereless Bodies: Micrometeorite Bombardment

Spectral reflectance and thermal emittance measurements provide much of what we know about the surface composition (i.e., mineralogy and chemistry) of planetary bodies. For robust interpretation of surface composition, it is necessary to understand processes that produce surface alterations, and thus modify reflectance and emittance spectra. Space weathering is a process that produces spectral change (e.g., Hapke, 2001; Bennett et al., 2013; Pieters & Noble, 2016). It is not a single process but a set of processes that include solar wind ions and electrons, UV photons, cosmic rays, and micrometeorite impacts. The continuous rain of charged particles and small impactors causes the physical and chemical state of a regolith to evolve with exposure. The physical products of space weathering include (1) the creation of thin (~100 nm), amorphous coatings that contain submicroscopic-size metallic iron (SMFe, also called nanophase iron or npFe^0), and Fe sulfide grains, which are produced by melt/vapor deposition from micrometeorite impacts and solar wind sputtering (e.g., Noguchi et al., 2001; Noble et al., 2011); (2) the formation of agglutinates, which are glass-welded agglomerates of minerals, lithic fragments, and pieces of earlier agglutinates (Duke et al., 1970; Basu, 1977); and (3) comminution of material to finer grain size (McKay et al., 1974, 1991). These physical changes affect the reflectance and emittance spectra of a surface, which consequently complicates the derivation of compositional information from such spectra.

Micrometeorites are of asteroidal and cometary origin and can produce melt and vapor (Brownlee, 1985; Bradley et al., 1988; Bradley, 1994; Nesvorný et al., 2010). The melt and vapor can form rims that encompass grains, agglutinates, and/or glass spherules. Agglutinates comprise about 25–30% of the particles in lunar soil on average, but a mature soil can contain as much as 65% agglutinates (McKay et al., 1991). Micron-sized micrometeorites only form enough melt to coat grains and form inclusions and mixed rims (Keller & McKay, 1997). The majority of agglutinates and glass spherules found in a lunar regolith are largely formed by mm and larger particles (see Chapter 18 for the spectral properties of the Moon). The continuous shower of micrometeorites means that agglutinates are continuously produced and agglutinate abundance in a soil increases with time and is proportional to the cumulative exposure age of the regolith. However, the rate at which agglutinates are produced is a function of solar distance and grain size of the target. The amount of melt production is enhanced by micrometeorite velocity, flux, and finer grain size of the regolith. Hence, an order of magnitude more impact melt and vapor is generated at Mercury (17–20 km/s) compared to the Moon (~12 km/s), and only minor amounts of melt and almost no vapor would be produced by dust in the Asteroid Main Belt (~ 5 km/s) (Cintala, 1992; Borin et al., 2009). The higher melt production would in turn enhance the concentration of glass in Mercury's regolith relative to the Moon or asteroids, which could be a major factor in explaining why Mercury is so dark and spectrally featureless (Blewett et al., 1997; Denevi & Robinson, 2008; Murchie et al., 2015). The important spectrum-altering property common to both agglutinates and inclusion-rich rims is the presence of SMFe.

SMFe is the main agent that alters UV-VNIR reflectance spectra (Hapke et al., 1975). As the abundance of SMFe increases with exposure to the space environment, the VNIR continuum slope increases, overall reflectance decreases, and the characteristic crystal field absorptions of mafic silicates in the regolith diminish in intensity (Adams & McCord, 1971a,b; Pieters et al., 1993, 2000; Fischer & Pieters, 1996; Hapke, 2001; Taylor et al., 2001; Noble and Pieters, 2003). SMFe is contained not only within rims surrounding grains (Keller & McKay, 1993, 1997) but also in agglutinates (Papike et al., 1981; Walker & Papike, 1981). In fact, agglutinitic glass contains much of the reduced Fe in lunar soils (Morris, 1980; Keller & McKay, 1994). Hence, agglutinates alter the spectral properties of a regolith to a greater extent because they obscure the spectral signature of comminuted rock fragments in lunar soils (Adams & McCord, 1971b; Charette et al., 1976; Fischer & Pieters, 1996). The slightly larger population of large SMFe (grain size of ~50 nm) versus small SMFe (<10 nm in size) in agglutinates than in the glassy rims serves as a darkening neutral opaque in the UV-VNIR spectral range (Lucey & Noble, 2008; Lucey & Riner, 2011).

The UV (<0.4 μm), and in particular the far-UV (<0.2 μm), is also affected by SMFe. However, the spectral behavior of SMFe in the UV is the opposite of that seen in the VNIR. With the addition of SMFe, spectra decrease in reflectance with increasing wavelength or become more blue in the UV rather than steeper and more red in the VNIR (Hendrix & Vilas, 2006; Hendrix et al., 2012; Denevi et al., 2014). Vilas and Hendrix (2015) concluded that the UV/blue reflectance characteristics allow earlier detection of the onset of space weathering compared to the VNIR continuum. Reversals in the reflectance continuum between UV and VNIR are also observed for nanophase iron embedded in silica gels (Noble et al., 2007). Results from this experimental work show that continuum slope is influenced by variations in abundance and size of the nanophase iron particles (Figure 13.3a).

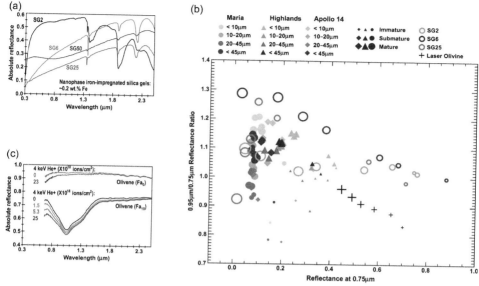

Figure 13.3 Spectral properties of Fe-impregnated and irradiated samples. (a) Reflectance spectra of nanophase iron-impregnated silica gels. Each sample contains ~0.2 wt.% abundance of Fe, but differs in the average grain size (in nm) of the nanophase iron particles as indicated on the figure (after Noble et al., 2007). (b) Reflectance at 0.75 μm versus 0.95/0.75 μm reflectance ratio for laser-irradiated olivine, nanophase iron-impregnated silica gels (after Noble et al., 2007), and lunar soil samples (after Gillis-Davis et al., 2017). (c) Reflectance spectra of <45 μm Fa_0 forsterite (Fe-free synthetic olivine) before (black, a) and after (red, b) irradiation with 23×10^{18} ions/cm^2, and of <45 μm Fa_{10} forsterite (limited Fe) with fluences of 4 keV He$^+$ as indicated (after Loeffler et al., 2009).

These spectrum-altering effects can be measured remotely using multispectral datasets, which enables the remote determination of space weathering as a tool for planetary mapping. For the Moon, the most popular method to determine maturity remotely uses a comparison of 0.95 μm/0.75 μm ratio versus 0.75 μm reflectance, termed the Optical Maturity parameter or OMAT (Lucey et al., 2000). In brief, OMAT decouples spectral trends produced by varying ferrous iron content in a regolith and spectral trends created by regolith maturity. For a given iron content, immature soils exhibit a higher 0.75 μm reflectance and a relatively lower 0.95 μm/0.75 μm reflectance ratio (Figure 13.3b). As that soil is exposed to the space environment and matures, the 0.75 μm reflectance decreases (darkens), and the 0.95 μm/0.75 μm ratio increases (spectrally reddens). All spectral trends on this plot appear to converge toward a theoretical, hypermature dark and red origin point given in Lucey et al. (2000). The quantified Euclidean distance from the theoretical hypermature endmember to a sample data point represents an index of maturity. OMAT does have some limitations (Wilcox et al., 2005; Garrick-Bethell et al., 2011; Nettles et al., 2011; Hemingway et al., 2015), and other methods for deriving soil maturity exist (Fischer & Pieters, 1994, 1996; Shkuratov et al., 1999; Nettles et al., 2011).

It is not possible, or at least highly uncertain, to compare OMAT degree of space weathering between other airless bodies such as Mercury and Vesta. Hence laboratory simulations can serve to expand our understanding of the complex process of space weathering by examining its multifaceted components individually and systematically. The spectrum-altering effects of space weathering are affected by multiple factors (i.e., target composition, rate of overturn, melt and vapor production, thickness of grain coatings, abundance of agglutinates, incidence angle, gravity, and factors related to solar distance such as irradiation fluence, micrometeorite velocity, photon irradiation, and surface temperature). Simulations of micrometeoroid impacts via pulsed laser experiments (Moroz et al., 1996; Yamada et al., 1999; Hiroi & Sasaki, 2001; Sasaki et al., 2001, 2003; Brunetto et al., 2006; Lazzarin et al., 2006; Loeffler et al., 2008, 2016; Gillis-Davis et al., 2017; Kaluna et al., 2017) as well as thermal space weathering (Hapke et al., 1975; Hiroi et al., 1996; Kohout et al., 2014) have provided insight into how these multiple factors interact. Continued work in this area will serve as a basis for examining how each of these factors contributes to the space weathering process as a system and potentially allow effects of each process (micrometeorites and solar wind), the alteration rates, and their dependence on surface properties, to be measured in a physical unit that could be intercomparable among airless bodies.

Space weathering effects on thermal emittance spectra are less well studied than the effects on VNIR reflectance. Because the mid-IR region is sensitive to other properties of geologic materials (such as silicate frameworks) and more representative of a bulk sample, it can probe different aspects of space weathering (Moroz et al., 2014). Mid-IR spectroscopic changes on simulated micrometeorite impacts include changes in emission band positions and shapes, and disappearance of bands (e.g., Moroz et al., 2014; Matsuoka et al., 2015; Lantz et al., 2017). These changes are due to the same processes that affect VNIR reflectance spectra, such as amorphization and impact melting. Processes that are important in shorter wavelength reflectance spectra, such as production of SMFe, have little effect on TIR emission spectra (Moroz et al., 2014). Mid-IR spectra are more affected by comminution and production of impact glass (Moroz et al., 2014).

13.9 "Space Weathering" of Atmosphereless Bodies: Ion Implantation

Energetic ions, such as solar wind ions, cosmic rays, or magnetospheric ions, are believed to be one of the main weathering agents active on the surfaces of airless bodies. These ions can induce chemical or structural changes in the surface minerals. For airless bodies, such as the Moon or near-Earth asteroids, laboratory studies have focused on simulating the 1 keV H^+ and 4 keV He^{2+} ions present in the solar wind, because the flux is so high compared with other components (Gosling, 2007), although other less abundant but more damaging ions could potentially contribute as well. One complication in these laboratory studies is that in practice it is difficult to obtain a pure beam of either 1 keV H^+ or 4 keV He^{2+} at a flux sufficient to study space weathering effects, and thus one has to adjust the experimental approach.

When laboratory simulations of solar wind bombardment on mineral grains are performed, knowing the environmental conditions under which the irradiation and analysis is performed is important (as is sample preparation and any possible sample chamber contaminants) (Nash, 1967; Dybwad, 1971). The question of the laboratory environment during analysis of irradiated samples has focused mainly on whether this analysis needs to be performed while the sample is still under vacuum (in situ) or if it could still accurately be performed after the sample has been exposed to atmosphere (ex situ), which is often a necessity. Unfortunately, the answer to this question depends on the analytical technique employed as well as possibly the sample being studied. As a brief example, San Carlos olivine (Fo_{90}), a mineral commonly used in irradiation studies, is most often analyzed by VNIR reflectance spectroscopy. Although this technique is not surface sensitive in that the probed depth is significantly larger than the penetration depth of the solar wind ions, the alteration produced in a thin rim on a grain can often be observed in the reflectance spectra (Figure 13.3c). It turns out that for Fo_{90} the ion-induced spectral changes are preserved even after the sample is exposed to atmosphere (Loeffler et al., 2009). This observation is fortunate because most ion irradiation experiments perform ex situ analysis of reflectance spectra (Hapke, 2001; Brunetto & Strazzulla, 2005). However, unlike reflectance spectroscopy, X-ray photoelectron spectroscopy (XPS), which is more surface sensitive, illustrates a case where in situ analysis is a necessity. It has been shown by XPS that simulations of solar wind bombardment can chemically reduce the iron present on the surface Fo_{90} (Dukes et al., 1999). However, these changes would be significantly lessened or possibly not even observed with ex situ measurements, as the chemically reduced surface quickly oxidizes when it is exposed to the atmosphere (Loeffler et al., 2009).

13.10 Summary and Conclusions

Laboratory studies of the effects of planetary surface conditions must often overcome formidable technical challenges. Technical advances are allowing us to construct and operate ever-increasingly sophisticated environmental chambers that allow for in situ acquisition of reflectance or emittance spectra. Even when in situ analysis is not possible, careful sample handling and independent analyses can determine whether exposure to ambient terrestrial conditions causes unwanted changes in sample optical, compositional, or structural properties.

Such laboratory studies have shown that solid planetary surface materials can be sensitive to endogenic and exogenic processes that may be complex or unexpected. Empirical laboratory studies of the effects of the environment on the optical properties of planetary surface materials will continue to play a pivotal role in interpretation of optical remote sensing data into the foreseeable future. Combinations of observational data, empirical studies, and theoretical modeling also allow insights into the relative importance of different planetary surface modification processes to be estimated (e.g., Shestopalov et al., 2013).

Acknowledgments

The authors wish to thank their colleagues who have contributed to our increasing understanding of how the planetary surface environment affects reflectance and emittance spectra, and to the funding agencies that have supported these efforts. We also wish to thank H. Kaluna for producing some of the figures used in this chapter, T. Roush for providing temperature spectra, and C. Stangarone for sharing data with us.

References

Adams J.B. & McCord T.B. (1971a) Alteration of lunar optical properties: Age and composition effects. *Science*, **171**, 567–571.

Adams J.B. & McCord T.B. (1971b) Optical properties of mineral separates, glass, and anorthositic fragments from Apollo mare samples. *Proceedings of the 2nd Lunar Sci. Conf.*, 2183–2195.

Basu A. (1977) Steady state, exposure age, and growth of agglutinates in lunar soils. *Proceedings of the 8th Lunar Planet. Sci. Conf.*, 3617–3632.

Bauch K.E., Hiesinger H., & Helbert J. (2011) Insolation and resulting surface temperatures of study regions on Mercury. *42nd Lunar Planet. Sci. Conf.*, Abstract #2257.

Beck P., Quirico E., Montes-Hernandez G., et al. (2010) Hydrous mineralogy of CM and CI chondrites from infrared spectroscopy and their relationship with low albedo asteroids. *Geochimica et Cosmochimica Acta*, **74**, 4881–4892.

Beck P., Schmitt B., Cloutis E.A., & Vernazza P. (2015) Low-temperature reflectance spectra of brucite and the primitive surface of 1-Ceres? *Icarus*, **257**, 471–476.

Bennett C.J., Pirim C., & Orlando T.M. (2013) Space weathering of Solar System bodies: A laboratory perspective. *Chemical Reviews*, **113**, 9086–9150.

Bishop J.L. & Pieters C.M. (1995) Low-temperature and low atmospheric pressure infrared reflectance spectroscopy of Mars soil analog materials. *Journal of Geophysical Research*, **100**, 5369–5379.

Blewett D.T., Lucey P.G., Hawke B.R., Ling G.G., & Robinson M.S. (1997) A comparison of Mercurian reflectance and spectral quantities with those of the Moon. *Icarus*, **129**, 217–231.

Blewett D.T., Vaughan W.M., Xiao Z., et al. (2013) Mercury's hollows: Constraints on formation and composition from analysis of geological setting and spectral reflectance. *Journal of Geophysical Research*, **118**, 1013–1032.

Borin P., Cremonese G., Marzari F., Bruno M., & Marchi S. (2009) Statistical analysis of micrometeoroids flux on Mercury. *Astronomy and Astrophysics*, **503**, 259–264.

Bradley J.P. (1994) Chemically anomalous, preaccretionally irradiated grains in interplanetary dust from comets. *Science*, **265**, 925–929.

Bradley J.P., Sandford S.A., & Walker R.M. (1988) Interplanetary dust particles. In: *Meteorites and the early Solar System* (J.F. Kerridge & M.S. Matthews, eds.). University of Arizona Press, Tucson, 861–895.

Brownlee D. (1985) Cosmic dust-collection and research. *Annual Review of Earth and Planetary Sciences*, **13**, 147–173.

Brunetto R. & Strazzulla G. (2005) Elastic collisions in ion irradiation experiments: A mechanism for space weathering of silicates. *Icarus*, **179**, 265–273.

Brunetto R., Romano F., Blanco A., et al. (2006) Space weathering of silicates simulated by nanosecond pulse UV excimer laser. *Icarus*, **180**, 546–554.

Burns R.G. (1993) *Mineralogical applications of crystal field theory*, 2nd edn. Cambridge University Press, Cambridge.

Charette M.P., Soderblom L.A., Adams J.B., Gaffey M.J., & McCord T.B. (1976) Age-color relationships in the lunar highlands. *Proceedings of the 7th Lunar Sci. Conf.*, 2579–2592.

Christensen P.R., Bandfield J.L., Hamilton V.E., et al. (2000) A thermal emission spectral library of rock-forming minerals. *Journal of Geophysical Research*, **105**, 9735–9739.

Ciddor P.E. (1996) Refractive index of air: New equations for the visible and near infrared. *Applied Optics*, **35**, 1566–1573.

Cintala M.J. (1992) Impact-induced thermal effects in the lunar and mercurian regoliths. *Journal of Geophysical Research*, **97**, 947–973.

Clark R.N. (1981) The spectral reflectance of water-mineral mixtures at low temperatures. *Journal of Geophysical Research*, **86**, 3074–3086.

Cloutis E.A., Craig M.A., Mustard J.F., et al. (2007) Stability of hydrated minerals on Mars. *Geophysical Research Letters*, **34**, L20202, DOI:10.1029/2007GL031267.

Cloutis E.A., Craig M.A., Kruzelecky R.V., et al. (2008) Spectral reflectance properties of minerals exposed to simulated Mars surface conditions. *Icarus*, **195**, 140–168.

Cooper J.F., Johnson R.E., Mauk B.H., Garrett H.B., & Gehrels N. (2001) Energetic ion and electron irradiation of the icy Galilean satellites. *Icarus*, **149**, 133–159.

Dalton J.B. & Pitman K.M. (2012) Low temperature optical constants of some hydrated sulfates relevant to planetary surfaces. *Journal of Geophysical Research*, **117**, E09001, DOI:10.1029/2011JE004036.

Dalton J.B., Prieto-Ballesteros O., Kargel J.S., Jamieson C.S., Jolivet J., & Quinn R. (2005) Spectral comparison of heavily hydrated salts with disrupted terrains on Europa. *Icarus*, **177**, 472–490.

De Angelis S., Carli C., Tosi F., et al. (2017) Temperature-dependent VNIS spectroscopy of hydrated Mg-sulfates. *Icarus*, **281**, 444–458.

De Sanctis M.C., Ammannito E., Raponi A., et al. (2015) Ammoniated phyllosilicates with a likely outer Solar System origin on (1) Ceres. *Nature*, **528**, 241–244.

De Sanctis M.C., Raponi A., Ammannito E., et al. (2016) Bright carbonate deposits as evidence of aqueous alteration on (1) Ceres. *Nature*, **536**, 54–57.

Delbo M., Mueller M., Emery J.P., et al. (2015) Asteroid thermophysical modeling. In: *Asteroids IV* (P. Michel, F. E. De Meo, & W.F. Bottke Jr., eds.). University of Arizona Press, Tucson, 107–128.

Denevi B.W. & Robinson M.S. (2008) Mercury's albedo from Mariner 10: Implication for the presence of ferrous iron. *Icarus*, **197**, 239–246.

Denevi B.W., Robinson M.S., Boyd A.K., Sato H., Hapke B.W., & Hawke B. (2014) Characterization of space weathering from Lunar Reconnaissance Orbiter Camera ultraviolet observations of the Moon. *Journal of Geophysical Research*, **119**, 976–997.

Duke M.B., Woo C.C., Bird M.L., Sellers G.A., & Finkelman R.B. (1970) Lunar soil: Size distribution and mineralogical constituents. *Science*, **167**, 648–650.

Dukes C.A., Baragiola R.A., & McFadden L.A. (1999) Surface modification of olivine by H^+ and He^+ bombardment. *Journal of Geophysical Research*, **104** (E1) 1865–1872.

Dybwad J. (1971) Radiation effects on silicates (5-keV H^+, D^+, He^+, N_2^+). *Journal of Geophysical Research*, **76**, 4023–4029.

Fama M., Loeffler M.J., Raut U., & Baragiola R.A. (2010) Radiation-induced amorphization of crystalline ice. *Icarus*, **207**, 314–319.

Ferrari S., Nestola F., Massironi M., et al. (2014) In-situ high-temperature emissivity spectra and thermal expansion of C2/c pyroxenes: Implications for the surface of Mercury. *American Mineralogist*, **99**, 786–792.

Fink U. & Sill G.T. (1982) The infrared spectral properties of frozen volatiles. In: *Comets* (L.L. Wilkening, ed.). University of Arizona Press, Tucson, 164–202.

Fischer E.M. & Pieters C.M. (1994) Remote determination of exposure degree and iron concentration of lunar soils using VIS-NIR spectroscopic methods. *Icarus*, **111**, 475–488.

Fischer E.M. & Pieters C.M. (1996) Composition and exposure age of the Apollo 16 Cayley and Descartes regions from Clementine data: Normalizing the optical effects of space weathering. *Journal of Geophysical Research*, **101**, 2225–2234.

Garenne A., Beck P., Montes-Hernandez G., et al. (2014) The abundance and stability of "water" in type 1 and 2 carbonaceous chondrites (CI, CM and CR). *Geochimica et Cosmochimica Acta*, **137**, 93–112.

Garenne A., Beck P., Montes-Hernandez G., et al. (2016) Bidirectional reflectance spectroscopy of carbonaceous chondrites: Implications for water quantification and primary composition. *Icarus*, **264**, 172–183.

Garrick-Bethell I., Head J.W., & Pieters C.M. (2011) Spectral properties, magnetic fields, and dust transport at lunar swirls. *Icarus*, **212**, 480–492.

Gillis-Davis J.J., Lucey P.G., Bradley J.P., et al. (2017) Incremental laser space weathering of Allende reveals non-lunar like space weathering effects. *Icarus*, **286**, 1–14.

Gosling J.T. (2007) The solar wind. In: *Encyclopedia of the Solar System* (L.A. McFadden, P.R. Weissman, & T. V. Johnson, eds.). Academic Press, Amsterdam, 99–116.

Gundlach B. & Blum J. (2012) Outgassing of icy bodies in the Solar System – II: Heat transport in dry, porous surface dust layers. *Icarus*, **219**, 618–629.

Hamilton V.E. (2010) Thermal infrared (vibrational) spectroscopy of Mg–Fe olivines: A review and applications to determining the composition of planetary surfaces. *Chemie der Erde*, **70**, 7–33.

Hapke B. (2001) Space weathering from Mercury to the asteroid belt. *Journal of Geophysical Research*, **106**, 10039–10073.

Hapke B.W., Cassidy W.A., & Wells E.N. (1975) Effects of vapor-phase deposition processes on the optical, chemical, and magnetic properties of the lunar regolith. *Moon*, **13**, 339–353.

Hapke B.W., Wells E., Wagner J., & Partlow W. (1981) Far-UV, visible, and near-IR reflectance spectra of frosts of H_2O, CO_2, NH_3 and SO_2. *Icarus*, **47**, 361–367.

Helbert J., Müller N., Kostama P., Marinangeli L., Piccioni G., & Drossart P. (2008) Surface brightness variations seen by VIRTIS on Venus Express and implications for the evolution of the Lada Terra region, Venus. *Geophysical Research Letters*, **35**, L11201.

Helbert J., Nestola F., Ferrari S., et al. (2013) Olivine thermal emissivity under extreme temperature ranges: Implication for Mercury surface. *Earth and Planetary Science Letters*, **371–372**, 252–257.

Hemingway D.J., Garrick-Bethell I., & Kreslavsky M.A. (2015) Latitudinal variation in spectral properties of the lunar maria and implications for space weathering. *Icarus*, **261**, 66–79.

Hendrix A.R. & Vilas F. (2006) The effects of space weathering at UV wavelengths: S-class asteroids. *The Astronomical Journal*, **132**, 1396–1404.

Hendrix A.R., Retherford K.D., Gladstone G.R., et al. (2012) The lunar far-UV albedo: Indicator of hydration and weathering. *Journal of Geophysical Research*, **117**, E12001, DOI:10.1029/2012JE004252.

Hinrichs J.L. & Lucey P.G. (2002) Temperature-dependent near-infrared spectral properties of minerals, meteorites, and lunar soil. *Icarus*, **155**, 169–180.

Hiroi T. & Sasaki S. (2001) Importance of space weathering simulation products in compositional modeling of asteroids: 349 Dembowska and 446 Aeternitas as examples. *Meteoritics and Planetary Science*, **36**, 1587–1596.

Hiroi T., Zolensky M.E., Pieters C.M., & Lipschutz M.E. (1996) Thermal metamorphism of the C, G, B, and F asteroids seen from the 0.7 μm, 3 μm, and UV absorption strengths in comparison with carbonaceous chondrites. *Meteoritics and Planetary Science*, **31**, 321–327.

Ip W.H., Williams D.J., McEntire R.W., & Mauk B.H. (1998) Ion sputtering and surface erosion at Europa. *Geophysical Research Letters*, **25**, 829–832.

Izawa M.R.M., Applin D.M., Mann P., et al. (2013) Reflectance spectroscopy (200–2500 nm) of highly-reduced phases under oxygen- and water-free conditions. *Icarus*, **226**, 1612–1617.

Kaluna H.M., Ishii H.A., Bradley J.P., Gillis-Davis J.J., & Lucey P.G. (2017) Simulated space weathering of Fe- and Mg-rich aqueously altered minerals using pulsed laser irradiation. *Icarus*, **292**, 245–258.

Keller L.P. & McKay D.S. (1993) Discovery of vapor deposits in the lunar regolith. *Science*, **261**, 1305–1307.

Keller L.P. & McKay D.S. (1994) The nature of agglutinitic glass in the fine-size fraction of lunar soil 10084. *25th Lunar Planet. Sci. Conf.*, Abstract, 685–686.

Keller L.P. & McKay D.S. (1997), The nature and origin of rims on lunar soil grains, *Geochimica et Cosmochimica Acta*, **61**, 2331–2341.

Kieffer H.H. (1969) A reflectance spectrometer/environmental chamber for frosts. *Applied Optics*, **8**, 2497–2500.

Kohout T., Cuda J., Filip J., et al. (2014) Space weathering simulations through controlled growth of iron nanoparticles on olivine. *Icarus*, **237**, 75–83.

Koike C., Chihara H., Tsuchiyama A., Suto H., Sogawa H., & Okuda H. (2003) Compositional dependence of infrared absorption spectra of crystalline silicate—II. Natural and synthetic olivines. *Astronomy and Astrophysics*, **399**, 1101–1107.

Lantz C., Brunetto R., Barucci M.A., et al. (2017) Ion irradiation of carbonaceous chondrites: A new view of space weathering on primitive asteroids. *Icarus*, **285**, 43–57.

Lazzarin M., Marchi S., Moroz L.V., et al. (2006) Space weathering in the main asteroid belt: The big picture. *Astrophysical Journal*, **647**, L179–L182.

Loeffler M., Baragiola R., & Murayama M. (2008) Laboratory simulations of redeposition of impact ejecta on mineral surfaces. *Icarus*, **196**, 285–292.

Loeffler M.J., Dukes C.A., & Baragiola R.A. (2009) Irradiation of olivine by 4 keV He$^+$: Simulation of space weathering by the solar wind. *Journal of Geophysical Research*, 114, E03003.

Loeffler M.J., Dukes C.A., Christoffersen R., & Baragiola R.A. (2016) Space weathering of silicates simulated by successive laser irradiation: In situ reflectance measurements of Fo$_{90}$, Fo$_{99+}$, and SiO$_2$. *Meteoritics and Planetary Science*, **51**, 261–275.

Logan L.M., Hunt G.R., Salisbury J.W., & Balsamo S.R. (1973) Compositional implications of Christiansen frequency maximums for infrared remote sensing applications. *Journal of Geophysical Research*, **78**, 4983–5003.

Lucey P.G. & Noble S K. (2008) Experimental test of a radiative transfer model of the optical effects of space weathering. *Icarus*, **197**, 348–353.

Lucey P.G. & Riner M.A. (2011) The optical effects of small iron particles that darken but do not redden: Evidence of intense space weathering on Mercury. *Icarus*, **212**, 451–462.

Lucey P.G., Keil K., & Whitely R. (1998) The influence of temperature on the spectra of the A-asteroids and implications for their silicate chemistry. *Journal of Geophysical Research*, **103**, 5865–5871.

Lucey P.G., Blewett D.T., Taylor G.J., & Hawke B.R. (2000) Imaging of lunar surface maturity. *Journal of Geophysical Research*, **105**, 20,377–20,386.

Mastrapa R.M.E. & Brown R.H. (2006) Ion irradiation of crystalline H$_2$O-ice: Effect on the 1.65 μm band. *Icarus*, **183**, 207–214.

Matsuoka M., Nakamura T., Kimura Y., et al. (2015) Pulse-laser irradiation experiments of Murchison CM2 chondrite for reproducing space weathering on C-type asteroids. *Icarus*, **254**, 135–143.

Maturilli A., Helbert J., St. John J.M., et al. (2014a) Komatiites as Mercury surface analogues: Spectral measurements at PEL. *Earth and Planetary Science Letters*, **398**, 58–65.

Maturilli A., Shiryaev A.A., Kulakova I.I., & Helbert J. (2014b) Infrared reflectance and emissivity spectra of nanodiamonds. *Spectroscopy Letters*, **47**, 446–450.

McKay D., Fruland R., & Heiken G. (1974) Grain size and the evolution of lunar soils. *Proceedings of the 5th Lunar Sci. Conf.*, 887–906.

McKay D.S., Heiken G., Basu A., et al. (1991) The lunar regolith. In: *Lunar sourcebook: A user's guide to the Moon* (G.H. Heiken, D.T. Vaniman, & B.M. French, eds.). Cambridge University Press, Cambridge, 285–365.

Michalski G., Böhlke J.K., & Thiemens M. (2004) Long term atmospheric deposition as the source of nitrate and other salts in the Atacama Desert, Chile: New evidence from mass-independent oxygen isotopic compositions. *Geochimica et Cosmochimica Acta*, **68**, 4023–4038.

Moroz L.V., Fisenko A.V., Semjonova L.F., Pieters C.M., & Korotaeva N.N. (1996) Optical effect of regolith processes on S-asteroids as simulated by laser shot on ordinary chondrites and other mafic materials. *Icarus*, **122**, 366–382.

Moroz L., Schade U., & Wasch R. (2000) Reflectance spectra of olivine-orthopyroxene-bearing assemblages at decreased temperatures: Implications for remote sensing of asteroids. *Icarus*, **147**, 79–93.

Moroz L.V., Starukhina L.V., Rout S.S., et al. (2014) Space weathering in silicate regoliths with various FeO contents: New insights from laser irradiation experiments and theoretical spectral simulations. *Icarus*, **235**, 187–206.

Morris R. (1980) Origins and size distribution of metallic iron particles in the lunar regolith. *Proceedings of the 11th Lunar Planet. Sci. Conf.*, 1697–1712.

Murchie S.L., Klima R.L., Denevi B.W., et al. (2015) Orbital multispectral mapping of Mercury with the MESSENGER Mercury Dual Imaging System: Evidence for the origins of plains units and low-reflectance material. *Icarus*, **254**, 287–305.

Nash D.B. (1967) Proton-irradiation darkening of rock powders – Contamination and temperature effects and applications to solar-wind darkening of Moon. *Journal of Geophysical Research*, **72**, 3089–3104.

Nestola F., Nimis P., Ziberna L., et al. (2011a) First crystal-structure determination of olivine in diamond: Composition and implications for provenance in the Earth's mantle. *Earth and Planetary Science Letters*, **305**, 249–255.

Nestola F., Pasqual D., Smyth J.R., et al. (2011b) New accurate elastic parameters for the forsterite-fayalite solid solution. *American Mineralogist*, **96**, 1742–1747.

Nesvorný D., Jenniskens P., Levison H.F., Bottke W.F., Vokrouhlický D., & Gounelle M. (2010) Cometary origin of the zodiacal cloud and carbonaceous micrometeorites: Implications for hot debris disks. *The Astrophysical Journal*, **713**, 816–836.

Nettles J.W., Staid M., Besse S., et al. (2011) Optical maturity variation in lunar spectra as measured by Moon Mineralogy Mapper data. *Journal of Geophysical Research*, **116**, E00G17, DOI:10.1029/2010JE003748.

Noble S.K. & Pieters C.M. (2003) Space weathering on Mercury: Implications for remote sensing. *Solar System Research*, **37**, 34–39.

Noble S.K., Pieters C.M., & Keller J. (2007) An experimental approach to understanding the optical effects of space weathering. *Icarus*, **192**, 629–642.

Noble S.K., Hiroi T., Keller L.P., Rahman Z., Sasaki S., & Pieters C.M. (2011) Experimental space weathering of ordinary chondrites by nanopulse laser: TEM results. *42nd Lunar Planet. Sci. Conf.*, Abstract #1382.

Noguchi T., Nakamura T., Kimura M., et al. (2001) Incipient space weathering observed on the surface of Itokawa dust particles. *Science*, **333**, 1121–1125.

Papike J.J., Simon S.B., White C., & Laul J.C. (1981) The relationship of the lunar regolith <10 μm fraction and agglutinates. Part I: A model for agglutinate formation and some indirect supportive evidence. *Proceedings of the 12th Lunar Planet. Sci. Conf.*, 409–420.

Peterson R.C., Nelson W., Madu B., & Shurvell H.F. (2007) Meridianiite: A new mineral species observed on Earth and predicted to exist on Mars. *American Mineralogist*, **92**, 1756–1759.

Pieters C.M. & Noble S.K. (2016) Space weathering on airless bodies. *Journal of Geophysical Research*, **121**, 1865–1884.

Pieters C.M., Fischer E.M., Rode O., & Basu A. (1993) Optical effects of space weathering: The role of the finest fraction. *Journal of Geophysical Research*, **98**, 20,817–20,824.

Pieters C.M., Taylor L.A., Noble S.K., et al. (2000) Space weathering on airless bodies: Resolving a mystery with lunar samples. *Meteoritics and Planetary Science*, **35**, 1101–1107.

Pommerol A., Schmitt B., Beck P., & Brissaud O. (2009) Water sorption on martian regolith analogs: Thermodynamics and near-infrared reflectance spectroscopy. *Icarus*, **204**, 114–136.

Roush T.L. & Singer R.B. (1986) Gaussian analysis of temperature effects on the reflectance spectra of mafic minerals in the 1-μm region. *Journal of Geophysical Research*, **91**, 10,301–10,308.

Roush T.L. & Singer R.B. (1987) Possible temperature variation effects on the interpretation of spatially resolved reflectance observations of asteroid surfaces. *Icarus*, **69**, 571–574.

Roush T.L., Pollack J.B., Witteborn F.C., & Bregman J.D. (1990) Ice and minerals on Callisto: A reassessment of the reflectance spectra. *Icarus*, **86**, 355–382.

Sasaki S., Nakamura K., Hamabe Y., Kurahashi E., & Hiroi T. (2001), Production of iron nanoparticles by laser irradiation in a simulation of lunar-like space weathering. *Nature*, **410**, 555–557.

Sasaki S., Kurahashi E., Yamanaka C., & Nakamura K. (2003) Laboratory simulation of space weathering: Changes of optical properties and TEM/ESR confirmation of nanophase metallic iron. *Advances in Space Research*, **31**, 2537–2542.

Schade U. & Wasch R. (1999) NIR reflectance spectroscopy of mafic minerals in the temperature range between 80 and 473 K. *Advances in Space Research*, **23**, 1253–1256.

Shannon R.D. (1976) Revised effective ionic radii and systematic studies of interatomic distances in halides and chalcogenides. *Acta Crystallographica A*, **32**, 751–767.

Sherman D.M. (1985) SCF-Xα-SW MO study of Fe-O and Fe-OH chemical bonds; applications to Mössbauer spectra and magnetochemistry of hydroxyl-bearing Fe^{3+} oxides and silicates. *Physics and Chemistry of Minerals*, **12**, 311–314.

Shestopalov D.I. Golubeva L.F., & Cloutis E.A. (2013) Optical maturation of asteroid surfaces. *Icarus*, **225**, 781–793.

Shkuratov Y.G., Kaydash V.G., & Opanasenko N.V. (1999) Iron and titanium abundance and maturity degree distribution on the lunar nearside. *Icarus*, **137**, 222–234.

Singer R.B. & Roush T.L. (1985) Effects of temperature on remotely sensed mineral absorption features. *Journal of Geophysical Research*, **90**, 12,434–12,444.

Singh S., Cornet T. Chevrier V.F., et al. (2016) Near-infrared spectra of liquid/solid acetylene under Titan relevant conditions and implications for *Cassini*/VIMS detections. *Icarus*, **270**, 429–434.

Smith P.H., Lemmon M.T., Lorenz R.D., Sromovsky L.A., Caldwell J.J. & Allison M.D. (1996) Titan's surface, revealed by HST imaging. *Icarus*, **119**, 336–349.

Strom R.G. & Sprague A. (2003) *Exploring Mercury: The iron planet*. Springer-Verlag, London.

Takir D., Emery J.P., McSween H.Y., et al. (2013) Nature and degree of aqueous alteration in CM and CI carbonaceous chondrites. *Meteoritics and Planetary Science*, **48**, 1618–1637.

Taylor L.A., Pieters C.M., Keller L.P., Morris R.V., & McKay D.S. (2001) Lunar mare soils: Space weathering and the major effects of surface-correlated nanophase Fe. *Journal of Geophysical Research*, **106**, 27,985–27,999.

Vaniman D.T., Bish D.L., Chipera S.J., Fialips C.I., Carey J.W., & Feldman W.C. (2004) Magnesium sulphate salts and the history of water on Mars. *Nature*, **431**, 663–665.

Vasavada A., Paige D.A., & Wood S.E. (1999) Near-surface temperatures on Mercury and the Moon and the stability of polar ice deposits. *Icarus*, **141**, 179–193.

Vilas F. & Hendrix A.R. (2015) The UV/blue effects of space weathering manifested in S-complex asteroids. I. Quantifying change with asteroid age. *The Astronomical Journal*, **150**, 64.

Vilas F., Domingue D., Helbert J., et al. (2016) Mineralogical indicators of Mercury's hollows composition in MESSENGER color observations. *Geophysical Research Letters*, **43**, 1450–1456.

Walker R.J. & Papike J.J. (1981) The relationship of the lunar regolith <10 μm fraction and agglutinates. Part II: Chemical composition of agglutinate glass as a test of the "fusion of the finest fraction" (F^3) model. *Proceedings of the 12th Lunar Planet. Sci. Conf.*, 421–432.

Wasiak F.C., Luspay-Kuti A., Welivitiya W.D.D.P., et al. (2013) A facility for simulating Titan's environment. *Advances in Space Research*, **51**, 1213–1220.

Wilcox B.B., Lucey P.G., & Gillis J.J. (2005) Mapping iron in the lunar mare: An improved approach. *Journal of Geophysical Research*, **110**, E11001, DOI:10.1029/2005JE002512.

Wu H.B., Chan M.N., & Chan C.K. (2007) FTIR characterization of polymorphic transformation of ammonium nitrate. *Aerosol Science and Technology*, **41**, 581–588.

Yamada M., Sasaki S., Nagahara H., et al. (1999) Simulation of space weathering of planet-forming materials: Nanosecond pulse laser irradiation and proton implantation on olivine and pyroxene samples. *Earth, Planets and Space*, **51**, 1255–1265.

14

Hyper- and Multispectral Visible and Near-Infrared Imaging Analysis

WILLIAM H. FARRAND, ERZSÉBET MERÉNYI, AND MARIO C. PARENTE

14.1 Introduction

Multispectral imaging enables the mapping of materials on planetary surfaces with unique spectral features recorded by the sensor's spectral channels. The added power of a hyperspectral sensor is uniquely *identifying* materials that have spectrally diagnostic absorption features within the wavelength range of the sensor. Processing approaches developed for multi- and hyperspectral sensors in the solar reflective region of the visible and near-infrared to short-wave infrared (VNIR–SWIR, 0.4–2.5 µm) take advantage of the higher spectral dimensionality of these datasets for comprehensive mapping of those materials. Methods for processing multi- and hyperspectral data fall into several categories including spectral matching methods, spectral parameterization, spectral mixture-based techniques, and machine learning approaches. Spectral parameterization for hyperspectral sensors is discussed in Chapter 23. Spectral matching, in the form of an expert system that utilizes spectral feature fitting is discussed in Chapter 11; however, several other spectral matching techniques are described here. We also discuss the concept of spectral endmembers, linear and nonlinear spectral unmixing, as well as partial unmixing that allows mapping of endmembers with superior background suppression. Finally, we discuss machine learning techniques including self-organizing maps, supervised artificial neural networks, and support vector machines. The methods discussed in this chapter provide complementarity with other chapters and give examples of analytics capabilities (the recovered detail from spectral imagery) versus computational cost and the expertise required.

14.2 Spectral Matching

The high spectral resolution of hyperspectral sensors means that the spectrum associated with each image pixel has the potential to show changes in shape produced by absorption bands and spectral slope differences. Spectral matching techniques ask the question, for each pixel in a scene, "How well does this pixel spectrum match that of the target material spectrum?"

A spectral matching technique that is widely used with hyperspectral data (since it takes advantage of the complete spectral shape of a target spectrum) is the Spectral Angle Mapper (SAM; Kruse et al., 1993). This method is invariant with regard to pixel brightness and is

Table 14.1 *Symbols used in this chapter*

Symbol	Quantity represented	Symbol	Quantity represented
θ	Spectral angle	**x** and **y**	Two spectra (feature vectors) $\in Rnb$
nb	Number of bands	p and q	Two probability mass functions
ne	Number of endmembers	e_j	residual between the measured and modeled values in band j
ρ	Reflectance	**z**	Vector operator, as for CEM
f_i	Fraction of endmember i	**μ**	Mean of background spectra
m	Intrinsic dimensionality	**C**	Sample covariance matrix
w	SOM prototype vector	**W** and **V**	Weight matrices used in SOM and SOM-hybrid classifier
f	SVM objective function	ξ_j	Parameters for handling nonseparable input data in SVM
v	SVM vector of coefficients	ϕ	Kernel applied to nonlinear boundaries in SVM

sensitive solely to spectral shape. The spectral angle (θ) between two spectra **x** and **y** is (in radians):

$$\theta = \arccos\left(\frac{\mathbf{x} \cdot \mathbf{y}}{\|\mathbf{x}\| \cdot \|\mathbf{y}\|}\right). \tag{14.1}$$

When this is used for classification, a threshold is set where a larger spectral angle threshold allows for more potential matches and a lower threshold is more restrictive. Note that symbols used in equations in this chapter are listed in Table 14.1.

Another valuable spectral matching approach uses the Spectral Information Divergence (SID) metric (Chang, 2000) that is based on the concept of divergence in information theory (e.g., Kullback, 1968). Thus for two spectra **x** and **y**, with probability mass functions $p = (p_1, p_2, \ldots, p_{nb})^T$ and $q = (q_1, q_2, \ldots, q_{nb})^T$ where the probability measure p_j for any band j of vector **x** (x_j) is $p_j = \frac{x_j}{\sum_{j=1}^{nb} x_j}$. The average discrepancy in the self-information of **y** relative to **x**, $D(\mathbf{x}\|\mathbf{y})$, is

$$D(\mathbf{x}\|\mathbf{y}) = \sum_{j}^{nb} p_j \log(p_j/q_j), \tag{14.2}$$

where nb is the number of bands and j is the band number. $D(\mathbf{x}\|\mathbf{y})$, also known as the Kullback–Leibler divergence between spectrum **y** relative to **x**, is nonnegative, nonsymmetric, and it is equal to 0 only if **x** = **y**. It can be similarly defined for $D(\mathbf{x}\|\mathbf{y})$, and the SID value between spectra **x** and **y** is calculated according to

$$\text{SID}(\mathbf{x},\mathbf{y}) = D(\mathbf{x}\|\mathbf{y}) + D(\mathbf{y}\|\mathbf{x}). \tag{14.3}$$

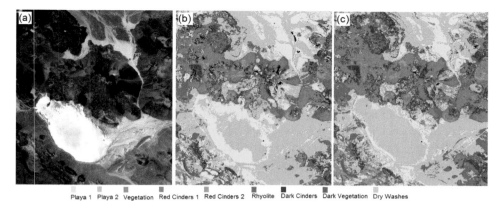

Figure 14.1 Subsection of AVIRIS scene f090819t01p00r06 over Lunar Crater Volcanic Field, Nevada. (a) Color composite of 1.7, 0.9, and 0.46 μm bands. (b) SAM class map of scene, map key at bottom. (c) SID class map of scene, map key at bottom.

Smaller SID values indicate spectra that are spectrally similar. Figure 14.1 compares classification maps from these two approaches.

A large number of other supervised classification approaches have been used with multispectral data including classic methods such as Minimum Distance, Mahalanobis Distance, and Maximum Likelihood classifiers. The use of many of these techniques is impractical with hyperspectral datasets that contain hundreds of spectral channels since maximum likelihood and other covariance-based classifiers require at least as many training samples per class as the number of bands plus one. This standard is often impossible for the hyperspectral mapping of poorly exposed materials. Merényi et al. (2014) describe the diminished performance of covariance-based classifiers, such as Maximum Likelihood and Mahalanobis Distance, on hyperspectral data containing many classes using AVIRIS data.

14.3 Spectral Mixture Analysis

14.3.1 Linear Spectral Mixture Analysis

The ground instantaneous field of view of a sensor, which captures the response of an area of ground recorded in an image pixel, almost always contains more than one material. This reality creates the mixed pixel problem. Singer and McCord (1979) referred to linear mixing as "areal" or "checkerboard" mixing because reflected radiance reaches the sensor from individual areas that can be visualized as checkerboard squares. To assess the fractional contribution of the materials whose combined spectra form the observed mixed pixel, a processing approach is needed that deconvolves the recorded multi- or hyperspectral pixel response into its component elements. At the longer wavelengths of the mid-IR, where minerals have higher absorption coefficients than in the VNIR, a simple linear combination of component material spectra has been shown to accurately model observed thermal emission spectra (Ramsey & Christensen, 1998; and Chapter 12). However, in the reflected solar region of the VNIR–SWIR, photons have a greater tendency to interact with more than one material before being scattered back toward the sensor. This results in a nonlinear or "intimate" mixing of signals. Models based on radiative transfer scattering theory have been developed that allow for determinations of the absolute abundances of endmember materials despite the effects of nonlinear mixing. However, these models require prior knowledge of materials in the scene and rely on the use of optical constants of component minerals. Linear Spectral Mixture Analysis (SMA) can be carried out using spectra extracted from the scene and is thus computationally more straightforward (e.g., Adams et al., 1993).

The spectra used, in varying fractional abundances, to form a linear combination for each pixel are the *spectral endmembers*. Spectral endmembers can consist of either "*image endmembers*" or "*reference endmembers.*" Image endmembers are individual pixel spectra, or averages of pixel spectra, that can be linearly combined to model the observed pixel spectra, within some acceptable level, for each spatial element of the scene. Reference endmembers are laboratory or field spectra of pure materials that correspond to the image endmembers and can ideally be used in their place.

The foundational equation of linear spectral mixture analysis is

$$\rho_j = \sum_{i=1}^{ne} f_i \rho_{i,j} + e_j, \qquad (14.4)$$

where ρ_j is the pixel reflectance in band j, f_i is the fraction of endmember i, $\rho_{i,j}$ is the reflectance of endmember i in band j, ne is the number of endmembers, and e_j is the residual between the measured and modeled values in band j. Across the scene, the mismatch between the modeled and actual scene is recorded in terms of root-mean-square error (RMSE) where

$$\text{RMSE} = \left[\frac{1}{nb}\sum_{j=1}^{nb} e_j^2\right]^{1/2}, \qquad (14.5)$$

where nb is the number of bands, and j and e_j are as defined earlier.

Critically important for the success of SMA in modeling a scene is the selection of the spectral endmembers that represent the spectral variability of the scene or its "intrinsic dimensionality," which is equal to the number of endmembers minus 1. Thus, three endmembers define a triangle existing on a two-dimensional (henceforth, 2-D) plane with intrinsic dimensionality of 2. A measure of the intrinsic dimensionality can be obtained from a plot of the eigenvalues of the dataset. A break in slope in a plot of eigenvalue versus band number indicates the transition from significant endmembers to minor spectral phases that do not significantly contribute to the spectral variability of the scene.

Several techniques have been developed to determine the identity of the spectral endmembers in any given hyperspectral scene. Boardman and Kruse (1994) developed a sequential method for the determination of spectral endmembers. First, to reduce the spectral dimensionality of the data, a Minimum Noise Fraction (MNF) transformation (Green et al., 1988) is applied. The MNF transform consists of two consecutive data transformations. First, the noise in the data is estimated, either using dark current information or from the data itself by a method such as a shift-difference operation (which assumes relative homogeneity in the image). A noise covariance matrix can then be formed and used to decorrelate and rescale the noise in the data. This results in transformed data in which the noise has unit variance in each of the bands and no band-to-band correlations. The second transformation is a standard principal components transformation (e.g., Richards, 2013) of the noise-whitened data. The combined MNF transformation results in a set of images ordered from first to last in terms of increasing levels of noise. Then, a Pixel Purity Index (PPI) is applied to the significant (those that are not dominated by noise) MNF bands. The PPI procedure determines the "purest" (those consisting mostly or entirely of a single spectral endmember) pixels in a hyperspectral scene. This is achieved by selecting a random vector through the center (the mean value) of the n-D cloud of significant MNF images and projecting each pixel in the image onto this random vector. Then, a histogram of the projected coordinates is formed in which the "pure" pixels are those that compose the tails of the histogram. This process is successively repeated with new random vectors. Those pixels that fall on the tails of the histogram are flagged as being the most "pure" pixels in the scene. Finally, the significant MNF values for these "spectrally pure" pixels are examined in an interactive n-D data visualization procedure. Boardman and Kruse (1994) adapted the projection pursuit algorithm (Friedman & Tukey, 1974) for n-D data visualization. This interactive n-D visualization approach allows for the selection of pixels defining vertices in n-D space that correspond to the image endmembers of the scene.

Programs for automated endmember determination have also been developed and widely used. N-FINDR (Winter, 1999) makes use of a concept described by Boardman (1995) where the m-D data cloud of a hyperspectral image (where m is the intrinsic dimensionality) is fitted by an m-D simplex. For example, a 2-D data cloud is fitted by a triangle and a 3-D data cloud would be fitted by a tetrahedron. However, a straightforward application of this "convex hull" fitting breaks down with higher numbers of endmembers due to the "curse of dimensionality." The N-FINDR algorithm approximates this methodology using a set of seed pixel spectra from the scene as initial endmembers, calculating the volume of the simplex formed by these initial endmembers, then testing each pixel in the scene to see if replacing the initial endmembers with the test pixel creates a larger simplex volume. If it does, the test pixel replaces the initial endmember and the process is repeated.

The Sequential Maximum Angle Convex Cone (SMACC) automatic endmember determination (Gruninger et al., 2004; Gilmore et al., 2011) algorithm models the dataset as a linear combination of unique spectra that define a convex cone containing mixed spectra in the dataset. The mixed spectra are modeled as a linear mixture of the unique, convex cone defining spectra. The algorithm proceeds by finding the most spectrally unique pixel, taken to be the one with the highest albedo, then finding the one that is most different from that first endmember based on a spectral angle metric (Section 14.2), then finding the pixel spectrum that is most dissimilar from the first two, and so on. The program proceeds until

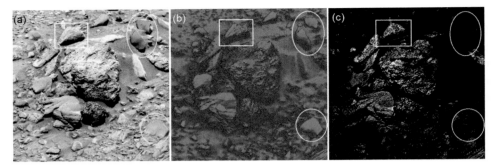

Figure 14.2 View of Spirit sol 663 target Goatgrass. (a) Composite of 673-, 535-, and 432-nm bands of rougher textured rocks, with a representative one noted within the white rectangle, that are spectrally distinct from smoother-textured rocks (represented by those in yellow circles). (b) Composite of fractions images derived from SMA with a bright dust endmember assigned to red, smooth-textured rock endmember assigned to green, and shade assigned to blue. (c) RMSE image. The smooth-textured rocks were the endmember mapped in the SMA; thus they have low RMSE and spectrally distinct rough-textured rocks have high RMSE values.

some preset number of endmembers is reached or until all pixels are within the spectral angle threshold of similarity.

Having determined the endmembers, the result of SMA (from Eq. 14.4) is a set of fraction images, with DN values ideally ranging between 0 and 1, one for each endmember plus an RMSE image that represents the goodness of fit between the modeled and measured scene. An example from Spirit rover Pancam data is shown in Figure 14.2. Bright dust, smooth textured rocks, and shade are shown in the color composite in Figure 14.2b. Rougher textured rocks are spectrally distinct from those with smoother textures, thus the former show up with high values in the RMSE image of Figure 14.2c. Coherent sets of pixels with high RMSE data numbers could either be a material distributed across the scene, in which case it could be used as an additional endmember, or if it represents a spatially isolated occurrence, it could be treated as an anomalous material mapped by the RMSE image or used as the target spectrum for one of the "partial unmixing" techniques discussed in Section 14.4.

"Shade" (the blue component in Figure 14.2b) is an example of a unique non-material endmember used in linear SMA with VNIR–SWIR data. The ideal shade image endmember would be that of pixel spectra extracted from a shaded black surface. Nonmaterial endmembers have also been successfully used in the Multiple-Endmember Linear Spectral Unmixing Model (MELSUM) of Combe et al. (2008).

14.3.2 Nonlinear Spectral Mixture Analysis

As noted earlier, while linear SMA is a useful approximation for use with VNIR–SWIR remote sensing data, mixing is very often nonlinear in this wavelength range. To account for nonlinear effects, various approaches that define the mixing in terms of the optical constants of the materials involved, or that deal with single scattering albedo (Chapter 2), have been used.

As described in Chapter 2, there are two major empirical models that address the scattering behavior of particulate surfaces. These are the radiative transfer theories developed by Hapke (1981) and a series of succeeding papers, summarized in Hapke (2012), and Shkuratov et al. (1999). Input parameters for the Hapke model include the viewing geometry (incidence, emittance, and phase angles), particle size, and single-scattering albedos of the component grains. For the Shkuratov model, viewing geometry is not a required input parameter; however, parameters that are required are the real and imaginary parts of the refractive index of the component particles, the average optical path length of the component grains (essentially grain size), and the volume filled by the particles.

An example of a nonlinear unmixing, based on the Hapke model, is presented in Figure 14.3 using the methodology of Liu et al. (2016). Those authors obtained the single-scattering albedo of minerals based on their optical constants using multiple phase-angle measurements and the method of Sklute et al. (2015). Those mineral spectra were then used as the potential endmember spectra in a spectral library of probable minerals and a range of grain sizes. Single-scattering albedo, derived from hyperspectral CRISM data atmospherically corrected using a radiative transfer code, could then be linearly unmixed to obtain mineral abundances. The example in Figure 14.3 is an unmixing of an Fe/Mg smectite-rich spectrum from CRISM data over the Mawrth Vallis region on Mars.

14.4 Partial Unmixing

"Partial unmixing" is a concept introduced by Smith et al. (1994) and Boardman et al. (1995), where the multi- or hyperspectral scene is decomposed into a set of "foreground"

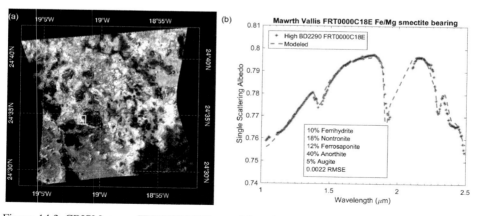

Figure 14.3 CRISM scene FRT0000C18E over Mawrth Vallis region of Mars. (a) Composite of bands centered at 2.5, 1.5, and 1.08 μm. Yellow box indicates ROI extraction site for high BD2290 (high 2290-nm band depth) pixels. (b) Measured and modeled ROI average spectrum with component fractions and RMSE as determined from nonlinear unmixing model described in Liu et al. (2016).

and "background" spectra. The foreground spectra are target materials of interest with spectra already known to the analyst or extracted from representative locations in the image. The background spectra represent "everything else," e.g., the undesired background materials. Smith et al. (1994) put this problem into the form of a "foreground/background analysis" algorithm. Chen and Reed (1987) and Stocker et al. (1990) described it as a linear matched filter. Harsanyi (1993) and Farrand and Harsanyi (1995) described an alternate matched filter formulation called "constrained energy minimization" or CEM. CEM and linear matched filter are functionally equivalent. As with the spectral matching methods of Section 14.2, these partial unmixing approaches can be considered as a form of supervised classification of the scene.

Schaum (2001) summarized the matched filter response, $mf(\mathbf{x})$ for each pixel vector, $\mathbf{x}$, as

$$mf(\mathbf{x}) = \frac{(\mathbf{x}_t - \boldsymbol{\mu})^T \mathbf{C}^{-1}(\mathbf{x} - \boldsymbol{\mu})}{\sqrt{(\mathbf{x}_t - \boldsymbol{\mu})^T \mathbf{C}^{-1}(\mathbf{x}_t - \boldsymbol{\mu})}}, \qquad (14.6)$$

where $\mathbf{x}_t$ is the target material spectrum, $\mathbf{x}$ is the pixel spectrum, μ denotes the mean of background spectra, and $\mathbf{C}$ is the covariance matrix of the scene.

CEM is based on the twin constraints that (1) when vector operator $\mathbf{z}$ is applied to a pixel spectrum, $\mathbf{x}$, that is identical to the target spectrum, $\mathbf{x}_t$, the result is 1, e.g., $\mathbf{x}^T * \mathbf{z} = 1$; and (2) the total energy of $\mathbf{x}^T * \mathbf{z}$ for other pixels different from the target signature is minimized. The CEM operator, $\mathbf{z}$, is obtained from

$$\mathbf{z} = \frac{\mathbf{C}^{-1}\mathbf{x}_t}{\mathbf{x}^T \mathbf{C}^{-1} \mathbf{x}_t}. \qquad (14.7)$$

The high background suppression of CEM is demonstrated in Figure 14.4b.

14.5 Machine Learning Approaches

14.5.1 Self-Organizing Maps for Clustering, Discovery, and for Aiding Supervised Classification

The Self-Organizing Map (SOM, by Kohonen, 1988), is the best-known member of the artificial neural map family. It is an unsupervised artificial neural network (ANN) – a neural machine learning paradigm that mimics the information summarization and organization of cortical areas in biological brains. SOMs can be used for unsupervised manifold learning and discovery of clusters (groups of similar spectra). An SOM can also serve as a hidden layer in a multilayer ANN to aid supervised classification. SOMs are especially well suited for multi- and hyperspectral data that contain many classes with subtle differences among classes and where the classes have widely varying or irregular statistical properties. They are applicable to high-dimensional data where the available training sets are small (too small to build a nonsingular covariance matrix) or unevenly distributed.

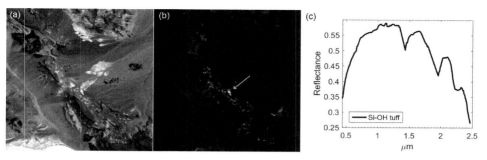

Figure 14.4 Portion of AVIRIS scene over Lunar Crater Volcanic Field in Nye County, Nevada. (a) 1.7-, 0.9-, 0.45-µm composite. (b) Fraction image from CEM using the spectrum of a tuff with an Si–OH absorption as the target spectrum. Arrows point to primary exposure of low abundance Si–OH tuff. (c) Reflectance spectrum extracted from AVIRIS data of tuff with 2.2 µm Si–OH absorption.

An SOM consists of a (usually) 2-D lattice A of N artificial neurons, where each neuron is connected to an input layer by an nb-D weight vector $\mathbf{w}_i \in \mathbf{R}^{nb}$ – also called a *prototype* vector. During learning the prototypes are moved in the data space to reflect the distribution of the nb-D input data space. After (typically random) initialization, SOM learning updates the prototype vectors $\mathbf{w}_i$ of neural units, i, iteratively through many time steps, t, as follows. First, a randomly selected input vector (spectrum) $\mathbf{x} \in \mathbf{R}^{nb}$ is compared to the prototypes of all SOM neurons, and the neuron whose prototype is closest to $\mathbf{x}$ (by some distance, e.g., Euclidean), say neuron c, "wins" the input (Eq. 14.8):

$$\|\mathbf{w}_c - \mathbf{x}\| \leq \|\mathbf{w}_i - \mathbf{x}\|, i \in A, \text{ for } i = 1, \ldots, N. \tag{14.8}$$

Second, the winning neuron, and neurons in its lattice neighborhood, adapt their prototypes to become more similar to the input vector.

$$\mathbf{w}_i(t+1) = \mathbf{w}_i(t) + \alpha(t) h_{c,i}(t) \Big(\mathbf{x} - \mathbf{w}_i(t) \Big), \text{ for } i = 1, \ldots, N. \tag{14.9}$$

Here, $0 \leq \alpha(t) \leq 1$ is the learning rate, $h_{c,i}(t)$ is a neighborhood function defining which neurons' prototypes get updated and to what extent as a function of the lattice distance between the winner neuron c and i. Frequently a 2-D Gaussian neighborhood is used, where all prototypes are updated but the extent of the update decreases with the lattice distance.

This learning process accomplishes two things. One is *adaptive vector quantization*: the weight vectors of SOM neurons become prototypes of similar input patterns and they follow the data distribution, i.e., more prototypes are allocated to dense regions than to sparse regions. Therefore, this data summarization captures the salient details of the data structure by a number of SOM prototypes magnitudes smaller than the number of data vectors. Simultaneously, the prototypes are organized on the SOM lattice according to their similarity relations in data space. Thus, neurons neighboring in the SOM lattice collectively represent groups of similar data vectors after sufficient learning in this *topology-preserving* mapping. The SOM expresses an intelligent summarization of both the statistics (the n-D density distribution) and the topology of the data manifold while preserving the spectral dimensionality (and details) in the prototypes.

SOM learning needs very little parameter tuning, and no prespecified number of clusters. Cluster discovery is done by segmenting the learned SOM into groups of similar prototypes (data points mapped to a prototype cluster make up a data cluster). For correct segmentation, the topological correctness of the manifold learning and the expressiveness of the representation of prototype similarities are critical. Many of these issues have been addressed by Merényi et al. (2009) and references therein. These include measures of the degree of topology preservation, a new prototype similarity measure, and cluster validity measure.

SOMs, and neural maps in general, offer other interesting properties for various advantages (e.g., Merényi et al., 2007, 2009). Importantly for the present context, SOMs provide complementarities and trade-offs to SMA, and to statistical learning. SOMs are nonparametric machine learning algorithms, and thus alleviate the problem of the unavailability of statistical models for multi- and hyperspectral data. They also support precise (supervised) classification when used as a hidden layer in a feed-forward multilayer ANN. The topology-preserving mapping and shared representation of inputs by neighbor neurons enables SOMs and SOM-hybrid neural architectures to separate classes whose spectra are very similar – in contrast to SMA, where endmembers need to be extremes of the data cloud. Therefore, they can handle many spectral species in a single model. They can achieve high classification accuracy with fewer labeled training samples than methods involving covariance matrices (e.g., Maximum Likelihood and Mahalonobis Distance classifiers), and are less sensitive to uneven distribution of training samples across classes. These aspects are detailed in a hyperspectral study classifying 23 spectral classes in the Lunar Crater Volcanic Field (same scene as in Figures 14.1 and 14.7) with high accuracy (Merényi et al., 2014).

SOMs have been used in remote analysis of astronomical and planetary objects for many years. Studies include analyses of laboratory reflectance and Mars Global Surveyor Thermal Emission Spectrometer (TES) data (Hogan & Roush, 2002; Roush & Hogan, 2007; Roush et al., 2007). Howell et al. (1994) used SOMs to discover olivine- and pyroxene-rich end-groups of asteroids and refine their taxonomy. Felder et al. (2014) derived 16 clusters from Moon Mineralogy Mapper imagery representing basic types of lunar surface materials. Simultaneous inference of temperature and grain size from spectra of icy volatiles is presented by Zhang et al. (2010).

Figure 14.5 shows clustering of a Pancam scene from the Mars Exploration Rover Spirit, which can be compared to band depth and CEM mapping of the same scene in Farrand et al. (2008a). SOM clustering and SOM-hybrid classification into many spectral classes helped detect new mineralogy trends at the Mars Pathfinder landing site (Farrand et al., 2008b).

SOM processing is more expensive computationally than linear spectral mixture analysis or many covariance-based statistical learning methods; therefore its best use is for data with complex structure where discovery depends on separation of many spectral species or subtle spectral variations. SOM learning is slow on conventional sequential computers but it can be greatly accelerated through parallel implementations. The most severe bottleneck

Figure 14.5 SOM cluster map of MER Spirit Pancam scene p2582 from Sol 608 near the summit of Husband Hill (after Taşdemir & Merényi, 2008). (a) The clusters (color coded at the bottom) were extracted using an advanced SOM visualization. (b) Composite of CEM fraction images of rocks with strong 900-nm absorption band (red, Bowline type rocks), shallow 900-nm band (green, Tenzing type rocks) and bright drift, blue (from Farrand et al., 2008a). Owing to the nonlinear separation capability, the SOM cluster map delineates more spectral types.

is interactive cluster extraction from the SOM, which has generally produced better results than automated procedures (e.g., clustering the SOM using K-means or hierarchical agglomerative methods). An automation approach that approximates the quality of interactive SOM segmentation for complex hyperspectral data is proposed in Merényi et al. (2016).

14.5.2 Support Vector Machines and Kernel Methods

Support Vector Machines (SVMs) (Boser et al., 1992; Cortes & Vapnik, 1995) is a classification approach based on the notion of the margin on either side of a hyperplane separating two data classes. The classifier builds the linear boundary between the classes by maximizing the margin and thereby creating the largest possible distance between the separating hyperplane and the data on either side of it. Once the optimum separating hyperplane is found, data points that lie on its margin are known as support vector points and the classifier assigns labels to unobserved points exclusively as a linear combination of the support vectors.

SVM training involves the minimization of the objective function:

$$f = \frac{1}{2}\mathbf{v}^T\mathbf{v} + c\sum_{j=1}^{nb} \xi_j, \tag{14.10}$$

subject to the constraints: $\mathbf{y}_j\left(\mathbf{v}^T \phi(\mathbf{x}_j) + b\right) \geq 1 - \xi_j, \xi_j \geq 0$, where c is a trade-off constant, $\mathbf{v}$ is a vector of coefficients, b is a constant, and ξ_j represents parameters for handling nonseparable input data. Note that $\mathbf{y}_j \in \pm 1$ represent the class labels and $\mathbf{X}_j$ represent the vector of input data points. The kernel, ϕ, is used to handle nonlinear boundaries and it

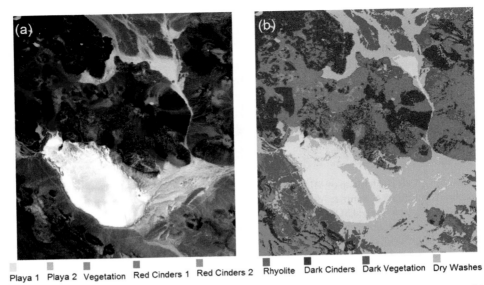

Figure 14.6 Subsection of AVIRIS scene f090819t01p00r06 over Lunar Crater Volcanic Field, Nevada. (a) Color composite of 1.7-, 0.9-, and 0.46-μm bands. (b) SVM classification map of the scene with map key at bottom.

transforms data from the input space to a higher dimensional feature space, so that the problem at hand becomes linear in this feature space. SVMs are inherently two-class classifiers. The traditional way to do multiclass classification with SVMs is to build several one-versus-rest classifiers and to choose the class that classifies the test datum with greatest margin. Another strategy is to build a set of one-versus-one classifiers, and to choose the class that is selected by the most classifiers.

SVMs have become a widely used method in hyperspectral image classification research (e.g., Gualtieri & Chettri, 2000; Melgani & Bruzzoni, 2004; Gilmore et al., 2008; Maulik and Chakraborty, 2017). The value of this method, vis-à-vis supervised spectral matching techniques discussed in Section 14.2, and shown in Figure 14.1, is presented in Figure 14.7 with an SVM classification of the same scene. While the SVM classification is more computationally intensive, comparison of Figure 14.7b with Figures 14.1b,c shows that the SVM was able to more fully map out the low albedo dark cinders (blue in the classification maps) and mixed vegetation and dark soil (the "dark vegetation" class) and also not confuse the two playa classes with dry wash deposits.

Several methods have exploited kernel constructions from the machine learning literature, including radial basis functions and reproducing kernels (e.g., Camps-Valls & Bruzzone, 2005). Brown et al. (2000), show that linear unmixing can be performed using SVMs. A number of pure pixels are selected from each spectral class, and an SVM is constructed for each class to separate it from all other classes. The support vectors are then considered to be the endmembers, and mixed pixels will lie in the margin separating the classes. Since the intersection of these margins will form a simplex with the support vectors as vertices, the classical linear mixing model (Section 14.3.1) is recovered. Nonlinear unmixing is accommodated by using a quadratic polynomial kernel.

Since the introduction of SVMs to remote sensing applications, researchers have quickly moved on from the use of general kernels toward techniques able to exploit the rich information contained in hyperspectral images, such as spatial context and unlabeled pixel spectra. Spatial information can be used to improve spectral classification (Fauvel et al., 2013). Camps-Valls et al. (2006) introduced spatial–spectral classifiers that combine dedicated kernels for the spectral and spatial information. The framework has been recently extended to deal with convex combinations of kernels through multiple-kernel learning (Tuia et al., 2010) and generalized composite kernels (Li et al., 2013). Other techniques define graph-based kernels capturing multiscale higher-order relations in a spatial neighborhood (Camps-Valls et al., 2010). Another alternative is to include spatial contextual information with Markov Random Fields (MRFs) (Moser et al., 2013 and references therein).

Recently, researchers have started to exploit the abundant unlabeled information contained in the image itself and have introduced semisupervised approaches in which one adds a regularization term $\Omega(f)$ to the SVM objective function f that enforces assumed similarities between labeled and unlabeled pixel spectra. One graph-based approach exploits the spectral correlation between neighboring pixels and attempts to assign similar labels to them (Gomez-Chova et al., 2008). Another technique exhibits a regularizer that enforces wide and empty SVM margins (Bruzzone et al., 2006). Other strategies learn a suitable kernel directly from the unlabeled spectra (Tuia & Camps-Valls, 2009).

Active learning techniques can rank the unlabeled pixels according to their expected importance for future labeling. The top-ranked pixels are then screened by a human operator, who provides labels, enlarging the training set. Such approaches applied to SVM use the intuition that a sample away from the decision boundary has a high confidence about its class assignment and is thus not interesting for future sampling. In other words, the approach tends to select the points lying within the margin of the SVM classifier as they are more likely to become support vectors (Tuia et al., 2011 and references therein).

14.6 Conclusions

This chapter reviewed several techniques for the analysis of multi- and hyperspectral data in the VNIR–SWIR spectral range (0.4–2.5 μm). We have reviewed supervised classification techniques that can rapidly match and map a set of known spectra including the approaches of Spectral Angle Mapper (SAM) and Spectral Information Divergence (SID).

Linear Spectral Mixture Analysis (SMA) is used for the characterization of spectral variability in a multi- or hyperspectral scene as well as for mapping the relative abundance of the endmembers (either derived from the scene or from library or field spectra) that account for the spectral variability.

In the VNIR–SWIR spectral range, mixing is more likely to be nonlinear than it is at longer wavelengths. Nonlinear unmixing requires a priori knowledge of materials in the scene as well as information on the optical constants of the constituent materials and thus are necessarily more analytically demanding than linear SMA.

Machine learning methods complement SMA and spectral matching. First, supervised classifiers map known spectral types when there is a sufficient number of known (labeled) samples in the image. In contrast to using single examples of spectral endmembers, as in SMA, a classification is produced by deriving the general characteristics of spectral types from a suite of spectra (the training sample set) for each class, representing the variability within that class. Further, the result is a class map, with well-defined (crisp) class boundaries. SMA works best with a small number of spectral endmembers that drive most of the spectral variability of the scene. Machine learning methods can map many classes with small spectral differences. Unsupervised clustering works without prior knowledge of spectral types in the scene and maps constituents in a scene purely based on spectral differences.

The techniques presented in this chapter also represent points in a trade-off space of computational cost versus information return. SMA is fast, inexpensive, and provides reliable and easy-to-interpret analysis of hyperspectral scenes. The spectral variability of a scene can generally be captured by a relatively small number of spectral endmembers, and effectively mapped using SMA. However, subtle spectral differences, represented by a larger number of classes, often need to be mapped. SVMs require considerably more computational cycles but they can produce more detail and better mapping accuracy than many other classifiers (one example is given by Figures 14.1 and 14.6). SOM clustering and SOM-hybrid supervised classification have the highest computational cost of the discussed methods. In return, they are capable of intricate discoveries, and simultaneous accurate mapping of many classes with subtle spectral differences, under statistically challenging circumstances (limited training sample size, unevenly represented classes, widely varying cluster/class properties).

Acknowledgments

We thank T. Glotch for spectral library data provided to demonstrate the nonlinear unmixing approach of Liu et al. (2016).

References

Adams J.B., Smith M.O., & Gillespie A.R. (1993) Imaging spectroscopy: Interpretation based on spectral mixture analysis. In: *Remote geochemical analysis: Elemental and mineralogical composition* (Pieters C.M. & Englert, eds.). Cambridge University Press, New York, 145–166.

Boardman J.W. (1993) Automating spectral unmixing of AVIRIS data using convex geometry concepts. *4th JPL Airborne Earth Science Workshop,* 11–14.

Boardman J.W. & Kruse F.A. (1994) Automated spectral analysis: A geologic example using AVIRIS data, north Grapevine Mountains, Nevada. *Proceedings of the 10th Thematic Conference on Geological Remote Sensing*, ERIM, Ann Arbor, MI, I-407–418.

Boardman J.W., Kruse F.A., & Green R.O. (1995) Mapping target signatures via partial unmixing of AVIRIS data. *5th Annual JPL Airborne Earth Science Workshop.*

Boser B.E., Guyon I.M., & Vapnik V.N. (1992) A training algorithm for optimal margin classifiers. *Proceedings of the Annual Workshop on Computational Learning Theory,* 144–152.

Brown M., Lewis H.G., & Gunn S.R. (2000) Linear spectral mixture models and support vector machines for remote sensing. *IEEE Transactions on Geoscience and Remote Sensing*, **38**, 2346–2360.

Bruzzone L., Chi M., & Marconcini M. (2006) A novel transductive SVM for semisupervised classification of remote sensing images. *IEEE Transactions on Geoscience and Remote Sensing*, **44**, 3363–3373.

Camps-Valls G. & Bruzzone L. (2005) Kernel-based methods for hyperspectral image classification, *IEEE Transactions on Geoscience and Remote Sensing*, **43**, 1351–1362.

Camps-Valls G., Gomez-Chova L., Muñoz-Marí J., Vila-Francs J., & Calpe-Maravilla J. (2006) Composite kernels for hyperspectral image classification. *IEEE Geoscience Remote Sensing Letters*, **3**, 93–97.

Camps-Valls G., Shervashidze N., & Borgwardt K.M. (2010) Spatio-spectral remote sensing image classification with graph kernels. *IEEE Geoscience Remote Sensing Letters*, **7**, 741–745.

Chandrasekhar S. (1960) *Radiative transfer.* Dover Publications, Mineola, NY.

Chang C.-I. (2000) An information theoretic-based approach to spectral variability, similarity and discriminability for hyperspectral image analysis. *IEEE Transactions on Information Theory*, **46**, 1927–1932.

Chen J.Y. & Reed I.S. (1987) A detection algorithm for optical targets in clutter. *IEEE Transactions on Aerospace Electronic Systems*, AES-23(1).

Combe J.P., Le Mouelic S., Sotin C., et al. (2008) Analysis of OMEGA/Mars express data hyperspectral data using a multiple-endmember linear spectral unmixing model (MELSUM): Methodology and first results. *Planetary and Space Science*, **56**, 951–975.

Cortes C. & Vapnik V.N. (1995) Support-vector networks. *Machine Learning*, **20**(3), 273–297.

Farrand W.H. & Harsanyi J.C. (1995) Discrimination of poorly exposed lithologies in imaging spectrometer data. *Journal of Geophysical Research*, **100**, 1565–1578.

Farrand W.H. & Harsanyi J.C. (1997) Mapping the distribution of mine tailings in the Coeur d'Alene River Valley, Idaho through the use of a Constrained Energy Minimization technique. *Remote Sensing of the Environment*, **59**, 64–76.

Farrand W.H., Bell J.F. III, Johnson J.R., et al. (2008a) Rock spectral classes observed by the Spirit rover's Pancam on the Gusev crater plains and in the Columbia Hills. *Journal of Geophysical Research*, **113**, E12S38, DOI:10.1029/2008JE003237.

Farrand W.H., Merényi E., Johnson J.R., & Bell J.F., III (2008b) Comprehensive mapping of spectral classes in the imager for Mars Pathfinder Super Pan. *Mars*, **4**, 33–55.

Fauvel M., Tarabalka Y., Benediktsson J.A., Chanussot J., & Tilton J.C. (2013) Advances in spectral-spatial classification of hyperspectral images. *Proceedings of the IEEE*, **101**, 652–675.

Felder M.P., Grumpe A., & Wöhler C. (2014) Automatic segmentation of spectrally similar lunar surface areas with emphasis on the spectral absorption features. *45th Lunar Planet. Sci. Conf.*, Abstract # 2537.

Friedman J.H. & Tukey J.W. (1974) A projection pursuit algorithm for exploratory data analysis. *IEEE Transactions on Computers*, **23**, 881–890.

Gilmore M.S., Bornstein B., Merrill M.D., Castaño R., & Greenwood J.P. (2008) Generation and performance of automated jarosite mineral detectors for visible/near-infrared spectrometers at Mars. *Icarus*, **195**, 169–183.

Gilmore M.S., Thompson D.R., Anderson L.J., Karamzadeh N., Mandrake L., & Castaño R. (2011) Superpixel segmentation for analysis of hyperspectral datasets, with application to CRISM data, M^3 data, and Ariadnes Chaos, Mars. *Journal of Geophysical Research*, **116**, E07001, DOI:10.1029/2010JE003763.

Gomez-Chova L., Camps-Valls G., Muñoz-Marí J., & Calpe J. (2008) Semi-supervised image classification with Laplacian support vector machines. *IEEE Geoscience Remote Sensing Letters*, **5**, 336–340.

Green A.A., Berman M., Switzer P., & Craig M.D. (1988) A transformation for ordering multispectral data in terms of image quality with implications for noise removal, *IEEE Transactions on Geoscience and Remote Sensing*, **26**, 65–74.

Gruninger J.H., Ratkowski A.J., & Hoke M.L. (2004) The sequential maximum angle convex cone (SMACC) endmember model. *Proceedings of SPIE 5425, Algorithms and Technologies for Multispectral, Hyperspectral, and Ultraspectral Imagery X*, **1**, DOI:10.1117/12.543794.

Gualtieri J.A. & Chettri S. (2000) Support vector machines for classification of hyperspectral data. *IGARSS 2000. IEEE 2000 International Geoscience and Remote Sensing Symposium, IEEE 2000 International Support vector machines for classification of hyperspectral data*, **2**, 813–815.

Hapke B. (1981) Bidirectional reflectance spectroscopy: 1. Theory. *Journal of Geophysical Research*, **86**, 3039–3054.

Hapke B. (2012) *Theory of reflectance and emittance spectroscopy.* Cambridge University Press, New York.

Harsanyi J.C. (1993) *Detection and classification of subpixel spectral signatures in hyperspectral image sequences.* PhD dissertation, Department of Electrical Engineering, University of Maryland.

Hogan R.C. & Roush T.L. (2002) SOM classification of martian TES data. *23rd Lunar Planet. Sci. Conf.*, Abstract #1693.

Howell E.S., Merényi E., & Lebofsky L.A. (1994) Using neural networks to classify asteroid spectra. *Journal of Geophysical Research*, **99**, 10,847–10,865.

Kohonen, T. (1988) *Self-organization and associative memory*. Springer-Verlag, New York.

Kruse F.A., Lefkoff A.B., Boardman J.W., et al. (1993) The spectral image processing system (SIPS)—interactive visualization and analysis of imaging spectrometer data. *Remote Sensing of Environment*, **44**, 145–163.

Kullback S. (1968) *Information theory and statistics*. Dover, Gloucester, MA.

Li J., Marpu P.R., Plaza A., Bioucas-Dias J., & Benediktsson J.A. (2013) Generalized composite kernel framework for hyperspectral image classification. *IEEE Transactions on Geoscience and Remote Sensing*, **51**, 4816–4829.

Liu Y., Glotch T.D., Scudder N., et al. (2016) End-member identification and spectral mixture analysis of CRISM hyperspectral data: A case study on southwest Melas Chasma, Mars. *Journal of Geophysical Research*, **121**, 2004–2036.

Maulik U. & Chakraborty D. (2017) Remote Sensing Image Classification: A survey of support-vector-machine-based advanced techniques. *IEEE Geoscience and Remote Sensing Magazine*, **5**(1), 33–52.

Melgani F. & Bruzzone L. (2004) Classification of hyperspectral remote sensing images with support vector machines. *IEEE Transactions on Geoscience and Remote Sensing*, **42**, 778–1790.

Merényi E., Farrand W.H., Brown R.H., Villmann Th., & Fyfe C. (2007) Information extraction and knowledge discovery from high-dimensional and high-volume complex data sets through precision manifold learning. *Proceedings of NASA Science Technology Conference (NSTC2007)*, College Park, MD, June 19–21, 2007.

Merényi E., Taşdemir K., & Zhang L. (2009) Learning highly structured manifolds: Harnessing the power of SOMs. *Similarity based clustering*. (M. Biehl, B. Hammer, M. Verleysen, & T. Villmann, eds.). Lecture Notes in Computer Science. Springer, Berlin and Heidelberg, 138–168.

Merényi E., Farrand W.H., Taranik J.V., & Minor T.B. (2014) Classification of hyperspectral imagery with neural networks: Comparison to conventional tools, *EURASIP Journal on Advances in Signal Processing*, **71**, DOI:10.1186/1687-6180-2014-71.

Merényi E., Taylor J., & Isella A. (2016) Mining complex hyperspectral ALMA cubes for structure with neural machine learning. *Proceedings of the IEEE Symposium Series of Computational Intelligence and Data Mining, SSCI 2016,* Athens, Greece, December 6–9, 2016, DOI:10.1109/SSCI.2016.7849952.

Moser G., Serpico S.B., & Benediktsson J.A. (2013) Land-cover mapping by Markov modeling of spatial-contextual information. *Proceedings of the IEEE*, **101**, 631–651.

Poulet F. & Erard S. (2004) Nonlinear spectral mixing: Quantitative analysis of laboratory mineral mixtures. *Journal of Geophysical Research*, **109**(E2), DOI:10.1029/2003JE002179.

Poulet F., Cuzzi J.N., Cruikshank D.P., Roush T., & Dalle Ore C.M. (2002) Comparison between the Shkuratov and Hapke scattering theories for solid planetary surfaces: Application to the surface composition of two Centaurs. *Icarus*, **160**, 313–324.

Poulet F., Bibring J.-P., Langevin Y., et al. (2009) Quantitative compositional analysis of martian mafic regions using MEx/OMEGA reflectance data: 1. Methodology, uncertainties and examples of application. *Icarus*, **201**, 69–83.

Ramsey M.S. & Christensen P.R. (1998) Mineral abundance determination: Quantitative deconvolution of thermal emission spectra, *Journal of Geophysical Research*, **103**, 577–596.

Reed I.S. & Yu X. (1990) Adaptive multiple-band CFAR detection of an optical pattern with unknown spectral distribution. *IEEE Transactions on Acoustics, Speech, and Signal Processing*, **38**, 1760–1770.

Ren H. & Chang C.I. (2000) A target-constrained interference-minimized filter for subpixel target detection in hyperspectral imagery. *IGARSS 2000. IEEE 2000 International Geoscience and Remote Sensing Symposium. IEEE 2000 International Support vector machines for classification of hyperspectral data*, **4**. 1545–1547.

Richards J.A. (2013) Supervised classification techniques. In: *Remote sensing digital image analysis*. Springer, Berlin and Heidelberg, 247–318.

Roush T.L. & Hogan R.C. (2007) Automated classification of visible and near-infrared spectra using self-organizing maps. *Proceedings of the IEEE Aerospace Conference 2007*, 1–10, DOI:10.1109/AERO.2007.352701.

Roush T.L., Helbert J., Hogan R.C., & Maturilli A. (2007) Classification of Mars analogue mixtures and end-member minerals using self-organizing maps. *38th Lunar Planet. Sci. Conf.*, Abstract #1291.

Schaum A.P. (2001) Spectral subspace matched filtering. *Proceedings of the SPIE 4381, Algorithms for Multispectral, Hyperspectral, and Ultraspectral Imagery VII*, **1** (August 20, 2001), DOI:10.1117/12.436996.

Schowengerdt R.A. (2012) *Techniques for image processing and classifications in remote sensing*. Academic Press, San Diego.

Shkuratov Y.G., Starukhina L., Hoffmann H., & Arnold G. (1999) A model of spectral albedo of particulate surfaces: Implications for optical properties of the Moon, *Icarus*, **137**, 235–246.

Singer R.B. & McCord T.B. (1979) Mars: Large scale mixing of bright and dark materials and implications for analysis of spectral bright regions on Mars. *Proceedings of the 10th Lunar Planet. Sci. Conf.*, 1825–1848.

Sklute E.C., Glotch T.D., Piatek J., Woerner W., Martone A., & Kraner M. (2015) Optical constants of synthetic potassium, sodium, and hydronium jarosite. *American Mineralogist*, **100**, 1110–1122.

Smith M.O., Roberts D.A., Hill J., et al. (1994) A new approach to quantifying abundances of materials in multispectral images. *Geoscience and Remote Sensing Symposium, 1994. IGARSS'94. Surface and Atmospheric Remote Sensing: Technologies, Data Analysis and Interpretation, International*, 2372–2374.

Stocker A., Reed I.S., & Yu X. (1990) Multidimensional signal processing for electro-optical target detection. *Proceedings of SPIE 1305, Signal and Data Processing of Small Targets 1990*, 218, DOI:10.1117/12.21593.

Taşdemir K. & Merényi E. (2008) Cluster analysis in remote sensing spectral imagery through graph representation and advanced SOM representation. *11th International Conference on Discovery Science*, DS-2008, Budapest, Hungary, October 13–16, 2008. Lecture Notes in Computer Science, Volume 5255/2008, 272–283.

Tuia D. & Camps-Valls G. (2009) Semi-supervised remote sensing image classification with cluster kernels. *IEEE Geoscience and Remote Sensing Letters*, **6**, 224–228.

Tuia D., Camps-Valls G., Matasci G., & Kanevski M. (2010) Learning relevant image features with multiple kernel classification. *IEEE Transactions on Geoscience and Remote Sensing*, **48**, 3780–3791.

Tuia D., Volpi M., Copa L., Kanevski M., & Muñoz-Marí J. (2011) A survey of active learning algorithms for supervised remote sensing image classification. *IEEE Journal of Selected Topics on Signal Processing*, **5**, 606–617.

Varshney P.K. & Arora M.K. (2004) *Advanced image processing techniques for remotely sensed hyperspectral data*. Springer Science+Business Media, New York.

Vieira E.F. & Ponz J.D (1998) Automated spectral classification using astronomical data analysis software and systems VII. A.S.P. Conference Series, **145**, 508.

Winter M.E. (1999) N-FINDR: An algorithm for fast autonomous spectral end-member determination in hyperspectral data. *SPIE's International Symposium on Optical Science, Engineering, and Instrumentation*. International Society for Optics and Photonics, 266–275.

Zhang L., Merényi E., Grundy W.M., & Young E.Y. (2010) Inference of surface parameters from near-infrared spectra of crystalline H_2O ice with neural learning. *Publications of the Astronomical Society of the Pacific*, **122** (893), 839–852.

15

Thermal Infrared Spectral Modeling

JOSHUA L. BANDFIELD AND A. DEANNE ROGERS

15.1 Introduction

Thermal infrared (TIR; defined here as ~5–50 μm) spectral measurements are sensitive to a large variety of surface and atmospheric characteristics and conditions, which can be used to better understand a wide range of present and past processes. The collection of large and systematic TIR datasets has driven the development of spectral models and techniques that can be used to help untangle the complex set of factors that influence these measurements. In this sense, these techniques increase the interpretability and utility of the data. We focus on spectral modeling techniques that have been applied primarily to the analysis of martian and lunar datasets. However, the techniques described here have a range of potential applications to a wealth of TIR spectral measurements to be returned from upcoming missions to other Solar System bodies.

The methods described in this chapter are by no means comprehensive, but we instead illustrate a sample of models and techniques that have been applied to the datasets. These methods are only a means to the more important end of understanding the processes that shape planetary surfaces throughout the Solar System. In that light, we also provide examples of studies where these techniques have been applied to derive useful compositional and physical properties.

15.2 Modal Mineralogy from Spectral Mixture Analysis

15.2.1 Introduction

Remote spectral measurements are typically acquired from surfaces consisting of physical mixtures, whether at the scale of an individual rock, outcrop, or landscape. Quantifying the spectral contributions from individual components to the measured spectrum permits determination of modal mineralogy that provides important information about the geologic evolution of a planetary surface. Types of physical mixtures include unconsolidated particulate materials (e.g., sands or regolith) and mineral grains/crystals in rocks, or can include multiple rock, soil, and/or land-cover types at larger spatial scales. Methods for estimating component abundance from TIR spectra, described in Sections 15.2.2 and

15.2.3, are applicable to individual spectra or spectral imagery. We note that additional spectral unmixing methods, which rely on models of radiative transfer, are described in Chapter 2.

15.2.2 Linear Spectral Mixture Analysis

15.2.2.1 Description and Methodology

Linear Spectral Mixture Analysis (LSMA) predicts the areal fraction of endmember components in a spectrum measured of a mixture. Mathematically, it is a form of multivariate linear regression, which is expressed in general form as

$$Y(j) = \sum_{i=1}^{n} \beta_i X(j)_i + e, \qquad (15.1)$$

where β_i is a set of coefficients that relates Y, the dependent variable, to X, the independent variables, and e is the residual.

In the case of LSMA, j is the spectral channel; Y is the mixed surface spectrum; X is a matrix of library spectra of candidate mixture components, of length n; β_i is the vector of coefficients for each library spectrum; and e is the residual error between the measured spectrum ($Y(j)$) and modeled spectrum ($\Sigma \beta_i X(j)_i$). If the number of library spectra (n) is less than or equal to the number of spectral channels, the system of equations is constrained or overdetermined, and β_i may be approximated using a least squares fit (e.g., Geladi & Kowalski, 1986). The residual error e can be plotted as a function of spectral channel and visually inspected to determine the quality of fit; the root-mean-square (RMS) of e for each spectral channel provides a useful measure of overall fit in a single parameter (e.g., Gillespie, 1992; Ramsey & Christensen, 1998) (Figure 15.1).

The spectral library can consist of laboratory-measured phases, synthetic spectra, model-derived spectra (e.g., Chapter 2), or scene-derived endmembers, among other options (e.g., Gillespie, 1992). A blackbody spectrum (where emissivity is equal to one at all wavelengths) also may be included to account for differences in spectral contrast between the library spectra and the mixed spectrum (Hamilton et al., 1997). The coefficients that are output from the model, once normalized to exclude the blackbody component, represent the areal abundance of each spectral library component in the mixture.

Note that Eq. (15.1) has no constraint on the sign of the modeled coefficients. Thus for surface applications a nonnegative constraint must be imposed on the least squares minimization process, such that the best-fit model uses only coefficients that are zero or positive. Two common variants of LSMA accomplish nonnegativity in different ways. Ramsey and Christensen (1998) used an iterative spectrum removal approach, in which library spectra with negative coefficients are removed from the library, reducing the size of X on each iteration until only spectra with positive coefficients remain. The nonnegative least squares method of Lawson and Hanson (1974) allows all spectra to remain available to the

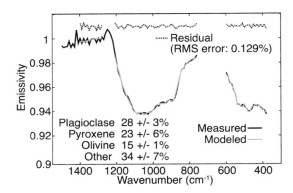

Figure 15.1 Basaltic sands spectrum derived from MER Opportunity Mini-TES spectra (after Glotch & Bandfield, 2006) using factor analysis and target transformation. The spectrum is best fit with plagioclase, pyroxene, olivine, and other minor components (Rogers & Aharonson, 2008). The residual error spectrum shows that the goodness of fit is relatively constant across all portions of the modeled wavelength range.

minimization algorithm through an iterative process, until the least squares solution is reached using only zero or positive coefficients. A comparison of the two algorithms by Rogers and Aharonson (2008) showed that both methods generally yield similar results, but that the iterative removal method of Ramsey and Christensen (1998) resulted in ejection of a true endmember in ~3% of cases examined. The accuracy of the iterative removal method is also more negatively affected by large library sizes than that of the method presented by Lawson and Hanson (1974).

The underlying assumptions of these LSMA techniques are that (1) the spectral library used to model the spectrum contains all of the endmembers present in the mixture and (2) the spectral emissivity of each component in the mixture contributes to the mixed spectrum in proportion to its true areal abundance, at each spectral channel. The first assumption calls for careful construction of a spectral library that contains all likely components that could be present in the mixture. The second assumption is generally true for optically thick grains/crystals, meaning, the photon path length is shorter than the thickness of the grain. For TIR wavelengths, this assumption is usually met for grain sizes >~60 µm, and mineral abundances may be estimated with accuracy within ~10–15% (absolute) of the known abundance (Ramsey & Christensen, 1998; Feely and Christensen, 1999; Hamilton and Christensen, 2000; Wyatt et al., 2001a,b; Thorpe et al., 2015; see Section 15.2.2.2 for exceptions). At smaller grain sizes (<~60 µm), modal abundances were retrieved with better than 15% accuracy for a binary mixture using a library of only the spectral endmembers of the same grain size (Ramsey et al., 1998). However, further work is needed to determine how well modal abundances could be recovered for more complex mixtures of fine particles and also using a blind library.

The size of the library in a single iteration is limited to the number of spectral channels (Lawson & Hanson, 1974; Ramsey & Christensen, 1998); however, some variants of the technique, such as multiple endmember spectral mixture analysis (MESMA), allow use of larger spectral libraries and do not require construction of a pretailored library (Roberts et al., 1998; Li & Mustard, 2003; Johnson et al., 2006; Huang et al., 2013). In MESMA, all possible linear combinations from a user-defined small number of components are attempted with the spectral library. Unused spectra from the library are then alternately added to the best fit model from the previous step, and the library spectrum that best improves the RMS error is kept as an endmember. This process continues until either RMS error is not improved through the addition of library spectra or the number of retained endmembers reaches a user-defined maximum number of components. This LSMA technique is particularly useful for multispectral imagery, where, though the number of components in an individual pixel could be low, the number of spectral endmembers in the scene might still exceed the number of spectral channels in the measurement.

These methods have permitted local, regional, and global mapping of mineral abundances on Mars with TES (Chapter 24), as well as detailed compositional analysis of rocks and soils at the Mars Exploration Rover landing sites with the Miniature Thermal Emission Spectrometer (Mini-TES; Chapter 25). In addition to deriving mineral abundance, LSMA may also be used to map the distribution and fractional abundance of spectrally distinctive units. Here, LSMA is performed on a pixel-by-pixel basis using spectral images (e.g., Gillespie, 1992; Bandfield et al., 2004). This is particularly useful for multispectral imagery, where poor spectral resolution prevents unequivocal mineral identifications, yet is sensitive to changes in bulk composition. For example, Ramsey et al. (2002) used this technique to determine the distribution of lithological units exposed in the ejecta of Meteor Crater, Arizona from Advanced Spaceborne Thermal Emission and Reflection Radiometer (ASTER) data. Similarly, Pan et al. (2015a) mapped the distribution of three spectrally distinctive units in the central peak of Jones crater, Mars using Thermal Emission Imaging System (THEMIS) data (Figure 15.2a). In this instance, one of the units is large enough to be spatially resolved by TES. LSMA of the TES spectrum revealed a pyroxene-rich subsurface lithology (Figure 15.2b).

15.2.2.2 Complicating Factors

The spectral library used for modeling should ideally consist of phases of similar grain size to one another. Because the depth of fundamental absorptions are affected by grain size (e.g., Chapters 2 and 3), there is the potential to over- or underestimate a phase if the library spectrum of that phase was measured from much larger or smaller grains than the other phases in the library. Weakly absorbing minerals such as chlorides, or very thin (<~10 μm) coatings, can exhibit high transparency in the TIR (Lane and Christensen, 1998; Baldridge et al., 2004) that precludes accurate retrieval of their abundance, and subsequently the abundances of other phases with which they are mixed. Last, systematic instrumental errors and/or anisothermality can affect the mixed spectrum in a way that alters the spectral response, thus adding potential uncertainty to derived mineral abundances.

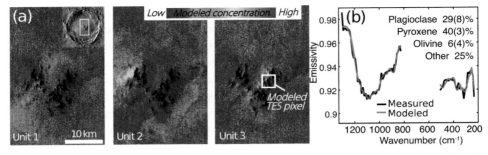

Figure 15.2 THEMIS spectral unit distributions (a) for the central peak of Jones crater (19.1°S, 340.3°E) (inset). Three distinctive units, varying in olivine, high-silica, and pyroxene abundance, are observed in the central peak (modified from Pan et al., 2015a). TES data coverage is available for Unit 3, the easternmost unit. The best-fit model shows a pyroxene-rich composition (b). A variety of surface studies have combined THEMIS and TES measurements in this manner to determine surface compositional properties at a higher spatial and spectral resolution than is possible with either instrument alone (e.g., Chapter 24).

15.2.3 Partial Least Squares Analysis

15.2.3.1 Description and Methodology

Partial Least Squares (PLS) analysis differs from LSMA in that it relies on a set of well-characterized sample mixtures with known mineral abundances and accompanying TIR spectra in order to predict the abundances from a spectrum of an unknown mixture. Like LSMA, PLS is also a form of multiple linear regression (Eq. 15.1), but uses a fundamentally different set of inputs. Here, X_i is a matrix of reference TIR spectra of mixtures (the "training" set), Y is the matrix of known mineral abundances for those spectra, j is the number of spectra in the training set, β is a set of regression coefficients, and i is the spectral channel.

An assumption of PLS analysis is that the system of observations can be described with a small number of latent variables, similar to factor analysis techniques (Section 15.3.1). Thus, prior to generating regression coefficients, X is transformed and reduced to a smaller number of variables that maximize both variance within X and covariance of X with Y. The smaller set of latent variables that are most predictive of mineral abundance in the set of sample spectra are used to predict the abundances for the unknown spectra. Practically, this means that, unlike LSMA, it does not require that all spectral channels be treated equally and removes the assumption of linear mixing across all channels. PLS is a widely used technique in chemometrics; for a detailed treatment of the mathematics, the reader is referred to Geladi and Kowalski (1986) and Wold et al. (2001).

In a study that applied PLS analysis to TIR spectra of coarse-grained granitoid rocks, Hecker et al. (2012) found mineral abundance accuracies that were similar to those of LSMA. However, PLS analyses are clearly superior when applied to fine-particulate mixtures. A comparison of PLS and LSMA for TIR spectra of very fine-grained (<~10 μm) compacted particulate mixtures showed that model accuracy for these mixtures was greatly improved using PLS, with abundances correctly predicted (within 10% absolute of known) for ~80–90% of samples with PLS, compared to ~40% of samples for LSMA (Pan et al., 2015b). The poor performance of LSMA was attributed to the fine-grained, optically thin nature of the grains in the mixture, resulting in transparency effects over portions of the spectral range and invalidating the assumption of linear mixing across all channels. However, because portions of the spectral range could be approximated by linear mixing, PLS was able to successfully predict the modal mineralogy of the mixtures (Pan et al., 2015b).

15.2.3.2 Complicating Factors

One of the practical limitations in using PLS is the need for a well-characterized set of mixtures to use as the training set, which ideally should encompass the range of compositions and particle sizes in the mixed spectrum of interest. Application of PLS to unknown TIR spectra first requires generating a range of relevant synthetic mixtures of interest, and/or detailed characterization of natural mixtures (e.g., rocks) through other techniques (e.g., X-ray diffraction or petrographic imaging), which commonly have their own challenges with accurate abundance retrievals. Unlike other remote geochemical techniques that utilize PLS, such as Laser-Induced Breakdown Spectroscopy (LIBS; Chapter 8), such training sets are generally rare at this time. However, the promise of major improvements in accuracy of routine TIR-retrievals from fine-grained mixtures (Section 15.2.3.1) suggests that building such training sets would be worthwhile.

Like LSMA, the variable spectral contrast of features within the reference library can negatively affect the accuracy of derived abundances, and minerals that are transparent across much of the spectral range might also reduce accuracy. A discussion of advantages and limitations of PLS applied to TIR spectral data is given in Pan et al. (2015b).

15.3 Target Transformation

15.3.1 Description and Methodology

Target transformation (Malinowski, 1991) enables a search for and recovery of independently variable spectral components in a set of mixed spectra in a semiautomated manner. This technique uses factor analysis of the derived set of eigenvectors and eigenvalues from a set of spectra to estimate the number of independently variable spectral components present. The significant higher order eigenvectors containing spectral information besides random noise can be used to reconstruct not only the original data, but also the independent spectral endmember components present in a system of mixed spectra, even if they are not present in the original data. As a result, target transformation can test for the presence of a spectral endmember component in the original mixed data. This methodology is described in detail by Malinowski (1991).

In practice, target transformation is a linear least squares fit of the significant eigenvectors to an endmember test spectrum (Figure 15.3). This test spectrum can be a laboratory

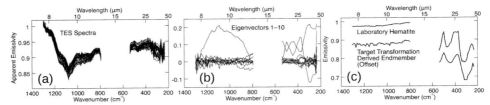

Figure 15.3 TES spectra of Mars. (a) A set of 366 TES apparent emissivity spectra from OCK (Orbit Counter Keeper) 3210 and ICKs (Incremental Counter Keeper) 1691–1760. (b) The first 10 eigenvectors derived from the mean-removed TES spectra shown in the top panel. Spectral features are concentrated in the lower order eigenvectors while higher order eigenvectors lack spectral information and are dominated by random noise. (c) The target transformation recovered the hematite spectral endmember from the series of TES spectra by fitting eigenvectors 1–10 to a laboratory spectrum of hematite. Small differences in the spectral shape reflect real differences, such as crystal orientation, between the martian and laboratory hematite.

spectrum, a spectrum derived from another remotely sensed dataset, or even a series of synthetic spectral shapes. If the test spectrum can be matched by a linear fit of the eigenvectors, then it is a possible independent endmember component present in the system. The target transformation leverages the reduction of the dimensionality of a dataset. By using only the significant eigenvectors, the dimensionality of the set of mixed data is greatly reduced from the number of spectral bands (commonly in the hundreds) to the number of significant eigenvectors (typically 5–15 for geologic materials). Linear combinations of the small number of eigenvectors are not typically able to reconstruct spectral endmembers unless they are present within the set of mixed spectra.

15.3.2 Example Applications

This methodology has been applied to laboratory and spacecraft TIR spectral data (Bandfield et al., 2000; Christensen et al., 2000; Bandfield et al., 2002; Glotch & Bandfield, 2006; Hamilton & Ruff, 2012; Glotch & Rogers, 2013; Geminale et al., 2015; Ruff & Hamilton, 2017). Target transformation was used to identify and derive the spectral shape of hematite on the martian surface using Thermal Emission Spectrometer data (TES; Christensen et al., 2000; Figure 15.3c). This is an ideal situation because of the distinctive spectral shape of hematite (an oxide) relative to other martian atmospheric and surface spectral components (e.g., silicates, ice, CO_2). The hematite spectral shape is cleanly isolated from the TES data, allowing for detailed analysis. As shown in Figure 15.3c, there are minor differences between spectral shape of the laboratory hematite and the target transformation derived endmember. These differences show a preferential [001] axis emission of the martian hematite, consistent with the hematite spherules observed at Meridiani Planum (Glotch et al., 2006a).

Factor analysis and target transformation have also been used to investigate compositional diversity in Mini-TES data at both Mars Exploration Rover landing sites (Glotch &

Bandfield, 2006; Hamilton & Ruff, 2012; Ruff & Hamilton, 2017). At Meridiani Planum, target transformation was applied to Mini-TES data to identify basalt, dust, hematite, and outcrop spectral endmembers. In particular, the target transformation-derived outcrop endmember was analyzed and shown to be dominated by sulfates and poorly crystalline silicate components (Glotch et al., 2006b).

Hamilton and Ruff (2012) used target transformation to identify olivine as an independently variable component in Mini-TES measurements of Adirondack Class basalts within the Gusev Plains and in the Columbia Hills. More recently, Ruff and Hamilton (2017) identified a continuum of compositions within Wishstone and Watchtower Class rocks, from plagioclase-rich volcaniclastic compositions to a rock dominated by an amorphous silicate alteration component.

15.4 Atmospheric Corrections

15.4.1 Introduction

TIR observations of the martian surface from orbit must contend with atmospheric effects on measured radiance. Aerosols (dust and water ice) and gases attenuate and scatter surface emitted radiance and emit their own radiance into the measurement field of view. Additional effects are also present, such as atmospheric emitted radiance that is reflected off the surface, passing again through the atmosphere into the field of view.

Even in cases where we have excellent knowledge of the state of the atmosphere, this can be an extraordinarily complex problem. Various strategies have been applied to retrieving surface radiance and emissivity. In cases with specialized observations, it is possible to apply more rigorous radiative transfer modeling. However, in most instances, there are far too many measurements and poorly constrained atmospheric conditions, making sophisticated models too computationally expensive and underconstrained to be practical.

15.4.2 Atmospheric Correction Using Multiple Emission Angle Observations

Specialized multiple emission angle measurements were acquired by TES periodically throughout the mission for characterization of atmospheric and surface compositional and physical properties. The predictably varying atmospheric path length over a fixed surface was used to derive atmospheric aerosol opacity and surface emissivity using radiative transfer modeling (Bandfield & Smith, 2003). Although spatial coverage was limited, this technique was used to determine two primary surface spectral units: (1) surface dust associated with moderate- and high-albedo surfaces, and (2) basaltic materials associated with low-latitude low-albedo surfaces (Figure 15.4). Accurate recovery of the spectral response of surface dust led to the identification of minor amounts of carbonate that is ubiquitous to martian moderate- and high-albedo regions (Bandfield et al., 2003).

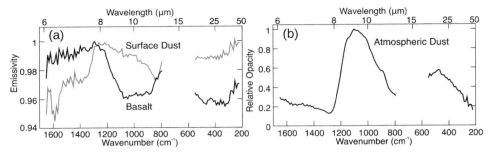

Figure 15.4 Surface (a) and atmospheric (b) spectra derived from TES multiple emission angle measurements. These specialized observations allow for a relatively robust separation of surface and atmospheric radiative contributions to the TIR measurements.

15.4.3 Atmospheric Correction of Nadir TES and THEMIS Measurements

Nadir-only measurements of the martian surface present a poorly constrained atmospheric correction problem. These observations contain a single observation with unknown abundances of dust and water-ice aerosols. There is no wavelength at which the aerosol abundances can be determined without first knowing surface emissivity, which is itself unknown and the goal of the atmospheric correction.

One method commonly employed to determine surface emissivity from relatively warm daytime TES nadir measurements uses a linear fit of atmospheric spectral shapes with a library of laboratory surface emissivity spectra (Smith et al., 2000). This takes advantage of the fact that TIR spectra of mixed mineral surfaces can be closely approximated by a linear combination of each mineral component (Section 15.2; e.g., Thomson & Salisbury, 1993; Ramsey & Christensen, 1998). In addition, it has been shown that martian and atmospheric fractional contributions to the measured spectra can be closely modeled assuming a linear combination of a small number of atmospheric spectral shapes (which are derived from target transformation; Section 15.3; Bandfield et al., 2000). Strictly, the atmospheric components combine in a nonlinear manner that is dictated by radiative transfer. However, using data from a limited range of temperatures and opacities allows for a close linear approximation. In practice, a spectral library of atmospheric and surface spectral shapes is fit in a least squares manner to TES apparent emissivity spectra. The atmospheric spectral shapes are multiplied by their weightings derived from the least squares fit and subtracted from the TES spectrum, producing surface emissivity (Smith et al., 2000).

Other atmospheric correction methods have also been applied under limited circumstances (Smith et al., 2000; Bandfield, 2008). Although most of these methods require simplifying assumptions, they have generally produced consistent results despite the widely varying approaches. In addition, the retrieved surface emissivity spectra are consistent with spectral ratios (e.g., Ruff & Christensen, 2002) and have also been confirmed by measurements of surface emissivity acquired by the Mars Exploration Rovers (e.g., Christensen et al., 2004; Rogers & Aharonson, 2008).

The limited wavelength range and spectral sampling of THEMIS data (Chapter 24) presents a different challenge for removing the atmospheric effects in these data. There is little spectral difference between atmospheric dust and coarse particulate silicates on the surface in the 7–12 μm spectral range. However, THEMIS data contain abundant spatial information that allows for a correction method that can be combined with the spectral information present in the TES data (Bandfield et al., 2004).

15.5 Thermophysical Properties from Spectroscopic Measurements

In addition to deriving a single kinetic surface temperature, spectroscopic TIR measurements can be used to determine the range of temperatures present within the measurement field of view that reflects properties such as subpixel slopes. In addition, spectroscopic observations can be used to identify thermal anomalies, such as volcanism on Io, even when the hot spots fill only a small fraction of the measurement field of view (Sinton, 1981). It is also important to understand the effects of surface anisothermality on TIR spectra even when focusing on surface compositions, because the effects can interfere with compositional interpretations (Bandfield, 2009; Greenhagen et al., 2010).

Rocks on planetary surfaces have a relatively narrow range of thermophysical properties, allowing for their temperature to be modeled and predicted. The two unknown factors, rock fraction and regolith temperature, can be determined using spectral radiance measurements at two or more wavelengths (Christensen, 1986; Bandfield et al., 2011). The methodology works well in practice using nighttime radiance measurements and has provided valuable data for landing site safety characterization and scientific studies (Christensen, 1986; Nowicki & Christensen, 2007; Bandfield et al., 2011; Golombek et al., 2012).

Christensen (1986) used multispectral Viking Infrared Thermal Mapper (IRTM) data to map the distribution of rocks on Mars. A similar analysis was completed by Nowicki and Christensen (2007) using TES data. More recently, multispectral Lunar Reconnaissance Orbiter Diviner Radiometer data have been used to derive rock abundance and rock-free regolith temperatures on the Moon (Bandfield et al., 2011). These data have been used to characterize the lunar regolith (e.g., Ghent et al., 2014).

Rough surfaces can exhibit subpixel anisothermality, especially at high angles of solar incidence, which produce both warm, strongly lit and cold, shadowed surfaces. TIR spectra of rough surfaces typically show prominent blue slopes in spectra (Figure 15.5b; Bandfield, 2009; Bandfield et al., 2015). The effects of surface roughness can dominate telescopic disk-integrated measurements of asteroids (e.g., Spencer, 1990; Lagerros, 1998). For the Moon, Diviner multispectral data show that most surfaces are rough (20–25° RMS slope distribution) at cm scales (Figure 15.5b; Bandfield et al., 2015). These effects can extend to wavelengths <3 μm (Figure 15.5a). Although less common, rough surfaces are also present on Mars, resulting in spectral slopes in apparent emissivity spectra (Bandfield, 2009).

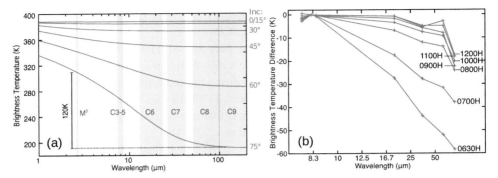

Figure 15.5 (a) Modeled lunar brightness temperatures for an albedo of 0.10 and RMS surface slope distribution of 20°. Higher angles of solar incidence increases surface anisothermality, resulting in large brightness temperature variations. Channels 3–9 indicate Diviner spectral coverage and M^3 indicates the wavelength used for derivation of surface temperature for thermal correction of Moon Mineralogy Mapper spectra. (b) Diviner spectra (channels 3–9) of an equatorial mare surface (304.65 to 307.86°E, –0.13 to 0.16°N) at local times of 0630–1200 H. Changes in spectral slope are due to surface roughness.

15.6 Summary

Spectral modeling techniques have been developed for the analysis of planetary surfaces using large TIR spacecraft datasets. These techniques can be applied to three main spectral analysis problems: (1) correction for atmospheric effects for the recovery of surface emissivity; (2) isolation and separation of surface spectral endmembers for the characterization of surface mineralogy; and (3) determination of surface anisothermality for the retrieval of surface physical properties and correction for thermal emission in near-infrared spectral data. These modeling techniques have been extensively applied to martian and lunar spacecraft datasets, forming a basis for the retrieval of surface physical and compositional properties.

Acknowledgment

Joshua Bandfield (1974-2019) was a kind person, brilliant scientist, selfless mentor, and very dear friend. Through his scientific contributions and generous collegiality, he positively influenced the lives of many people in multiple corners of the planetary science community. His absence will be deeply felt by all who knew him, for many years to come. –D. R.

References

Baldridge A.M., Farmer J.D., & Moersch J.E. (2004) Mars remote-sensing analog studies in the Badwater Basin, Death Valley, California. *Journal of Geophysical Research*, **109**, E12006, 1–18.

Bandfield J.L. (2008) High-silica deposits of an aqueous origin in western Hellas Basin, Mars. *Geophysical Research Letters*, **35**, DOI:10.1029/2008GL033807.

Bandfield J.L. (2009) Effects of surface roughness and graybody emissivity on martian thermal infrared spectra. *Icarus*, **202**, 414–428.

Bandfield J.L. & Smith M.D. (2003) Multiple emission angle surface-atmosphere separations of thermal emission spectrometer data. *Icarus*, **161**, 47–65.

Bandfield J.L., Hamilton V.E., & Christensen P.R. (2000) A global view of martian surface compositions from MGS-TES. *Science*, **287**, 1626–1630.

Bandfield J.L., Edgett K.S., & Christensen P.R. (2002) Spectroscopic study of the Moses Lake dune field, Washington: Determination of compositional distributions and source lithologies. *Journal of Geophysical Research*, **107**, DOI:10.1029/2000JE001469.

Bandfield J.L., Glotch T.D., & Christensen P.R. (2003) Spectroscopic identification of carbonate minerals in the martian dust. *Science*, **301**, 1084–1087.

Bandfield J.L., Rogers D., Smith M.D., & Christensen P.R. (2004) Atmospheric correction and surface spectral unit mapping using Thermal Emission Imaging System data. *Journal of Geophysical Research*, **109**, DOI:10.1029/2004JE002289.

Bandfield J.L., Ghent R.R., Vasavada A.R., Paige D.A., Lawrence S.J., & Robinson M.S. (2011) Lunar surface rock abundance and regolith fines temperatures derived from LRO Diviner Radiometer data. *Journal of Geophysical Research*, **116**, DOI:10.1029/2011JE003866.

Bandfield J.L., Hayne P.O., Williams J.-P., Greenhagen B.T., & Paige D.A. (2015) Lunar surface roughness derived from LRO Diviner Radiometer observations. *Icarus*, **248**, 357–372.

Christensen P.R. (1986) The spatial distribution of rocks on Mars. *Icarus*, **68**, 217–238.

Christensen P.R., Bandfield J.L., Clark R.N., et al. (2000) Detection of crystalline hematite mineralization on Mars by the Thermal Emission Spectrometer: Evidence for near-surface water. *Journal of Geophysical Research*, **105**, 9623–9642.

Christensen P.R., Ruff S.W., Fergason R.L., et al. (2004) Initial results from the Mini-TES Experiment in Gusev crater from the Spirit rover. *Science*, **305**, 837–842.

Feely K.C. & Christensen P.R. (1999) Quantitative compositional analysis using thermal emission spectroscopy: Application to igneous and metamorphic rocks. *Journal of Geophysical Research*, **104**, 24,195–24,210.

Geladi P. & Kowalski B.R. (1986) Partial least-squares regression: A tutorial. *Analytica Chimica Acta*, **185**, 1–17.

Geminale A., Grassi D., Altieri F., et al. (2015) Removal of atmospheric features in near infrared spectra by means of principal component analysis and target transformation on Mars: I. Method. *Icarus*, **253**, 51–65.

Ghent R.R., Hayne P.O., Bandfield J.L., et al. (2014) Constraints on the recent rate of lunar ejecta breakdown and implications for crater ages. *Geology*, **42**, 1059–1062.

Gillespie A. (1992) Spectral mixture analysis of multispectral thermal infrared images. *Remote Sensing of Environment*, **42**, 137–145.

Glotch T.D. & Bandfield J.L. (2006) Determination and interpretation of surface and atmospheric Miniature Thermal Emission Spectrometer spectral end-members at the Meridiani Planum landing site. *Journal of Geophysical Research*, **111**, E12S06, 1507–1509.

Glotch T.D. & Rogers A.D. (2013) Evidence for magma-carbonate interaction beneath Syrtis Major, Mars. *Journal of Geophysical Research*, **118**, 126–137.

Glotch T.D., Christensen P.R., & Sharp T.G. (2006a) Fresnel modeling of hematite crystal surfaces and application to martian hematite spherules. *Icarus*, **181**, 408–418.

Glotch T.D., Bandfield J.L., Christensen P.R., et al. (2006b) Mineralogy of the light-toned outcrop at Meridiani Planum as seen by the Miniature Thermal Emission Spectrometer and implications for its formation. *Journal of Geophysical Research*, **111**, E12S03, DOI:10.1029/2005JE002672.

Golombek M., Huertas A., Kipp D., & Calef F. (2012) Detection and characterization of rocks and rock size-frequency distributions at the final four Mars Science Laboratory landing sites. *International Journal of Mars Science and Exploration*, **7**, 1–22.

Greenhagen B.T., Lucey P.G., Wyatt M.B., et al. (2010) Global silicate mineralogy of the Moon from the Diviner Lunar Radiometer. *Science*, **329**, 1507–1509.

Hamilton V.E. & Christensen P.R. (2000) Determining the modal mineralogy of mafic and ultramafic igneous rocks using thermal emission spectroscopy. *Journal of Geophysical Research*, **105**, 9717–9733.

Hamilton V.E. & Ruff S.W. (2012) Distribution and characteristics of Adirondack-class basalt as observed by Mini-TES in Gusev crater, Mars and its possible volcanic source. *Icarus*, **218**, 917–949.

Hamilton V.E., Christensen P.R., & McSween H.Y. (1997) Determination of martian meteorite lithologies and mineralogies using vibrational spectroscopy. *Journal of Geophysical Research*, **102**, 25593–25604.

Hecker C., Dilles J.H., van der Meijde M., & van der Meer F.D. (2012) Thermal infrared spectroscopy and partial least squares regression to determine mineral modes of granitoid rocks. *Geochemistry Geophysics Geosystems*, **13**, Q03021, DOI:10.1029/2011GC004004.

Huang J., Edwards C.S., Ruff S.W., Christensen P.R., & Xiao L. (2013) A new method for the semiquantitative determination of major rock-forming minerals with thermal infrared multispectral data: Application to THEMIS infrared data. *Journal of Geophysical Research*, **118**, 2146–2152.

Johnson J.R., Staid M.I., Titus T.N., & Becker K. (2006) Shocked plagioclase signatures in Thermal Emission Spectrometer data of Mars. *Icarus*, **180**, 60–74.

Lagerros J.S. (1998) Thermal physics of asteroids. IV. Thermal infrared beaming. *Astronomy and Astrophysics*, **332**, 1123–1132.

Lane M.D. & Christensen P.R. (1998) Thermal infrared emission spectroscopy of salt minerals predicted for Mars. *Icarus*, **135**, 528–536.

Lawson C.L. & Hanson R.J. (1974) *Solving least squares problems*. Prentice-Hall, Englewood Cliffs, NJ.

Li L. & Mustard J.F. (2003) Highland contamination in lunar mare soils: Improved mapping with multiple endmember spectral mixture analysis (MESMA). *Journal of Geophysical Research*, **108**, DOI:10.1029/2002JE001917.

Malinowski E.R. (1991) *Factor analysis in chemistry*, 2nd edn. John Wiley & Sons, New York.

Nowicki S. & Christensen P. (2007) Rock abundance on Mars from the thermal emission spectrometer. *Journal of Geophysical Research*, **112**, DOI:10.1029/2006JE002798.

Pan C., Rogers A., & Michalski J. (2015a) Thermal and near-infrared analyses of central peaks of martian impact craters: Evidence for a heterogeneous martian crust. *Journal of Geophysical Research*, **120**, 662–688.

Pan C., Rogers A., & Thorpe M. (2015b) Quantitative compositional analysis of sedimentary materials using thermal emission spectroscopy: 2. Application to compacted fine-grained mineral mixtures and assessment of applicability of partial least squares methods. *Journal of Geophysical Research*, **120**, 1984–2001.

Ramsey M.S. (2002) Ejecta distribution patterns at Meteor Crater, Arizona: On the applicability of lithologic endmember deconvolution for spaceborne thermal infrared data of Earth and Mars. *Journal of Geophysical Research*, **107**, DOI:10.1029/2001JE001827.

Ramsey M.S. & Christensen P.R. (1998) Mineral abundance determination: Quantitative deconvolution of thermal emission spectra. *Journal of Geophysical Research*, **103**, 577–596.

Roberts D.A., Gardner M., Church R., Ustin S., Scheer G., & Green R. (1998) Mapping chaparral in the Santa Monica Mountains using multiple endmember spectral mixture models. *Remote Sensing of Environment*, **65**, 267–279.

Rogers A. & Aharonson O. (2008) Mineralogical composition of sands in Meridiani Planum determined from Mars Exploration Rover data and comparison to orbital measurements. *Journal of Geophysical Research*, **113**, DOI:10.1029/2007JE002995.

Ruff S.W. & Christensen P.R. (2002) Bright and dark regions on Mars: Particle size and mineralogical characteristics based on Thermal Emission Spectrometer data. *Journal of Geophysical Research*, **107**, 5119, DOI:10.1029/2001JE001580.

Ruff S.W. & Hamilton V.E. (2017) Wishstone to Watchtower: Amorphous alteration of plagioclase-rich rocks in Gusev crater, Mars. *American Mineralogist*, **102**, 235–251.

Sinton W.M. (1981) The thermal emission spectrum of Io and a determination of the heat flux from its hot spots. *Journal of Geophysical Research*, **86**, 3122–3128.

Smith M.D., Bandfield J.L., & Christensen P.R. (2000) Separation of atmospheric and surface spectral features in Mars Global Surveyor Thermal Emission Spectrometer (TES) spectra. *Journal of Geophysical Research*, **105**, 9589–9607.

Spencer J.R. (1990) A rough-surface thermophysical model for airless planets. *Icarus*, **83**, 27–38.

Thomson J.L. & Salisbury J.W. (1993) The mid-infrared reflectance of mineral mixtures (7–14 µm). *Remote Sensing of Environment*, **45**, 1–13.

Thorpe M.T., Rogers A.D., Bristow T.F., & Pan C. (2015) Quantitative compositional analysis of sedimentary materials using thermal emission spectroscopy: 1. Application to sedimentary rocks. *Journal of Geophysical Research*, **120**, 1956–1983.

Wold S., Sjöström M., & Eriksson L. (2001) PLS-regression: A basic tool of chemometrics. *Chemometrics and Intelligent Laboratory Systems*, **58**, 109–130.

Wyatt M.B., Hamilton V.E., McSween H.Y., Christensen P.R., & Taylor L.A. (2001a) Analysis of terrestrial and martian volcanic compositions using thermal emission spectroscopy: 1. Determination of mineralogy, chemistry, and classification strategies. *Journal of Geophysical Research*, **106**, 14711–14732.

Wyatt M.B., Hamilton V.E., McSween H.Y., Jr., Christensen P.R., & Taylor L.A. (2001b) Analysis of terrestrial and martian volcanic compositions using thermal emission spectroscopy, 1. Determination of mineralogy, chemistry, and classification strategies. *Journal of Geophysical Research*, **106**, 14,711–14,732.

16

Geochemical Interpretations Using Multiple Remote Datasets

SUNITI KARUNATILLAKE, LYNN M. CARTER, HEATHER B. FRANZ,
LYDIA J. HALLIS, AND JOEL A. HUROWITZ

16.1 Introduction

Both Mars (e.g., Bell, 2008) and the Moon (e.g., Heiken et al., 1991; Jolliff et al., 2006) have sufficient remote sensing data to enable working across "regional," "local," and "in situ" lateral spatial scales at a variety of wavelengths in the electromagnetic spectrum. This chapter often uses Mars as an example because of the extent of complementary data available.

Most planetary studies today involve multiple types of data. Here we focus on several case studies in which the synthesis of the different observation types lead to unique inferences. The case studies include compositional comparisons between radar, infrared (IR), and optical observations; oxygen and iron isotopic variation analyses; molar ratio analyses of key major, minor, and aqueously mobile elements; chemical index of alteration (CIA) under variable water to rock volumetric ratios (W/R); and principal component analysis (PCA) with chemical and thermophysical observations.

16.2 Radar and Compositional Studies of Mare Serenitatis, the Moon

To date, the Moon is the only planetary object for which multiple wavelengths of high-resolution radar imaging can be compared with IR, UV, and optical data of the surface. Figure 16.1a is a P-band (70-cm wavelength) radar image of Mare Serenitatis. This image, acquired using Arecibo Observatory and the Green Bank Telescope (GBT), reveals morphological features of large lava flows, including channels and flow lobes (Campbell et al., 2009, 2014). Similar features have been mapped in Mare Nubium (Morgan et al., 2016; Carter et al., 2017). The radar wave can penetrate through regolith in places where the dielectric losses are low. Radar observations are described in more detail in Chapters 10 (theory) and 31 (applications to planetary exploration). Figure 16.1b is an S-band (12.6-cm wavelength) image of the same area, acquired using Arecibo Observatory and GBT. The lava flows are no longer visible; the shorter wavelength radar does not penetrate through as much regolith and also scatters from small surface and subsurface cm-sized rocks such as impact ejecta. A Clementine Ultraviolet/Visible

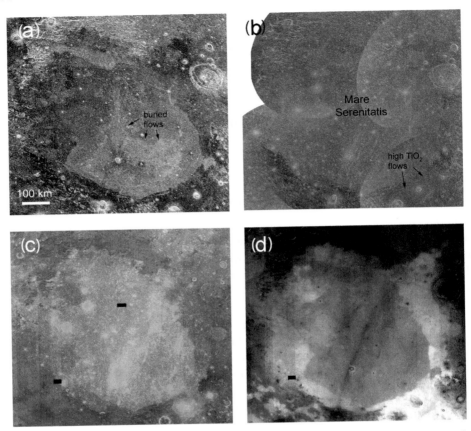

Figure 16.1 Remote sensing techniques used together can provide a more detailed and accurate view of surface composition, subsurface structure, and subsurface composition. As an example, multiple wavelengths of radar data plus IR and optical remote sensing reveal areas of differing composition and surface texture in Mare **Serenitatis.** (a) P-band (70-cm wavelength) radar image (Campbell et al., 2014). (b) S-band (12.6-cm wavelength) image (Carter et al., 2009). (c) Clementine UV/VIS data. (d) Derived TiO_2 image.

(UV/VIS) false color image of Mare Serenitatis (Figure 16.1c with red as the 750/415-nm ratio, green as the 750/950-nm ratio, and blue as the 415/750-nm ratio), illustrates compositional and maturity differences. Iron-rich, higher Ti basalt units are mapped in blue tones, whereas the low Ti units have a yellow-red color. The oldest highland regions appear red due to the presence of mature soils. A Clementine-data derived TiO_2 image using the method of Lucey et al. (2000) shows compositional differences among basalt mare units (Figure 16.1d).

The separate observations can be combined to provide deeper insights than those available from only one source. The lunar radar backscatter at 70 cm is greatest in broad areas with 2–3% derived TiO_2 abundance (Campbell et al., 2014). In both

P-band and S-band data, the radar backscatter is very low from high-TiO_2 basalts near the southern edge of Serenitatis, suggesting that the radar wave is strongly attenuated by TiO_2 (Campbell et al., 2014; Morgan et al., 2016). At shorter radar wavelengths (S-band image), the buried flows are invisible even in low-TiO_2 regions, which requires them to be buried below the sensing depth of the radar. Campbell et al. (2014) suggest that the reflectivity differences in low-TiO_2 regions of central Serenitatis are caused by small, 1–2%, variations in TiO_2 content of the upper part of the flows. For the lunar mare, radar may provide an alternative and precise TiO_2 discrimination method, allowing us to map additional flow compositional units. Returned samples provide much needed details for such remote sensing datasets; dielectric measurements of Apollo-collected basalts show that TiO_2 has large dielectric losses at the radar frequencies (Carrier et al., 1991).

16.3 Isotopic and Geochemical Characterization across Apollo Landing Sites, the Moon

The Moon also provides a geologic context for synthesizing isotopic and geochemical data, which, while sampled at in situ spatial scales, enables analyses that apply regionally, across Apollo 11, 12, 15, and 17 landing sites. These are proximal to Mare Tranquillitatis, Oceanus Procellarum, and Mare Serenitatis, respectively (Figure 16.2). Ultimately, the synthesis of oxygen and iron isotopic data with elemental trends yields insight into multiple mantle sources of basalts in these regions and an estimate of the oxygen and iron isotopic composition of the lunar mantle, which constrains models of lunar formation and differentiation (Hallis et al., 2010; Liu et al., 2010).

We consider trends in oxygen and iron isotopes of lunar basalts. Both oxygen and iron isotopes can provide valuable insight into geochemical processes on terrestrial planets. Isotope ratios for those elements are typically expressed as delta values, representing deviations in parts per thousand (i.e., per mil, or ‰) from the corresponding isotope ratios of relevant reference standards:

$$\delta^{18}O = \left[\frac{\left(^{18}O/^{16}O\right)_{sample}}{\left(^{18}O/^{16}O\right)_{reference}} - 1\right] \cdot 1000 \qquad (16.1)$$

and

$$\delta^{56}Fe = \left[\frac{\left(^{56}Fe/^{54}Fe\right)_{sample}}{\left(^{56}Fe/^{54}Fe\right)_{reference}} - 1\right] \cdot 1000, \qquad (16.2)$$

where the reference for oxygen is the Vienna Standard Mean Ocean Water (V-SMOW) and that for iron is established by the Institute for Reference Materials and Measurements (IRMM-014).

Figure 16.2 illustrates results of coordinated $\delta^{18}O$ and $\delta^{56}Fe$ measurements of multiple lunar basalts, where diamonds show data from Liu et al. (2010) and triangles show data from Spicuzzo et al. (2007). Note that Fe isotopic data are not available for every sample. Liu et al. (2010) used samples returned by four different Apollo missions, with landing sites as indicated in Figure 16.2a. On a plot of $\delta^{18}O$ versus Mg# (i.e., MgO/(MgO + FeO) as molar fractions), both low-Ti and high-Ti basalts display trends of increasing $\delta^{18}O$ with decreasing Mg#, consistent with crystal fractionation that favors incorporation of Mg over Fe in early precipitates (Figure 16.2b). Liu et al. (2010) interpreted trends in $\delta^{18}O$ as evidence for crystal fractionation rather than magma mixing or assimilation due to the

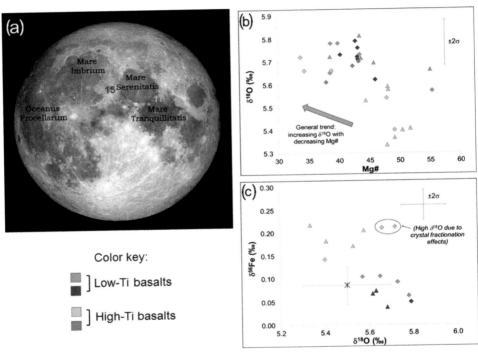

Figure 16.2 Trends in oxygen and iron isotopes of lunar basalts: diamonds show data from Liu et al. (2010) and triangles show data from Spicuzzo et al. (2007). (a) Landing sites of Apollo missions where samples were collected. (b) On a plot of $\delta^{18}O$ versus Mg#, both low-Ti and high-Ti basalts display trends of increasing $\delta^{18}O$ with decreasing Mg#, indicating crystal fractionation. Iron isotopes do not display such a trend, reflecting minimal iron isotopic fractionation expected in highly reducing lunar magmas. (c) Low-Ti and high-Ti basalts show distinct groupings in $\delta^{56}Fe$–$\delta^{18}O$ space, reflecting lower $\delta^{18}O$ of ilmenite compared to silicates. Two high-Ti basalt samples showed higher-than-typical $\delta^{18}O$ due to crystal fractionation effects, as indicated. Asterisk shows average lunar upper mantle composition calculated by Liu et al. (2010) from isotopic mass balance between low-Ti and high-Ti basalts, constrained by their estimated global proportions at the lunar surface from Clementine remote sensing data.

relationship between $\delta^{18}O$ and La/Sm that indicates a single mantle source within each subgroup.

Iron isotopes behave in an opposite manner, with ilmenite nominally characterized by heavier $\delta^{56}Fe$ than silicates (Craddock et al., 2010). However, iron isotopes in lunar basalts do not display trends analogous to those observed in $\delta^{18}O$, in contrast to observations of terrestrial ocean island basalts from Kilauea, for example. These display fractionations of up to 0.2‰ in $\delta^{56}Fe$ between early olivine crystals and relatively enriched residual melt (Teng et al., 2008; Liu et al., 2010). Liu et al. (2010) suggest that minimal iron isotopic fractionation in lunar basalts is consistent with highly reducing conditions expected for lunar magmas, as the Fe isotopic fractionation observed in terrestrial basalts is driven by preferential partitioning of heavy Fe into Fe^{3+} as opposed to more compatible Fe^{2+} during mantle melting (Dauphas et al., 2009). As shown in Figure 16.2c, low-Ti and high-Ti basalts show distinct groupings in $\delta^{56}Fe$-$\delta^{18}O$ space, reflecting lower $\delta^{18}O$ of ilmenite compared to silicates. Two high-Ti basalt samples indicated in the figure showed higher-than-typical $\delta^{18}O$ due to late-stage crystallization (Liu et al., 2010).

Liu et al. (2010) combined the isotopic data with Clementine remote sensing data, estimated the global proportions of low-Ti and high-Ti lunar basalts, and subsequently estimated the oxygen and iron isotopic composition of the lunar upper mantle. Analysis of Clementine UV/VIS camera data suggests that the lunar surface contains proportions of ~90% low-Ti basalt and ~10% high-Ti basalt (Giguere et al., 2000). From the average isotopic compositions for these groups of basalts and mass balance based on the Clementine proportions, Liu et al. (2010) calculated average $\delta^{18}O$ and $\delta^{56}Fe$ for the lunar upper mantle of 5.5 ± 0.2‰ and 0.085 ± 0.04‰, respectively. This composition is indicated by the asterisk in Figure 16.2c. Possible fractionation mechanisms proposed to explain these findings ultimately place constraints on the formation and evolution of the Moon (Liu et al., 2010).

16.4 Insight from the Chemical Index of Alteration, Mars

As with mixing and unmixing models, CIA (Figures 16.3a,b) also highlights the utility of chemical observations. Chemical weathering of silicate rocks via hydrolysis leads to an exchange of the cations Na^+, K^+, Ca^{2+}, and Mg^{2+} for H^+, as well as the possible loss of Si^{4+} (Kramer, 1968). Na^+, K^+, and Ca^{2+} are commonly supplied by the weathering of feldspar and volcanic glass. Mg^{2+} is derived from glasses, sheet silicates, and mafic minerals and resides in chloritic and smectitic clays (Nesbitt & Young, 1984; Pettijohn et al., 1987). The CIA is defined using molar fractions as

$$\text{CIA} = \left(Al_2O_3 / (Al_2O_3 + NaO + K_2O + CaO) \right) * 100,$$

where CaO represents the CaO content of silicate minerals only (Fedo et al., 1995). In general, hydrolytic weathering causes a progressive transformation of affected components into clay minerals, ultimately kaolinite. Kaolinite has a CIA value of 100 and represents the highest degree of weathering. The CIA for illite is between 75 and ~90, for muscovite it is at 75, and for feldspar it is at 50. Fresh basalts have values between 30 and 45, and fresh granites and granodiorites at 45–55 (Nesbitt & Young, 1982; Fedo et al., 1995) (Figure 16.3a).

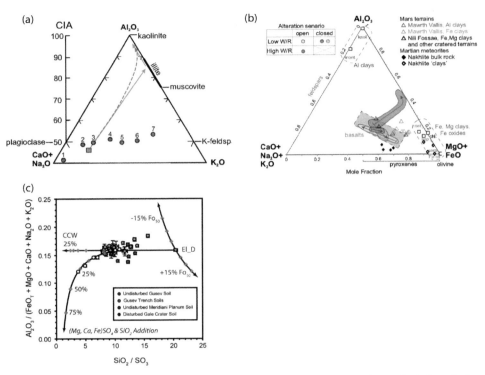

Figure 16.3 Chemical alteration trends in martian rocks and soils. (a) CIA analyses of whole-rock data provide a ratio of predominantly immobile Al_2O_3 versus the mobile cations Na^+, K^+, and Ca^{2+} given as oxides for igneous rocks. Gabbro (1), tonalite (2), granodiorite (3), granite (4), A-type granite (5), charnokite (6), and potassic granite (7) are represented by the red circles. The ideal weathering trends of terrestrial UCC-type source lithologies (blue arrows) would be parallel to the predicted weathering trend of Nesbitt and Young (1984). This is similar to the trend shown in green for Average Proterozoic Shale (Condie, 1993). Plot adapted from Arnaud et al. (2009). (b)The chemical and mineralogical changes observed during aqueous alteration of basalt are most relevant to the martian surface. In open, low W/R systems olivine dissolution and water evaporation (producing salts) are the two most important processes, shown by the yellow envelope (Hurowitz & McLennan, 2007). In closed, low-W/R systems (orange and green envelopes by Cann & Vine [1966] and Ehlmann et al. [2011a], respectively) primary minerals are replaced by Mg, Fe-rich clays and oxides, with products such as silica and zeolites precipitated in pore spaces. The Nakhlite basaltic martian meteorites contain both closed system (open diamonds) and open system alteration products (evaporates and salts). Plot adapted from Ehlmann et al. (2011a). (c) Plot of $Al_2O_3/(FeO_T + MgO + CaO + Na_2O + K_2O)$ vs. SiO_2/SO_3, with all data as mole percentages of the oxides. Trends are illustrated that mark the CCW process, the effects of olivine fractionation, and the addition of sulfate salts and silica, which influenced the chemical composition of subsurface ("trench") regolith analyzed by the Spirit rover at Gusev crater (from Hurowitz and Fischer, 2014).

For the lithologies of Earth's upper continental crust (UCC), ideal weathering trends would be parallel to the predicted weathering trend of Nesbitt and Young (1984), which is similar to the observed trend for Average Proterozoic Shale (Condie, 1993). Note that Figures 16.3a,b do not show the lower part of the diagram (Al_2O_3 <40). This low-Al region is where most martian lithologies, including the basaltic martian meteorites, would plot.

Specifically considering the CIA for primary igneous rocks, whole-rock data can be used to give a ratio of predominantly immobile Al_2O_3 vs. the mobile cations Na^+, K^+, Ca^{2+}, Mg^{2+}, and Fe^{2+} (Figure 16.3b). The chemical and mineralogical changes observed during aqueous alteration of basalt are most relevant to our martian surface case study. Under differing conditions, rocks of similar compositions can alter in chemically distinct ways. The two key parameters are the W/R and the closed or open nature of the system. In terrestrial rocks, high-W/R open system weathering leads to substantial chemical fractionation (purple envelope in Figure 16.3b) (Nesbitt & Wilson, 1992), with alkali and magnesium cations being removed, and primary minerals being replaced by Al-rich phyllosilicates and Fe oxides. Liberated cations may be redeposited as salts. Low-W/R alteration results in little chemical fractionation and transport, although the mineralogy of the rock is changed. In open, low-W/R systems olivine dissolution and water evaporation (producing salts) are the two most important processes (yellow envelope in Figure 16.3b). In closed, low-W/R systems (orange and green envelopes in Figure 16.3b) primary minerals are replaced by Mg, Fe-rich phyllosilicates and oxides, with products such as silica and zeolites precipitated in pore spaces.

With the exception of the Al-rich phyllosilicate bearing units of Mawrth Vallis (e.g., Bishop et al., 2008; Wray et al., 2008), phyllosilicate-bearing units on Mars detected via remote sensing lie approximately on the mixing line between basalt and Mg, Fe-rich phyllosilicates, indicating low-W/R system alteration (Ehlmann et al., 2011b). The open or closed nature of the units is more difficult to determine. The Fe, Mg-rich phyllosilicates of Nili Fossae contain wide compositional variation, roughly following the open system low-W/R compositional trend (Poulet et al., 2008; Ehlmann et al., 2011a). In contrast, Mawrth Vallis Fe-rich smectites show limited compositional variation (Bishop et al., 2008), indicative of a closed system (Ehlmann et al., 2011a).

16.5 Compositional Mixing and Unmixing Calculations, Mars

Next, we focus on the insight that results from relative variations in the ratios of oxides using in situ observations at Gusev crater, Mars (Figure 16.3c). Chapter 28 by Gellert and Yen discusses Alpha Particle X-ray Spectrometer (APXS) instrumentation and underlying principles needed to derive the in situ oxide abundances. The Gusev observations by the Spirit rover highlight weathering processes at low W/R, which strongly influence the chemical composition of the martian regolith (Hurowitz & McLennan, 2007; Hurowitz & Fischer, 2014). Both also contrast with terrestrial chemical weathering that occurs at several orders of magnitude higher W/R and many tens of K higher temperatures, except at hyperarid cryospheric locations such as the Atacama Desert or the Antarctic Dry Valleys.

Figure 16.3c highlights how mathematical mixing and unmixing calculations can shed light on some of the geochemical processes affecting regolith deposits on the surface of Mars. In this diagram, we seek to explain chemical and physical modification processes through mole ratios of $Al_2O_3/(FeO_T + MgO + CaO + Na_2O + K_2O)$ against SiO_2/SO_3. These processes have impacted the composition of soils from the Gusev crater, Meridiani Planum, and Gale crater landing sites. The former ratio is informative because it juxtaposes the behavior of an element (Al) that is relatively immobile during chemical weathering processes, against a suite of elements (Fe, Mg, Ca, Na, K) that can be relatively mobile during such processes. The ratio can therefore be used to evaluate changes in composition relative to an unweathered parent (or protolith) that occur as a result of chemical weathering. The SiO_2/SO_3 ratio tracks the degree to which S-bearing volatiles (e.g., gases, aerosols, water) have impacted the chemical composition of the regolith relative to an S-poor protolith composition. In this diagram, we model a "cation conservative weathering (CCW)" process in which Fo_{50} olivine (($Fe_{0.5}$, $Mg_{0.5}$)$_2SiO_4$) is dissolved by an S-bearing fluid (e.g., sulfuric acid), followed by the precipitation of divalent metal sulfates and silica, producing a horizontal trend on the diagram.

The regolith target "El Dorado" (El_D) is the starting composition for this calculation where for every mole of MgO, FeO, and SiO_2 subtracted, a mole of $MgSO_4$, $FeSO_4$, and SiO_2 was added back into the calculated residual chemical composition. El Dorado was part of a field of soil "ripples" encountered on Husband Hill in Gusev crater (Sullivan et al., 2008); it represents an endmember composition in SiO_2/SO_3 space as one of the least chemically weathered regolith samples encountered at either Mars Exploration Rover landing site (Ming et al., 2008; Morris et al., 2008). As shown, the CCW trend matches the horizontal scatter exhibited by most of the remaining regolith analyses, suggesting that a process of chemical alteration, promoted by small quantities of S-rich fluid, has played an important role in the chemical modification of soil compositions on Mars. The lack of physical fractionation of altered substrate from alteration product strongly suggests that this chemical alteration process took place at low W/R.

16.6 Characterizing Bulk Soil across Mars

Our PCA example combines regional scale chemical observations with computed areal abundance of surficial particles at up to few tens of μm grain size, which we term the Dust Cover Index (DCI) for brevity. While both chemical map and DCI data are regional to global in extent, they use fundamentally distinct observation methods: Mars Odyssey Gamma-Ray Spectrometer (GRS) for the former and thermal properties inferred from IR emission spectra for the latter. Chapters 9 and 30 describe the spectra-to-elemental abundance derivation. The general dominance of soil across the planet at regional scales also makes this case study representative of bulk soil processes (Karunatillake et al., 2014, 2016), despite the dramatic difference in sampling depth between the chemical observations (few dm) and IR spectra (tens of μm).

Figure 16.4a shows the second (PC2) versus first (PC1) principal components of multivariate data in mid-latitudinal martian soil. This PC combination incorporates chemical maps (i.e., Al, Ca, Cl, Fe, H, K, S, Si, and Th mass fractions with H reported

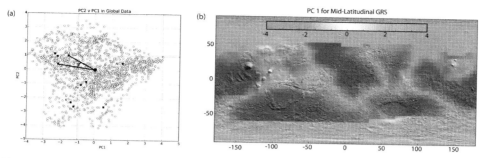

Figure 16.4 Principal components analyses for Mars. (a) Plot of the second versus first principal components of multivariate data in mid-latitudinal martian soil. The origin is marked by a black disk, while black squares show the axes of variables projected into the space of first (PC1) and second principal components (PC2). Circles show data projected onto the same space. (b) Variations in PC1 overlain on MOLA shaded relief.

stoichiometrically as H_2O) and the DCI from Ruff et al. (2002). Summarily, the DCI, computed as the average thermal emissivity value of the surface between 1350 and 1400 cm^{-1}, corresponds to the areal fraction covered by fine-grained silicates, generally finer than 40 µm (Putzig et al., 2005). The associated bivariate plot (Figure 16.4a) represents the synoptic insight possible from PCA, using just PC1 and PC2, which dominate by collectively accounting for ~45% of the total variance in the data.

Our plot allows the correlation of variables to be represented in multivariate space from pairwise angles, where smaller angles indicate stronger correlation. The example of H_2O and DCI highlights this with an angle of 14.8°, indicating a strong correlation (Figure 16.4a). This pairwise analysis also enables identifying groups of spatially coupled variables across complementary data such as chemical and thermophysical. In this particular case, Cl, H_2O, and S define a visually evident cluster. Fe and Ca form another distinct cluster. Also as a validation of the methodology, K and Th define a cluster, consistent with prior observations (Taylor et al., 2006, 2010).

The results enable us to establish consistency with – though not causation by – global scale soil processes, due in part to the areal dominance of soil at regional scales (e.g., Christensen, 1986; Nowicki & Christensen, 2007). For example, the cluster of volatiles standing apart from the DCI (Figure 16.4a) is consistent with chemical association between S and H in soil, likely as hydrous sulfates, as the primary influence on H variations. Adsorption on dust grains is disfavored. For geographic context, we also display variations in PC1 overlain on MOLA shaded relief (Figure 16.4b), comparable with observations by Gasnault et al. (2010). This illustrates a relationship between composition and topography on Mars.

16.7 Conclusions

In this chapter we summarize optimal analytical methods to synthesize complementary remote sensing observations for insight into geochemical processes, using Mars and the

Moon as example terrestrial bodies of the Solar System. The observation methods, including instruments and underlying principles, are described in corresponding chapters. Some of our case studies involve in situ and meteoritic data that provide necessary references to understand the trends seen from satellite observations.

The Moon provides the context to synthesize radar and mineralogical data in one instance, and isotopic-geochemical data in another. Insight from samples returned by the Apollo missions (see Chapter 19) (Figure 16.1) further highlight the utility of sample return missions to Solar System bodies. Mixing–unmixing trends, from in situ observations at Gusev crater, Mars, illustrates insight into geochemical processes. This includes olivine loss from either a physical (hydrodynamic) or a chemical (dissolution) removal mechanism. As an equally informative analysis, CIA can constrain the W/R, and the degree of isolation from the surroundings (i.e., open vs. closed system). Complementing such methods, PCA enables us to assess the likelihood of distinct compositional trends in multivariate data space. Such analyses can be advanced further pending the availability of future sample return from Mars and beyond. The continuation of complementary instrumentation on space missions, such as visible/near-IR/thermal IR and gamma-ray/neutron spectrometers, are essential for such progress.

Acknowledgments

The authors thank graduate student D. R. Hood (through a LaSPACE GSRA award) for assistance with the PCA results and undergraduate student D. L. Laczniak (through NASA's PGGURP program) for figure editing. Additional support was provided by NASA MDAP grant 80NSSC18K1375 and a Louisiana EPSCoR Research Award Program (RAP) grant LEqSF-EPS(2017)-RAP-22 to SK. Funding from the People Programme (Marie Curie Actions) of the European Union's Seventh Framework Programme (FP7/2007–2013) under REA grant agreement No. 624137 was provided to LH. JAH acknowledges support from NASA Award NNX13AR09G.

References

Bell J. (2008) *The martian surface: Composition, mineralogy, and physical properties.* Cambridge University Press, New York.

Bishop J.L., Noe Dobrea E.Z., McKeown N.K., et al. (2008) Phyllosilicate diversity and past aqueous activity revealed at Mawrth Vallis, Mars. *Science*, **321**, 830–833.Campbell B.A., Hawke B.R., Carter L.M., Ghent R.R., & Campbell D.B. (2009) Rugged lava flows on the Moon revealed by Earth-based radar. *Geophysical Research Letters*, **36**, 1–5.

Campbell B.A., Ray Hawke B., Morgan G.A., Carter L.M., Campbell D.B., & Nolan M. (2014) Improved discrimination of volcanic complexes, tectonic features, and regolith properties in Mare Serenitatis from Earth-based radar mapping. *Journal of Geophysical Research*, **119**, 313–330.

Cann J.R. & Vine F.J. (1966) An area on the crest of the Carlsberg Ridge: Petrology and magnetic survey. *Philosophical Transactions of the Royal Society of London A: Mathematical and Physical Sciences*, **259**, 198 LP–217.

Carrier W.D., Olhoeft G.R., & Mendell W. (1991) Physical properties of the lunar surface. In: *Lunar sourcebook: A user's guide to the Moon.* Cambridge University Press, New York, 475–594.

Carter L.M., Campbell B.A., Hawke B.R., Campbell D.B., & Nolan M.C. (2009) Radar remote sensing of pyroclastic deposits in the southern Mare Serenitatis and Mare Vaporum regions of the Moon. *Journal of Geophysical Research*, **114**, 2156–2202.

Carter L.M., Petro N.E., Campbell B.A., Baker D.M.H., & Morgan G.A. (2017) Earth-based radar and orbital remote sensing observations of mare basalt flows and pyroclastic deposits in Mare Nubium. *37th Lunar Planet. Sci. Conf.*, Abstract #1736.

Christensen P.R. (1986) The spatial distribution of rocks on Mars. *Icarus*, **68**, 217–238.

Condie K.C. (1993) Chemical composition and evolution of the upper continental crust: Contrasting results from surface samples and shales. *Chemical Geology*, **104**, 1–37.

Craddock P.R., Dauphas N., & Clayton R.N. (2010) Mineralogical control on iron isotopic fractionation during lunar differentiation and magmatism. *41st Lunar Planet. Sci. Conf.*, Abstract #1230.

Dauphas N., Pourmand A., & Teng F.-Z. (2009) Routine isotopic analysis of iron by HR-MC-ICPMS: How precise and how accurate? *Chemical Geology*, **267**, 175–184.

Ehlmann B.L., Mustard J.F., Clark R.N., Swayze G.A., & Murchie S.L. (2011a) Evidence for low-grade metamorphism, hydrothermal alteration, and diagenesis on Mars from phyllosilicate mineral assemblages. *Clays and Clay Minerals*, **59**, 359–377.

Ehlmann B.L., Mustard J.F., Murchie S.L., et al. (2011b) Subsurface water and clay mineral formation during the early history of Mars. *Nature*, **479**, 53–60.

Fedo C.M., Nesbitt H.W., & Young G.M. (1995) Unravelling the effects of potassium metasomatism in sedimentary rocks and paleosols, with implications for paleoweathering conditions and provenance. *Geology*, **23**, 921–924.

Gasnault O., Jeffrey Taylor G., Karunatillake S., et al. (2010) Quantitative geochemical mapping of martian elemental provinces. *Icarus*, **207**, 226–247.

Giguere T.A., Taylor G.J., Hawke B.R., & Lucey P.G. (2000) The titanium contents of lunar mare basalts. *Meteoritics and Planetary Science*, **35**, 193–200.

Hallis L.J., Anand M., Greenwood R.C., Miller M.F., Franchi I.A., & Russell S.S. (2010) The oxygen isotope composition, petrology and geochemistry of mare basalts: Evidence for large-scale compositional variation in the lunar mantle. *Geochimica et Cosmochimica Acta*, **74**, 6885–6899.

Heiken G.H., Vaniman D.T., & French B.M., eds. (1991) *Lunar sourcebook: A user's guide to the Moon*. Cambridge University Press, New York.

Hurowitz J.A. & Fischer W.W. (2014) Contrasting styles of water–rock interaction at the Mars Exploration Rover landing sites. *Geochimica et Cosmochimica Acta*, **127**, 25–38.

Hurowitz J.A. & McLennan S.M. (2007) A 3.5 Ga record of water-limited, acidic weathering conditions on Mars. *Earth and Planetary Science Letters*, **260**, 432–443.

Jolliff B.L., Wieczorek M.A., Shearer C.K., & Neal C.R., eds. (2006) *New views of the Moon*. Reviews in Mineralogy and Geochemistry Series, **60**. Mineralogical Society of America.

Karunatillake S., Wray J.J., Gasnault O., et al. (2014) Sulfates hydrating bulk soil in the martian low and middle latitudes. *Geophysical Research Letters*, **41**, 7987–7996.

Karunatillake S., Wray J.J., Gasnault O., et al. (2016) The association of hydrogen with sulfur on Mars across latitudes, longitudes, and compositional extremes. *Journal of Geophysical Research*, **121**, 1–29.

Kramer J.R. (1968) Mineral-water equilibria in silicate weathering. *International Geological Congress, 23rd session*, 149–160.

Liu Y., Spicuzza M.J., Craddock P.R., et al. (2010) Oxygen and iron isotope constraints on near-surface fractionation effects and the composition of lunar mare basalt source regions. *Geochimica et Cosmochimica Acta*, **74**, 6249–6262.

Lucey P.G., Blewett D.T., & Jolliff B.L. (2000) Lunar iron and titanium abundance algorithms based on final processing of Clementine ultraviolet-visible images. *Journal of Geophysical Research*, **105**, 20297–20305.

Ming D.W., Gellert R., Morris R.V., et al. (2008) Geochemical properties of rocks and soils in Gusev crater, Mars: Results of the Alpha Particle X-Ray Spectrometer from Cumberland Ridge to Home Plate. *Journal of Geophysical Research*, **113**, E12S39, DOI:10.1029/2008JE003195.

Morgan G.A., Campbell B.A., Campbell D.B., & Hawke B.R. (2016) Investigating the stratigraphy of Mare Imbrium flow emplacement with Earth-based radar. *Journal of Geophysical Research*, **121**, 1498–1513.

Morris R.V., Klingelhöfer G., Schröder C., et al. (2008) Iron mineralogy and aqueous alteration from Husband Hill through Home Plate at Gusev crater, Mars: Results from the Mössbauer instrument on the Spirit Mars Exploration Rover. *Journal of Geophysical Research*, **113**, E12S42, DOI:10.1029/2008JE003201.

Nesbitt H.W. & Wilson R.E. (1992) Recent chemical weathering of basalts. *American Journal of Science*, **292**, 740–777.

Nesbitt H.W. & Young G.M. (1982) Early Proterozoic climates and plate motions inferred from major element chemistry of lutites. *Nature*, **299**, 715–717.

Nesbitt H.W. & Young G.M. (1984) Prediction of some weathering trends of plutonic and volcanic rocks based on thermodynamic and kinetic considerations. *Geochimica et Cosmochimica Acta*, **48**, 1523–1534.

Nowicki S.A. & Christensen P.R. (2007) Rock abundance on Mars from the Thermal Emission Spectrometer. *Journal of Geophysical Research*, **112**, E05007, DOI:10.1029/2006JE002798.

Pettijohn F.J., Potter P.E., & Siever R. (1987) *Sand and sandstone*. Springer-Verlag, New York.

Poulet F., Mangold N., Loizeau D., et al. (2008) Abundance of minerals in the phyllosilicate-rich units on Mars. *Astronomy and Astrophysics*, **487**, L41–U193, DOI:10.1051/0004-6361:200810150.

Putzig N., Mellon M., Kretke K., & Arvidson R. (2005) Global thermal inertia and surface properties of Mars from the MGS mapping mission. *Icarus*, **173**, 325–341.

Ruff S.W. & Christensen P.R. (2002) Bright and dark regions on Mars: Particle size and mineralogical characteristics based on Thermal Emission Spectrometer data. *Journal of Geophysical Research*, **107**, 5127, DOI:10.1029/2001JE001580.

Spicuzza M.J., Day J.M.D., Taylor L.A., & Valley J.W. (2007) Oxygen isotope constraints on the origin and differentiation of the Moon. *Earth and Planetary Science Letters*, **253**(1–2), 254–265.

Sullivan R., Arvidson R., Bell J.F., et al. (2008) Wind-driven particle mobility on Mars: Insights from Mars Exploration Rover observations at "El Dorado" and surroundings at Gusev crater. *Journal of Geophysical Research*, **113**, E06S07, DOI:10.1029/2008JE003101.

Taylor G.J., Boynton W.V., Brückner J., et al. (2006) Bulk composition and early differentiation of Mars. *Journal of Geophysical Research*, **112**. E03S10, DOI:10.1029/2005JE002645.

Taylor G.J., Martel L.M.V., Karunatillake S., Gasnault O., & Boynton W.V. (2010) Mapping Mars geochemically. *Geology*, **38**, 183–186.

Teng F.-Z., Dauphas N., & Helz R.T. (2008) Iron isotope fractionation during magmatic differentiation in Kilauea Iki Lava Lake. *Science*, **320**, 1620–1622.

Wray J.J., Ehlmann B.L., Squyres S.W., Mustard J.F., & Kirk R.L. (2008) Compositional stratigraphy of clay-bearing layered deposits at Mawrth Vallis, Mars. *Geophysical Research Letters*, **35**, L12202, DOI:10.1029/2008GL034385.

Part IV

Applications to Planetary Surfaces

17

Spectral Analyses of Mercury

SCOTT L. MURCHIE, NOAM R. IZENBERG, AND RACHEL L. KLIMA

17.1 Introduction to Spectral Reflectance of Mercury

Spectral reflectance has been a primary tool for investigating the composition and geology of Mercury. In the 1970s, Mariner 10 images provided the first moderate-resolution views of the surface, revealing the heavily cratered plains and color differences associated with different geologic units. Over the next four decades telescopic reflectance spectra at ultraviolet (UV) through short-wave infrared (SWIR) wavelengths (~0.1–2.5 µm) revealed a featureless reflectance spectrum lacking evidence for absorptions due to ferrous iron in silicate minerals, though thermal emission spectra suggested the presence of various silicate phases. The most comprehensive UV through SWIR spectral measurements have been acquired by the Mercury Dual Imaging System (MDIS) wide-angle camera (WAC) and the Mercury Atmospheric and Surface Composition Spectrometer (MASCS) Visible and Infrared Spectrograph (VIRS) and Ultraviolet and Visible Spectrometer (UVVS) on the MESSENGER spacecraft. From mid-UV through SWIR wavelengths (0.2–2.5 µm), reflection of sunlight by a silicate planetary regolith like that of Mercury is controlled by the partitioning of radiation between absorbance and reflectance, and by the scattering of reflected light in different directions. The behavior of each factor provides information on the composition and texture of the surface (Chapter 2).

Key variables that affect the balance of absorbance and reflectance include the amount and oxidation state of transition metals, principally iron, in silicates and oxides; the presence of opaque components, including ilmenite and carbon-bearing phases; and the degree of space weathering, that is, modification of the optical properties of the surface by sustained interaction with the space environment. Ferrous iron in minerals creates crystal-field absorptions near 0.85–1.05 µm wavelength that are particularly important in distinguishing key Fe-bearing phases, including oxides, olivine, pyroxene, and glasses (Chapters 1 and 4). Opaque minerals typically darken and flatten the reflectance spectrum of a mixture in which they occur. Space weathering occurs on the Moon (Chapter 18) and near-Earth asteroids (Chapter 19), and is widely thought to occur on Mercury, by the formation of submicroscopic amorphous rims on regolith grains, which contain nm-scale inclusions of Fe or FeS formed from impact-generated vapor and solar wind sputtering (e.g., Hapke et al., 1975; Pieters, 1993; Hapke, 2001; Gaffey, 2010; Domingue et al., 2014). Space weathering on those bodies has the effect of darkening and reddening (i.e., steepening the slope of reflectance vs. wavelength in) silicate spectra at visible to SWIR wavelengths (McCord & Adams, 1972a,b; Fischer & Pieters, 1994) and brightening and bluing (i.e., reducing the slope of) the spectrum in the UV (Hendrix & Vilas, 2006; Hendrix et al., 2012). Both opaque minerals and space weathering subdue mineralogical absorptions.

The manner in which a surface scatters reflected light in different directions is known as the surface's photometric behavior. The key angles defining the scattering geometry are incidence angle i (between incoming sunlight and the surface normal), emergence angle e (between outgoing scattered light and the surface normal), and phase angle g (between incoming and outgoing rays). A substantial literature exists on photometric functions of planetary regoliths and is reviewed by Domingue et al. (2016). That subject is not treated in detail here; rather, photometric modeling is accepted as a critical tool to provide a photometric normalization of image and spectral data acquired at differing geometries to a common illumination and observation geometry. This normalization is essential for construction of mosaics in which variations in reflectance ideally result only from inherent differences in surface properties. Similarly, normalization to an illumination geometry like that of laboratory measurements is critical to interpreting composition in the context of analog materials. The most recent normalization of MESSENGER data described in this chapter is that of Domingue et al. (2016).

17.1.1 The Moon as an Analog to Mercury

Prior to MESSENGER, the Moon had widely been assumed to foreshadow Mercury's surface composition. On the Moon (Chapter 18), there are three major causes for variations in spectral reflectance: (1) the modal abundance of iron-containing pyroxene, which varies between the basaltic maria and anorthositic highlands, resulting in a lower reflectance and stronger 1-μm absorption in the maria (Adams & McCord, 1970; Pieters, 1993); (2) variations in UV to visible reflectance and spectral slope due to differences in the abundance of the opaque mineral ilmenite ($FeTiO_3$) (Charette et al., 1974; Lucey et al., 1995, 1998; Blewett et al., 1997a); and (3) "optical maturation" of regolith by space weathering (e.g., Fischer & Pieters, 1994; Lucey et al., 1995, 1998, 2000; Blewett et al., 1997a; Hapke, 2001). As explained in the text that follows, the sources of spectral heterogeneity on Mercury are quite different than those on the Moon, except possibly for space weathering: opaque minerals have a dominant effect, and differences in major element composition have little spectral effect because Mercury's silicates lack a ferrous iron component.

17.1.2 Mariner 10 and Earth-Based Studies

Prior to MESSENGER, Earth-based telescopic studies were the main constraint on mineralogy of mercurian surface material. The UV-SWIR spectrum is red sloped with no 1-μm iron crystal field absorption detectable above the level of noise in measurements (e.g., Vilas

& McCord, 1976; McCord & Clark, 1979; Vilas et al., 1984). In contrast, on the Moon, a 1-μm absorption occurs in both mare and most highland materials (Chapter 18). Mercury's lack of a 1-μm feature had been interpreted to indicate that the amount of ferrous iron in surface silicates is much lower than in most lunar soils, as low as <2 wt.% (McCord & Adams, 1972a; Hapke, 1977; McCord & Clark, 1979; Blewett et al., 1997b; Warell & Blewett, 2004; Warell et al., 2006). The lack of a 1-μm feature required pre-MESSENGER mineralogical interpretations to be made based on albedo at UV-SWIR wavelengths and mid- to thermal-IR emission spectra (Vilas & McCord, 1976; Vilas et al., 1984; Sprague et al., 1994, 1995, 2002; Emery et al., 1998; Warell et al., 2006). Although most hypothesized compositions were low in iron, the proposed silicate mineralogies ranged from plagioclase-rich through pyroxene-rich.

The Mariner 10 imaging system (Murray et al., 1974) acquired images covering about 40% of Mercury's surface through 0.355- and 0.575-μm wavelength filters using a vidicon detector. At this wavelength range, spectral slope and reflectance can be used to estimate content of opaque minerals and differences in optical maturity (Lucey et al., 1998, 2000). However, the Mariner 10 image data do not cover the 1-μm absorption, precluding additional constraints on silicate composition. Difficulties in calibrating Mariner 10 images also resulted in an initial focus on qualitative analyses (e.g., Hapke et al., 1975; Rava & Hapke, 1987). In the early 1990s, Mariner 10 images of Mercury were recalibrated (Robinson & Lucey, 1997), yielding new findings about Mercury's surface: (1) there are two spectral trends, one inferred to result from differences in opaque content and the other from differences in optical maturity, and (2) craters and basins exhume materials with both higher and lower opaque mineral contents, indicating vertical heterogeneity in composition (Robinson & Lucey, 1997; Robinson & Taylor, 2001; Blewett et al., 1997b).

17.1.3 Spectral Measurements by MESSENGER

The MESSENGER spacecraft (Solomon et al., 2001) measured Mercury's spectral reflectance at UV-SWIR wavelengths using three instruments. The MDIS/WAC (Hawkins et al., 2007) sampled the spectrum through 11 discrete filters centered at 0.43–1.02 μm (Figure 11 in Hawkins et al., 2007; see also Figure 17.4a), with a pixel instantaneous field of view (IFOV) of 179 μrad and a $10.5° \times 10.5°$ field of view (FOV). MDIS was mounted on a pivot, enabling nadir pointing and data to be collected in Mercury orbit at low emergence and phase angles, with a goal of obtaining images at the lowest possible solar incidence angle. MASCS/VIRS (McClintock & Lankton, 2007) was a point spectrometer with a 400-μrad (0.023°) IFOV sampling the wavelength range 0.3–1.44 μm at 5 nm per channel. VIRS was fixed to the spacecraft and constrained by spacecraft pointing restrictions to view the surface at phase angles of 78° to 102°. MASCS/UVVS was a separate point spectrometer coaligned with VIRS, but with distinct operating modes for the surface and atmosphere. The surface mode had a $0.05° \times 0.04°$ IFOV with detectors sampling the wavelength in ranges 0.115–0.19 μm at 0.3 nm channel^{-1}, 0.16–0.32 μm at 0.7 nm channel^{-1}, and 0.25–0.6 μm at 0.6 nm channel^{-1}.

Three MESSENGER flybys of Mercury in 2008–09 provided >90% coverage of the surface by MDIS/WAC with 11-color multispectral imaging at a few km per pixel, plus sparse measurements of representative surface units by MASCS/VIRS. The flyby imaging showed that Mercury's surface contains widespread, smooth volcanic plains (Head et al., 2008, 2009; Murchie et al., 2008; Denevi et al., 2009) that vary in spectral character from lower reflectance and a less red spectral slope through higher reflectance with a redder slope. Reflectance of optically mature surfaces at 0.75 μm (normalized to $i = 30°$, $e = 0°$, $g = 30°$) ranges from ~0.8 to 1.25 that of the average surface (Denevi et al., 2009). Both MDIS and MASCS/VIRS data confirmed absence of a 1-μm crystal field absorption, and thus a low ferrous iron content in silicates (McClintock et al., 2008; Robinson et al., 2008; Blewett et al., 2009).

During operations in orbit about Mercury from March 2011 through April 2015, three MDIS multispectral mapping campaigns were conducted (Denevi et al., 2018): a global 8-color map at ~1000 m pixel^{-1} (Figure 17.1; Denevi et al., 2018), a 3-color map of northern and equatorial latitudes at an improved spatial sampling of ~400 m pixel^{-1}, and a 5-color map of the northern smooth plains at a uniform phase angle to search for subtle spectral units within the very obliquely illuminated plains. In addition, 11-color image sets were specifically targeted at selected features. The 8-color map's spectral sampling every ~110 nm could detect and map Fe^{2+}-bearing olivine, pyroxene, or glasses if Fe^{2+} is present locally in sufficient quantity.

Globally distributed mapping with MASCS/VIRS was accomplished by measuring along the track of the instrument FOV formed by the spacecraft's orbital motion, with

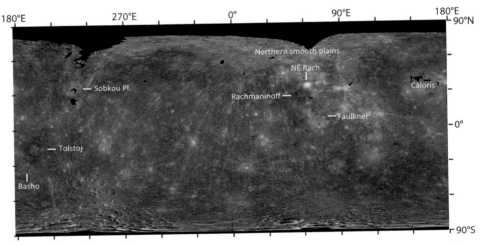

Figure 17.1 Synoptic view of MDIS/WAC normalized spectral reflectance of Mercury, showing parameterized spectral variations and illumination and viewing geometries in simple cylindrical projection. Red (R), green (G), and blue (B) image planes are R = principal component 2 (PC2), stretched 0.01–0.045; G = principal component 1 (PC1), stretched 0.072–0.372; and B = 0.43-μm/0.99-μm reflectance ratio, stretched 0.349–0.717. Key geographic features are indicated. Figure from Murchie et al. (2019), used with permission from Cambridge University Press.

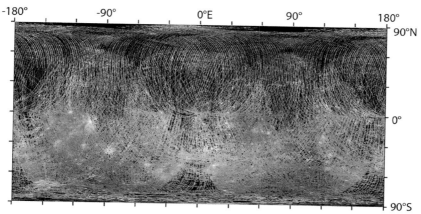

Figure 17.2 Synoptic view of MASCS/VIRS normalized spectral reflectance of Mercury showing parameterized spectral variations in simple cylindrical projection. RGB image with R = reflectance at 0.575 μm, stretched 0.03–0.10; G = 0.415-μm/0.75-μm reflectance ratio, stretched 0.48–0.65; and B = 0.31-μm/0.39-μm reflectance ratio, stretched 0.57–0.71. Figure from Murchie et al. (2019), used with permission from Cambridge University Press.

spectra covering all wavelengths simultaneously collected typically at a 1-s cadence. Pointing perpendicular to the tracks was offset in some orbits to distribute coverage as evenly as possible (Figure 17.2). In addition, specific features of interest were targeted with single profiles or small raster scans. Key results of MASCS/VIRS global mapping were described by Izenberg et al. (2014).

MASCS/UVVS measurements were acquired from Mercury orbit separately from VIRS spectra, because of inherent design differences between the two spectrometers. The UVVS scanned its grating to build wavelength coverage, so that measurements of differing wavelengths were nonsimultaneous. To ensure that all wavelengths of a spectrum sampled the same region of the surface, while the grating scanned through wavelengths, the MESSENGER spacecraft pointed and tracked a fixed point on the surface, and collected spectra of a grid of such points over equatorial and southern latitudes (see Figure 17.3).

17.2 Mercury's Reflectance Compared with Other Bodies

Some aspects of Mercury's surface composition may be inferred by comparing its normal albedo – reflectance normalized to $i = 0°$, $e = 0°$, $g = 0°$ – with that of other planetary bodies at 0.55–0.60 μm (Murchie et al., 2015, 2019). Mercury's normal albedo is intermediate to those of surfaces dominated optically by ferrous iron–containing silicates (e.g., 4 Vesta, 433 Eros, the Moon) and those of low-albedo C- and D-type asteroids (Chapter 19). C- and D-type asteroids are both low in albedo; C-types are spectrally neutral, displaying little slope at visible wavelengths, whereas D-types are strongly red sloped. Both asteroid types

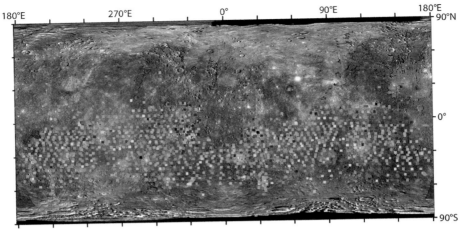

Figure 17.3 Synoptic view of MASCS/UVVS grid of spectral reflectance measurements of Mercury in a simple cylindrical projection. Each measurement was taken with all wavelengths measuring the same geographic location. The color scale represents the ratio of reflectances at 0.22 μm and 0.29 μm, on a rainbow scale where violet to red represents values from 0.25 to 0.45. The base is MDIS reflectance normalized to $i = 30°$, $e = 0°$, $g = 30°$. Figure from Murchie et al. (2019), used with permission from Cambridge University Press.

are thought to be darkened by up to several or more percent opaque phases, including magnetite and carbon-containing phases such as organics and graphite, which are present in carbonaceous chondrite meteorites (e.g., Cloutis et al., 2011).

On average, the normal albedo of Mercury is lower than that of the lunar nearside (Denevi & Robinson, 2008) and approximates that of the maria, although ferrous iron in silicates does not darken Mercury's surface (Domingue et al., 2014). This comparison indicates the occurrence of one or more pervasive opaque phases on Mercury that are either absent from or much less abundant on the Moon. Using Mariner 10 data, Denevi and Robinson (2008) and Braden and Robinson (2013) compared average reflectance properties of the two bodies as well as those of fresh crater ray materials, which are least affected by space weathering. Mercury is ~15% lower in reflectance than the lunar nearside, and mercurian fresh crater ray material is ~45% lower in reflectance than fresh crater ray material in the lunar highlands, further supporting a greater abundance of opaque minerals in mercurian than lunar surface material.

17.3 Mercury's Spectral Properties and Spectral Variability

17.3.1 Spectral Units and Relationship to Morphologic Features

The surface of Mercury is dominated by three geomorphic units: smooth plains, intercrater plains, and the rims, walls, and ejecta of impact craters and basins (Trask & Guest, 1975;

Denevi et al., 2013a,b; Whitten et al., 2014). MDIS multispectral imaging and MASCS/VIRS spectral measurements show that variations in reflectance and spectral slope divide those geomorphic units into four spatially extensive spectral units (Robinson et al., 2008; Denevi et al., 2009; Murchie et al., 2015, 2019), whose type spectra are shown in Figure 17.4a, and type locations are described by Murchie et al. (2019). Three of the spectral units form plains: low-reflectance blue plains (LBP), intermediate plains (IP), and high-reflectance red plains (HRP), and occur in both the smooth and intercrater plains morphologic units. The fourth spectral unit, low-reflectance material (LRM; Robinson et al., 2008), is darker than LBP and is concentrated in floors, rims, and ejecta of impact craters and basins; it also occurs as numerous small patches in the most heavily cratered regions (Klima et al., 2016; Denevi et al., 2019). (Acronyms for Mercury's spectral units and other terms are summarized in Table 17.1.) A broad, shallow absorption-like upward curvature in the spectrum of LRM, centered near 0.6 µm, is particularly evident when the spectrum is ratioed to spectrally bland HRP (Figures 17.4b,c). This curvature originally led to an interpretation that LRM may contain a large fraction of ilmenite (which has such a spectral feature) as on the Moon (Robinson et al., 2008; Denevi et al., 2009). Subsequently, after MESSENGER's X-ray and gamma-ray spectrometers found that Ti content of Mercury's surface is too low for large amounts of ilmenite to be present (Nittler et al., 2011; Weider et al., 2012, 2014; Evans et al., 2012; Murchie et al., 2015), the 0.6-µm feature was reinterpreted as originating from graphite (which has a similar spectral feature), a mixture of grain sizes of metallic iron, or some combination of the two (Murchie et al., 2015). MESSENGER elemental abundance measurements support graphite as the source of this feature (Peplowski et al., 2016), though metallic iron may occur locally and/or in low abundances.

Three additional high-reflectance spectral units have limited geographic distributions (Figure 17.4). (1) Fresh crater materials are brighter and less red than plains or LRM, consistent with having experienced less space weathering (Robinson et al., 2008; Blewett et al., 2009). (2) Bright materials within and forming haloes around the central depressions of hollows (Blewett et al., 2011) are less red and brighter than surrounding materials. Most hollows are closely associated with LRM though they also occur in the red unit (Blewett et al., 2013; Thomas et al., 2014a; Izenberg et al., 2015). Subtle absorption features seen in bright hollow materials at Dominici and Hopper Craters have been interpreted to originate from MgS or graphite (Vilas et al., 2016). (3) A "red unit" consisting of pyroclastic deposits (Robinson et al., 2008; Goudge et al., 2014) is brighter and slightly redder than HRP. The red unit (as well as hollows formed on it) have a much more prominent falloff in reflectance below 0.4 µm than do all other materials (see Figure 8.4c), consistent with the edge of a more well-defined oxygen–metal charge transfer (OMCT) band that results from an oxidized transition metal in silicates. Given measured elemental abundances on Mercury, the OMCT band most likely results from a small concentration of Fe^{2+} in one or more silicate minerals (McClintock et al., 2008; Goudge et al., 2014; Izenberg et al., 2014). The abundance of Fe^{2+} required to create the observed moderate OMCT band, possibly 0.1 wt.% or less (Cloutis et al., 2008), is much lower than the ≥1 wt.% needed to create a 1-µm absorption detectable above the noise in MASCS and MDIS data (Klima et al., 2011).

Table 17.1 *Definition of Mercury- and MESSENGER-related acronyms*

Acronym	Term
MESSENGER	MErcury Surface, Space ENvironment, GEochemistry, and Ranging
MDIS	Mercury Dual Imaging System
WAC	Wide-Angle Camera
MASCS	Mercury Atmospheric and Surface Composition Spectrometer
VIRS	Visible and Infrared Spectrograph
UVVS	Ultraviolet and Visible Spectrometer
LRM	Low-Reflectance Material
LBP	Low-reflectance Blue Plains
IP	Intermediate Plains
HRP	High-reflectance Red Plains
OMCT	Oxygen–Metal Charge Transfer
UV	Ultraviolet
SWIR	Short-Wave Infrared

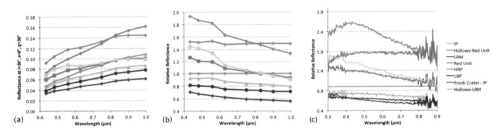

Figure 17.4 Type spectra of major and minor spectral units measured by MASCS/VIRS and MDIS/WAC. Unit nomenclature follows the conventions of Murchie et al. (2015). Locations of the type areas are given in Table 8.2 of Murchie et al. (2015). (a) MDIS spectra corrected to $i = 30°$, $e = 0°$, $g = 30°$. Symbols show the effective centers of the MDIS filters. (b) The same spectra ratioed to HRP to highlight differences. (c) MASCS/VIRS spectra ratioed to HRP to highlight differences. Note: The standard photometric correction for MASCS/VIRS data is to $i = 45°$, $e = 45°$, $g = 90°$, which is within the range of photometric angles actually sampled by the data set; those data are not routinely shown on the same scale as MDIS spectra normalized to $i = 30°$, $e = 0°$, $g = 30°$. Figure from Murchie et al. (2019), used with permission from Cambridge University Press.

MASCS/UVVS spectra of most spectral units are compared in Figure 17.5 with MASCS/VIRS spectra of the same spots. UVVS spectra reproduce the differences in relative strengths of a UV turndown into the OMCT band found in VIRS data. The red unit exhibits the strongest turndown and a relatively well-defined band minimum; HRP, IP, and LBP lack well defined band minima (Maxwell et al., 2016). Mid-UV reflectance in LRM is higher than in IP or LBP, an important observation to constrain the opaque phase in LRM (Section 17.4.3).

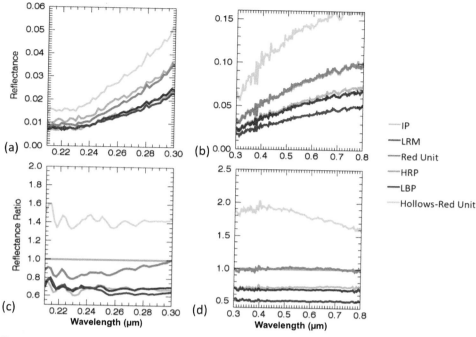

Figure 17.5 Representative spectra of all major and most minor spectral units measured by MASCS/UVVS and MASCS/VIRS. Unit nomenclature follows the conventions of Murchie et al. (2015). Locations of the areas whose spectra are shown are given in Table 8.3 of Murchie et al. (2015). Locations of these spectra are not type locations, so spectral contrast is less than in Figure 17.4, for example, the lesser UV turndown. (a) UVVS spectra corrected to $i = 45°$, $e = 45°$, $g = 90°$. (b) VIRS spectra of the same locations corrected similarly. (c) UVVS spectra ratioed to the spectrum of HRP to highlight differences. (d) VIRS spectra ratioed to the spectrum of HRP. Note that the spots sampled by the two spectrometers were measured at different times and geometries, so data may disagree slightly at the overlap wavelength of 0.3 μm.

17.3.2 Primary and Secondary Spectral Trends

Mercury's spectral variations define two major spectral trends, each characterized by variations in spectral slope and reflectance at UV to SWIR wavelengths (Figure 17.6a). These trends were revealed with the assistance of principal components analysis, in which data are transformed into the coordinate system of the eigenvectors of the data set. The first principal component, PC1, corresponds to the first eigenvector and represents overall variations in reflectance. The primary spectral trend, defined by PC2, is between the endmembers LRM (having a low value of PC2) and the red unit (having a high value of PC2), with LBP, IP, and HRP intermediate to the endmembers and becoming respectively more like the red unit (Murchie et al., 2015, 2019; Klima et al., 2018). In the 0.4–1.0 μm wavelength range, the trend is between lower reflectance and shallower spectral slope in LRM, and a higher

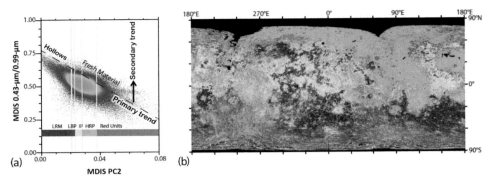

Figure 17.6 Principle Components Analyses. (a) Scatterplot of MDIS PC2 versus 0.43-μm/0.99-μm reflectance ratio, color coded by density of points and with coordinates given by the values of PC2 and ratio of reflectances measured at 0.43 μm and 0.99 μm. Warmer colors have more points in the global map. Approximate PC2 boundaries of the major spectral units forming the primary trend are shown. The PC2 values in spectral units are different from those listed by Murchie et al. (2015) as a result of updated data calibration. The primary and secondary spectral trends are indicated; fresh crater materials and hollows are offset vertically from the primary trend and plot above the dashed line. (b) Approximate geographic distribution of spectral units forming the primary trend, shown on a base map of MDIS reflectance at 0.75 μm. Colors correspond to those in the scatterplot. Figure from Murchie et al. (2019), used with permission from Cambridge University Press.

reflectance and slightly steeper spectral slope in the red unit. Regions of LRM with the most well-defined 0.6-μm absorption-like feature have the lowest PC2 values (Murchie et al., 2015, 2019; Klima et al., 2018). In MASCS/VIRS data, a similar trend is defined by a 0.31-μm/0.39-μm reflectance ratio, which parameterizes the UV turndown into the OMCT band: lower values represent a stronger band and higher values a weaker band (Izenberg et al., 2014). The red unit has the lowest UV reflectance ratio and the most well-defined OMCT band, and LRM has the highest UV reflectance ratio and least well-defined OMCT band.

The secondary spectral trend is characterized by increasing reflectance and a progressively less red ("bluer") spectral slope, with the spectral slope defined in MDIS data by higher values of the 0.43-μm/0.99-μm reflectance ratio (Murchie et al., 2015), and in MASCS/VIRS data by a higher 0.415-μm/0.75-μm reflectance ratio (Izenberg et al., 2014). For the most part, higher reflectance ratios correspond with optically immature fresh crater materials, and low reflectance ratios with mature, space-weathered materials. In Figure 17.6a, fresh crater materials are offset positively in the plot; those that plot above the dashed line are the freshest materials. Hollows are also offset from their background terrains along the secondary spectral trend (Murchie et al., 2015). Izenberg et al. (2015) noted that where hollows occur in the red unit, they retain the red unit's low UV 0.31-μm/0.39-μm reflectance ratio and stronger turndown into the OMCT band (compare hollows spectra in Figure 17.4c), thus retaining the signature of substrate. On this basis, Izenberg et al. suggested that bright hollow materials are less space weathered equivalents of their substrate, whether the red unit or LRM, exposed when the hollows were formed.

17.3.3 Spatial Distribution of Spectral Units

Figure 17.6b shows the spatial distribution of different spectral units. Materials at the "red" end of the primary spectral trend are unevenly distributed across Mercury's surface. The two largest contiguous areas of HRP are the northern smooth plains (Head et al., 2011) and smooth plains interior to the Caloris basin (Murchie et al., 2008; Ernst et al., 2015), two of the three largest occurrences of smooth plains mapped by Denevi et al. (2013a). Other occurrences of smooth HRP include interior plains of the Rachmaninoff, Rembrandt, and Tolstoj impact basins, and scattered smooth plains within and around other smaller basins. There are scattered exposures of HRP material throughout the IP within impact crater rims, walls, and ejecta, which are interpreted as HRP buried by overlying volcanic IP and excavated by the impacts (Ernst et al., 2010). Large areas of HRP also form intercrater plains at southern mid-latitudes from 220°E to 260°E longitude, and in northern equatorial to mid-latitudes from 60°E to 120°E longitude. Most of Mercury's surface consists of IP, which forms both smooth and intercrater plains. LBP forms smooth plains southwest of Caloris, and the major occurrences are as intercrater plains over the region 60°S–50°N, 290°–350°E, within and to the south and east of the "high-Mg" region identified from MESSENGER X-ray spectrometer measurements (Weider et al., 2015). LRM occurs in rims and ejecta of impact craters and basins off all relative ages: LRM in the relatively old Tolstoj and Rembrandt basins, the intermediate age Rachmaninoff basin, and the relatively young Basho crater are closely comparable in reflectance and spectral slope (Murchie et al., 2015). Large annuli also define ejecta of highly degraded and infilled impact basins (Fassett et al., 2012). LRM composed of numerous small patches occurs at southern mid-latitudes at 310°E–30°E and 120°E–180°E, in areas with the highest density of large craters that may represent Mercury's oldest preserved crust (Klima et al., 2018; Denevi et al., 2019).

17.4 Insights into Mercury's Geologic History from Reflectance Spectroscopy

17.4.1 Relation of Reflectance Spectra and Elemental Composition

None of the spatially resolved spectral units exhibits an identifiable 1-μm crystal field absorption attributable to Fe^{2+} in silicates, placing an important constraint on chemistry of mercurian silicates (Izenberg et al., 2014; Murchie et al., 2015). MESSENGER's X-ray and gamma-ray spectrometers found a low total Fe abundance, with a global range of ~1–2 wt.% (1.3–2.6 wt.% expressed as FeO) (Nittler et al., 2011; Evans et al., 2012; Weider et al., 2012, 2014). Sulfur is abundant, with a global range of ~1–4 wt.% S (Nittler et al., 2011; Evans et al., 2012; Weider et al., 2012). From these observations and geochemical modeling, Klima et al. (2013) and Zolotov et al. (2013) concluded that FeO in mercurian silicates must be <1 wt.%, and that the majority of the iron exists as free metal or sulfides. The low Fe and high S abundances are consistent with a reducing environment, 4.5–7.3 log10 units below the iron–wustite buffer.

The reduced chemistry of mercurian surface materials prevents a lunar-like correlation of major element chemistry and spectral reflectance properties. Without ferrous iron, the major rock-forming minerals – plagioclase, olivine occurring as forsterite, pyroxene as enstatite and diopside – are all high in reflectance. Instead, a dominant control on spectral reflectance is relative abundance of a few percent or less of graphite (Murchie et al., 2015, 2019).

17.4.2 Structure of the Upper Crust

Spectral variations among deposits that form smooth and intercrater plains act as tracers with which to reconstruct the structure of Mercury's upper crust, using crater scaling relations and the many occurrences of one spectral unit excavated from beneath another. Ernst et al. (2010) examined several regions imaged multispectrally by MDIS during MESSENGER's first two flybys of Mercury. They concluded that much of the upper 5 km of Mercury's crust consists of layers of HRP, IP, and LBP, with LRM forming the bottom of the stratigraphic column. By applying similar methods to the Caloris interior plains, Ernst et al. (2015) inferred that >2.5 km of HRP buries basin floor material consisting of LRM, and that LRM extends to at least 11 km depth. Denevi et al. (2013b) examined a highly infilled Caloris-sized basin ("b30" of Fassett et al., 2012) and inferred that >500 m of intercrater IP buries at least 3.5 km of older HRP burying LRM. Modeling of overlapping ejecta of multiple impacts suggests that, to account for the observed areal extent of LRM, the source depth from which it originates is on average 30 km, but that depth to subsurface LRM is laterally heterogeneous (Rivera-Valentin & Barr, 2014).

17.4.3 Formation of the Earliest Crust

The stratigraphy, spectral reflectance, and elemental composition of LRM provide evidence for the formation of an exotic primary flotation crust on Mercury. The similarity of LRM regardless of exposure age, and its position at the bottom of Mercury's stratigraphic column, indicates that its darkening agent is intrinsic to deeply buried crustal rocks. Given the low abundance of Ti, the opaque phases plausibly present at abundances sufficiently high to account for the low reflectance of LRM are graphite present at 1–5 wt.%, a mixture of nanophase and microphase metallic iron or iron-sulfide, or some combination (Murchie et al., 2015). Modeling of UV spectral reflectance by Trang et al. (2016) found that iron alone would produce a lower UV reflectance than is observed, but that an iron–carbon mixture can accurately reproduce the UV spectrum of LRM. Murchie et al. (2015) summarized three possible origins for a layer of a carbon- and/or iron-bearing LRM excavated from depth: a primary graphite flotation crust predicted by geochemical modeling of an early magma ocean (Vander Kaaden & McCubbin, 2015), a late-accreting carbonaceous veneer (Wänke, 1981; Wänke & Dreibus, 1994), or a layer that was shock-darkened during and before the late heavy bombardment (Gillis-Davis et al., 2013). These hypotheses make distinct predictions about abundances of carbon and iron in LRM compared to plains: C-enriched, C- and Fe-enriched, or similar, respectively.

The iron and carbon contents of LRM were bounded by measurements by MESSENGER's X-ray and neutron spectrometers, acquired at low altitudes late in the mission (Peplowski et al., 2016). Carbon and iron are detectable with neutron spectroscopy because carbon is a thermal neutron scatterer that increases thermal neutron flux, whereas iron is a neutron absorber that decreases thermal neutron flux. The neutron spectrometer detected thermal neutron enhancements highly correlated with occurrences of LRM, consistent with an increase in carbon content of 1–3 wt.% above the average abundance outside the LRM, in good agreement with estimates from spectral mixture modeling. In contrast, iron content was determined independently by the X-ray spectrometer to be low. Based on these results, Peplowski et al. (2016) proposed that LRM incorporates a primary graphite flotation crust mixed with igenous materials, by impacts or by being assimilated into ascending magmas.

17.4.4 Constraints on Volcanology

Mercury's red unit is distinct both in its spectral properties and in its style of emplacement. Whereas most of Mercury's volcanic materials are plains-forming and thought to have been emplaced in flood volcanic eruptions (Head et al., 2008, 2009, 2011), the red unit occurs in diffuse-edged blankets typically ~20–130 km in radius, interpreted as pyroclastic deposits. Under Mercury's reducing conditions, plausible volatiles to have driven the explosive eruptions include CO, CS_2, COS, and S_2 (Zolotov, 2011). To produce the observed size range of deposits, Kerber et al. (2011) and Thomas et al. (2014b) estimated a volatile content of 1–10 wt.%.

Several observations suggest the volatiles and their origin. The red unit is the brightest material on Mercury and contains the only apparent absorption attributable to oxidized iron, the OMCT band. In the largest deposit, northeast of Rachmaninoff, the MESSENGER X-ray and neutron spectrometers showed that S is depleted by 1.7 wt.% compared to average terrains (Weider et al., 2016) and C by 1 wt.% (almost entirely; Peplowski et al., 2016). Zolotov (2011) suggested that highly reduced mercurian magmas could have been slightly oxidized by FeO and SiO_2 assimilated from country rock. Absence of a 1-μm absorption limits Fe^{2+} in silicates to <1 wt.%, and more than a few tenths of a percent Fe^{2+} creates an OMCT band so strong that it saturates with no obvious turndown of reflectance into the UV; thus typical crustal material may have ~0.3–1 wt.% Fe^{2+}. Weider et al. (2016) proposed that this Fe^{2+} in crustal rocks oxidized C and S in graphite and sulfides, supplying volatile gases to drive pyroclastic volcanism, in the process creating the red unit's spectral characteristics. Its strongest observed UV turndown on Mercury is consistent with an Fe^{2+} content being lowered to 0.1 wt.% or less by reduction to Fe^0; the higher reflectance is consistent with carbon having been oxidized and escaped; and the depleted C and S contents are consistent with graphite and sulfides having been oxidized and C and S removed. Typical amounts of S and C available for oxidation correspond with the exsolved volatile content required to account for observed halo sizes.

17.5 Summary

Reflectance spectroscopy of Mercury is an exercise in recovering geochemical information based on the fewest of clues: low but spatially variable albedo, the edge of an OMCT band present in the brightest deposits, and curvature of the visible-wavelength spectrum in the darkest materials. Taken together with evidence from superposition relations evident from imaging, and elemental abundance measurements from nuclear spectroscopy, the limited spectral information about Mercury has yielded insights into crustal structure, evolution, composition, and even volcanologic processes. This synergy shows how spectral reflectance can best be interpreted on other planetary bodies: in a geologic context, and in conjunction with complementary information on composition, morphology, and stratigraphy.

References

Adams J.B. & McCord T.B. (1970) Remote sensing of lunar surface mineralogy: Implications from visible and near-infrared reflectivity of Apollo 11 samples. *Proceedings of the Apollo 11 Lunar Sci. Conf.*, **3**, 1937–1945.

Blewett D.T., Lucey P.G., Hawke B.R., & Jolliff B.L. (1997a) Clementine images of the lunar sample-return stations: Refinement of FeO and TiO_2 mapping techniques. *Journal of Geophysical Research*, **102**, 16319–16325.

Blewett D.T., Lucey P.G., Hawke B.R., Ling G.G., & Robinson M.S. (1997b) A comparison of mercurian reflectance and spectral quantities with those of the Moon. *Icarus*, **129**, 217–231.

Blewett D.T., Robinson M.S., Denevi B.W., et al. (2009) Multispectral images of Mercury from the first MESSENGER flyby: Analysis of global and regional color trends. *Earth and Planetary Science Letters*, **285**, 272–282.

Blewett D.T., Chabot N.L., Denevi B.W., et al. (2011) Hollows on Mercury: Evidence from MESSENGER for geologically recent volatile-related activity. *Science*, **333**, 1856–1859.

Blewett D.T., Vaughan W.V., Xiao Z., et al. (2013) Mercury's hollows: Constraints on formation and composition from analysis of geological setting and spectral reflectance. *Journal of Geophysical Research*, **118**, 1013–1032.

Braden S.E. & Robinson M.S. (2013) Relative rates of optical maturation of regolith on Mercury and the Moon. *Journal of Geophysical Research*, **118**, 1903–1914.

Charette M.P., McCord T.B., Pieters C.M., & Adams J.B. (1974) Application of remote spectral reflectance measurements to lunar geology classification and determination of titanium content of lunar soils. *Journal of Geophysical Research*, **79**, 1605–1613.

Cloutis E.A., McCormack K.A., Bell J.F., et al. (2008) Ultraviolet spectral reflectance properties of common planetary minerals. *Icarus*, **197**, 321–347.

Cloutis E.A., Hudon P., Hiroi T., Gaffey M.J., & Mann P. (2011) Spectral reflectance properties of carbonaceous chondrites: 2. CM chondrites. *Icarus*, **216**, 309–346.

Denevi B.W. & Robinson M.S. (2008) Mercury's albedo from Mariner 10: Implications for the presence of ferrous iron. *Icarus*, **197**, 239–246.

Denevi B.W., Robinson M.S., Solomon S.C., et al. (2009) The evolution of Mercury's crust: A global perspective from MESSENGER. *Science*, **324**, 613–618.

Denevi B.W., Ernst C.M., Meyer H.M., et al. (2013a) The distribution and origin of smooth plains on Mercury. *Journal of Geophysical Research*, **118**, 891–907.

Denevi B.W., Ernst C.M., Whitten J.L., et al. (2013b) The volcanic origin of a region of intercrater plains on Mercury. *44th Lunar Planet. Sci. Conf.*, Abstract #1218.

Denevi B.W., Chabot N.L., Murchie S.L., et al. (2018), Calibration, projection, and final image products of MESSENGER's Mercury Dual Imaging System, *Space Science Reviews*, **214**, 1–52.

Denevi B.W., Ernst C.M., Prockter L.M., & Robinson M.S. (2019) The geologic history of Mercury. In: *Mercury: The view after MESSENGER* (S.C. Solomon, L.R. Nittler, & B. J. Anderson, eds.). Cambridge University Press, Cambridge.

Domingue D.L., Chapman C.R., Killen R.M., et al. (2014) Mercury's weather-beaten surface: Understanding Mercury in the context of lunar and asteroidal space weathering studies. *Space Science Reviews*, **181**, 121–214.

Domingue D.L., Denevi B.W., Murchie S.L., & Hash C. (2016) Application of multiple photometric models to disk-resolved measurements of Mercury's surface: Insights into Mercury's regolith characteristics. *Icarus*, **268**, 172–203.

Emery J.P., Sprague A.L., Witteborn F.C., Colwell J.E., Kozlowski R.W.H., & Wooden D.H. (1998) Mercury: Thermal modeling and mid-infrared (5–12 μm) observations. *Icarus*, **136**, 104–123.

Ernst C.M., Murchie S.L., Barnouin O.S., et al. (2010) Exposure of spectrally distinct material by impact craters on Mercury: Implications for global stratigraphy. *Icarus*, **209**, 210–223.

Ernst C.M., Denevi B.W., Barnouin O.S., et al. (2015) Stratigraphy of the Caloris basin, Mercury: Implications for volcanic history and basin impact melt. *Icarus*, **250**, 413–429.

Evans L.G., Peplowski P.N., Rhodes E.A., et al. (2012) Major-element abundances on the surface of Mercury: Results from the MESSENGER Gamma-Ray Spectrometer. *Journal of Geophysical Research*, **117**, E00L07, DOI:10.1029/2012JE004178.

Fassett C.I., Head J.W., Baker D.M.H., et al. (2012) Large impact basins on Mercury: Global distribution, characteristics, and modification history from MESSENGER orbital data. *Journal of Geophysical Research*, **117**, E00L08, DOI:10.1029/2012JE004154.

Fischer E.M. & Pieters C.M. (1994) Remote determination of exposure degree and iron concentration of lunar soils using VIS-NIR spectroscopic methods. *Icarus*, **111**, 475–488.

Gaffey M.J. (2010) Space weathering and the interpretation of asteroid reflectance spectra. *Icarus*, **209**, 564–574.

Gillis-Davis J.J., van Niekerk D., Scott E.R.D., McCubbin F.M., & Blewett D.T. (2013) Impact darkening: A possible mechanism to explain why Mercury is spectrally dark and featureless. Abstract P11A–07, presented at 2013 Fall Meeting, American Geophysical Union, San Francisco, December 9–13.

Goudge T.A., Head J.W., Kerber L., et al. (2014) Global inventory and characterization of pyroclastic deposits on Mercury: New insights into pyroclastic activity from MESSENGER orbital data. *Journal of Geophysical Research*, **119**, 635–658.

Hapke B. (1977) Interpretations of optical observations of Mercury and the Moon. *Physics of the Earth and Planetary Interiors*, **15**, 264–274.

Hapke B. (2001) Space weathering from Mercury to the asteroid belt. *Journal of Geophysical Research*, **106**, 10,039–10,073.

Hapke B., Danielson G.E., Klaasen K., & Wilson L. (1975) Photometric observations of Mercury from Mariner 10. *Journal of Geophysical Research*, **80**, 2431–2443.

Hawkins S.E. III, Boldt J.D., Darlington E.H., et al. (2007) The Mercury Dual Imaging System on the MESSENGER spacecraft. *Space Science Reviews*, **131**, 247–338.

Head J.W., Murchie S.L., Prockter L.M., et al. (2008) Volcanism on Mercury: Evidence from the first MESSENGER flyby. *Science*, **321**, 69–72.

Head J.W., Murchie S.L., Prockter L.M., et al. (2009) Volcanism on Mercury: Evidence from the first MESSENGER flyby for extrusive and explosive activity and the volcanic origin of plains. *Earth and Planetary Science Letters*, **285**, 227–242.

Head J.W., Chapman C.R., Strom R.G., et al. (2011) Flood volcanism in the high northern latitudes of Mercury revealed by MESSENGER. *Science*, **333**, 1853–1856.

Hendrix A.R. & Vilas F. (2006) The effects of space weathering at UV wavelengths: S-class asteroids. *Astronomical Journal*, **132**, 1396–1404.

Hendrix A.R., Retherford K.D., Gladstone G.R., et al. (2012) The lunar far-UV albedo: Indicator of hydration and weathering. *Journal of Geophysical Research*, **117**, E12001, DOI:10.1029/2012JE004252.

Izenberg N.R., Klima R.L., Murchie S.L., et al. (2014) The low-iron, reduced surface of Mercury as seen in spectral reflectance by MESSENGER. *Icarus*, **228**, 364–374.

Izenberg N.R., Thomas R.J., Blewett D.T., & Nittler L.R. (2015) Are there compositionally different types of hollows on Mercury? *46th Lunar Planet. Sci. Conf.*, Abstract #1344.

Kerber L., Head J.W., Blewett D.T., et al. (2011) The global distribution of pyroclastic deposits on Mercury: The view from MESSENGER flybys 1–3. *Planetary and Space Science*, **59**, 1895–1909.

Klima R.L., Dyar M.D., & Pieters C.M. (2011) Near-infrared spectra of clinopyroxenes: Effects of calcium content and crystal structure. *Meteoritics and Planetary Science*, **42**, 235–253.

Klima R.L., Izenberg N.R., Murchie S.L., et al. (2013) Constraining the ferrous iron content of minerals in Mercury's crust. *44th Lunar Planet. Sci. Conf.*, Abstract #1602.

Klima R.L., Denevi B.W., Ernst C.M., Murchie S.L., & Peplowski P.N. (2018) Global distribution and spectral properties of low-reflectance material on Mercury. *Geophysical Research Letters*, **45**, 2945–2953.

Lucey P.G., Taylor G.J., & Malaret E. (1995) Abundance and distribution of iron on the Moon. *Science*, **268**, 1150–1153.

Lucey P.G., Blewett D.T., & Hawke B.R. (1998) Mapping the FeO and TiO_2 content of the lunar surface with multispectral imaging. *Journal of Geophysical Research*, **103**, 3679–3699.

Lucey P.G., Blewett D.T., Taylor G.J., & Hawke B.R. (2000) Imaging of lunar surface maturity. *Journal of Geophysical Research*, **105**, 20377–20386.

Maxwell R.E., Izenberg N.R., & Holsclaw G.M. (2016) Implications for iron and carbon in Mercury surface materials from ultraviolet reflectance. *47th Lunar Planet. Sci. Conf.*, Abstract #1606.

McClintock W.E. & Lankton M.R. (2007) The Mercury Atmospheric and Surface Composition Spectrometer for the MESSENGER mission. *Space Science Reviews*, **131**, 481–522.

McClintock W.E., Izenberg N.R., Holsclaw G.M., et al. (2008) Spectroscopic observations of Mercury's surface reflectance during MESSENGER's first Mercury flyby. *Science*, **321**, 62–65.

McCord T.B. & Adams J.B. (1972a) Mercury: Surface composition from the reflection spectrum. *Science*, **178**, 745–747.

McCord T.B. & Adams J.B. (1972b) Mercury: Interpretation of optical observations. *Icarus*, **17**, 585–588.

McCord T.B. & Clark R.N. (1979) The Mercury soil: Presence of Fe^{2+}. *Journal of Geophysical Research*, **84**, 7664–7668.

Murchie S.L., Watters T.R., Robinson M.S., et al. (2008) Geology of the Caloris basin, Mercury: A view from MESSENGER. *Science*, **321**, 73–77.

Murchie S.L., Klima R.L., Denevi B.W., et al. (2015) Orbital multispectral mapping of Mercury with the MESSENGER Mercury Dual Imaging System: Evidence for the origins of plains units and low-reflectance material. *Icarus*, **254**, 287–305.

Murchie S.L, Klima R.L., Domingue D.L., Izenberg N.R., Blewett D.T., & Helbert J. (2019) Spectral reflectance constraints on the composition and evolution of Mercury's surface. In: *Mercury: The view after MESSENGER* (S.C. Solomon, L.R. Nittler, & B.J. Anderson, eds.). Cambridge University Press, Cambridge.

Murray B.C., Belton M.J.S., Danielson G.E., et al. (1974) Venus: Atmosphere motion and structure from Mariner 10 pictures. *Science*, **183**, 1307–1315.

Nittler L.R., Starr R.D., Weider S.Z., et al. (2011) The major-element composition of Mercury's surface from MESSENGER X-ray spectrometry. *Science*, **333**, 1847–1850.

Peplowski P.N., Klima R.L., Lawrence D.J., et al. (2016) Remote sensing evidence for an ancient carbon-bearing crust on Mercury. *Nature Geoscience*, **9**, 273–276.

Pieters C.M. (1993) Compositional diversity and stratigraphy of the lunar crust derived from reflectance spectroscopy. In: *Remote geochemical analysis: Elemental and mineralogic composition* (C.M. Pieters & P.A. J. Englert, eds.). Cambridge University Press, Cambridge, 309–340.

Rava B. & Hapke B. (1987) An analysis of the Mariner 10 color ratio map of Mercury. *Icarus*, **71**, 397–429.

Rivera-Valentin E.G. & Barr A.C. (2014) Impact-induced compositional variations on Mercury. *Earth and Planetary Science Letters*, **391**, 234–242.

Robinson M.S. & Lucey P.G. (1997) Recalibrated Mariner 10 color mosaics: Implications for mercurian volcanism. *Science*, **275**, 197–200.

Robinson M.S. & Taylor G.J. (2001) Ferrous oxide in Mercury's crust and mantle. *Meteoritics and Planetary Science*, **36**, 841–847.

Robinson M.S., Murchie S.L., Blewett D.T., et al. (2008) Reflectance and color variations on Mercury: Regolith processes and compositional heterogeneity. *Science*, **321**, 66–69.

Solomon S.C., McNutt R.L. Jr., Gold R.E., et al. (2001) The MESSENGER mission to Mercury: Scientific objectives and implementation. *Planetary and Space Science*, **49**, 1445–1465.

Sprague A.L., Kozlowski R.W.H., Witteborn F.C., Cruikshank D.P., & Wooden D.H. (1994) Mercury: Evidence for anorthosite and basalt from mid-infrared (7.3–13.5 micrometers) spectroscopy. *Icarus*, **109**, 156–167.

Sprague A.L., Hunten D.M., & Lodders K. (1995) Sulfur at Mercury, elemental at the poles and sulfides in the regolith. *Icarus*, **118**, 211–215.

Sprague A.L., Emery J.P., Donaldson K.L., Russell R.W., Lynch D.K., & Mazuk A.L. (2002) Mercury: Mid-infrared (3–13.5 µm) observations show heterogeneous composition, presence of intermediate and basic soil types, and pyroxene. *Meteoritics and Planetary Science*, **37**, 1255–1268.

Thomas R.J., Rothery D.A., Conway S.J., & Anand M. (2014a) Hollows on Mercury: Materials and mechanisms involved in their formation. *Icarus*, **229**, 221–235.

Thomas R.J., Rothery D.A., Conway S.J., & Anand M. (2014b) Mechanisms of explosive volcanism on Mercury: Implications from its global distribution and morphology. *Journal of Geophysical Research*, **119**, 2239–2254.

Trang D., Lucey P.G., & Izenberg N.R. (2016) Mapping of submicroscopic carbon and iron on Mercury with radiative transfer modeling of MESSENGER VIRS reflectance spectra. *47th Lunar Planet. Sci. Conf.*, Abstract #1396.

Trask N.J. & Guest J.E. (1975) Preliminary geologic terrain map of Mercury. *Journal of Geophysical Research*, **80**, 2461–2477.

Vander Kaaden K.E., & McCubbin F.M. (2015) Exotic crust formation on Mercury: Consequences of a shallow, FeO-poor mantle. *Journal of Geophysical Research*, **120**, 195–209.

Vilas F. & McCord T.B. (1976) Mercury: Spectral reflectance measurements (0.33–1.06 μm) 1974/75. *Icarus*, **28**, 593–599.

Vilas F., Leake M.A., & Mendell W.W. (1984) The dependence of reflectance spectra of Mercury on surface terrain. *Icarus*, **59**, 60–68.

Vilas F., Domingue D.L., Helbert J., et al. (2016) Mineralogical indicators of Mercury's hollows composition in MESSENGER color observations. *Geophysical Research Letters*, **43**, 1450–1456, DOI:10.1002/2015GL067515.

Wänke H. (1981) Constitution of terrestrial planets. *Philosophical Transactions of the Royal Society of London A*, **303**, 287–302.

Wänke H. & Dreibus G. (1994) Water abundance and accretion history of terrestrial planets. *Conference on Deep Earth and Planetary Volatiles*, Lunar and Planetary Institute, Houston, TX, 46.

Warell J. & Blewett D.T. (2004) Properties of the hermean regolith: V. New optical reflectance spectra, comparison with lunar anorthosites, and mineralogical modeling. *Icarus*, **168**, 257–276.

Warell J., Sprague A.L., Emery J.P., Kozlowski R.W.H., & Long A. (2006) The 0.7–5.3 μm spectra of Mercury and the Moon: Evidence for high-Ca pyroxene on Mercury. *Icarus*, **180**, 281–291.

Weider S.Z., Nittler L.R., Starr R.D., et al. (2012) Chemical heterogeneity on Mercury's surface revealed by the MESSENGER X-Ray Spectrometer. *Journal of Geophysical Research*, **117**, E00L05, DOI:10.1029/2012JE004153.

Weider S.Z., Nittler L.R., Starr R.D., McCoy T.J., & Solomon S.C. (2014) Variations in the abundance of iron on Mercury's surface from MESSENGER X-Ray Spectrometer observations. *Icarus*, **235**, 170–186.

Weider S.Z., Nittler L.R., Starr R.D., et al. (2015) Evidence for geochemical terranes on Mercury: Global mapping of major elements with MESSENGER's X-Ray Spectrometer. *Earth and Planetary Science Letters*, **416**, 109–120.

Weider S.Z., Nittler L.R., Murchie S.L., et al. (2016) Evidence from MESSENGER for sulfur- and carbon-driven explosive volcanism on Mercury. *Geophysical Research Letters*, **43**, 3653–3661, DOI:10.1002/2016GL068325.

Whitten J.L., Head J.W., Denevi B.W., & Solomon S.C. (2014) Intercrater plains on Mercury: Insights into unit definition, characterization, and origin from MESSENGER datasets. *Icarus*, **241**, 97–113.

Zolotov M. (2011) On the chemistry of mantle and magmatic volatiles on Mercury. *Icarus*, **212**, 24–41.

Zolotov M., Sprague A.L., Hauck S.A. II, Nittler L.R., Solomon S.C., & Weider S.Z. (2013) The redox state, FeO content, and origin of sulfur-rich magmas on Mercury. *Journal of Geophysical Research*, **118**, 138–146.

18

Compositional Analysis of the Moon in the Visible and Near-Infrared Regions

CARLÉ M. PIETERS, RACHEL L. KLIMA, AND ROBERT O. GREEN

18.1 Introduction

Spectroscopic studies of the Moon have long provided the foundation for remote compositional analysis of our Solar System. As our nearest neighbor, the Moon is the only evolved geologic body, other than Earth, for which we have samples that were collected with their geologic context intact. Laboratory studies of these samples, coupled with spectroscopic studies of the same materials, have provided insight into such varied topics as planetary dynamics, planetary magmatic evolution, and regolith evolution. Initial access to lunar material has raised awareness of a myriad of phenomena, including radiative transfer through regolith mixtures and space weathering, that must be considered when studying airless bodies. For decades since the Apollo samples were returned, the Moon has continued to serve as the cornerstone for developing and testing our understanding of data returned from other remote sensing missions, from Mercury (Chapter 17) through the asteroid belt (Chapter 19).

Shortly after their return by Apollo 11, rock and soil samples from the Moon were measured in the laboratory using reflected visible through near-infrared (NIR) wavelength light (Adams & McCord, 1970; Adams & Jones, 1970; McCord & Johnson, 1970). Electronic mineral absorption bands (Chapter 1), predicted by crystal field theory and measured previously through transmitted light measurements of terrestrial mineral crystals, were recognized in spectra of all lunar samples, with soils exhibiting the weakest bands and crystalline materials exhibiting the strongest bands. These measurements directly validated that remote telescopic spectra could be used to characterize the mineral properties of the Moon and other bodies. Telescopic data of the Apollo 11 landing sites most closely resembled spectra of soil, but it was recognized that similar measurements of rocky areas had the potential to provide detailed mineralogic information about other locations across the Moon.

Laboratory spectra of example lunar soil, rock, glass, and mineral samples (Figure 18.1) exhibit prominent diagnostic absorption bands, principally due to electronic transitions in transition metal ions with incompletely filled d orbitals (e.g., Fe^{2+}) in a crystal structure. Light is absorbed, resulting in a transition from a lower-energy to a higher-energy orbital state as described by crystal field theory (Burns, 1993), and the energy (wavelength) of the

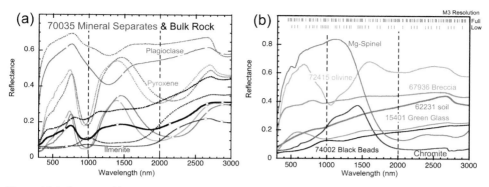

Figure 18.1 Spectra of lunar rock and mineral samples measured in the RELAB. (a) Apollo 17 lunar basalt (70035) and its mineral separates (including two types of pyroxene), each prepared as coarse (solid) and fine (dotted) particles (Isaacson et al., 2011). The samples of bulk basalt are shown in black. (b) Additional mineral separates identified in lunar materials. Lunar samples are from Apollo 15, 16, or 17 (numbered) measured in RELAB, but the Mg-spinel and chromite are terrestrial samples (Cloutis et al., 2004). The lunar olivine sample contains trace amounts of Cr-spinel. Spectral resolution for the two modes of measurement by M^3 are shown along the top. Vertical dashed lines are provided at 1 and 2 μm for ease of comparison with other spectra.

absorption is dependent on the coordination and symmetry of anions and transition metals in the crystal structure (Chapter 1). Distinctive spectral signatures of olivine, pyroxenes, feldspar, spinel, ilmenite, glass, and their mixtures (Chapters 1 and 4) enable remote assessment of lunar crustal composition. Evaluation of pyroxene composition (Ca^{2+}, Fe^{2+}, and Mg^{2+}) is particularly important for crustal lithologies of the relatively dry Moon (e.g., Adams, 1974; Hazen et al., 1978; Cloutis & Gaffey 1991; Sunshine & Pieters, 1993; Klima et al., 2007, 2008, 2011a). Lunar samples document that the primary lunar crust (highlands) is dominated by anorthositic plagioclase. Although crystalline plagioclase typically does not contain iron, trace amounts of Fe^{2+} incorporated into the structure of lunar plagioclase result in a broad absorption near 1250 nm (Conel & Nash, 1970), the strength of which has been shown to be linearly related to the total amount of iron in the plagioclase (Bell & Mao, 1973; Cheek et al., 2011a).

In addition to the major rock-forming silicate minerals (olivine, pyroxene, and plagioclase), opaque minerals and glasses are both present in lunar samples and can be identified spectrally on the lunar surface. Ilmenite, though optically dark or opaque, exhibits an increase in reflectance around 1000 nm and toward 2500 nm that can be most clearly distinguished when present as extremely fine particles suspended in a transparent matrix, for example, in lunar pyroclastic glasses (Adams et al., 1974). Chromite and Mg-spinel both exhibit extremely strong absorption bands near 2000 nm, which are evident even when only small amounts of these minerals are present. For example, olivine sample 72415 (Figure 18.1) contains tiny inclusions of chromite, resulting in the broad absorption near 2000 nm in this spectrum. The absorption spectra of amorphous or processed material (often just referred to as "glass") depend on the degree of crystallinity. Volcanic quench glasses such as Apollo 15 green glass (15401) primarily exhibit characteristic broad absorptions that can be attributed to Fe and Ti in short-range coordination (Bell et al., 1976), and unusual fully melted and rapidly quenched impact melts exhibit similar broad absorption features, the strength of which are dependent on composition (largely Fe and Ti). Natural impact melts and breccias, however, exhibit a range of spectral properties and are usually complex and partially crystallized, exhibiting characteristic absorptions of fine-grained minerals dispersed within them (Tompkins & Pieters, 2010).

18.2 Historic Context

The spectral properties of the lunar nearside were studied well before the Apollo and Luna landings on the Moon using visible wavelength instruments on telescopes. As NIR detectors expanded the capabilities of spectrometers, compositional analyses became a successful focus of many lunar telescopic analyses, summarized in McCord et al. (1981) and later reviewed in Pieters (1993). It wasn't until the last decade of the twentieth century, however, that sensors were sent to orbit the Moon and for the first time enabled information sensitive to lunar compositional properties to be systematically acquired and mapped across the entire globe. Multispectral visible/near-infrared (VNIR) cameras were first sent on one small spacecraft, and a little later in the same decade a separate small spacecraft carrying a gamma-ray and neutron spectrometer was also sent to orbit the Moon to collect global data. The combined data from these two small missions, Clementine and Lunar Prospector respectively, astounded and invigorated the science community with dramatic and previously unseen compositional information about this small planetary companion of Earth. These "New Views of the Moon" (Figure 18.2) sparked fresh avenues of scientific investigations across the planetary science community (e.g., Jolliff et al., 2006).

Highlights from this initial "New Views" era include the following: (1) The lunar farside is quite different from the nearside that was viewed with telescopes and contains the most extreme topography, the largest and oldest basin on the Moon (South Pole-Aitken), and the most iron-poor as well as the most iron-rich nonmare regions of the Moon. (2) The largest concentration of heat-producing elements, however, were found to occur on the nearside and are correlated with a wide diversity of mare basalt ages and compositions. (3) The polar areas are special and exhibit enhanced concentrations of hydrogen, presumed to be OH or H_2O trapped or buried in the cold shadowed areas. All of these new insights stimulate further exploration with more modern and capable sensors.

To advance our understanding of the composition of the Moon, two key activities are associated with the turn of the millennium. First, sensors evolved and improved, driven by increasing commercial, scientific, and strategic applications of remote sensing. Second, the "New Views" of the Moon stimulated the expanding international space exploration community to recognize that there was *much* yet to learn from Earth's nearest neighbor. Before the end of the first decade of the new millennium, spacecraft were prepared and

Figure 18.2 Summary of global lunar data returned by the Clementine and Lunar Prospector missions that orbited the Moon 1994–98. Integrated analyses of the data are discussed in Jolliff et al. (2006). The Clementine color composite uses spectral parameters to highlight variations in the nature of the continuum (blue to red) and the magnitude of the ferrous absorption near 1 μm (green or yellow). The RGB display is enhanced R = 750/415, G = 750/950, B = 415/750 nm.

successfully launched to the Moon by Europe, Japan, China, India, and then the United States, acquiring a diverse and highly valuable range of new data. Sophisticated and capable optical spectroscopic instruments that acquired global data available to the community were flown by the Japanese (Multiband Imager [MI] and Spectral Profiler [SP]) on the Kaguya spacecraft and by the Indians (a US instrument, Moon Mineralogy Mapper, M^3) on the Chandrayaan-1 spacecraft. These instruments and their initial data were analyzed jointly for cross-comparison (Ohtake et al., 2013; Pieters et al., 2013). Ongoing analyses of these data continue to be highly productive, and several scientific results and important discoveries from this generation of spectral measurements are itemized in Section 18.4.

A key aspect of successful modern sensors is the ability to provide both spatial and spectral information seamlessly in scientific analyses. One approach is to integrate independent instruments, with one emphasizing spatial information using few spectral channels and another emphasizing spectral information with lower spatial information. Analyses of data from the instruments on Kaguya provide an excellent example of this approach, acquiring high spatial information in nine MI channels and high spectral resolution for a single spatial element producing a profile along the orbit using SP. Together they create global coverage over time. A second approach is enabled by advanced sensors that build on significant improvements in both optics and detectors. Imaging spectrometers simultaneously collect detailed spectra for each spatial element (pixel or instantaneous field of view [IFOV]) within an image. M^3 on Chandrayaan-1 was the first imaging spectrometer flown to the Moon. The principal data product is an "image cube" of data comprising two spatial dimensions (an image) and one spectral dimension (a spectrum for each image element). Because it generates large amounts of data in a single pass, M^3 was designed to work in two modes: one at full spatial and spectral resolution for targeted areas, and the other at low-resolution mode (12× data reduction) with more global coverage (see Figure 18.1b). Owing to significant spacecraft operational anomalies, however, most M^3 data were acquired in the low-resolution mode (Boardman et al., 2011). These three instruments (MI, SP, M^3) obtained global spectroscopic data that are publicly available at http://l2db .selene.darts.isas.jaxa.jp/ and https://pds-imaging.jpl.nasa.gov/volumes/m3.html. The general properties of the instruments that obtained global lunar spectral data (to date) are summarized in Tables 18.1 and 18.2 for lunar multispectral cameras (including Clementine cameras and LROC-WAC) and modern NIR spectrometers respectively.

18.3 Constraints from "Space Weathering"

With the first return of samples, it became readily apparent that lunar soils were not just broken and pulverized local lunar rocks (e.g., Adams & McCord, 1970). Although the bulk composition of soils generally reflects that of the local terrain, lunar soils in fact have been highly processed and altered during their residence on the lunar surface as discussed in Heiken et al. (1991). A variety of small-scale but pervasive processes are active in the harsh environment at 1 AU where the Moon resides. These include continuous micrometeorite bombardment, interaction with radiation and energetic particles from the Sun and galactic sources, thermal cycling, etc. Their combined effects over time have been termed lunar "space weathering" and create the distinctive and complex lunar soils analyzed in Earth-based laboratories and observed remotely (Pieters et al., 2000; Hapke, 2001; Pieters & Noble, 2016). The optical effects of the cumulative products on spectra of lunar samples from a basaltic terrain are shown in Figure 18.3a. Compared to fresh rock powders, well-developed (mature) lunar soils are generally darker at visible wavelengths, have weaker diagnostic absorption bands, and exhibit a characteristic "red-sloped" lunar continuum into the NIR.

NIR spectroscopic data acquired remotely exhibit the same trends across the lunar surface as seen in the laboratory with returned rock and soil samples. In a directly parallel manner, freshly exposed (unweathered) material at a young crater (Figure 18.3b) is typically brighter than surrounding more developed mature soil. Similarly, spectra moving along a short traverse away from this fresh crater illustrate variations of a mixing trend as the recently exposed fresh material is incorporated into and mixes with the local

Table 18.1 *Bandpass information of lunar multispectral cameras that provide global data*

Clem UVVIS[a,b]	Clem NIR[a,c]	MI VIS[d]	MI NIR[d]	LROC-WAC[e]
Effective spatial resolution/pixel (IFOV)				
>115 m	>178 m	~20 m	~62 m	~400 m
Nominal orbit				
425 × 2900 km	425 × 2900 km	100 km	100 km	50 km
Bandpass wavelength of each filter (nm)				
415 ± 20	1102 ± 30	415 ± 20	1000 ± 27	320 ± 32
750 ± 5	1248 ± 30	750 ± 12	1050 ± 28	360 ± 15
900 ± 10	1499 ± 30	900 ± 21	1250 ± 33	415 ± 36
950 ± 15	1996 ± 31	950 ± 30	1550 ± 48	565 ± 20
1000 ± 15	2620 ± 30	1000 ± 42		605 ± 20
	2792 ± 146			645 ± 23
				690 ± 39

[a] Nozette et al. (1994); McEwen and Robinson (1998); IFOV is *minimum* (elliptical orbit).
[b] McEwen et al. (1998); Eliason et al. (1999).
[c] Lucey et al. (2000); Eliason et al. (2003); Cahill et al. (2004).
[d] Ohtake et al. (2010, 2013); Pieters et al. (2013).
[e] Robinson et al. (2010); Boyd et al. (2012); Denevi et al. (2014). The effective spatial resolution is limited by the UV channels; VIS channels have higher spatial resolution.

Table 18.2 *Instrument properties for lunar NIR spectrometers that provide global data*

Instrument and mode	Spatial resolution (IFOV, m)	Swath width /no. of elements	No. of spectral channels	Spectral range (nm)	Spectral sampling (nm)
SP VIS-NIR1[a]	500^2	500 m/1	162	500–1700	6–8
SP NIR2[b]	500^2	500 m/1	112	1700–2600	8
M³ Full[c]					
100-km orbit	70^2	40 km/600	260	460–3000	10
200-km orbit	70 × 140	80 km/600	260	460–3000	10
M³ Low[c]					
100-km orbit	140^2	40 km/300	85	460–3000	20–40
200-km orbit	140 × 280	80 km/300	85	460–3000	20–40

[a] Yamamoto et al. (2011); Ohtake et al. (2013); Pieters et al. (2013).
[b] Yamamoto et al. (2011); Yamamoto et al. (2014).
[c] Boardman et al. (2011); Green et al. (2011); Lundeen et al. (2011); Pieters et al. (2013). Because of spacecraft issues, most M³ data were obtained in low-resolution mode with about half from a 200-km orbit.

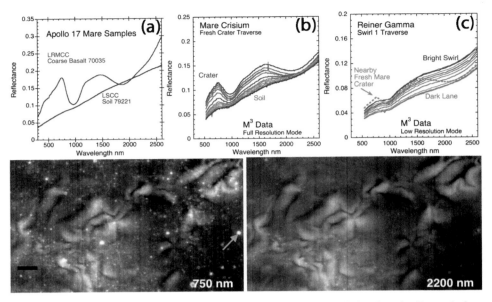

Figure 18.3 (Top) Reflectance spectra of (a) mare basalt rock and a well-developed soil sample from Apollo 17. The Apollo basalt is the same as in Figure 18.1. (b) Traverse of full-resolution M^3 data across a fresh crater into surrounding mature soil illustrating the same optical properties as Apollo samples. (c) Traverse of low-resolution M^3 data across an unusual albedo feature (Reiner Gamma SW swirl) compared to a local fresh mare crater. (Bottom) M^3 reflectance images at 750 and 2200 nm for the swirl region containing the spectra shown in (c). On the 750-nm image the swirl traverse location is shown as a small red line in the upper right, the example fresh crater is indicated with a green arrow, and the scale bar is 5 km. Fresh craters of all sizes disappear in the M^3 image at 2200 nm due to significant pyroxene absorptions in immature mare soils.

background mature soil. Even though the returned samples and remote measurements of Figures 18.3a and b are from completely different basaltic locations on the Moon, the spectra illustrate the same fundamental relation between freshly broken basaltic rocks and well-developed soils that have accumulated lunar space weathering products.

Following several decades of investigation, the ubiquitous presence of nanophase metallic iron (npFe^0) was identified as the principal product associated with the optical effects of space weathering on lunar materials (Pieters et al., 2000; Hapke, 2001). This optically active component has been shown to accumulate on the rims of grains and is also incorporated into grain interiors during soil processing and reworking (e.g., Keller & McKay, 1993, 1997; Sasaki et al., 2001; Noble et al., 2007; Keller & Zhang, 2015; Pieters & Noble, 2016). Detailed petrographic and compositional analyses of a suite of soils with different degrees of lunar space weathering were undertaken along with coordinated spectroscopic measurements by a Lunar Soil Characterization Consortium (LSCC) (Taylor et al., 2001, 2010; Pieters & Taylor, 2003; Noble et al., 2006). For all soils studied an increase in the abundance of npFe^0 weathering products was observed with decreasing particle size, independently confirming the surficial nature of npFe^0 on soil particles. All LSCC spectroscopic and compositional data are available online for integration or further analyses (www.planetary.brown.edu/relabdocs/LSCCsoil.html).

Not all materials on the Moon that are unusually bright at visible wavelengths are immature or unweathered soils, however. The most enigmatic are the lunar "swirls," wispy high albedo markings with no topographic relief that are commonly associated with prominent magnetic anomalies detected from orbit (e.g., Blewett et al., 2011). An example swirl region is shown in Figures 18.3c and d for a basaltic area SW of Reiner Gamma, a major lunar magnetic anomaly. It has been hypothesized that these unusual albedo features represent immature soils protected from ionized particle radiation (e.g., Kramer et al., 2011). Although the relation between local fresh basaltic craters (green spectrum) and background soils (red) in this area are comparable to those observed elsewhere (Figure 18.3b), the traverse of spectra from bright to dark areas across the swirl (Figure 18.3c) bears no resemblance to the well documented space weathering properties illustrated in Figures 18.3a–b. Instead, swirl soils exhibit an almost uniform brightening at all wavelengths. As illustrated in Figures 18.3d–e, all swirl terrains in the region exhibit the same properties as those documented in Figure 18.3c traverse. Freshly exposed basalt (prominent as small bright craters in the 750-nm image) are almost invisible in the 2200-nm image due to the strong pyroxene absorption near 2200-nm characteristic of immature soil. In contrast, swirl albedo features show little evidence of unaltered soils and are bright at all wavelengths across the scene. It cannot be overemphasized that space weathering involves several processes. The high-albedo markings at swirls indeed have been shown to exhibit less interaction with solar wind H (in the form of lower surficial OH: e.g., Pieters & Garrick-Bethell, 2015). Nevertheless, the detailed spectroscopic data also show that the overall brightening pattern of swirls (Figure 18.3c–e) is not simply due to immature soils. Instead, these lovely enigmatic albedo features scattered across the Moon highlight our incomplete understanding of how multiple processes are interwoven during surface exposure and alteration, especially concerning effects of a local strong magnetic field on the evolution (and perhaps mobility) of soil particles.

18.4 Principal Results from Modern Instruments

Lunar samples have been invaluable not only for their documentation of the remarkable character and composition of Earth's nearest planetary neighbor, but also because they provide unique constraints (ground truth) for interpreting global remotely acquired data. In addition to the mineral, rock, and soil samples mentioned earlier, a large fraction of returned lunar samples are breccias, reflecting the long and continuous history of impact processes suffered by crustal materials. Through the decades, cumulative analyses have provided a general understanding of lunar crustal evolution that involves an early magma ocean, formation of a mafic mantle and feldspathic cumulate crust, probable early mantle overturn and creation of "Mg-suite" materials, and later internal heating resulting in basaltic volcanism (Heiken et al., 1991; Jolliff et al., 2006).

Major scientific questions remain to be addressed for the Moon that are fundamental to planetary science (NRC, 2007). It is in this context that modern spectroscopic instruments are being used to test and significantly expand our knowledge of the character of the Moon and lunar crustal evolution. Along the way, several unexpected but important discoveries have also been made including the identification and global occurrence of exceptionally pure anorthosite, recognition of a completely new rock type consisting of Mg-spinel and anorthosite, and, of course, the surprise of finding much of the upper lunar surface is host to OH species.

18.4.1 Mineralogy of the Highland Crust

18.4.1.1 Upper Crust: PAN and "Featureless" Plagioclase

Prior to sending high spatial and spectral resolution instruments to study the lunar surface, remote identification of the principal component in the highland crust, namely plagioclase feldspar, was indirect and based on a *lack* of observed diagnostic absorptions due to ferrous iron-bearing minerals (e.g., Pieters, 1993; Hawke et al., 2003). The logic was (and still is) that since these areas were high albedo, found throughout the highlands, and exhibited no detectable mafic minerals, the only known common lunar material that meets those criteria is plagioclase. However, as seen in Figure 18.1a, crystalline lunar plagioclase, even with tiny amounts of FeO (0.1–0.3%), is relatively transparent and exhibits the well-defined crystal field absorption near 1.25 µm (Burns, 1993). One (as yet unproven) hypothesis for extensive "featureless" plagioclase on the Moon is that the extensive impact history of the lunar feldspathic crust created "diaplectic glass" (maskelynite), a shocked form of plagioclase that has lost its original crystal structure (Stöffler, 1972, 1974) and cannot produce the normal crystal field absorption at 1.25 µm.

The paradigm changed when two independent spacecraft with different instruments clearly detected the plagioclase diagnostic 1.25 µm absorption and began mapping crystalline plagioclase across the Moon (Ohtake et al., 2009; Pieters et al., 2009; Donaldson Hanna et al., 2014). Some crystalline plagioclase exposures such as the Inner Rook Mountains (inner ring of Orientale) are enormous in scale (Cheek et al., 2013), strongly suggesting they represent the predicted thick plagioclase cumulate product of a magma ocean. Mineral mixing experiments demonstrated that the remote spectroscopic measurements limit the amount of mafic minerals (e.g., pyroxene) that can be present with the crystalline plagioclase to be less than ~2% (e.g., Cheek & Pieters, 2014). This created another dilemma because most lunar anorthositic samples in our collections from Apollo contain a small amount of ferrous minerals and are often called ferroan anorthosites (FAN). To produce anorthosites with the purity observed remotely on such a massive scale is not easily achieved with standard magma ocean fractional crystallization models. Thus, these crystalline lunar anorthosites observed across the highlands are affectionately known as Pure Anorthosites, or PAN (Ohtake et al., 2009). Example remote measurements of crystalline PAN along with (abundant) neighboring "featureless" plagioclase are shown in Figure 18.4a.

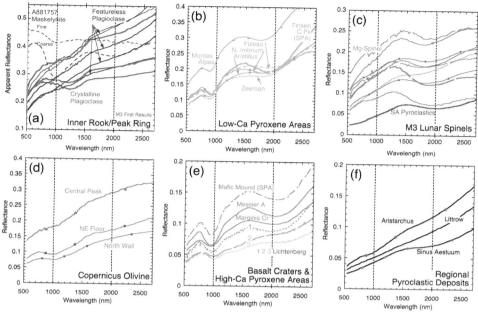

Figure 18.4 Example NIR spectra of lunar areas obtained from orbit. (a) Crystalline and "featureless" plagioclase found along the Orientale Inner Rook mountains obtained by M^3 (after Pieters et al., 2009). Shown for comparison are laboratory spectra of clean maskelynite glass separated from the gabbroic lunar meteorite A881757 (Pieters, 1996). (b) Example spectra from low-Ca pyroxene-bearing rocks found across the lunar surface. The Montes Alpes spectrum was taken from a large massif in the mountains northeast of Mare Imbrium near Vallis Alpes. Aristillus crater is located in Mare Imbrium, but penetrates through the mare and excavates LCP-rich material. The North Imbrium Norite spectrum is collected from a small, relatively fresh, unnamed crater. Finsen, Fizeau, and Zeeman craters are found within the South Pole-Aitken Basin. (c) Comparison of M^3 spectra of spinel found in diverse lunar settings (after Pieters et al., 2014). Areas top to bottom at 2100 nm are: Endymion1, Endymion2, ThomsonS1, WernerWall1, Montes Teneriffe1, Moscoviense 1, ThomsonS2, TheophilusS1, TychoSp3, Sinus Aestuum2. (d) Olivine exposures in Copernicus crater (after Dhingra et al., 2015). The prominent olivine found in the north wall is associated with impact melt whereas the central peak olivine is excavated from depth. The few olivine-bearing blocks found on the floor have an indeterminate origin. (e) Spectra of lunar basalt craters and areas containing high-Ca pyroxene obtained by M^3. Of the three small numbered fresh craters near Lichtenberg (1) is from the older low-Ti basalt whereas (2 and 3) are from the young relatively Ti-rich picritic basalt (Staid et al., 2011). (f) Spectra of regional pyroclastic (dark mantle) deposits. Aristarchus plateau deposits exhibit properties containing quench glass, whereas many large regional pyroclastic deposits such as Littrow near Apollo 17 appear to be of the more slowly cooled (micro) crystalline form. (Sinus Aestuum contains a form of spinel and the spectrum also appears in Figure18.4c.)

The relation of "featureless" plagioclase to PAN, both common in remote observations, has not been fully resolved. The maskelynite hypothesis for "featureless" plagioclase is not well supported because maskelynite itself (lunar or experimental sample) results in a glass absorption near 1 μm in laboratory measurements (Pieters, 1996; Johnson & Hörz, 2003), although no comparable absorption is observed remotely for highland areas. On the other hand, no lunar sample of anorthosite composition (with or without shock features) has yet been measured with a "featureless" spectrum. PAN and "featureless" plagioclase are nevertheless often observed contiguously on the Moon, arguing against an especially low FeO content accounting for the common featureless spectra. In some situations, spatial resolution may be responsible for not detecting crystalline plagioclase (Ohtake et al., 2009, 2013), where PAN is more readily detected at high spatial resolution. Differential space weathering (Yamamoto et al., 2015), extensive sample fracturing and internal scattering of transparent media (Pieters, 2017), or development of a thick amorphous rind on feldspathic grains (e.g., Keller et al., 2016) might be important for "featureless" plagioclase, but remain untested.

18.4.1.2 Lower Crust: Low-Ca pyroxene, Mg-Spinel, and Olivine

Low-Ca pyroxenes are found ubiquitously across the lunar highlands to different degrees at all scales, with the most prominent localities concentrated in and/or around major impact basins, including the South Pole–Aitken Basin and the Imbrium Basin (Pieters et al., 2001; Isaacson & Pieters, 2009; Klima et al., 2011b; Nakamura et al., 2012; Lucey et al., 2014). The presence of low-Ca pyroxene both above and below the FAN/PAN anorthosite of the upper crust (Pieters et al., 2001; Hawke et al., 2003; Moriarty & Pieters, 2016a, 2018) suggest a lower crust origin which provided extensive low-Ca pyroxene-bearing debris from the basin impact period of early lunar history. In low-Ca pyroxenes, the band positions of the 1 and 2 µm absorptions are controlled primarily by crystallographic effects determined by the ratio of iron to magnesium in the mineral, with both absorption bands shifting to longer wavelengths as iron substitutes for magnesium. Thus, band position can be used to assess the overall iron content of the pyroxenes. Several examples of low-Ca (Mg-rich) pyroxenes found distributed across the surface of the Moon are presented in Figure 18.4b. Based on M^3 data, the band positions of most low-Ca pyroxenes on the Moon suggest an average pyroxene with moderate amounts of iron (Mg50–75), while a few select locations exhibit shorter wavelength bands consistent with an average higher magnesium content (>Mg80) (Klima et al., 2011b). One of the most prominent of such high-Mg exposures is found in the Montes Alpes region, where large massifs associated with the Imbrium impact are located. In addition to high-Mg, low-Ca pyroxenes, some of these blocks exhibit gabbroic and olivine-rich compositions, suggesting that these lithologies sometimes coexist at depth in the lunar crust, potentially as layered mafic intrusions at the base of the anorthositic crust (e.g., Klima et al., 2017).

The presence of abundant *Mg-Spinel* in association with plagioclase was not recognized remotely or in the lunar samples prior to its first detection by the M^3 imaging spectrometer at Moskoviense (Pieters et al., 2011). The original detection of Mg-spinel was found as a discrete ~4 km mega-clast covered with regolith along the inner ring of the basin. It was not detectable in visible wavelength images and provided no morphological clue (such as a crater) of its existence. Only NIR spectroscopy revealed its presence through its unusual (for the Moon) character. The area exhibits no detectible ferrous absorption near 1 µm, but a pronounced broad absorption is observed near 2 µm as well as an inflection near 2.5 µm that indicates the presence of the longer wavelength spinel absorption (near 3 μm). The pair of absorption bands at 2 and 3 μm characteristic of Mg-spinel (Figure 18.1) are now readily identified at several locations on the Moon as illustrated in Figure 18.4c. Their geologic settings across the highland crust are diverse but also include several prominent basins such as Moscoviense, Nectaris, Imbrium, and SPA (see summary in Pieters et al., 2014). This low-Fe, high-Mg form of spinel ("pink" spinel) has been observed in a few lunar samples, but it is found only in low abundance and also in association with other mafic minerals such as olivine or pyroxene. Since the prominent remote observations of Mg-spinel occurs entirely in feldspathic terrain, this new rock type has been informally termed pink spinel anorthosite (PSA). Although local concentrations of PSA have been now been found globally and at multiple sites on the Moon (e.g., Pieters et al., 2014), no sample of PSA has been identified in returned samples. A current model for the origin of this newly recognized rock type is interaction of an early "Mg-suite" mafic liquid with the cumulate anorthosite crust (Prissel et al., 2014).

Clasts of *olivine*-rich materials (troctolite and possibly dunite) are readily recognized as important rock types in returned highland samples, and olivine is also a common component in several Apollo basalts (e.g., Heiken et al., 1991; Jolliff et al., 2001). In lunar telescopic measurements, enhanced olivine was suspected of being present in the young unsampled basalts of the western nearside (Pieters et al., 1980; Staid et al., 2001) while exposures of olivine-rich material were first unambiguously detected remotely in the central peaks of Copernicus (Pieters, 1982). The higher spatial resolution and broad spectral range of recent orbital spectrometers (M^3 and SP) confirmed that olivine is indeed abundantly present in the young Ti-rich basalt of western Oceanus Procellarum (Staid et al., 2011). In the highlands several additional areas across the Moon and around basins were identified where distinctly olivine-rich materials are exposed (Yamamoto et al., 2010). Most are associated with highly feldspathic materials suggesting troctolite as the principal rock type. However, diversity also occurs in close proximity; NIR spectra of three different exposures of olivine-dominated materials at Copernicus are illustrated in Figure 18.4d. In addition, olivine-dominated materials are sometimes found coexisting with other lithologies as discrete areas along the walls of medium-size craters (8–25 km) such as Proclus and an unnamed crater in Schrödinger (Kramer et al., 2013; Donaldson Hanna et al., 2014). Although the composition of olivine assessed with NIR spectra is generally Mg-rich, current data are insufficient to accurately deconvolve all components for a precise determination (e.g., Isaacson et al., 2011). The origin of olivine found in highland areas is currently ambiguous. Although some exposures are associated with basins and are thus clearly derived from a relatively deep source (Yamamoto et al., 2010), all such exposures are blocks embedded within a highly feldspathic matrix, suggesting a crustal affinity.

18.4.1.3 Character of the Megaregolith

Many individual minerals comprising the lunar highland crust are identified and well characterized from lunar samples and also studied remotely at local craters or outcrops.

Nevertheless, the vast majority of highland samples studied as returned samples or meteorites are complex, multicomponent breccias formed as a result of the intense impact history suffered during the first ~500 My of crustal evolution. This period of heavy bombardment in the early period of Solar System evolution included formation of the giant basins and resulted in a heavily fractured, broken, and mixed outer layer of variable crust ~10 km thick referred to as a "megaregolith" (Heiken et al., 1991). The basin-scale impacts excavated into the lower crust and perhaps upper mantle, mixed all components, and redistributed them across the Moon, to be remixed by other later impacts. Although several highland exposures might be considered uplifted bedrock (e.g., the Inner Rook Mountains, or Peak Ring, of Orientale), most of what has been observed remotely represents mega-blocks redistributed during major impact events. Reconstructing the possible crustal origin of crystalline component found in the walls or central peaks of subsequent craters continues to be a challenge. In addition, poorly defined systematics of melting and recrystallization of components (Tompkins & Pieters, 2010; Pieters & Taylor, 2003) complicate interpretations of available observations of impact products on the lunar surface.

18.4.2 Mineralogy of Lunar Basaltic Volcanism

18.4.2.1 Dominant role of Hi-Ca pyroxene

Mare basalts are clearly distinguishable from other material on the Moon not only by their distinctive smooth morphology, but also by their prominent pyroxene absorption bands, detected principally near 960–1000 nm and 2100–2300 nm. The longer wavelength position of these absorption bands compared to most highland pyroxene-bearing materials indicates that substantially more Ca is present in the average pyroxene structure. The absorption band widths are broader than what is observed for pure terrestrial clinopyroxenes in the laboratory (e.g., Klima et al., 2011a). This suggests that a range of pyroxene compositions are present (like returned mare samples that have been measured, e.g., Figure 18.1a) due to extensive zoning in the pyroxenes and/or due to mechanical mixtures of different pyroxenes in the regolith (Moriarty & Pieters, 2016b). Measurement of lunar optical properties near 2200 nm is also influenced by thermal emission, complicating the assessment of specific pyroxene composition (see Section 4.4). In addition to spectra of the basaltic terrain shown in Figure 18.3, example lunar spectra of areas containing abundant high-Ca pyroxene are shown in Figure 18.4e.

Although pyroxene dominates the spectra of most lunar basalts, other minerals (plagioclase, olivine, ilmenite) are also present and often have a measurable effect on the bulk rock or soil spectrum (e.g., Isaacson et al., 2011). The presence of olivine in small amounts (<10%) with pyroxene is normally not detectible. However, picritic basalts where olivine abundance is equal to or greater than pyroxene have been identified at the unsampled young basalts of Oceanus Procellarum (Staid et al., 2011). Such examples from small fresh craters near Lichtenberg are also shown in Figure 18.4e.

18.4.2.2 Pyroclastic Material

Basalts are not the only form of volcanism on the Moon. Pyroclastic (fire fountain) deposits have been known since Apollo, and regional and local deposits have been identified remotely (e.g., Gaddis et al., 2003). Most are very dark and are often called "dark mantle material," or DMM. The regional deposits are believed to consist largely of different portions of quench glass and a (micro) crystallized equivalent that (to a first order) are comparable to the orange glass and black beads from Apollo 17 but with a range of mafic compositions. The geochemistry of the Apollo orange and black beads indicates a mantle origin, but their relation to nearby basalts is not clear. In contrast, the smaller localized pyroclastic deposits are unsampled, but typically occur around vents or fractures in floor fractured craters and are more difficult to characterize because they often contain abundant local country rock/breccia. Example spectra illustrating the diversity found at the large regional dark pyroclastic deposits are Figure 18.4f. As highlighted in the text that follows, currently only one area of high-albedo and highly unusual pyroclastic deposits has been definitively identified in the highlands at Compton Belkovich (Jolliff et al., 2011).

18.4.3 OH/H_2O on the Moon

One of the most surprising new discoveries from recent lunar spectroscopic data came in 2009, when a distinct absorption band near 3 μm that is diagnostic of OH^- and or H_2O-bearing material was observed in data collected by M^3, as well as by the Deep Impact and *Cassini* NIR spectrometers (Clark, 2009; Pieters et al., 2009b; Sunshine et al., 2009). Though Lunar Prospector neutron spectrometer data had shown evidence for enhanced H at both lunar poles (Feldman et al., 1998), and a possible presence of water ice collecting in permanently shadowed polar regions had been hypothesized earlier (Watson et al., 1961), any form of water was not expected to be broadly distributed on the illuminated portions of the Moon. Nevertheless, hydrated material was detected across the Moon, with absorptions generally strengthening toward high latitudes (Pieters et al., 2009b). Example M^3 spectra for areas exhibiting OH/H_2O feature near 3 μm are shown in Figure 18.5a.

The long-wavelength cutoff of M^3 at 3 μm complicates analysis of the specific depth and shape of this feature, particularly on warmer surfaces where thermal emission contributes to the signal measured by the detector and can effectively "mask" the absorption band. The NIR spectrometers on both Deep Impact and *Cassini* fortunately extended to longer wavelengths allowing the thermal component to be constrained, and provided information about the character of the OH band, albeit over a much lower spatial resolution. These data suggest that the broadly distributed 3 μm band exhibits a minimum near 2.8 μm, but is broad, extending to as long as 3.5 μm (Sunshine et al., 2009). Furthermore, there is the suggestion that this absorption may wax and wane over the course of a lunar day (Sunshine et al., 2009; McCord et al., 2011; Li & Milliken, 2017).

McCord et al. (2011) suggested the combination of two absorptions, mapping the global distribution of a band centered at 2.8 μm and one at longer wavelengths. They concluded that

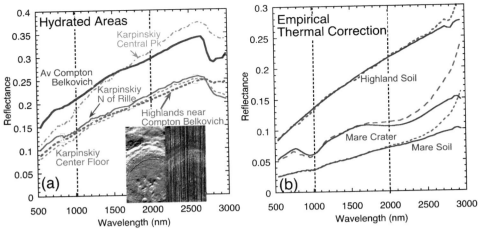

Figure 18.5 M^3 spectra of hydrated areas. (a) M^3 spectra of lunar areas exhibiting distinct OH/H_2O feature near 3 μm (solid lines) compared with background (dashed). The hydration feature is inherently smooth (see Figure 18.1a) and deviations are residual measurement artifacts (see Lundeen et al., 2011). Compton Belkovich is an unusual feature at 61°N, 100°E that is interpreted to result from silicic volcanism (Jolliff et al., 2011). Karpinskiy is a floor-fractured crater found at high northern latitudes (73°N, 167°E) and exhibits regional OH as well as additional concentration north of a major fracture rille (Cheek et al., 2011b). (Inset) M^3 images of Karpinskiy at 1489 nm (brightness) and hydration band strength parameter (illustrating variations in OH near concentric rille). (b) Example Level 1 M^3 spectra before thermal correction (red dashed) compared with Level 2 initial calibrated data (solid). Since Level 2 data also incorporate photometric corrections, the Level 1 reflectance data are scaled to Level 2 at 1509 nm for comparison.

the 2.8-μm band is present to some extent at all latitudes, while that near 3 μm is present only poleward of about 45°, with both bands increasing in depth toward the poles. The broad and possibly variable nature of the absorption band(s) near 3 μm has led to the conclusion that this signal is dominated by adsorbed OH$^-$ (and potentially H_2O), likely produced by interaction of solar wind H$^+$ with oxygen in the regolith silicates (Clark, 2009; Pieters et al., 2009b; Sunshine et al., 2009; McCord et al., 2011). There has been evidence that this adsorbed OH$^-$/H_2O more readily bonds with anorthosite than mafic minerals, and may prefer freshly fractured surfaces containing oxygen with a higher abundance of defects or dangling bonds (Pieters et al., 2009b; Dyar et al., 2010; McCord et al., 2011). However, some of this observed lithological preference may simply be variations in apparent band strength resulting from albedo and transparency differences of the host or the result of inadequate thermal removal in the darker, warm, mare materials (e.g., Li & Milliken, 2017). Initial analyses were performed using data that was thermally corrected using a standard M^3-derived empirical approximation (Clark et al., 2011; see Section 4.4.2). One alternative thermal model that incorporates Diviner data to empirically correct much of M^3 data (Li & Milliken, 2016a) results in a similar latitudinal 3 μm band depth trend to that observed using the initial calibration, but shows less distinction between mare and highland material at equatorial latitudes.

In addition to the global veneer of solar-implanted OH⁻/H$_2$O, there is evidence for minor hydrated species in local terrain such as OH⁻/H$_2$O trapped in material native to the Moon or even exogenic H$_2$O-enriched material deposited on the Moon by impacts such as comets. Using data that were processed with the standard M^3 thermal correction, a localized enhancement in the 2.8 μm OH⁻ band was observed in the central peak of Bullialdus crater, and interpreted as native lunar water, excavated with materials from depth (Klima et al., 2013). Similarly, a very prominent local OH⁻/H$_2$O anomaly was detected on the lunar surface associated with the presumed silicic pyroclastic complex (Jolliff et al., 2011) Compton Belkovich on the farside (Bhattacharya et al., 2013; Petro et al., 2013; Klima & Petro, 2017). Furthermore, using the thermal correction derived by Li and Milliken (2016a) significant endogenic associated hydration is identified and mapped across some, but not all, of the dark regional pyroclastic deposits (Li & Milliken, 2016b; Milliken & Li, 2017).

18.4.4 Analysis Issues and Challenges

Even though spectroscopic data continue to provide remarkable insight into the mineral character and evolution of the Moon, the data are not perfect or without issues and challenges. Some issues are coupled to a specific instrument, its capabilities, calibration, and operation. Others are inherent in the nature of the data itself. We mention a few of these potential pitfalls here. None are insurmountable, but neither should any be neglected.

18.4.4.1 Mixtures, Breccia, and Soil

As illustrated in previous sections, most lunar rock types consist of a small number of relatively anhydrous minerals that exhibit highly diagnostic absorption bands that are accessible through NIR measurements. Lunar materials are thus well suited for mapping and analysis with modern remote spectroscopic techniques. Nevertheless, most things in nature are ultimately a mixture, and on the scale of returned samples, typical lunar breccias and soils are extremely complex, with much of the mass often in the form of petrographically amorphous or poorly defined matrix material. It is thus essential that the geologic context of materials identified be used to constrain the interpretation of minerals detected. Estimating the abundance of minerals from bulk NIR reflectance spectra also requires independent knowledge of the optical properties of components suspected to be present as well as a variety of analytical methods and models to "deconvolve" or un-entwine the individual parts (see Chapter 2). Because of the (sometimes extreme) nonlinearity of optical effects from different components in a naturally complex mixture, accurate mineral abundance measurements remain a challenge (but continue to improve) (Lucey et al., 2004; Crites et al., 2015). More commonly, valuable relative abundance estimates and comparisons between different terrains are made using the reasonable assumption that processes which alter and mix materials over time (such as impact cratering and space weathering) are approximately the same for a particular region or feature, and direct comparisons of absorption properties can be made for terrain with similar geologic context or morphologic characteristics, i.e., affected by similar degree of structural or alteration history. No specific approach guideline has been developed, but application examples using this assumption include Tompkins and Pieters (1999), Kramer et al. (2011b), and Sun and Li (2015).

18.4.4.2 Thermal Component

Though NIR measurements primarily detect reflected light, thermal emission from warm surfaces begins to contribute to the signal at wavelengths beyond ~2 µm at 1 AU. The warmer the surface, the greater the contribution from the blackbody thermal emission curve (Clark, 1979). This thermal emission has little effect on analysis of the 1 µm band region, so minerals such as olivine and plagioclase, and to some extent pyroxene (particularly low-Ca pyroxene) can be characterized at those wavelengths without issue. However, quantitative analyses of basalts and of other mixtures including high-Ca pyroxenes, spinel, or OH^-/H_2O are contingent on accurate thermal removal. For the case of M^3, mapping and analysis of OH^-/H_2O is severely affected, as thermal emission on dark, warm surfaces may completely mask the absorptions near 3 µm. In principle, if the temperature of a surface is known, a blackbody curve at that temperature could be simply subtracted from the measured signal. However, in practice, determining the true temperature of the material measured, either by measuring it with another sensor or by modeling from first principles, is nontrivial, particularly when only a small portion of the blackbody curve is measured. For M^3 data, thermal removal was initially conducted using an iterative approach to fit the excess thermal component measured at each M^3 surface pixel (Clark et al., 2011). Examples of M^3 spectra from a mare soil, highland soil, and a high-calcium pyroxene (HCP)-rich mare crater before and after thermal removal are shown in Figure 18.5b. This purely empirical method is computationally fast, relative to detailed thermal modeling, and results in a useful corrected spectrum as well as an estimate of surface temperature for each pixel measured on the surface. However, when compared to Diviner thermal measurements, this standard M^3 method tends to underestimate the temperature. This is particularly important when a lunar absorption band overlaps the thermal emission component. For example, when the material contains high-Ca pyroxene, errors can be on the order of 10 K (Clark et al., 2011). An alternative method takes advantage of an apparent power law relationship between the reflectance of many lunar soils at 1.55 µm and at 2.54 µm to derive a thermal correction for the M^3 data (Li & Milliken, 2016a). They show that for eight sample locations at which Diviner temperatures were measured at near the same time of day that the M^3 spectra were obtained, their empirically corrected data produce reflectance data similar to those corrected using the Diviner temperatures. However, their method is largely calibrated using lunar soils and glasses, and may not properly predict the reflectance in fresh, rocky regions with strong mineral absorption features (Li & Milliken, 2016a). With the improvement of our understanding of the thermal behavior of the Moon, new thermophysical models (e.g., Bandfield et al., 2018) are showing promise for thermally correcting M^3 data without having to rely on empirically derived relationships. The next generation of lunar sensors can greatly improve lunar hydration analyses by simultaneous measurement of the thermal properties.

18.4.4.3 Lunar Continuum

As seen in the various spectra of the natural lunar surface in Figures 18.3, 18.4, and 18.5, all spectra exhibit a continuum sloped toward longer wavelengths to different degrees, and the diagnostic features of minerals present are combined with this continuum. There is no simple method to remove or account for the lunar continuum, largely because it has multiple causes that include (1) the remnants (wings or tails) of very strong absorptions elsewhere such as those in the ultraviolet (oxygen–metal) and mid-IR (silicate); (2) wavelength dependent scattering of various nanophase opaque particles; (3) broad charge-transfer absorptions involving Fe and Ti; (4) changing portions of body (internal) and Fresnel (surface) scattering with opacity and wavelength. Each mineral, glass, and matrix material of a rock or soil contribute to the continuum of the measured bulk sample or area, further increasing the complexity of its origin.

Thus, although the properties of an individual lunar continuum appear simple (smoothly varying toward longer wavelengths), in practice approximations must be made, and while some approaches are better than others *all* currently have limitations. A common practice in lunar spectroscopy is to estimate and "remove" the continuum in order to better examine the character of diagnostic absorptions present. This typically involves fitting a continuum shape (straight line or curved) to a part of the measured spectrum at wavelengths believed to be the least affected by the absorption band being studied and dividing the spectrum by the continuum fit (e.g., Pieters, 1993). Such a procedure can be refined to work reasonably well for relatively strong absorption bands of olivine and pyroxene (e.g., see discussion in Moriarty & Pieters, 2016b). It is much less reliable for weak absorptions or absorptions where the long wavelength part of the absorption is not measured (e.g., spinel, OH).

Although a continuum may mask or distort diagnostic features of minerals present, the lunar continuum may also contain valuable information because some of its components are linked to composition as well as weathering (e.g., Hiroi et al., 1997). An empirical general correlation of the visible continuum slope (300–700 nm) of lunar basaltic soils with TiO_2 content has been noted for decades (Charette et al., 1974; Pieters, 1978, 1993; Lucey et al., 1998; Giguere et al., 2000). Although limited to the scope of returned samples, the correlation appears partially linked to the opacity of the bulk soil. However, caution is recommended: the precision of continuum measurements made remotely with modern sensors now far exceeds the accuracy of the correlation itself, and other variables certainly also contribute to patterns observed.

18.5 Expectations for the Future

Serious exploration of the Moon in the decades ahead will benefit enormously from continuing improvement of spectroscopic instruments and associated technologies. With its close proximity to Earth, the Moon can take advantage of capabilities not achievable elsewhere in the Solar System. Two intertwined elements will enable major progress for exploration and utilization of lunar materials: (1) Advanced imaging spectrometers with improved spatial resolution, spectral resolution, and spectral range. Enabling technologic improvements are in detector arrays, optics, electronics, thermal control, and miniaturization. (2) Enhanced downlink (communication) capabilities as well as on-board artificial intelligence (AI) including optical laser communication and high-power system-on-a-chip on-board processing. These capabilities are advancing with investments in space infrastructure for objectives throughout the Solar System. As communication and AI improve, the latency period of delivered results should drop from years to near-real time, opening a new era of exploration. We look forward to quickly identifying a valuable lunar outcrop (from orbit) or a rock in the distance (on the surface) that has just the mineral components sought for scientific inquiry or resource utilization.

Imaging spectrometers, which provide both spatial and spectral information simultaneously, have evolved considerably through recent decades (e.g., Vane et al., 1993; Green et al., 2011). Key measurement requirements of modern imaging spectrometers are specified in terms of (1) spectral, (2) radiometric, (3) spatial, and (4) uniformity characteristics. Modern designs span spectral ranges from the ultraviolet to the thermal IR and are matched to specific radiometric and spatial requirements. Nevertheless, two orthogonal uniformity characteristics that are *essential* for modern spectroscopic analysis are cross-track spectral uniformity (aka "spectral smile") across the field of view (FOV) and the spectral instantaneous field of view (IFOV) uniformity (aka "keystone"). These are illustrated in Figure 18.6 for a pushbroom imaging spectrometer configuration. In the desired high-uniformity system, all spectra measured simultaneously across the FOV have the *same* line-to-line spectral calibration *and* all wavelengths measured for an in-line spectrum are derived from the *same* element on the surface. This latter property is akin to band-to-band registration in multispectral cameras and is critical to achieve the identical footprint for analysis of spectroscopic measurements that may contain hundreds of sampled wavelengths. Achieving excellent measurement uniformity (within 0.1 pixel across an array) is a critical property for imaging spectrometer design and implementation.

State-of-the-art imaging spectrometers take combined advantage of the latest telescope, spectrometer, signal chain, and thermal designs and technologies that together result in desired improved capabilities and products. Two examples of modern imaging spectrometer implementations appropriate for exploration of the Moon and in situ resource discovery are based on Offner and Dyson design forms (Mouroulis et al., 2000), as they provide a high-throughput, compact instrument with excellent spectroscopic uniformity.

One example is an all reflective Offner design (Van Gorp et al., 2014), the Ultra Compact Imaging Spectrometer (UCIS). The UCIS design is scalable for spectral ranges from 300 to 5000 nm and spectral sampling from 3 to 30 nm for landed or orbital experiments. The optical design is based on a compact two-mirror telescope that provides a 30° FOV, 1.4 milliradian IFOV, and a diffraction grating on a convex

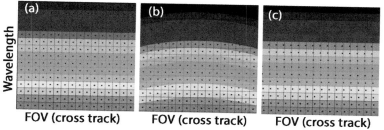

Figure 18.6 Schematic illustration of uniformity requirements across an array detector for a pushbroom type imaging spectrometer in which one spatial dimension and full spectral information are measured simultaneously. An image cube (two spatial and one spectral dimensions) is obtained through a timed sequence (after Green et al., 2011). The dots represent individual locations on the surface. The squares represent the detector array elements. The color bands correspond to different wavelengths of light. (a) The desired high uniformity of an imaging spectrometer: all cross-track elements have the same spectral calibration and all wavelengths of a spectrum are measured simultaneously for the same surface element. (b) An imaging spectrometer with poor cross-track spectral uniformity: spectral measurements (and calibration) varies for each cross track spatial element. (c) An imaging spectrometer with poor IFOV uniformity: the spatial location on the surface varies with the wavelength measured, violating a key requirement for spectroscopy.

surface with a structured blaze as with M^3 (Green et al., 2011) to enable optimized performance over the desired spectral range. The resource needs are ≤3 kg and ≤3 W for implementation on a lander or rover. A second example is the Snow and Water Imaging Spectrometer (SWIS) (Bender et al., 2015), a Dyson imaging spectrometer based on a cubesat form developed for Earth science. The SWIS design is scalable for spectral ranges from 300 to 14,000 nm and sampling from 3 to 30 nm. Excellent radiometric performance is achieved through a high F/1.8 optical speed. This is particularly important for imaging spectrometers in the range from 5000 to 14,000 nm, such as in an airborne F/1.6 thermal IR Dyson imaging spectrometer developed recently (Hook et al., 2013). The optical design of the SWIS instrument uses a slit and a refractive element to direct light to the concave grating, where it is dispersed and reflected back to the refractive element that images the dispersed light on the detector array. High-throughput Dyson spectrometers can be adapted for a range of orbital and surface missions for exploration and resources assessment of the Moon.

Next-generation imaging spectrometers under development provide essential measurements needed for understanding surface composition, resources, and time-variable properties of the Moon. Such instruments deliver high-quality spectra from the ultraviolet to the thermal IR by matching spectral, radiometric, and spatial characteristics to mission requirements from orbit or on the surface. In addition, most modern instruments require less mass, power, and volume resources than their predecessors. With recent and ongoing advances in design and technologies, advanced imaging spectrometer instruments are ready to enable the next, exciting, and important steps in exploration of the Moon from orbit and on the surface.

Acknowledgments

This integration was enabled through support from NASA SSERVI contract # NNA14AB01A. We thank D. Dhingra, M. Staid, K. Donaldson Hanna, and J. Mustard for contributing example spectra.

References

Adams J. & McCord T. (1970) Remote sensing of lunar surface mineralogy: Implications from visible and near-infrared reflectivity of Apollo 11 samples. *Geochimica et Cosmochimica Acta Supplement*, 1, 1937.

Adams J.B. (1974) Visible and near-infrared diffuse reflectance spectra of pyroxenes as applied to remote sensing of solid objects in the Solar System. *Journal of Geophysical Research*, 79, 4829–4836.

Adams J.B. & Jones R.L. (1970) Spectral reflectivity of lunar samples. *Science*, 167, 737–739.

Adams J.B., Pieters C., & McCord T.B. (1974) Orange glass: Evidence for regional deposits of pyroclastic origin on the moon. *Proceedings of the 5th Lunar Planet. Sci. Conf.*, 171–186.

Bandfield J.L., Poston M.J., Klima R.L., & Edwards C.S. (2018) Widespread distribution of OH/H_2O on the lunar surface inferred from spectral data. *Nature Geoscience*, 11(3), 173–177.

Bell P., Mao H., & Weeks R. (1976) Optical spectra and electron paramagnetic resonance of lunar and synthetic glasses: A study of the effects of controlled atmosphere, composition, and temperature. *Proceedings of the Lunar Planet. Sci. Conf.*, 2543–2559.

Bell P.M. & Mao H.K. (1973) Optical and chemical analysis of iron in Luna 20 plagioclase. *Geochimica et Cosmochimica Acta*, 37, 755–759.

Bender H.A., Mouroulis P., Smith C.D., et al. (2015) Snow and water imaging spectrometer (SWIS): Optomechanical and system design for a CubeSat-compatible instrument. In: *Imaging Spectrometry XX* (T.S. Pagano & J.F. Silny, eds.). *SPIE Proceedings*, 9611.

Besse S., Sunshine J., & Gaddis L. (2014) Volcanic glass signatures in spectroscopic survey of newly proposed lunar pyroclastic deposits. *Journal of Geophysical Research*, 119, 355–372.

Bhattacharya S., Saran S., Dagar A., et al. (2013) Endogenic water on the Moon associated with non-mare silicic volcanism: Implications for hydrated lunar interior. *Current Science*, 105, 685–691.

Blewett D.T., Coman E.I., Hawke B.R., Gillis-Davis J.J., Purucker M.E., & Hughes C.G. (2011) Lunar swirls: Examining crustal magnetic anomalies and space weathering trends. *Journal of Geophysical Research*, 116, E02002, DOI:10.1029/2010JE003656.

Boardman J.W., Pieters C.M., Green R.O., et al. (2011) Measuring moonlight: An overview of the spatial properties, lunar coverage, selenolocation, and related Level 1B products of the Moon Mineralogy Mapper. *Journal of Geophysical Research*, 116, E00G14, DOI:10.1029/2010JE003730.

Boyd A.K., Robinson M.S., & Sato H. (2012) Lunar Reconnaissance Orbiter wide angle camera photometry: An empirical solution. *43rd Lunar Planet. Sci. Conf.*, Abstract # 2795.

Burns R.G. (1993) *Mineralogical applications of crystal field theory.* Cambridge University Press, Cambridge.

Cahill J.T., Lucey P.G., Gillis J.J., & Steutel D. (2004) Verification of quality and compatibility for the newly calibrated Clementine NIR data set. *35th Lunar Planet. Sci. Conf.*, Abstract #1469.

Charette M.P., McCord T.B., Pieters C., & Adams J.B. (1974) Application of remote spectral reflectance measurements to lunar geology classification and determination of titanium content of lunar soils. *Journal of Geophysical Research*, 79, 1605–1613.

Cheek L., Pieters C.M., Parman S., Dyar M.D., Speicher E.A., & Cooper R.F. (2011a) Spectral characteristics of plagioclase with variable iron content: Applications to remote sensing of the lunar crust. *42nd Lunar Planet. Sci. Conf.*, Abstract #1617.

Cheek L.C. & Pieters C.M. (2014) Reflectance spectroscopy of plagioclase-dominated mineral mixtures: Implications for characterizing lunar anorthosites remotely. *American Mineralogist*, 99, 1871–1892.

Cheek L.C., Pieters C.M., Boardman J.W., et al. (2011) Goldschmidt crater and the Moon's north polar region: Results from the Moon Mineralogy Mapper (M^3). *Journal of Geophysical Research*, 116, E00G02, DOI:10.1029/2010je003702.

Cheek L.C., Donaldson H.K.L., Pieters C.M., Head J.W., & Whitten J.L. (2013) The distribution and purity of anorthosite across the Orientale Basin: New perspectives from Moon Mineralogy Mapper data. *Journal of Geophysical Research*, 118, 1805–1820.

Clark R.N. (1979) Planetary reflectance measurements in the region of planetary thermal emission. *Icarus*, 40, 94–103.

Clark R.N. (2009) Detection of adsorbed water and hydroxyl on the Moon. *Science*, 326, 562–564.

Clark R.N., Pieters C.M., Green R.O., Boardman J.W., & Petro N.E. (2011) Thermal removal from near-infrared imaging spectroscopy data of the Moon. *Journal of Geophysical Research*, **116**, E00G16, DOI:10.1029/2010JE003751.

Cloutis E.A. & Gaffey M.J. (1991) Spectral-compositional variations in the constituent minerals of mafic and ultramafic assemblages and remote sensing implications. *Earth, Moon, and Planets*, **53**, 11–53.

Cloutis E.A., Sunshine J.M., & Morris R.V. (2004) Spectral reflectance–compositional properties of spinels and chromites: Implications for planetary remote sensing and geothermometry. *Meteoritics and Planetary Science*, **39**, 545–565.

Conel J.E. & Nash D.B. (1970) Spectral reflectance and albedo of Apollo 11 lunar samples: Effects of irradiation and vitrification and comparison with telescopic observations. *Apollo 11 Lunar Sci. Conf.*, 2013–2023.

Crites S.T., Lucey P.G., & Taylor G.J. (2015) The mafic component of the lunar crust: Constraints on the crustal abundance of mantle and intrusive rock, and the mineralogy of lunar anorthosites. *American Mineralogist*, **100**, 1708–1716.

Denevi B.W., Robinson M.S., Boyd A.K., Sato H., Hapke B.W., & Hawke B.R. (2014) Characterization of space weathering from Lunar Reconnaissance Orbiter Camera ultraviolet observations of the Moon. *Journal of Geophysical Research*, **119**, 976–997.

Dhingra D., Pieters C.M., & Head J.W. (2015) Multiple origins for olivine at Copernicus crater. *Earth and Planetary Science Letters*, **420**, 95–101.

Donaldson Hanna K.L., Cheek L.C., Pieters C.M., et al. (2014) Global assessment of pure crystalline plagioclase across the Moon and implications for the evolution of the primary crust. *Journal of Geophysical Research*, **119**, 1516–1545.

Dyar M.D., Hibbitts C.A., & Orlando T.M. (2010) Mechanisms for incorporation of hydrogen in and on terrestrial planetary surfaces. *Icarus*, **208**, 425–437.

Eliason E., Isbell C., Lee E., et al. (1999) Mission to the Moon: The Clementine UVVIS Global Mosaic. PDS CL_4001–4078. www.lpi.usra.edu/lunar/tools/clementine/instructions/UVVIS_DIM_Info.html

Eliason E.M., Lee E.M., Becker T.L., et al. (2003) A near-infrared (NIR) global multispectral map of the Moon from Clementine. *34th Lunar Planet. Sci. Conf.*, Abstract #2093.

Feldman W.C., Maurice S., Binder A.B., Barraclough B.L., Elphic R.C., & Lawrence D.J. (1998) Fluxes of fast and epithermal neutrons from lunar Prospector: Evidence for water ice at the lunar poles. *Science*, **281**, 1496–1500.

Gaddis L.R., Staid M.I., Tyburczy J.A., Hawke B.R., & Petro N.E. (2003) Compositional analyses of lunar pyroclastic deposits. *Icarus*, **161**, 262–280.

Giguere T.A., Taylor G.J., Hawke B.R., & Lucey P.G. (2000) The titanium contents of lunar mare basalts. *Meteoritics and Planetary Science*, **35**, 193–200.

Green R. (2016) 30 years of thermally controlled imaging spectrometers for Earth and planetary science. *46th International Conference on Environmental Systems*.

Green R., Pieters C.M., Mouroulis P., et al. (2011) The Moon Mineralogy Mapper (M^3) imaging spectrometer for lunar science: Instrument description, calibration, on-orbit measurements, science data calibration and on-orbit validation. *Journal of Geophysical Research*, **116**, E00G19, DOI:10.1029/2011JE003797.

Hapke B. (2001) Space weathering from Mercury to the asteroid belt. *Journal of Geophysical Research*, **106**, 10039–10073.

Hawke B.R., Peterson C.A., Blewett D.T., et al. (2003) Distribution and modes of occurrence of lunar anorthosite. *Journal of Geophysical Research*, **108**, 5050, DOI:10.1029/2002JE001890.

Hazen R.M., Bell P.M., & Mao H.K. (1978) Effects of compositional variation on absorption spectra of lunar pyroxenes. *Proceedings of the 9th Lunar Planet. Sci. Conf.*, 2919–2934.

Heiken G., Vaniman D., & French B.M. (1991) *Lunar sourcebook: A user's guide to the Moon*. Cambridge University Press, New York.

Hiroi T., Pieters C., & Morris R. (1997) New considerations for estimating lunar soil maturity from VIS-NIR reflectance spectroscopy. *28th Lunar Planet. Sci. Conf.*, Abstract #1152.

Hook S.J., Johnson W.R., & Abrams M.J. (2013) NASA's Hyperspectral Thermal Emission Spectrometer (HyTES). In: *Thermal infrared remote sensing: Sensors, methods, applications* (C. Kuenzer & S. Dech, eds.). Springer, Dordrecht, 93–115.

Isaacson P.J. & Pieters C.M. (2009) Northern Imbrium noritic anomaly. *Journal of Geophysical Research*, **114**, E09007, DOI:10.1029/2008JE003293.

Isaacson P.J., Sarbadhikari A.B., Pieters C.M., et al. (2011a) The lunar rock and mineral characterization consortium: Deconstruction and integrated mineralogical, petrologic, and spectroscopic analyses of mare basalts. *Meteoritics and Planetary Science*, **46**, 228–251.

Isaacson P.J., Pieters C.M., Besse S., et al. (2011b) Remote compositional analysis of lunar olivine rich lithologies with Moon Mineralogy Mapper (M^3) spectra. *Journal of Geophysical Research*, **116**, E00G11, DOI:10.1029/2010JE003731.

Johnson J.R. & Hörz F. (2003) Visible/near-infrared spectra of experimentally shocked plagioclase feldspars. *Journal of Geophysical Research*, **108**, 5120, DOI:10.1029/2003JE002127.

Jolliff B.I., Wieczorek M.A., Shearer C.K., & Neal C.R. (2006) *New views of the Moon*. Reviews in Mineralogy and Geochemistry, **60**. Geochemical Society.

Jolliff B.L., Wiseman S.A., Lawrence S.J., et al. (2011) Non-mare silicic volcanism on the lunar farside at Compton–Belkovich. *Nature Geoscience*, **4**, 566–571.

Keller L. & Zhang S. (2015) Rates of space weathering in lunar soils. *Space Weathering of Airless Bodies: An Integration of Remote Sensing Data, Laboratory Experiments and Sample Analysis Workshop*, Abstract #2056.

Keller L., Berger E., Christoffersen R., & Zhang S. (2016) Direct determination of the space weathering rates in lunar soils and Itokawa regolith from sample analyses. *47th Lunar Planet. Sci. Conf.*, Abstract #2525.

Keller L.P. & McKay D.S. (1993) Discovery of vapor deposits in the lunar regolith. *Science*, **261**, 1305–1307.

Keller L.P. & McKay D.S. (1997) The nature and origin of rims on lunar soil grains. *Geochimica et Cosmochimica Acta*, **61**, 2331–2341.

Klima R., Cahill J., Hagerty J., & Lawrence D. (2013) Remote detection of magmatic water in Bullialdus crater on the Moon. *Nature Geoscience*, **6**, 737–741.

Klima R., Buczkowski D., Ernst C., & Greenhagen B. (2017) Geological and spectral analysis of low-calcium pyroxenes around the Imbrium Basin on the Moon. *48th Lunar Planet. Sci. Conf.*, Abstract #2502.

Klima R.L. & Petro N.E. (2017) Remotely distinguishing and mapping endogenic water on the Moon. *Philosophical Transactions of the Royal Society A*, **375**, 20150391.

Klima R.L., Pieters C.M., & Dyar M.D. (2007) Spectroscopy of synthetic Mg-Fe pyroxenes I: Spin-allowed and spin-forbidden crystal field bands in the visible and near-infrared. *Meteoritics and Planetary Science*, **42**, 235–253.

Klima R.L., Pieters C.M., & Dyar M.D. (2008) Characterization of the 1.2 micrometer M1 pyroxene band: Extracting cooling history from near-IR spectra of pyroxenes and pyroxene-dominated rocks. *Meteoritics and Planetary Science*, **43**, 1591–1604.

Klima R.L., Dyar M.D., & Pieters C.M. (2011a) Near-infrared spectra of clinopyroxenes: Effects of calcium content and crystal structure. *Meteoritics and Planetary Science*, **46**, 379–395.

Klima R.L., Pieters C.M., Boardman J.W., et al. (2011b) New insights into lunar petrology: Distribution and composition of prominent low Ca pyroxene exposures as observed by the Moon Mineralogy Mapper (M^3). *Journal of Geophysical Research*, **116**, DOI:10.1029/2010JE003719.

Kramer G.Y., Besse S., Dhingra D., et al. (2011a) M^3 spectral analysis of lunar swirls and the link between optical maturation and surface hydroxyl formation at magnetic anomalies. *Journal of Geophysical Research*, **116**, E00G18, DOI:10.1029/2010JE003729.

Kramer G.Y., Besse S., Nettles J., et al. (2011b) Newer views of the Moon: Comparing spectra from Clementine and the Moon Mineralogy Mapper. *Journal of Geophysical Research*, **116**, E00G04, DOI:10.1029/2010JE003728.

Kramer G.Y., Kring D.A., Nahm A.L., & Pieters C.M. (2013) Spectral and photogeologic mapping of Schrödinger Basin and implications for post-South Pole-Aitken impact deep subsurface stratigraphy. *Icarus*, **223**, 131–148.

Li S. & Milliken R.E. (2016a) An empirical thermal correction model for Moon Mineralogy Mapper data constrained by laboratory spectra and Diviner temperatures. *Journal of Geophysical Research*, **121**, 2081–2107.

Li S. & Milliken R.E. (2016b) Heterogeneous water content in the lunar interior: Insights from orbital detection of water in pyroclastic deposits and silicic domes. *47th Lunar Planet. Sci. Conf.*, Abstract #1568.

Li S. & Milliken R.E. (2017) Water on the surface of the Moon as seen by the Moon Mineralogy Mapper: Distribution, abundance, and origins. *Science Advances*, **3**, e1701471.

LSCC data. Spectra: www.planetary.brown.edu/relabdocs/LSCCsoil.html; Composition: pgi.utk.edu/lunar-soil-characterization-consortium-lscc-data/

Lucey P.G. (2004) Planets-L08701. Mineral maps of the Moon. *Geophysical Research Letters*, **31**, L08701, DOI:10.1029/2003GL019406.

Lucey P.G., Blewett D.T., & Hawke B.R. (1998) Mapping the FeO and TiO_2 content of the lunar surface with multispectral imagery. *Journal of Geophysical Research*, **103**, 3679–3699.

Lucey P.G., Blewett D.T., Eliason E.M., et al. (2000) Optimized calibration constants for the Clementine NIR camera. *31st Lunar Planet. Sci. Conf.*, Abstract #1273.

Lucey P.G., Norman J.A., Crites S.T., et al. (2014) A large spectral survey of small lunar craters: Implications for the composition of the lunar mantle. *American Mineralogist*, **99**, 2251–2257.

Lundeen S., McLaughlin S., & Alanis R. (2011) Moon Mineralogy Mapper Data Product software interface specification. PDS document Version 9.10. Jet Propulsion Laboratory, JPL D-39032, Pasadena, CA.

McCord T.B. & Johnson T.V. (1970) Lunar spectral reflectivity (0.30 to 2.50 microns) and implications for remote mineralogical analysis. *Science*, **169**, 855–858.

McCord T.B., Clark R.N., Hawke B.R., et al. (1981) Moon: Near-infrared spectral reflectance, a first good look. *Journal of Geophysical Research*, **86**, 10883–10892.

McCord T.B., Taylor L.A., Combe J.P., et al. (2011) Sources and physical processes responsible for OH/H_2O in the lunar soil as revealed by the Moon Mineralogy Mapper (M^3). *Journal of Geophysical Research*, **116**, E00G05, DOI:10.1029/2010JE003711.

McEwen A.S. & Robinson M. (1997) Mapping of the Moon by Clementine. *Advances in Space Research*, **19**(10), 1523–1533.

McEwen A.S., Eliason E., Lucey P., et al. (1998) Summary of radiometric calibration and photometric normalization steps for the Clementine UVVIS images. *29th Lunar Planet. Sci. Conf.*, Abstract 1466–1467.

Milliken R.E. & Li S. (2017) Remote detection of widespread indigenous water in lunar pyroclastic deposits. *Nature Geoscience*, **10**, 561–565.

Moriarty III D.P. & Pieters C.M. (2016a) South Pole–Aitken Basin as a probe to the lunar interior. *47th Lunar Planet. Sci. Conf.*, Abstract #1763.

Moriarty III D.P. & Pieters C.M. (2016b) Complexities in pyroxene compositions derived from absorption band centers: Examples from Apollo samples, HED meteorites, synthetic pure pyroxenes, and remote sensing data. *Meteoritics and Planetary Science*, **51**, 207–234.

Moriarty III D.P. & Pieters C.M. (2018) The character of South Pole-Aitken Basin: Patterns of surface and subsurface composition. *Journal of Geophysical Research*, **123**, 729–747.

Mouroulis P., Green R.O., & Chrien T.G. (2000) Design of pushbroom imaging spectrometers for optimum recovery of spectroscopic and spatial information. *Applied Optics*, **39**, 2210–2220.

Nakamura R., Yamamoto S., Matsunaga T., et al. (2012) Compositional evidence for an impact origin of the Moon's Procellarum basin. *Nature Geoscience*, **5**(11),775–778, DOI:10.1038/ngeo1614.

National Research Council, Space Studies Board. (2007) *Scientific context for exploration of the Moon*. National Academies Press, Washington, DC.

Noble S.K., Pieters C.M., Hiroi T., & Taylor L.A. (2006) Using the modified Gaussian model to extract quantitative data from lunar soils. *Journal of Geophysical Research*, **111**, DOI:10.1029/2006JE002721.

Noble S.K., Pieters C.M., & Keller L.P. (2007) An experimental approach to understanding the optical effects of space weathering. *Icarus*, **192**, 629–642.

Nozette S., Rustan P., Pleasance L.P., et al. (1994) The Clementine Mission to the Moon: Scientific overview. *Science*, **266**, 1835–1839.

Ohtake M., Matsunaga T., Haruyama J., et al. (2009) The global distribution of pure anorthosite on the Moon. *Nature*, **461**, 236–240.

Ohtake M., Matsunaga T., Yokota Y., et al. (2010) Deriving the absolute reflectance of lunar surface using SELENE (Kaguya) multiband imager data. *Space Science Reviews*, **154**, 57–77.

Ohtake M., Pieters C., Isaacson P., et al. (2013) One Moon, many measurements 3: Spectral reflectance. *Icarus*, **226**, 364–374.

Petro N.E., Isaacson P.J., Pieters C.M., Jolliff B.L., Carter L.M., & Klima R.L. (2013) Presence of OH/H_2O associated with the lunar Compton-Belkovich volcanic complex identified by the Moon Mineralogy Mapper (M^3). *44th Lunar Planet. Sci. Conf.*, 2688.

Pieters C.M. (1978) Mare basalt types on the front side of the moon: A summary of spectral reflectance data. *Proceedings of the 9th Lunar Sci. Conf. (Suppl. 10, Geochimica et Cosmochimica Acta)*, 2825–2849.

Pieters C.M. (1982) Copernicus crater central peak: Lunar mountain of unique composition. *Science*, **215**, 59–61.

Pieters C.M. (1993) Compositional diversity and stratigraphy of the Lunar crust derived from reflectance spectroscopy. In: *Remote geochemical analysis: Elemental and mineralogical composition* (C. Pieters & P. Englert, eds.). Cambridge University Press, Cambridge, 309–339.

Pieters C.M. (1996) Plagioclase and maskelynite diagnostic features. *27th Lunar Planet. Science Conf.*, Abstract #1031.

Pieters C.M. (2017) Origin and importance of "featureless" plagioclase on the Moon. *5th Eur. Lunar Symp.*, Munster, Germany.

Pieters C.M. & Garrick-Bethell I. (2015) Hydration variations at lunar swirls. *46th Lunar Planet. Sci. Conf.*, Abstract #2120.

Pieters C.M. & Noble S.K. (2016) Space weathering on airless bodies. *Journal of Geophysical Research*, **121**, 1865–1884.

Pieters C.M. & Taylor L.A. (2003) Systematic global mixing and melting in lunar soil evolution. *Geophysical Research Letters*, **30**, 2048, DOI:10.1029/2003GL018212.

Pieters C.M., Head J.W., Adams J.B., McCord T.B., Zisk S.H., & Whitford Stark J.L. (1980) Late high titanium basalts of the western maria: Geology of the Flamsteed region of Oceanus Procellarum. *Journal of Geophysical Research*, **85**, 3913–3938.

Pieters C.M., Taylor L.A., Noble S.K., et al. (2000) Space weathering on airless bodies: Resolving a mystery with lunar samples. *Meteoritics and Planetary Science*, **35**, 1101–1107.

Pieters C.M., Head III J.W., Gaddis L.R., Jolliff B.L., & Duke M. (2001) Rock types of South Pole-Aitken Basin and extent of basaltic volcanism. *Journal of Geophysical Research*, **106**, 28,001–28,022.

Pieters C.M., Boardman J., Buratti B., et al. (2009a) Mineralogy of the lunar crust in spatial context: First results from the Moon Mineralogy Mapper (M^3). *40th Lunar Planet. Sci. Conf.*, Abstract #2052.

Pieters C.M., Goswami J.N., Clark R., et al. (2009b) Character and spatial distribution of OH/H_2O on the surface of the Moon seen by M^3 on Chandrayaan-1. *Science*, **326**, 568–572.

Pieters C.M., Besse S., & Boardman J. (2011) Mg spinel lithology: A new rock type on the lunar farside. *Journal of Geophysical Research*, **116**, E00G08, DOI:10.1029/2010JE003727.

Pieters C.M., Boardman J.W., Ohtake M., et al. (2013) One Moon, many measurements 1: Radiance values. *Icarus*, **226**, 951–963.

Pieters C.M., Hanna K.D., Cheek L., et al. (2014) The distribution of Mg-spinel across the Moon and constraints on crustal origin. *American Mineralogist*, **99**, 1893–1910.

Prissel T., Parman S., Jackson C., et al. (2014) Pink Moon: The petrogenesis of pink spinel anorthosites and implications concerning Mg-suite magmatism. *Earth and Planetary Science Letters*, **403**, 144–156.

Robinson M., Brylow S., Tschimmel M., et al. (2010) Lunar Reconnaissance Orbiter Camera (LROC) instrument overview. *Space Science Reviews*, **150**, 81–124.

Sasaki S., Nakamura K., Hamabe Y., Kurahashi E., & Hiroi T. (2001) Production of iron nanoparticles by laser irradiation in a simulation of lunar-like space weathering. *Nature*, **410**, 555–557.

Staid M.I. & Pieters C.M. (2001) Mineralogy of the last lunar basalts: Results from Clementine. *Journal of Geophysical Research*, **106**, 27887–27900.

Staid M.I., Pieters C.M., Besse S., et al. (2011) The mineralogy of late stage lunar volcanism as observed by the Moon Mineralogy Mapper on Chandrayaan 1. *Journal of Geophysical Research*, **116**, E00G10, DOI:10.1029/2010JE003735.

Stöffler D. (1972) Deformation and transformation of rock-forming minerals by natural and experimental shock processes. 1.Behavior of minerals under shock compression. *Fortschritte der Mineralogie*, **49**, 50–113.

Stöffler D. (1974) Deformation and transformation of rock-forming minerals by natural and experimental shock processes: II. Physical properties of shocked minerals. *Fortschritte der Mineralogie*, **51**, 256–289.

Sun Y.S. & Li L.L. (2015) Characterization of lunar crust mineralogy with M^3 data. *46th Lunar Planet. Sci. Conf.*, Abstract #2941.

Sunshine J.M. & Pieters C.M. (1993) Estimating modal abundances from the spectra of natural and laboratory pyroxene mixtures using the modified Gaussian model. *Journal of Geophysical Research*, **98**, 9075–9087.

Sunshine J.M., Farnham T.L., Feaga L.M., et al. (2009) Temporal and spatial variability of lunar hydration as observed by the Deep Impact spacecraft. *Science*, **326**, 565–568.

Taylor L.A., Pieters C.M., Keller L.P., Morris R.V., & McKay D.S. (2001) Lunar mare soils: Space weathering and the major effects of surface correlated nanophase Fe. *Journal of Geophysical Research*, **106**, 27,985–27,999.

Taylor L.A., Pieters C., Patchen A., et al. (2010) Mineralogical and chemical characterization of lunar highland soils: Insights into the space weathering of soils on airless bodies. *Journal of Geophysical Research*, **115**, E02002, DOI:10.1029/2009JE003427.

Tompkins S. & Pieters C.M. (1999) Mineralogy of the lunar crust: Results from Clementine. *Meteoritics and Planetary Science*, **34**, 25–41.

Tompkins S. & Pieters C.M. (2010) Spectral characteristics of lunar impact melts and inferred mineralogy. *Meteoritics and Planetary Science*, **45**, 1152–1169.

Van Gorp B., Mouroulis P., Blaney D.L., Green R.O., Ehlmann B.L., & Rodriguez J.I. (2014) Ultra-compact imaging spectrometer for remote, in situ, and microscopic planetary mineralogy. *Journal of Applied Remote Sensing*, **8**, 084988.

Vane G. (1993) Imaging spectrometry of the Earth and other Solar System bodies. In: *Remote geochemical analysis: Elemental and mineralogical composition* (C.M. Pieters & P.A.J. Englert, eds.). Cambridge University Press, Cambridge, 121–143.

Watson K., Murray B.C., & Brown H. (1961) The behavior of volatiles on the lunar surface. *Journal of Geophysical Research*, **66**, 3033–3045.

Yamamoto S., Nakamura R., Matsunaga T., et al. (2010) Possible mantle origin of olivine around lunar impact basins detected by SELENE. *Nature Geoscience*, **3**, 533–536.

Yamamoto S., Matsunaga T., Ogawa Y., et al. (2011) Preflight and in-flight calibration of the spectral profiler on board SELENE (Kaguya). *IEEE Transactions on Geoscience and Remote Sensing*, **49**, 4660–4676.

Yamamoto S., Matsunaga T., Ogawa Y., et al. (2014) Calibration of NIR 2 of spectral profiler onboard Kaguya/SELENE. *IEEE Transactions on Geoscience and Remote Sensing*, **52**, 6882–6898.

Yamamoto S., Nakamura R., Matsunaga T., et al. (2015) Featureless spectra on the Moon as evidence of residual lunar primordial crust. *Journal of Geophysical Research*, **120**, 2190–2205.

19
Spectral Analyses of Asteroids

JOSHUA P. EMERY, CRISTINA A. THOMAS, VISHNU REDDY, AND NICHOLAS A. MOSKOVITZ

19.1 Introduction to Asteroids

The Solar System is chock-full of a terrific diversity of planets and planetary bodies, and astronomers are now finding an even greater variety around other stars. How did these planets come to be? We cannot answer this question by only studying the planets themselves, as they are the product of complex formation processes and long histories of evolution. The geologic evolution, particularly melting and subsequent geochemical processing, significantly alters the original constituents from which the planets were built.

The Solar System is, fortunately, full of many smaller bodies that record various stages of its formation and evolution. Asteroids are relatively small (less than a few hundred km), mostly rocky (though some are metallic) bodies that orbit the Sun (Figure 19.1a). The largest concentration is in the Main Belt, between about 2.2 and 3.3 AU (Figure 19.1b), where there are estimated to be ~1 million asteroids larger than 1 km (Gladman et al., 2009; O'Brien & Sykes, 2011). Those asteroids with perihelion distances smaller than 1.3 AU are defined as near-Earth asteroids (NEAs). A few groups also reside beyond the Main Belt, including the Cybeles (near 3.5 AU, between the 2:1 and 5:3 mean motion resonances with Jupiter), the Hildas (near 4 AU, in the 2:3 mean motion resonance), and the Trojans (5.2 AU, in a 1:1 resonance with Jupiter, in the stable Lagrange regions).

Asteroids are a mixture of primitive material and material that has been physically and geochemically processed to varying degrees. So-called "primitive" asteroids are those that have not melted or been chemically altered since accreting from the solar nebula. These primitive asteroids provide a direct window into the compositions out of which planets formed and the thermochemical conditions in different regions of the nebula. Though not melted, there is evidence in the meteorite record and from telescopic spectroscopy that some otherwise primitive asteroids experienced gentle heating. These primitive bodies therefore conceal within them clues to the earliest stages of planetesimal heating and alteration. Some asteroids show evidence that they have undergone partial melting, as inferred from petrologic analyses of spectrally analogous meteorites. These asteroids therefore record a further stage in the alteration process, where planetesimals heated sufficiently to begin melting some of the component minerals. Large planetesimals did fully melt and differentiate. Many did so with compositions (e.g., H_2O content) different

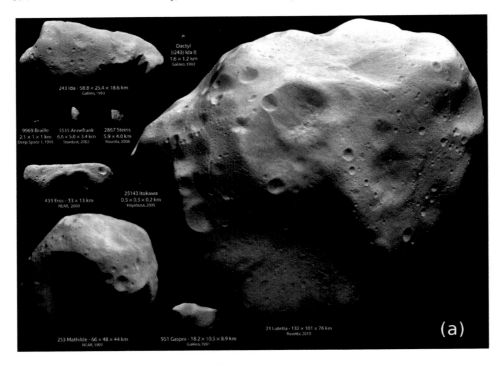

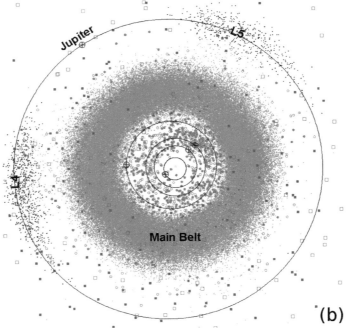

Figure 19.1 Overview of asteroids. (a) Images of several asteroids visited by spacecraft (modified from collage produced by E. Lakdawalla of the Planetary Society). (b) Overview of asteroid populations of the inner Solar System. Green points are Main Belt asteroids, red points are near-Earth asteroids, black dots are Trojan asteroids (labeled L4 and L5), and blue symbols are comets (modified from a plot produced by the Minor Planet Center, www.minorplanetcenter.net/).

from the planets (e.g., McSween et al., 2002), so these evolved bodies provide a laboratory for investigating the differentiation process.

In addition to representing many of the alteration processes experienced by planets and their precursors, asteroid surfaces have been exposed to space for up to 4.5 Gyr. Their surfaces are therefore excellent laboratories for investigating the interactions of matter with the space environment. Asteroids are much less massive than the planets (the entire mass of the current Main Belt is only about 4% the mass of the Moon, or 0.05% the mass of Earth). Owing to their diminutive size and large numbers, the population records gravitational sculpting during the long dynamic history of the Solar System. Combining compositional and dynamical studies of asteroids is a powerful tool for unraveling that history. Asteroids also contain potentially valuable resources. When humans eventually become a spacefaring civilization, knowledge of the distributions of those resources, as can be determined by spectral observations, will be a valuable commodity.

The Solar System contains other groups of small bodies as well, most notably comets, irregular satellites, and Kuiper Belt objects. These groups are generally defined based on their current orbital properties, but they likely represent a continuum of compositions. Investigating links among different categories of small bodies provides deep insight into Solar System evolution. For instance, simulations of the effects of giant planet migration on small-body populations (i.e., the Nice model) suggest that Trojan asteroids likely originated in the Kuiper Belt (Morbidelli et al., 2005; Nesvorný et al., 2013), and Hilda group asteroids may have as well (Levison et al., 2009). When comets lose a sufficient fraction of their ice, they remain inactive, thereby masquerading as asteroids (e.g., Hartmann et al., 1987; Weissman et al., 2002). Nevertheless, this chapter will focus on the objects traditionally defined as asteroids, as described earlier. Comets are discussed in Chapter 20.

19.2 Asteroid Surfaces

Primitive bodies are composed of the material that condensed from the solar nebula in the region in which they accreted. In the region that asteroids formed, crystalline olivine and pyroxene comprised the bulk of the rocky portion of these condensates (e.g., Grossman, 1972; Henning, 2010). Amorphous silicates of olivine and pyroxene composition were also present and increased in abundance with increasing distance from the Sun. At some distance, nebular temperatures decreased sufficiently for H_2O to condense (e.g., Lewis, 1972). This distance is commonly known as the snow line or frost line, and models suggest that it occurred in what is now the Main Belt (though it likely moved around radially by up to a few AU over the nebular lifetime; e.g., Dodson-Robinson et al., 2009). Asteroids that accreted beyond the snow line contained a significant fraction of H_2O. If the interior of an ice-rich asteroid achieves high enough temperatures to melt ice, the liquid H_2O drives hydration reactions, forming hydrated silicates and other hydrothermal products (e.g., carbonates). Complex organic molecules were not a significant component of the condensate fraction in the inner nebula (e.g., Prinn & Fegley, 1989). Farther from the Sun, nebular temperatures may have been low enough that interstellar organic material may have been incorporated into those distant small bodies (e.g., Kuiper Belt objects). From this (simplified) view of the material available to accrete into primitive asteroids, the materials that characterize primitive body compositions include crystalline and amorphous olivine and pyroxene, phyllosilicates (e.g., serpentine-like minerals) from reactions after accretion, carbonates, organic molecules, and H_2O. Nebular models also predict, and meteorite analyses show, the presence of metal, oxides, and sulfides (Section 19.3).

Remote spectral characterization of olivine and pyroxene chemistry and mineralogy on asteroid surfaces provides significant insight into geochemical processing of more evolved bodies. When a rock (asteroids in this case) experiences temperatures above the melting point of one of its minerals, or a mineral is heated above the melting point of one of its solid-solution endmembers, partial melting occurs. For chondritic (primitive) compositions, crystalline olivine and pyroxene still dominate the products of partial melting, though with different chemistries. When sufficiently heated, phyllosilicates lose their water and revert to olivine or pyroxene mineralogies. Full melting of the asteroid leads to significant compositional alteration, the products of which depend on the pressure, temperature, and oxygen fugacity as well as the starting compositions. Nevertheless, crystalline olivine and pyroxene are still important products over a wide range of conditions, though other silicate groups, such as feldspars, may also be abundant. Spectral characterization of phyllosilicate mineralogy is similarly useful for bodies that accreted with a significant fraction of H_2O.

The texture of asteroid surfaces also contributes to their spectral properties. Comminution of the surfaces over 4.5 Gyr forms fine-grained regoliths on the oldest surfaces. Smaller asteroids generally have younger surfaces, and thermal inertia measurements indicate that many asteroids smaller than ~10 km have somewhat coarser grained regoliths (Delbó et al., 2015), as was seen first-hand by the Hayabusa images of the asteroid Itokawa. The very low gravity regime on asteroid surfaces may enable formation of very underdense, extreme "fairy-castle" structures, which in turn could affect spectra (Emery et al., 2006; Vernazza et al., 2012).

Irradiation of asteroid surfaces by high-speed solar wind ions and bombardment by micrometeorites – space weathering – can also affect measured spectra of asteroid surfaces. On the Moon, space weathering causes glassy silicate rims with embedded nm-scale Fe-metal particles to form on surface regolith grains (Chapters 13 and 18; Hapke, 2001; Pieters & Noble, 2016). These rims reduce spectral contrast of absorption bands in the visible and near-infrared (NIR), lower the albedo, and redden the spectra (cause reflectance to increase more steeply with increasing wavelength). Absorption band centers do not appear to shift (e.g., Loeffler et al., 2009), so the underlying mineralogy and chemistry are still accessible to spectroscopy. At mid-infrared (mid-IR) wavelengths, effects are less pronounced, but the rims cause emissivity features to "soften" and decrease somewhat in spectral contrast (Brucato et al., 2004). For S-type asteroids, space weathering effects are expected to be similar to lunar-style, though starting composition may affect the relative expression of each effect. Because of their low albedos and weaker-to-absent absorptions at NIR wavelengths, C-type asteroids show much weaker effects from space weathering; slight reddening and bluing each occur for different samples (Brunetto et al., 2015). In irradiation experiments of hydrated meteorites, Lantz et al. (2017) found that the sharp 2.71-μm OH-stretch absorption broadened and moved to longer wavelengths (~2.80 μm), suggesting that space weathering may affect the mineral structure of phyllosilicates. Another effect of solar wind irradiation observed on the Moon is formation of OH from implanted H^+. This effect appears spectrally as a weak (few %) absorption feature near 2.8 to 3.0 μm (Clark, 2009; Pieters et al., 2009; Sunshine et al., 2009). Rivkin et al. (2018) report similar weak absorptions on the two largest near-Earth asteroids, Eros and Ganymed, presumably from a similar process.

Remote spectroscopy probes only the uppermost portion of the regolith (mm-scale at most). It is therefore natural to wonder whether measured spectra of asteroids are truly representative of their near-surface composition, or if space weathering and infall of exogenous material completely masks the true composition. Comminution and space weathering are surficial processes; the surface properties resulting from those processes are not perfectly representative of unaltered material. There is no evidence, however, that asteroid regoliths are different compositions than the asteroid below. Lunar soils are similar compositions to lunar rocks (Chapter 18). Except for space weathering effects, telescopic spectra of asteroids agree well with laboratory spectra of meteorites. Spacecraft missions to asteroids have not revealed any spots that would suggest the surface is masking some fundamentally different near-surface composition. The Dawn mission (see Chapter 20) uncovered spectral evidence of exogenous carbonaceous material on 4 Vesta, but that material is a minor component of the surface and does not mask the fundamentally basaltic nature revealed in telescopic and spacecraft spectra (De Sanctis et al., 2012; McCord et al., 2012; Reddy et al., 2012). We therefore fully expect that telescopic spectra of asteroids are generally representative of the asteroids' true composition. An exception to this rule may be icy asteroids in the outer Main Belt and beyond. Ice in the regolith of asteroids that orbit ~3–5.2 AU from the Sun will sublimate to form an ice-free layer cm to m thick (depending on distance and obliquity; Guilbert-Lepoutre, 2014; Schorghofer, 2016). The Dawn mission detected some ice exposures on 1 Ceres (Combe et al., 2016), and it will be interesting to see whether NASA's Lucy mission will see any ice-rich outcrops on the Trojan asteroids it flies past.

19.3 Observing Asteroids

Planetary astronomers obtain spectra of asteroid surfaces through a combination of telescopic (ground- and space-based) and spacecraft observations. Depending on the wavelength of the observation and temperature of the body, the recorded flux may be dominated by sunlight reflected off the surface, heat thermally emitted by the surface, or a combination of the two (Figure 19.2). Reflectance spectroscopy focuses on analysis of sunlight reflected from the surface (Chapters 4 and 11), whereas thermal emission spectroscopy focuses on analysis of radiated heat (Chapter 3).

Most asteroids appear as point sources, since their angular size is smaller than the angular resolutions of current telescopic observations. Such observations are referred as disk-integrated because they observe the entire visible hemisphere. Disk-resolved spectra are

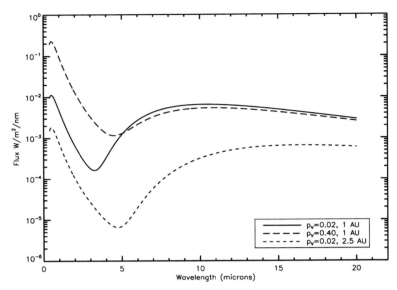

Figure 19.2 Spectral Energy Distributions (SEDs) for two near-Earth asteroids with different albedos and a Main Belt asteroid. The reflected sunlight dominates the SEDs at wavelengths shorter than ~3.5 μm and thermal emission dominates longward of ~5.5 μm. The crossover between these two regimes depends on the temperature of the body and, therefore, the albedo and distance from the Sun.

available for a small number of asteroids, mostly from spacecraft observations, but it is possible to spatially resolve a limited number of objects using ground-based adaptive optics systems and space-based telescopes.

19.3.1 Reflectance Spectroscopy

Reflectance spectroscopy covers a wide range of wavelengths (~0.3–4.0 μm) that contains many diagnostic spectral features (see Chapters 1 and 4), and much of this wavelength range is observable from the ground. Observations are often separated into discrete wavelength ranges: ultraviolet (UV; e.g., Cochran & Vilas, 1997; Li et al., 2011), visible (0.3–0.8 μm; e.g., Xu et al., 1995; Bus & Binzel, 2002), NIR (0.8–2.5 μm; e.g., Emery et al., 2011; Thomas et al., 2014), and the 3-μm region (2.8–4.0 μm; e.g., Rivkin et al., 2002).

A generalized data acquisition plan includes sequences of observed spectra of the target asteroid and corresponding sequences of spectra of solar analog stars. Observations of solar analog standard stars enable the removal of the solar spectral signature and telluric absorption effects for ground-based observations. Astronomers have identified several stars as suitable standards, including 16 Cyg B and Hyades 64 (e.g., Xu et al., 1995; Bus & Binzel, 2002). One can also identify suitable standards using the SIMBAD astronomical database (Wenger et al., 2000) to select G-dwarfs with solar-like color indices (B-V, V-K) near the position of the object (e.g., Takir & Emery, 2012; Thomas et al., 2014). Observations of standard stars near in time and space (less than a few degrees) to the observed asteroids provide the best telluric correction. If the standard stars do not provide a good telluric correction, atmospheric transmission can be modeled with a program such as ATRAN (Lord, 1992). The final reflectance spectrum is computed by dividing the calibrated asteroid spectrum by the calibrated standard star spectrum. Telescopic spectra are generally normalized to unity at a specified wavelength (nominally at 0.55 μm).

Observing procedures vary slightly for different wavelength regions. Visible wavelength observations are less sensitive to telluric effects and more sensitive to scattered light than NIR and 3-μm observations. Therefore, visible wavelength observations should avoid bright skies (e.g., twilight and near the Moon). At NIR wavelengths, the atmospheric emission contributes significantly to the background flux and varies on a timescale of a few minutes. Nodding the object between two positions along the slit enables background flux to be measured. Because scattering in Earth's atmosphere is wavelength dependent, the seeing disk (apparent size) and angle of refraction are also wavelength dependent (differential refraction; Filippenko, 1982). Differential refraction in particular can lead to wavelength-dependent light loss out of the slit and therefore incorrect spectral slopes. The best observing practice is to align the slit along the angle of refraction for a given position in the sky – the parallactic angle – so that all the light is captured in the slit.

The 3-μm region is within the crossover area of the SED for most asteroids and has varying degrees of thermal contamination. Thermal and molecular line emission from Earth's astmosphere lead to brighter background fluxes and more rapid variation than at shorter wavelengths. Therefore, individual exposure times are very short. The thermal flux from the asteroid needs to be modeled and removed before the spectrum is analyzed.

UV, visible, and NIR spectrometers on spacecraft are optimized for the expected spectral properties of the mission target. The scientific requirements of the mission determine the wavelengths and the spectral and spatial resolutions. These spectrometers are calibrated prior to launch, in flight, and have individualized calibration procedures.

19.3.2 Thermal Emission Spectroscopy

Thermal emission spectra (mid-IR range) of asteroids have been measured for wavelengths from ~5 to ~38 μm. Much of this wavelength range cannot be observed from the ground because of low atmospheric transmission. Observations have been taken using ground-based (e.g., Lim et al., 2005), airborne (e.g., Cohen et al., 1998; Vernazza et al., 2017), and space-based (e.g., Dotto et al., 2000; Emery et al., 2006) telescopes. In addition, the OSIRIS-REx mission includes a thermal emission spectrometer that is observing the NEA Bennu from ~6 to 50 μm (Hamilton et al., 2019).

Ground-based observations generally focus on the 8–13-μm atmospheric transmission window. Again, a generalized data acquisition plan includes sequences of observed spectra of the target asteroids and corresponding standard stars. At mid-IR wavelengths, the sky and the telescope emit their own thermal radiation. The sky background varies rapidly (time scale of seconds) and must be measured frequently. This is often done by "chopping" (tilting the secondary mirror slightly) to empty sky adjacent to the target asteroid many times during each nod position. The standards should be bright stars whose IR spectra are well known. Wavelength calibration uses prominent telluric lines in the observations.

Space-based observations can cover the full mid-IR wavelength range, and here the zodiacal cloud and interstellar medium are the major sources of background emission. This background emission contains spectral structure and could lead to spurious features if not removed. Background flux is generally removed by nodding along the slit and subtracting spectral images of the two nod positions from each other.

Thermal emission dominates the observed flux from an asteroid in the mid-IR wavelengths. The thermal flux spectrum is a combination of the thermal continuum flux that depends on the temperature of the body and the emissivity spectrum. The best fit thermal continuum is computed by minimizing the chi-squared between a thermal model of the surface and the observation. The final emissivity spectrum is calculated by subtracting any reflectance component from the SED and then dividing the result by the modeled thermal continuum.

Kirchhoff's law defines the complementary relationship between emissivity and reflectance at a given wavelength (e.g., Hapke, 2012; see Chapter 3). There are different measures of reflectance and emission, and one must be careful to use appropriate complementary quantities. Telescopic reflectance and emissivity spectra are *not* complementary quantities. Telescopic reflectance is bidirectional, whereas telescopic emissivity is a hemispherical-directional measurement. The correct complementary quantity to a telescopic emissivity measurement is a hemispherical-directional reflectance (r_{hd}).

The hemispherical-directional emissivity of a particulate surface of isotropic scatterers given the hemispherical-directional reflectance of the surface is

$$\varepsilon_d(e) = 1 - r_{hd}(e), \qquad (19.1)$$

where e is the emission angle. Bidirectional reflectance depends on both incidence (i) and emission angles. For isotropic scatterers, Hapke (2012) finds

$$r_{hd}(e) = 1 - \gamma H(\mu), \qquad (19.2)$$

and

$$r_{bi}(i,e) = \frac{w}{4\pi} \frac{\mu_o}{\mu_o + \mu} H(\mu_o) H(\mu). \qquad (19.3)$$

In these relations, w is the single-scattering albedo, $\gamma = \sqrt{1-w}$, μ_o and μ are the cosines of the incidence and emission angles, respectively, and $H(x)$ is the Chandrasekhar H function. No simple relation exists between r_{hd} and r_{bi}. One could, for a given single-scattering albedo, compute the ratio r_{hd}/r_{bi} as a function of incidence and emission angles and use this for a proper implementation of Kirchhoff's law for telescopic spectra.

19.4 Analysis of Asteroid Spectra

Wavelength-dependent reflectance data can be obtained *spectroscopically* (dispersing flux in wavelength using an optical element such as a prism or dispersion grating) or *spectrophotometrically* (employing multiple filters to measure flux in discrete wavelength regions). Techniques for analyzing these spectral data include (1) classification into taxonomic bins (Tholen, 1984; Bus & Binzel, 2002; Carvano et al., 2010), (2) curve matching to either laboratory spectra (e.g., Fornasier et al., 2010) or to linear combinations of compositional end members (e.g., Yang & Jewitt, 2010), (3) parametric analysis of absorption features (Cloutis et al., 1986; Gaffey et al., 1993; Dunn et al., 2010), and (4) physically motivated models to determine detailed surface mineralogy (e.g., Sunshine & Pieters, 1993; Hapke, 2012). Examples of asteroid spectra from visible through mid-IR wavelengths are given in Figure 19.3.

Early taxonomic systems adopted hybrid schemes, classifying asteroids into groups that shared similar properties based on multiple data products such as spectral slope, albedo, and polarization phase curves (e.g., Chapman et al., 1975; Tholen, 1984). As the number of known asteroids rapidly expanded, it became increasingly difficult to obtain a diverse and complete set of data products for a representative fraction of the asteroid population. As such, taxonomic schemes shifted to purely spectroscopy-based systems at visible wavelengths (Xu et al., 1995; Bus & Binzel, 2002; Lazzaro et al., 2004), NIR wavelengths

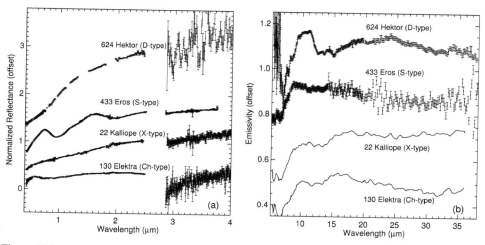

Figure 19.3 Example asteroid spectra from visible through mid-IR wavelengths including four major taxonomic complexes. (a) Visible and NIR reflectance spectra, normalized at 0.55 μm and offset for clarity (Hektor: +0.5, Kalliope: –0.5, Elektra: –0.75). (b) Emissivity spectra, scaled to a bolometric emissivity of 0.9 and offset for clarity (Hektor: +0.2, Kalliope: –0.2, Elektra: –0.4). The spectral data are from previous studies: Hektor reflectance (Emery & Brown, 2003), Hektor emissivity (Emery et al., 2006), Eros reflectance (Thomas et al., 2014; Wigton, 2015), Eros emissivity (Vernazza et al., 2010), Kalliope reflectance (Hardersen et al., 2011), Elektra reflectance (Takir & Emery, 2012), and Kalliope and Elektra emissivity (Marchis et al., 2012).

(Burbine & Binzel, 2002; DeMeo et al., 2009), and around 3 μm (Takir & Emery, 2012). Recently, taxonomic systems based on large datasets of broadband photometric colors have been developed (e.g., Carvano et al., 2010; Popescu et al., 2016). As current and future surveys such as Gaia (Mignard et al., 2007), WISE (Mainzer et al., 2011b), and LSST (Ivezić et al., 2008) increasingly produce larger and larger catalogs of physical data, we can expect a shift back to hybrid classification systems (e.g., Mainzer et al., 2011a; DeMeo & Carry, 2013).

In many cases, spectral taxonomic systems have been defined based on principal component analysis (PCA), a statistical method that reduces the dimensionality of a spectral or spectrophotometric dataset into a basis set of uncorrelated values that reflect the maximum possible variance in the original data. As the variance in asteroid spectral data is in part attributable to composition, the use of PCA to classify data can distinguish broad compositional groups. Coarse compositional assignment via taxonomy can be achieved for large numbers of objects with data of relatively low signal-to-noise ratio (S/N; ~10). Such population-level statistics provides important insight into the overall distribution of asteroid types and serves as a tracer of fundamental processes of Solar System formation and evolution (e.g., Gradie & Tedesco, 1982; Walsh et al., 2012). Deriving detailed, quantitative mineralogy for such a large number of objects is currently prohibitive because of observing time and data quality requirements. However, taxonomic classification is not a substitute for quantitative compositional or mineralogic analysis. Similarly, curve-matching techniques are not necessarily quantitatively diagnostic, but instead offer insight into the plausibility of compositional analogs.

Quantitative composition can be determined using empirically derived parametric descriptions of spectral absorption features. This technique is based on deriving relationships between measured spectral properties (e.g., absorption band centers and integrated areas) and corresponding compositional details (e.g., pyroxene mineralogy, pyroxene to olivine ratio). Fundamental laboratory spectroscopy of pyroxenes (Adams, 1974) and olivines (King & Ridley, 1987) paved the way for parametric band analysis of visible through NIR spectra of pyroxene–olivine compositions applicable to ordinary chondrite meteorites (e.g., Cloutis et al., 1986). Refinements to encompass a wider range of compositions (Reddy et al., 2015), alternative wavelength regimes (Lim et al., 2011), corrections to account for differences in temperature between the laboratory and space environments (Burbine et al., 2009), and effects of observational viewing geometry (e.g., Sanchez et al., 2012) have since been implemented.

Two primary sources of uncertainty contribute to the accuracy of compositional band analysis of asteroid spectra. The first is uncertainty in the absolute calibration of the parametric equations. Comparisons of derived compositions to known compositions of laboratory samples (e.g., Dunn et al., 2010) suggest that compositions derived from band analysis are accurate to within ~5%. The second source of uncertainty is related to S/N and is more difficult to quantify, particularly when nonrandom noise sources (e.g., residual telluric absorption features) affect spectral regions of interest. Furthermore, S/N errors are associated with flux or reflectance values and thus do not directly quantify uncertainty associated with band parameters such as absorption band centers. Full Monte Carlo simulations to quantify the effects of random and nonrandom noise on the accuracy of band parameters are necessary to quantify this source of uncertainty.

Physically motivated spectral modeling techniques can also be used to derive quantitative compositional information (Chapter 2). Reddy et al. (2015) review the Modified Gaussian Model (MGM; Sunshine & Pieters, 1993) and radiative transfer techniques (e.g., Shkuratov et al., 1999; Hapke, 2012). In short, these techniques use physically motivated functions to represent spectral data. In the case of the MGM, modified Gaussians functions are employed to describe electronic transition absorption strengths based on crystal field theory. MGM has proven particularly useful in characterizing the compositions of S- and V-type asteroids based on visible and NIR data (e.g., Sunshine et al., 2004; Mayne et al., 2011).

Radiative transfer models solve for the reflectivity of a particulate surface by using optical constants of known minerals and by making assumptions about surface scattering, particle size distributions, and surface mineralogy. Such models have been exploited to constrain the mineralogy of S- and V-type asteroids (e.g., Lawrence & Lucey, 2007; Vernazza et al., 2008). These codes have also been widely applied to primitive compositions, whose spectra are generally devoid of major visible and NIR absorption features. In these cases, it is often possible to only place upper limits on the presence of various constituent minerals (e.g., Emery & Brown, 2004). Radiative transfer codes have also been applied in the 3-μm region (e.g., Rivkin and Emery, 2010). Such modeling has yet to be widely applied to emissivity spectra of asteroids in the mid-IR, mostly because of the complexities of scattering at those wavelengths.

Looking ahead, more sophisticated spectral modeling schemes will likely emerge as computational capabilities continue to grow. For example, radiative transfer equations can now be solved for irregularly shaped particles thanks to discrete dipole approximation (DDA) codes (e.g., Zubko et al., 2008; Muinonen & Pieniluoma, 2011). Future efforts to further develop a DDA formalism for simultaneously constraining both composition and regolith grain properties (size and shape) of asteroidal surfaces may prove to be extremely fruitful.

19.5 Overview of Asteroid Compositions: Connecting Ground-Based and Spacecraft Data

The past three decades have seen major advances in physical characterization of asteroids. Flyby and rendezvous of several Main Belt and near-Earth asteroids by spacecraft have provided us with "ground truth" measurements to validate and refine our ground-based spectroscopic observations of these bodies. In this section, we briefly review the properties of several taxonomic types and the spacecraft data that have provided "ground truth." We refer the reader to DeMeo et al. (2009, 2015) for more in-depth discussion of asteroid spectral classes, the mineralogy inferred for each class, and the inferred meteorite analogs.

19.5.1 S-Complex Asteroids

S-type asteroid spectra are characterized by 1-μm and 2-μm absorption bands due to the minerals olivine and pyroxene (e.g., Eros in Figure 19.3). S-type asteroids can be further divided into seven subtypes depending on the relative ratio of olivine and pyroxene (Gaffey et al., 1993). The S(IV) subtypes have surface composition analogous to ordinary chondrite meteorites, which make up about 87% of all the meteorites that fall on Earth. Ordinary chondrites are further divided into three different types, H-, L-, and LL-chondrites, depending on their metal content and silicate chemistry.

S-type asteroids visited by spacecraft span the entire range of ordinary chondrite types we have in our terrestrial collection. Spectra of 243 Ida and 951 Gaspra, which the Galileo spacecraft flew by, are consistent with H- and LL-chondrites, respectively (Kelley et al., 2017). 433 Eros was the target of NEAR-Sheomaker spacecraft. Data from multiple instruments on NEAR found that the surface composition is consistent with a space-weathered L6 ordinary chondrite (McCoy et al., 2001; McFadden et al., 2001; Nittler et al., 2001; Bell et al., 2002; Izenberg et al., 2003). Reddy et al. (2015) compared the olivine (fayalite, Fa) and pyroxene (ferrosilite, Fs) values for Eros from ground-based spectra and NEAR NIS instrument and found them to be consistent.

The ultimate ground-truth for S-type asteroid spectral calibrations came from the Japanese Hayabusa mission to near-Earth asteroid 25143 Itokawa (Fujiwara et al., 2006). Hayabusa brought back microscopic samples of the S-type NEA (Nakamura et al., 2011). The Fa and Fs values of these particles match those derived from ground-based and spacecraft telescopic spectra (Abe et al., 2006; Binzel et al., 2001).

19.5.2 C-Complex Asteroids

C-complex asteroids are characterized by shallowly sloped visible and NIR spectra and generally have low albedos. Ch-type spectra, a subset of the C-complex, are defined by absorptions at 0.7 μm (Vilas & Gaffey, 1989; Bus & Binzel, 2002) and contain features near 3 μm (Jones et al., 1990; Rivkin et al., 2015) that are attributed to phyllosilicates (e.g., Elektra in Figure 19.3). Two C-complex asteroids have been the targets of spacecraft rendezvous and flyby missions. Main Belt asteroid (253) Mathilde was visited by the NEAR Shoemaker spacecraft (Veverka et al., 1997), but the NIR spectrometer was not turned on for the flyby (due to power constraints). Clark et al. (1999) report no significant color variations across the surface, and note that ground-based observations in the visible (Binzel et al., 1996) and 3-μm (Rivkin et al., 1997) regions show no evidence of hydration.

Ceres was the second target of NASA's Dawn mission and has been the focus of intense ground-based observing campaigns (e.g., Rivkin et al., 2006). Ceres has one of the richest spectra among low-albedo asteroids, showing UV features indicative of graphitized carbon and sulfur (Hendrix et al., 2016) and NIR features attributed to magnetite, carbonates, and other products of aqueous alteration (e.g., Milliken & Rivkin, 2009). Mid-IR (8–13 μm) observations have inconsistently found evidence for carbonates, but on the whole are consistent with the 3–4 μm observations (Cohen et al., 1998; Lim et al., 2005; Vernazza et al., 2017). Dawn spacecraft observations of Ceres (Chapter 20) through the VIR spectrometer show similar global spectra (e.g., De Sanctis et al., 2015). In addition, the better spectral and spatial coverage of Dawn/VIR data revealed a rich array of geologic materials on the surface, including phyllosilicates, ammoniated clays, salts (sodium carbonate), H_2O, and organics (De Sanctis et al., 2015, 2017; Combe et al., 2016). See Chapter 20 for more details.

19.5.3 X-Complex

X-Complex asteroids are characterized by moderately red spectral slopes and no (or very weak) absorption features in visible and NIR spectra (Bus & Binzel, 2002). They are considered to include asteroids of a very wide range of compositions. When albedo is known, the X-complex separates into three classes (Tholen, 1984): E-type (high albedo, p_v >30%), M-type (moderate albedo, 10%< p_v <30%), and P-type (low albedo, p_v <10%).

E-Type Asteroids: E-type asteroids were initially linked to enstatite achondrites, but ground-based observations suggest that the type includes several silicate mineralogies (e.g., Clark et al., 2004). The European Space Agency's Rosetta spacecraft flew by the small E-type Main Belt asteroid 2867 Šteins in 2008. Ground-based spectra had revealed an absorption feature at 0.49 μm in spectra of Šteins (Fornasier et al., 2008), attributed to oldhamite or Ti-bearing pyroxenes (Shestopalov et al., 2010), and Rosetta data are consistent with these ground-based results (e.g., Markus et al., 2013). Mid-IR emissivity spectra from Spitzer are consistent with an enstatite-rich surface (Barucci et al., 2008; Groussin et al., 2011).

M-Type Asteroids: M-type asteroids (e.g., Kalliope in Figure 19.3) were originally associated with iron meteorites, but astronomical observations suggest that many are silicate-rich. The only M-type asteroid that has been visited by spacecraft is 21 Lutetia, which, like Šteins, was a flyby target for Rosetta in 2010. Visible and NIR spectra of Lutetia are featureless and have been associated with either enstatite chondrites or carbonaceous chondrites (Barucci et al., 2012, 2015; Moyano-Cambero et al., 2016). Mid-IR emissivity spectra from Spitzer are most consistent with carbonaceous chondrites, as are polarimetric observations (Barucci et al., 2008). Ground-based 3-μm spectra of the southern hemisphere contain an absorption ascribed to phyllosilicates (Rivkin et al., 2011), but ground-based and VIRTIS-M spectra of the northern hemisphere show no absorptions. It seems that Lutetia's surface may be heterogeneous on a global scale. NASA's Psyche mission will explore a different M-type world.

P-Type Asteroids: P-type asteroids are primitive objects that are likely organic-rich and do not have analogs in the meteorite population. No spacecraft have visited any P-type objects, but NASA's Lucy mission will give us our first glimpse of these asteroids.

19.5.4 V-Type Asteroids

V-type asteroids have spectra that are dominated by nearly pure pyroxene absorption bands. (4) Vesta was the first target of NASA's Dawn mission (Russell et al., 2012; see Chapter 20). Earth-based telescopes have observed Vesta for almost a century (e.g., Bobrovnikoff, 1929), revealing a basaltic (pyroxene-dominated) surface with a weak 3-μm absorption feature. The Dawn mission confirmed many of these observations, including the albedo variations on the two hemispheres of Vesta that were shown to be due to in fall of carbonaceous chondrite impactors (McCord et al., 2012; Reddy et al., 2012), the detection of hydrated minerals associated with these terrains (Hasegawa et al., 2003), and the detection of olivine on the surface. Dawn observations of Vesta have provided an opportunity to verify more than a century of ground-based color and spectral observations of Vesta.

19.5.5 Other Asteroid Spectral Types

Several other asteroid spectral types have not yet been visited by spacecraft. Most notable of these are the D-types. D-types are characterized by spectra that are steeply red-sloped and featureless at visible and NIR wavelengths (e.g., Hektor in Figure 19.3). There are no meteorite analogs of the D-types, and compositions are uncertain, though mid-IR features are consistent with the presence of fine-grained silicates at the surfaces (Emery et al., 2006; Vernazza et al., 2012). NASA's Lucy mission to the Trojan asteroids will give us our first up-close look at D-types.

A-type objects have a broad 1-μm absorption band and are thought to be remnants of olivine-rich mantle material for one or more differentiated objects. A-type objects are rare, but they are important for understanding differentiation in the early Solar System. These objects have not been visited by spacecraft and are not included in any upcoming mission plans.

19.6 Summary and Future Prospects

Planetary astronomers have made great strides since the 1960s and over the last couple of decades with improved ground-based and spacecraft spectroscopy to enhance understanding of Solar System formation and evolution. Advances have come both from sheer volume of spectra acquired and from improvements in technology and analysis techniques. As is generally true of science, however, these advances have led to the ability to ask, and expect to be able to answer, more sophisticated questions. DeMeo et al. (2015) provide an excellent perspective on the current state of knowledge of the compositional structure of the asteroid belt, and they highlight questions at the forefront of asteroid science. The future advances required to address these questions (and undoubtedly open new ones) will likely come in four general forms.

Sample Return. Meteorite analogs have been a critical component of asteroid spectral calibrations, with the Vesta-HED connection being the shining success. Unfortunately, meteorites generally have uncertain origins, and no other meteorite to (single) parent body connection as robust as the Vesta-HED connection has been established. Furthermore, several asteroid taxonomic groups have no spectral analogs among the meteorites. By using spacecraft to return samples to Earth, both of these deficiencies can be overcome. JAXA's Hayabusa mission was spectacularly successful in linking S-type asteroids and ordinary chondrites, despite returning a tiny fraction of the amount expected, and two additional asteroid sample return missions are underway (OSIRIS–REx and Hayabusa 2). Targeting asteroids for which we have no meteorite analogs should be high priority for future sample return missions.

Flyby and Orbital Missions. Improved spatial resolution and spectral coverage enabled by spacecraft missions provide critical compositional and geologic information. Spacecraft missions have sampled only a very small fraction of the diversity of the asteroid belt. The upcoming Psyche and Lucy missions, to 16 Psyche and several Trojan asteroids, respectively, will improve that sampling somewhat. Nevertheless, many asteroid types remain a mystery. Missions that focus on remote characterization can sample diversity (in terms of composition, size, and other factors) more economically than sample return, and these missions will be a critical element of the exploration portfolio.

Improved Wavelength Coverage. Spectral analysis of some asteroid types and important processes are hindered by lack of spectra in certain wavelength ranges. Ground-based spectroscopy at $\lambda > 2.5$ µm is difficult in general, and it is impossible to get spectra from the ground at certain wavelengths. Knowledge of the phyllosilicate mineralogy of hydrated asteroids, for instance, would be significantly advanced by spectral capabilities in the 2.5–2.9 µm spectral region. Characterization of complex organics, sulfates, and carbonates would similarly be improved with better capabilities in the 3- to 5-µm region. The James Webb Space Telescope will have capabilities over this wavelength range, but, given time pressures, will probably not be able to observe a large fraction of the asteroid population. The UV is similarly underexploited for investigating silicate mineralogy, space weathering, and organic materials. Building future capabilities at wavelengths that are inaccessible, or very difficult to access, from the ground should be a priority.

Increasing Database of Asteroid Spectra. Asteroids display a very wide range of spectral characteristics – far wider than observed in the meteorite population. As the collective database of asteroid spectra, in multiple spectral regimes, continues to grow, scientists will gain a deeper, richer understanding of the compositional complexity of the inner Solar System. This increased level of detail is necessary for addressing the increasingly sophisticated questions we now know to ask about the Solar System.

References

Abe M., Takagi Y., Kitazato K., et al. (2006) Near-infrared spectral results of asteroid Itokawa from the Hayabusa spacecraft. *Science*, **312**, 1334–1338.

Adams J.B. (1974) Visible and near-infrared diffuse reflectance spectra of pyroxenes as applied to remote sensing of solid objects in the Solar System. *Journal of Geophysical Research*, **79**, 4829–4836.

Barucci M., Fornasier S., Dotto E., et al. (2008) Asteroids 2867 Steins and 21 Lutetia: Surface composition from far infrared observations with the Spitzer space telescope. *Astronomy and Astrophysics*, **477**, 665–670.

Barucci M., Belskaya I., Fornasier S., et al. (2012) Overview of Lutetia's surface composition. *Planetary and Space Science*, **66**, 23–30.

Barucci M.A., Fulchignoni M., Ji J., Marchi S., & Thomas N. (2015) The flybys of asteroids (2867) Šteins, (21) Lutetia, and (4179) Toutatis. In: *Asteroids IV* (P. Michel, F.E. DeMeo, & W.F. Bottke, eds.). University of Arizona Press, Tucson, 433–450.

Bell J. III, Izenberg N., Lucey P., et al. (2002) Near-IR reflectance spectroscopy of 433 Eros from the NIS instrument on the NEAR mission: I. Low phase angle observations. *Icarus*, **155**, 119–144.

Binzel R.P., Burbine T.H., & Bus S.J. (1996) Groundbased reconnaissance of asteroid 253 Mathilde: Visible wavelength spectrum and meteorite comparison. *Icarus*, **119**, 447–449.

Binzel R.P., Rivkin A.S., Bus S.J., Sunshine J.M., & Burbine T.H. (2001) MUSES-C target asteroid (25143) 1998 SF36: A reddened ordinary chondrite. *Meteoritics and Planetary Science*, **36**, 1167–1172.

Bobrovnikoff N.T. (1929) The spectra of minor planets. *Lick Observatory Bulletin*, **14**, 18–27.

Brucato J.R., Strazzulla G., Baratta G., & Colangeli L. (2004) Forsterite amorphisation by ion irradiation: Monitoring by infrared spectroscopy. *Astronomy and Astrophysics*, **413**, 395–401.

Brunetto R., Loeffler M.J., Nesvorný D., Sasaki S., & Strazzulla G. (2015) Asteroid surface alteration by space weathering processes. In: *Asteroids IV* (P. Michel, F.E. DeMeo, & W.F. Bottke, eds.). University of Arizona Press, Tucson, 597–616.

Burbine T.H. & Binzel R.P. (2002) Small main-belt asteroid spectroscopic survey in the near-infrared. *Icarus*, **159**, 468–499.

Burbine T.H., Buchanan P.C., Dolkar T., & Binzel R.P. (2009) Pyroxene mineralogies of near-Earth vestoids. *Meteoritics and Planetary Science*, **44**, 1331–1341.

Bus S.J. & Binzel R.P. (2002) Phase II of the small main-belt asteroid spectroscopic survey: The observations. *Icarus*, **158**, 106–145.

Carvano J.M., Hasselmann P.H., Lazzaro D., & Mothé-Diniz T. (2010) SDSS-based taxonomic classification and orbital distribution of main belt asteroids. *Astronomy and Astrophysics*, **510**, A43.

Chapman C.R., Morrison D., & Zellner B. (1975) Surface properties of asteroids: A synthesis of polarimetry, radiometry, and spectrophotometry. *Icarus*, **25**, 104–130.

Clark B.E., Veverka J., Helfenstein P., et al. (1999) NEAR photometry of Asteroid 253 Mathilde. *Icarus*, **140**, 53–65.

Clark B.E., Bus S.J., Rivkin A.S., et al. (2004) E-type asteroid spectroscopy and compositional modeling. *Journal of Geophysical Research*, **109**, DOI:10.1029/2003JE002200.

Clark R.N. (2009) Detection of adsorbed water and hydroxyl on the Moon. *Science*, **326**, 562–564.

Cloutis E.A., Gaffey M.J., Jackowski T.L., & Reed K.L. (1986) Calibrations of phase abundance, composition, and particle size distribution for olivine-orthopyroxene mixtures from reflectance spectra. *Journal of Geophysical Research*, **91**, 11641–11653.

Cochran A.L. & Vilas F. (1997) The McDonald Observatory serendipitous UV/blue spectral survey of asteroids. *Icarus*, **127**, 121–129.

Cohen M., Witteborn F.C., Roush T., Bregman J., & Wooden D. (1998) Spectral irradiance calibration in the infrared. VIII. 5–14 micron spectroscopy of the asteroids Ceres, Vesta, and Pallas. *The Astronomical Journal*, **115**, 1671–1679.

Combe J.-P., McCord T.B., Tosi F., et al. (2016) Detection of local H_2O exposed at the surface of Ceres. *Science*, **353**, aaf3010.

Delbó M., Mueller M., Emery J.P., Rozitis B., & Capria M.T. (2015) Asteroid thermophysical modeling. In: *Asteroids IV* (P. Michel, F.E. DeMeo, & W.F. Bottke, eds.), University of Arizona Press, Tucson, 107–128.

DeMeo F.E. & Carry B. (2013) The taxonomic distribution of asteroids from multi-filter all-sky photometric surveys. *Icarus*, **226**, 723–741.

DeMeo F.E., Binzel R.P., Slivan S.M., & Bus S.J. (2009) An extension of the Bus asteroid taxonomy into the near-infrared. *Icarus*, **202**, 160–180.

DeMeo F.E., Alexander C.M.O., Walsh K.J., Chapman C.R., & Binzel R.P. (2015) The compositional structure of the asteroid belt. In: *Asteroids IV* (P. Michel, F.E. DeMeo, & W.F. Bottke, eds.). University of Arizona Press, Tucson, 13–41.

De Sanctis M.C., Combe J.-P., Ammannito E., et al. (2012) Detection of widespread hydrated materials on Vesta by the VIR imaging spectrometer on board the Dawn mission. *The Astrophysical Journal Letters*, **758**, L36.

De Sanctis M.C., Ammannito E., Raponi A., et al. (2015) Ammoniated phyllosilicates with a likely outer Solar System origin on (1) Ceres. *Nature*, **528**, 241–244.

De Sanctis M.C., Ammannito E., McSween H.Y., et al. (2017) Localized aliphatic organic material on the surface of Ceres. *Science*, **355**, 719–722.

Dodson-Robinson S.E., Willacy K., Bodenheimer P., Turner N.J., & Beichman C.A. (2009) Ice lines, planetesimal composition and solid surface density in the solar nebula. *Icarus*, **200**, 672–693.

Dotto E., Müller T., Barucci M., et al. (2000) ISO results on bright Main Belt asteroids: PHT-S observations. *Astronomy and Astrophysics*, **358**, 1133–1141.

Dunn T.L., McCoy T.J., Sunshine J., & McSween H.Y. Jr. (2010) A coordinated spectral, mineralogical, and compositional study of ordinary chondrites. *Icarus*, **208**, 789–797.

Emery J.P. & Brown R.H. (2003) Constraints on the surface composition of Trojan asteroids from near-infrared (0.8–4.0 μm) spectroscopy. *Icarus*, **164**, 104–121.

Emery J.P. & Brown R.H. (2004) The surface composition of Trojan asteroids: Constraints set by scattering theory. *Icarus*, **170**, 131–152.

Emery J.P., Cruikshank D.P., & Van Cleve J. (2006) Thermal emission spectroscopy (5.2–38 μm) of three Trojan asteroids with the Spitzer Space Telescope: Detection of fine-grained silicates. *Icarus*, **182**, 496–512.

Emery J.P., Burr D.M., & Cruikshank D.P. (2011) Near-infrared spectroscopy of Trojan asteroids: Evidence for two compositional groups. *The Astronomical Journal*, **141**, 25.

Filippenko A.V. (1982) The importance of atmospheric differential refraction in spectrophotometry. *Publications of the Astronomical Society of the Pacific*, **94**, 715–721.

Fornasier S., Migliorini A., Dotto E., & Barucci M. (2008) Visible and near infrared spectroscopic investigation of E-type asteroids, including 2867 Steins, a target of the Rosetta mission. *Icarus*, **196**, 119–134.

Fornasier S., Clark B., Dotto E., Migliorini A., Ockert-Bell M., & Barucci M. (2010) Spectroscopic survey of M-type asteroids. *Icarus*, **210**, 655–673.

Fujiwara A., Kawaguchi J., Yeomans D., et al. (2006) The rubble-pile asteroid Itokawa as observed by Hayabusa. *Science*, **312**, 1330–1334.

Gaffey M.J., Bell J.F., Brown R.H., et al. (1993) Mineralogical variations within the S-type asteroid class. *Icarus*, **106**, 573–602.

Gladman B.J., Davis D.R., Neese C., et al. (2009) On the asteroid belt's orbital and size distribution. *Icarus*, **202**, 104–118.

Gradie J. & Tedesco E. (1982) Compositional structure of the asteroid belt. *Science*, **216**, 1405–1407.

Grossman L. (1972) Condensation in the primitive solar nebula. *Geochimica et Cosmochimica Acta*, **36**, 597–619.

Groussin O., Lamy P., Fornasier S., & Jorda L. (2011) The properties of asteroid (2867) Steins from Spitzer Space Telescope observations and OSIRIS shape reconstruction. *Astronomy and Astrophysics*, **529**, A73.

Guilbert-Lepoutre A. (2014) Survival of water ice in Jupiter Trojans. *Icarus*, **231**, 232–238.

Hamilton V.E., Simon A.A., Christensen P.R., et al. (2019) Evidence for widespread hydrated minerals on asteroid (101955) Bennu. *Nature Astronomy*, **3**, 332–340.

Hapke B. (2001) Space weathering from Mercury to the asteroid belt. *Journal of Geophysical Research*, **106**, 10,039–10,073.

Hapke B. (2012) *Theory of reflectance and emittance spectroscopy*. Cambridge University Press, Cambridge.

Hardersen P.S., Cloutis E.A., Reddy V., Mothé-Diniz T., & Emery J.P. (2011) The M-/X-asteroid menagerie: Results of an NIR spectral survey of 45 main-belt asteroids. *Meteoritics and Planetary Science*, **46**, 1910–1938.

Hartmann W.K., Tholen D.J., & Cruikshank D.P. (1987) The relationship of active comets, "extinct" comets, and dark asteroids. *Icarus*, **69**, 33–50.

Hasegawa S., Murakami K., Ishiguro M., et al. (2003) Evidence of hydrated and/or hydroxylated minerals on the surface of asteroid 4 Vesta. *Geophysical Research Letters*, **30**, DOI:10.1029/2003GL018627.

Hendrix A.R., Vilas F., & Li J.Y. (2016) Ceres: Sulfur deposits and graphitized carbon. *Geophysical Research Letters*, **43**, 8920–8927.

Henning T. (2010) Cosmic silicates. *Annual Review of Astronomy and Astrophysics*, **48**, 21–46.

Ivezić Ž., Axelrod T., Brandt W., et al. (2008) Large Synoptic Survey Telescope: From science drivers to reference design. *Serbian Astronomical Journal*, **176**, 1–13.

Izenberg N.R., Murchie S.L., Bell J.F. III, et al. (2003) Spectral properties and geologic processes on Eros from combined NEAR NIS and MSI data sets. *Meteoritics and Planetary Science*, **38**, 1053–1077.

Jones T.D., Lebofsky L.A., Lewis J.S., & Marley M.S. (1990) The composition and origin of the C, P, and D asteroids: Water as a tracer of thermal evolution in the outer belt. *Icarus*, **88**, 172–192.

Kelley M.S., Sanchez J.A., & Reddy V. (2017) Characterization of spacecraft targets: Ida and Gaspra. *Conference on Asteroids, Comets, Meteors*, Montevideo, Uruguay, poster 2.e.67.

King T.V. & Ridley W.I. (1987) Relation of the spectroscopic reflectance of olivine to mineral chemistry and some remote sensing implications. *Journal of Geophysical Research*, **92**, 11,457–11,469.

Lantz C., Brunetto R., Barucci M., et al. (2017) Ion irradiation of carbonaceous chondrites: A new view of space weathering on primitive asteroids. *Icarus*, **285**, 43–57.

Lawrence S.J. & Lucey P.G. (2007) Radiative transfer mixing models of meteoritic assemblages. *Journal of Geophysical Research*, **112**, DOI:10.1029/2006JE002765.

Lazzaro D., Angeli C., Carvano J., Mothé-Diniz T., Duffard R., & Florczak M. (2004) S 3 OS 2: The visible spectroscopic survey of 820 asteroids. *Icarus*, **172**, 179–220.

Levison H.F., Bottke W.F., Gounelle M., Morbidelli A., Nesvorný D., & Tsiganis K. (2009) Contamination of the asteroid belt by primordial trans-Neptunian objects. *Nature*, **460**, 364–366.

Lewis J.S. (1972) Low temperature condensation from the solar nebula. *Icarus*, **16**, 241–252.

Li J.-Y., Bodewits D., Feaga L.M., et al. (2011) Ultraviolet spectroscopy of asteroid (4) Vesta. *Icarus*, **216**, 640–649.

Lim L.F., McConnochie T.H., Bell J.F. III, & Hayward T.L. (2005) Thermal infrared (8–13 μm) spectra of 29 asteroids: The Cornell mid-infrared asteroid spectroscopy (MIDAS) survey. *Icarus*, **173**, 385–408.

Lim L.F., Emery J.P., & Moskovitz N.A. (2011) Mineralogy and thermal properties of V-type Asteroid 956 Elisa: Evidence for diogenitic material from the Spitzer IRS (5–35 μm) spectrum. *Icarus*, **213**, 510–523.

Loeffler M., Dukes C., & Baragiola R. (2009) Irradiation of olivine by 4 keV He+: Simulation of space weathering by the solar wind. *Journal of Geophysical Research*, **114**, DOI:10.1029/2008JE003249.

Lord S.D. (1992) *A new software tool for computing Earth's atmospheric transmission of near- and far-infrared radiation. NASA TM-103957*. NASA Ames Research Center, Moffett Field, CA.

Mainzer A., Masiero J., Grav T., et al. (2011a) NEOWISE studies of asteroids with Sloan photometry: Preliminary results. *The Astrophysical Journal*, **745**, 7.

Mainzer A., Bauer J., Grav T., et al. (2011b) Preliminary results from NEOWISE: An enhancement to the wide-field infrared survey explorer for Solar System science. *The Astrophysical Journal*, **731**, 53.

Marchis F., Enriquez J., Emery J., et al. (2012) Multiple asteroid systems: Dimensions and thermal properties from Spitzer Space Telescope and ground-based observations. *Icarus*, **221**, 1130–1161.

Markus K., Arnold G., Hiesinger H., et al. (2013) Comparison of ground-based and VIRTIS-M/ROSETTA reflectance spectra of asteroid 2867 Steins with laboratory reflectance spectra in the VIS and IR. *EGU General Assembly*, Abstract #EGU2013-11287.

Mayne R., Sunshine J., McSween H. Jr., Bus S., & McCoy T.J. (2011) The origin of Vesta's crust: Insights from spectroscopy of the Vestoids. *Icarus*, **214**, 147–160.

McCord T.B., Li J.-Y., Combe J.-P., et al. (2012) Dark material on Vesta from the infall of carbonaceous volatile-rich material. *Nature*, **491**, 83–86.

McCoy T.J., Burbine T., McFadden L., et al. (2001) The composition of 433 Eros: A mineralogical—chemical synthesis. *Meteoritics and Planetary Science*, **36**, 1661–1672.

McFadden L.A., Wellnitz D.D., Schnaubelt M., et al. (2001) Mineralogical interpretation of reflectance spectra of Eros from NEAR near-infrared spectrometer low phase flyby. *Meteoritics and Planetary Science*, **36**, 1711–1726.

McSween H. Y. Jr., Ghosh A., Grimm R.E., Wilson L., & Young E.D. (2002) Thermal evolution models of asteroids. In: *Asteroids III* (W. Bottke, Cellino Paolicchi, & R.P. Binzel, eds.). University of Arizona Press, Tucson, 559–571.

Mignard F., Cellino A., Muinonen K., et al. (2007) The Gaia mission: Expected applications to asteroid science. *Earth, Moon, and Planets*, **101**, 97–125.

Milliken R.E. & Rivkin A.S. (2009) Brucite and carbonate assemblages from altered olivine-rich materials on Ceres. *Nature Geoscience*, **2**, 258–261.

Morbidelli A., Levison H.F., Tsiganis K., & Gomes R. (2005) Chaotic capture of Jupiter's Trojan asteroids in the early Solar System. *Nature*, **435**, 462–465.

Moyano-Cambero C.E., Trigo-Rodríguez J.M., Llorca J., Fornasier S., Barucci M.A., & Rimola A. (2016) A plausible link between the asteroid 21 Lutetia and CH carbonaceous chondrites. *Meteoritics and Planetary Science*, **51**, 1795–1812.

Muinonen K. & Pieniluoma T. (2011) Light scattering by Gaussian random ellipsoid particles: First results with discrete-dipole approximation. *Journal of Quantitative Spectroscopy and Radiative Transfer*, **112**, 1747–1752.

Nakamura T., Noguchi T., Tanaka M., et al. (2011) Itokawa dust particles: A direct link between S-type asteroids and ordinary chondrites. *Science*, **333**, 1113–1116.

Nesvorný D., Vokrouhlický D., & Morbidelli A. (2013) Capture of Trojans by jumping Jupiter. *The Astrophysical Journal*, **768**, 45.

Nittler L.R., Starr R.D., Lim L., et al. (2001) X-ray fluorescence measurements of the surface elemental composition of asteroid 433 Eros. *Meteoritics and Planetary Science*, **36**, 1673–1695.

O'Brien D.P. & Sykes M.V. (2011) The origin and evolution of the asteroid belt—Implications for Vesta and Ceres. *Space Science Reviews*, **163**, 41–61.

Pieters C.M. & Noble S.K. (2016) Space weathering on airless bodies. *Journal of Geophysical Research*, **121**, 1865–1884.

Pieters C.M., Goswami J.N., Clark R.N., et al. (2009) Character and spatial distribution of OH/H_2O on the surface of the Moon seen by M3 on Chandrayaan-1. *Science*, **326**, 568–572.

Popescu M., Licandro J., Morate D., et al. (2016) Near-infrared colors of minor planets recovered from Vista-VHS survey (MOVIS). *Astronomy and Astrophysics*, **591**, A115.

Prinn R.G., & Fegley B. Jr. (1989) Solar nebula chemistry: Origins of planetary, satellite and cometary volatiles. In: *Origin and evolution of planetary and satellite atmospheres* (S.K. Atreya, J.B. Pollack, & M.S. Matthews, eds.). University of Arizona Press, Tucson, 78–136.

Reddy V., Le Corre L., O'Brien D.P., et al. (2012) Delivery of dark material to Vesta via carbonaceous chondritic impacts. *Icarus*, **221**, 544–559.

Reddy V., Dunn T., Thomas C.A., Moskovitz N., & Burbine T. (2015) Mineralogy and surface composition of asteroids. In: *Asteroids IV* (P. Michel, F.E. DeMeo, & W.F. Bottke, eds.). University of Arizona Press, Tucson, 65–87.

Rivkin A.S. & Emery J.P. (2010) Detection of ice and organics on an asteroidal surface. *Nature*, **464**, 1322–1323.

Rivkin A.S., Clark B.E., Britt D.T., & Lebofsky L.A. (1997) Infrared spectrophotometry of the NEAR flyby target 253 Mathilde. *Icarus*, **127**, 255–257.

Rivkin A.S., Howell E.S., Vilas F., & Lebofsky L.A. (2002) Hydrated minerals on asteroids: The astronomical record. *Asteroids III*, **1**, 235–253.

Rivkin A.S., Volquardsen E.L., & Clark B.E. (2006) The surface composition of Ceres: Discovery of carbonates and iron-rich clays. *Icarus*, **185**, 563–567.

Rivkin A.S., Clark B.E., Ockert-Bell M., et al. (2011) Asteroid 21 Lutetia at 3 μm: Observations with IRTF SpeX. *Icarus*, **216**, 62–68.

Rivkin A.S., Thomas C.A., Howell E.S., & Emery J.P. (2015) The Ch-class asteroids: Connecting a visible taxonomic class to a 3 μm band shape. *The Astronomical Journal*, **150**, 198.

Rivkin A.S., Howell E.S., Emery J.P., & Sunshine J. (2018) Evidence for OH or H_2O on the surface of 433 Eros and 1036 Ganymed. *Icarus*, **304**, 74–82.

Russell C.T., Raymond C.A., Coradini A., et al. (2012) Dawn at Vesta: Testing the protoplanetary paradigm. *Science*, **336**, 684–686.

Sanchez J.A., Reddy V., Nathues A., Cloutis E.A., Mann P., & Hiesinger H. (2012) Phase reddening on near-Earth asteroids: Implications for mineralogical analysis, space weathering and taxonomic classification. *Icarus*, **220**, 36–50.

Schorghofer N. (2016) Predictions of depth-to-ice on asteroids based on an asynchronous model of temperature, impact stirring, and ice loss. *Icarus*, **276**, 88–95.

Shestopalov D.I., Golubeva L.F., McFadden L.A., Fornasier S., & Taran M.N. (2010) Titanium-bearing pyroxenes of some E asteroids: Coexisting of igneous and hydrated rocks. *Planetary and Space Science*, **58**, 1400–1403.

Shkuratov Y., Starukhina L., Hoffmann H., & Arnold G. (1999) A model of spectral albedo of particulate surfaces: Implications for optical properties of the Moon. *Icarus*, **137**, 235–246.

Sunshine J.M. & Pieters C.M. (1993) Estimating modal abundances from the spectra of natural and laboratory pyroxene mixtures using the modified Gaussian model. *Journal of Geophysical Research*, **98**, 9075–9087.

Sunshine J.M., Bus S.J., McCoy T.J., Burbine T.H., Corrigan C.M., & Binzel R.P. (2004) High-calcium pyroxene as an indicator of igneous differentiation in asteroids and meteorites. *Meteoritics and Planetary Science*, **39**, 1343–1357.

Sunshine J.M., Farnham T.L., Feaga L.M., et al. (2009) Temporal and spatial variability of lunar hydration as observed by the Deep Impact spacecraft. *Science*, **326**, 565–568.

Takir D. & Emery J.P. (2012) Outer main belt asteroids: Identification and distribution of four 3-μm spectral groups. *Icarus*, **219**, 641–654.

Tholen D.J. (1984) *Asteroid taxonomy from cluster analysis of photometry*. PhD thesis, University of Arizona, Tucson.

Thomas C.A., Emery J.P., Trilling D.E., Delbó M., Hora J.L., & Mueller M. (2014) Physical characterization of Warm Spitzer-observed near-Earth objects. *Icarus*, **228**, 217–246.

Vernazza P., Binzel R., Thomas C., et al. (2008) Compositional differences between meteorites and near-Earth asteroids. *Nature*, **454**, 858–860.

Vernazza P., Carry B., Emery J., et al. (2010) Mid-infrared spectral variability for compositionally similar asteroids: Implications for asteroid particle size distributions. *Icarus*, **207**, 800–809.

Vernazza P., Delbo M., King P., et al. (2012) High surface porosity as the origin of emissivity features in asteroid spectra. *Icarus*, **221**, 1162–1172.

Vernazza P., Castillo-Rogez J., Beck P., et al. (2017) Different origins or different evolutions? Decoding the spectral diversity among C-type asteroids. *The Astronomical Journal*, **153**, 72.

Veverka J., Thomas P., Harch A., et al. (1997) NEAR's flyby of 253 Mathilde: Images of a C asteroid. *Science*, **278**, 2109–2114.

Vilas F. & Gaffey M.J. (1989) Phyllosilicate absorption features in main-belt and outer-belt asteroid reflectance spectra. *Science*, **246**, 790–792.

Walsh K.J., Morbidelli A., Raymond S.N., O'Brien D., & Mandell A. (2012) Populating the asteroid belt from two parent source regions due to the migration of giant planets—"The Grand Tack." *Meteoritics and Planetary Science*, **47**, 1941–1947.

Weissman P.R., A'Hearn M.F., McFadden L., & Rickman H. (2002) Evolution of comets into asteroids. *Asteroids III*, 669–686.

Wenger M., Ochsenbein F., Egret D., et al. (2000) The SIMBAD astronomical database-The CDS reference database for astronomical objects. *Astronomy and Astrophysics Supplement Series*, **143**, 9–22.

Wigton N.R. (2015) *Near-infrared (2–4 micron) spectroscopy of near-Earth asteroids: A search for OH/H_2O on small planetary bodies.* MS thesis, University of Tennessee.

Xu S., Binzel R.P., Burbine T.H., & Bus S.J. (1995) Small Main-Belt Asteroid Spectroscopic Survey: Initial results. *Icarus*, **115**, 1–35.

Yang B. & Jewitt D. (2010) Identification of magnetite in B-type asteroids. *The Astronomical Journal*, **140**, 692–698.

Zubko E., Shkuratov Y., Mishchenko M., & Videen G. (2008) Light scattering in a finite multi-particle system. *Journal of Quantitative Spectroscopy and Radiative Transfer*, **109**, 2195–2206.

20

Visible and Near-Infrared Spectral Analyses of Asteroids and Comets from Dawn and Rosetta

M. CRISTINA DE SANCTIS, FABRIZIO CAPACCIONI,
ELEONORA AMMANNITO, AND GIANRICO FILACCHIONE

20.1 Introduction

Comets and some asteroids are believed to be primitive bodies, remnants of the formation of the Solar System. Comets, including the Jupiter family comet 67P/CG, are considered the prototypes of the volatile-rich planetesimals formed beyond the snow line, while most asteroids, orbiting between Mars and Jupiter, formed inside the snow line. 1 Ceres and 4 Vesta are two of the largest minor planets (see Chapter 19), apparent survivors from the earliest days of the formation of the Solar System. They are the key main belt objects for exploring the state of the early Solar System across the snow line. The Dawn mission was designed to explore these two small bodies. Dawn's goal was to orbit 4 Vesta and 1 Ceres to obtain measurements that provide an understanding of the conditions and processes acting during the Solar System's earliest epoch. The Rosetta mission's primary objective has been to study the evolution of the Jupiter family comet 67P/CG during its orbit around the Sun. As a secondary goal, Rosetta performed two flybys of asteroids 2867 Šteins, classified as E-type, and 21 Lutetia, which although classified as M-type, has characteristics typical of E- and C-type asteroids.

Determination of the mineral composition of surface materials in their geologic context is a primary objective of the Rosetta and Dawn missions. The solid compounds (silicates, oxides, salts, organics, and ices) of the bodies investigated by Rosetta and Dawn can be identified through visible and near-infrared (VNIR) spectroscopy (see Chapters 1, 4, and 5) using high spatial resolution imaging to map the heterogeneity of surfaces and high spectral resolution spectroscopy to unambiguously determine the composition.

20.2 Instruments and Mission Description

Designed by ESA as the cornerstone cometary science mission of the Horizon 2000 plan, the Rosetta spacecraft was launched on March 2, 2004, carrying the surface landing module Philae. During the 10-year-long journey to the comet, the spacecraft had gravitational assist fly-bys at Earth and Mars. After a hibernation of 2.5 years, the mission escorted the comet 67P/CG from August 2014 until September 30, 2016, when the spacecraft was purposely crashed in a wide pit located in the Ma'at region. The spacecraft followed the comet as it approached the Sun on the inbound part of the trajectory (perihelion at 1.27 AU on August 13, 2015) and then during the outbound trajectory.

The Dawn mission to Vesta and Ceres was launched in September 2007 with a scheduled 2011 arrival at Vesta and 2015 arrival at Ceres (Russell et al., 2007). The Dawn spacecraft spent ~14 months in orbit around Vesta, and it is still orbiting Ceres at this time. Dawn is carrying a suite of instruments that includes two framing cameras, a VNIR spectrometer, and a gamma-ray and neutron detector, coupled with radio tracking to measure the gravity field. This instrument suite revealed geologically and geochemically complex worlds (Russell et al., 2012, 2016).

20.2.1 Rosetta/OSIRIS

The Optical, Spectroscopic, and Infrared Remote Imaging System (OSIRIS) (Figure 20.1) was a system of two imaging cameras (a narrow-angle and one wide-angle) on board the Rosetta mission (Keller et al., 2007). The Narrow-Angle Camera (NAC) with 2.20° by 2.22° field of view (FOV) and 18.6 μrad instantaneous field of view (IFOV) was designed to acquire high-resolution images of comet 67P/Churyumov-Gerasimenko's nucleus. The Wide-Angle Camera (WAC) with 11.35° by 12.11° FOV and 101 μrad IFOV was designed to map the gas and dust composing the coma near the comet. Both channels use a 2048 × 2048 charge-

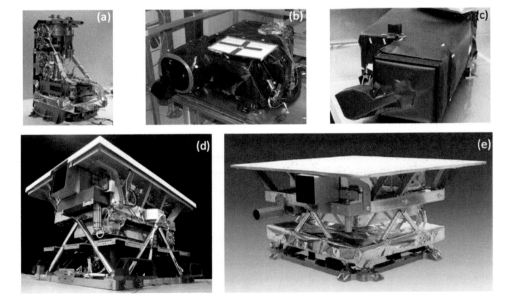

Figure 20.1 Cameras and VIS-IR spectrometers on board Dawn and Rosetta. (a) Dawn Framing Camera (FC). (b) Rosetta OSIRIS Narrow-Angle Camera (NAC). (c) Rosetta OSIRIS Wide-Angle Camera (WAC). (d) Dawn Visible and Infrared Mapping Spectrometer (VIR-MS). (e) Rosetta Visible InfraRed Thermal Imaging Spectrometer (VIRTIS).

coupled device (CCD) as detectors, and two filter wheels with 16 filters each are employed to acquire color images. The NAC filters are centered at various wavelengths between 269.3 and 989.3 nm, while the WAC filters are sensitive between 246.2 and 631.6 nm.

20.2.2 Rosetta/VIRTIS

The Visible, Infrared, and Thermal Imaging Spectrometer (VIRTIS) on board the Rosetta mission (Figure 20.1) was a dual-channel spectrometer (Coradini et al., 2007) designed to infer composition, spectrophotometric, and thermal properties of comet 67P/CG. The mapping channel (VIRTIS-M) used a Shafer telescope joined to an Offner relay to illuminate the entrance slit of the Offner spectrometer. By means of a dual-zone convex grating the optical beam was split between the VIS (0.25–1.0 μm) and IR (1.0–5.0 μm) ranges. A CCD and a Mercury Cadmium Telluride (MCT) detector were used as focal planes. Both channels acquired 256 spatial samples (IFOV = 250 μrad, FOV = 64 mrad) and 432 bands (spectral sampling 1.9 nm/band and 9.7 nm/band for the VIS and IR channels, respectively). The instrument used a mixed thermal control, with an external passive radiator to cool down the optical head structure and CCD to temperatures between 130 and 140 K and an active cooler to maintain MCT detector at about 85 K. The instrument built real-time hyperspectral images by moving a scan mirror placed at the entrance of the telescope. For each scan step equal to the IFOV, the two detectors acquired simultaneously a line of the image. Alternatively, VIRTIS-M operated in pushbroom mode, by maintaining the scan mirror at a fixed position and taking advantage of the spacecraft movement to build the image. Depending on the observation strategy defined for the different phases of the mission, one of the two modes was selected.

For the high spectral resolution channel (VIRTIS-H) the light was collected by an off-axis parabola and then collimated by a second off-axis parabola before entering a cross-dispersing prism made of lithium fluoride. After exiting the prism, the light was diffracted by a flat reflective grating, which dispersed the light in a direction perpendicular to the prism dispersion. The prism produced very high spectral resolution by separating orders 9 through 13 across a two-dimensional detector array with a low groove density echelle grating design: the spectral resolution varies in each order between $\lambda/\Delta\lambda$ = 1200 and 3500. The instrument used an MCT detector similar to the one previously described for the M channel. A second active cooler was used to bring the detector down to $T < 85$ K. As VIRTIS-H was not an imaging channel, it was geometrically aligned to the FOV's center of the VIRTIS-M channel. The instrument IFOV was rectangular and equal to 1.75 by 0.58 mrad. Finally, VIRTIS-H completely relied on the spacecraft attitude for pointing during the acquisitions.

20.2.3 Dawn FC

The Dawn FC (Figure 20.1) comprises a refractive lens system, a set of seven band-pass filters (from UV to 1 μm) and a clear filter with a wheel mechanism, a baffle with a door in front of the optics tube, a CCD at the focal plane, a thermal stabilization system, and supporting electronics. For redundancy, two identical cameras were provided, both located side by side on the +Z-deck of the spacecraft. Each camera has a mass of 5.5 kg. The FC consists of a stack of two major components, the Camera Head (CH) and the Electronics Box (E-Box). The E-Box contains the electronics for the Data Processing Unit (DPU), Power Converter Unit (PCU), and Mechanism Controller Unit (MCU). The CH is thermally insulated from the E-Box. The optical system has an FOV of 5.5° × 5.5° and an IFOV of 93.7 μrad. The camera head contains the CCD, the electronics for CCD operation and readout, and the filter wheel mechanism. The lens system is located in the lens barrel mounted on the CH, and atop the lens barrel is the baffle with the door mechanism attached. For further details see Sierks et al. (2011a).

20.2.4 Dawn VIR

The Dawn Visible and Infrared Mapping Spectrometer (VIR-MS or VIR) is a hyperspectral imaging spectrometer (De Sanctis et al., 2011). VIR (Figure 20.1) covers the range from the near UV (0.25 μm) to the near IR (5.0 μm) and has high spectral resolution and imaging capabilities. VIR combines two data channels in one compact instrument. It is inherited from the VIRTIS-M mapping spectrometer on board the ESA Rosetta mission (see above). VIR has the same imaging and spectral capabilities as VIRTIS-M, but it has improved electronic and grating performances. For details see De Sanctis et al. (2011).

20.3 Mission Targets: Asteroids, Dwarf Planets, and Comets

20.3.1 21 Lutetia and 2867 Šteins

The composition and nature of 21 Lutetia have long been perplexing (Barucci et al., 2007 and references therein). Its spectral classification, on the basis of ground spectroscopic observations, included the possibilities of a C-type or an M-type asteroid in the Tholen taxonomy (Tholen & Barucci, 1989). The more recent classification scheme of DeMeo et al. (2009) confirmed the spectral ambiguity classifying 21 Lutetia as an object of the Xc class with a flat and nearly featureless spectrum that is compatible with some carbonaceous chondrites and Enstatite chondrites (Birlan et al., 2006; Ockert-Bell et al., 2010). Ground observations (Rivkin et al., 2011b) provided some indication of the presence of a hydration feature that led to an estimate of a maximum water abundance of 2%.

The knowledge of 2867 Šteins prior to the Rosetta fly-by was considerably more scarce, the asteroid being a small one with few known properties. In particular, the large albedo of 0.45 ± 0.1, an estimated diameter of 4.6 km (Fornasier et al., 2007) and the spectral properties were compatible with that of E-type asteroids (Barucci et al., 2007 and references therein).

20.3.2 4 Vesta

4 Vesta is one of the largest objects in the asteroid belt, with a mean diameter of 525 km. The surface of 4 Vesta is covered by pyroxene-bearing basaltic materials suggesting a thermal evolution that led to differentiation and formation of an iron core (Pieters et al., 2011 and references therein). This early information indicated 4 Vesta as the possible source of the HED (Howardites, Eucrites, and Diogenites) meteorites (McCord et al., 1970). Later ground based observations greatly improved the spectra and confirmed this association (McFadden et al., 1977; Feierberg et al., 1980). Further observations indicated that Vesta's surface is not uniform (Binzel et al., 1997; Cochran and Vilas, 1998): the observed variations were believed to reflect impact excavations and lava flows in different regions. The asteroid has a huge impact basin in the southern hemisphere that exhibits inhomogeneous composition, linked to the layering of the lower crust and upper mantle (Thomas et al., 1997).

20.3.3 1 Ceres

1 Ceres is the largest and most massive body of the main belt. Its diameter is ~945 km and the general nature of the composition of Ceres' surface was known, from early ground-based telescopic observations, to be hydroxylated with mineralogy similar to carbonaceous chondrites (Gaffey & McCord, 1979). The mean spectrum of Ceres observed from ground is relatively flat shortward of 2.5 µm, with a strong feature at 3.07 µm, that has been interpreted as due to phyllosilicates, brucite (Rivkin et al., 2011a), other hydrated materials, or NH_4-bearing clays (King et al., 1992). An Fe charge transfer feature at 0.7 µm is also present in Ceres spectrum. Phyllosilicates, originated from aqueous alteration processes, display this signature (Vilas & Gaffey, 1989; Vilas, 1994). The dwarf planet density also indicated that Ceres could have retained its primordial water, in the form of liquid or ice (Rivkin et al., 2011a). Only small variations of spectral features with rotation were found, suggesting a globally homogeneous surface composition (Rivkin et al., 2011a).

20.3.4 67P/CG

Although discovered in 1960 by the two Ukrainian researchers that gave it their names, the comet 67P/CG was quite unknown until its selection as the target of the Rosetta mission. From 2004 the comet has been the subject of repeated observations (see Snodgrass et al., 2013) to build up the best understanding possible of the behavior of 67P/CG before Rosetta's arrival. The rotation period of 67P/CG was determined as 12.76137 ± 0.00006 hr, and the thermal inertia was estimated to be <15 Jm^{-2} K^{-1} $s^{-1/2}$, implying a surface regolith composed of fine and porous particles (Lowry et al., 2012). Moreover, from the thermal inertia the authors derived an effective radius of 1.97 ± 0.04 km and an albedo of 0.060 ± 0.005. During its last perihelion passage prior to the Rosetta encounter the comet became active in November, 2007 at a pre-perihelion distance from the Sun of 4.3 AU. The major findings were that the dust brightness could be described well by $Af\rho \propto r^{-3.2}$ pre-perihelion and $Af\rho \propto r^{-3.4}$

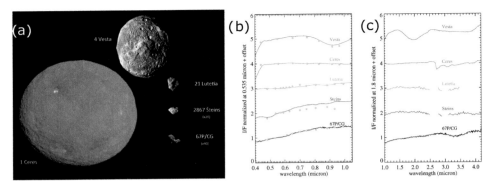

Figure 20.2 Comparison of the five minor bodies discussed in this chapter. (a) Images of 1 Ceres, 4 Vesta, 21 Lutetia, 2867 Šteins (enlarged by a factor 20) and 67P/CG (enlarged by a factor 40). Images courtesy NASA and ESA. (b) I/F in VIS range (Camera filters + VIRTIS/VIR) for all targets (with offsets). (c) I/F in the IR range (VIRTIS/VIR) for all targets (with offsets).

post-perihelion, and that the comet had a higher dust-to-gas ratio than average, with log(Af $\rho/Q(H_2O)$) = –24.94 ± 0.22 cm s molecule^{-1} at r <2 AU (Snodgrass et al., 2013). Afρ is a quantity designed by A'Hearn et al. (1984) for comparing the dust continuum under different observing conditions, where A is the albedo, f is the filling factor of grains within the field of view, and ρ is the radius of the coma. A model fit to the photometric data suggested that only a small fraction (1.4%) of the surface was active (Snodgrass et al., 2013).

20.4 Spectral Analysis Methods

20.4.1 Data Treatment in the Reflectance/Emission Superposition Region

VNIR imaging spectroscopy is a powerful technique to infer composition and physical properties of surfaces; however, an accurate data treatment is required to extract quantitative information. Following the calibration of the data in terms of its radiance factor *I/F* (reflected radiance over incident flux density) additional post-processing is necessary to derive spectral reflectance from *I/F*: (1) the removal of the surface thermal emission at wavelengths >3 μm for targets with temperatures above ~200 K; and (2) the correction with respect to the photometric response, necessary to remove the effects associated with illumination and observation geometry from the data. A detailed discussion on the photometric modeling and corrections can be found in Ciarniello et al. (2015). Although the thermal emission removal from the VIRTIS data is discussed at length in several papers, for instance in Keihm et al. (2012) and in Barucci et al. (2016), a brief description is provided here. On an atmosphereless body of the Solar System, depending on the surface properties (thermal conductivity, porosity, albedo, etc.) and on the heliocentric distance, the thermal emission contribution can be predominant in the 3–5 μm range respect to the solar reflected flux. Thus, the thermal emission contribution needs to be removed for a proper interpretation of the spectral reflectance data. The spectral radiance received from a surface is given by the sum of the solar input reflected by the surface and of the thermal flux emitted by the surface. The contribution of the reflected component is derived by means of the photometric modeling (Ciarniello et al., 2015) up to the wavelength where thermal emission starts to contribute; for longer wavelengths up to 5 μm the reflected components is extrapolated as a continuum with a fixed slope. This method can introduce some uncertainty in the spectral slope of the retrieved spectrum but it does not alter the capability to retrieve spectral features in the thermal emission range. The thermal emission is modeled as a graybody with temperature (T) and emissivity (ε_{eff}) as free parameters that can be minimized by means of a Levenberg–Marquardt least squares optimization algorithm. The modeled thermal emission is then subtracted from the measured radiance spectrum and then the reflected radiance in the full spectral range is converted to spectral reflectance. In general, the signal-to-noise ratio in the thermal range is considerably lower than at shorter wavelengths because the detector's dynamics is degraded by the instrument's thermal background. A similar approach has been used by Sunshine et al. (2007) to model Deep Impact data of comet 9P/Tempel 1.

20.4.2 Spectral Indicators Suited for Mineralogy and Detection of Ices (Spectral Slopes, Absorption Band Parameters, Spectral Ratios, etc.)

Hyperspectral imagers, including those described in this chapter, measure a large number of channels and not all of these spectral channels carry relevant information (spectral redundancy). Spectral analysis can thus take advantage of the reduction of the dimensionality of the datasets by using different methods to describe specific spectral properties. For instance, spectral slopes (the angular coefficient of the linear best fit to the reflectance within a specific spectral range) are quite sensitive to color changes and can be used to trace both composition and grain size changes (Filacchione et al., 2012), or to trace surface weathering (Pieters et al., 2012). In Section 20.4.4 are shown some spectral slope maps for 67P/CG, which have been used to characterize the abundance of water ice on the surface of the cometary nucleus. Absorption bands can be characterized by a number of indicators, e.g., band depths, band areas, and band center wavelengths (Clark, 1999) that are computed after normalizing the bands' reflectance with respect to the local continuum. Band depth is determined using the wavelength of the band center after continuum removal. The continuum is usually computed as a straight line by fitting reflectance data between left and right band wings. The band depth and area are correlated with the absorbing species' abundance, grain size, and solar phase. The wavelength of the band center is linked mainly to composition and in some cases temperature. Other band indicators, including asymmetry, can be used in specific cases when the principal band is skewed by a secondary one, as in the case of the olivine signature at about 1 μm.

20.4.3 Spectral Fitting Technique

Spectral fitting by means of Hapke's radiative transfer theory (Hapke, 2005) is a quantitative approach that allows for the derivation of the surface composition by identifying the endmembers, their physical properties (regolith grain size, surface roughness) and mixing modalities (areal, intimate) (see Chapter 2). While this approach is much more rigorous with respect to spectral indicators, it is limited by the availability of the optical constants necessary to perform the calculation and by the intrinsic limitations of an analytical theory applied to remote sensing data. This technique has been widely applied by adopting the method described by Ciarniello et al. (2011) to Rosetta and Dawn datasets to infer surface composition of the 67P/CG nucleus (De Sanctis et al., 2015a; Barucci et al., 2016; Filacchione et al., 2016b,c; Raponi et al., 2016) and Ceres (De Sanctis et al., 2015b, 2016). Using Hapke (2005) theory it is possible to calculate the ratio of the bidirectional reflectance $r(i, e, g, \lambda)$ (correlated to the radiance factor by a simple multiplication by π) of a semi-infinite medium at wavelengths λ and λ_0 as

$$\frac{r(i,e,g,\lambda)}{r(i,e,g,\lambda_0)} = \frac{\frac{w(\lambda)K}{4\pi} \frac{\mu_{0e}}{\mu_{0e}+\mu_e}[B_{SH}(g)p(g,\lambda)+H(w(\lambda),\mu_{0e}/K)+H(w(\lambda),\mu_e/K)-1]S(i,e,g,\theta)B_{CB}(g,\lambda)}{\frac{w(\lambda_0)K}{4\pi} \frac{\mu_{0e}}{\mu_{0e}+\mu_e}[B_{SH}(g)p(g,\lambda_0)+H(w(\lambda_0),\mu_{0e}/K)+H(w(\lambda_0),\mu_e/K)-1]S(i,e,g,\theta)B_{CB}(g,\lambda)}.$$

(20.1)

where i, e, g are the incidence, emission and phase angles, respectively; $w(\lambda)$ is the single scattering albedo (SSA); K is the porosity of the medium; $p(g, \lambda)$ is the single particle phase function; μ_{0e}, μ_e are the effective cosines of the incidence and emission angles, differing from μ_0 and μ (the cosines of the incidence and emission angles) for the effect of surface roughness; $H(w, \mu_e/K)$ is the Chandrasekhar function giving the multiple scattering effect; $B_{SH}(g)$ describe the shadow-hiding opposition effect; $B_{CB}(g, \lambda)$ the coherent back-scattering opposition effect; $S(i, e, g, \theta)$ is the shadow function required to describe the large-scale roughness and θ is the average surface slope. All terms that do not have a wavelength-dependence cancel out in Eq. 20.1).

In the case of areal mixing between different components of the surface material, the medium is modeled as patches of separate endmembers, with each solar photon interacting only with one endmember. In the case of intimate mixtures, the particles of two endmember materials are in contact with each other and both components participate in the scattering of a single solar photon. While in an intimate mixture the single-scattering albedo of the mixture is the average weighted through abundance, in the areal mixing the albedo of the two components are computed independently and then averaged with weighting by abundance. As an example, in Figure 20.3 is shown the best-fit solution of a water-ice–rich area on 67P/CG's surface as discussed by Filacchione et al. (2016b).

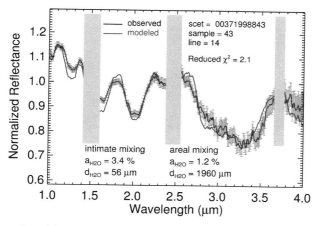

Figure 20.3 A water-ice–rich VIRTIS spectrum of 67P/CG's nucleus. The debris field in the Imothep region is displayed as a red curve with error bars; at each wavelength the error corresponds to the inverse of the signal-to-noise ratio compared with a best-fit synthetic model (blue curve). The modeled spectrum is obtained through an areal mixing of pure water ice covering 1.2% of the pixel area with a grain size distribution peaked around $d_{H2O} = 2$ mm in diameter, while the remaining 98.8% of the pixel area is made of an intimate mixture of water ice and the ubiquitous dark material in proportions 3.4% and 95.4%, respectively. The water ice has a size distribution peaked at $d_{H2O} = 56$ μm in diameter for the intimate mixture (from Filacchione et al., 2016b).

20.4.4 Compositional Maps

Compositional maps derived from imaging spectrometers are widely used to correlate surface composition with geologic and morphologic features resolved by cameras. A similar data fusion allows improving our understanding of the formation processes, evolution, and active mechanisms occurring on Solar System bodies. Without entering into the technical details concerning the rendering of irregularly shaped body maps (Filacchione et al., 2016a), we can associate a given spectral parameter (albedo, spectral slope, band parameter, spectral modeling best-fit result) with a cell placed at a given position in a longitude–latitude grid (De Sanctis et al., 2012a; Ammannito et al., 2013a). While this process seems to be quite straightforward, many uncertainties (quality of the digital shape model, lack in accuracy of the reconstruction of navigation data causing errors in the retrieval of the IFOV with respect to the target) can affect the final results. Moreover, in the case of very irregular bodies, such as for 67P/CG, the effect of shadows and rough local topography is relevant and must be taken into account in the rendering of the compositional maps. Examples of compositional maps associated with spectral band centers (Figure 20.4a) or spectral bands depths (Figure 20.5b), respectively, for the Dawn targets 4 Vesta and 1 Ceres or 0.5–0.8 μm spectral slope and organic material band depth (Figure 20.5) for 67P/CG are discussed in Section 20.5.

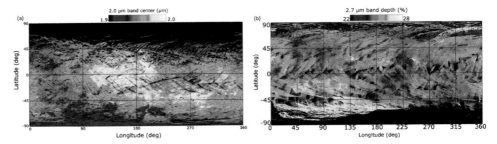

Figure 20.4 Distribution maps for spectral features. (a) Distribution of the center wavelength of the 2 μm pyroxene band on Vesta as mapped by VIR. The position of the band center characterizes the mineralogy of pyroxenes on Vesta (after De Sanctis et al., 2012a). (b) Distribution of the depth of the 2.7 μm phyllosilicate band on Ceres as mapped by VIR. This band depth is an indication of the relative abundance of phyllosilicates on Ceres (after Ammannito et al., 2016).

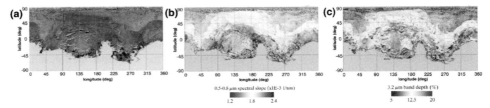

Figure 20.5 Compositional maps of comet 67P/CG, in simple cylindrical projection, obtained in the early mission phase at heliocentric distances varying between 3.6 and 3.2 AU. (a) Albedo in visible colors (B at 0.44 μm, G at 0.55 μm, R at 0.7 μm). (b) 0.5–0.8 μm spectral slope. (c) 3.2 μm organic material band depth (after Filacchione et al., 2016a).

20.5 Science Highlights

We summarize here the main results from the instruments on board the Rosetta and Dawn missions, focusing on the derived composition of the bodies visited.

20.5.1 Results from Rosetta

The spectral observations of 21 Lutetia performed by VIRTIS on board Rosetta on July 10, 2010 did not detect any notable absorption features of either silicates or hydrated minerals in the spectral range from 0.4–3.5 μm (Coradini et al., 2011). Spectral signatures of surface alteration resulting from space weathering were also not observed. Moreover, the surface is characterized by extreme uniformity with reflectance variations <5% in the different regions (Magrin et al., 2012). No indication of the presence of water (as adsorbed or in hydrated minerals) has been found; this is not compatible with ground observation data by Rivkin et al. (2011b), who derived a 2–3% OH band depth. This discrepancy could be partially explained by the difference in latitudes of the ground and Rosetta observations, although the global surface uniformity observed by VIRTIS seems to rule out this possibility. The surface temperature reached a maximum value of 245 K and correlated well with topographic features. The thermal inertia is in the range of 20–30 $Jm^{-2} K^{-1} s^{-1/2}$, comparable to that of a lunar-like powdery regolith. Lutetia is likely a remnant of the primordial planetesimal population, unaltered by differentiation processes and composed of chondritic materials of enstatitic or carbonaceous origin, dominated by iron-poor minerals that have not suffered aqueous alteration (Coradini et al., 2011).

VIRTIS and OSIRIS spectral observations of 2867 Šteins, and its fairly large albedo of 0.4 (Keller, 2010), essentially confirmed 2867 Šteins as a member of the E-type class. E-type asteroids are commonly identified as potential parent bodies of Aubrite achondritic meteorites (Fornasier et al., 2007). These are breccias of igneous cumulates formed in high temperature environments with Fe end Ni mainly observed as metals or as sulfides (troilite, oldhamite) and a Mg-rich silicate composition (orthopyroxenes, enstatites, etc.). VIRTIS observed the 490-nm band (Figure 20.3), tentatively attributed to sulfides (Burbine et al., 2002), and OSIRIS images showed a very homogeneous color (variation <1%), suggesting compositional homogeneity throughout (Keller, 2010).

Comet 67P/CG's nucleus appears bilobate where the larger lobe (aka the "body") has a size of about 4.1 km × 3.3 km × 1.8 km, and the smaller lobe (the "head") is 2.6 km × 2.3 km × 1.8 km (Sierks et al., 2015). The two parts are connected by a transition region named "neck." See Figure 20.2, where 67P/CG's nucleus is shown in the context of the other objects discussed in this chapter. For a detailed description of 67P/CG's morphologic properties see Sierks et al. (2015), Thomas et al. (2015), and El-Maarry et al. (2016). The surface of 67P/CG appears very dark, with a visible albedo of 6.2% (Ciarniello et al., 2015; Fornasier et al., 2015), and mainly dehydrated with a widespread distribution of organic materials (Capaccioni et al., 2015), probably produced by a complex mixture of dark disordered poly-aromatic compounds, opaque refractories including Fe-sulfides and several chemical species containing –COOH, NH_4^+, CH_2/CH_3, or –OH (alcohols) (Quirico et al., 2016). The maps shown in Figure 20.5 report the distribution of water ice at the surface. The northern regions display a bluer spectrum (flatter slopes in the VNIR spectra) and a deeper 3.2-μm band depth, where water ice contaminates this band at the left shoulder (Filacchione et al., 2016a). In some localized areas, the occurrence of landslides has caused the exposure of deeper layers where water-ice–rich material has been identified (Pommerol et al., 2015; Barucci et al., 2016; Ciarniello et al., 2016; Filacchione et al., 2016a,b; Raponi et al., 2016). Clear evidence of diurnal and seasonal evolution of surface ices, respectively water and carbon dioxide ices, have been detected and characterized (De Sanctis et al., 2015a; Filacchione et al., 2016c). Their presence has been interpreted as a process acting within the thermal skin depth that is correlated with the vapor transport in the subsurface of the nucleus with subsequent recondensation at the surface.

67P/CG is the only comet observed so far by a spacecraft to show clear spectral evidence of the presence of organic compounds on the surface. Analysis of coma gases and grains confirm the presence of organic material within the nucleus as well.

20.5.2 Results from Dawn

Dawn observed asteroid 4 Vesta for about 14 months, and the comparison between the data from the VIR Dawn imaging spectrometer and the different classes of HED indicates that the average spectrum of Vesta resembles that of the howardites (De Sanctis et al., 2013). However, at high spatial resolution, the surface of Vesta shows multiple, different HED lithologies (Ammannito et al., 2013a). A very small percentage of the surface is covered by diogenite, and also the basaltic eucrites are relatively scarce with respect to the meteoritic collection. The largest abundance of diogenitic material is found in the southern huge basin, indicating that diogenite clearly occurs below a basaltic upper crust (De Sanctis et al., 2012a; Ammannito et al., 2013a). However, diogenite is also found elsewhere, and its distribution does not provide unambiguous constraints for the magma ocean formation model. Olivine was expected to be present in the Vesta interior: in the mantle of a vertically layered body as required by the magma ocean models, or at the base of (or within) the mantle–crust boundary as claimed by fractionation models. Contrary to this expectation, olivine was first detected by VIR-Dawn in two wide areas located in the northern hemisphere (Ammannito et al., 2013b; see Figure 20.4), far from the southern basins that were expected to have excavated the crust down to reach the mantle. A few other smaller occurrences of olivine are present elsewhere and are not only concentrated in the southern basins (Ruesch et al., 2014; Palomba et al., 2015). Both the diogenite and the olivine distribution mapped by VIR on board Dawn require a revision of the planetesimal petrogenetic models (Mandler & Elkins-Tanton, 2013).

Hydrated material, possibly carbonaceous chondrites that constitute the most abundant inclusion type found in howardites, has also been identified on Vesta (De Sanctis et al., 2012b). The 2.8-μm OH absorption is distributed across Vesta's surface and shows areas enriched and depleted in hydrated materials. The uneven distribution indicates ancient processes that differ from those believed to be responsible for the occurrence of OH on other airless bodies, such as the Moon. The origin of vestan OH provides new insight into the delivery of hydrous materials not only in the main belt, but also in the inner Solar System, suggesting processes that may have played a role in the formation of terrestrial planets (De Sanctis et al., 2012b).

Parts of asteroid 1 Ceres' surface are heavily cratered, but the largest expected craters are absent (Marchi et al., 2016). Ceres appears gravitationally relaxed at only the longest wavelengths, implying a mechanically strong lithosphere with a weaker deep interior (Park et al., 2016). Ceres' surface displays hydroxylated silicates, including ammoniated clays, carbonates, and a dark component not yet spectrally identifiable (De Sanctis et al., 2015b; Ammannito et al., 2016; Ciarniello et al., 2016). The surface of Ceres observed by Dawn is dark and punctuated by small, bright regions (Nathues et al., 2015; Li et al., 2016; Russell et al., 2016), with one large complex of very bright areas on the floor of the Occator crater (239° longitude and 22°N latitude). The location and morphology of the bright material in Occator indicate the presence of subsurface conduits of bright material. Infrared spectra of the bright material in Occator show the presence of abundant sodium carbonates and ammonia salts and no measureable water-ice bands (De Sanctis et al., 2016). These properties are consistent with recent hydrothermal activity and cryovolcanism delivering materials produced at depth to the surface. The ubiquitous distribution of ammoniated phyllosilicate across Ceres indicates a global mixing process and pervasive aqueous alteration (Ammannito et al., 2016) as shown in Figure 20.4b. The possibility of abundant volatiles at depth is corroborated by geomorphologic features such as flat crater floors with pits, lobate flows, domes, and water ice in some craters (Combe et al., 2016). All of this evidence is consistent with a geologically differentiated body with a silicate core and an ice-rich mantle. In addition, the presence of ammonia-rich compounds indicates the large (in view of the relevant amount of water in the interior of Ceres) contribution of outer Solar System materials to the planetesimals that formed Ceres.

References

A'Hearn M.F., Schleicher D.G., Feldman P.D., Millis R.C. & Thompson D.T. (1984) Comet Bowell 1980b. *The Astronomical Journal*, **89**, 579–591.

Ammannito E., De Sanctis M.C., Capaccioni F., et al. (2013a) Vestan lithologies mapped by the visual and infrared spectrometer on Dawn. *Meteoritics and Planetary Science*, **48**, 2185–2198.

Ammannito E., De Sanctis M., Palomba E., et al. (2013b) Olivine in an unexpected location on Vesta's surface. *Nature*, **504**, 122–125.

Ammannito E., DeSanctis M., Ciarniello M., et al. (2016) Distribution of phyllosilicates on the surface of Ceres. *Science*, **353**, aaf4279.

Barucci M.A., Fulchignoni M., & Rossi A. (2007) Rosetta asteroid targets: 2867 Steins and 21 Lutetia. *Space Science Reviews*, **128**, 67–78.

Barucci M.A., Filacchione G., Fornasier S., et al. (2016) Detection of exposed H_2O ice on the nucleus of comet 67P/Churyumov-Gerasimenko-as observed by Rosetta OSIRIS and VIRTIS instruments. *Astronomy and Astrophysics*, **595**, A102.

Binzel R.P., Gaffey M.J., Thomas P.C., Zellner B.H., Storrs A.D., & Wells E.N. (1997) Geologic mapping of Vesta from 1994 Hubble space telescope images. *Icarus*, **128**, 95–103.

Birlan M., Vernazza P., Fulchignoni M., et al. (2006) Near infra-red spectroscopy of the asteroid 21 Lutetia-I. New results of long-term campaign. *Astronomy and Astrophysics*, **454**, 677–681.

Burbine T.H., McCoy T.J., Nittler L.R., Benedix G.K., Cloutis E.A., & Dickinson T.L. (2002) Spectra of extremely reduced assemblages: Implications for Mercury. *Meteoritics and Planetary Science*, **37**, 1233–1244.

Capaccioni F., Coradini A., Filacchione G., et al. (2015) The organic-rich surface of comet 67P/Churyumov-Gerasimenko as seen by VIRTIS/Rosetta. *Science*, **347**, aaa0628.

Ciarniello M., Capaccioni F., Filacchione G., et al. (2011) Hapke modeling of Rhea surface properties through Cassini-VIMS spectra. *Icarus*, **214**, 541–555.

Ciarniello M., Capaccioni F., Filacchione G., et al. (2015) Photometric properties of comet 67P/Churyumov-Gerasimenko from VIRTIS-M onboard Rosetta. *Astronomy and Astrophysics*, **583**, A31.

Ciarniello M., Raponi A., Capaccioni F., et al. (2016) The global surface composition of 67P/Churyumov-Gerasimenko nucleus by Rosetta/VIRTIS. II) Diurnal and seasonal variability. *Monthly Notices of the Royal Astronomical Society*, **462**, S443–S458.

Ciarniello M., De Sanctis M.C., Ammannito E., et al. (2017) Spectrophotometric properties of dwarf planet Ceres from the VIR spectrometer on board the Dawn mission. *Astronomy and Astrophysics*, **598**, A130.

Clark R.N. (1999) Spectroscopy of rocks and minerals, and principles of spectroscopy. *Manual of Remote Sensing*, **3**, 2–2.

Cochran A.L. & Vilas F. (1998) The changing spectrum of Vesta: Rotationally resolved spectroscopy of pyroxene on the surface. *Icarus*, **134**, 207–212.

Combe J.-P., McCord T.B., Tosi F., et al. (2016) Detection of local H_2O exposed at the surface of Ceres. *Science*, **353**, aaf3010.

Coradini A., Capaccioni F., Drossart P., et al. (2007) VIRTIS: An imaging spectrometer for the Rosetta mission. *Space Science Reviews*, **128**, 529–559.

Coradini A., Capaccioni F., Erard S., et al. (2011) The surface composition and temperature of asteroid 21 Lutetia as observed by Rosetta/VIRTIS. *Science*, **334**, 492–494.

De Sanctis M.C., Raponi A., Ammannito E., et al. (2016) Bright carbonate deposits as evidence of aqueous alteration on (1) Ceres. *Nature*, **536**, 54–57.

De Sanctis M.C. (2011) The VIR spectrometer. *Space Science Reviews*, **163**, 329–369.

De Sanctis M.C., Ammannito E., Capria M., et al. (2012a) Spectroscopic characterization of mineralogy and its diversity across Vesta. *Science*, **336**, 697–700.

De Sanctis M.C., Combe J.-P., Ammannito E., et al. (2012b) Detection of widespread hydrated materials on Vesta by the VIR imaging spectrometer on board the Dawn mission. *The Astrophysical Journal Letters*, **758**, L36.

De Sanctis M.C., Ammannito E., Capria M.T., et al. (2013) Vesta's mineralogical composition as revealed by the visible and infrared spectrometer on Dawn. *Meteoritics and Planetary Science*, **48**, 2166–2184.

De Sanctis M.C., Capaccioni F., Ciarniello M., et al. (2015a) The diurnal cycle of water ice on comet 67P/Churyumov–Gerasimenko. *Nature*, **525**, 500–503.

De Sanctis M.C., Ammannito E., Raponi A., et al. (2015b) Ammoniated phyllosilicates with a likely outer Solar System origin on (1) Ceres. *Nature*, **528**, 241–244.

El-Maarry M.R., Thomas N., Gracia-Berná A., et al. (2016) Regional surface morphology of comet 67P/Churyumov-Gerasimenko from Rosetta/OSIRIS images: The southern hemisphere. *Astronomy and Astrophysics*, **593**, A110.

Feierberg M.A., Larson H.P., Fink U., & Smith H.A. (1980) Spectroscopic evidence for two achondrite parent bodies: Asteroids 349 Dembowska and 4 Vesta. *Geochimica et Cosmochimica Acta*, **44**, 513–524.

Filacchione G., Capaccioni F., Clark R., et al. (2010) Saturn's icy satellites investigated by *Cassini*–VIMS: II. Results at the end of nominal mission. *Icarus*, **206**, 507–523.

Filacchione G., Capaccioni F., Ciarniello M., et al. (2012) Saturn's icy satellites and rings investigated by *Cassini*–VIMS: III–Radial compositional variability. *Icarus*, **220**, 1064–1096.

Filacchione G., Capaccioni F., Ciarniello M., et al. (2016a) The global surface composition of 67P/CG nucleus by Rosetta/VIRTIS. (I) Prelanding mission phase. *Icarus*, **274**, 334–349.

Filacchione G., De Sanctis M., Capaccioni F., et al. (2016b) Exposed water ice on the nucleus of comet 67P/Churyumov–Gerasimenko. *Nature*, **529**, 368–372.

Filacchione G., Raponi A., Capaccioni F., et al. (2016c) Seasonal exposure of carbon dioxide ice on the nucleus of comet 67P/Churyumov-Gerasimenko. *Science*, **354**, aag3161.

Fornasier S., Marzari F., Dotto E., Barucci M., & Migliorini A. (2007) Are the E-type asteroids (2867) Steins, a target of the Rosetta mission, and NEA (3103) Eger remnants of an old asteroid family? *Astronomy and Astrophysics*, **474**, L29–L32.

Fornasier S., Hasselmann P., Barucci M., et al. (2015) Spectrophotometric properties of the nucleus of comet 67P/Churyumov-Gerasimenko from the OSIRIS instrument onboard the ROSETTA spacecraft. *Astronomy and Astrophysics*, **583**, A30.

Fornasier S., Mottola S., Keller H.U., et al. (2016) Rosetta's comet 67P/Churyumov-Gerasimenko sheds its dusty mantle to reveal its icy nature. *Science*, aag2671.

Gaffey M. & McCord T. (1979) Mineralogical and petrological characterizations of asteroid surface materials. *Asteroids*, 688–723.

Hapke B. (2005) *Theory of reflectance and emittance spectroscopy*. Cambridge University Press, Cambridge.

Keihm S., Tosi F., Kamp L., et al. (2012) Interpretation of combined infrared, submillimeter, and millimeter thermal flux data obtained during the Rosetta fly-by of Asteroid (21) Lutetia. *Icarus*, **221**, 395–404.

Keller H.U., Barbieri C., Lamy P., et al. (2007) OSIRIS: The scientific camera system onboard Rosetta. *Space Science Reviews*, **128**, 433–506.

Keller H., Barbieri C., Koschny D., et al. (2010) E-type asteroid (2867) Steins as imaged by OSIRIS on board Rosetta. *Science*, **327**, 190–193.

King T.V., Clark R., Calvin W., Sherman D.M., & Brown R. (1992) Evidence for ammonium-bearing minerals on Ceres. *Science*, **255**, 1551–1553.

Li J.-Y., Reddy V., Nathues A., et al. (2016) Surface albedo and spectral variability of Ceres. *The Astrophysical Journal Letters*, **817**, L22.

Lowry S., Duddy S., Rozitis B., et al. (2012) The nucleus of Comet 67P/Churyumov-Gerasimenko. A new shape model and thermophysical analysis. *Astronomy and Astrophysics*, **548**, A12.

Magrin S., La Forgia F., Pajola M., et al. (2012) (21) Lutetia spectrophotometry from Rosetta-OSIRIS images and comparison to ground-based observations. *Planetary and Space Science*, **66**, 43–53.

Mandler B.E. & Elkins-Tanton L.T. (2013) The origin of eucrites, diogenites, and olivine diogenites: Magma ocean crystallization and shallow magma chamber processes on Vesta. *Meteoritics and Planetary Science*, **48**, 2333–2349.

Marchi S., Ermakov A., Raymond C., et al. (2016) The missing large impact craters on Ceres. *Nature Communications*, **7**, 12257.

McCord T.B., Adams J.B., & Johnson T.V. (1970) Asteroid vesta: Spectral reflectivity and compositional implications. *Science*, **168**(3938), 1445–1447.

McFadden L.A., McCord T.B., & Pieters C. (1977) Vesta: The first pyroxene band from new spectroscopic measurements. *Icarus*, **31**, 439–446.

Nathues A., Hoffmann M., Schaefer M., et al. (2015) Sublimation in bright spots on (1) Ceres. *Nature*, **528**, 237–240.

Ockert-Bell M., Clark B.E., Isaacs M.E., Cloutis R., Fornasier E.A., & Bus S. (2010) The composition of M-type asteroids: Synthesis of spectroscopic and radar observations. *Icarus*, **210**, 674–692.

Palomba E., Longobardo A., De Sanctis M.C., et al. (2015) Detection of new olivine-rich locations on Vesta. *Icarus*, **258**, 120–134.

Park R.S., Konopliv A.S., Bills B.G., et al. (2016) A partially differentiated interior for (1) Ceres deduced from its gravity field and shape. *Nature*, **537**, 515–517.

Pieters C.M., McFadden L.A., Prettyman T., et al. (2011) Surface composition of Vesta: Issues and integrated approach. *Space Science Reviews*, **163**, 117–139.

Pieters C., Ammannito E., Blewett D., et al. (2012) Distinctive space weathering on Vesta from regolith mixing processes. *Nature*, **491**, 79–82.

Pommerol A., Thomas N., El-Maarry M.R., et al. (2015) OSIRIS observations of meter-sized exposures of H_2O ice at the surface of 67P/Churyumov-Gerasimenko and interpretation using laboratory experiments. *Astronomy and Astrophysics*, **583**, A25.

Quirico E., Moroz L., Schmitt B., et al. (2016) Refractory and semi-volatile organics at the surface of comet 67P/Churyumov-Gerasimenko: Insights from the VIRTIS/Rosetta imaging spectrometer. *Icarus*, **272**, 32–47.

Raponi A., Ciarniello M., Capaccioni F., et al. (2016) The temporal evolution of exposed water ice-rich areas on the surface of 67P/Churyumov–Gerasimenko: Spectral analysis. *Monthly Notices of the Royal Astronomical Society*, **462**, S476–S490.

Rivkin A.S., Li J.-Y., Milliken R.E., et al. (2011a) The surface composition of Ceres. *Space Science Reviews*, **163**, 95–116.

Rivkin A.S., Clark B.E., Ockert-Bell M., et al. (2011b) Asteroid 21 Lutetia at 3 μm: Observations with IRTF SpeX. *Icarus*, **216**, 62–68.

Ruesch O., Hiesinger H., De Sanctis M.C., et al. (2014) Detections and geologic context of local enrichments in olivine on Vesta with VIR/Dawn data. *Journal of Geophysical Research*, **119**, 2078–2108.

Russell C.T., Capaccioni F., Coradini A., et al. (2007) Dawn mission to Vesta and Ceres: Symbiosis between terrestrial observations and robotic exploration. *Earth, Moon, and Planets*, **101**, 65–91.

Russell C., Raymond C., Coradini A., et al. (2012) Dawn at Vesta: Testing the protoplanetary paradigm. *Science*, **336**, 684–686.

Russell C., Raymond C., Ammannito E., et al. (2016) Dawn arrives at Ceres: Exploration of a small, volatile-rich world. *Science*, **353**, 1008–1010.

Sierks H., Keller H.U., Jaumann R., et al. (2011a) The Dawn Framing Camera. *Space Science Reviews*, **163**, 263–327.

Sierks H., Lamy P., Barbieri C., et al. (2011b) Images of asteroid 21 Lutetia: A remnant planetesimal from the early Solar System. *Science*, **334**, 487–490.

Sierks H., Barbieri C., Lamy P.L., et al. (2015) On the nucleus structure and activity of comet 67P/Churyumov-Gerasimenko. *Science*, **347**, aaa1044.

Snodgrass C., Tubiana C., Bramich D., Meech K., Boehnhardt H., & Barrera L. (2013) Beginning of activity in 67P/Churyumov-Gerasimenko and predictions for 2014–2015. *Astronomy and Astrophysics*, **557**, A33.

Sunshine J.M., Groussin O., Schultz P.H., et al. (2007) The distribution of water ice in the interior of Comet Tempel 1. *Icarus*, **190**, 284–294.

Tholen D.J. & Barucci M.A. (1989) Asteroid taxonomy. In: *Asteroids II* (R. Binzel, T. Gehrels, & M.S. Matthews, eds.). University of Arizona Press, Tucson, 298–315.

Thomas P.C., Binzel R.P., Gaffey M.J., Zellner B.H., Storrs A.D., & Wells E. (1997) Vesta: Spin pole, size, and shape from HST images. *Icarus*, **128**, 88–94.

Thomas N., Sierks H., Barbieri C., et al. (2015) The morphological diversity of comet 67P/Churyumov-Gerasimenko. *Science*, **347**, aaa0440.

Vilas F. (1994) A cheaper, faster, better way to detect water of hydration on Solar System bodies. *Icarus*, **111**, 456–467.

Vilas F. & Gaffey M.J. (1989) Phyllosilicate absorption features in main-belt and outer-belt asteroid reflectance spectra. *Science*, **246**, 790–792.

21

Spectral Analyses of Saturn's Moons Using the *Cassini* Visual Infrared Mapping Spectrometer

BONNIE J. BURATTI, ROBERT H. BROWN, ROGER N. CLARK,
DALE P. CRUIKSHANK, AND GIANRICO FILACCHIONE

21.1 Introduction

The moons of Saturn represent several classes of unique objects, including at least two – Enceladus and Titan – with possibly habitable environments and complex chemistry. These bodies range from the planet-like Titan, with a substantial atmosphere and a hydrological cycle based on methane, to the regular medium-sized moons that include active Enceladus, to the odd families of small, inner irregular moons, to the outer irregular moons that are probably captured objects, possibly from the Kuiper Belt. The small moons include the ring moons Pan and Daphnis; the co-orbitals Janus and Epimetheus; the shepherding moons Prometheus, Pandora, and Atlas; and the Tethys Lagrangians Telesto and Calypso. Figure 21.1 is a montage of a representative sample of Saturn's family of moons, provided by the Visual and Infrared Mapping Spectrometer (VIMS) and the Imaging Science Subsystem (ISS) on the *Cassini* spacecraft. This chapter describes the spectral properties of these bodies based on the visible to near-infrared (VNIR) spectra of volatiles (Chapter 5).

Ground-based observations identified crystalline water ice as the major constituent of the surfaces of the main moons (see review in Cruikshank et al., 2005), and their low densities, ranging from 0.98 for Tethys to 1.61 for Enceladus, implied a bulk composition dominated by water ice. Some low-albedo opaque contaminant was also present: models that included tholins and others with nanophases of metallic iron plus nano-iron oxides gave good fits to their spectra and explained the reddish slope exhibited by some moons in the visible. Based on likely formation scenarios for the moons and their current surface temperatures, other expected but not detected constituents included hydrates of ammonia, CH_4, and CO, although ammonia hydrate may have been detected by ground-based observers (Emery et al., 2005; Verbiscer et al., 2006). A possible detection by VIMS of ammonia on Dione was also made (Clark et al., 2008).

VIMS was the first infrared instrument designed to study the surface composition of the Saturnian moons as part of a deep-space mission. The specifics of VIMS are outlined in Brown et al. (2004). The wavelength range, from 0.35 µm to 5.1 µm, covers 99% of the reflected solar spectrum in 352 spectral channels, with spectral resolution ranging from 1.46 nm in the visible region (0.35–1.05 µm) to 16.6 nm in the NIR (0.85–5.1 µm). Each

Figure 21.1 Montage of a representative group of Saturn's moons. (Upper row) Mimas, Enceladus, Helene, and Dione. (Lower row) Titan, Hyperion, Iapetus, Phoebe. The Titan image was obtained by VIMS; the others by ISS. The moons are not to scale in size, albedo or color. Their mean radii are (from top to bottom, left to right, in km): 198; 252; 18; 561; 2575.5; 135; 734.5; 106.5. NASA/JPL-Caltech.

image of up to 64 × 64 pixels is constructed by the motion of a mirror that is capable of articulation in two dimensions. The basic data unit is the three-dimensional image cube, with two spatial dimensions and one spectral dimension. The field of view of a full 64 × 64 image is 32 × 32 mrad, offering 0.5 mrad spatial resolution for each pixel. A Nyquist sampling mode that offered double the spatial resolution in one dimension was also available.

Mission planners for *Cassini* designed a series of targeted flybys in which the spacecraft trajectory was altered to approach specific moons to within a distance typically of 1000 km or less, focusing on Enceladus and Titan (Table 21.1). A targeted flyby of Phoebe was conducted before Saturn orbit insertion. Mimas and Tethys were not included in this plan, but a serendipitous close flyby of Tethys on September 24, 2005 was transformed into a targeted, even closer, flyby. Additional untargeted flybys, in which the spacecraft happened to approach a moon without a specific trajectory maneuver designed to do so, also yielded key data. For Titan and Enceladus, each flyby was centered on a major goal such as determination of the gravity field and internal structure, radar properties, optical and infrared remote sensing for geology and composition, and fields and particles. In addition to these encounters with the main moons, approaches to the small moons were sufficiently close so that VIMS spectra could be obtained on at least nine of the moons. Figure 21.2 shows typical spectra obtained for each of the major moons of Saturn plus the co-orbital and shepherd moons.

Table 21.1 *Summary of* Cassini's *major flybys of Saturn's main moons, June 1, 2004–April 22, 2017*

Object	Number of flybys	Distance (km)	Comments
Mimas	1	9500	Not a targeted flyby
Enceladus	22	20–4999	Flybys dedicated to specific disciplines
Tethys	1	1503	Converted serendipitous flyby
Dione	5	96–516	Three gravity flybys; two fields and particles
Rhea	4	100–5750	Gravity; remote sensing (2); fields and particles
Hyperion	1	522	
Titan	126	953–31,130	Flybys dedicated to specific disciplines
Iapetus	1	1229	Additional flyby at 123 000 km altitude
Phoebe	1	2068	Prior to Saturn orbit insertion; flyby was June 11, 2004.

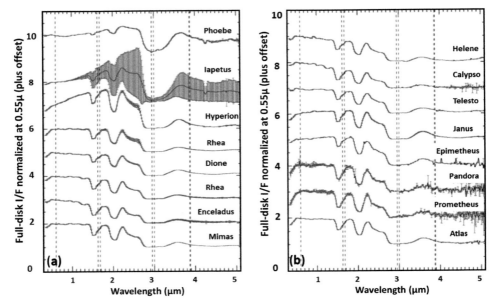

Figure 21.2 VIMS spectra (based on Filacchione et al., 2012). (a) Typical spectra of Saturn's main moons. (b) Typical spectra of the co-orbitals Janus and Epimetheus, and the shepherd moons Pandora, Prometheus, and Atlas. The spectra of the moons are dominated by the water ice absorption bands at 1.6, 2.0, and 3 μm. The range in spectra of Iapetus is shown by the shaded area. The dashed lines delineate the locations of the VIMS blocking filters, which remove signal from unwanted orders; these regions are considered to have poorer data.

21.2 Enceladus

Enceladus is the most significant of the five regular inner moons of Saturn because it is currently active. Its south polar region is dominated by active fissures (dubbed "tiger stripes") approaching temperatures of 200 K and jets composed of μm-sized water-ice particles expelled from a liquid ocean (Porco et al., 2006; Spencer et al., 2006; Goguen et al., 2013). The preliminary *Cassini* VIMS findings for Enceladus are summarized in Brown et al. (2006). The early remote sensing flybys of this moon showed that its spectrum was dominated by crystalline water ice, with some amorphous ice possibly existing in the regions between the tiger stripes (Brown et al., 2006). The amorphous ice interpretation was made before the spectral effects of sub-μm ice grains were known (Clark et al., 2012); new models that include these effects are needed to search for possible amorphous ice. CO_2 was discovered on Enceladus, and seemed to be globally prevalent (Brown et al., 2006). The spectral signature of CO_2 in the active south polar region, with the placement of the band at 4.26 μm, suggested that the ice was complexed with other molecules, most likely water ice, with more free ice in the cold regions outside the tiger stripes. Possible organics are also present on the surface of Enceladus (Figure 21.3), but the dearth of laboratory data prevents the detection of specific compounds. No evidence for CO was found on Enceladus, even though it was expected to be there. The position of spectral bands of water ice indicates the size of the component particles. For the surface of Enceladus, the dominant grain sizes are 50–150 μm, with increased sizes of up to 300 μm in the tiger stripe area. Both VIMS and ground-based observations showed a coloring agent on the moon that caused a slight reddening of its visible spectrum (Verbiscer et al., 2006).

With the presence of possible but unconfirmed organics, a liquid ocean in contact with a mineral-rich mantle (Postberg et al., 2011), and a source of energy that drives the moon's activity, Enceladus presents a potentially habitable environment commensurate with our current understanding of such environments for primitive bacterial life.

21.3 Mimas, Tethys, Dione, Rhea

Mimas and Tethys had no planned targeted flybys, but one serendipitous flyby of Tethys was transformed into a targeted flyby (see Table 21.1). As for Enceladus, the surfaces of the four additional regular inner moons of Saturn are dominated by water ice. CO_2 has been discovered on all five of the inner main moons, as well as on Phoebe, Iapetus, and Hyperion (Buratti et al., 2005; Clark et al., 2005, 2008, 2012; Cruikshank et al., 2007, 2010 see Figure 21.4a). NH_3 frost has been tentatively identified on Dione (Clark et al., 2008). They also identified a feature at 2.42 μm on Dione that could be attributed to a cyanide compound, although no fundamental band was found, or to trapped H_2 (Clark et al., 2008). More recent calibrations of the VIMS data (Clark et al., 2016) appear to have weakened the strength of any H_2 absorption, implying that less is present on Dione than initially thought.

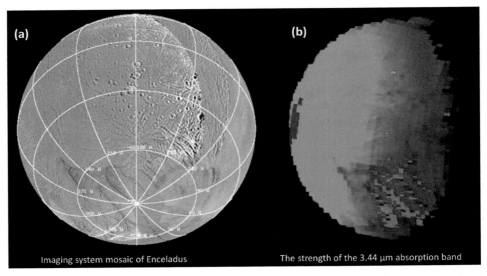

Figure 21.3 Images of Enceladus. (a) An ISS mosaic. (b) The strength of the 3.44 μm absorption band characteristic of light organics; redder areas indicate stronger absorptions (based on Brown et al., 2006. NASA/JPL-Caltech). Laboratory data are not yet sufficiently extensive to identify specific compounds.

The surface composition of the moons is determined by exogenous processes including the accretion of particles from the E-ring onto their surfaces, which result in higher albedos for moons closer to Enceladus and brighter leading sides for moons exterior to Enceladus (Buratti et al., 1990; Verbiscer et al., 2007) and magnetospheric effects. In the latter case, patterns of high thermal inertia and increased UV reflectivity in the ISS UV3 filter centered at 0.34 μm coincide with regions of enhanced impact and annealing by high-energy magnetospheric electrons (Howett et al., 2011, 2012, 2014; Schenk et al., 2011; Paranicas et al., 2012). The patterns, nicknamed "pacmen" are especially prominent on Mimas and Tethys, with fainter presentations on Dione and Rhea.

One of the most intriguing observations concerning these moons is the discovery of red streaks of unknown origin on Tethys (Schenk, 2015). They are arcuate, lower albedo features centered on the anti-Saturn meridian consisting of parallel lineations a few km wide and 50–250 km long. Figure 21.4b is a ratio of VIMS spectra of these streaks to the surrounding areas. The comparison shows stark spectral differences: there is more water ice absorption in the red streaks, as well as an enhanced Fresnel peak, all characteristic of recent outgassing and freezing of water particles. The spectral shape between 3.5 and 5.1 μm shows possible enhancement of organic compounds in the red streaks. This shape is also characteristic of the sublimation of water ice (Brown et al., 2012). Some of these spectral contrasts may be due to differences in particle sizes; in any case they are suggestive of some sort of ongoing outgassing activity from the red streaks.

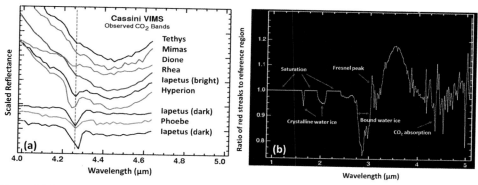

Figure 21.4 VIMS spectra of Saturn's moons. (a) Representative spectra from Saturn's main moons, including from both bright and dark terrains on Iapetus, showing the CO_2 absorption band (adapted from Clark et al., 2008). Error bars are about twice the width of the lines. (b) A ratio of the spectrum of the red streaks on Tethys to the surrounding areas, showing key spectral differences between the two. Up to 4 μm the error bars are about the size of the line but increase to about ±0.01 at longer wavelengths.

21.4 Iapetus and Hyperion

Iapetus represented one of the key conundrums in planetary science with its extreme albedo dichotomy: on its leading hemisphere, normal reflectances as low as 0.02 contrasted with the trailing side with albedos typical of the other icy moons of Saturn, up to 0.7 and characteristic of water ice (Squyres et al., 1984). The discovery of the Phoebe ring (Verbiscer et al., 2009) and the elucidation of the dynamical behavior of the resulting dust (Tamayo et al., 2011) provided a definite source for low-albedo particles impinging and collecting on its leading side. CO_2 was found on Iapetus during an untargeted flyby just prior to the Huygens probe separation (Buratti et al., 2005), and spectra with km-scale resolution were obtained during a targeted flyby on September 10, 2007. Clark et al. (2012) identified additional key components of the surface of Iapetus in addition to CO_2 and H_2O, including bound water, H_2, OH-bearing minerals, trace organics, and possibly ammonia. The CO_2 on Iapetus was enriched in the low-albedo areas, reaching a maximum at the apex of motion where the albedo is lowest. One key component of the Clark et al. analysis is the presence of nanophase metallic iron particles and an iron oxide, probably hematite. These particles as well as sub-μm ice grains could all contribute the observed Rayleigh scattering peak in the visible part of the spectrum. Furthermore, these small nanophase iron particles could act as a coloring agent throughout the Saturnian system, providing an explanation for the consistent reddening into the visible seen on the icy moons and the rings. Figure 21.5 is a summary of the features in the spectrum of Iapetus.

An analysis of the 2.7–4.0 μm spectral region by Cruikshank et al. (2008, 2014) identified the presence of both aromatic and aliphatic hydrocarbons in the dark material on Iapetus. The aromatic band near the C–H stretching modes of aromatic hydrocarbons

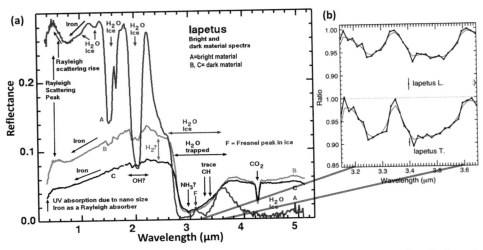

Figure 21.5 **Iapetus spectra.** (a) Spectra of bright and dark material on Iapetus (after Clark et al., 2012). Features due to H_2O ice and other components are labeled on the spectra. (b) The identification of aromatic hydrocarbons in the bright spectrum. This plot provides the residuals resulting from a best-fit Hapke compositional model to the spectra of the leading and trailing sides of Iapetus (after Cruikshank et al., 2014). Absorption bands at 3.28 μm are characteristic of aromatic hydrocarbons, while those between 3.35 and 3.6 μm are characteristic of aliphatic hydrocarbons.

at ~3.28 μm is especially strong, and is likely due to aromatic hydrocarbons. Because surfaces rich in aromatic hydrocarbons are readily carbonized via UV radiation, these substances are not stable on the surface of Iapetus, and thus the process of accretion of Phoebe ring particles must be ongoing. Denk et al. (2010) noted that there are two distinct color regimes on Iapetus: the low-albedo material on the leading side is consistently redder than that on the trailing side. This difference could be explained by the accumulation of greater amounts of the exogenously created red chromophore or by different abundances of the iron particles.

Hyperion was observed closely during one targeted flyby on September 26, 2005. An initial analysis by Cruikshank et al. (2007) identified two primary compositional terrains: a relatively bright unit covering most of the surface that is dominated by the spectral signature of water ice, and a low-albedo unit with a spectral signature similar to that of the low-albedo material on Iapetus. This result suggests that Hyperion is being similarly coated with material from the Phoebe ring, but because the moon is in chaotic rotation, it is not found preferentially on one hemisphere. Rather, it tends to be located in the bottoms of craters or pits, similar to sun cups on Earth, in which dark material is concentrated in warmer areas through a process of thermal segregation. This material is rich in organics, including possibly aromatics. CO_2 was also found on Hyperion (see Figure 21.4a), probably combined in some way with water ice. Dalton et al. (2012) identify aromatic and aliphatic hydrocarbons in the low-albedo material, and possibly

H_2. As for Iapetus and the other moons, nanophase iron is thought to act as a coloring agent (Clark et al., 2012).

The moons are darkened and reddened by a coloring agent that may also be exogenous in origin. One idea advanced by Clark et al. (2012) has sub-0.5-µm diameter particles composed of nanophase hematite and metallic iron contaminating Dione, Hyperion, Epimetheus, Iapetus, Phoebe, parts of the ring system, and perhaps other moons, with an enhanced amount on the low-albedo hemisphere of Iapetus. Cruikshank et al. (2014) posit the existence of aromatic and aliphatic hydrocarbons as at least a component of the elusive dark red material on Iapetus, Hyperion, and Phoebe (see Section 21.5). The highest concentration of dark material occurs on the leading hemisphere of Iapetus. There, the dark material's spectrum displays a relatively unique 3 µm absorption that is matched well by adsorbed water in nanophase iron oxides (Clark et al., 2012). The absorption is significantly shifted from the 3-µm absorption observed in spectra of tholins. Clark et al. (2005) also showed that the gray spectral color of Phoebe was matched by the same nano-iron + nano-iron oxide as the red material on Iapetus but in different proportions and with grains mixed with ice.

Both Iapetus and Hyperion have spectral bands that have not been identified (Clark et al., 2012; Dalton et al., 2012). They may be due to higher order hydrocarbons, some of which have not been measured in the laboratory.

21.5 Phoebe

Phoebe is the largest example of an outer irregular moon of a giant planet. Its inclined retrograde orbit suggests it is a captured object. Besides water ice, initial spectral mapping by VIMS identified abundant CO_2 as well as spectral bands for organics, nitriles, cyanide compounds, and ferrous iron–bearing minerals (Clark et al., 2005), though Clark et al. (2012) found H_2 is a better explanation for the spectrum than nitriles and cyanides. The absorption feature at 2.42 µm for H_2 was observed for the first time on Phoebe, although the new calibration makes this identification less certain, as is the case for Iapetus. The aromatic hydrocarbon band is also present at the same wavelength as Iapetus (Figure 21.5), although the band is weaker (Cruikshank et al., 2008). In general, the spectrum of Phoebe is similar to that of Iapetus (Clark et al., 2012; see Figure 21.5), with the same phenomenon of Rayleigh scattering in the visible and the same general scheme of absorption bands. In the visible Phoebe is more spectrally neutral, possibly because of additional native carbon on its surface.

The materials discovered by VIMS on Phoebe are typical of those formed in low-temperature environments and thus indicate that it formed in a region exterior to the Saturnian system, in a colder region of the proto-solar nebula. Phoebe's composition, which is similar to that of comets and interstellar dust, coupled with its high rock-to-volatile ratio, suggests that the moon may have formed in the outer solar system, perhaps in the Kuiper Belt (Clark et al., 2005; Johnson & Lunine, 2005).

21.6 The Small Moons

The small inner moons of Saturn fall into several unique categories: the co-orbital moons, Janus and Epimetheus, which probably resulted from a collision; the ring shepherds Atlas, Prometheus and Pandora; Pan and Daphnis, which dwell within the rings (and clear the Encke and Keeler gaps, respectively); and the lagrangians, Calypso and Telesto for Tethys, and Polydeuces and Helene for Dione. These moons offer clues to the evolution of the Saturnian system, including collisional processes and the origin of Saturn's ring systems. The geology and dynamics of the small moons is reviewed in Thomas et al. (2013).

So far VIMS spectra have been analyzed for Atlas, Epimetheus, Janus, Pandora, Telesto, Calypso, and Helene (Buratti et al., 2010; Filacchione et al., 2012; see Figure 21.2). None of these moons shows absorption bands other than those for water ice, although they show color differences among themselves, probably due to contamination by Saturn's inner ring system. The inner shepherd moons, Atlas and Pandora, tend to be redder than the co-orbitals and the Tethys lagrangians in the visible, again due to contamination by the main ring system. The lagrangians are bluer because of contamination by particles of ice from the E-ring. The lagrangians are also similar in color to their "parent" bodies, mainly because they are subjected to the same exogenous processes at the same distance from Saturn (see Figure 21.6).

The small inner moons also show color and possible compositional differences across their surfaces. The nearly round Helene is covered with unique striations that indicate some type of downslope movement. The lowest lying material is substantially bluer, with an Imaging Science Subsystem IR3/UV3 (0.93 µm/0.34 µm) ratio that is almost 15% lower than that of the higher-elevation materials (Thomas et al., 2013). One explanation for this color difference is smaller sizes of the low-lying material. Telesto and Calypso also show spatially resolved color changes in the visible and near-IR, but they are not as distinct as those of Helene. All of these moons are in synchronous rotation, and spectra obtained during the final phase of the *Cassini* mission in 2017 (the F-ring and proximal orbits) will enable a comparison of leading/trailing compositional differences on some of the moons. "Best ever" closest approaches of Atlas, Daphnis, Epimetheus, Pan, and Pandora occurred during this period. Spectra of the ring moons have not been obtained, but their presence in the midst of Saturn's main ring system suggests they are coated with ice and the same contaminants that are found in the main ring system.

A large family of small irregular moons orbits Saturn at great distances (~0.1 AU), falling into four dynamical families and possibly corresponding compositional families (Gladman et al., 2001). Little is known about the composition of these objects, but given that their colors vary, the bodies must be compositionally diverse, reflecting multiple captures into orbit around Saturn. Their colors range from gray to red, although they are not as a red as Kuiper Belt Objects (KBOs), which have V–R indices (defined as the astronomical magnitude in the visible filter minus that of the red filter) approaching 0.9, compared to a maximum of 0.65 for the outer Saturnian moons (Grav & Bauer, 2007; Jewitt & Haghighipour, 2007). Grav and Bauer (2007) showed that the colors of the dynamical

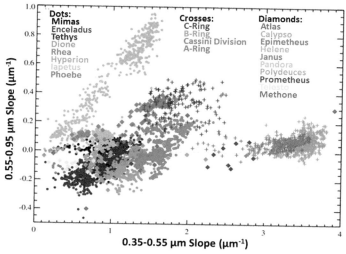

Figure 21.6 A comparison of the visible colors of the Saturnian moons from VIMS. The three "ecological zones" form clear groups in the graph. The right grouping contains the inner moons and the A and B rings (CD is the *Cassini* Division). The middle group consists of Mimas, Enceladus, Tethys, Dione, and Rhea, which are progressively redder as their distance from the E-ring increases. The upper left group contains the outer moons of Hyperion and Iapetus, which are contaminated by the Phoebe ring. Phoebe itself is more neutral in color. Adapted from Filacchione et al. (2012).

families are correlated in at least three out of four cases, offering proof in addition to dynamical arguments that the families may have originated from a common body. VIMS observations of several moons were obtained, but no spectra have yet been extracted because of their faintness.

21.7 Titan

Titan is a complex geologic world, with features ranging from lakes and fluvial systems, to massive equatorial dune fields, and possibly cryovolcanoes, as well as a weather system with clouds and hydrocarbon rain driven by a methanologic cycle. The composition of Titan is elusive because its thick nitrogen–methane atmosphere (1.6 bars) and haze obscures the surface. In the IR region of the spectrum, there are several windows of clarity that correspond to methane transmission bands: the most significant are at 2.01, 2.73, and 4.94 μm. Furthermore, the optical depth of the haze decreases as the wavelength increases. Exploiting these bands of clarity in the IR where the surface can be viewed, Griffith et al. (2003) showed that the albedo of the surface of Titan measured though these windows is most consistent with water ice and the presence of some darkening agent, but this study assumed that the decreasing reflectance with increasing wavelength was a signature of water ice. Many materials, including most organic compounds, also display this property because of increasing absorption strengths (Clark et al., 2009, 2010; Kokaly et al., 2017a,b). Early results from *Cassini* VIMS seemed to confirm this view (Buratti et al., 2006; McCord et al., 2006) although the same assumption regarding spectral slope was made. Some CO_2 was present as well (McCord et al., 2008). Later work by Clark et al. (2010) presented a more complex picture in which a substrate of water ice was covered by a panoply of organic compounds, some of which formed in Titan's haze layer. Clark et al. found no evidence for any exposure in the optical surface for water ice. Among the molecules detected were abundant benzene, ethane, methane, and hydrogen cyanide; the latter was also found in Titan's cloud system. Acetylene was not detected, and there is an unidentified spectral line at 5.01 μm, as well as many other lines that can be identified only with additional laboratory spectral measurements. Later work incorporating a radiative transfer model for the atmosphere based on the results of stellar occultations suggested there are areas of Titan covered with exposed water ice (Hayne et al., 2014), but this study did not include the effect of reddening by small particles in the atmosphere. Sections of the equatorial regions of Titan are covered with low-albedo dune fields, most likely composed of organic-rich sand grains (Barnes et al., 2008). Lakes filled with hydrocarbons, primarily ethane and methane, and possibly additional low-molecular-mass hydrocarbons, span the northern polar regions (Brown et al., 2008) and seem in some cases to be rimed with an as-yet unidentified evaporate deposit. Barnes et al. (2011) suggest that dry lake beds in Titan's north pole, and possibly the 5-μm bright regions of Regio Hui (28°S, 125°W) and Regio Hotei (28°S, 80°W), are such deposits rich in hydrocarbons.

21.8 Summary and Conclusions

The main composition of the moons of Saturn is water ice, with a patina of admixtures placed primarily by exogenous processes. There are three basic "ecological zones" in the system, with Saturn's rings controlling each of these zones (Buratti et al., 2010; Filacchione et al., 2010, 2013). The inner small moons are contaminated by the A and B main ring system, with the same red chromophore – possibly nano-iron, or tholins (or both) – causing the visible spectral slope observed on these objects. Further out, the E-ring of Saturn, which is composed of μm-sized particles expelled from the interior of Enceladus through its plume, coats the main medium-sized moons to form bright icy objects. An additional red chromophore, which is also present in the C-ring, colors the surfaces of the moons not coated with the E-ring, particularly as the distance from Enceladus increases. Finally, the Phoebe ring, which is composed of low-albedo dust expelled from Phoebe by collisional processes, containing nano-iron and hydrated nano-iron oxides plus a component of organic molecules, coats the leading side of Iapetus and all of Hyperion. The nano-iron could be a component of space-weathered silicates (Clark et al., 2012), although the dearth of iron particles detected by the *Cassini* Cosmic Dust Analyzer in the Saturnian system (e.g., McBride et al., 2007) means the identification of nano-iron is only a hypothesis at present. The spectral differences on these bodies is determined by the amount of "Phoebe dust" present. Figure 21.6 shows a summary of how the three types of contamination work in the system.

Titan is a unique world, with its thick atmosphere; photochemical processes in its haze layer; and system of liquid, solid, and gaseous hydrocarbons. The consensus view of its composition is that it possesses a bedrock of water ice, which is apparently not visible through a layer of haze comprising both complex and simple hydrocarbons, at least at the few km resolution of the VIMS data. The uncertainty over water ice detection rests in spectral models that are incomplete and do not include the spectral effects of the many organic compounds on Titan's surface and in the atmosphere. Yet to be determined is the thickness of the organic layer covering the ice.

Acknowledgments

Part of this research was carried out at the Jet Propulsion Laboratory, California Institute of Technology under contract with NASA. This project was funded by the *Cassini* mission.

References

Barnes J.W., Brown R.H., Soderblom L., et al. (2008) Spectroscopy, morphometry, and photoclinometry of Titan's dunefields from *Cassini*/VIMS. *Icarus*, **195**, 400–414.
Barnes J.W., Bow J., Schwartz J., et al. (2011) Organic sedimentary deposits in Titan's dry lakebeds: Probable evaporite. *Icarus*, **216**, 136–140.
Brown R.H., Baines K.H., Bellucci G., et al. (2004) The *Cassini* visual and infrared mapping spectrometer (VIMS) investigation. *Space Science Reviews*, **115**, 111–168.
Brown R.H., Clark R.N., Buratti B.J., et al. (2006) Composition and physical properties of Enceladus' surface. *Science*, **311**, 1425–1428.
Brown R.H., Soderblom L.A., Soderblom J.M., et al. (2008) The identification of liquid ethane in Titan's Ontario Lacus. *Nature*, **454**, 607–610.
Brown R.H., Lauretta D.S., Schmidt B., & Moores J. (2012) Experimental and theoretical simulations of ice sublimation with implications for the chemical, isotopic, and physical evolution of icy objects. *Planetary and Space Science*, **60**, 166–180.
Buratti B.J., Mosher J.A., & Johnson T.V. (1990) Albedo and color maps of the saturnian satellites. *Icarus*, **87**, 339–357.
Buratti B.J., Cruikshank D.P., Brown R.H., et al. (2005) *Cassini* visual and infrared mapping spectrometer observations of Iapetus: Detection of CO_2. *The Astrophysical Journal Letters*, **622**, L149–L152.
Buratti B.J., Sotin C., Brown R.H., et al. (2006) Titan: Preliminary results on surface properties and photometry from VIMS observations of the early flybys. *Planetary and Space Science*, **54**, 1498–1509.
Buratti B.J., Bauer J.M., Hicks M.D., et al. (2010) *Cassini* spectra and photometry 0.25–5.1 μm of the small inner satellites of Saturn. *Icarus*, **206**, 524–536.
Clark R.N., Brown R.H., Jaumann R., et al. (2005) Compositional maps of Saturn's moon Phoebe from imaging spectroscopy. *Nature*, **435**, 66–69.
Clark R.N., Curchin J.M., Jaumann R., et al. (2008) Compositional mapping of Saturn's satellite Dione with *Cassini* VIMS and implications of dark material in the Saturn system. *Icarus*, **193**, 372–386.
Clark R.N., Curchin J.M., Barnes J.W., et al. (2010) Detection and mapping of hydrocarbon deposits on Titan. *Journal of Geophysical Research*, **115**, E10005, DOI:10.1029/2009JE003369.
Clark R.N., Cruikshank D.P., Jaumann R., et al. (2012) The surface composition of Iapetus: Mapping results from *Cassini* VIMS. *Icarus*, **218**, 831–860.
Clark R.N., Brown R.H., & Lytle D.M. (2016) The VIMS wavelength and radiometric calibration. *NASA Planetary Data System, The Planetary Atmospheres Node*, https://atmos.nmsu.edu/data_and_services/atmospheres_data/Cassini/vims_2.html.
Cruikshank D.P., Owen T.C., Dalle Ore C., et al. (2005) A spectroscopic study of the surfaces of Saturn's large satellites: H_2O ice, tholins, and minor constituents. *Icarus*, **175**, 268–283.
Cruikshank D.P., Dalton J.B., Dalle Ore C.M., et al. (2007) Surface composition of Hyperion. *Nature*, **448**, 54–56.
Cruikshank D.P., Wegryn E., Dalle Ore C., et al. (2008) Hydrocarbons on Saturn's satellites Iapetus and Phoebe. *Icarus*, **193**, 334–343.

Cruikshank D.P., Meyer A.W., Brown R.H., et al. (2010) Carbon dioxide on the satellites of Saturn: Results from the *Cassini* VIMS investigation and revisions to the VIMS wavelength scale. *Icarus*, **206**, 561–572.

Cruikshank D.P., Dalle Ore C.M., Clark R.N., & Pendleton Y.J. (2014) Aromatic and aliphatic organic materials on Iapetus: Analysis of *Cassini* VIMS data. *Icarus*, **233**, 306–315.

Dalton J.B. III, Cruikshank D.P., & Clark R.N. (2012) Compositional analysis of Hyperion with the *Cassini* Visual and Infrared Mapping Spectrometer. *Icarus*, **220**, 752–776.

Denk T., Neukum G., Roatsch T., et al. (2010) Iapetus: Unique surface properties and a global color dichotomy from *Cassini* imaging. *Science*, **327**, 435–439.

Emery J.P., Burr D.M., Cruikshank D.P., Brown R.H., & Dalton J.B. (2005) Near-infrared (0.8–4.0 µm) spectroscopy of Mimas, Enceladus, Tethys, and Rhea. *Astronomy and Astrophysics*, **435**, 353–362.

Filacchione G., Capaccioni F., Clark R., et al. (2010) Saturn's icy satellites investigated by *Cassini*–VIMS: II. Results at the end of nominal mission. *Icarus*, **206**, 507–523.

Filacchione G., Capaccioni F., Ciarniello M., et al. (2012) Saturn's icy satellites and rings investigated by *Cassini*–VIMS: III–Radial compositional variability. *Icarus*, **220**, 1064–1096.

Filacchione G., Capaccioni F., Clark R.N., et al. (2013) The radial distribution of water ice and chromophores across Saturn's system. *The Astrophysical Journal*, **766**, 76.

Gladman B., Kavelaars J., Holman M., et al. (2001) Discovery of 12 satellites of Saturn exhibiting orbital clustering. *Nature*, **412**, 163–166.

Goguen J.D., Buratti B.J., Brown R.H., et al. (2013) The temperature and width of an active fissure on Enceladus measured with *Cassini* VIMS during the 14 April 2012 South Pole flyover. *Icarus*, **226**, 1128–1137.

Grav T. & Bauer J. (2007) A deeper look at the colors of the saturnian irregular satellites. *Icarus*, **191**, 267–285.

Griffith C.A., Owen T., Geballe T.R., Rayner J., & Rannou P. (2003) Evidence for the exposure of water ice on Titan's surface. *Science*, **300**, 628–630.

Hayne P.O., McCord T.B., & Sotin C. (2014) Titan's surface composition and atmospheric transmission with solar occultation measurements by *Cassini* VIMS. *Icarus*, **243**, 158–172.

Howett C.J.A., Spencer J.R., Schenk P., et al. (2011) A high-amplitude thermal inertia anomaly of probable magnetospheric origin on Saturn's moon Mimas. *Icarus*, **216**, 221–226.

Howett C.J.A., Spencer J.R., Hurford T., Verbiscer A., & Segura M. (2012) PacMan returns: An electron-generated thermal anomaly on Tethys. *Icarus*, **221**, 1084–1088.

Howett C.J.A., Spencer J.R., Hurford T., Verbiscer A., & Segura M. (2014) Thermophysical property variations across Dione and Rhea. *Icarus*, **241**, 239–247.

Jewitt D. & Haghighipour N. (2007) Irregular satellites of the planets: Products of capture in the early Solar System. *Annual Review of Astronomy and Astrophysics*, **45**, 261–295.

Johnson T.V. & Lunine J.I. (2005) Saturn's moon Phoebe as a captured body from the outer Solar System. *Nature*, **435**, 69–71.

Kokaly R.F., Clark R.N., Swayze G.A., et al. (2017) USGS spectral library version 7, https://dx.doi.org/10.5066/F7RR1WDJ. https://speclab.cr.usgs.gov/spectral-lib.html. US Geological Survey.

McBride N., Hillier J., Green S., et al. (2007) *Cassini* cosmic dust analyser: Composition of dust at Saturn. *Workshop on Dust in Planetary Systems*, 107–110.

McCord T.B., Hansen G.B., Buratti B.J., et al. (2006) Composition of Titan's surface from *Cassini* VIMS. *Planetary and Space Science*, **54**, 1524–1539.

McCord T.B., Hayne P., Combe J.-P., et al. (2008) Titan's surface: Search for spectral diversity and composition using the *Cassini* VIMS investigation. *Icarus*, **194**, 212–242.

Paranicas C., Roussos E., Krupp N., et al. (2012) Energetic charged particle weathering of Saturn's inner satellites. *Planetary and Space Science*, **61**, 60–65.

Porco C.C., Helfenstein P., Thomas P.C., et al. (2006) *Cassini* observes the active south pole of Enceladus. *Science*, **311**, 1393–1401.

Postberg F., Schmidt J., Hillier J., Kempf S., & Srama R. (2011) A salt-water reservoir as the source of a compositionally stratified plume on Enceladus. *Nature*, **474**, 620–622.

Schenk P., Hamilton D.P., Johnson R.E., et al. (2011) Plasma, plumes and rings: Saturn system dynamics as recorded in global color patterns on its midsize icy satellites. *Icarus*, **211**, 740–757.

Schenk P.M., Buratti B., Byrne P., McKinnon W.B., Nimmo F., & Scipioni F. (2015) Blood stains on Tethys: Evidence of recent activity. *American Geophysical Union, Fall Meeting 2015*, Abstract #P21B-02.

Spencer J.R., Pearl J.C., Segura M., et al. (2006) *Cassini* encounters Enceladus: Background and the discovery of a south polar hot spot. *Science*, **311**, 1401–1405.

Squyres S.W., Buratti B., Veverka J., & Sagan C. (1984) Voyager photometry of Iapetus. *Icarus*, **59**, 426–435.

Tamayo D., Burns J.A., Hamilton D.P., & Hedman M.M. (2011) Finding the trigger to Iapetus' odd global albedo pattern: Dynamics of dust from Saturn's irregular satellites. *Icarus*, **215**, 260–278.

Thomas P.C., Burns J.A., Hedman M., et al. (2013) The inner small satellites of Saturn: A variety of worlds. *Icarus*, **226**, 999–1019.

Verbiscer A.J., Peterson D.E., Skrutskie M.F., et al. (2006) Near-infrared spectra of the leading and trailing hemispheres of Enceladus. *Icarus*, **182**, 211–223.
Verbiscer A.J., French R., Showalter M., & Helfenstein P. (2007) Enceladus: Cosmic graffiti artist caught in the act. *Science*, **315**, 815–817.
Verbiscer A.J., Skrutskie M.F., & Hamilton D.P. (2009) Saturn's largest ring. *Nature*, **461**, 1098–1100.

22

Spectroscopy of Pluto and Its Satellites

DALE P. CRUIKSHANK, WILLIAM M. GRUNDY, DONALD E. JENNINGS,
CATHERINE B. OLKIN, SILVIA PROTOPAPA, DENNIS C. REUTER,
BERNARD SCHMITT, AND S. ALAN STERN

22.1 Introduction and Early Observations of Pluto

The first component of Pluto's surface to be detected spectroscopically was solid methane, seen in the near-infrared (NIR) wavelength region (1-2.5 µm) with ground-based telescopes beginning in 1980 (Cruikshank & Silvaggio, 1980; Soifer et al., 1980). Subsequently, the observations were extended to 5 µm (Olkin et al., 2007; Protopapa et al., 2008). The presence of methane had been deduced from photometric observations of Pluto in three wavelength bands selected to discriminate among the most likely components of an icy surface in the outer Solar System (Cruikshank et al., 1976), following the reasoning of Lewis (1972). Methane has several strong absorption bands across the NIR spectrum accessible to instruments available in the 1970s and 1980s. Resolution and signal precision improved as instrumental capabilities advanced, enabling Owen et al. (1993) to report the detection of solid N_2 and CO at 2.15 µm and 2.35 µm, respectively. Nitrogen ice had been found on Triton a few years earlier because Triton is typically brighter than Pluto, and because its N_2 band is significantly stronger than that of Pluto. Nitrogen ice occurs in two crystalline phases. The cubic or α phase exists at $T < 35.6$ K, while above this temperature the hexagonal β-phase occurs (Scott, 1976). The two phases have different spectral band characteristics, and on Pluto and Triton the shape of the 2.15-µm band indicates that the β-phase is present, consistent with the temperatures of the surfaces derived from vapor pressure considerations. The trajectory of these studies and discoveries is described in the fully referenced review by Cruikshank et al. (2015) and in narrative form with historical context in Cruikshank and Sheehan (2018).

On Pluto (and on Triton), solid N_2 and CH_4 are found in solid solutions having spectral characteristics dependent on the abundance of these two molecules. They are completely miscible in one another only over two narrow abundance ranges at the two extremes: N_2-rich ice with a small amount of substituted CH_4 on one side of the binary phase diagram (Prokhvatilov & Yantsevich, 1983), and on the other side CH_4-rich ice with a few percent N_2 molecules. At Pluto's surface temperature of ~40 K, for example, in N_2-rich ice the saturation limit of CH_4 is ~0.06, while in CH_4-rich ice the saturation limit of N_2 is ~0.035 (see Chapter 5). The CH_4 bands first found in ground-based spectra were observed to be shifted toward shorter wavelengths, indicating that CH_4 is dissolved in N_2. Detailed

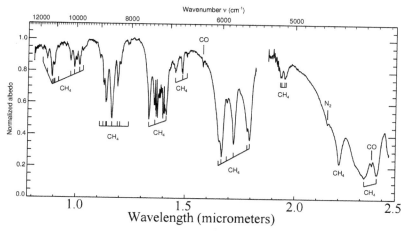

Figure 22.1 The NIR spectrum of Pluto, 0.8–2.5 μm (a global average that includes the light from Charon), with the absorption bands identified. This is an average of the best spectra from 65 datasets obtained from 2001 to 2013 in a monitoring program. For an assessment of the variations of individual species with the disk-averaged spectrum at a full range of central longitudes on the surface, see Grundy et al. (2013), the original source of this average spectrum.

radiative transfer modeling of the NIR spectrum of Pluto, using these shifts and the presence of a weak band at 1.69 μm, demonstrated the presence of both CH_4 diluted in N_2-rich ice and of CH_4-rich ice segregated either spatially or vertically (Douté et al., 1999). The spectroscopic characteristics of CH_4 and N_2 mixtures have been explored in laboratory studies by Quirico et al. (1996), Quirico and Schmitt (1997a), and Protopapa et al. (2015). In the latter reference, the complex refractive indices of several mixtures relevant to Pluto and other icy planetary bodies are presented. Carbon monoxide (CO) also occurs in solid solution in N_2, with the width and intensity of the 2.35-μm band varying with temperature and with the phase of the N_2 in which the CO occurs (Quirico & Schmitt, 1997b).

These spectral characteristics of N_2, CH_4, and CO separately and in combination enabled some preliminary conclusions about the spatial distribution of volatile ices and their secular evolution on Pluto by observing the planet when different hemispheres were visible from Earth, even though the disk could not be resolved (Douté et al., 1999; Grundy et al., 2013, 2014). This set the stage for the close-up spectral mapping of Pluto (and Charon) with the instruments on board the *New Horizons* spacecraft (see Section 22.3).

While spectroscopy of Pluto's surface was moving forward with improvements in astronomical instrumentation, the first spectral indications of the planet's atmosphere were obtained by Young (1994) and Young et al. (1997), by the detection of several rotational lines in the $2\upsilon_3$ band of CH_4 near 1.66 μm. The presence of the atmosphere (with a suspected haze layer) had been established in 1988 from observations of a stellar occultation by Pluto (Elliot et al., 1989), and was initially thought to consist mainly of CH_4. The discovery of N_2 ice on the surface by Owen et al. (1993) showed that the principal gaseous component was N_2 in vapor-pressure equilibrium with the surface, but details of the contribution of CH_4 to the atmosphere were lacking until Young's spectroscopic detection of the gas-phase bands. As noted in the text that follows, atmospheric haze was confirmed by the *New Horizons* spacecraft, and has been recognized to play an important role in the evolution of the surface as well as in atmospheric chemistry.

22.2 Spectroscopy of Charon

Pluto's largest satellite was discovered in 1978. While early observations during the epoch of mutual transits and occultations with Pluto in the 1980s, as viewed from Earth, indicated the presence of H_2O ice on its surface (Buie et al., 1987; Marcialis et al., 1987), high-quality NIR spectra became available only after the introduction of advanced optical systems in large ground-based telescopes. Water ice was confirmed by its NIR spectral signature, and to the surprise of the investigators was shown to occur in the hexagonal crystalline state, despite Charon's low temperature and the expectation that the ice would be in a disordered amorphous state (Brown & Calvin, 2000; Buie & Grundy, 2000). Furthermore, a weak but persistent absorption band at ~2.2 µm indicated the presence of another molecule, now generally thought to arise from a hydrate (or possibly a salt) of ammonia in the surface ice. The $NH_3 \cdot nH_2O$ band is not distributed uniformly over Charon's surface, and there are suspected variations in the band shape with longitude (Cook et al., 2007, 2017). The actual state or states of hydration are unknown.

22.3 Spectral Mapping of Pluto with *New Horizons*

The *New Horizons* spacecraft, launched on January 20, 2006, flew past Pluto and Charon on July 14, 2015, at a distance of 12 500 km from the planet's surface and 17 900 km from the satellite. The initial results of the multifaceted investigation of Pluto and its satellites are described in Stern et al. (2015), and in many subsequent publications. Spectral mapping in the wavelength region 0.4–2.5 µm was accomplished with the Ralph instrument package, consisting of the Multispectral Visible Imaging Camera (MVIC) and the Linear Etalon Imaging Spectral Array (LEISA) (Reuter et al., 2008). LEISA afforded a spatial resolution of the full hemisphere of Pluto visible at the encounter of ~6.2–7 km per pixel, and as sharp as ~2.8 km per pixel over a portion of the planet during the closest approach phase. MVIC provided images in four color filters at spatial resolutions up to 0.7 km per pixel.

22.3.1 Ralph Instrument Description

The Ralph visible and infrared remote sensing package on *New Horizons* consists of two focal planes fed by a shared telescope. It obtains multiwavelength coverage by having different wavelengths sensed by different pixels, avoiding the need for moving parts within the instrument. Using MVIC, this is achieved by means of four separate 5024 × 32 pixel charge-coupled device (CCD) arrays, each with a different interference filter covering the detector. MVIC is operated in time-delay integration (TDI) mode in which the image is read out along the short axis of each chip at a rate corresponding to the scan rate, so the effective exposure time is 32 times the time needed to scan across a single pixel. In LEISA, a pair of filters is affixed to the 256 × 256 pixel HgCdTe array, but the LEISA filters are wedged interference filters, so that each row of pixels is sensitive to a different wavelength. One of the LEISA filters covers wavelengths from 1.25 to 2.5 μm at a resolution ($\lambda/\Delta\lambda$) 240 while the other covers 2.1 to 2.25 μm at resolution 560. The whole LEISA array is repeatedly read out and saved at a frame rate corresponding to the time needed to shift the field of view by one LEISA pixel.

To collect a multispectral dataset with Ralph, the field of view is scanned across the target scene by slewing the spacecraft. During such a scan, each point in the scene is eventually imaged at each wavelength, but because the spacecraft is moving relative to the target, the geometry changes over the course of the scan. The result is a point cloud of individual footprints, each with a distinct wavelength and geometry. These data sets are archived and publicly accessible through the Small Bodies Node of the Planetary Data System (PDS). With suitable software, such as the Integrated Software for Imagers and Spectrometers (ISIS3, available from the US Geological Survey), they can be reprojected to create multispectral maps.

22.3.2 Characterization of Pluto with LEISA Data

With the spatial resolution achieved in the LEISA spectral images, it became possible to discern regions of the surface in which the H_2O ice spectral signature could be mapped; H_2O could not be reliably identified in Earth-based full-disk spectra, largely due to the dominant contribution of Charon's water to the integrated spectrum. Because the vapor pressure of H_2O at ~40 K is exceedingly low, water ice is regarded as bedrock underlying the more volatile N_2, CH_4, and CO ices in most regions of the surface, and providing mechanical strength to support the varied topography of Pluto, which from stereo imagery is seen to range between about −3.5 and +2.5 km with respect to the mean datum (Schenk et al., 2018).

At least one exposure of H_2O ice may represent material ejected from a subsurface reservoir in a cryovolcanic event. Water ice bearing the spectral signature of NH_3 or $NH_3 \cdot nH_2O$ covers several hundred square km centered near the main trough of Virgil Fossae, a graben complex resulting from extensional tectonic forces in Pluto's crust (Dalle Ore et al., 2019; Cruikshank et al. 2019a,b). The topography of several nearby small craters and other structural features is muted, possibly by a mantle of cryoclastic material ejected from one or more vents in or near Virgil Fossae. The NH_3-bearing water ice in this exposure also carries a unique red color suggestive of the presence of an organic tholin (see below) that may have existed in the subsurface fluid reservoir from which the material now seen on the surface emanated. Ammonia in H_2O has the effect of lowering the freezing point of the mixture by as much as 100 K, and, especially in combination with organic molecules, is also critically important in terms of the prebiotic chemistry of Pluto's interior and surface. The water ice exposed around Pulfrich crater (see Figure 22.4, reference point and spectrum *a*) and in a few other locations may also be found to carry the NH_3 signature in the continued investigation of Pluto's surface with the spectral imaging data from *New Horizons*.

In addition to the H_2O-covered regions, Schmitt et al. (2017) have identified two different types of ice on Pluto, based on the positions and strengths of the spectral bands seen in the LEISA data in different geographical regions (Figure 22.2). In Sputnik Planitia and at medium northern latitudes nitrogen is prominent in an N_2-rich mix with CH_4 and CO as minor components; in this mixture the 2.15-μm N_2 band is generally prominent and the CH_4 bands are matrix-shifted, with the 2.35-μm CO band also present. In another distinct surface compositional unit, mostly characteristic of the whole north polar region, the ice is the CH_4-rich phase, with the possible inclusion of N_2 and CO; in this unit the CH_4 bands are not matrix shifted and the N_2 band itself is absent or very weak. Using maps of the distribution of these compositional units, transitional trends can be identified as the strengths and shifts of the spectral bands change from one region to another. For example, the progressive increase of the CH_4-rich phase signature, concurrent with a decrease of the CO/CH_4 ratio in an N_2-rich ice indicates that a fractionation sublimation sequence transforms one kind of ice to the other, resulting in either a N_2-rich:CH_4-rich binary mixture or a layer of CH_4-rich ice that might obscure an underlying N_2-rich layer (Schmitt et al., 2017).

In an analysis of the same dataset, but using a different analytical approach, Protopapa et al. (2017) calculated, through the Hapke radiative transfer model (see Chapter 2), the abundances and textural (grain size) properties of the icy (Figure 22.2) and non-icy materials (Figure 22.3) across Pluto's surface. Their analysis was focused primarily on the constraints that the surface volatile distribution could place on models of the transport of volatiles on Pluto in response to the changing insolation patterns induced by short- and long-term seasonal changes known to occur. This analysis of the large-scale latitudinal variations of CH_4 and N_2 ices revealed three major swaths across the surface, one enriched in CH_4 and extending from the north pole to 55°N, one dominated by N_2 and extending southward to 35°N, and a third narrower band, rich in CH_4, that reaches latitude 20°N (Figure 22.2). This pattern is consistent with the change in insolation over a few decades (less than one Pluto orbit around the Sun, which is 248 years). The great convecting N_2 ice sheet called Sputnik Planitia, lying at the longitude and in the equatorial region corresponding to the Charon antipode, interrupts this pattern. It appears to serve as a cold trap for the condensation of atmospheric N_2 (Bertrand & Forget, 2016) with some evidence for the migration of N_2 across Sputnik Planitia driven by sublimation and subsequent condensation (Figure 22.2; Protopapa et al., 2017; Bertrand et al., 2018). Outside of Sputnik Planitia the distribution of Pluto's highly volatile ices appears to be variable on seasonal time scales. The climate, and consequently the distribution of volatile ices on Pluto's surface, is also expected to be responsive to insolation on the much longer Milanković-like time scale (~3 My), induced by changing obliquity, eccentricity, and longitude of perihelion (Binzel

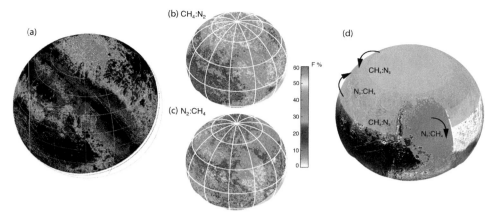

Figure 22.2 Maps of ices at Pluto. (a) Map of the CH_4-rich ice dominated area (red to green) and N_2-rich–dominated regions (light to dark blue). Black is for no CH_4 or N_2 ices, mostly Cthulhu (lower left) and around Pulfrich crater (see also Figure 22.4), where H_2O ice is exposed (from Schmitt et al., 2017). (b) Map of Pluto superposed on a reprojected version of the base map derived from *New Horizons* Long Range Reconnaissance Imager (LORRI) images from pixel-by-pixel modeling to show the spatial distribution and fractional abundance (F%) of the CH_4-rich component. (c) Fractional abundance of the N_2-rich surface ice. (d) Schematic view of the large-scale variations identified across the surface of Pluto. The arrows indicate the direction of the N_2 sublimation transport discussed by Protopapa et al. (2017). The flow of N_2 evaporating from the north polar region is condensing at lower latitudes, an N_2 sublimation front is likely expanding northward from Cthulhu, while there is a migration of N_2 from the northern part toward the equatorial part of Sputnik Planitia. Panels (b) and (c) from Protopapa et al. (2017).

et al., 2017; Bertrand et al., 2018, 2019). Indeed, Pluto's surface retains signatures of this long-period climate variation, including glacial flows east of Sputnik Planitia, mountains in the north eroded by glacial flow, and the bladed terrain (Moore et al., 2018). Stern et al. (2017) have pointed out a number of geologic features that qualitatively support the view of major variations in Pluto's climate over long periods of time.

There is a non-ice component of Pluto's surface; it is characterized by a range of colors from dark brown to light tan or yellow, and it has the effect of not only adding color to the surface, but also strongly affecting the albedo of nearly every region the surface. The range of albedo across the surface is a factor of 10, greater than on any other planetary body studied so far, except for Iapetus (Buratti et al., 2017; Figure 22.4). The very low albedo, reaching a minimum of <0.1, affects the local surface temperature, hence the ability to retain deposits of the volatile ices, N_2 in particular, but also CO and CH_4. Although diagnostic spectral absorption bands attributed to the colored material have not yet been identified, there is a consensus view that this surface component is a macromolecular mix of refractory organic solids that have precipitated from the atmosphere (Grundy et al., 2018) or have formed in the surface by energetic processing of the native ices, N_2, CH_4, and CO. An additional source of colored complex organics that have erupted onto Pluto's surface from subsurface fluid reservoirs through cryovolcanic activity has been proposed by Cruikshank et al. (2019a,b).

All three sources are plausible, and at different epochs in Pluto's history one or the other could dominate. In the current epoch, atmospheric photochemical processes transform CH_4 into other hydrocarbons and nitriles, among other possible molecules. Atmospheric CH_4 currently shields the surface from Ly-α radiation from the Sun and the galaxy, but extreme ultraviolet (UV) may be acting to process the surface ices. In other epochs when Pluto's atmosphere might have been absent or significantly less opaque to UV, direct processing of the surface ices may have been the dominant factor in their production.

The non-ice component is commonly referenced as a tholin, which can be defined in the general sense as a relatively refractory (stable at $T > 100$ K) mix of organic molecules produced in the laboratory by energetic processing of simpler carbon-bearing and other molecules in the gas or solid phase. Molecules created in the laboratory by the energetic processing can consist of single and multiple aromatic units linked by aliphatic branching structures; the molecular masses of the components of a tholin may reach several hundred or even thousands of Da. Starting molecules typically include small hydrocarbons, but the carbon source can also be carbon oxides (CO, CO_2, etc.), and other initial components can include N_2 and H_2O. Depending on the initial components, the degree and kind of processing, as well as other factors, tholins produced in the laboratory have a range of colors, generally very similar to those of Pluto's surface. Tholins produced by UV and electron bombardment of a Pluto mix of ices (N_2:CH_4:CO = 100:1:1) have a rich organic chemical makeup (Materese et al., 2014, 2015), as well as colors that match some regions of Pluto as measured with the *New Horizons* MVIC (Spencer et al., 2016).

The details of the spectral mapping presented above pertain to the hemisphere in sunlight at the time of the encounter, and which is opposite the direction toward Charon. On the approach to Pluto, the Charon-directed hemisphere was imaged at low to moderate spatial resolution at three broadband wavelengths with the MVIC. These data enable making color images that can be compared quantitatively with similar data for the encounter hemisphere and for which spectral mapping correlation of color with composition exists. A collage of 16 MVIC images covering all longitudes is given by Olkin et al. (2017), together with a full analysis of the color differences and similarities.

New Horizons revealed a complex series of haze layers in Pluto's atmosphere (Cheng et al., 2017). Particles comprising the hazes, as well as a number of hydrocarbons and nitriles, are produced in the atmosphere by photochemical processes (Gao et al., 2017), and on further processing and aggregation, fall to the planet's surface as solid particles. To first order, the particulate fallout and accumulation should occur isotropically, at least on the hemisphere in sunlight, but seasonal effects remain to be clarified. Grundy et al. (2018) have considered the geologic effects of an accumulated layer of refractory particles (about 20 m deep over the age of the Solar System, if the current rate is typical) that are expected to have the general molecular and microstructural properties as some kinds of tholin produced in the laboratory, although specific spectral features have not been found or identified. The nonuniformity of Pluto's coloration suggests that after haze particles precipitate to the surface they are further affected by chemical interactions with the local host surface or by additional processing by some other mechanism.

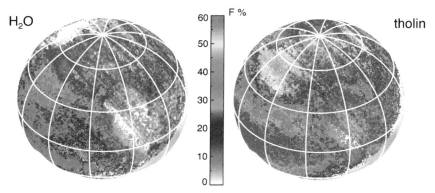

Figure 22.3 H_2O ice and tholin at Pluto. The abundances (F%) of H_2O ice (A) and tholin (B) are projected on the LORRI base map of Pluto (Figure 22.4). These maps were derived from pixel-by-pixel modeling of the LEISA data and are adapted from Protopapa et al. (2017).

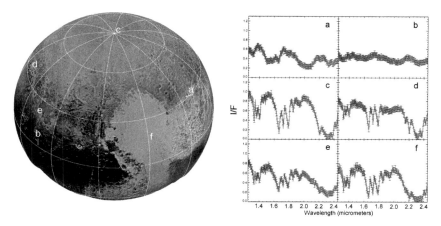

Figure 22.4 Spectra of Pluto. (Left) Base map of Pluto, showing the wide range of albedo (normal albedo range 0.08–1.0; Buratti et al., 2017) across the surface of the hemisphere observed at the *New Horizons* encounter. The map identifies six regions for which the extracted spectra are shown in the adjacent panels. Spectrum (a) shows the characteristic signatures of H_2O ice at the location of Pulfrich crater, while (b) comes from a region in Cthulhu Macula where there is almost no discernible signature of any ice. The signature absorption band of N_2 at 2.15 μm, though intrinsically weak, is strongest in regions (d) and (f). These and other regions show variable strength of the CH_4 bands. The wavelength positions of the CH_4 and N_2 bands are illustrated in Figure 22.1 (Pluto spectral data are from Protopapa et al., 2017).

22.4 Spectral Mapping of Charon with *New Horizons*

Spatially resolved spectral images of the encounter hemisphere of Charon obtained with LEISA demonstrate the ubiquitous presence of (mostly) crystalline H_2O ice, as well as the spotty distribution of the spectral signature attributed to $NH_3 \cdot nH_2O$ (Cook et al., 2017,

2018; Dalle Ore et al., 2018). The north polar region of Charon is mantled with a brown-colored material similar to some colored regions on Pluto. Grundy et al. (2016) has proposed that CH_4 gas continuously streaming from Pluto's atmosphere is preferentially frozen on the pole of Charon that is the coldest region on the satellite because it is in long-term darkness during that phase of the seasonal cycle. Lyman-alpha ultraviolet sunlight, back-scattered by interplanetary neutral hydrogen, processes the frozen CH_4 to produce tholin-like material on the cold, dark pole, followed by additional processing by direct solar ultraviolet radiation when that polar region returns to sunlight. Thus, in this view the colored north polar region acquired its coloration largely during the polar nighttime.

22.5 Pluto's Small Satellites

Pluto has four small satellites that were discovered with the Hubble Space Telescope. Nix and Hydra were found in 2005, Kerberos in 2011, and Styx in 2012; the latter two were discovered while *New Horizons* was in flight. All four are irregular in shape, ranging from Styx at $16 \times 9 \times 8$ km, to Hydra at $65 \times 45 \times 25$ km, and they spin rapidly in chaotic rotation (Showalter & Hamilton, 2015). An overview of *New Horizons* observations of the small satellites is given by Weaver et al. (2016).

Observations of Kerberos, Nix, and Hydra were obtained with LEISA (Cook et al., 2018), and all three show the clear spectral signature of H_2O ice. High-quality spectra of Nix and Hydra show the 1.65-μm H_2O ice band indicating the presence of the crystalline phase. Another species has an absorption band near 2.21 μm, coincident with NH_3 hydrate, although the corresponding 1.99-μm band in the hydrate is not detected. Other ammoniated species are under consideration, awaiting improved laboratory data for a firm identification. In the case of Nix, which was spatially well resolved, the ammoniated material is distributed nonuniformly across the surface, and is not associated with a red-brown patch seen in MVIC images of the satellite.

22.6 Summary

Pluto's surface composition is varied and complex, resulting from a combination of factors that include interaction with the atmosphere, changes over long- and short-term seasonal cycles, the migration of volatiles, chemical and physical interactions among the molecular components, a complex geologic history and present, the presence of internal heat, and other processes not yet identified. Episodes and events of cryovolcanic activity in tectonically active regions may have ejected fluids from the interior onto the surface, with the composition of those materials providing clues to Pluto's internal chemistry. In many respects Pluto may be unique among the small icy bodies of the outer Solar System, but it shares some similarities to Triton, and probably to other dwarf planets of comparable size that also host large amounts of N_2 and/or CH_4 ice (e.g., Eris, Sedna, Makemake, Quaoar), yet to be investigated. Charon has its own properties and history, and may be a template for other smaller icy bodies without atmospheres in the Kuiper Belt population, while Pluto's four small satellites may be a window on the characteristics of a vast number of fragments from collisions among larger bodies in the Kuiper Belt. The *New Horizons* mission to Pluto and beyond has opened extraordinary new vistas on bodies in the third zone of the Solar System, and stands as a tribute to the scientists, engineers, technicians, and managers who made it possible.

References

Bertrand T. & Forget F. (2016) Observed glacier and volatile distribution on Pluto from atmosphere–topography processes. *Nature*, **540**, 86–89.

Bertrand T., Forget F., Umurhan O., et al. (2018) The nitrogen cycles on Pluto over seasonal and astronomical timescales. *Icarus*, **309**, 277–296.

Bertrand T., Forget F., Umurhan O.M., et al. (2019) The methane cycles on Pluto over seasonal and astronomical timescales. *Icarus*, **329**, 148–165.

Binzel R.P., Earle A.M., Buie M.W., et al. (2017) Climate zones on Pluto and Charon. *Icarus*, **287**, 30–36.

Brown M.E. & Calvin W.M. (2000) Evidence for crystalline water and ammonia ices on Pluto's satellite Charon. *Science*, **287**, 107–109.

Buie M.W. & Grundy W.M. (2000) The distribution and physical state of H_2O on Charon. *Icarus*, **148**, 324–339.

Buie M.W., Cruikshank D.P., Lebofsky L.A., & Tedesco E.F. (1987) Water frost on Charon. *Nature*, **329**, 522–523.

Buratti B.J., Hofgartner J.D., Hicks M.D., et al. (2017) Global albedos of Pluto and Charon from LORRI *New Horizons* observations. *Icarus*, **287**, 207–217.

Cheng A.F., Summers M.E., Gladstone G.R., et al. (2017) Haze in Pluto's atmosphere. *Icarus*, **290**, 112–133.

Cook J.C., Desch S.J., Roush T.L., Trujillo C.A., & Geballe T.R. (2007) Near-infrared spectroscopy of Charon: Possible evidence for cryovolcanism on Kuiper belt objects. *The Astrophysical Journal*, **663**, 1406–1419.

Cook J.C., Dalle Ore C.M., Binzel R.P., et al. (2017) Mapping Charon at 2.21 microns. *48th Lunar Planet. Sci. Conf.*, Abstract #2236.

Cook J.C., Dalle Ore C.M., Protopapa S., et al. (2018) Composition of Pluto's small satellites: Analysis of *New Horizons* spectral images. *Icarus*, **315**, 30–45.

Cruikshank D.P. & Sheehan W. (2018) *Discovering Pluto: Exploration at the edge of the Solar System*. University of Arizona Press, Tucson.

Cruikshank D.P. & Silvaggio P.M. (1980) The surface and atmosphere of Pluto. *Icarus*, **41**, 96–102.

Cruikshank D.P., Pilcher C.B., & Morrison D. (1976) Pluto: Evidence for methane frost. *Science*, 835–837.

Cruikshank D.P., Grundy W.M., DeMeo F.E., et al. (2015) The surface compositions of Pluto and Charon. *Icarus*, **246**, 82–92.

Cruikshank D.P., Materese C.K., Pendleton Y.J., et al. (2019a) Prebiotic chemistry of Pluto. *Astrobiology*, **17**(7).

Cruikshank D.P., Umurhan O.M., Beyer R.A., et al. (2019b) Recent cryovolcanism in Virgil Fossae on Pluto. *Icarus*, **330**, 155–168.

Dalle Ore C.M., Protopapa S., Cook J.C., et al. (2018) Ices on Charon: Distribution of H_2O and NH_3 from *New Horizons* LEISA observations. *Icarus*, **300**, 21–32.

Dalle Ore C.M., Cruikshank D.P., Protopapa S., et al. (2019) Detection of ammonia on Pluto's surface in a region of geologically recent tectonism. *Science Advances*, **5**, eaav5731.

Douté S., Schmitt B., Quirico E., et al. (1999) Evidence for methane segregation at the surface of Pluto. *Icarus*, **142**, 421–444.

Elliot J.L., Dunham E., Bosh A., et al. (1989) Pluto's atmosphere. *Icarus*, **77**, 148–170.

Gao P., Fan S., Wong M.L., et al. (2017) Constraints on the microphysics of Pluto's photochemical haze from *New Horizons* observations. *Icarus*, **287**, 116–123.

Grundy W.M., Olkin C.B., Young L.A., Buie M.W., & Young E.F. (2013) Near-infrared spectral monitoring of Pluto's ices: Spatial distribution and secular evolution. *Icarus*, **223**, 710–721.

Grundy W.M., Olkin C.B., Young L.A., & Holler B.J. (2014) Near-infrared spectral monitoring of Pluto's ices II: Recent decline of CO and N_2 ice absorptions. *Icarus*, **235**, 220–224.

Grundy W.M., Cruikshank D.P., Gladstone G.R., et al. (2016) The formation of Charon's red poles from seasonally cold-trapped volatiles. *Nature*, **539**, 65–68.

Grundy W.M., Bertrand T., Binzel R.P., et al. (2018) Pluto's haze as a surface material. *Icarus*, **314**, 232–245.

Lewis J.S. (1972) Low temperature condensation from the solar nebula. *Icarus*, **16**, 241–252.

Marcialis R.L., Rieke G.H., & Lebofsky L.A. (1987) The surface composition of Charon: Tentative identification of water ice. *Science*, **237**, 1349–1351.

Materese C.K., Cruikshank D.P., Sandford S.A., Imanaka H., Nuevo M., & White D.W. (2014) Ice chemistry on outer Solar System bodies: Carboxylic acids, nitriles, and urea detected in refractory residues produced from the UV-photolysis of N_2:CH_4:CO-containing ices. *Astrophysical Journal*, **788**, 111.

Materese C.K., Cruikshank D.P., Sandford S.A., Imanaka H., & Nuevo M. (2015) Ice chemistry on outer Solar System bodies: Electron radiolysis of N_2-CH_4- and CO- containing ices. *Astrophysical Journal*, **812**, 150.

Moore J.M., Howard A.D., Umurhan O.M., et al. (2018) Bladed terrain on Pluto: Possible origins and evolution. *Icarus*, **300**, 129–144.

Olkin C.B., Young E.F., Young L.A., et al. (2007) Pluto's spectrum from 1.0–4.2µm: Implications for surface properties. *Astronomical Journal*, **133**, 420–431.

Olkin C., Spencer J.R., Grundy W.M., et al. (2017) The global color of Pluto from *New Horizons*. *Astronomical Journal*, **154**, 258, DOI:10.3847/1538-3881/aa965b.

Owen T.C., Roush T.L., Cruikshank D.P., et al. (1993) Surface ices and the atmospheric composition of Pluto. *Science*, **261**, 745–748.

Prokhvatilov A. & Yantsevich L. (1983) X-ray investigation of the equilibrium phase diagram of CH_4-N_2 solid mixtures. *Soviet Journal of Low Temperature Physics*, **9**, 94–98.

Protopapa S., Boehnhardt H., Herbst T., et al. (2008) Surface characterization of Pluto and Charon by L and M band spectra. *Astronomy & Astrophysics*, **490**, 365–375.

Protopapa S., Grundy W., Tegler S., & Bergonio J. (2015) Absorption coefficients of the methane–nitrogen binary ice system: Implications for Pluto. *Icarus*, **253**, 179–188.

Protopapa S., Grundy W.M., Reuter D.C., et al. (2017) Pluto's global surface composition through pixel-by-pixel Hapke modeling of *New Horizons* Ralph/LEISA data. *Icarus*, **287**, 218–228.

Quirico E. & Schmitt B. (1997a) A spectroscopic study of CO diluted in N_2 ice: Applications for Triton and Pluto. *Icarus*, **128**, 181–188.

Quirico E. & Schmitt B. (1997b) Near-infrared spectroscopy of simple hydrocarbons and carbon oxides diluted in solid N_2 and as pure ices: Implications for Triton and Pluto. *Icarus*, **127**, 354–378.

Quirico E., Schmitt B., Bini R., & Salvi P.R. (1996) Spectroscopy of some ices of astrophysical interest: SO_2, N_2 and N_2: CH_4 mixtures. *Planetary and Space Science*, **44**, 973–986.

Reuter D.C., Stern S.A., Scherrer J., et al. (2008) Ralph: A visible/infrared imager for the *New Horizons* Pluto/Kuiper Belt mission. *Space Science Reviews*, **140**, 129–154.

Schenk P., Beyer R.A., McKinnon W.B., et al. (2018) Basins, fractures and volcanoes: Global cartography and topography of Pluto from *New Horizons*. *Icarus*, **314**, 400–433.

Schmitt B., Philippe S., Grundy W., et al. (2017) Physical state and distribution of materials at the surface of Pluto from *New Horizons* LEISA imaging spectrometer. *Icarus*, **287**, 229–260.

Scott T.A. (1976) Solid and liquid nitrogen. *Physics Reports*, **27**, 89–157.

Showalter M. & Hamilton D. (2015) Resonant interactions and chaotic rotation of Pluto's small moons. *Nature*, **522**, 45–49.

Soifer B.T., Neugebauer G., & Matthews K. (1980) The 1.5–2.5 µm spectrum of Pluto. *Astronomical Journal*, **85**, 166–167.

Spencer J.R., Stern A., Olkin C., et al. (2016) The colors of Pluto: Clues to its geological evolution and surface/atmospheric interactions. *AGU Fall Meeting*, Abstract #P54A-01.

Stern S.A., Bagenal F., Ennico K., et al. (2015) The Pluto system: Initial results from its exploration by *New Horizons*. *Science*, **350**, aad1815.

Stern S.A., Binzel R.P., Earle A.M., et al. (2017) Past epochs of significantly higher pressure atmospheres on Pluto. *Icarus*, **287**, 47–53.

Weaver H.A., Buie M.W., Buratti B.J., et al. (2016) The small satellites of Pluto as observed by *New Horizons*. *Science*, **351**, aae0030.

Young L.A. (1994) *Bulk properties and atmospheric structure of Pluto and Charon*. PhD thesis, Massachusetts Institute of Technology.

Young L.A., Elliot J., Tokunaga A., de Bergh C., & Owen T. (1997) Detection of gaseous methane on Pluto. *Icarus*, **127**, 258–262.

23

Visible to Short-Wave Infrared Spectral Analyses of Mars from Orbit Using CRISM and OMEGA

SCOTT L. MURCHIE, JEAN-PIERRE BIBRING, RAYMOND E. ARVIDSON,
JANICE L. BISHOP, JOHN CARTER, BETHANY L. EHLMANN, YVES LANGEVIN,
JOHN F. MUSTARD, FRANCOIS POULET, LUCIE RIU, KIMBERLY D. SEELOS,
AND CHRISTINA E. VIVIANO

23.1 Introduction to Orbital VSWIR Spectral Reflectance of Mars

The Observatoire pour la Mineralogie, l'Eau, les Glaces et l'Activité (OMEGA) (Bibring et al., 2004a) and the Compact Reconnaissance Imaging Spectrometer for Mars (CRISM) (Murchie et al., 2007) have revolutionized knowledge of the primary and secondary mineralogic compositions of the martian surface and the history of aqueous environments recorded by its secondary mineralogy. In this chapter we review the historical context for these two investigations, their key findings, and the geologic history of the shallow martian crust revealed by analyses of their returned data. Both OMEGA and CRISM also investigated Mars' permanent and seasonal polar H_2O and CO_2 ice caps, the composition and vertical structure of clouds and aerosols, and seasonal and interannual variability of the atmosphere. The latter topics are not treated in detail here, but the interested reader is referred to key papers on polar ices by Bibring et al. (2004b), Langevin et al. (2005, 2007), Brown et al. (2008, 2010a, 2012, 2014, 2016), Calvin et al. (2009), and Cull et al. (2010), and to papers on atmospheric trace gases and aerosols by Vincendon et al. (2007, 2011), Wolff et al. (2009), Smith et al. (2009, 2013), Clancy et al. (2012, 2013, 2017), and Guzewich et al. (2014).

23.1.1 Physical Factors Controlling Martian VSWIR Spectral Reflectance

The VSWIR spectral range contains absorption features diagnostic of a broad range of primary and secondary minerals present in the crust and altered surface layer of Mars. These include electronic transition absorptions due to Fe^{2+} in olivine, pyroxene, and glass, Fe^{3+} in oxides and oxyhydroxides (Chapter 1; Adams, 1974; Sherman et al., 1982; Sunshine et al., 1990; Cloutis & Gaffey, 1991), and vibrational absorptions due to H_2O-, OH-, SO_4-, and CO_3-bearing alteration products including phyllosilicates, hydrated or hydroxylated sulfates, oxyhydroxides, and carbonates (Chapter 4; Hunt & Salisbury, 1971a,b; Rossman, 1976; Clark et al., 1990; Cloutis et al., 2006). Briefly stated, key absorptions include 1- and 2-μm composite absorptions in pyroxene and Fe^{2+}-bearing glass and a broad ~0.9–1.3-μm composite absorption in olivine, whose relative strengths and positions are indicative of metal cation composition; a 1.3-μm absorption in plagioclase feldspars containing Fe^{2+}; combination and overtone metal–OH absorptions at ~1.4 µm and 2.2–2.35 µm, whose positions and shapes are diagnostic of OH bonds with Si, Al, Fe, and Mg in a variety of hydroxylated silicates; combination and overtone H_2O absorptions at ~1.4 and ~1.9 µm, whose shapes and positions help distinguish various hydrated minerals including silicates, sulfates, and carbonates; a sulfate overtone absorption at 2.4 µm; and carbonate overtone absorptions at 2.3 µm, 2.5 µm, 3.3–3.4 µm, and 3.8–4.0 µm, whose positions are sensitive to the cation present.

The spectral signature of surface components is filtered and overprinted by absorptions due to atmospheric gases (primarily CO_2, CO, and H_2O), and by scattering and absorption by dust and (primarily H_2O) ice aerosols. The gas absorptions are strongest near 1.4, 1.6, 1.9–2.1, and 2.7 µm, wavelengths close to those of H_2O and OH overtone absorptions. Under typical atmospheric aerosol opacities, scattering by dust and H_2O ice aerosols can contribute as much as 20–30% of martian radiance measured from space over the short-wavelength end of the VSWIR spectral range (e.g., Clancy et al., 1995; Erard, 2001). Depending on aerosol opacity, absorptions due to Fe^{3+} and H_2O ice in aerosols, and reflectance of surface materials, aerosols can either increase or decrease radiance measured from orbit and modulate both the continuum and depths of absorptions. Due to the slow change in their radiance with wavelength, aerosols have the least effect on detection of narrow absorptions such as those of hydroxylated silicates and hydrated sulfates; they introduce larger perturbations to shapes and depths of broad, shallow absorptions due to olivine, pyroxene, and glass.

23.1.2 Historical Context from Telescopic Measurements and Phobos 2/ISM

VSWIR spectroscopy using Earth-based telescopes, acquired from the 1960s to 1990s, provided basic constraints on the composition of the martian regolith over regions as small as hundreds of km scale. All regions outside of the seasonal polar caps exhibit a strong red slope below 0.8 µm, resulting from the edge of a strong charge transfer band due to nanophase ferric oxide in Mars' pervasive bright, mobile dust (Morris et al., 1989). In some regions, weak absorptions near 0.66 and 0.89 µm indicate a small amount of crystalline hematite (Bell et al., 1990). Dark regions exhibit broad, shallow 1- and 2-µm absorptions consistent with pyroxene and a pervasive negative SWIR spectral slope (decreasing reflectance with increasing wavelength), consistent with dark regions having a predominantly basaltic composition with a thin coating of dust (Singer et al., 1979; Singer & McSween, 1993). Bright, dusty regions lack similarly well-defined iron silicate absorptions and in telescopic data resemble basaltic volcanic glass that has been altered by hydration and oxidation (Singer, 1982; Bell et al., 1993; Morris et al., 1993). A strong 3-µm absorption in both bright and dark regions indicates a pervasive component of molecular H_2O within regolith particles or adsorbed on their surfaces (Moroz, 1964; Houck et al., 1973; Pimental et al., 1974).

The first Mars orbital VSWIR investigation was the Imaging Spectrometer for Mars (ISM), a French-built instrument on board the Soviet Phobos 2 mission. ISM was a whisk-broom imaging spectrometer that measured 128 spectral channels from 0.77 to 3.16 μm (Bibring et al., 1989). ISM operated during February–March 1989 and acquired 11 hyperspectral images of the martian equatorial region. Its improved spatial resolution compared to Earth-based telescopic measurements led to the discovery of heterogeneity of pyroxene composition within dark regions (Mustard et al., 1997) and a stronger 3-μm absorption due to molecular H_2O associated with interior layered deposits of Valles Marineris and unconformable sedimentary deposits in the western part of Arabia (Murchie et al., 1993, 2000). From these data, the existence of surface components other than basalt and dust – a hydrated salt and a dark, crystalline ferric mineral – was inferred (Murchie et al., 2000).

ISM whetted the Mars science community's appetite for groundbreaking new compositional information, but it was the OMEGA instrument that delivered major new discoveries.

23.1.3 Overview of OMEGA and Its Key Findings

OMEGA is a combined push-broom (0.4–1.0 μm) and whisk-broom (1.0–5.1 μm) style imaging spectrometer on Mars Express that measures 352 contiguous channels covering 0.35–5.1 μm over a field of view (FOV) 8.8° in width with an instantaneous field of view (IFOV) of 1.2 mrad. Spectral sampling is 7, 14, and 20 nm/channel from 0.35–1.0, 1.0–2.5, and 2.5–5.1 μm, respectively. Depending on altitude in the elliptical orbit, infrared spatial sampling can be 0.3–5 km and averages ~1 km. It began operations in Mars' orbit in late 2003.

OMEGA data extended earlier findings about igneous materials and revolutionized understanding of the nature of aqueous alteration of Mars' crust as recorded by its secondary mineralogy. Stratigraphically older volcanic materials are enriched in olivine and low-Ca pyroxenes compared to younger volcanics, suggesting evolution of magma composition over time (Mustard et al., 2005; Poulet et al., 2007), possibly due to higher temperatures or greater incorporation of H_2O into older magmas leading to greater partial melting (Poulet et al., 2009a; Baratoux et al., 2013). Extensive deposits of olivine-rich rocks in Nili Fossae could be mantle material excavated as impact melt from the Isidis basin-forming impact or picritic volcanics emplaced immediately after the Isidis event 3.98 Gyr ago (Mustard et al., 2007). Depending on which interpretation is correct, Nili's olivine-rich rock could signify lower crustal composition or an early period of lower-silica volcanism (see Section 23.2.2). Materials excavated from depths ≥2 km were recognized at dozens of locations to contain a variety of Fe/Mg-phyllosilicates, plus other hydrated minerals that could not be identified at OMEGA's spatial and spectral resolution (Bibring et al., 2005; Poulet et al., 2005). In addition, both the Nili Fossae (Mangold et al., 2007) and Mawrth Vallis regions (Loizeau et al., 2007) contain tens of thousands km^2 of clay deposits – a lower layer of Fe/Mg-rich smectite clay locally with an upper layer rich in Al-clay (Michalski & Noe Dobrea, 2007).

Younger, late Noachian to Hesperian-aged sedimentary deposits superposed on the Noachian crust, including layered deposits in Valles Marineris (Nedell et al., 1987;

Komatsu et al., 1993) and in western Arabia (Malin & Edgett, 2000), were revealed by OMEGA to contain a distinctly different assemblage of secondary minerals – hydrated sulfates plus ferric minerals including hematite (Arvidson et al., 2005, 2006; Gendrin et al., 2005a,b; Bibring et al., 2007; Griffes et al., 2007). In some layered deposits, dominance of either polyhydrated or monohydrated sulfates in different layers suggests temporal changes in depositional environments, diagenesis, or weathering to different hydration states (Mangold et al., 2008). The temporal sequence of martian secondary mineralogies, early Noachian phyllosilicates followed by late Noachian to Hesperian sulfates plus hematite, followed by late Hesperian and Amazonian formations that exhibit little evidence for aqueous alteration, was inferred by Bibring et al. (2006) to result from a change in the surface environment from higher water activity and neutral pH, to acidic with lower water activity, and later to completely dry. The transition from neutral to acidic conditions is discussed in more detail in Sections 23.2.3.4, 23.2.3.5, and 23.2.3.7.

The new view of Mars that emerged from OMEGA served as the guide to planning the first few years of CRISM's investigation of Mars, principally selection of the first several thousand sites to be covered by high spatial resolution targeted observations.

23.1.4 Introduction to CRISM and Its Key Findings

CRISM, on the Mars Reconnaissance Orbiter (Murchie et al., 2007, 2009a), is a pushbroom imaging spectrometer on a gimbal that covers the wavelength range 0.36–3.92 μm at 6.55 nm/channel. It observes Mars' surface in two main modes. "Targeted observations" use gimbal motion to compensate for ground track motion and cover an area up to ~10 × 10 km in size at 18 m pixel^{-1} (or up to 10 × 20 km at 36 m pixel^{-1}) with contiguous, hyperspectral coverage. The continuously varying viewing geometry introduces variations in emergence and phase angles within each scene. Targeted observations acquired through late 2012 included up to 10 overlapping, reduced-spatial resolution images taken before and after the main scene at larger emergence angles, to support fitting and normalization of the effects of the varying photometric geometry of the main scene. CRISM targeted observations commonly are interpreted using simultaneous ("coordinated") images from MRO's High-Resolution Imaging Science Experiment (HiRISE) (McEwen et al., 2007) and Context Camera (Malin et al., 2007).

CRISM "multispectral mapping" and "hyperspectral mapping" use nadir-pointed, 10-km wide strips hundreds of km in length, with data binned spatially to 90 or 180 m/pixel. In multispectral mode, 72 selected wavelengths cover the most common mineral absorptions. This mode was intended originally to create a map with 3–10 times improved spatial resolution compared with OMEGA, to provide context for targeted observations and to find new regions of interest. It was augmented with hyperspectral mapping modes, having up to 190 additional channels (262 total) to provide contiguous, hyperspectral coverage at key wavelengths to improve discrimination of spectrally similar secondary minerals.

CRISM data show that aqueous alteration of the Noachian crust is far more extensive than was evident at OMEGA spatial resolution; that mineral assemblages are far more diverse than had been recognized; that knowledge of small-scale compositional heterogeneity dramatically improves understanding of past environments; and that the previously supposed transition from neutral to acidic environments is gradational in space and time. Hydrated silicates occur in thousands of exposures of Noachian crust (Mustard et al., 2008; Murchie et al., 2009b), and >30 secondary minerals have been identified that form over a range of temperatures from just above the freezing point to 350°C (Ehlmann et al., 2009, 2010, 2011a,b; Buczkowski et al., 2010; Carter et al., 2013; Viviano-Beck et al., 2014). Their exposure from below ~2 km of basalt suggests that much of the alteration of Noachian crust occurred in aquifers at conditions ranging from just above freezing to low-grade metamorphic/hydrothermal. Remnants of Al-phyllosilicate-bearing layered clay deposits like those surrounding Mawrth Vallis identified in OMEGA data were also found to occur throughout the southern Noachian highlands and are widely interpreted to be remnants of a pedogenic layer formed during a late Noachian period of subaerial aqueous alteration, i.e., pedogenesis (Bishop et al., 2008; Wray et al., 2008, 2009a; Ehlmann et al., 2009; McKeown et al., 2009; Murchie et al., 2009b; Le Deit et al., 2012; Carter et al., 2015). Other thin layers of secondary mineral-bearing formations suggest a variety of other aqueous surface environments in the late Noachian period: (a) Mg-carbonate formed by alteration of olivine-rich rocks (Ehlmann et al., 2008a; Brown et al., 2010b); (b) deltaic bottomset beds formed in late Noachian open-basin lakes, that are enriched in clay and carbonate of probable detrital origin (Ehlmann et al., 2008b; Grant et al., 2008; DeHouck et al., 2010; Milliken & Bish, 2010; Goudge et al., 2012) and hydrated silica interpreted as authigenic because it extends into the lake basins beyond other obviously detrital sediments (Carter et al., 2013); and (c) closed-basin lake deposits on the west flank of Tharsis that contain acid sulfates and kaolinite, suggesting an acidic, precipitate-dominated environment (Wray et al., 2011; Ehlmann et al., 2016). Hesperian- to Amazonian-aged deposits within and surrounding Valles Marineris contain hydrated silica (opal), jarosite, and fluvial channels, suggesting formation by acidic waters (Milliken et al., 2008; Weitz et al., 2008a; Roach et al., 2010; Thollot et al., 2012). Thick, late Noachian to Hesperian-aged hydrated sulfate-bearing layered deposits are more widespread in Noachian highlands than had previously been apparent and occur within craters and other topographic lows throughout Arabia (Wiseman et al., 2008, 2010) and in Gale crater (Milliken et al., 2010; Thomson et al., 2011). Gale was selected as the Mars Science Laboratory landing site based on its stratigraphic section containing secondary mineral assemblages spanning the Noachian into the Hesperian period. In many of the sulfate-bearing layered deposits, there is compositional stratification into layers whose sulfates have different hydration states ranging from anhydrous to ~4 (or more) water molecules bound to each sulfate molecule, possibly recording environmental changes during the deposits' emplacement (Bishop et al., 2009; Murchie et al., 2009c; Lichtenberg et al., 2010; Liu et al., 2012, 2016).

23.1.5 Isolating and Visualizing the Martian Surface

For the reasons described in Section 23.1.1, separating surface and atmospheric contributions to radiance measured by orbital instruments is critical to accurate determination and interpretation of surface composition. Analyses of OMEGA and CRISM data employ several methods to accomplish that separation. The most basic step is dividing measured radiance by the cosine of the solar incidence angle to correct for illumination geometry and by a scaled, empirical atmospheric transmission spectrum (Bibring et al., 2005; Mustard et al., 2008; Murchie et al., 2009a) to remove absorptions due to atmospheric gases. For CRISM targeted observations, sampling of different emergence angles by gimbal motion provides a dataset from which to derive a normalization of the aerosol contribution throughout the scenes to that at the lowest emergence angle sampled (Seelos et al., 2016a). Such a procedure makes targeted observations internally more comparable. For generation of mosaicked CRISM mapping data, where there are large strip-to-strip variations in aerosol opacity, radiative transfer modeling is required to simultaneously remove atmospheric gas absorptions and model the surface reflectance that would be expected at uniformly low aerosol opacities (McGuire et al., 2008, 2009, 2013). That modeling uses databases of empirical measurements of seasonal and latitudinal variance in atmospheric dust and ice aerosol opacity. The most sophisticated analyses of OMEGA and CRISM data have used radiative transfer modeling to retrieve single-scattering albedo spectra (Fraeman et al., 2013; Liu et al., 2016; Arvidson et al., 2014; Fox et al., 2016; Wiseman et al., 2016).

Analysis and visualization of the large volumes of data returned by OMEGA and CRISM (>60 billion CRISM spectra) have utilized "summary parameters," spectral indices that represent band depths (Clark & Roush, 1984) and spectral slopes that show likely locations of different mineral phases and regolith textural variations (e.g., coatings indicated by spectral slope). The original formulae for summary parameters were based on absorptions in minerals detected in OMEGA data (Pelkey et al., 2007), and later updated to reduce false-positive and -negative detections and to include minerals that had not been recognized at the spatial resolution of OMEGA data (Viviano-Beck et al., 2014).

23.2 Minerals Identified by OMEGA and CRISM in their Geologic Context

As of 2017, >30 different minerals or mineral groups had been identified on Mars by using CRISM or OMEGA data. Type spectra of the most common minerals in the martian geologic record are shown in Figure 23.1, compared with laboratory spectra. Absorptions diagnostic of these minerals – and that are easily detectable through Mars' CO_2-rich atmosphere – occur throughout the wavelength range 0.5–3.9 μm. Viviano-Beck et al. (2014) provided type spectra and type locations for these minerals, and detailed their diagnostic absorptions and geologic settings; those data are archived at the Planetary Data System's Geosciences Node. Viviano-Beck et al. (2014) discuss the diagnostic absorptions of identified minerals in detail. Here we focus on the minerals' geologic settings in the context of the martian time stratigraphic system (Tanaka et al., 2014), and

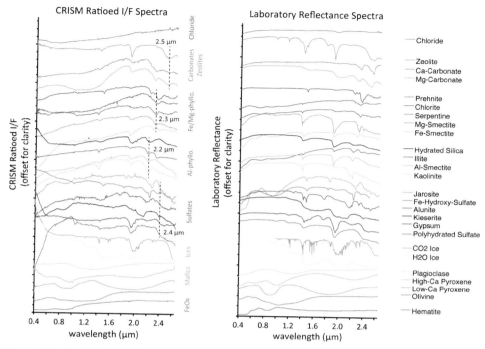

Figure 23.1 Observed CRISM 0.4–2.6 μm spectra of common minerals on Mars. (Left) Ratioed to martian dust to highlight diagnostic absorptions. (Right) Compared with laboratory analog materials shown in units of reflectance, scaled and offset for clarity.

their occurrences as parts of mineral assemblages. This contextual information provides evidence for the evolution of martian geologic processes, and the ancient environments in which secondary minerals formed. The minerals and their settings are summarized in this section, and the geologic history inferred from them is summarized in Section 23.3.

23.2.1 The Surface Layer

The surface layer of Mars consists mostly of regolith, sand, and dust decoupled physically from underlying intact bedrock, that together represent most of what orbital VSWIR spectroscopy measures. Intact bedrock whose mineralogy is detectable from orbit – and must thus have no more than several μm of adhering dust (Fischer & Pieters, 1993) – is restricted in most areas of the planet to steep slopes and areas undergoing active eolian erosion. OMEGA and CRISM data, with their higher spatial resolutions than those of previous VSWIR data, reveal the surface layer to be more heterogeneous and closely related to bedrock than had been thought. Bright, fine-grained dust does not typically have an absorption near 0.89 μm because of crystalline ferric minerals, as had been thought from analysis of telescopic (Bell et al., 1990) and ISM spectra (e.g., Erard & Calvin, 1997;

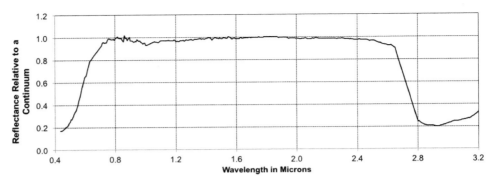

Figure 23.2 Example CRISM spectrum of martian dust. The data are from CRISM FRT00007901 (summit of Olympus Mons), corrected for atmospheric gas absorptions as described in Section 23.1.5, and divided by an upper hull continuum to accentuate weak absorption features.

Murchie et al., 1993, 2000). Instead, dust has a 1-μm absorption consistent with olivine plus some amount of pyroxene. Wavelengths <0.8 μm have a steeply red spectral slope due to nanophase ferric oxide (Figure 23.2; compare with olivine in Figure 23.1). These spectral signatures are consistent with in situ mineralogic measurements from MER/Spirit's Mössbauer spectrometer indicating that roughly half of the Fe in soil occurs as olivine and pyroxene, and the remainder as nanophase oxide (Morris et al., 2006). The retention of half of the Fe in finely powdered crystalline primary phases indicates that dust is not primarily a result of palagonitization (i.e., hydration and oxidation of glass) but instead of disaggregation of lightly oxidized rock.

The sand component of the surface layer is heterogeneous and exhibits compositional affinity to nearby bedrock, suggesting that sand sources are local to regional. Rippled sand deposits near olivine-rich bedrock in Nili Fossae are rich in olivine (Mustard et al., 2009); rippled sands near sulfate-rich layered deposits in Candor Chasma exhibit a strong sulfate spectral signature (Murchie et al., 2009c). In a study of sand in Valles Marineris, Chojnacki et al. (2014a,b) showed that sand is mineralogically diverse, related to discrete source deposits, and subject to eolian fractionation.

23.2.2 Primary Silicate Minerals in Bedrock

OMEGA and CRISM data show that most Noachian- to Hesperian-aged bedrock in the southern highlands exhibits a VSWIR spectral signature dominated by pyroxene (Figures 23.3 and 23.4), consistent with a predominantly basaltic crustal composition (e.g., Mustard et al., 2005; Poulet et al., 2007; Ody et al., 2012; Riu et al., 2019). Spectra are dominated optically by other primary mafic phases only in specific settings. Low-Ca pyroxene dominates the most ancient igneous rock in the southern highlands (Mustard et al., 2005), much of which formed during the pre-Noachian period (before formation of Hellas basin (Frey et al., 2003; Carr & Head, 2010), a period that some authors group with the early

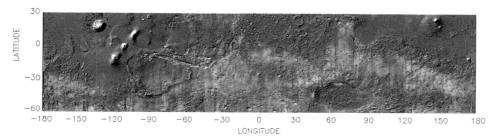

Figure 23.3 CRISM summary parameter map generated from OMEGA data, showing distribution in the equatorial region of major primary phases. Olivine is shown using OLINDEX3 in red, low-Ca pyroxene using LCPINDEX2 in green, and high-Ca pyroxene and Fe^{2+}-bearing glass using HCPINDEX2 in blue. The data are shown on a shaded relief MOLA elevation base. Uncolored areas are obscured by dust; the reddish tone along their margins is due to the olivine signature in dust.

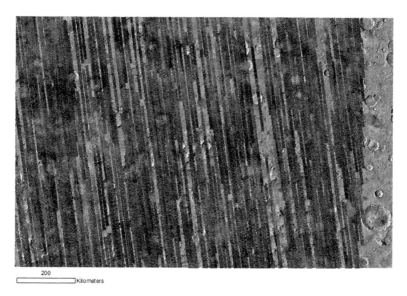

Figure 23.4 Map of key minerals in the Tyrrhena Terra region of the Noachian highlands, generated from CRISM multispectral data. High-Ca pyroxene is shown in blue using the summary parameter HCPINDEX2, low-Ca pyroxene in green using LCPINDEX2, and Fe/Mg phyllosilicates and/or olivine in red using OLINDEX3. The latter summary parameter is sensitive to Fe-phyllosilicates as well as olivine. The data are overlain on a THEMIS dayside IR mosaic.

Noachian). These exposures include the rim materials of the Isidis (Mustard et al., 2005, 2009), Argyre (Buczkowski et al., 2010), and Hellas (Riu et al., 2019) basins; rim and central peak materials of large craters (Mustard et al., 2005; Skok et al., 2012); and the lower walls and floor of eastern Valles Marineris (Mustard et al., 2005; Murchie et al., 2009b; Flahaut et al., 2012; Quantin et al., 2012; Viviano-Beck et al., 2017). Rocks dominated spectrally by high-Ca pyroxene form younger volcanic units that are Hesperian to Amazonian in age, indicating a shift in mafic mineral composition in igneous rocks around ~3.5 Ga (Viviano et al., 2019). Materials dominated by olivine have a greater range of age, and are found in five geologic settings (Poulet et al., 2007; Ody et al., 2013). Three settings correlate spatially and stratigraphically with those of low-Ca pyroxene: rim and ejecta materials of Argyre, Hellas, and Isidis; outcrops in the lower walls of Valles Marineris (and sand on the chasma floor derived from those outcrops) that formed as dikes and sills (Flahaut et al., 2011) or ancient extrusions that were buried by subsequent basalts (Wilson & Mustard, 2013); and buttes near Hellas. Olivine-rich rock is also exposed in the northern plains where superposed craters penetrated both the spectrally bland northern plains and the underlying basaltic crust, exposing Noachian and pre-Noachian material like that in the three settings described in the previous sentence (Carter et al., 2010; Pan et al., 2017). Finally, an enhanced spectral signature of olivine is observed in some intercrater plains in the southern highlands (Ody et al., 2013; McBeck et al., 2014). Non-basaltic primary mineralogies are conspicuously absent in materials exposed from depth in the Tharsis plateau by craters and escarpments (including chasma walls of the western part of Valles Marineris), indicating that a thick accumulation of basalt forms the Tharsis plateau (Flauhaut et al., 2012; Quantin et al., 2012; Viviano-Beck et al., 2017).

Plagioclase is a widespread igneous mineral on Mars, but it is not detected directly by orbital VSWIR techniques where the 1- and 2-μm absorptions due to >5% mafic phases obscure the 1.25-μm plagioclase absorption (Cheek & Pieters, 2012). Landed CheMin measurements show that in the soils of Gale crater it is the single most abundant phase (Bish et al., 2013). Plagioclase is detected directly from its 1.25-μm absorption in probable dacite forming Hesperian-aged volcanic materials in Nili Patera on the Syrtis Major Planum volcanic shield, and in probable anorthosite in Noachian-aged plutons exhumed by impacts; both materials are thought to result from igneous differentiation (Carter & Poulet, 2013; Wray et al., 2013). In agreement with CheMin results, plagioclase is inferred to be the most abundant mineral in typical martian volcanic materials based on the results of spectral mixture modeling, in which observed spectra can be fit by 40–60 vol.% plagioclase with lesser high- and low-Ca pyroxene and olivine (Chapter 15; Bandfield et al., 2000; Poulet et al., 2009a,b).

Dark material in the northern plains is spectrally distinct from that in the highlands, as recognized previously in Thermal Emission Spectrometer (TES) data (Chapter 24). Thermal emission spectra of the Acidalia region have been interpreted alternatively to indicate a glassy andesitic composition (Bandfield et al., 2000) or altered basalt (Wyatt & McSween, 2002; Wyatt et al., 2004). At VSWIR wavelengths the dark materials exhibit a negative spectral slope at 0.8–2.6 μm without strong absorptions due to Fe^{2+} or OH in minerals, consistent with altered coatings but inconsistent with a high degree of aqueous alteration of basalt (Mustard et al., 2005; Poulet et al., 2007). Horgan and Bell (2012) showed that removal of the spectral continuum reveals weak absorptions consistent with Fe^{2+}-bearing glass, occurring either as leached silica-enriched rinds or as volcanic glass. Exposure by impact craters of volcanic rock indistinguishable from olivine basalts in the southern highlands shows that glassy material in Acidalia is of order only hundreds of meters in thickness (Salvatore et al., 2010).

23.2.3 Secondary Mineral Assemblages in Bedrock and the Surface Layer

The most surprising result from CRISM is the diversity and widespread occurrence of secondary minerals whose formation involved aqueous alteration, only some of which had been detected previously by OMEGA, Mars Global Surveyor/TES (Christensen et al., 2001), or Mars Odyssey/THEMIS (Christensen et al., 2004). The phases are found in several different geologic settings, in approximate order of decreasing age: (1) hydrated and hydroxylated silicates exhumed from depth in the Noachian highlands crust, mostly by impact craters and basins; (2) carbonates of differing composition and stratigraphic position; (3) detrital and chemical sediments filling craters and other topographic lows, commonly with horizontal bedding and inflowing valley networks providing evidence for a past lacustrine environment; (4) a globally distributed, shallow layer of compositionally stratified material altered to clays, silica, and sulfates; (5) scattered, eroded remnants of late Noachian and younger sulfate-rich layered deposits superposed unconformably on older eroded surfaces; (6) silica and weathered material associated with volcanic constructs; and (7) thin beds of silica and acid sulfates within and adjacent to Valles Marineris.

23.2.3.1 Deep Crustal Hydrous Silicates

Hydrated and hydroxylated silicates exhumed from depth in the Noachian crust (Figure 23.4) are the predominant secondary mineral assemblage that formed under aqueous conditions (Murchie et al., 2009b; Ehlmann et al., 2011a,b; Carter et al., 2013; Ehlmann & Edwards, 2014). The most common minerals are Fe/Mg-smectite clays, which dominate at depths of origin from 0 to 5 km, and chlorites, which dominate at greater depths to ≥7 km (Sun & Milliken, 2015). Modeling of mineral formation kinetics suggests that preservation of smectite at 0.3–5 km depth without alteration to chlorite or illite may have resulted from limited availability of water (Tosca & Knoll, 2009).

There is evidence for both radiogenic heating on early Mars and impact heating throughout martian history having supported aqueous alteration to form hydroxylated and hydrated silicates. Evidence for radiogenic heating includes regional occurrence of minerals that form at higher temperatures than smectite and chlorite, which typically form at temperatures of <175°C (McSween et al., 2015). Zeolite is observed in many exposures, suggesting temperatures ranging from ambient to 225°C (Ehlmann et al., 2009, 2011b). Detection of serpentine and prehnite, which form at temperatures of ambient to 400°C and 200°–400°C respectively, point to even higher temperature hydrothermal environments. Minerals formed at these higher temperatures are more localized in their occurrence than are chlorite or zeolite; prehnite occurs, for example, near Nili Fossae [Ehlmann et al., 2009] and near Hellas (Buczkowski et al., 2010; Clark et al., 2015), but not at all in the walls of Valles Marineris (Viviano-Beck et al., 2017). Serpentine is found in the Nili Fossae and Claritas

Rise tectonic zones, in a few highlands impact craters, and associated with olivine-rich material surrounding the Isidis basin (Ehlmann et al., 2010). Epidote, whose occurrence implies the highest formation temperatures of 250°–600°C, has been identified only where exposed from beneath Syrtis Major lavas (Carter et al., 2013; Viviano-Beck et al., 2014). The observed secondary minerals indicate metamorphic facies ranging from diagenetic through zeolite facies through sub-greenschist (prehnite-pumpellyite) facies (Ehlmann et al., 2011b).

The walls of Valles Marineris expose evidence for more localized variations in alteration temperatures. Pre-Noachian rock is altered dominantly to Fe/Mg-smectite, but a distinct, higher-temperature alteration assemblage (chlorite + zeolite + carbonate ± serpentine) occurs in a confined longitudinal band where the Claritas Rise intersects Valles Marineris. The higher-temperature secondary minerals are concentrated within a heavily fractured band bearing olivine-rich dikes, implying hydrothermal alteration by endogenic heating (Viviano-Beck et al., 2017).

Comparison of CRISM data with laboratory and field analogs, coupled with consideration of early martian environments, led to the hypothesis that hydrated and hydroxylated silicates excavated from depth typically represent several percent alteration of a basaltic protolith (Ehlmann et al., 2011a). This hypothesis is supported by in situ investigation of early to pre-Noachian smectite-bearing rock in the rim of Endeavour crater (Wray et al., 2009b) by MER/Opportunity, which suggests alteration of several percent of a basaltic protolith by hydrothermal fluids (Arvidson et al., 2014; Fox et al., 2016).

Other evidence supports a major – even dominant – role for impact heating in alteration of the Noachian crust. Alteration at depth is discontinuous, occurring over domains km in extent that are not traceable for hundreds or even several tens of km, consistent with alteration driven by localized impact heating (Ding et al., 2015; Viviano-Beck et al., 2017) and/or limited by a discontinuous supply of water (Sun & Milliken, 2015; Sun & Milliken, 2018). The discontinuously altered materials in the walls of Valles Marineris are typically brecciated (Viviano-Beck et al., 2017), consistent with impact ejecta or megaregolith. Glass and hydrated silica, which can devitrify to phyllosilicate, are widespread in Noachian craters, most conspicuously in central peaks but also in floors and ejecta (Tornabene et al., 2013; Cannon & Mustard, 2015; Sun & Milliken, 2018). At Toro and Ritchey craters, Marzo et al. (2010) and Ding et al. (2015) showed that distribution of secondary minerals in the crater interior is consistent with an impact-generated hydrothermal system. Sun & Milliken (2015) studied hydrated silica and smectites within craters and found examples that are both ancient and exhumed, and younger and draped on topography suggesting impact melts, demonstrating that impacts sometimes drive localized hydrothermal alteration even in relatively young craters. Tornabene et al. (2013) and Osinski et al. (2013) synthesized evidence that impact melting and impact-generated hydrothermal systems must have been an important driver for formation of hydrated and hydroxylated silicates: the likelihood of hydrothermal systems in impact craters larger than a few km in size, the widespread distribution of hydrated silica in craters, and the tendency of hydrated impact melt to devitrify into crystalline hydrated silicates even at low temperatures over geologic time. Regional differences in metamorphic grade could even be due to implanted heat of impact basin-forming events as an alternative to endogenic heating.

23.2.3.2 Carbonate-Bearing Rocks

Carbonate-bearing rocks occur in two geologic settings. First, Noachian Mg-rich carbonates are concentrated in altered, olivine-rich materials surrounding the Isidis basin (Ehlmann et al., 2008a; Bishop et al., 2013a), which formed possibly as Isidis impact melt (Mustard et al., 2007) or as volcanic materials (Hamilton & Christensen, 2005). They are thought to have sequestered 1–12 mbar of CO_2 from the Noachian atmosphere (Edwards & Ehlmann, 2015). Second, Fe/Ca carbonates are part of a distinct clay- and carbonate-bearing mineral assemblage that typically is exhumed from 2 to 5 km depth through overlying Noachian to Hesperian surface units, suggesting formation in a pre-Noachian environment. Known exposures are concentrated in Tyrrhena Terra and Margaritifer Terra, but also occur elsewhere (Michalski & Niles, 2010; Michalski et al., 2013a; Wray et al., 2016). The close association of Fe/Ca carbonates with clay minerals is compatible with alteration of a protolith with limited removal of cations that precipitated as carbonate (Bultel et al., 2015; Jain & Chauhan, 2015). Although alteration by subsurface fluids may play a role in carbonate formation, layering present in some deposits suggests deposition in surface water followed by burial (Wray et al., 2016). Modeling of surface–atmosphere interactions by van Berk and Fu (2011) and Zolotov (2015) under different atmospheric chemistries predicts that the observed assemblage of clay minerals plus Fe/Ca carbonate could have formed by weathering of a basaltic protolith under an early, denser atmosphere (possibly with hundreds of mbar of CO_2) in contact with surface water. Thus, if the deep Fe/Ca carbonates formed at the surface during the pre-Noachian period, depending of the spatial extent of the now-buried deposits, the amount of sequestered CO_2 could approach or exceed the limited amount sequestered in Noachian Mg-carbonates (Wray et al., 2016).

23.2.3.3 Lacustrine Sediments

Lacustrine detrital and chemical sediments occur both in crater interiors and in intercrater topographic lows. Compositionally, there are three endmember secondary mineral assemblages, each occurring in deposits with distinct morphologies. The first is intracrater deposits of detrital sediments bearing secondary minerals (Goudge et al., 2012 and references therein). Approximately 79 open crater lake basins (having outlets) have incoming valley networks and contain deposits of probable lacustrine origin, including deltaic deposits proximal to the incoming valley and layered deposits in more distal locations on the crater floor. Of these, 10 craters contain recognized aqueous minerals in the layered deposits or deltaic bottomset beds, including Fe/Mg-smectite, kaolinite, and carbonate; Fe/Mg-smectite is the most commonly occurring mineral. In all cases, the secondary minerals appear to have eroded from the drainage basin. In some cases, hydrated silica also occurs on the delta or beyond it on the crater floor; based on location of the silica, it may have formed as a precipitate (Carter et al., 2013). The low abundance of secondary minerals in these settings is probably a consequence of the basalt-dominated composition of average highlands crust, plus minimal formation of authigenic minerals (Goudge et al., 2012; Poulet et al., 2014).

The second endmember is chloride-rich sediments within closed topographic basins, also into which valley networks typically debouch. The chloride was originally inferred from low emissivity at thermal IR wavelengths (Osterloo et al., 2008). Approximately 640 occurrences have been identified in closed basin craters (lacking outflowing channels) and intercrater plains (Osterloo et al., 2010). The deposits share a distinctive fine-scale fractured surface, a relatively gray color at wavelengths <0.8 μm, and a weak 3-μm band suggesting a relatively anhydrous composition (Murchie et al., 2009b; Glotch et al., 2010; Jensen & Glotch, 2011). Spectral modeling suggests that chlorides comprise 10–25 vol.% of a mixture with detrital sediments, consistent with formation in playa environments (Jensen & Glotch, 2011; Glotch et al., 2016). The chlorides are generally not associated with other hydrated minerals except for Fe/Mg-rich clays, an association that is especially noteworthy in northwestern Terra Sirenum (Ruesch et al., 2012). Stratigraphic relations suggest two separate depositional episodes, first phyllosilicates and then chlorides (Murchie et al., 2009b; Glotch et al., 2010), separated by a period of erosion.

The third endmember is found in a few closed basin lake deposits on the western flank of Tharsis, including in Columbus and Cross craters. These deposits are concentrated on the inner crater rims and exhibit interbedded kaolinite, hydrated silica, acid sulfates including alunite and jarosite, hydrated sulfates (Wray et al., 2011; Ehlmann et al., 2016), and possibly hydrated chlorine salts (Hanley & Horgan, 2016). The lack of inflowing valley networks and the mineral assemblage present suggest that these lakes were fed by acidic, probably sulfurous groundwater (Wray et al., 2011; Ehlmann et al., 2016).

23.2.3.4 Layered Clays

"Layered clays" occur as a late Noachian, meters to hundreds of meters thick layer of Fe/Mg smectite, commonly overlain by Al-smectite, kaolinite, and/or hydrated silica, with those secondary mineralogies capped by an erosion-resistant glassy or mafic layer (Bishop et al., 2008). The type location is near Mawrth Vallis (Poulet et al., 2005; Loizeau et al., 2007; Wray et al., 2008), where layered clays cover ~900000 km^2 (Noe Dobrea et al., 2010). Smaller exposures occur along and high in the walls of Valles Marineris (Murchie et al., 2009b; Weitz et al., 2014), in Margaritifer Terra (Le Deit et al., 2012; Seelos et al., 2016b), in Nili Fossae (Ehlmann et al., 2009; Ehlmann & Dundar, 2015), and scattered throughout the Noachian highlands (Wray et al., 2009a; Le Deit et al., 2012). Figure 23.5 shows detailed stratigraphy of the deposit near Mawrth Vallis described by Bishop et al. (2013b, 2016). There, a basal layer of Fe-smectite (nontronite) transitions upward to material with a doublet absorption at 2.23 and 2.29 μm that is interpreted to be acid-leached clay. Above that, Al-smectite (montmorillonite) and opal are capped by the amorphous aluminosilicates allophane and imogolite (Bishop & Rampe, 2016). Hypotheses for formation of these widely occurring deposits include alteration of volcanic ash and/or pedogenesis (Ehlmann et al., 2009; McKeown

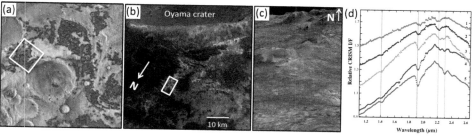

Figure 23.5 Stratigraphy of layered clays near Mawrth Vallis. (a) Regional view of Mawrth Vallis showing Fe/Mg-phyllosilicates in red and Al-phyllosilicates, hydrated silica, and amorphous phases in blue, overlain on THEMIS dayside IR imagery. Mineral outcrops were identified by thresholding summary parameter maps derived from OMEGA data (redrawn from Bishop et al., 2017). (b) Close-up view of inset in (a) with CRISM false-color data overlain on CTX imagery over an HRSC stereo terrain model. (c) HiRISE stereo terrain model with CRISM false-color data showing inset in (b). (d) Spectra 1–5 represent the stratigraphic units illustrated in (b) and (c) and described by Bishop et al. (2016). Nontronite (spectrum 5) is mapped in red at the bottom of the stratigraphic sequence, overlain by ferrous clays in purple (spectrum 4), a unit containing possible acid-leached clay in yellow (spectrum 3), Al-phyllosilicates and opal in blue (spectrum 2), and poorly crystalline aluminosilicates such as allophane and imogolite in green (spectrum 1) at the top of the profile. All of the spectra are ratioed to a nearby spectrally bland region to highlight weak absorptions.

et al., 2009; Michalski et al., 2013b) contemporaneously with formation of valley networks (Ehlmann et al., 2011a; Carter et al., 2015). Occurrence of allophane and imogolite is consistent with immature soils formed on volcanic ash (Bishop & Rampe, 2016). Acid sulfates (e.g., bassanite, alunite, jarosite) also occur in association with the layered clays (Farrand et al., 2009; Wray et al., 2010; Ehlmann & Dundar, 2015), and the implied high SO_4 concentrations suggest a transition during this period of martian history from a neutral to an acidic surface weathering environment (Bibring et al., 2006, 2007; Zolotov, 2015; Zolotov & Mironenko, 2016).

23.2.3.5 Sulfate-Bearing Layered Deposits

Unconformable, sulfate-bearing deposits are the most studied of martian secondary mineral-bearing deposits, due largely to MER/Opportunity's in situ investigation of the areally extensive occurrence in Meridiani Planum, known as the "Burns Formation" (e.g., Squyres et al., 2004a,b; Arvidson et al., 2005, 2006, 2015). Other occurrences are in Valles Marineris (Gendrin et al., 2005a; Weitz et al., 2008b, 2012; Murchie et al., 2009c; Roach et al., 2009; Liu et al., 2012); outlying chasmata not directly connected to Valles Marineris such as Juventae Chasma (Figure 23.6; Bishop et al., 2009; Noel et al., 2015) and Aram Chaos (Lichtenberg et al., 2010); scattered impact craters in the Noachian highlands including in Arabia (Wiseman et al., 2008, 2010) and Gale crater (Milliken et al., 2010; Thomson et al., 2011); and in Olympia Undae and other dune fields surrounding the north polar layered deposits (Horgan et al., 2009; Massé et al.,

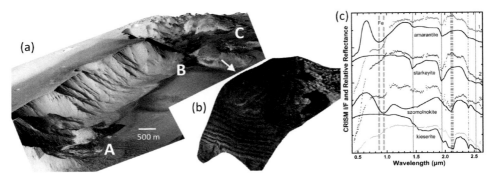

Figure 23.6 Unconformable, hydrated sulfate-bearing layered deposits in Juventae Chasma (after Noel et al., 2015). (a) Three-dimensional (3D) view inside the chasma showing three separate mounds of the deposit, using HRSC and CTX imagery overlain by CRISM mineral maps with yellow-green indicating monohydrated sulfates and magenta-purple indicating polyhydrated sulfates. (b) Zoomed and rotated view of inset from (a), showing CRISM image FRT00009C0A with summary product BD920 in the red plane showing ferric minerals, BD2100 in green showing monohydrated sulfates, and SINDEX in blue showing polyhydrated sulfate. Both 3D views use 2× vertical exaggeration. (c) CRISM corrected reflectance spectra from the mound in (b) compared with laboratory measurements of polyhydrated and monohydrated sulfates. Gray lines mark diagnostic spectral features. All of the spectra are ratioed to a nearby region to highlight weak absorptions; the slight concavities near 2.2 μm in the upper two spectra are an artifact of absorptions in the denominator spectrum.

2012). The Gale crater occurrence, which sits atop a complete stratigraphic section including older clay-bearing lacustrine deposits, is being investigated in situ by the Mars Science Laboratory Curiosity rover (Grotzinger et al., 2012). Sulfate-bearing unconformable deposits are massive to layered, depending on their location and position in the stratigraphic column. All of the deposits are highly sculpted by wind, and those in Valles Marineris are also sculpted by outflow channels, suggesting a poorly indurated lithology; deposits in Olympia Undae are reworked into dunes. The assemblage of minerals detected includes hematite plus sulfate minerals, which have been argued to indicate an acidic weathering environment (Bibring et al., 2007). The sulfate cations appear generally to be dominated by Mg and Fe, and the hydration state of sulfates is variable from deposit to deposit and within the stratigraphic column. Minerals detected include polyhydrated sulfates, possibly starkeyite ($MgSO_4 \cdot 4H_2O$) and/or amarantite ($FeOHSO_4 \cdot 3H_2O$), monohydrated sulfates (kieserite, $MgSO_4 \cdot H_2O$ and szmolonokite, $FeSO_4 \cdot H_2O$), and $FeOHSO_4$, which is not a named mineral. All of these sulfates are indicative of desiccating and/or high-temperature conditions. In some deposits, interbedded clay-bearing beds suggest that arid, acidic conditions were punctuated by wetter and/or less acidic periods (Flahaut et al., 2015; Powell et al., 2017). The north polar deposit is distinct in its sulfates being dominated by gypsum ($CaSO_4 \cdot 2H_2O$).

Hypotheses for formation of sulfate-rich unconformable deposits in Meridiani Planum and in Valles Marineris include trapping of eolian sediment in evaporites formed by discharge of saline groundwater fed by precipitation (Andrews-Hanna et al., 2007) or, alternatively, by melting of high-elevation ice in ice-covered late Noachian highlands (Kite et al., 2013; Scanlon et al., 2013; Fastook & Head, 2015; Wordsworth et al., 2015). A very different genetic mechanism has also been proposed, acid weathering along boundaries of dust grains embedded inside massive ancient ice deposits (Niles & Michalski, 2009; Massé et al., 2012). Whether or not the latter mechanism is applicable to the Valles Marineris and Meridani Planum deposits, sub-ice alteration may explain gypsum-bearing north polar deposits. Variations in hydration state within the stratigraphic column of sulfate-rich unconformable deposits have been interpreted as recording changes in the depositional environment (Gendrin et al., 2005a; Mangold et al., 2008; Bishop et al., 2009; Murchie et al., 2009c; Lichtenberg et al., 2010; Wiseman et al., 2010; Liu et al., 2012; and others). Modeling and laboratory studies suggest that both temperature and chemistry of aqueous solutions could have affected the hydration state of precipitated sulfates. Al-Samir et al. (2017) modeled formation of deposits in Juventae Chasma as having formed by leaching of mafic bedrock by acidic groundwater, followed by discharge and evaporation of that water at the surface. They found that the observed sequence of monohydrated sulfate overlain by polyhydrated sulfate could have arisen from precipitation in a standing water body filling the chasma, with monohydrated sulfates precipitating from hydrothermal fluids at 100°–200°C and polyhydrated sulfates from cooler water at <75°C. Alternatively, Wang et al. (2016) showed that coprecipitation of halides and oxyhydroxides could facilitate dehydration of "fresh" amorphous sulfates to a monohydrate such as kieserite, whereas deposits most dominated by sulfate anions would tend instead to form starkeyite.

23.2.3.6 Hydrous Minerals in Volcanic Settings

Secondary mineral assemblages associated with volcanic constructs are rare. Skok et al. (2010) described mounds of hydrated silica on the flanks of a volcanic cone within Nili Patera, and interpreted them as hot spring deposits possibly comparable to the silica-rich deposits at "Home Plate" that were investigated in situ by MER/Spirit (Squyres et al., 2008; Arvidson et al., 2008, 2010; Ruff et al., 2011). Ackiss et al. (2018) described occurrences of Fe oxide, zeolite, and gypsum on Sisyphi Montes, table mountains interpreted to have formed by subglacial volcanism (Ghatan & Head, 2002), and interpreted these minerals to have formed by lava–water interactions.

23.2.3.7 Opal-Bearing Deposits

The youngest surface environments containing spatially extensive, aqueously formed secondary mineral assemblages occur surrounding and within Valles Marineris. Tens of meters thick, light-toned layered deposits on the surrounding plateau exhibit variations between beds in brightness, color, mineralogy, and erosional properties that are atypical of the layered deposits described in Section 23.2.3.5. Reflectance spectra indicate opaline silica and jarosite, consistent with low-temperature, acidic aqueous alteration of basaltic materials. The correspondence between fluvial landforms, silica, and jarosite supports sustained, possibly acidic precipitation, surface runoff, and fluvial deposition during the Hesperian and possibly into the Amazonian period (Milliken et al., 2008; Weitz et al., 2008a, 2010). Possibly related deposits occur inthe interiors of several chasmata – notably Noctis Labyrinthus, Melas Chasma, Coprates Catena, and Ophir Chasma – where deposits that contain opaline silica, jarosite, Al-phyllosilicates, Fe/Mg smectite clays (some of which are acid- leached), and hydrated sulfates infill topographic lows, form mounds, or are draped on eroded chasma walls and floors (Roach et al., 2010; Thollot et al., 2012; Grindrod et al., 2012; Weitz et al., 2013, 2015; Weitz & Bishop, 2016). These deposits indicate that aqueous environments persisted into the Hesperian and even Amazonian periods. Their association with the most tectonically modified region of Mars would be consistent with a hydrothermal source for the waters forming some of these deposits.

23.2.3.8 Young Secondary Mineral Assemblages?

The most recently formed secondary mineral assemblages may be perchlorates, which have been detected by landed missions (e.g., Hecht et al., 2009; Martín-Torres et al., 2015), and are powerful freezing point depressants that could allow brine flow even under modern martian conditions (e.g., Dickson et al., 2013; Massé et al., 2014; Fischer et al., 2016). Flow of brine has been suggested possibly to play a role in formation of recurring slope lineae (RSL), dark streaks that form on sunward-facing slopes during seasons of peak daytime temperature at equatorial and mid-latitudes, grow downslope over months, and fade in autumn (McEwen et al., 2011, 2014; Ojha et al., 2014). The spectral signature of RSL suggests depletion of fines (Ojha et al., 2013). Detection of the spectral signature of hydrated oxychlorine salts including perchlorates near RSL has been reported (Ojha et al., 2015). However subsequent analysis has suggested that most or all such detections of small patches of oxychlorine salts instead represent a data artifact that mimics the spectrum of this mineral (Leask et al., 2018).

Fresh shallow valleys (FSVs) of fluvial origin, and lake basins that the FSVs fed and connected, are of late Hesperian to middle Amazonian age possibly overlapping the ages of RSL. They are similar in latitudinal distribution to RSL and particularly conspicuous in northern Arabia (Wilson et al., 2016). As of 2018, aqueously formed secondary minerals have not been recognized in association with FSVs.

23.3 Martian Geologic History Revealed by VSWIR Reflectance Spectroscopy

Figure 23.7 displays a schematic timeline, modified from that of Ehlmann et al. (2011a), of key processes during the geologic history of Mars for which there is evidence from VSWIR mineral detections. Newer findings largely support Bibring et al.'s (2006) theory that Mars progressed through three environmental stages each characterized by a distinct type of alteration of mafic protolith to secondary minerals: the pre- to late Noachian "phyllosian" stage during which rocks were altered primarily to phyllosilicates in relatively neutral pH environments; the late Noachian to Hesperian "theiikian" stage characterized by acidic

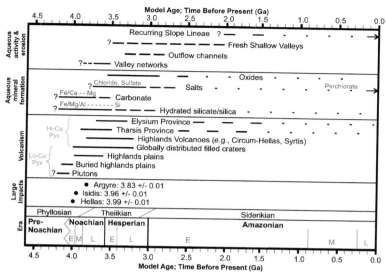

Figure 23.7 Schematic history of martian alteration environments as indicated by secondary mineralogy, compared with time stratigraphic periods and the ages of water-related surface features and geologic units rich in alteration morphologies. (Adapted from Ehlmann et al., 2011a.) Additional content comes from references cited in this chapter and from Werner (2008, 2009) and Werner and Tanaka (2011).

alteration to oxides and sulfates; and the Hesperian to Amazonian "siderikian" stage characterized by minimal and dominantly oxidative, anhydrous alteration under cold, anhydrous conditions. Refinements to that concept are that "phyllosian" alteration occurred mostly in the subsurface, with widespread surface water relatively limited in time and commonly acidic; that "phyllosian" and "theiikian" conditions overlapped in time; and that surface water, rare though it was, did not disappear during the "siderikian" stage.

The oldest identified primary minerals form pre-Noachian igneous rocks rich in low-Ca pyroxene and olivine, that overlie differentiated plutons within which anorthosite formed by flotation of plagioclase and pyroxenite and dunite formed by settling of low-Ca pyroxene and olivine (Skok et al., 2012). Secondary minerals formed during this time include hydrated and hydroxylated silicates that were subsequently exhumed from depth, and assemblages of Fe/Ca-carbonates and clays that may have formed at the pre-Noachian surface and been buried by later volcanism. Both endogenic and impact heating appear to have played roles in driving alteration.

During the early to middle Noachian period, formation of deposits containing secondary minerals punctuated the emplacement of basaltic volcanic materials rich in low-Ca pyroxene and in some places olivine. Alteration at depth continued, driven both by endogenic and impact heating; examples were subsequently exposed in impact craters and escarpments such as the walls of Valles Marineris.

The late Noachian period featured a peak in aqueous activity and chemical alteration at the martian surface, including formation of Mg-rich carbonates by alteration of olivine surrounding the Isidis basin. Crater-density measurements indicate that valley networks, layered clays of possible pedogenic origin, and lacustrine accumulations of detrital sediments were roughly coeval (Ehlmann et al., 2011a). Chloride-rich lacustrine sediments have slightly younger relative ages, as indicated by crater densities (Hynek et al., 2015) and superposition on older, clay-bearing material in the same lake basins (Murchie et al., 2009b; Glotch et al., 2010). This geologically limited period of widespread surface water marked the transition from phyllosian to theiikian environmental conditions, with acidic conditions clearly dominating the latest stage of formation of layered clays. Around this time, composition of volcanic materials transitioned to mafic minerals being dominated by high-Ca pyroxene.

Younger spatially extensive deposits containing secondary minerals record acidic environments, and are restricted to special locations. Sulfate- and hematite-bearing unconformable layered deposits formed in the late Noachian to Hesperian period in Meridiani Planum, scattered highlands craters, and Valles Marineris. These deposits exhibit compositional layering that may record changes in water chemistry and/or temperature. Some studies interpreted the Valles Marineris and Meridiani Planum deposits to have formed by trapping of eolian sediment in evaporites precipitated from groundwater discharge. Alternatively, some deposits may have formed by acid alteration of sediments trapped in ice, such as the north polar gypsum-bearing deposits. Geographically localized alteration by acidic surface water, probably partly of hydrothermal origin, persisted into the Amazonian period within and around Valles Marineris. During the remainder of the Amazonian period through modern times, formation of altered material continues mostly by wind abrasion and oxidative weathering to form dust, and only minor aqueous activity continues.

23.4 Future Directions

What is needed to improve the state of knowledge of the martian surface from VSWIR spectroscopy beyond OMEGA and CRISM? This question has been considered by the Mars Exploration Program Analysis Group (MEPAG), and the recommendation was twofold: improved VSWIR spatial resolution, and accompaniment by thermal IR data at a spatial resolution several times better than that provided by THEMIS. The authors of this chapter believe that recommendation is supported by the literature cited herein.

The evolution of knowledge about martian surface composition using VSWIR spectroscopy, from telescopic studies through ISM and then OMEGA followed by CRISM, illustrates an important point that is almost intuitive in hindsight. As resolution has improved from regional to geologic unit, to outcrop scale, more of the diversity of martian surface composition is resolvable through windows in dust and sand cover and in smaller and smaller outcrops. For example, the occurrence of layered deposits rich in hydrated sulfates was near the limit of interpretability of ISM data, but unambiguous in OMEGA data. Hydrated silica-rich deposits near Valles Marineris were unresolved by OMEGA, but easily resolved by CRISM. There is clear evidence that further improvements in spatial resolution – to meters scale – would continue to yield new discoveries. Since 2012, CRISM has used a modified observing technique whereby certain targeted observations employ spatial oversampling in the along-track direction. In ground processing, the oversampling allows rendering of the data at an improved resolution with a pixel scale of 12 or even 9 m/pixel, versus the 18 m/pixel native spatial sampling (Kreisch et al., 2016). Several studies that have used this approach demonstrated detection of small exposures that are not resolvable at 18 m/pixel, providing a more detailed view of compositional stratigraphy (e.g., Fraeman et al., 2013; Fox et al., 2016; Powell et al., 2017). Thus, expanding the state of knowledge of martian surface composition beyond OMEGA and CRISM would benefit strongly from improvements in spatial resolution.

Recent analyses show that a second improvement in the state of knowledge will come from high spatial resolution measurements covering both VSWIR and thermal IR wavelengths. In their analysis of carbonate-olivine assemblages in the Nili Fossae region, Edwards and Ehlmann (2015) derived improved constraints on carbonate abundance from simultaneously fitting both wavelength ranges. An ideal complement to VSWIR data at meters scale would thus be thermal IR data at CRISM-like pixel scales.

References

Ackiss S., Horgan B., Seelos F., Farrand W., & Wray J. (2018) Mineralogical evidence for subglacial volcanoes in the Sisyphi Montes region of Mars. *Icarus*, **311**, 357–370.

Adams J. (1974) Visible and near-infrared diffuse reflectance spectra of pyroxenes as applied to remote sensing of solid objects in the solar system. *Journal of Geophysical Research*, **79**, 4829–4836.

Al-Samir M., Nabhan S., Fritz J., et al. (2017) The paleolacustrine evolution of Juventae Chasma and Maja Valles with implications for the formation of interior layered deposits on Mars. *Icarus*, **292**, 125–143.

Andrews-Hanna J.C., Phillips R.J., & Zuber M.T. (2007) Meridiani Planum and the global hydrology of Mars. *Nature*, **446**, 163–166.

Arvidson R.E., Poulet F., Bibring J.-P., et al. (2005) Spectral reflectance and morphologic correlations in eastern Terra Meridiani, Mars. *Science*, **307**, 1591–1594.

Arvidson R.E., Poulet F., Morris R.V., et al. (2006) Nature and origin of the hematite-bearing plains of Terra Meridiani based on analyses of orbital and Mars Exploration Rover data sets. *Journal of Geophysical Research*, **111**, E12S08, DOI:10.1029/2006JE002728.

Arvidson R.E., Ruff S.W., Morris R.V., et al. (2008) Spirit Mars rover mission to the Columbia Hills, Gusev crater: Mission overview and selected results from the Cumberland Ridge to Home Plate. *Journal of Geophysical Research*, **113**, E12S33, DOI:10.1029/2008JE003183.

Arvidson R.E., Bell J.F. III, Bellutta P., et al. (2010) Spirit Mars rover mission: Overview and selected results from the northern Home Plate Winter Haven to the side of Scamander crater. *Journal of Geophysical Research*, **115**, E00F03, DOI:10.1029/2010JE003633.

Arvidson R.E., Squyres S.W., Bell J.F. III, et al. (2014) Ancient aqueous environments at Endeavour crater, Mars. *Science*, **343**, 1248097, DOI:10.1126/science.1248097.

Arvidson R.E., Bell J.F. III, Catalano J.G., et al. (2015) Mars Reconnaissance Orbiter and Opportunity observations of the Burns formation: Crater hopping at Meridiani Planum. *Journal of Geophysical Research*, **120**, 429–451.

Bandfield J.L., Hamilton V.E., & Christensen P.R. (2000) A global view of martian surface compositions from MGS-TES. *Science*, **287**, 1626–1630.

Baratoux D., M. Toplis J., Monnereau M., & Sautter V. (2013) The petrological expression of early Mars volcanism. *Journal of Geophysical Research*, **118**, 59–64.

Bell J.F. III, McCord T.B., & Owensby P.D. (1990) Observational evidence of crystalline iron oxides on Mars. *Journal of Geophysical Research*, **95**, 14447–14461.

Bell J.F. III, Morris R.V., & Adams J.B. (1993) Thermally altered palagonitic tephra: A spectral and process analog to the soil and dust of Mars. *Journal of Geophysical Research*, **98**, 3373–3385.

Bibring J.-P., Langevin Y., Soufflot A., et al. (1989) Results from the ISM experiment. *Nature*, **341**, 591–593.

Bibring J.-P., Soufflot A., Berthé M., et al. (2004a) OMEGA: Observatoire pour la Minéralogie, l'Eau, les Glaces et l'Activité. In: *Mars Express: The scientific payload* (A. Wilson, ed.). ESA SP-1240. ESA Publications Division, Noordwijk, Netherlands, 37–49.

Bibring J.-P., Langevin Y., Poulet F., et al. (2004b) Perennial water ice identified in the south polar cap of Mars. *Nature*, **428**, 627–630.

Bibring J.-P, Langevin Y., Gendrin A., et al. & OMEGA team (2005) Mars surface diversity as revealed by the OMEGA/Mars Express observations. *Science*, **307**, 1576–1581.

Bibring J.-P, Langevin Y., Mustard J.F., et al. (2006) Global mineralogical and aqueous Mars history derived from OMEGA/Mars Express data. *Science*, **312**, 400–404, DOI:10.1126/science.1122659.

Bibring J.-P., Arvidson R.E., Gendrin A., et al. (2007) Coupled ferric oxides and sulfates on the martian surface. *Science*, **317**, 1206–1209.

Bish D.L., Blake D.F., Vaniman D.T., et al. & MSL Science Team (2013) X-ray diffraction results from Mars Science Laboratory: Mineralogy of Rocknest at Gale crater. *Science*, **341**, 1238932.

Bishop J.L. & Rampe E.B. (2016) Evidence for a changing martian climate from the mineralogy at Mawrth Vallis. *Earth and Planetary Science Letters*, **448**, 42–48.

Bishop J.L., Noe Dobrea E.Z., McKeown N.K., et al. (2008) Phyllosilicate diversity and past aqueous activity revealed at Mawrth Vallis, Mars. *Science*, **321**, 830–833.

Bishop J.L., Parente M., Weitz C.M., et al. (2009) Mineralogy of Juventae Chasma: Sulfates in the light-toned mounds, mafic minerals in the bedrock, and hydrated silica and hydroxylated ferric sulfate on the plateau. *Journal of Geophysical Research*, **114**, E00D09, DOI:10.1029/2009JE003352.

Bishop J.L., Tirsch D., Tornabene L.L., et al. (2013a) Mineralogy and morphology of geologic units at Libya Montes, Mars: Ancient aqueous outcrops, mafic flows, fluvial features and impacts. *Journal of Geophysical Research*, **118**, 487–513.

Bishop J.L., Loizeau D., McKeown N.K., et al. (2013b) What the ancient phyllosilicates at Mawrth Vallis can tell us about possible habitability on early Mars. *Planetary and Space Science*, **86**, 130–149.

Bishop J.L., Gross C., Rampe E.B., et al. (2016) Mineralogy of layered outcrops at Mawrth Vallis and implications for early aqueous geochemistry on Mars. *47th Lunar Planet. Sci. Conf.*, Abstract #1332.

Bishop J.L., Michalski J.R., & Carter J. (2017) Remote detection of clay minerals. In: *Infrared and Raman spectroscopies of clay minerals* (W.P. Gates, J.T. Kloprogge, J. Madejová, & F. Bergaya, eds.). Elsevier, the Netherlands, 482–514.

Brown A.J., Byrne S., Tornabene L.L., & Roush T. (2008) Louth crater: Evolution of a layered water ice mound. *Icarus*, **196**, 433–445.

Brown A.J., Calvin W.M., McGuire P.C., & Murchie S.L. (2010a) Compact Reconnaissance Imaging Spectrometer for Mars (CRISM) south polar mapping: First Mars year of observations. *Journal of Geophysical Research*, **115**, E00D13, DOI:10.1029/2009JE003333.

Brown A.J., Hook S.J., Baldridge A.M., et al. (2010b) Hydrothermal formation of clay-carbonate alteration assemblages in the Nili Fossae region of Mars. *Earth and Planetary Science Letters*, **297**, 174–182.

Brown A.J., Calvin W.M., & Murchie S.L. (2012) Compact Reconnaissance Imaging Spectrometer for Mars (CRISM) north polar springtime recession mapping: First 3 Mars years of observations. *Journal of Geophysical Research*, **117**, E00J20, DOI:10.1029/2012JE004113.

Brown A.J., Piqueux S., & Titus T.N. (2014) Interannual observations and quantification of summertime H_2O ice deposition on the martian CO_2 ice south polar cap. *Earth and Planetary Science Letters*, **406**, 102–109.

Brown A.J., Calvin W.M., Becerra P., & Byrne S. (2016) Martian north polar cap summer water cycle. *Icarus*, **277**, 401–415.

Buczkowski D.L., Murchie S., Clark R., et al. (2010) Investigation of an Argyre basin ring structure using Mars Reconnaissance Orbiter/Compact Reconnaissance Imaging Spectrometer for Mars. *Journal of Geophysical Research*, **115**, E12011, DOI:10.1029/2009JE003508.

Bultel B., Quantin-Nataf C., Andréani M., Clénet H., & Lozac'h L. (2015) Deep alteration between Hellas and Isidis basins. *Icarus*, **260**, 141–160.

Calvin W.M., Roach L.H., Seelos F.P., et al. (2009) Compact Reconnaissance Imaging Spectrometer for Mars observations of northern martian latitudes in summer. *Journal of Geophysical Research*, **114**, E00D11, DOI:10.1029/2009JE003348.

Cannon K.M. & Mustard J.F. (2015) Preserved glass-rich impactites on Mars. *Geology*, 43, 635–638.

Carr M.H. & Head J.W. (2010) Geologic history of Mars. *Earth and Planetary Science Letters*, **294**, 185–203.

Carter J. & Poulet F. (2013) Ancient plutonic processes on Mars inferred from the detection of possible anorthositic terrains. *Nature Geoscience*, **6**, 1008–1012.

Carter J., Poulet F., Bibring J.-P., & Murchie S. (2010) Discovery of hydrated silicates in crustal outcrops in the northern plains of Mars. *Science*, **328**, 1682–1686.

Carter J., Poulet F., Bibring J.-P., Mangold N., & Murchie S. (2013) Hydrous minerals on Mars as seen by the CRISM and OMEGA imaging spectrometers: Updated global view. *Journal of Geophysical Research*, **118**, 831–858.

Carter J., Loizeau D., Mangold N., Poulet F., & Bibring J.-P. (2015) Widespread surface weathering on early Mars: A case for a warmer and wetter climate. *Icarus*, **248**, 373–382.

Cheek L.C. & Pieters C.M. (2012) Variations in anorthosite purity at Tsiolkovsky crater on the Moon. *43rd Lunar Planet. Sci. Conf.*, Abstract #2624.

Chojnacki M., Burr D.M., & Moersch J.E. (2014a) Valles Marineris dune fields as compared with other martian populations: Diversity of dune compositions, morphologies, and thermophysical properties. *Icarus*, **230**, 96–142.

Chojnacki M., Burr D.M., Moersch J.E., & Wray J.J. (2014b) Valles Marineris dune sediment provenance and pathways. *Icarus*, **232**, 187–219.

Christensen P.R., Bandfield J.L., Hamilton V.E., et al. (2001) Mars Global Surveyor Thermal Emission Spectrometer experiment: Investigation description and surface science results. *Journal of Geophysical Research*, **106**, 23,823–23,871.

Christensen P.R., Jakosky B.M., Kieffer H.H., et al. (2004) The Thermal Emission Imaging System (THEMIS) for the Mars 2001 Odyssey mission. *Space Science Reviews*, **110**, 85–130.

Clancy R.T., Lee S.W., Gladstone G.R., McMillan W.W., & Roush T. (1995) A new model for Mars atmospheric dust based upon analysis of ultraviolet through infrared observations from Mariner 9, Viking, and Phobos. *Journal of Geophysical Research*, **100**, 5251–5263.

Clancy R.T., Sandor B.J., Wolff M.J., et al. (2012) Extensive MRO CRISM observations of 1.27 μm O_2 airglow in Mars polar night and their comparison to MRO MCS temperature profiles and LMD GCM simulations. *Journal of Geophysical Research*, **117**, E00J10, DOI:10.1029/2011JE004018.

Clancy R.T., Sandor B.J., García-Muñoz A., et al. (2013) First detection of Mars atmospheric hydroxyl: CRISM Near-IR measurement versus LMD GCM simulation of OH Meinel band emission in the Mars polar winter atmosphere. *Icarus*, **226**, 272–281.

Clancy R.T., Smith M.D., Lefèvre F., et al. (2017) Vertical profiles of Mars 1.27 μm O_2 dayglow from MRO CRISM limb spectra: Seasonal/global behaviors, comparisons to LMDGCM simulations, and a global definition for Mars water vapor profiles. *Icarus*, **293**, 132–156.

Clark R.N. & Roush T.L. (1984) Reflectance spectroscopy: Quantitative analysis techniques for remote sensing applications. *Journal of Geophysical Research*, **89**, 6329–6340.

Clark R.N., King T.V.V., Klejwa M., Swayze G.A., & Vergo N. (1990) High spectral resolution reflectance spectroscopy of minerals. *Journal of Geophysical Research*, **95**, 12,653–12,680.

Clark R.N., Swayze G.A., Murchie S.L., Seelos F.P., Seelos K., & Viviano-Beck C.E. (2015) Mineral and other materials mapping of CRISM data with Tetracorder 5. *46th Lunar Planet. Sci. Conf.*, Abstract #2410.

Cloutis E. & Gaffey M. (1991) Pyroxene spectroscopy revisited: Spectral-compositional correlations and relationship to geothermometry. *Journal of Geophysical Research*, **96**, 22,809–22,826.

Cloutis E.A., Hawthorne F.C., Mertzman S.A., et al. (2006) Detection and discrimination of sulfate minerals using reflectance spectroscopy. *Icarus*, **184**, 121–157.

Cull S., Arvidson R.E., Morris R.V., Wolff M., Mellon M.T., & Lemmon M.T. (2010) The seasonal ice cycle at the Mars Phoenix landing site: II. Post-landing CRISM and ground observations. *Journal of Geophysical Research*, **115**, E00E19, DOI:10.1029/2009JE003410.

Dehouck E., Mangold N., Le Mouélic S., Ansan V., & Poulet F. (2010) Ismenius Cavus, Mars: A deep paleolake with phyllosilicate deposits. *Planetary and Space Science*, **58**, 941–946.

Dickson J.L., Head J.W., Levy J.S., & Marchant D.R. (2013) Don Juan Pond, Antarctica: Near-surface $CaCl_2$-brine feeding Earth's most saline lake and implications for Mars. *Scientific Reports*, **3**, 1166, DOI:10.1038/srep01166.

Ding N., Bray V.J., McEwen A.S., et al. (2015) The central uplift of Ritchey crater, Mars. *Icarus*, **252**, 255–270.

Edwards C.S. & Ehlmann B.L. (2015) Carbon sequestration on Mars. *Geology*, **43**, 863–866.

Ehlmann B.L. & Dundar M. (2015) Are Noachian/Hesperian acidic waters key to generating Mars' regional-scale aluminum phyllosilicates? The importance of jarosite co-occurrences with Al-phyllosilicate units. *46th Lunar Planet. Sci. Conf.*, Abstract #1635.

Ehlmann B.L. & Edwards C.S. (2014) Mineralogy of the martian surface. *Annual Review of Earth Planetary of Science*, **42**, 291–315.

Ehlmann B.L., Mustard J., Murchie S., et al. (2008a) Orbital identification of carbonate-bearing rocks on Mars. *Science*, **322**, 1828–1832.

Ehlmann B.L., Mustard J.F., Fassett C.I., et al. (2008b) Clay-bearing minerals and organic preservation potential in sediments from a martian delta environment, Jezero crater, Nili Fossae, Mars. *Nature Geoscience*, **1**, 355–358.

Ehlmann B.L., Mustard J.F., Swayze G.A., et al. (2009) Identification of hydrated silicate minerals on Mars using MRO-CRISM: Geologic context near Nili Fossae and implications for aqueous alteration. *Journal of Geophysical Research*, **114**, E00D08, DOI:10.1029/2009JE003339.

Ehlmann B.L., Mustard J.F., & Murchie S.L. (2010) Geologic setting of serpentine deposits on Mars. *Geophysical Research Letters*, **37**, 610, DOI:10.1029/2010GL042596.

Ehlmann B.L., Mustard J.F., Murchie S.L., et al. (2011a) Aqueous environments during Mars' first billion years: Evidence from the clay mineral record. *Nature*, **479**, 53–60.

Ehlmann B.L., Mustard J.F., Clark R.N., Swayze G.A., & Murchie S.L. (2011b) Evidence for low-grade metamorphism, hydrothermal alteration, and diagenesis on Mars from phyllosilicate mineral assemblages. *Clays & Clay Minerals*, **59**, 359–377.

Ehlmann B.L., Swayze G.A., Milliken R.E., et al. (2016) Discovery of alunite in Cross crater, Terra Sirenum, Mars: Evidence for acidic, sulfurous waters. *American Mineralogist*, **101**, 1527–1542.

Erard S. (2001) A spectrophotometric model of Mars in the near-infrared. *Geophysical Research Letters*, **28**, 1291–1294.

Erard S. & Calvin W. (1997) New composite spectra of Mars, 0.4–5.7 μm. *Icarus*, **130**, 449–460.

Farrand W.H., Glotch T.D., Rice J.W., Hurowitz J.A., & Swayze G.A. (2009) Discovery of jarosite within the Mawrth Vallis region of Mars: Implications for the geologic history of the region. *Icarus*, **204**, 478–488.

Fastook J.L. & Head J.W. (2015) Glaciation in the Late Noachian Icy Highlands: Ice accumulation, distribution, flow rates, basal melting, and top-down melting rates and patterns. *Planetary and Space Science*, **106**, 82–98.

Fischer E. & Pieters C. (1993) The continuum slope of Mars: Bidirectional reflectance investigations and applications to Olympus Mons. *Icarus*, **102**, 185–202.

Fischer E., Martínez G.M., & Rennó N.O. (2016) Formation and persistence of brine on Mars: Experimental simulations throughout the diurnal cycle at the Phoenix landing site. *Astrobiology*, **16**, 937–948.

Flahaut J., Mustard J.F., Quantin C., Clenet H., Allemand P., & Thomas P. (2011) Dikes of distinct composition intruded into Noachian-aged crust exposed in the walls of Valles Marineris. *Geophysical Research Letters*, **38**, L15202, DOI:10.1029/2011GL048109.

Flahaut J., Quantin C., Clenet H., Allemand P., Mustard J., & Thomas P. (2012) Noachian crust and key geologic transitions in the lower walls of Valles Marineris: Insights into early igneous processes on Mars. *Icarus*, **221**, 420–435.

Flahaut J., Carter J., Poulet F., et al. (2015) Embedded clays and sulfates in Meridiani Planum, Mars. *Icarus*, **248**, 269–288.

Fox V.K., Arvidson R.E., Guinness E.A., et al. (2016) Smectite deposits in Marathon Valley, Endeavour crater, Mars, identified using CRISM hyperspectral reflectance data. *Geophysical Research Letters*, **43**, 4885–4892.

Fraeman A.A., Arvidson R.E., Catalano J.G., et al. (2013) A hematite-bearing layer in Gale crater, Mars: Mapping and implications for past aqueous conditions. *Geology*, **41**, 1103–1106.

Frey H.V., Frey E.L., Hartmann W.K., & Tanaka K.L. (2003) Evidence for buried "pre-Noachian" crust pre-dating the oldest observed surface units on Mars. *34th Lunar Planet. Sci. Conf.*, Abstract #1848.

Gendrin A., Mangold N., Bibring J.-P., et al. (2005a) Sulfates in martian layered terrains: The OMEGA/Mars Express view. *Science*, **307**, 1587–1591.

Gendrin A., Bibring J.-P., Mustard J.F., et al. & OMEGA Team (2005b) Identification of predominant ferric signatures in association to the martian sulfate deposits. *36th Lunar Planet. Sci. Conf.*, Abstract #1378.

Ghatan G.J. & Head J.W. III (2002) Candidate subglacial volcanoes in the south polar region of Mars: Morphology, morphometry, and eruption conditions. *Journal of Geophysical Research*, **107**, 5048, DOI:10.1029/2001JE001519.

Glotch T.D., Bandfield J.L., Tornabene L.L., Jensen H.B., & Seelos F.P. (2010) Distribution and formation of chlorides and phyllosilicates in Terra Sirenum, Mars. *Geophysical Research Letters*, **37**, L16202, DOI:10.1029/2010GL044557.

Glotch T.D., Bandfield J.L., Wolff M.J., Arnold J.A., & Che C. (2016) Constraints on the composition and particle size of chloride salt-bearing deposits on Mars. *Journal of Geophysical Research*, **121**, 454–471.

Goudge T.A., Head J.W., Mustard J.F., & Fassett C.I. (2012) An analysis of open-basin lake deposits on Mars: Evidence for the nature of associated lacustrine deposits and post-lacustrine modification processes. *Icarus*, **219**, 211–229.

Grant J.A., Irwin R.P. III, Grotzinger J.P., et al. (2008) HiRISE imaging of impact megabreccia and sub-meter aqueous strata in Holden crater, Mars. *Geology*, **36**, 195–198.

Griffes J.L., Arvidson R.E., Poulet F., & Gendrin A. (2007) Geologic and spectral mapping of etched terrain deposits in northern Meridiani Planum. *Journal of Geophysical Research*, **112**, E08S09, DOI:10.1029/2006JE002811.

Grindrod P.M., West M., Warner N.H., & Gupta S. (2012) Formation of an Hesperian-aged sedimentary basin containing phyllosilicates in Coprates Catena, Mars. *Icarus*, **218**, 178–195.

Grotzinger J.P., Crisp J., Vasavada A.R., et al. (2012) Mars Science Laboratory mission and science investigation. *Space Science Reviews*, **170**, 5–6.

Guzewich S.D., Smith M.D., & Wolff M.J. (2014) The vertical distribution of martian aerosol particle size. *Journal of Geophysical Research*, **119**, 2694–2708.

Hamilton V.E. & Christensen P.R. (2005) Evidence for extensive, olivine-rich bedrock on Mars. *Geology*, **33**, 433–436.

Hanley J. & Horgan B. (2016) A novel method to remotely sense martian chlorine salts. *47th Lunar Planet. Sci. Conf.*, Abstract #2983.

Hecht M.H., Kounaves S.P., Quinn R.C., et al. (2009) Detection of perchlorate and the soluble chemistry of martian soil at the Phoenix lander site. *Science*, **325**, 64–67.

Horgan B.H. & Bell J.F. III (2012) Widespread weathered glass on the surface of Mars. *Geology*, **40**, 391–394.

Horgan B.H., Bell J.F. III, Noe Dobrea E.Z., et al. (2009) Distribution of hydrated minerals in the north polar region of Mars. *Journal Geophysical Research*, **114**, E01005, DOI:10.1029/2008JE003187.

Houck J., Pollack J., Sagan C., Schaak D., & J. Decker (1973) High altitude spectroscopic evidence for bound water on Mars. *Icarus*, **18**, 470–480.

Hunt G.R. & Salisbury J.W. (1971a) Visible and near infrared spectra of minerals and rocks. II. Carbonates. *Modern Geology*, **2**, 23–30.

Hunt G. & Salisbury J. (1971b) Visible and infrared spectra of minerals and rocks. IV: Sulphides and sulphates. *Modern Geology*, **3**, 1–14.

Hynek B.M., Osterloo M.K., & Kierein-Young K.S. (2015) Late-stage formation of martian chloride salts through ponding and evaporation. *Geology*, **43**, 787–790.

Jain N. & Chauhan P. (2015) Study of phyllosilicates and carbonates from the Capri Chasma region of Valles Marineris on Mars based on Mars Reconnaissance Orbiter-Compact Reconnaissance Imaging Spectrometer for Mars (MRO-CRISM) observations. *Icarus*, **250**, 7–17.

Jensen H.B. & Glotch T.D. (2011) Investigation of the near-infrared spectral character of putative martian chloride deposits. *Journal of Geophysical Research*, **116**, E00J03, DOI:10.1029/2011JE003887.

Kite E.S., Halevy I., Kahre M.A., Wolff M.J., & Manga M. (2013) Seasonal melting and the formation of sedimentary rocks on Mars, with predictions for the Gale crater mound. *Icarus*, **223**, 181–210.

Komatsu G., Geissler P.E., Strom R.G., & Singer R.B. (1993) Stratigraphy and erosional landforms of layered deposits in Valles Marineris, Mars. *Journal of Geophysical Research*, **98**, 11,105–11,121.

Kreisch C.D., O'Sullivan J.A., Arvidson R.E., et al. (2016) Regularization of Mars Reconnaissance Orbiter CRISM along-track oversampled hyperspectral imaging observations of Mars, *Icarus*, **282**, 136–151.

Langevin Y., Poulet F., Bibring J.-P., Schmitt B., Douté S., & Gondet B. (2005) Summer evolution of the north polar cap of Mars as observed by OMEGA/Mars Express. *Science*, **307**, 1581–1584.

Langevin Y., Bibring J.-P., Montmessin F., et al. (2007) Observations of the south seasonal cap of Mars during recession in 2004–2006 by the OMEGA visible/near-infrared imaging spectrometer on board Mars Express. *Journal of Geophysical Research*, **112**, E08S12, DOI:10.1029/2006JE002841.

Leask E., Ehlmann B., Dundar M., Murchie S., & Seelos F. (2018) Challenges in the search for perchlorate and other hydrated minerals with 2.1-μm absorptions on Mars. *Geophysical Research Letters*, **45**, 12,180–12,189.

Le Deit L., Flahaut J., Quantin C., et al. (2012) Extensive surface pedogenic alteration of the martian Noachian crust suggested by plateau phyllosilicates around Valles Marineris. *Journal of Geophysical Research*, **117**, E00J05, DOI:10.1029/2011JE003983.

Lichtenberg K., Arvidson R., Morris R., et al. (2010) Stratigraphy of hydrated sulfates in the sedimentary deposits of Aram Chaos, Mars. *Journal of Geophysical Research*, **115**, E00D17, DOI:10.1029/2009JE003353.

Liu Y., Arvidson R.E., Wolff M.J., et al. (2012) Lambert albedo retrieval and analyses over Aram Chaos from OMEGA hyperspectral imaging data. *Journal of Geophysical Research*, **117**, E00J11, DOI:10.1029/2012JE004056.

Liu Y., Glotch T.D., Scudder N.A., et al. (2016) End-member identification and spectral mixture analysis of CRISM hyperspectral data: A case study on southwest Melas Chasma, Mars. *Journal of Geophysical Research*, **121**, 2004–2036.

Loizeau D., Mangold N., Poulet F., et al. (2007) Phyllosilicates in the Mawrth Vallis region of Mars. *Journal of Geophysical Research*, **112**, DOI:10.1029/2006JE002877.

Loizeau D., Carter J., Bouley S., et al. (2012) Characterization of hydrated silicate-bearing outcrops in Tyrrhena Terra, Mars: Implications to the alteration history of Mars. *Icarus*, **219**, 476–497.

Malin M.C. & Edgett K.S. (2000) Sedimentary rocks of early Mars. *Science*, **290**, 1927–1937.

Malin M.C., Bell J.F. III, Cantor B.A., et al. (2007) Context camera investigation on board the Mars Reconnaissance Orbiter. *Journal of Geophysical Research*, **112**, E05S04, DOI:10.1029/2006JE002808.

Mangold N., Poulet F., Mustard J.F., et al. (2007) Mineralogy of the Nili Fossae region with OMEGA/Mars Express data: 2. Aqueous alteration of the crust. *Journal of Geophysical Research*, **112**, E08S04, DOI:10.1029/2006JE002835.

Mangold N., Gendrin A., Gondet B., et al. (2008) Spectral and geological study of the sulfate-rich region of West Candor Chasma, Mars. *Icarus*, **194**, 519–543.

Martín-Torres F.J., Zorzano M.-P., Valentín-Serrano P., et al. (2015) Transient liquid water and water activity at Gale crater on Mars. *Nature Geoscience*, **8**, 357–361.

Marzo G.A., Davila A.F., Tornabene L.L., et al. (2010) Evidence for Hesperian impact-induced hydrothermalism on Mars. *Icarus*, **208**, 667–683.

Massé M., Bourgeois O., Le Mouélic S., Verpoorter C., Spiga A., & Le Deit L. (2012) Wide distribution and glacial origin of polar gypsum on Mars. *Earth and Planetary Science Letters*, **317**, 44–55.

Massé M., Beck P., Schmitt B., et al. (2014) Spectroscopy and detectability of liquid brines on Mars. *Planetary and Space Science*, **92**, 136–149.

McBeck J., Seelos K.D., Ackiss S.E., & Buczkowski D. (2014) Using CRISM and THEMIS to characterize high thermal inertia terrains in the northern Hellas region of Mars. *American Geophysical Union, Fall Meeting 2014*, Abstract #P41B-3900.

McEwen A.S., Eliason E.M., Bergstrom J.W., et al. (2007) Mars Reconnaissance Orbiter's High Resolution Imaging Science Experiment (HiRISE). *Journal of Geophysical Research*, **112**, E05S02, DOI:10.1029/2005JE002605.

McEwen A.S., Ojha L., Dundas C.M., et al. (2011) Seasonal flows on warm martian slopes. *Science*, **333**, 740–743.

McEwen A., Dundas C.M., Mattson S.S., et al. (2014) Recurring slope lineae in equatorial regions of Mars. *Nature Geoscience*, **7**, 53–58.

McGuire P.C., Wolff M.J., Smith M.D., et al. & CRISM Team (2008) MRO/CRISM retrieval of surface Lambert albedos for multispectral mapping of Mars with DISORT-based rad. transfer modeling: Phase 1 – Using historical climatology for temperatures, aerosol opacities, & atmosheric Pressures. *IEEE Transactions on Geoscience and Remote Sensing*, **46**, 4020–4040.

McGuire P.C., Bishop J.L., Brown A.J., et al. (2009) An improvement to the volcano-scan algorithm for atmospheric correction of CRISM and OMEGA spectral data. *Planetary and Space Science*, **57**, 809–815.

McGuire P.C., Arvidson R.E., Bishop J.L., et al. (2013) Mapping minerals on Mars with CRISM: Atmospheric and photometric correction for MRDR map tiles, version 2, and comparison to OMEGA. *44th Lunar Planet. Sci. Conf.*, Abstract #1581.

McKeown N., Bishop J., Noe Dobrea E., et al. (2009) Characterization of phyllosilicates observed in the central Mawrth Vallis region, Mars, their potential formational processes, and implications for past climate. *Journal of Geophysical Research*, **114**, E00D10, DOI:10.1029/2008JE003301.

McSween H.Y., Labotka T.C., & Viviano-Beck C.E. (2015) Metamorphism in the martian crust. *Meteoritics and Planetary Science*, **50**, 590–603.

MEPAG NEX-SAG Report (2015) Report from the Next Orbiter Science Analysis Group (NEX-SAG), chaired by B. Campbell and R. Zurek, posted December, 2015 by the Mars Exploration Program Analysis Group (MEPAG) at http://mepag.nasa.gov/reports.cfm

Michalski J.R. & Niles P.B. (2010) Deep crustal carbonate rocks exposed by meteor impact on Mars. *Nature Geoscience*, **3**, 751–755.

Michalski J.R. & Noe Dobrea Eldar Z. (2007) Evidence for a sedimentary origin of clay minerals in the Mawrth Vallis region, Mars. *Geology*, **35**, 951–954.

Michalski J.R., Cuadros J., Niles P.B., Parnell J., Rogers A.D., & Wright S.P. (2013a) Groundwater activity on Mars and implications for a deep biosphere. *Nature Geoscience*, **6**, 133–138.

Michalski J.R., Niles P.B., Cuadros J., & Baldridge A.M. (2013b) Multiple working hypotheses for the formation of compositional stratigraphy on Mars: Insights from the Mawrth Vallis region. *Icarus*, **226**, 816–840.

Milliken R., Swayze G., Arvidson R., et al. (2008) Opaline silica in young deposits on Mars. *Geology*, **36**, 847–850.

Milliken R.E. & Bish D.L. (2010) Sources and sinks of clay minerals on Mars. *Philosophical Magazine*, **90**, 2293–2308.

Milliken R.E., Grotzinger J.P., & Thomson B.J. (2010) Paleoclimate of Mars as captured by the stratigraphic record in Gale crater. *Geophysical Research Letters*, **37**, L04201, DOI:10.1029/2009GL041870.

Moroz V. (1964) The infrared spectrum of Mars (1.1–4.1 μm). *Soviet Astronomy*, **8**, 273–281.

Morris R.V., Agresti D.G., Lauer H.V. Jr., Newcomb J.A., Shelfer T.D., & Murali A.V. (1989) Evidence for pigmentary hematite on Mars based on optical, magnetic, and Mossbauer studies of superparamagnetic (nanocrystalline) hematite. *Journal of Geophysical Research*, **94**, 2760–2778.

Morris R.V., Golden D.C., Bell J.F. III, Lauer H.V. Jr., & Adams J.B. (1993) Pigmenting agents in martian soils: Inferences from spectral, Mössbauer, and magnetic properties of nanophase and other iron oxides in Hawaiian palagonitic soil PN-9. *Geochimica Cosmochimica Acta*, **57**, 4597–4609.

Morris R.V., Klingelhöfer G., Schröder C., et al. (2006) Mössbauer mineralogy of rock, soil, and dust at Gusev crater, Mars: Spirit's journey through weakly altered olivine basalt on the plains and pervasively altered basalt in the Columbia Hills. *Journal of Geophysical Research*, **111**, E02S13, DOI:10.1029/2005JE002584.

Murchie S., Mustard J., Bishop J., Head J., Pieters C., & Erard S. (1993) Spatial variations in the spectral properties of bright regions on Mars. *Icarus*, **105**, 454–468.

Murchie S., Kirkland L., Erard S., Mustard J., & Robinson M. (2000) Near-infrared spectral variations of martian surface materials from ISM imaging spectrometer data. *Icarus*, **147**, 444–471.

Murchie S., Arvidson R., Bedini P., et al. (2007) Compact Reconnaissance Imaging Spectrometer for Mars (CRISM) on Mars Reconnaissance Orbiter (MRO). *Journal of Geophysical Research*, **112**, E05S03, DOI:10.1029/2006JE002682.

Murchie S.L, Seelos F.P., Hash C.D., et al. & CRISM Team (2009a) The CRISM investigation and data set from the Mars Reconnaissance Orbiter's Primary Science Phase. *Journal of Geophysical Research*, **114**, E00D07, DOI:10.1029/2009JE003344.

Murchie S.L., Mustard J.F., Ehlmann B.L., et al. (2009b) A synthesis of martian aqueous mineralogy after one Mars year of observations from the Mars Reconnaissance Orbiter. *Journal of Geophysical Research*, **114**, E00D06, DOI:10.1029/2009JE003342.

Murchie S., Roach L., Seelos F., et al. (2009c) Compositional evidence for the origin of layered deposits in Candor Chasma, Mars. *Journal of Geophysical Research*, **114**, E00D05, DOI:10.1029/2009JE003343.

Mustard J.F., Murchie S., Erard S., & Sunshine J. (1997) In situ compositions of martian volcanics: Implications for the mantle. *Journal of Geophysical Research*, **102**, 25,605–25,615.

Mustard J.F., Poulet F., Gendrin A., et al. (2005) Olivine and pyroxene diversity in the crust of Mars. *Science*, **307**, 1594–1597.

Mustard J.F., Poulet F., Head J.W., et al. (2007) Mineralogy of the Nili Fossae region with OMEGA/Mars Express data: 1. Ancient impact melt in the Isidis Basin and implications for the transition from the Noachian to Hesperian. *Journal of Geophysical Research*, **112**, E08S03, DOI:10.1029/2006JE002834.

Mustard J., Murchie S., Pelkey S.M., et al. (2008) Hydrated silicate minerals on Mars observed by the CRISM instrument on MRO. *Nature*, 454, 305–309.

Mustard J., Ehlmann B., Murchie S., et al. (2009) Composition, morphology, and stratigraphy of Noachian/Phyllosian Crust around the Isidis basin. *Journal of Geophysical Research*, **114**, E00D12, DOI:10.1029/2009JE003349.

Nedell S., Squyres S., & Andersen D. (1987) Origin and evolution of the layered deposits in the Valles Marineris, Mars. *Icarus*, **70**, 409–441.

Niles P.B. & Michalski J. (2009) Meridiani Planum sediments on Mars formed through weathering in massive ice deposits. *Nature Geoscience*, **2**, 215–220.

Noe Dobrea E., Bishop J., McKeown N., et al. (2010) Mineralogy and stratigraphy of phyllosilicate-bearing and dark mantling units in the greater Mawrth Vallis / west Arabia Terra area: Constraints on geological origin. *Journal of Geophysical Research*, **115**, E00D19, DOI:10.1029/2009JE003351.

Noel A., Bishop J.L., Al-Samir M., et al. (2015) Mineralogy, morphology and stratigraphy of the light-toned interior layered deposits at Juventae Chasma. *Icarus*, **251**, 315–331.

Ody A., Poulet F., Langevin Y., et al. (2012) Global maps of anhydrous minerals at the surface of Mars from OMEGA/MEx. *Journal of Geophysical Research*, **117**, E00J14, DOI:10.1029/2012JE004117.

Ody A., Poulet F., Bibring J.-P., et al. (2013) Global investigation of olivine on Mars: Insights into crust and mantle compositions. *Journal of Geophysical Research*, **118**, 234–262.

Ojha L., Wray J.J., Murchie S.L., McEwen A.S., Wolff M.J., & Karunatillake S. (2013) Spectral constraints on the formation mechanism of recurring slope lineae. *Geophysical Research Letters*, **40**, 5621–5626.

Ojha L., McEwen A., Dundas C., et al. (2014) HiRISE observations of Recurring Slope Lineae (RSL) during southern summer on Mars. *Icarus*, **231**, 365–376.

Ojha L., Wilhelm M.B., Murchie S.L., et al. (2015) Spectral evidence for hydrated salts in seasonal brine flows on Mars. *Nature Geoscience*, **8**, 829–832.

Osinski G.R., Tornabene L.L., Banerjee N.R., et al. (2013) Impact-generated hydrothermal systems on Earth and Mars. *Icarus*, **224**, 347–363.

Osterloo M.M., Hamilton V.E., Bandfield J.L., et al. (2008) Chloride-bearing materials in the southern highlands of Mars. *Science*, **319**, 1651–1654.

Osterloo M.M., Anderson F.S., Hamilton V.E., & Hynek B.M. (2010) Geologic context of proposed chloride-bearing materials on Mars. *Journal of Geophysical Research*, **115**, E10012, DOI:10.1029/2010JE003613.

Pan L., Ehlmann B.L., Carter J., & Ernst C.M. (2017) The stratigraphy and history of Mars' northern lowlands through mineralogy of impact craters: A comprehensive survey. *Journal of Geophysical Research*, **122**, 1824–1854.

Pelkey S.M., Mustard J.F., Murchie S., et al. (2007) CRISM multispectral summary products: Parameterizing mineral diversity on Mars from reflectance. *Journal of Geophysical Research*, **112**, E08S14, DOI:10.1029/2006JE002831.

Pimental G., Forney P., & Herr K. (1974) Evidence about hydrate and solid water in the martian surface from the 1969 Mariner infrared spectrometer. *Journal of Geophysical Research*, **79**, 1623–1634.

Poulet F., Bibring J.-P., Mustard J.F., et al. (2005) Phyllosilicates on Mars and implications for early martian climate. *Nature*, **438**, 623–627

Poulet F., Gomez C., Bibring J.-P., et al. (2007) Martian surface mineralogy from Observatoire pour la Minéralogie, l'Eau, les Glaces et l'Activité on board the Mars Express spacecraft (OMEGA/MEx): Global mineral maps. *Journal of Geophysical Research*, **112**, E08S02, DOI:10.1029/2006JE002840.

Poulet F., Mangold N., Loizeau D., et al. (2008) Abundance of minerals in the phyllosilicate-rich units on Mars. *Astronomy and Astrophysics*, **487**, L41–L44.

Poulet F., Mangold N., Platevoet B., et al. (2009a) Quantitative compositional analysis of martian mafic regions using the MEx/OMEGA reflectance data. 2. Petrological implications. *Icarus*, **201**, 84–101.

Poulet F., Bibring J.-P., Langevin Y., et al. (2009b) Quantitative compositional analysis of martian mafic regions using the MEx/OMEGA reflectance data 1. Methodology, uncertainties and examples of application. *Icarus*, **201**, 69–83.

Poulet F., Carter J., Bishop J.L., Loizeau D., & Murchie S.L. (2014) Mineral abundances at the final four Curiosity study sites and implications for their formation. *Icarus*, **231**, 65–76.

Powell K.E., Arvidson R.E., Zanetti M., Guinness E.A., & Murchie S.L. (2017) The structural, stratigraphic, and paleoenvironmental record exposed on the rim and walls of Iazu crater, Mars. *Journal of Geophysical Research*, **122**, 1138–1156.

Quantin C., Flahaut J., Clenet H., Allemand P., & Thomas P. (2012) Composition and structures of the subsurface in the vicinity of Valles Marineris as revealed by central uplifts of impact craters. *Icarus*, **221**, 436–452.

Riu L., Poulet F., Carter J., et al. (2019) The M3 project: 1– A global hyperspectral image-cube of the martian surface. *Icarus*, **319**, 281–292.

Roach L., Mustard J., Murchie S., et al. (2009) Testing evidence of recent hydration state change in sulfates on Mars. *Journal of Geophysical Research*, **114**, E00D02, DOI:10.1029/2008JE003245.

Roach L.H., Mustard J.F., Swayze G., et al. (2010) Hydrated mineral stratigraphy of Ius Chasma, Valles Marineris. *Icarus*, **206**, 253–268.

Rossman G. (1976) Spectroscopic and magnetic studies of ferric iron hydroxysulfates: The series $Fe(OH)SO_4 \cdot nH_2O$ and jarosite. *American Mineralogist*, **61**, 398–401.

Ruesch O., Poulet F., Vincendon M., et al. (2012) Compositional investigation of the proposed chloride-bearing materials on Mars using near-infrared orbital data from OMEGA/MEx. *Journal of Geophysical Research*, **117**, E00J13.

Ruff S.W., Farmer J.D., Calvin W.M., et al. (2011) Characteristics, distribution, origin, and significance of opaline silica observed by the Spirit rover in Gusev crater, Mars. *Journal of Geophysical Research*, **116**, E00F23, DOI:10.1029/2010JE003767.

Salvatore M.R., Mustard J.F., Wyatt M.B., & Murchie S.L. (2010) Definitive evidence of Hesperian basalt in Acidalia and Chryse planitiae. *Journal of Geophysical Research*, **115**, E07005, DOI:10.1029/2009JE003519.

Scanlon K.E., Head J.W., Madeleine J.-B., Wordsworth R.D., & Forget F. (2013) Orographic precipitation in valley network headwaters: Constraints on the ancient martian atmosphere. *Geophysical Research Letters*, **40**, 4182–4187.

Seelos F.P., Viviano-Beck C.E., Morgan M.F., Romeo G., Aiello J.J., & Murchie S.L. (2016a) CRISM hyperspectral targeted observation PDS product sets – TERs and MTRDRs. *47th Lunar Planet. Sci. Conf.*, Abstract #1783.

Seelos K.D., Seelos F.P., Buczkowski D.L., & Viviano-Beck C.E. (2016b) Mapping laterally extensive phyllosilicates in west Margaritifer Terra, Mars. *47th Lunar Planet. Sci. Conf.*, Abstract #7043.

Sherman D., Burns R., & Burns V. (1982) Spectral characteristics of the iron oxides with application to the martian bright region mineralogy. *Journal of Geophysical Research*, **87**, 10,169–10,180.

Singer R.B. (1982) Spectral evidence for the mineralogy of high-albedo soils and dust on Mars. *Journal of Geophysical Research*, **87**, 10,159–10,168.

Singer R.B. & McSween H.Y. Jr. (1993) The igneous crust of Mars: Compositional evidence from remote sensing and the SNC meteorites. In: *Resources of near-Earth space* (J.S. Lewis, M.S. Matthews, & M.L. Guerrieri, eds.). ARI, Heidelberg, 709–736.

Singer R.B., McCord T.B., Clark R.N., Adams J.B., & Huguenin R.L. (1979) Mars surface composition from reflectance spectroscopy: A summary. *Journal of Geophysical Research*, **84**, 8415–8426.

Skok J.R., Mustard J.F., Ehlmann B.L., Milliken R.E., & Murchie S.L. (2010) Silica deposits in the Nili Patera caldera on the Syrtis Major volcanic complex on Mars. *Nature Geoscience*, **3**, 838–841.

Skok J.R., Mustard J.F., Tornabene L.L., Pan C., Rogers D., & Murchie S.L. (2012) A spectroscopic analysis of martian crater central peaks: Formation of the ancient crust. *Journal of Geophysical Research*, **117**, E00J18, DOI:10.1029/2012JE004148.

Smith M.D., Wolff M.J., Clancy R.T., & Murchie S.L. (2009) Compact Reconnaissance Imaging Spectrometer observations of water vapor and carbon monoxide. *Journal of Geophysical Research*, **114**, E00D03, DOI:10.1029/2008JE003288.

Smith M.D., Wolff M.J., Clancy R.T., Kleinböhl A., & Murchie S.L. (2013) Vertical distribution of dust and water ice aerosols from CRISM limb-geometry observations. *Journal of Geophysical Research*, **118**, 321–334.

Squyres S.W., Grotzinger J.P., Arvidson R.E., et al. (2004a) In situ evidence for an ancient aqueous environment at Meridiani Planum, Mars. *Science*, **306**, 1709–1714.

Squyres S.W., Arvidson R.E., Bell J.F., III, et al. (2004b) The Opportunity rover's Athena science investigation at Meridiani Planum, Mars. *Science*, **306**, 1698–1703.

Squyres S.W., Arvidson R.E., Ruff S., et al. (2008) Detection of silica-rich deposits on Mars. *Science*, **320**, 1063–1067.

Sun V.Z. & Milliken R.E. (2015) Ancient and recent clay formation on Mars as revealed from a global survey of hydrous minerals in crater central peaks. *Journal of Geophysical Research*, **120**, 2293–2332.

Sun V.Z. & Milliken R.E. (2018) Distinct geologic settings of opal-A and more crystalline hydrated silica on Mars. *Geophysical Research Letters*, **45**, 10,221–10,228.

Sunshine J., Pieters C., & Pratt S. (1990) Deconvolution of mineral absorption bands: An improved approach. *Journal of Geophysical Research*, **95**, 6955–6966.

Tanaka K.L., Robbins S.J., Fortezzo C.M., Skinner J.A., & Hare T.M. (2014) The digital global geologic map of Mars: Chronostratigraphic ages, topographic and crater morphologic characteristics, and updated resurfacing history. *Planetary and Space Science*, **95**, 11–24.

Thollot P., Mangold N., Ansan V., et al. (2012) Most Mars minerals in a nutshell: Various alteration phases formed in a single environment in Noctis Labyrinthus. *Journal of Geophysical Research*, **117**, E00J06, DOI:10.1029/2011JE004028.

Thomson B.J., Bridges N.T., Milliken R., et al. (2011) Constraints on the origin and evolution of the layered mound in Gale crater, Mars using Mars Reconnaissance Orbiter data. *Icarus*, **214**, 413–432.

Tornabene L.L., Osinski G.R., McEwen A.S., et al. (2013) An impact origin for hydrated silicates on Mars: A synthesis. *Journal of Geophysical Research*, **118**, 994–1012.

Tosca N.J. & Knoll A.H. (2009) Juvenile chemical sediments and the long term persistence of water at the surface of Mars. *Earth and Planetary Science Letters*, **286**, 379–386.

van Berk W. & Y. Fu (2011) Reproducing hydrogeochemical conditions triggering the formation of carbonate and phyllosilicate alteration mineral assemblages on Mars (Nili Fossae region). *Journal of Geophysical Research*, **116**, E10006, DOI:10.1029/2011JE003886.

Vincendon M., Langevin Y., Poulet F., Bibring J.-P., & Gondet B. (2007) Recovery of surface reflectance spectra and evaluation of the optical depth of aerosols in the near-IR using a Monte Carlo approach: Application to the OMEGA observations of high-latitude regions of Mars. *Journal of Geophysical Research*, **112**, E08S13, DOI:10.1029/2006JE002845.

Vincendon M., Pilorget C., Gondet B., Murchie S., & Bibring J.-P. (2011) New near-IR observations of mesospheric CO_2 and H_2O clouds on Mars. *Journal of Geophysical Research*, **116**, E00J02, DOI:10.1029/2011JE003827.

Viviano C., Murchie S., Daubar I., Morgan M., Seelos F., & Plescia J. (2019) Composition of Amazonian volcanic materials in Tharsis and Elysium, Mars, from MRO/CRISM reflectance spectra. *Icarus*, **328**, 274–286.

Viviano-Beck C.E., Seelos F.P., Murchie S.L., et al. (2014) Revised CRISM spectral parameters and summary products based on the currently detected mineral diversity on Mars. *Journal of Geophysical Research*, **119**, 1403–1431.

Viviano-Beck C.E., Murchie S.L., Beck A.W., & Dohm J.M. (2017) Compositional and structural constraints on the geologic history of eastern Tharsis Rise, Mars. *Icarus*, **284**, 43–58.

Wang A., Jolliff B.L., Liu Y., & Connor K. (2016) Setting constraints on the nature and origin of the two major hydrous sulfates on Mars: Monohydrated and polyhydrated sulfates. *Journal of Geophysical Research*, **121**, 678–694.

Weitz C.M. & Bishop J.L. (2016) Stratigraphy and formation of clays, sulfates, and hydrated silica within a depression in Coprates Catena, Mars. *Journal of Geophysical Research*, **121**, 805–835.

Weitz C.M., Milliken R.E., Grant J.A., McEwen A.S., Williams R.M.E., & Bishop J.L. (2008a) Light-toned strata and inverted channels adjacent to Juventae and Ganges chasmata, Mars. *Geophysical Research Letters*, **35**, L19202, DOI:10.1029/2008GL035317.

Weitz C.M. Lane M.D., Staid M., & Noe Dobrea E. (2008b) Gray hematite distribution and formation in Ophir and Candor chasmata. *Journal of Geophysical Research*, **113**, E02016, DOI:10.1029/2007JE002930.

Weitz C.M., Milliken R.E., Grant J.A., et al. (2010) Mars Reconnaissance Orbiter observations of light-toned layered deposits and associated fluvial landforms on the plateaus adjacent to Valles Marineris. *Icarus*, **205**, 73–102.

Weitz C.M., Noe Dobrea E.Z., Lane M.D., & Knudson A.T. (2012) Geologic relationships between gray hematite, sulfates, and clays in Capri Chasma. *Journal of Geophysical Research*, **117**, E00J09, DOI:10.1029/2012JE004092.

Weitz C.M., Bishop J.L., & Grant J.A. (2013) Gypsum, opal, and fluvial channels within a trough of Noctis Labyrinthus, Mars: Implications for aqueous activity during the Late Hesperian to Amazonian. *Planetary and Space Science*, **87**, 130–145.

Weitz C.M., Bishop J.L., Baker L.L., & Berman D.C. (2014) Fresh exposures of hydrous Fe-bearing amorphous silicates on Mars. *Geophysical Research Letters*, **41**, 8744–8751.

Weitz C.M., Noe Dobrea E., & Wray J.J. (2015) Mixtures of clays and sulfates within deposits in western Melas Chasma, Mars. *Icarus*, **251**, 291–314.

Werner S.C. (2008) The early martian evolution: Constraints from basin formation ages. *Icarus*, **195**, 45–60.

Werner S.C. (2009) The global martian volcanic evolutionary history. *Icarus*, **201**, 44–68.

Werner S.C. & Tanaka K.L. (2011) Redefinition of the crater-density and absolute-age boundaries for the chronostratigraphic system of Mars. *Icarus*, **215**, 603–607.

Wilson J.H. & Mustard J.F. (2013) Exposures of olivine-rich rocks in the vicinity of Ares Vallis: Implications for Noachian and Hesperian volcanism. *Journal of Geophysical Research*, **118**, 916–929.

Wilson S.A., Howard A.D., Moore J.M., & Grant J.A. (2016) A cold-wet middle-latitude environment on Mars during the Hesperian-Amazonian transition: Evidence from northern Arabia valleys and paleolakes. *Journal of Geophysical Research*, **121**, 1667–1694.

Wiseman S.M., Arvidson R.E., Andrews-Hanna J.C., et al. (2008) Phyllosilicate and sulfate-hematite deposits within Miyamoto crater in southern Sinus Meridiani, Mars. *Geophysical Research Letters*, **35**, L19204, DOI:10.1029/2008GL035363.

Wiseman S., Arvidson R., Morris R., et al. (2010) Spectral and stratigraphic mapping of hydrated sulfate and phyllosilicate-bearing deposits in northern Sinus Meridiani, Mars. *Journal of Geophysical Research*, **115**, E00D18, DOI:10.1029/2009JE003354.

Wiseman S.M., Arvidson R.E., Wolff M.J., et al. (2016) Characterization of artifacts introduced by the empirical volcano-scan atmospheric correction commonly applied to CRISM and OMEGA near-infrared spectra. *Icarus*, **269**, 111–121.

Wolff M.J., Smith M.D., Clancy R.T., et al. (2009) Wavelength dependence of dust aerosol single scattering albedo as observed by the Compact Reconnaissance Imaging Spectrometer. *Journal of Geophysical Research*, **114**, E00D04, DOI:10.1029/2009JE003350.

Wordsworth R.D., Kerber L., Pierrehumbert R.T., Forget F., & Head J.W. (2015) Comparison of "warm and wet" and "cold and icy" scenarios for early Mars in a 3-D climate model. *Journal of Geophysical Research*, **120**, 1201–1219.

Wray J.J., Ehlmann B.L., Squyres S.W., Mustard J.F., & Kirk R.L. (2008) Compositional stratigraphy of clay-bearing layered deposits at Mawrth Vallis, Mars. *Geophysical Research Letters*, **35**, L12202, DOI:10.1029/2008GL034385.

Wray J.J., Murchie S.L., Squyres S.W., Seelos F.P., & Tornabene L.L. (2009a) Diverse aqueous environments on ancient Mars revealed in the southern highlands. *Geology*, **37**, 1043–1046.

Wray J.J., Noe Dobrea E.Z., Arvidson R.E., et al. (2009b) Phyllosilicates and sulfates at Endeavour crater, Meridiani Planum, Mars. *Geophysical Research Letters*, **36**, L21201, DOI:10.1029/2009GL040734.

Wray J.J., Squyres S.W., Roach L.H., Bishop J.L., Mustard J.F., & Noe Dobrea E.Z. (2010) Identification of the Ca-sulfate bassanite in Mawrth Vallis, Mars. *Icarus*, **209**, 416–421.

Wray J.J., Milliken R.E., Dundas C.M. (2011) Columbus crater and other possible groundwater-fed paleolakes of Terra Sirenum, Mars. *Journal of Geophysical Research*, **116**, E01001, DOI:10.1029/2010JE003694.

Wray J.J., Hansen S.T., Dufek J., et al. (2013) Prolonged magmatic activity on Mars inferred from the detection of felsic rocks. *Nature Geoscience*, 6, 1013–1017.

Wray J.J., Murchie S.L., Bishop J.L., et al. (2016) Orbital evidence for more widespread carbonate-bearing rocks on Mars. *Journal of Geophysical Research*, **121**, 652–677.

Wyatt M.B. & McSween H.Y. (2002) Spectral evidence for weathered basalt as an alternative to andesite in the northern lowlands of Mars. *Nature*, **417**, 263–266.

Wyatt M.B., McSween H.Y. Jr., Tanaka K.L., & Head J.W. III (2004) Global geologic context for rock types and surface alteration on Mars. *Geology*, **32**, 645–648.

Zolotov M.Y. (2015) What solutions caused Noachian Weathering on Mars? *American Geophysical Union, Fall Meeting 2015*, Abstract #P33A-2118.

Zolotov M. Yu. & Mironenko M.V. (2016) Chemical models for martian weathering profiles: Insights into formation of layered phyllosilicate and sulfate deposits. *Icarus*, **275**, 203–220.

24

Thermal Infrared Spectral Analyses of Mars from Orbit Using the Thermal Emission Spectrometer and Thermal Emission Imaging System

VICTORIA E. HAMILTON, PHILIP R. CHRISTENSEN, JOSHUA L. BANDFIELD,
A. DEANNE ROGERS, CHRISTOPHER S. EDWARDS, AND STEVEN W. RUFF

24.1 Introduction

Thermal infrared (TIR) spectroscopy is a laboratory and remote sensing technique that is sensitive to the fundamental vibrational modes in geologic materials (see also Chapters 2 and 3). The TIR vibrational modes for each mineral or amorphous phase are based on their chemical composition and structure, enabling the identification of virtually all phases.

The Mars Global Surveyor (MGS) Thermal Emission Spectrometer (TES) was designed to address a wide range of science objectives, including the determination of surface mineralogy, volatile abundance and history, and atmospheric dynamics (Christensen et al., 1992, 2001). The TES instrument included a Michelson interferometer ($\sim$1650–200 cm^{-1}) and bore sighted solar reflectance and thermal radiance channels for the measurement of albedo and thermophysical properties at a spatial resolution of $\sim$3 × 6 km (Christensen et al., 2001). In this chapter, we focus on the results of TES martian surface mineralogy investigations using the interferometer. Methods for subtraction of atmospheric contributions and identification of compositional information are described in numerous publications (see also Chapter 15) and are not discussed here (Ramsey & Christensen, 1998; Bandfield et al., 2000, 2002, 2004; Smith et al., 2000; Ruff & Christensen, 2002, 2007; Ruff, 2004; Koeppen & Hamilton, 2008; Rogers & Aharonson, 2008; Rogers & Bandfield, 2009; Rogers & Fergason, 2011; Baldridge et al., 2013; Lane & Christensen, 2013; Smith et al., 2013; Rogers & Nekvasil, 2015; Hanna et al., 2016).

The 2001 Mars Odyssey Thermal Emission Imaging System (THEMIS) was designed to address similar science objectives as TES but at complementary, higher spatial resolution in a multispectral, imaging format (Christensen et al., 2004). THEMIS is composed of a visible subsystem with five channels centered between 0.42 and 0.86 μm having a maximum resolution of $\sim$18 m/pixel, and an uncooled, infrared subsystem consisting of nine, $\sim$1-μm wide channels from 6.8 to 14.9 μm having a spatial resolution of $\sim$100 m/pixel. As with TES, THEMIS measures emitted thermal radiance, which is converted to apparent emissivity after calibration (Bandfield et al., 2004; Christensen et al., 2004). Similar analytical methods are applied to THEMIS data as described in the preceding text for TES data.

24.2 Key Science Results

24.2.1 Global Mapping

24.2.1.1 Mineral/Phase Maps

We use the term "geologic phase" to encompass both minerals and amorphous materials, where the latter have been demonstrated by TES and instruments on subsequent missions (e.g., Bandfield et al., 2000; Bish et al., 2013) to be a significant component of martian surface materials. One of the first global maps derived from TES showed the distribution of coarse crystalline (gray) hematite (Christensen et al., 2001) in close association with layered, sedimentary units. These hematite deposits likely formed by a process involving chemical precipitation from aqueous fluids, under either ambient or hydrothermal conditions. The hematite-bearing region in Meridiani Terra was selected as the landing site for the Mars Exploration Rover Opportunity, and the hematite was found to be present in the form of small concretions within complex sediments recording a history that includes periodic inundation with surface water (e.g., Squyres et al., 2004).

The first full suite of global mineral maps using TES data were published by Bandfield (2002) and showed that elevated concentrations of plagioclase feldspar and high-Ca pyroxene are consistent with surfaces having basaltic compositions in low-albedo regions north of 45°S. Significant concentrations of plagioclase and high-silica (sheet silicates and/or amorphous) phases with low concentrations of high-Ca pyroxenes are concentrated in both southern and northern high-latitude, low-albedo regions. The spectral library used by Bandfield (2002) contained only two olivine compositions (forsterite and fayalite), and these were identified only in a small number of locations. Notable phases that were not identified include quartz, amphibole, basaltic glass, carbonate, and sulfate. Most of these phases or their spectral equivalents have been identified since and characterized at local scales using TES, THEMIS, and Mini-TES data (e.g., Bandfield et al., 2004; Bandfield, 2006; Glotch & Bandfield, 2006; Squyres et al., 2008; Lane et al., 2008; Morris et al., 2010; Ruff et al., 2011; Hamilton & Ruff, 2012; Ruff & Hamilton, 2017).

Subsequent global mapping studies commonly focused on specific minerals and generally sought to expand the representation of solid solution compositions and increase the spatial resolution of the maps. Koeppen and Hamilton (2008) searched for an expanded range of olivine solid solution spectra and found a diversity of compositions across the martian surface (at typical abundances of 10–20 vol.%) that are dominated by ~Fo_{68}, but with a broader range than represented by martian meteorites, including confident identifications of Fo_{91}-, Fo_{53}-, and Fo_{39} -bearing materials and one isolated detection of a fayalitic (Fo_1) composition. Milam et al. (2010) investigated the distribution and variation in plagioclase feldspar compositions and found that the average plagioclase composition is ~An_{60} and does not vary significantly or in correlated fashion between specific geologic terrains. X-ray diffraction analysis of the Rocknest basaltic sand shadow by the Mars Science Laboratory's Curiosity provided local confirmation of a plagioclase feldspar having a solid solution composition of ~$An_{57\pm3}$ and olivine with $Fo_{61\pm3}$ (Bish et al., 2013; Blake et al., 2013).

Global analyses of TES data that utilize spectral indices have sought to identify or discriminate between specific minerals and amorphous phases of interest in spectra of the martian surface. Spectral indices may not be uniquely interpretable, but when used cautiously can be very informative. Ruff (2004) examined the global distribution of a spectral feature at 1630 cm^{-1} associated with water in zeolites and concluded that they are a feasible component of the martian dust. To distinguish between an amorphous, alkali-rich component and spectrally similar dioctahedral smectite clay minerals in the global-scale "surface type 2" (ST2, Section 24.2.1.2) materials identified by TES, Ruff and Christensen (2007) developed two spectral indices. No extensive, global correlations were identified in that study, indicating that high-silica amorphous phases remain the best candidate components of ST2 material.

24.2.1.2 Global Rock-Type Maps

Regionally contiguous spatial variability in basaltic mineral assemblage is present at a global scale, reflecting processes related to crustal and surface evolution (Figure 24.1). The most striking differences in surface composition are between the northern lowlands and the southern highlands, with Acidalia and Chryse Planitia exhibiting elevated abundances of noncrystalline silicate phase(s) and decreased abundances of olivine and pyroxene compared to the highlands (Bandfield et al., 2000; Bandfield, 2002; Rogers & Christensen, 2007). The southern highlands exhibit regional variations in composition that are broadly correlated with crustal provinces (Rogers & Christensen, 2007; Rogers & Hamilton, 2015).

The global mapping efforts described in the preceding text did not isolate surfaces based on average particle size or morphology, thus the reported regional mineralogic assemblages

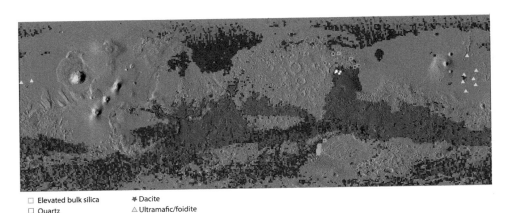

Figure 24.1 Distribution of TES-identified compositional classes (after Rogers & Christensen, 2007). Finer-scale variability is reported by Rogers and Hamilton (2015). Projection, boundaries, and base map are as in Rogers and Christensen (2007). Localized THEMIS detections of silica, quartz, dacitic, and ultramafic materials are shown as colored symbols; these are reported detections and additional occurrences may be present.

are likely to represent mixtures of regolith, impact ejecta, bedrock, and active sand deposits (e.g., Rogers and Christensen, 2007). Higher spatial resolution THEMIS data reveal that small, compositionally distinctive outcrops contribute to some of the regional differences apparent in lower resolution TES global mineral assemblage maps. However, some regions have few such outcrops, and the regionally derived mineralogic assemblage likely represents more of a bulk, near-surface regolith composition.

Most of the regions that are distinctive in mineral assemblage also show distinctions in Gamma-Ray Spectrometer (GRS) elemental mass fractions (e.g., Boynton et al., 2007). Because GRS has a deeper depth of sensitivity, the agreement between TES and GRS suggests that the compositional differences are not entirely due to surficial alteration (see also Section 24.2.2). Though thin alteration surfaces probably contribute to the measured assemblage, this alteration must be controlled by the underlying primary mineralogy. Global differences in mineral assemblage likely reflect large scale processes related to igneous crustal evolution, regional sedimentation, and glacial/periglacial activity (Rogers & Hamilton, 2015).

24.2.1.3 Martian Meteorite-Like Terrains

A long-standing objective in Mars remote sensing has been to identify terrains on the surface that are spectrally similar to martian meteorites so that these valuable martian samples can be placed into geologic context. Hamilton et al. (2003) searched the TES dataset for meteorite-like signatures on Mars. Basaltic lithologies represented by Zagami and Los Angeles were not identified with confidence, consistent with many martian dark regions being dominated by basalts having plagioclase > pyroxene (e.g., Bandfield et al., 2000; Rogers & Christensen, 2007), whereas shergottites have mineralogies with pyroxene > plagioclase (e.g., McSween, 2002 and references therein). It is also likely that the relatively young shergottites are derived from Amazonian-aged surfaces covered by dust. Local-scale (100s–1000s km^2) concentrations of olivine- and orthopyroxene-bearing materials such as ALH A77005, Chassigny, and ALH 84001 were found. Olivine-bearing materials are so common on the martian surface that identifying one or more candidate source regions is particularly challenging. Pyroxene-rich materials resembling Nakhla were identified near the detection limit throughout the eastern Valles Marineris region and portions of Syrtis Major. Importantly, martian meteorite-like TIR spectra represent only a minor portion of the dust-free surface of Mars and do not appear to be representative of the bulk composition of the ancient crust. The more recent discovery of the ancient polymict breccia Northwest Africa 7034 (e.g., Agee et al., 2013; Humayun et al., 2013) revealed new lithologies and TIR analyses of these lithologies may provide better matches to TIR spectra of dust-free, Noachian surfaces.

24.2.2 Comparison with Elemental/Chemical Techniques

The chemistries of phases identified in TES spectra can be used in combination with their derived abundances to calculate the effective bulk chemistry, thus enabling a direct comparison to data obtained from elemental/chemical techniques. Early conversions of individual TES "surface type 1" (ST1) and "surface type 2" (ST2) mineralogies to chemical data were presented by (McSween, 2002; McSween et al., 2003, 2004, 2006), and the first global data conversion was given by McSween et al. (2009). Early analyses suggested TES data are consistent with basaltic andesitic, calc-alkaline compositions that are typically distinct from the fields encompassed by the other datasets; this result was interpreted as indicating chemical alteration of the TES-measured materials as opposed to crustal recycling. Modeling of the global TES dataset by Koeppen and Hamilton (2008) used a refined library that had been tailored to the diversity of olivine-bearing, igneous materials based on more recent examinations of local-scale surface mineralogy. Hamilton and Rogers (2011) used the resulting mineralogic data to produce a new analysis of the global chemistry and found that if all possible weathering phases are excluded, the field of TES analyses shifts considerably toward basaltic, tholeiitic compositions that are more consistent with the other data sets (Figure 24.2) and represent variations in and/or the evolution of igneous compositions over time (see also Section 24.2.1).

24.2.3 Clues to Mars' History from the Distribution of Olivine-Rich Rocks

Complementing lower resolution TES data, THEMIS data have been used to identify local-scale, olivine-enriched materials over large sections of the planet's surface (Figure 24.3; Christensen et al., 2003; Hamilton & Christensen, 2005; Rogers et al., 2005; Edwards et al., 2008; Koeppen & Hamilton, 2008; Rogers et al., 2009; Rogers & Fergason, 2011; Edwards et al., 2014; Hanna et al., 2016). By comparing olivine content and composition with the elevation of outcrops in Ganges and Eos Chasmata, Edwards et al. (2008) and Edwards and

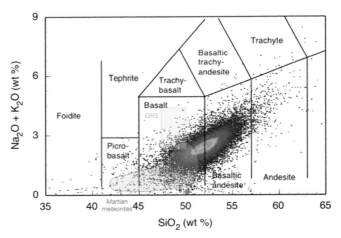

Figure 24.2 Silica versus alkalis igneous classification diagram (after LeBas et al., 1986) showing the chemistries of martian meteorites and Mars as measured by the Gamma-Ray Spectrometer (modified from McSween et al., 2009). The colored cloud of points represents TES-derived chemistries calculated by Hamilton and Rogers (2011) from data in Rogers and Hamilton (2015).

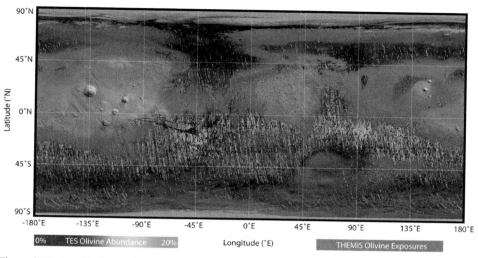

Figure 24.3 Distribution and abundance of olivine from TES overlaid on albedo map with detections from THEMIS shown in red. Data are from Koeppen and Hamilton (2008) and Bandfield et al. (2011).

Christensen (2011) determined that these outcrops were likely the exposures of a larger (~100 000 km^3), laterally extensive unit of olivine-enriched basalt that is representative of early volcanic flows on Mars, potentially associated with the initiation of Tharsis volcanism in the mid-Noachian. This unit represents a significant eruptive volume of lava (~100 000 km^3) and is potentially the largest continuous compositional unit discovered on Mars.

Some deeply infilled craters on Mars have been identified as a key host for isolated olivine-enriched exposures (McDowell & Hamilton, 2007; Edwards et al., 2014). THEMIS and TES data commonly show these materials to be enriched in olivine and pyroxene relative to the surrounding units (Edwards et al., 2014). Formation of these more mafic, high thermal inertia materials in the bottom of crater floors remains enigmatic but two prevailing hypotheses for their formation exist, including impact-induced decompression melting (Edwards et al., 2014) and/or the presence of an early, thin lithosphere that served as a lid for preexisting melt, with impact-generated fractures serving as a conduit for the melt to reach the surface. Any formation mechanism for these olivine-enriched crater floor materials must address the (1) nearly global distribution, as they are observed in nearly all low-albedo regions up to moderate latitudes in the ancient cratered highlands; (2) the heavy degree of infilling; (3) the rocky nature of the infilling material; and (4) the unique composition compared to surrounding regolith. Two prevailing hypotheses for their formation exist. The first is an impact-induced decompression melting hypothesis (Edwards et al., 2014) where the removal of overlying crustal material deforms the martian lithosphere/asthenosphere boundary sufficiently to generate extra melt that can be mobilized to the surface via impact-generated fractures. However, these infilled craters are generally <100 km in diameter, and the deformation of the boundary is unlikely to be sufficient to generate significant melt, unless the crust was thin and the asthenosphere was hot (Edwards et al., 2014). Alternatively, it is possible that an early, thin lithosphere served as a lid for preexisting melt in the hot asthenosphere. Impact-generated fractures would then serve as a conduit for this preexisting melt to reach the surface, resulting in concentration of olivine-bearing lavas on crater floors.

Additional isolated exposures in the martian highlands (Hamilton & Christensen, 2005; Bandfield et al., 2011; Edwards & Ehlmann, 2015) have been altered variably to contain carbonate, likely through a low-temperature in situ carbonation process that drew ~tens of mbars of CO_2 out of the early atmosphere. Additional olivine-enriched plains units in the southern highlands are observed and have been mapped in detail (e.g., Rogers et al., 2009; Rogers & Fergason, 2011; Rogers & Nazarian, 2013) having been interpreted as volcanic in origin. Many of the units are associated with large-impact basins, including Hellas (Rogers & Nazarian, 2013) and Isidis (Bandfield et al., 2011), suggesting these large basin-forming impacts and their pervasive fracture systems provided the means to erupt these materials onto the surface.

24.2.4 Salt Minerals: Carbonates and Chlorides

Carbonates have been identified in martian high albedo (>0.20) regions using TES data (Bandfield et al., 2003). Because of their highly contrasting absorption coefficients, small amounts (as little as 0.5% by weight) of carbonates can be detected in silicate-dominated fine particulates. Carbonates appear to be ubiquitous at low levels in the dust and bright soils, though it has been suggested these spectral features are instead due to the presence of hydrous iron sulfates (Lane et al., 2004). Carbonates have since been identified using a variety of orbital and in situ measurements (Boynton et al., 2009; Ehlmann et al., 2009; Palomba et al., 2009; Morris et al., 2010; Leshin et al., 2013; Glotch & Rogers, 2013; Wray et al., 2016; Ruff et al., this volume and Section 24.2.3).

A key discovery from THEMIS, supported by TES, was the detection of extensive chloride salt deposits throughout the ancient highlands (Osterloo et al., 2008, 2010). These deposits lack emissivity features over the THEMIS spectral region, as well as in the visible to near-infrared (VNIR), which suggests that they are anhydrous (Murchie et al., 2009; Glotch et al., 2010). These deposits are observed primarily in local lows and erosional windows (Osterloo et al., 2010), which is generally consistent with erosion down to an older unit in what is already an ancient terrain. THEMIS-based estimates put chloride abundances at 10–25% (Glotch et al., 2016). Although the chloride deposits are typically small (1–15 km), their widespread distribution throughout the southern highlands (Figure 24.4) suggests an early, global-scale formation process (Osterloo et al., 2010), such as evaporation following ponding of surface brines in playa/lake environments, or groundwater upwelling (Osterloo et al., 2010; Glotch et al., 2016). An outstanding problem for modeling chloride deposit formation is the general lack of associated carbonate and sulfate minerals, which are predicted to precipitate along with chlorides from late-stage brines (Tosca & McLennan, 2006). Brine fractionation, low

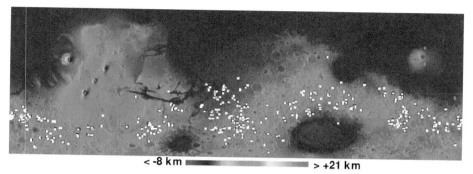

Figure 24.4 Global distribution of chloride-bearing deposits (white squares) overlaid on MOLA topography (from Osterloo et al., 2010).

abundances, or burial of the carbonate and sulfate materials could explain their nondetection, but this is a problem that must be explained.

24.2.5 Igneous Diversity

The martian crust appears to be dominated by basaltic compositions (Christensen et al., 2000; Rogers and Christensen, 2007; Rogers & Hamilton, 2015) (Section 24.2.1) with picritic basalts (Hoefen et al., 2003; Hamilton & Christensen, 2005; Edwards et al., 2008) (Section 24.2.3), providing an example of the development of a planetary crust very different from that of Earth and Earth's moon. A relatively SiO_2-enriched unit concentrated in the northern lowlands could have an igneous origin, or it may be the result of aqueous alteration (Section 24.2.7). The variety of igneous compositions known to exist on Mars indicates a planet with a complex set of igneous processes, though one without a well-understood paradigm.

TES and THEMIS data have been used to identify local exposures with a variety of igneous compositions (Christensen et al., 2005). Lava flows in the Nili Patera caldera of Syrtis major show spectral signatures consistent with a glass-rich dacite. Other sites in Syrtis Major (and Noachis Terra) show feldspar spectral features in NIR datasets (Carter & Poulet, 2013; Wray et al., 2013) and are possibly composed primarily of Ca-rich pyroxene and coarse feldspar (Rogers & Nekvasil, 2015).

Exposures composed primarily of plagioclase and quartz are present in northwest Syrtis Major and extend into Antoniadi crater (Bandfield et al., 2004; Christensen et al., 2005; Bandfield, 2006). The presence of hydrated silica indicates that at least some alteration has occurred (Ehlmann et al., 2009; Smith & Bandfield, 2012), but the bulk composition indicates the presence of a highly enriched source rock. The quartz-bearing surfaces are exposed within windows to the preexisting Noachian crust over an area covering 230 × 125 km (Bandfield, 2006). This suggests the possibility of extensive regions of SiO_2-enriched igneous compositions on Mars, possibly through partial melting of basaltic materials.

24.2.6 High-Silica Aqueous Phases

TES and THEMIS data show evidence for localized exposures of high-silica phases, indicating the presence of these materials in significant concentrations. In the few cases where exposures are large enough (km-scale), TES data have been used to identify the specific mineralogy (Bandfield et al., 2004; Bandfield, 2008) but high SiO_2 contents can also be readily identified using THEMIS data (Smith et al., 2013; Amador & Bandfield, 2016).

The only known occurrence of quartz on Mars to date (Figure 24.5; Section 24.2.6; Bandfield et al., 2004; Bandfield, 2006) has spatially coincident hydrated silica and nearby phyllosilicates and zeolites (determined from NIR measurements), indicating the presence of aqueous alteration (Ehlmann et al., 2009; Smith & Bandfield, 2012). By contrast, surfaces along the base of western Hellas Basin show evidence for the presence of up to 80 vol. % poorly crystalline high-silica materials sporadically exposed along a 650-km stretch (Bandfield, 2008; Bandfield et al., 2013). Surfaces with similar spectra have been identified in Mawrth Vallis, Nilosyrtis Mensae and Nili Fossae (Figure 24.5; Michalski & Fergason, 2009; Rogers & Bandfield, 2009; Amador & Bandfield, 2016; Bandfield & Amador, 2016). In these cases, phyllosilicates and other aqueous compositions are also commonly present locally. The elevated abundances of high-silica materials indicate that high water to rock ratios were present in these regions, but without elevated temperatures or for durations necessary for quartz diagenesis (Tosca & Knoll, 2009). Amorphous and poorly crystalline silica-rich phases appear to be common on Mars, based

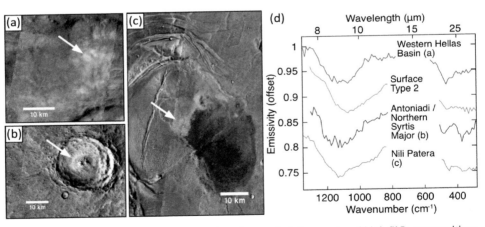

Figure 24.5 TES spectra of Mars. (Left) TES surface emissivity spectra of high SiO_2 compositions, including poorly crystalline silica surfaces in western Hellas Basin (green), global surface type 2 (cyan) composed primarily of plagioclase and an amorphous silica phase, quartz and plagioclase surfaces in Syrtis Major and Antoniadi crater (purple), and a dacitic composition in Nili Patera (orange). (Right) THEMIS bands 8–7–5 decorrelation stretch images show the locations of the high SiO_2 surfaces (arrows), which are yellow in the images.

on both TES/THEMIS and other spacecraft observations (e.g., Bandfield, 2002; Milliken et al., 2008; Ruff et al., 2011; Rice et al., 2013; Smith et al., 2013), although ambiguity remains in the interpretation of some of these phases (e.g., Kraft et al., 2003; Michalski et al., 2005).

24.2.7 The Distribution of Dusty Surfaces and Detection of Water

TIR spectra are sensitive to particle size (Hapke, 1981), transitioning from surface scattering-dominated at coarse particle sizes to volume scattering-dominated at finer size fractions (e.g., Vincent & Hunt, 1968; Salisbury et al., 1991) (see also Chapter 3). Empirically, the transition between these regimes typically occurs at particle sizes of roughly 65–100 μm in silicates, where below this size volume scattering results in a significant reduction in emissivity in the interband regions that can complicate compositional interpretation. Ruff and Christensen (2002) took advantage of this characteristic to define a martian Dust Cover Index (DCI) that is based on the average emissivity of TES spectra from 1400 to 1350 cm^{-1} (~7.14 to 7.41 μm). The DCI complements studies of the physical character of surfaces that also use albedo, thermal inertia, and geomorphology (Ruff et al., 2001; Ruff & Christensen, 2002; Lang et al., 2009).

Volume scattering also results in peaks in the emissivity spectrum where absorption coefficients are high (see also Section 24.2.5 and Chapter 3). A peak at ~1630 cm^{-1} in TES spectra of martian dust is likely attributable to the fundamental bending mode of H_2O (Ruff, 2004), consistent with observations of the fundamental stretching mode at higher wavenumbers (e.g., Audouard et al., 2014 and references therein). The origin of this feature is not well constrained but has been proposed by Ruff (2004) as possible evidence for zeolite in the dust, or could result from adsorbed water on small particles (e.g., Bishop et al., 1994; Yen et al., 1998). Further analysis is needed to definitively assign this specific feature to a mineral phase(s), adsorbed water, or both, although there is evidence from other data sets that H_2O is present in mineral or amorphous phases (e.g., Leshin et al., 2013; Meslin et al., 2013; Audouard et al., 2014).

24.3 Summary

TIR data collected by the Thermal Emission Spectrometer (TES) and Thermal Emission Imaging System (THEMIS) instruments have significantly impacted the understanding of martian surface mineralogy. Spatial/temporal variations in igneous lithologies have been identified, as well as laterally extensive olivine-enriched basalts representing what may be the largest continuous compositional unit on Mars. Quartz, carbonate, and chloride minerals, along with amorphous, silica-enriched phases have been identified in low-albedo regions. Based on these data, it is evident that Mars has experienced a diversity of primary and secondary geologic processes including igneous crustal evolution, regional sedimentation, aqueous alteration, and glacial/periglacial activity.

Acknowledgments

Our co-author Josh Bandfield passed away unexpectedly in June 2019. He was not just a colleague but also a dear friend and he will be missed greatly by all of us. His contributions to planetary infrared spectroscopy and thermophysics have been significant (as evidenced in this chapter and others) and his untimely passing also is a great loss to our community.

References

Agee C.B., Wilson N.V., McCubbin F.M., et al. (2013) Unique meteorite from early Amazonian Mars: Water-rich basaltic breccia Northwest Africa 7034. *Science*, **339**, 780–785.

Amador E.S. & Bandfield J.L. (2016) Elevated bulk-silica exposures and evidence for multiple aqueous alteration episodes in Nili Fossae, Mars. *Icarus*, **276**, 39–51.

Arvidson R.E. (1974) Wind-blown streaks, splotches, and associated craters on Mars: Statistical analysis of Mariner 9 photographs. *Icarus*, **21**, 12–27.

Audouard J., Poulet F., Vincendon M., et al. (2014) Water in the martian regolith from OMEGA/Mars Express. *Journal of Geophysical Research*, **119**, 1969–1989.

Baldridge A.M., Lane M.D., & Edwards C.S. (2013) Searching at the right time of day: Evidence for aqueous minerals in Columbus crater with TES and THEMIS data. *Journal of Geophysical Research*, **118**, 179–189.

Bandfield J.L. (2002) Global mineral distributions on Mars. *Journal of Geophysical Research*, **107**, DOI:10.1029/2001JE001510.

Bandfield J.L. (2006) Extended surface exposures of granitoid compositions in Syrtis Major, Mars. *Geophysical Research Letters*, **33**, DOI:L0620310.1029/2005GL025559.

Bandfield J.L. (2008) High-silica deposits of an aqueous origin in western Hellas Basin, Mars. *Geophysical Research Letters*, **35**, DOI:L1220510.1029/2008GL033807.

Bandfield J.L. & Amador E.S. (2016) Extensive aqueous deposits at the base of the dichotomy boundary in Nilosyrtis Mensae, Mars. *Icarus*, **275**, 29–44.

Bandfield J.L., Christensen P.R., & Smith M.D. (2000a) Spectral data set factor analysis and end-member recovery: Application to analysis of martian atmospheric particulates. *Journal of Geophysical Research*, **105**, 9573–9587.

Bandfield J.L., Hamilton V.E., & Christensen P.R. (2000b) A global view of martian surface compositions from MGS-TES. *Science*, **287**, 1626–1630.

Bandfield J.L., Edgett K.S., & Christensen P.R. (2002) Spectroscopic study of the Moses Lake dune field, Washington: Determination of compositional distributions and source lithologies. *Journal of Geophysical Research*, **107**, 5092, DOI:5010.1029/2000JE001469.

Bandfield J.L., Glotch T.D., & Christensen P.R. (2003) Spectroscopic identification of carbonate minerals in the martian dust. *Science*, **301**, 1084–1087.

Bandfield J.L., Hamilton V.E., Christensen P.R., & McSween H.Y. (2004) Identification of quartzofeldspathic materials on Mars. *Journal of Geophysical Research*, **109**, DOI:E1000910.1029/2004JE002290.

Bandfield J.L., Rogers A.D., & Edwards C.S. (2011) The role of aqueous alteration in the formation of martian soils. *Icarus*, **211**, 157–171.

Bandfield J.L., Amador E.S., & Thomas N.H. (2013) Extensive hydrated silica materials in western Hellas Basin, Mars. *Icarus*, **226**, 1489–1498.

Bish D.L., Blake D., Vaniman D., et al. (2013) X-ray diffraction results from Mars Science Laboratory: Mineralogy of Rocknest at Gale crater. *Science*, **341**, DOI:123893210.1126/science.1238932.

Bishop J.L., Pieters C.M., & Edwards J.O. (1994) Infrared spectroscopic analyses on the nature of water in montmorillonite. *Clays and Clay Minerals*, **42**, 702–716.

Blake D.F., Morris R.V., Kocurek G., et al. (2013) Curiosity at Gale crater, Mars: Characterization and analysis of the Rocknest sand shadow. *Science*, **341**, 1239505.

Boynton W.V., Taylor G.J., Evans L.G., et al. (2007) Concentration of H, Si, Cl, K, Fe, and Th in the low-and mid-latitude regions of Mars. *Journal of Geophysical Research*, **112**, DOI:10.1029/2007JE002887.

Boynton W.V., Ming D.W., Kounaves S.P., et al. (2009) Evidence for calcium carbonate at the Mars Phoenix landing site. *Science*, **325**, 61–64.

Carter J. & Poulet F. (2013) Ancient plutonic processes on Mars inferred from the detection of possible anorthositic terrains. *Nature Geoscience*, **6**, 1008–1012.

Christensen P.R. (1983) Eolian intracrater deposits on Mars: Physical properties and global distribution. *Icarus*, **56**, 496–518.

Christensen P.R., Bandfield L., Hamilton V.E., et al. (1992) Thermal Emission Spectrometer experiment: Mars Observer mission. *Journal of Geophysical Research*, **97**, 7719–7734.

Christensen P.R., Bandfield J.L., Smith M.D., Hamilton V.E., & Clark R.N. (2000) Identification of a basaltic component on the martian surface from Thermal Emission Spectrometer data. *Journal of Geophysical Research*, **105**, 9609–9621.

Christensen P.R., Morris R.V., Lane M.D., Bandfield J.L., & Malin M.C. (2001a) Global mapping of martian hematite mineral deposits: Remnants of water-driven processes on early Mars. *Journal of Geophysical Research*, **106**, 23873–23885.

Christensen P.R., Bandfield J.L., Hamilton V.E., et al. (2001b) Mars Global Surveyor Thermal Emission Spectrometer experiment: Investigation description and surface science results. *Journal of Geophysical Research*, **106**, 23823–23871.

Christensen P.R., Bandfield J.L., Bell J.F. III (2003) Morphology and composition of the surface of Mars: Mars Odyssey THEMIS results. *Science*, **300**, 2056–2061.

Christensen P.R., Jakosky B.M., Kieffer H.H., et al. (2004) The thermal emission imaging system (THEMIS) for the Mars 2001 Odyssey Mission. *Space Science Reviews*, **110**, 85–130.

Christensen P.R., McSween H.Y. Jr., Bandfield J.L., et al. (2005) Evidence for magmatic evolution and diversity on Mars from infrared observations. *Nature*, **436**, 504–509.

Edwards C.S. & Christensen P.R. (2011) Evidence for a widespread olivine-rich layer on Mars: Identification of a global impact ejecta deposit? *42nd Lunar Planet. Sci. Conf.*, Abstract #2560.

Edwards C.S. & Ehlmann B.L. (2015) Carbon sequestration on Mars. *Geology*, **43**, 863–866.

Edwards C.S. & Piqueux S. (2016) The water content of recurring slope lineae on Mars. *Geophysical Research Letters*, **43**, 8912–8919.

Edwards C.S., Christensen P., & Hamilton V. (2008) Evidence for extensive olivine-rich basalt bedrock outcrops in Ganges and Eos chasmas, Mars. *Journal of Geophysical Research*, **113**, E11003, DOI:10.1029/2008je003091.

Edwards C.S., Bandfield J.L., Christensen P.R., & Fergason R.L. (2009) Global distribution of bedrock exposures on Mars using THEMIS high-resolution thermal inertia. *Journal of Geophysical Research*, **114**, E11001, DOI:10.1029/2009JE003363.

Edwards C.S., Bandfield J.L., Christensen P.R., & Rogers A.D. (2014) The formation of infilled craters on Mars: Evidence for widespread impact induced decompression of the early martian mantle? *Icarus*, **228**, 149–166.

Ehlmann B.L. & Edwards C.S. (2014) Mineralogy of the martian surface. *Annual Review of Earth and Planetary Sciences*, **42**, 291–315.

Ehlmann B.L., Mustard J.F., Murchie S.L., et al. (2008) Orbital identification of carbonate-bearing rocks on Mars. *Science*, **322**, 1828–1832.

Ehlmann B.L., Mustard J.F., Swayze G.A., et al. (2009) Identification of hydrated silicate minerals on Mars using MRO-CRISM: Geologic context near Nili Fossae and implications for aqueous alteration. *Journal of Geophysical Research*, **114**, DOI:E00D0810.1029/2009JE003339.

Gillespie A.R. (1992) Enhancement of multispectral thermal infrared images: Decorrelation contrast stretching. *Remote Sensing of Environment*, **42**, 147–155.

Gillespie A.R., Kahle A.B., & Walker R.E. (1986) Color enhancement of highly correlated images. I. Decorrelation and HSI contrast stretches. *Remote Sensing of Environment*, **20**, 209–235.

Glotch T.D. & Bandfield J.L. (2006) Determination and interpretation of surface and atmospheric Miniature Thermal Emission Spectrometer spectral end-members at the Meridiani Planum landing site. *Journal of Geophysical Research*, **111**, E12S06, DOI:10.1029/2005JE002671.

Glotch T.D. & Rogers A.D. (2013) Evidence for magma-carbonate interaction beneath Syrtis Major, Mars. *Journal of Geophysical Research*, **118**, 126–137.

Glotch T.D., Bandfield J.L., Tornabene L.L., Jensen H.B., & Seelos F.P. (2010) Distribution and formation of chlorides and phyllosilicates in Terra Sirenum, Mars. *Geophysical Research Letters*, **37**, DOI:10.1029/2010GL044557.

Glotch T.D., Bandfield J.L., Wolff M.J., Arnold J.A., & Che C. (2016) Constraints on the composition and particle size of chloride salt-bearing deposits on Mars. *Journal of Geophysical Research*, **121**, 454–471.

Gooding J.L. (1992) Soil mineralogy and chemistry on Mars: Possible clues from salts and clays in SNC meteorites. *Icarus*, **99**, 28–41.

Hamilton V.E. & Christensen P.R. (2005) Evidence for extensive olivine-rich bedrock in Nili Fossae, Mars. *Geology*, **33**, 433–436.

Hamilton V.E. & Rogers A.D. (2011) A new view of martian surface geochemistry. *42nd Lunar Planet. Sci. Conf.*, Abstract #1273.

Hamilton V.E. & Ruff S.W. (2012) Distribution and characteristics of Adirondack-class basalt as observed by Mini-TES in Gusev crater, Mars and its possible volcanic source. *Icarus*, **218**, 917–949.

Hamilton V.E., Christensen P.R., McSween H.Y. Jr., & Bandfield J.L. (2003) Searching for the source regions of martian meteorites using MGS TES: Integrating martian meteorites into the global distribution of igneous materials on Mars. *Meteoritics and Planetary Science*, **38**, 871–885.

Hanna R.D., Hamilton V.E., & Putzig N.E. (2016) The complex relationship between olivine abundance and thermal inertia on Mars. *Journal of Geophysical Research*, **121**, 1293–1320.

Hapke B. (1981) Bidirectional reflectance spectroscopy: 1. Theory. *Journal of Geophysical Research*, **86**, 3039–3054.

Hoefen T.M., Clark R.N., Bandfield J.L., Smith M.D., Pearl J.C., & Christensen P.R. (2003) Discovery of olivine in the Nili Fossae region of Mars. *Science*, **302**, 627–630.

Humayun M., Nemchin A., Zanda B., et al. (2013) Origin and age of the earliest martian crust from meteorite NWA 7533. *Nature*, **503**, 513–516.

Kahn R. (1985) The evolution of CO_2 on Mars. *Icarus*, **62**, 175–190.

Koeppen W.C. & Hamilton V.E. (2008) Global distribution, composition, and abundance of olivine on the surface of Mars from thermal infrared data. *Journal of Geophysical Research*, **113**, E05001, DOI:10.1029/2007JE002984.

Kraft M.D., Michalski J.R., & Sharp T.G. (2003) Effects of pure silica coatings on thermal emission spectra of basaltic rocks: Considerations for martian surface mineralogy. *Geophysical Research Letters*, **30**, DOI:228810.1029/2003GL018848.

Lane M.D. & Christensen P.R. (2013) Determining olivine composition of basaltic dunes in Gale crater, Mars, from orbit: Awaiting ground truth from Curiosity. *Geophysical Research Letters*, **40**, 3517–3521.

Lane M.D., Dyar M.D., & Bishop J.L. (2004) Spectroscopic evidence for hydrous iron sulfate in the martian soil. *Geophysical Research Letters*, **31**, L1970210.1029/2004GL021231.

Lane M.D., Bishop J.L., Darby Dyar M., King P.L., Parente M., & Hyde B.C. (2008) Mineralogy of the Paso Robles soils on Mars. *American Mineralogist*, **93**, 728–739.

Lang N.P., Tornabene L.L., McSween H.Y. Jr., & Christensen P.R. (2009) Tharsis-sourced relatively dust-free lavas and their possible relationship to martian meteorites. *Journal of Volcanology and Geothermal Research*, **185**, 103–115.

Leshin L.A., Mahaffy P.R., Webster C.R., et al. (2013) Volatile, isotope, and organic analysis of martian fines with the Mars Curiosity rover. *Science*, **341**, 1238937.

McDowell M.L. & Hamilton V.E. (2007) Geologic characteristics of relatively high thermal inertia intracrater deposits in southwestern Margaritifer Terra, Mars. *Journal of Geophysical Research*, **112**, E12001, DOI:10.1029/2007JE002925.

McEwen A.S., Dundas C.M., Mattson S.S., et al. (2014) Recurring slope lineae in equatorial regions of Mars. *Nature Geoscience*, **7**, 53–58.

McFadden L.A. & Cline T.P. (2005) Spectral reflectance of martian meteorites: Spectral signatures as a template for locating source region on Mars. *Meteoritics and Planetary Science*, **40**, 151–172.

McSween H.Y. Jr. (2002) The rocks of Mars, from far and near. *Meteoritics and Planetary Science*, **37**, 7–25.

McSween H.Y., Grove T.L., & Wyatt M.B. (2003) Constraints on the composition and petrogenesis of the martian crust. *Journal of Geophysical Research*, **108**, DOI:10.1029/2003JE002175.

McSween H.Y., Arvidson R.E., Bell J., et al. (2004) Basaltic rocks analyzed by the Spirit rover in Gusev crater. *Science*, **305**, 842–845.

McSween H.Y., Ruff S., Morris R., et al. (2006) Alkaline volcanic rocks from the Columbia Hills, Gusev crater, Mars. *Journal of Geophysical Research*, **111**, DOI:E09S9110.1029/2006JE002698.

McSween H.Y., Taylor G.J., & Wyatt M.B. (2009) Elemental composition of the martian crust. *Science*, **324**, 736–739.

Meslin P.-Y., Gasnault O., Forni O., et al. (2013) Soil diversity and hydration as observed by ChemCam at Gale crater, Mars. *Science*, **341**, 1238670.

Michalski J.R. & Fergason R.L. (2009) Composition and thermal inertia of the Mawrth Vallis region of Mars from TES and THEMIS data. *Icarus*, **199**, 25–48.

Michalski J.R., Kraft M.D., Sharp T.G., Williams L.B., & Christensen P.R. (2005) Mineralogical constraints on the high-silica martian surface component observed by TES. *Icarus*, **174**, 161–177.

Milam K.A., McSween H.Y., Moersch J., & Christensen P.R. (2010) Distribution and variation of plagioclase compositions on Mars. *Journal of Geophysical Research*, **115**, E09004, DOI:10.1029/2008JE003495.

Milliken R.E., Swayze G.A., Arvidson R.E., et al. (2008) Opaline silica in young deposits on Mars. *Geology*, **36**, 847–850.

Mitchell J.L. & Christensen P.R. (2016) Recurring slope lineae and chlorides on the surface of Mars. *Journal of Geophysical Research*, **121**, 1411–1428.

Morris R.V., Ruff S.W., Gellert R., et al. (2010) Identification of carbonate-rich outcrops on Mars by the Spirit rover. *Science*, **329**, 1189667.

Murchie S.L., Mustard J.F., Ehlmann B.L., et al. (2009) A synthesis of martian aqueous mineralogy after 1 Mars year of observations from the Mars Reconnaissance Orbiter. *Journal of Geophysical Research*, **114**, DOI:10.1029/2009JE003342.

Niles P.B., Catling D.C., Berger G., et al. (2013) Geochemistry of carbonates on Mars: Implications for climate history and nature of aqueous environments. *Space Science Reviews*, **174**, 301–328.

Osterloo M., Hamilton V., Bandfield J., et al. (2008) Chloride-bearing materials in the southern highlands of Mars. *Science*, **319**, 1651–1654.

Osterloo M.M., Anderson F.S., Hamilton V.E., & Hynek B.M. (2010) Geologic context of proposed chloride-bearing materials on Mars. *Journal of Geophysical Research*, **115**, E10012: DOI:10.1029/2010JE003613.

Palomba E., Zinzi A., Cloutis E.A., D'Amore M., Grassi D., & Maturilli A. (2009) Evidence for Mg-rich carbonates on Mars from a 3.9 μm absorption feature. *Icarus*, **203**, 58–65.

Ramsey M.S. & Christensen P.R. (1998) Mineral abundance determination: Quantitative deconvolution of thermal emission spectra. *Journal of Geophysical Research*, **103**, 577–596.

Rice M.S., Cloutis E.A., Bell J.F. III, et al. (2013) Reflectance spectra diversity of silica-rich materials: Sensitivity to environment and implications for detections on Mars. *Icarus*, **223**, 499–533.

Rogers A.D. & Christensen P.R. (2007) Surface mineralogy of martian low-albedo regions from MGS-TES data: Implications for upper crustal evolution and surface alteration. *Journal of Geophysical Research*, **112**, E01003, DOI:10.1029/2006JE002727.

Rogers A.D. & Aharonson O. (2008) Mineralogical composition of sands in Meridiani Planum determined from Mars Exploration rover data and comparison to orbital measurements. *Journal of Geophysical Research*, **113**, E06S14, DOI:10.1029/2007JE002995.

Rogers A.D. & Bandfield J.L. (2009) Mineralogical characterization of Mars Science Laboratory candidate landing sites from THEMIS and TES data. *Icarus*, **203**, 437–453.

Rogers A.D. & Fergason R.L. (2011) Regional-scale stratigraphy of surface units in Tyrrhena and Iapygia Terrae, Mars: Insights into highland crustal evolution and alteration history. *Journal of Geophysical Research*, **116**, E08005, DOI:10.1029/2010JE003772.

Rogers A.D. & Hamilton V.E. (2015) Compositional provinces of Mars from statistical analyses of TES, GRS, OMEGA and CRISM data. *Journal of Geophysical Research*, **120**, 62–91.

Rogers A.D. & Nazarian A.H. (2013) Evidence for Noachian flood volcanism in Noachis Terra, Mars, and the possible role of Hellas impact basin tectonics. *Journal of Geophysical Research*, **118**, 1094–1113.

Rogers A.D. & Nekvasil H. (2015) Feldspathic rocks on Mars: Compositional constraints from infrared spectroscopy and possible formation mechanisms. *Geophysical Research Letters*, **42**, 2619–2626.

Rogers A.D., Christensen P.R., & Bandfield J.L. (2005) Compositional heterogeneity of the ancient martian crust: Analysis of Ares Vallis bedrock with THEMIS and TES data. *Journal of Geophysical Research*, **110**, E05010, DOI:10.1029/2005JE002399.

Rogers A.D., Aharonson O., & Bandfield J.L. (2009) Geologic context of bedrock exposures in Mare Serpentis, Mars: Implications for crust and regolith evolution in the cratered highlands. *Icarus*, **200**, 446–462.

Ruff S.W. (2004) Spectral evidence for zeolite in the dust on Mars. *Icarus*, **168**, 131–143.

Ruff S.W. & Christensen P.R. (2002) Bright and dark regions on Mars: Particle size and mineralogical characteristics based on Thermal Emission Spectrometer data. *Journal of Geophysical Research*, **107**, 5127, DOI:5110.1029/2001JE001580.

Ruff S.W. & Christensen P.R. (2007) Basaltic andesite, altered basalt, and a TES-based search for smectite clay minerals on Mars. *Geophysical Research Letters*, **34**, L10204, DOI:10.1029/2007GL029602.

Ruff S.W. & Hamilton V.E. (2017) Wishstone to Watchtower: Amorphous alteration of plagioclase-rich rocks in Gusev crater, Mars. *American Mineralogist*, **102**, 235–251.

Ruff S.W., Christensen P.R., Clark R.N., et al. (2001) Mars' "White Rock" feature lacks evidence of an aqueous origin: Results from Mars Global Surveyor. *Journal of Geophysical Research*, **106**, 23,921–23,927.

Ruff S.W., Christensen P.R., Blaney D., et al. (2006) The rocks of Gusev crater as viewed by the Mini-TES instrument. *Journal of Geophysical Research*, **111**, DOI:E12S1810.1029/2006JE002747.

Ruff S.W., Farmer J.D., Calvin W.M., et al. (2011) Characteristics, distribution, origin, and significance of opaline silica observed by the Spirit rover in Gusev crater, Mars. *Journal of Geophysical Research*, **116**, DOI: E00F2310.1029/2010JE003767.

Salisbury J.W. (1991) *Infrared (2.1–25 μm) spectra of minerals*. The Johns Hopkins University Press, Baltimore, MD.

Salisbury J.W., D'Aria D.M., & Jarosewich E. (1991) Midinfrared (2.5–13.5 μm) reflectance spectra of powdered stony meteorites. *Icarus*, **92**, 280–297.

Smith M.D., Bandfield J.L., & Christensen P.R. (2000) Separation of atmospheric and surface spectral features in Mars Global Surveyor Thermal Emission Spectrometer (TES) spectra. *Journal of Geophysical Research*, **105**, 9589–9607.

Smith M.R. & Bandfield J.L. (2012) Geology of quartz and hydrated silica-bearing deposits near Antoniadi crater, Mars. *Journal of Geophysical Research*, **117**, DOI:E0600710.1029/2011JE004038.

Smith M.R., Bandfield J.L., Cloutis E.A., & Rice M.S. (2013) Hydrated silica on Mars: Combined analysis with near-infrared and thermal-infrared spectroscopy. *Icarus*, **223**, 633–648.

Squyres S.W., Grotzinger J.P., Arvidson R.E., et al. (2004) In situ evidence for an ancient aqueous environment at Meridiani Planum, Mars. *Science*, **306**, 1709–1714.

Squyres S.W., Arvidson R.E., Ruff S., et al. (2008) Detection of silica-rich deposits on Mars. *Science*, **320**, 1063–1067.

Tosca N.J. & Knoll A.H. (2009) Juvenile chemical sediments and the long term persistence of water at the surface of Mars. *Earth and Planetary Science Letters*, **286**, 379–386.

Tosca N.J. & McLennan S.M. (2006) Chemical divides and evaporite assemblages on Mars. *Earth and Planetary Science Letters*, **241**, 21–31.

Vincent R.K. & Hunt G.R. (1968) Infrared reflectance from mat surfaces. *Applied Optics*, **7**, 53–59.

Wray J.J., Hansen S.T., Dufek J., et al. (2013) Prolonged magmatic activity on Mars inferred from the detection of felsic rocks. *Nature Geoscience*, **6**, 1013–1017.

Wray J.J., Murchie S.L., Bishop J.L., et al. (2016) Orbital evidence for more widespread carbonate-bearing rocks on Mars. *Journal of Geophysical Research*, **121**, 652–677.

Yen A.S., Murray B.C., & Rossman G.R. (1998) Water content of the martian soil: Laboratory simulations of reflectance spectra. *Journal of Geophysical Research*, **103**, 11,125–11,133.

25

Thermal Infrared Remote Sensing of Mars from Rovers Using the Miniature Thermal Emission Spectrometer

STEVEN W. RUFF, JOSHUA L. BANDFIELD, PHILIP R. CHRISTENSEN,
TIMOTHY D. GLOTCH, VICTORIA E. HAMILTON, AND A. DEANNE ROGERS

25.1 Introduction to the Miniature Thermal Emission Spectrometer

As described by Christensen et al. (2003), the Miniature Thermal Emission Spectrometer (Mini-TES) instrument was designed and flown on the two Mars Exploration Rovers (MER) to provide information about the geology and atmosphere of Mars through measurements of naturally emitted thermal infrared radiance (TIR; 5–29 μm; 339.50–1997.06 cm^{-1}; 167 channels at 9.99 cm^{-1}/channel). Mini-TES served as a remote sensing tool for surveying the mineralogy of rocks and soils immediately surrounding the rover and out to the horizon (Squyres et al., 2003). These observations in some cases were used to direct the rover to interesting science targets for detailed study with the rovers' robotic-arm-mounted instruments and also to provide broader context for the sparse measurements from these instruments. Mini-TES also provided temperatures of the lower atmospheric boundary layer and gathered information about atmospheric dust optical depth and water vapor column abundance. We focus on surface mineralogy results in this chapter, so the reader is referred to the work of Smith et al. (2004, 2006) for results of the atmospheric measurements. Surface temperatures obtained from Mini-TES spectra were used to derive thermal inertia, from which particle size information of the regolith was determined. The work of Fergason et al. (2006) presents those results.

Mini-TES is a Fourier transform interferometer/spectrometer housed within the Warm Electronics Box on each rover. It viewed the ground and sky using a 6.35-cm-diameter Cassegrain telescope attached to the periscope-like Pancam Mast Assembly (PMA) (Figure 25.1). A rotating elevation scan mirror within the PMA head directed scene radiance first to a fixed fold mirror, then down to the primary and secondary mirrors of the telescope, and into the flat-plate Michelson interferometer and uncooled deuterated triglycine sulfate (DTGS) pyroelectric detector. Combined with the 360° azimuth rotation of the PMA, Mini-TES had views of the sky up to 30° above the nominal horizon and ~50° below, with a spatial resolution of 20 mrad or optionally, 8 mrad using an actuated field stop.

The majority of Mini-TES observations of the surface were made as targeted "stares" in which a rock or soil was observed for an extended period of time (dwell) to enhance the signal-to-noise ratio of the data. Each scan of the interferometer, taking just under 2 s, is

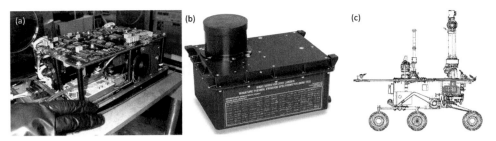

Figure 25.1 Images of the Mini-TES instrument (a) with cover off, (b) with cover on, and (c) its location on the rover shown diagrammatically in red.

called an incremental counter keeper (ICK) and produced a single spectrum. A typical targeted stare was 200 ICKs long and included blackbody calibration target observations acquired before and after the stare, requiring about 15 minutes total. Rastered observations also were made using the azimuth motions of the PMA and elevation motions by the Mini-TES moving mirror in the PMA head. Rasters typically involved many fewer ICKS per spot than a stare but covered more terrain. Key science results mostly came from targeted stares, as shown in this chapter.

Spirit's Mini-TES operated through sol 2174, after which power loss due to dust buildup on the rover's solar arrays and the onset of the martian winter while embedded in fine soil ended the mission (Arvidson et al., 2010). Opportunity's Mini-TES operated through sol 2243, after which degradation due to extreme thermal cycling precluded further use. An unanticipated power draw from a stuck heater on the rover's arm forced the use of a "deep sleep" mode to conserve power early in the mission, disabling the survival heater intended to maintain temperatures $> -50°C$.

25.2 Background: Calibration of Mini-TES Data

25.2.1 Standard Calibration

The output of an interferometric spectrometer such as Mini-TES is an interferogram, a time-varying voltage from the detector. This is converted to a signal as a function of wavenumber via Fourier transformation, then calibrated to radiance using a blackbody target at two different temperatures. Planck radiance at the highest brightness temperature is then divided out to produce the desired temperature-independent, unitless emissivity spectrum (e.g., Ruff et al., 1997). The two-point calibration of Mini-TES data was intended to use a solar-heated, rover-deck-mounted blackbody target and an unheated second blackbody target inside the PMA head. However, this scheme could not be used because of the failure of a temperature sensor on the deck-mounted blackbody target of each rover after the first night on the surface. Instead, a single-point scheme was developed and demonstrated to be sufficient for the Mini-TES science objectives (Christensen et al., 2004a,b).

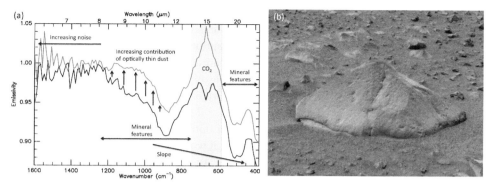

Figure 25.2 Characteristics of Mini-TES spectra. (a) Two Adirondack Class basalt spectra display a combination of features due to mineral components plus those due to atmosphere, surface dust, temperature inhomogeneity, and instrument noise as a function of target radiance. Red spectrum is from sol 55, sequence P3174, and black spectrum is from sol 41, sequence P3110. (b) The type example Adirondack basalt is shown in this approximate true color (ATC) Pancam view (Savransky & Bell, 2004; Bell et al., 2006) from sol 14 (P2542).

25.2.2 Spectral Characteristics

Every Mini-TES observation includes features due to atmospheric CO_2 centered on 667 cm^{-1} (15 μm). These features are displayed as either peaks or troughs depending on whether the temperature of the atmosphere along the Mini-TES line of sight was warmer or cooler than the surface target, respectively (Figure 25.2; Ruff et al., 2006). Components with different temperatures in the Mini-TES field of view (FOV; minimum diameter of ~10 cm), for example, sunlit and shadowed facets of a rock, or rock and soil in the same view, led to a continuum slope in the final emissivity spectrum (Figure 25.2). This can be accounted for using spectral modeling (Christensen et al., 2003) as described in Section 25.3.

Dust accumulation on Mini-TES optical surfaces was minimized by rotating the scan mirror into a closed position inside the PMA head. However, both instruments experienced dust accumulation over time scales many times greater than the originally planned 90 sols (one martian sol is ~24.6 Earth hours). For Spirit's Mini-TES, mirror-dust contamination occurred abruptly on sol 420 of the mission following the onset of dust devil activity (Ruff et al., 2006; Smith et al., 2006), and then again beginning with global dust storm activity on about sol 1220 (Ruff et al., 2011). A correction was first developed and implemented for use with Mini-TES atmospheric observations (Smith et al., 2006) and then validated for surface observations from Spirit (Ruff et al., 2011), but not after sol 1220 due to additional dust accumulation on multiple mirror surfaces. Opportunity's Mini-TES mirror slowly accumulated dust beginning early in the mission and then accelerated around sol 325 when atmospheric dust activity suddenly increased (Smith et al., 2006). The mirror-dust correction became ineffective sometime after sol 350 (Glotch & Bandfield, 2006), although measurements were still acquired through sol 2243.

An unanticipated spectral artifact occurred in conditions where optically thin airfall dust was out of thermal equilibrium with the underlying rock or soil substrate. As described by Hamilton and Ruff (2012), the resulting spectral character is similar to that of atmospheric dust and mirror-dust contamination. When the dust is warmer than the substrate, it contributes radiance most recognizable as a convex-up shape spanning ~900 to 1200 cm^{-1} (Figure 25.2). This feature is inverted when the dust is cooler than the substrate (Rivera-Hernandez et al., 2015). Because the mineralogy of the dust is essentially that of atmospheric dust, and because both optically thin surface dust and atmospheric dust have very similar spectral properties, it was found that a Mini-TES spectrum of the sky could be used as a spectral endmember to account for the contribution from optically thin surface dust (Glotch & Bandfield, 2006; Hamilton & Ruff, 2012).

25.3 Methods: Spectral Corrections and Analyses

Mini-TES emissivity spectra of rocks and coarse particulate soil (>~40 μm), like those from other TIR instruments, can be modeled using a library of spectral endmembers and a linear least squares algorithm to identify individual components and their areal abundance (e.g., Ramsey & Christensen, 1998; Rogers & Aharonson, 2008). Because the contributions of spectral slope and optically thin surface dust described earlier generally can be modeled as linear additions, these spectra can be added to the library (e.g., Christensen et al., 2004a,b; Glotch et al., 2006; Ruff et al., 2006; Rogers & Aharonson, 2008; Hamilton & Ruff, 2012). Mirror-dust artifacts must be removed prior to applying linear spectral mixture analysis.

The use of an iterative strategy for modeling of Mini-TES spectra is a common approach wherein different combinations of endmembers are applied, and the lowest-error models are deemed the most robust. This approach is used because the number of spectral end members is algebraically limited to one less than the number of data points in the spectrum to be modeled (Mini-TES spectra have 167 data points). In practice, this number is less than the algebraic limit, and for Mini-TES spectra, libraries with <60 spectra are typical. By varying the different mineral spectra in a library and evaluating the resulting spectral fits both visually and quantitatively with the root-mean-square error output from the algorithm, it is possible to converge on a more robust solution than with a single library. In addition, because Mini-TES spectra are in some cases accompanied by mineralogic and geochemical results from the rovers' Mössbauer spectrometer (MB) and Alpha Particle X-ray Spectrometer (APXS) respectively, it is possible to refine the iterative results to obtain better agreement with these other datasets (Glotch et al., 2006; Rogers & Aharonson, 2008).

Another approach to the analysis of Mini-TES spectra is known as Factor Analysis and Target Transformation (FATT), which is applied to a set of spectra (usually 10s to 100s) to identify spectral endmembers that represent the range of variation within the dataset. The methodology was originally applied to TIR spectra from the TES instrument (Bandfield et al., 2000) but has been used to good effect for identifying spectral classes and individual components among Mini-TES spectra (Glotch & Bandfield, 2006; Hamilton & Ruff, 2012;

Ruff & Hamilton, 2017). More detailed descriptions of the FATT technique and spectral modeling are presented in Chapter 15.

25.4 Results

In the following sections we highlight the mineralogic results obtained with Mini-TES observations from each of the two rovers. The Spirit Rover was sent to Gusev crater to test the hypothesis that it once hosted a lake (Squyres et al., 2004b). Opportunity explored Meridiani Planum to understand the origin of hematite deposits (Squyres et al., 2004a) that were identified years before with spectra from the orbiting TES instrument (Christensen et al., 2000).

25.4.1 Results from Spirit in Gusev Crater

25.4.1.1 Adirondack Class Basalt

On the floor of Gusev crater, Spirit encountered eroded basalt flows rather than sedimentary rocks indicative of a lake. Dubbed Adirondack Class after the name of the first rock investigated using the rover's arm instruments, these rocks are dominated by spectral features attributable to olivine (Figure 25.2), consistent with their chemical classification as picritic basalt (e.g., Christensen et al., 2004b; McSween et al., 2004). They have varying amounts of a loose dust coating that clearly resembles the globally homogenous dust on Mars, which is distributed by local, regional, and global atmospheric processes. Although Adirondack Class basalt is the dominant rock type of the Gusev plains, rare examples were encountered in the petrologically diverse Columbia Hills ~2 km east of the Spirit landing site. Here they are anomalous and thus were interpreted as exotic rocks delivered by meteorite impacts onto the plains (Grant et al., 2006; Ruff et al., 2006). A total of 104 unique rock targets observed with Mini-TES were recognized as Adirondack Class basalt, having an average olivine composition of ~Fo_{45} as determined with linear least squares modeling, and olivine abundance that varies independently of a basaltic matrix as determined using the FATT technique (Hamilton & Ruff, 2012).

25.4.1.2 Wishstone–Watchtower Alteration

The Columbia Hills are noteworthy for their diversity of igneous and volcaniclastic rocks spanning a range of composition, texture, and degree of alteration (e.g., Arvidson et al., 2008). The widespread plagioclase-rich clastic rocks of the Wishstone Class are the relatively unaltered endmembers of a suite of rocks that displays progressively more intense alteration, culminating in the Watchtower Class with its enrichments in Mg, Zn, S, Br, and Cl (e.g., Hurowitz et al., 2006). The MB instrument, which measures Fe-bearing mineral phases, revealed increasing values of nanophase ferric oxide (npOx), ferric to total iron ratio (Fe^{3+}/Fe_T), and mineralogic alteration index (MAI; the sum of npOx, hematite, goethite, and sulfate abundance) (Morris et al., 2008). Mini-TES observed nearly 200

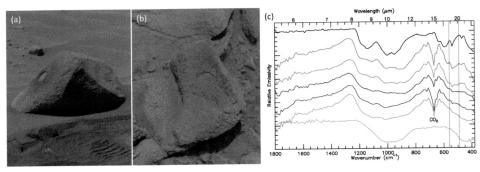

Figure 25.3 Alteration endmembers Wishstone and Watchtower and Mini-TES spectra that span the range of alteration. (a) Pancam ATC images of Wishstone (sol 342, P2571 subframe) and (b) Watchtower (sol 419, P2574 subframe) show ~4 cm diameter RAT grind. (c) Wishstone Class is rich in plagioclase, as shown by the laboratory spectrum of labradorite in black and vertical lines highlighting its prominent features. These features transition to those more similar to basaltic glass (laboratory spectrum in cyan) in Watchtower Class, an indication of alteration rather than a primary component.

rocks that span the range of alteration, with three intermediate subclasses represented by 13 or more examples in each case (Ruff & Hamilton, 2017). The plagioclase-rich character of Wishstone Class is clearly evident in Mini-TES spectra, in which features attributable to plagioclase composition in the range of ~An_{30} to ~An_{65} are present. These features are progressively diminished with increasing alteration intensity (Figure 25.3), which correlates with the increasing Fe^{3+}/Fe_T ratio and npOx component obtained from MB measurements.

A Mini-TES spectral component resembling basaltic glass accompanies the increasing alteration, akin to that described for the coating on the Adirondack Class rock called Mazatzal (Ruff & Hamilton, 2017). As in that case, the appearance and modeling of a basaltic glass component are interpreted as a proxy for one or more amorphous phases that are not present in available spectral libraries. One possible explanation for the apparent amorphous phase(s) is a water-limited alteration process in the extreme aridity and cold of Mars that is sufficient to depolymerize silicate tetrahedral networks but insufficient to allow significant cation mobility (Ruff & Hamilton, 2017). The apparent absence of any phyllosilicate or opaline silica phases among the alteration products supports the concept of limited cation mobility and is perhaps consistent with a form of "cation conservative" alteration as has been described for other rocks in Gusev crater (Hurowitz & Fischer, 2014).

25.4.1.3 Comanche Carbonate-Rich Outcrops

The identification of carbonate in the Columbia Hills outcrops dubbed Comanche and Comanche Spur was made nearly four years after Spirit observed them in late 2005. At that time, the Mini-TES mirror-dust correction had not been implemented, so the spectra were difficult to interpret. Following the correction, Mini-TES spectra combined with elemental data from APXS and spectral data from MB identified 16–34 wt.% Mg-Fe carbonate within these outcrops, along with Mg-rich olivine, and an amorphous silicate component (Morris et al., 2010). The proximity to the Home Plate hydrothermal system (described later) and the Mg-rich composition of the carbonate led to a hydrothermal origin hypothesis.

A study by Ruff et al. (2014) incorporated additional observations and modeling, resulting in an alternative hypothesis for the origin of Comanche carbonate involving an ephemeral lake. The nearby Algonquin outcrops show the same olivine features in Mini-TES spectra and a texture consistent with volcanic tephra like that of Comanche outcrops, indicating that Algonquin-type rocks may be the host of the carbonate. Modeling of a Mini-TES spectrum from Comanche clearly shows an Algonquin spectral component and also demonstrates that multiple carbonate phases likely are present rather than just a single Mg-Fe phase (Figure 25.4). Geochemical modeling using APXS data of Algonquin also supports the idea that rocks with its composition could generate fluids, through water-limited leaching, with a composition that could produce carbonates like those of Comanche. Furthermore, the morphology and thermal characteristics of Algonquin Class outcrops were recognized elsewhere in the Columbia Hills and beyond using orbital data, suggestive of a formerly more extensive deposit of Algonquin-like tephra. A hypothesis incorporating these results was proposed in which evaporation of an ephemeral lake, perhaps recharged by multiple floods, led to the occurrence and significant abundance of carbonates in Comanche (Ruff et al., 2014).

25.4.1.4 Home Plate Opaline Silica

Spirit observed multiple outcrops and an isolated occurrence of soil rich in silica (65–92 wt. % SiO_2) adjacent to Home Plate (Squyres et al., 2008). The silica phase is opal-A (hydrated

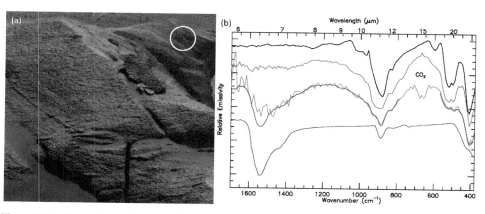

Figure 25.4 Comanche carbonate-rich outcrop and spectral analysis. (a) A portion of the Comanche outcrop is seen in this cropped Pancam scene (ATC) from sol 695 (sequence P2422). Circle represents ~15 cm diameter Mini-TES FOV. (b) Measured Mini-TES Comanche spectrum in green (sol 701, P3340) and modeled (blue) with a combination of olivine-rich Algonquin Class average (red; olivine shown in black) and a mixture of Mg and Fe carbonates (purple).

Figure 25.5 Spectral characteristics and morphology of Home Plate silica outcrops. (a) Mini-TES spectra of Home Plate silica (black, sol1168, P3968; green, sol 1116, P3857) strongly resemble spectra of halite-encrusted silica sinter from El Tatio, Chile (red and blue). (b) An example of Home Plate nodular silica from which the black Mini-TES spectrum in (a) was acquired; circle represents ~14 cm diameter FOV. Pancam ATC view spans ~50 cm (sol 1174, P2588). (c) Halite-encrusted nodular silica sinter in a hot spring discharge channel with microbial mats at El Tatio resembles Home Plate nodular silica at the same scale as in (b).

amorphous SiO_2) as determined with spectra from Mini-TES. Abundant evidence supports a hydrothermal origin for the silica, with important implications for astrobiology (Squyres et al., 2008). Hydrothermal systems produce silica both through acid leaching by fumarolic vapors and chemical sedimentary processes via hot springs and geysers. Stratigraphic, morphologic, textural, and spectral features of Home Plate silica occurrences favor the latter process, based on field analog studies and supporting laboratory investigations (Ruff et al., 2011; Ruff & Farmer, 2016).

The distinctive nodular and digitate morphology and spectral characteristics of the Home Plate silica outcrops are strikingly similar to those aspects of silica sinter deposits produced in active hot spring and geyser discharge channels found at El Tatio, Chile (Figure 25.5) (Ruff & Farmer, 2016). The high elevation (~4300 m) silica sinter deposits of El Tatio at the edge of Chile's Atacama Desert provide the best spectral and morphologic match yet to Home Plate silica, likely due to more Mars-like conditions compared with lower and wetter hydrothermal sites on Earth. The digitate silica nodules at El Tatio are complex sedimentary structures produced from a combination of biotic and abiotic processes. Aeolian erosion or other fully abiotic processes have not been ruled out for the martian silica structures, but because biotic processes cannot be ruled out, they have been described as potential biosignatures (Ruff & Farmer, 2016).

25.4.2 Results from Opportunity at Meridiani Planum

25.4.2.1 TES-Observed Components

Crystalline hematite (α-Fe_2O_3) covering thousands of km^2 of Meridiani Planum was identified from orbit using TES spectra, based on a set of features below ~600 cm^{-1} (Christensen et al., 2000). Also present in these spectra are the features of basaltic material and dust. The Opportunity Rover instruments confirmed each of these spectral components, providing ground truth for the TES spectra. From the surface, hematite was found to occur in spherical masses a few mm in diameter, colloquially referred to as blueberries (Squyres et al., 2004a). The spherules were interpreted as sedimentary concretions formed by precipitation from groundwater within a sulfate-rich host rock (Burns Formation described in the text that follows) and then eroded from this softer rock, forming an extensive lag deposit mixed with basaltic sand and dust. These are the components recognized in TES spectra from orbit, with the exception of the poorly exposed sulfate host rock.

Spectrally, the hematite composing the spherules was recognized as having arisen from the transformation of goethite, an aqueous pathway for the formation of hematite (Glotch & Bandfield, 2006), confirming earlier work based on the comparison of TES spectra of Meridiani Planum with laboratory spectra (Glotch et al., 2004). As shown in Figure 25.6, a ratio of spectra from hematite-rich versus hematite-poor surfaces produces well-resolved, goethite-derived hematite spectral features both from orbit-acquired and surface-acquired observations.

The basaltic component in TES spectra, when isolated from the hematite component, is notably similar to the basaltic sand of Meridiani Planum observed with Mini-TES (Yen et al., 2005). Combining complementary compositional datasets from Opportunity (APXS and MB), the mineralogy of the sand was determined to be ~10–15% olivine with about twice as much Fe-rich pyroxene, probably in the form of pigeonite. Plagioclase feldspar at 20–25% is the other dominant component, with sulfate and amorphous phases each accounting for ~10% of the sand, and other phases below accepted detection limits composing the remainder (Rogers & Aharonson, 2008). The agreement between these results and those from TES spectra lends credibility to mineralogic results derived from TES spectra elsewhere on Mars.

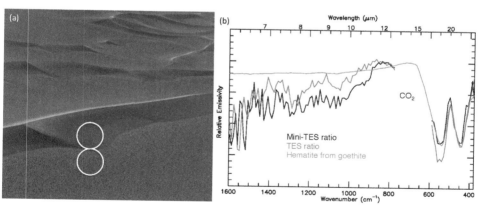

Figure 25.6 Spectral characteristics of Meridiani Planum hematite. (a) Cropped Pancam ATC mosaic from sol 211 (P2424) with Mini-TES FOVs (circles ~15 cm diameter). (b) A ratio of Mini-TES spectra (black) from a hematite-rich surface between aeolian bedforms in Endurance crater (lower circle) to a hematite-poor bedform (upper circle) is well matched by a ratio of orbital TES spectra (red) from hematite-rich Meridiani Planum versus a hematite-poor terrain to the south. Both resemble a laboratory spectrum of hematite produced from dehydroxylation of goethite (green). Mini-TES spectra are from sol 195 (P3834). TES spectra are from OCK 5411 and ICKs 1709-1711 and 1716-1717.

25.4.2.2 Burns Formation

The sulfate-rich sedimentary rock that hosts the hematite spherules is known as the Burns formation. Prior to the arrival of Opportunity, these outcrops were not recognized from orbital observations because of their limited exposures, mostly around heavily eroded impact craters. Opportunity fortuitously arrived in a small (~20 m) crater dubbed Eagle in which the finely layered outcrops were exposed and investigated with the full payload of instruments. The presence of abundant S was documented with the APXS (up to 25 wt.%), which is contained in S-bearing minerals including the Fe sulfate mineral jarosite, as identified with the MB instrument (Squyres et al., 2004a). Combined with the morphologic and textural features observed with the imaging instruments, it was suggested that these rocks resulted from a combination of aeolian and aqueous sedimentary processes in shallow ponds or lakes with repeated cycles of wetting and drying associated with a rising and falling groundwater table (Grotzinger et al., 2005; McLennan et al., 2005).

Mini-TES observations of Burns formation outcrops typically include a mixture of rock, hematite spherules, basaltic sand, and dust (Christensen et al., 2004b). However, use of the FATT technique allows the spectral character of the rock to be isolated, enhancing the spectral contrast and interpretability (Glotch & Bandfield, 2006) (Figure 25.7). By combining the complementary datasets from APXS and MB with iterative spectral modeling, a refined bulk mineralogy of Burns formation outcrop was identified (Glotch et al., 2006). It is composed primarily of Al-rich opaline silica; Mg-, Ca-, and Fe-bearing sulfates; plagioclase feldspar; nontronite; and hematite, with one or more hydrated phases evident from a feature in Mini-TES spectra of rock tailings produced by the Rock Abrasion Tool. Although these modeled results provide the best fits to the available data, some of the components are more robust than others. For example, the nontronite identification is not as

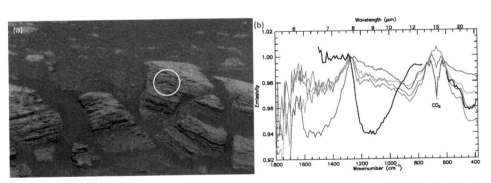

Figure 25.7 Example of Burns formation outcrop and representative Mini-TES spectra. (a) A portion of a Pancam ATC mosaic from sol 37 (P2392) shows layered Burns formation outcrop. (b) Mini-TES spectra of Burns formation include orange spectrum (sol 39, P3201) from white circle (~15 cm) in **a** and one acquired from the grind tailings (blue; sol 36, P3185) that displays a prominent peak near 1650 cm^{-1} due to molecular water bound in one or more minerals, and the FATT-derived version of Burns formation (black) that provides a clearer look at the spectral characteristics without the various contributions from surface dust.

robust as the amorphous silica component, which itself could include different combinations of silica phases (Glotch et al., 2006).

25.4.2.3 Meteorites

An abundance of meteorites was discovered on the plains of Meridiani Planum, beginning with a large cobble near the spacecraft's jettisoned heat shield. Dubbed Heat Shield Rock, it was identified as an iron meteorite using the full complement of rover instruments and ultimately renamed Meridiani Planum in recognition of its official status as an IAB-complex meteorite on Mars (Connolly et al., 2006; Schröder et al., 2008). The initial indication of its metallic Fe composition came from Mini-TES observations that showed notably high-contrast spectral features comparable to those in upward-looking Mini-TES spectra of the sky, i.e., strong features due to the emission of dust and CO_2 in the atmosphere (Figure 25.8). Such features evident in the spectra of a rock indicated that it was reflecting the downwelling radiance of the sky, consistent with a low emissivity metallic composition (Ruff et al., 2008). Although other meteorites were encountered by Opportunity, including additional irons (Ashley et al., 2011), these encounters came later in the mission when Mini-TES mirror-dust contamination was sufficient to render the spectra uninterpretable. However, the Spirit rover observed two adjacent cobbles on Low Ridge near Home Plate that had essentially the same morphology and Mini-TES spectral character as that of Heat Shield Rock (Figure 25.8). Although these rocks were not investigated by the arm-mounted instruments, their spectral and morphologic characteristics strongly suggest that they also are iron meteorites (Schröder et al., 2008).

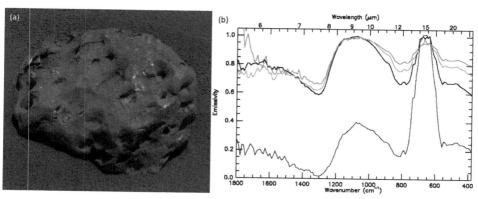

Figure 25.8 Confirmed and likely iron meteorites on Mars. (a) Heat Shield Rock, an iron meteorite formally renamed Meridiani Planum, is shown in this Pancam ATC image from sol 346 (P2591). (b) The Mini-TES spectra of Heat Shield Rock (black; sol342, P3929) and two other likely iron meteorites (red, sol 858, P3448; and green, sol 1029, P3443) near Home Plate in Gusev crater, resemble the spectrum of an upward-looking observation of the sky (blue; sol 342, P3577) indicative of their TIR reflective character.

25.5 Summary

Mini-TES proved to be a robust scientific instrument capable of providing a range of observations that greatly enhanced the science return from the Spirit and Opportunity rovers. As a reconnaissance tool, Mini-TES made >50 observations across the plains of Gusev crater that documented the nearly homogeneous occurrence of Adirondack Class basalt. Rare examples of Adirondack Class were readily identified exclusively with Mini-TES once Spirit arrived in the Columbia Hills, revealing the role of impact processes in redistributing rocks on Mars. A perhaps unique-to-Mars style of alteration among the Wishstone–Watchtower suite of rocks on Husband Hill was recognized with the benefit of hundreds of observations that greatly expanded the sparse coverage by the rover's arm-mounted instruments. In contrast, largely unaltered volcanic igneous rocks also were documented, led by observations from Mini-TES in the case of pyroxene-rich Backstay and Irvine Classes. Magmatic evolution is apparent among olivine-rich outcrops of Algonquin Class volcanic tephra in which spectral changes in olivine features associated with changes in Mg–Fe composition are evident in Mini-TES spectra. Among these tephra deposits, Mini-TES provided unambiguous evidence of a substantial carbonate component that perhaps is the only indication observed by Spirit of an ancient lake in Gusev crater. The identification of opaline silica deposits adjacent to Home Plate was made possible with Mini-TES spectra, revealing a hydrothermal system that produced outcrops of rocks with textures consistent with hot spring sinter and morphology that suggests the possibility of microbial mediation.

Opportunity was sent to Meridiani Planum to investigate the possibility that an ancient aqueous environment produced widespread deposits of the hematite identified with TES from orbit. All of the spectral components recognized from orbit were confirmed on the ground with the full instrument payload, including Mini-TES, thus providing the first ground truth of orbit-obtained martian mineralogy. An ancient aqueous environment was also confirmed, although spanning a range of conditions and processes not originally hypothesized. The hematite was found in mm-scale spherules eroding from sulfate-rich sandstones, likely produced from Fe-rich solutions percolating as groundwater through the rocks. Spectral evidence recognized in TES data and confirmed on the ground with Mini-TES showed that goethite likely was the original iron phase that transformed to hematite following dehydroxylation. The Burns formation bedrock, largely obscured by basaltic sand and a hematite lag deposit, is basaltic sandstone significantly altered to sulfate phases and, as shown by Mini-TES spectra, amorphous silica phases. Other significant discoveries include meteorites at both landing sites.

Acknowledgments

The untimely death of our co-author and friend Josh Bandfield in June 2019 came as a great shock to all who knew him. He accomplished much in his nearly 45 years on Earth, making substantial contributions to our understanding of worlds beyond it. Josh was the first to recognize features in TES spectra attributable to hematite in Meridiani Planum, Mars that ultimately led to exploration by the Opportunity rover, which is documented in part in this chapter. His diligence, conscientiousness, and scientific integrity produced a body of work that will endure, as will the memories of him that we now hold dear.

References

Arvidson R.E., Ruff S.W., Morris R.V., et al. (2008) Spirit Mars rover mission to the Columbia Hills, Gusev crater: Mission overview and selected results from the Cumberland Ridge to Home Plate. *Journal of Geophysical Research*, **113**, DOI:10.1029/2008JE003183.

Arvidson R.E., Bell J.F. III, Bellutta P., et al. (2010) Spirit Mars rover mission: Overview and selected results from the northern Home Plate winter haven to the side of Scamander crater. *Journal of Geophysical Research*, **115**, DOI:10.1029/2008JE003183.

Ashley J.W., Golombek M.P., Christensen P.R., et al. (2011) Evidence for mechanical and chemical alteration of iron-nickel meteorites on Mars: Process insights for Meridiani Planum. *Journal of Geophysical Research*, **116**, DOI:10.1029/2010JE003672.

Bandfield J.L., Christensen P.R., & Smith M.D. (2000) Spectral data set factor analysis and end-member recovery: Application to analysis of martian atmospheric particulates. *Journal of Geophysical Research*, **105**, 9573–9587.

Bell J.F. III, Joseph J., Sohl-Dickstein J.N., et al. (2006) In-flight calibration and performance of the Mars Exploration Rover Panoramic Camera (Pancam) instruments. *Journal of Geophysical Research*, **111**, DOI:10.1029/2005JE002444.

Christensen P.R., Bandfield J.L., Clark R.N., et al. (2000) Detection of crystalline hematite mineralization on Mars by the Thermal Emission Spectrometer: Evidence for near-surface water. *Journal of Geophysical Research*, **105**, 9623–9642.

Christensen P.R., Mehall G.L., Silverman S.H., et al. (2003) Miniature Thermal Emission Spectrometer for the Mars Exploration Rovers. *Journal of Geophysical Research*, **108**, DOI:10.1029/2003JE002117.

Christensen P.R., Wyatt M.B., Glotch T.D., et al. (2004a) Mineralogy at Meridiani Planum from the Mini-TES experiment on the Opportunity rover. *Science*, **306**, 1733–1739.

Christensen P.R., Ruff S.W., Fergason R.L., et al. (2004b) Initial results from the Mini-TES experiment in Gusev crater from the Spirit rover. *Science*, **305**, 837–842.

Connolly H.C.J., Zipfel J., Grossman J.N., et al. (2006) The Meteoritical Bulletin No. 90. *Meteoritics and Planetary Science*, **41**, 1383–1418.

Fergason R.L., Christensen P.R., Bell J.F. III, Golombek M.P., Herkenhoff K.E., & Kieffer H.H. (2006) Physical properties of the Mars Exploration Rover landing sites as inferred from Mini-TES–derived thermal inertia. *Journal of Geophysical Research*, **111**, DOI:10.1029/2005JE002583.

Glotch T.D. & Bandfield J.L. (2006) Determination and interpretation of surface and atmospheric Mini-TES spectral end-members at the Meridiani Planum landing site. *Journal of Geophysical Research*, **111**, DOI:10.1029/2005JE002671.

Glotch T.D., Morris R.V., Christensen P.R., & Sharp T.G. (2004) Effect of precursor mineralogy on the thermal infrared emission spectra of hematite: Application to martian hematite mineralization. *Journal of Geophysical Research*, **109**, DOI:10.1029/2003JE002224.

Glotch T.D., Bandfield J.L., Christensen P.R., et al. (2006) Mineralogy of the light-toned outcrop at Meridiani Planum as seen by the Miniature Thermal Emission Spectrometer and implications for its formation. *Journal of Geophysical Research*, **111**, DOI:10.1029/2005JE002672.

Grant J.A., Wilson S.A., Ruff S.W., Golombek M.P., & Koestler D.L. (2006) Distribution of rocks on the Gusev Plains and on Husband Hill, Mars. *Geophysical Research Letters*, **33**, DOI:10.1029/2006GL026964.

Grotzinger J.P., Arvidson R.E., Bell J.F. III, et al. (2005) Stratigraphy and sedimentology of a dry to wet eolian depositional system, Burns formation, Meridiani Planum, Mars. *Earth and Planetary Science Letters*, **240**, 11–72.

Hamilton V.E. & Ruff S.W. (2012) Distribution and characteristics of Adirondack-class basalt as observed by Mini-TES in Gusev crater, Mars and its possible volcanic source. *Icarus*, **218**, 917–949.

Hurowitz J.A. & Fischer W.W. (2014) Contrasting styles of water–rock interaction at the Mars Exploration Rover landing sites. *Geochimica et Cosmochimica Acta*, **127**, 25–38.

Hurowitz J.A., McLennan S.M., McSween H.Y. Jr., DeSouza P.A. Jr., & Klingelhöfer G. (2006) Mixing relationships and the effects of secondary alteration in the Wishstone and Watchtower Classes of Husband Hill, Gusev crater, Mars. *Journal of Geophysical Research*, **111**, DOI:10.1029/2006JE002795.

McLennan S.M., Bell J.F. III, Calvin W.M., et al. (2005) Provenance and diagenesis of the evaporate-bearing Burns formation, Meridiani Planum, Mars. *Earth and Planetary Science Letters*, **240**, 95–121.

McSween H.Y., Jr., Arvidson R.E., Bell J.F. III, et al. (2004) Basaltic rocks analyzed by the Spirit rover in Gusev crater. *Science*, **305**, 842–845.

Morris R.V., Klingelhofer G., Schroder C., et al. (2008) Iron mineralogy and aqueous alteration from Husband Hill through Home Plate at Gusev crater, Mars: Results from the Mössbauer instrument on the Spirit Mars Exploration Rover. *Journal of Geophysical Research*, **113**, DOI:10.1029/2008JE003201.

Morris R.V., Ruff S.W., Gellert R., et al. (2010) Identification of carbonate-rich outcrops on Mars by the Spirit rover. *Science*, **329**, 421–424.

Ramsey M.S. & Christensen P.R. (1998) Mineral abundance determination: Quantitative deconvolution of thermal emission spectra. *Journal of Geophysical Research*, **103**, 577–596.

Rivera-Hernandez F., Bandfield J.L., Ruff S.W., & Wolff M.J. (2015) Characterizing the thermal infrared spectral effects of optically thin surface dust: Implications for remote-sensing and in situ measurements of the martian surface. *Icarus*, **262**, 173–186.

Rogers A.D. & Aharonson O. (2008) Mineralogical composition of sands in Meridiani Planum determined from Mars Exploration Rover data and comparison to orbital measurements. *Journal of Geophysical Research*, **113**, DOI:10.1029/2007JE002995.

Ruff S.W. & Farmer J.D. (2016) Silica deposits on Mars with features resembling hot spring biosignatures at El Tatio in Chile. *Nature Communications*, **7**, 13554.

Ruff S.W. & Hamilton V.E. (2017) Wishstone to Watchtower: Amorphous alteration of plagioclase-rich rocks in Gusev crater, Mars. *American Mineralogist*, **102**, 235–251.

Ruff S.W., Christensen P.R., Barbera P.W., & Anderson D.L. (1997) Quantitative thermal emission spectroscopy of minerals: A laboratory technique for measurement and calibration. *Journal of Geophysical Research*, **102**, 14,899–14,913.

Ruff S.W., Christensen P.R., Blaney D.L., et al. (2006) The rocks of Gusev crater as viewed by the Mini-TES instrument. *Journal of Geophysical Research*, **111**, DOI:10.1029/2006JE002747.

Ruff S.W., Christensen P.R., Glotch T.D., Blaney D.L., Moersch J.E., & Wyatt M.B. (2008) The mineralogy of Gusev crater and Meridiani Planum derived from the Miniature Thermal Emission Spectrometers on the Spirit and Opportunity rovers. In: *The martian surface: Composition, mineralogy, and physical properties* (J. Bell, ed.). Cambridge University Press, Cambridge, 315–338.

Ruff S.W., Farmer J.D., Calvin W.M., et al. (2011) Characteristics, distribution, origin, and significance of opaline silica observed by the Spirit rover in Gusev crater. *Journal of Geophysical Research*, **116**, DOI:10.1029/2010JE003767.

Ruff S.W., Niles P.B., Alfano F., & Clarke A.B. (2014) Evidence for a Noachian-aged ephemeral lake in Gusev crater, Mars. *Geology*, **42**, 359–362.

Savransky D. & Bell J.F. III (2004) True color and chromaticity of the martian surface and sky from Mars Exploration Rover Pancam observations. *Eos, Transactions American Geophysical Union*, Abstract P21A-0197.

Schröder C., Rodionov D.S., McCoy T.J., et al. (2008) Meteorites on Mars observed by the Mars Exploration Rovers. *Journal of Geophysical Research*, DOI:10.1029/2007JE002990.

Smith M.D., Wolff M.J., Lemmon M.T., et al. (2004) First atmospheric science results from the Mars Exploration Rovers Mini-TES. *Science*, **306**, 1750–1753.

Smith M.D., Wolff M.J., Spanovich N., et al. (2006) One martian year of atmospheric observations using MER Mini-TES. *Journal of Geophysical Research*, **111**, DOI:10.1029/2006JE002770.

Squyres S.W., Arvidson R.E., Baumgartner E.T., et al. (2003) Athena Mars rover science investigation. *Journal of Geophysical Research*, **108**, DOI:10.1029/2003JE002121.

Squyres S.W., Arvidson R.E., Bell J.F. III, et al. (2004a) The Opportunity rover's Athena science investigation at Meridiani Planum, Mars. *Science*, **306**, 1698–1703.

Squyres S.W., Arvidson R.E., Bell J.F. III, et al. (2004b) The Spirit rover's Athena science investigation at Gusev crater, Mars. *Science*, **305**, 794–799.

Squyres S.W., Arvidson R.E., Ruff S.W., et al. (2008) Detection of silica-rich deposits on Mars. *Science*, **320**, 1063–1067.

Yen A.S., Gellert R., Schroder C., et al. (2005) An integrated view of the chemistry and mineralogy of martian soils. *Nature*, **436**, 49–54.

26

Compositional and Mineralogic Analyses of Mars Using Multispectral Imaging on the Mars Exploration Rover, Phoenix, and Mars Science Laboratory Missions

JAMES F. BELL III, WILLIAM H. FARRAND, JEFFREY R. JOHNSON, KJARTAN M. KINCH, MARK LEMMON, MARIO C. PARENTE, MELISSA S. RICE, AND DANIKA WELLINGTON

26.1 Introduction and Background

Imaging is a fundamental part of terrestrial and planetary exploration, enabling basic reconnaissance of the geologic properties and context of surfaces and of the meteorologic properties and temporal variability of atmospheres. The ability to acquire spectrally resolved (color) images further enhances the capability of remote sensing observations to provide additional information on composition, mineralogy, and/or physical properties, while still being able to use spatial contiguity as a guide to interpretation of surface/atmospheric processes and evolution. Constraints on imaging cadence, resolution, and/or collected/transmitted data volume force trade-offs, however, between the amount of spatial versus spectral coverage that can be achieved by any particular sensor or mission. At one extreme vertex of this trade space might be a high spectral resolution spectrometer that can observe only a small field of view ("point spectrometer"), while at another extreme might be a high spatial resolution 2-D imager that can acquire data in hundreds to thousands of wavelengths ("imaging spectrometer"). Intermediate modalities include instruments that opt to maintain high spatial resolution imaging capability but sample data in only a small number (typically 3 to ~12) of wavelengths; these are commonly known as "multispectral imaging" systems.

Multispectral visible to near-infrared (VNIR; ~400–1000 nm) imaging with silicon-based charge-coupled devices (CCDs) has played a major role in the exploration of Mars, both from orbital sensors (e.g., Chapter 23) as well as from camera systems on surface landers and rovers. The first modern 2-D digital imaging systems deployed on the surface of Mars were carried by the NASA Mars Pathfinder lander and rover in 1997 (Matijevic et al., 1997; Smith et al., 1997), and multispectral imaging results from that mission were reviewed by Farrand et al. (2008a).

Based on the success of that mission and its multispectral imaging capabilities, expanded sensors with similar multispectral sampling capabilities were deployed on the NASA Mars Exploration Rovers Spirit and Opportunity Panoramic Camera (Pancam) investigation beginning in 2004 (Bell et al., 2003). Initial Pancam multispectral imaging results were summarized by Bell et al. (2008), though that review did not capture the full extent of the results from either the Spirit (which returned data for more than 6 years) or Opportunity (which returned data for more than 14 years) missions.

A similar sensor and multispectral imaging capability were used for the NASA Phoenix Mars lander mission's Surface Stereo Imager (SSI; Lemmon et al., 2008), augmented by red, green, and blue (RGB) color imaging capability on the Phoenix Robotic Arm Camera (RAC; Keller et al., 2008) and Optical Microscope (OM; Hecht et al., 2008); those results are summarized here. Also summarized here are the initial results from the further-enhanced multispectral imaging capabilities deployed on the NASA Mars Science Laboratory Curiosity rover's Mast Camera (Mastcam) investigation (Bell et al., 2017; Malin et al., 2017) beginning in 2012, which are also augmented in terms of color capabilities by RGB imaging from the Curiosity Mars Hand Lens Imager (MAHLI; Edgett et al., 2012) and Mars Descent Imager (MARDI; Malin et al., 2017) investigations. Many more details about these and other Mars multispectral imaging systems can be found in the review by Gunn and Cousins (2016).

26.2 Methodology

The multispectral imaging investigations focused on here have used filter wheels to provide measurements at multiple diagnostic wavelengths (summarized in Table 26.1). Mars surface imaging investigations to date have included the goal of obtaining so-called "true color" imaging at or near the same RGB wavelengths as the human visual system (e.g., Bell et al., 2006). Prior to the Mars Science Laboratory mission, RGB images from Mars had been created by downlinking and merging separate images acquired RGB filters. Mastcam (as well as the Curiosity MAHLI and MARDI imagers) acquires RGB imaging using Bayer pattern filters similar to those used in commercial digital cameras; a JPEG color image is created within the camera electronics and then downlinked (e.g., Bell et al., 2017).

Most of the other filters used on Mars surface multispectral imaging investigations are narrower-band filters at wavelengths chosen to sample key segments of the 400–1100 nm spectrum, based on either previous telescopic and/or orbital studies of the reflectance properties of Mars, or the laboratory reflectance properties of a suite of minerals that were known or suspected to occur there. For example, telescopic and early Mars orbital reflectance spectra (such as those from the Hubble Space Telescope and the Phobos 2 ISM and Mars Express OMEGA orbital investigations; Bell & Ansty, 2007; Murchie et al., 2000; Bibring et al., 2005) revealed solid evidence for a variety of ferric (Fe^{3+}) oxides and ferrous (Fe^{2+}) silicates in VNIR spectra of Mars, as well as speculative evidence on the presence of hydrates, sulfates, clays, and carbonates (see Chapters 1, 2, 4, and 5, and the review by Bell, 1996). Figure 26.1 shows examples of reflectance spectra of some of these kinds of materials over the wavelength regions sampled by Mars surface multispectral imaging systems. One of the goals of narrowband filters flown on these systems has been sampling of these various absorption features, reflectance maxima, and spectral slope variations that could maximize the potential to map the spatial distribution of potential variations in composition, mineralogy, and/or physical properties (e.g., Bell et al., 2000;

Table 26.1 *Multispectral imaging wavelengths for Mars surface color imaging investigations*[a]

Viking Landers[b]	MPF IMP[c]	MER Pancam[d]	Phoenix SSI[e]	MSL Mastcam[f]
	443 ± 26	432 ± 32	447 ± 23	445 ± 20
480 ± 85	480 ± 27	482 ± 30	485 ± 21	495 ± 76[g]
	531 ± 30	535 ± 20	532 ± 28	527 ± 14
550 ± 60				554 ± 78[g]
	600 ± 21	601 ± 17	604 ± 15	
				640 ± 88[g]
675 ± 125	671 ± 20	673 ± 16	673 ± 19	676 ± 20
	752 ± 19	753 ± 20	754 ± 21	751 ± 20
	802 ± 21	803 ± 20	802 ± 22	805 ± 20
			833 ± 28	
875 ± 95	858 ± 34	864 ± 17	864 ± 37	867 ± 20
	898 ± 41	904 ± 26	900 ± 45	908 ± 22
	931 ± 27	934 ± 25	931 ± 25	937 ± 22
950 ± 70	967 ± 30		969 ± 30	
990 ± 95	1003 ± 29	1009 ± 38	1002 ± 30	1012 ± 42

[a] Bandwidth reported here is effective band center ± FWHM. All values are in nm. Shaded cells indicate stereo imaging wavelengths. Panchromatic filters, solar filters, or other special purpose narrowband filters not used for routine multispectral science are not listed here (see Gunn & Cousins, 2016 for complete lists).

[b] Huck et al. (1977); [c]Smith et al. (1997); [d]Bell et al. (2003).

[e] Lemmon et al. (2008). In addition, the Phoenix mission's RAC imaging system acquired three-color RGB multispectral data by illuminating the scene with red (634 nm), green (524 nm), or blue (464 nm) LEDs and imaging onto a panchromatic CCD (Keller et al., 2008). The Phoenix OM imaging system operated similarly, but acquired four-color data by also illuminating the scene with a UV (375 nm) LED (Hecht et al., 2008).

[f] Bell et al. (2017).

[g] These Bayer pattern RGB wavelengths were also used for three-color surface multispectral imaging on the MSL MAHLI (Edgett et al., 2012) and MARDI (Malin et al., 2017) imaging systems.

Farrand et al., 2006). Additional details and examples of common modern multispectral image analysis methods were described in Chapter 14.

Finally, multispectral imaging filters that span the maximum practical wavelength sensitivity of the system's photodetector can also be used to infer certain physical properties of surface materials or atmospheric aerosols (dust, water ice) using observations made

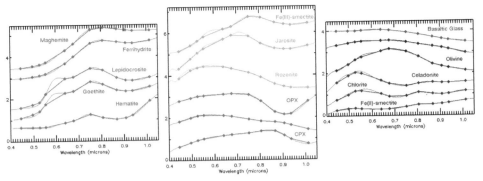

Figure 26.1 Examples of 400–1100 nm (0.4–1.1 μm) reflectance spectra of laboratory mineral samples representing the kinds of minerals that have previously been detected or inferred to exist on Mars. Solid lines are full resolution spectra, and the data points represent an example of a way to sample the full-resolution spectrum with 12 unique bandpasses designed to maximize the detectability of these kinds of phases. In this case, the data points show the multispectral sampling for the Mastcam multispectral imaging system on NASA's Mars Science Laboratory (Curiosity) rover (Table 26.1). Mineral classes are grouped by color. Red spectra: iron oxides/oxyhydroxides. Yellow spectra: sulfates. Green spectra: ferrous pyroxenes (OPX = orthopyroxene. CPX = clinopyroxene). Light blue: Fe-bearing glass. Dark blue: ferrous olivines. Purple: clay silicates. The y-axis is relative reflectance, and spectra are offset for clarity. (Modified from Horgan et al., 2017.)

over a wide range of diagnostic viewing angles. Because the manner in which VNIR light reflects from a surface depends on the viewing geometry and solar illumination angle, observations acquired under different phase angles can be used to quantify the albedo and scattering characteristics of different geologic materials.

26.3 Results

Early mission Pancam multispectral imaging results from the Mars Exploration Rovers (MER) Spirit and Opportunity were summarized in Bell et al. (2008). Here we provide a summary of the overall Pancam multispectral results from the completed Spirit rover and Phoenix lander missions, and an update and summary of results from the Opportunity and Curiosity rover missions. The summaries provided here are divided into results related to fine-grained soil components, rocky materials, and physical properties gleaned from multicolor photometric observations.

26.3.1 Multispectral Imaging from the Spirit Rover in Gusev Crater

Multispectral imaging on the Spirit rover provided the ability to remotely detect spectral variability associated with the presence of certain classes of materials that could not be studied as broadly, or at all, with other rover instruments, as well as to map that spectral variability around the rover and along its traverse. The resulting data sets from more than 2200 sols of operation between 2004 and 2010 significantly complement and enhance the results from the rovers' more spatially limited or in situ elemental chemistry (Chapter 28), infrared and Mössbauer spectroscopy (Chapters 25 and 27), and microscopic imaging investigations.

26.3.1.1 Fine-Grained Regolith Materials in Gusev Crater

Multispectral imaging on the Spirit rover was used to study a variety of fine-grained materials along the traverse in Gusev crater (mission details are summarized in Arvidson et al., 2006, 2008, 2010). Here, "fine-grained materials" include soil, dust, sand, and drift, where "soil" is a generic term for unconsolidated rock and mineral matter, "dust" is the finest component of the soil that can easily become airborne (typically < 5 µm), "sand" includes potentially mobilized silt- to sand-sized grains (typically 5–200 µm), and "drift" includes sand and dust that have been concentrated in aeolian bedforms (Bell et al., 2004a, 2008). Pancam spectra of typical bright and dark soils are consistent with previous average bright and dark region telescopic spectra (Bell et al., 1990; Mustard & Bell, 1994) and Pathfinder IMP multispectral data (Bell et al., 2000). The visible-wavelength spectral properties of dust observed by Pancam are dominated by the presence of fine-grained (nanophase to poorly crystalline) ferric oxides (e.g., Kinch et al., 2006). Dark sand and drift deposits (such as the dunes at "El Dorado") exhibit a shallow, broad absorption near 900 nm and a negative 900–1000 nm spectral slope attributed to the presence of ferrous-iron–bearing silicates (e.g., pyroxene, olivine) (Bell et al., 2008). These spectra are similar to some of the least-dusty basaltic rock surfaces at Gusev, and thus the dark soils at Gusev likely contain a significant component of less altered mafic material than the dust and other soil deposits.

Serendipitously, Spirit encountered a number of subsurface soils with whitish, yellowish, and reddish hues within the Columbia Hills (e.g., Figure 26.2) that were observed by Pancam at a total of 18 locations (e.g., Johnson et al., 2007; Wang et al., 2008; Arvidson et al., 2010; Rice et al., 2011; Farrand et al., 2016). Most of these deposits were brought up from depths of ~10 cm by the dragging motion of Spirit's inoperative right front wheel, with SO_3 concentrations measured by the Alpha Particle X-ray Spectrometer (APXS) of up to 38 wt.% (Ming et al., 2006, 2008). Mössbauer (MB) spectrometer data indicated ferric sulfates (Morris et al., 2008), and Pancam spectra (Figures 26.2a,d) were consistent with a heterogeneous mixture of hydrated ferric sulfate phases (e.g., Lane et al., 2008; Parente et al., 2009; Arvidson et al., 2011) such as ferricopiapite and fibroferrite (e.g., Johnson et al., 2007; Parente et al., 2009; Arvidson et al., 2011), Fe^{3+}-bearing phosphates (Lane et al., 2008), hydrated silica (Wang et al., 2008), and/or elemental sulfur (Morris et al., 2007). At the high-albedo soil exposure called Gertrude Weise, APXS measurements indicated a nearly pure silica composition (~98 wt.% SiO_2 when corrected for dust contamination), with minor TiO_2 (Squyres et al., 2008). Mid-IR spectroscopic measurements made by the Miniature Thermal Emission Spectrometer (Mini-TES) were consistent with hydrated amorphous silica (Squyres et al., 2008; Ruff et al., 2011), and Pancam spectra contained

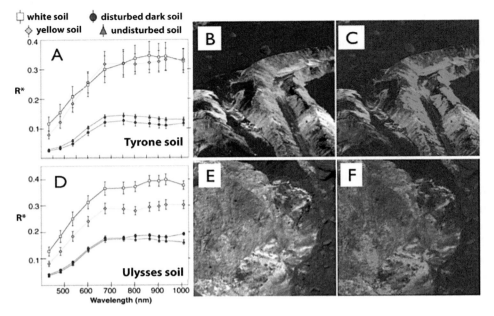

Figure 26.2 Examples of Gusev soil color endmembers. (a) Pancam spectra, (b) false color composite (432, 535, 753 nm), and (c) decorrelation stretch (DCS) image using the same filters of "Tyrone" from sol 790 (sequence P2531); field of view is ~65 cm. (d) Pancam spectra, (e) false color, and (f) DCS image of "Ulysses" from sol 1888 (sequence P2559), using the same filters as (b) and (c). Field of view is ~24 cm. (Modified from Rice et al., 2011.)

a downturn at 1009 nm indicative of the presence of H_2O and/ or OH (Wang et al., 2008; Rice et al., 2010).

The mineralogy, geochemistry, spatial variability, and geologic setting of the bright subsurface soils in Gusev crater suggest that they formed either in a hydrothermal environment from fumarolic condensates, by precipitation from geothermal waters, and/or from leaching of local basaltic rocks (e.g., Morris et al., 2008; Squyres et al., 2008; Yen et al., 2008; Ruff et al., 2011). While the time of hydrothermal activity within Gusev crater when the sulfate- and silica-rich soils formed is not well constrained, the processes that sorted, transported, and modified the bright soil material may be ongoing. Indeed, the layered structure observed by Pancam in the wall of the Ulysses soil trench suggested an ongoing pedogenic modification from downward migration of soluble materials by gravity-driven water (Arvidson et al., 2010).

26.3.1.2 Rocky Materials in Gusev Crater

The Gusev plains where Spirit landed have been interpreted as physically eroded, Hesperian-aged basaltic lava flows (e.g., Squyres et al., 2004; McSween et al., 2006a) and rocks observed on it were generally dark toned, with low 535-nm band depths, low to intermediate 904-nm band depths, and a fitted relative reflectance maximum near 673 nm (Squyres et al., 2006a). Rocks observed in the Columbia Hills were more diverse both chemically (Squyres et al., 2006a) and spectrally (Farrand et al., 2006; Ruff et al., 2006). For example, Farrand et al. (2008b) used several endmember determination and clustering approaches to define a set of rock spectral classes (Figure 26.3) that also roughly corresponded to different geographic locations along Spirit's traverse path.

Light-toned rocks observed on the West Spur of Husband Hill, as well as those observed higher on Husband Hill and on its south side, were softer than the plains basalts (as determined by the specific grind energy required to grind into them with Spirit's RAT) and appeared to be clastic in nature. Details on these classes are provided in Farrand et al. (2008b), but briefly

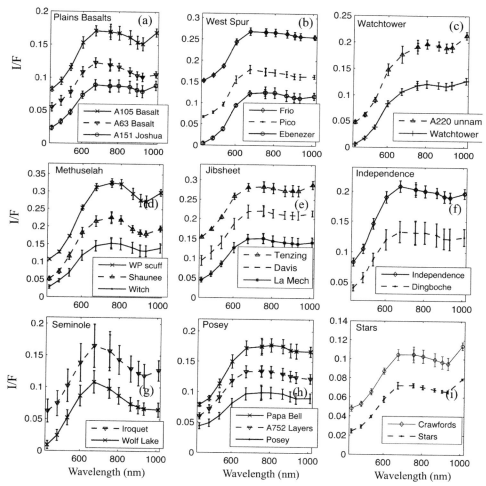

Figure 26.3 Representative Pancam 11-point radiance factor (I/F) spectra of the nine spectral classes of rocks from Spirit's traverse in Gusev crater (based on Farrand et al., 2008b).

stated, the light-toned clastic rocks of Husband Hill were more oxidized (as indicated by deeper 535-nm band depths; Figures 26.3b,c) and had a range of NIR spectral features from relatively flat NIR spectral slopes (Figures 26.3b,e,f,h) to those with significant 904-nm bands indicative of ferric oxides or ferrous silicates such as low-Ca pyroxene (Figures 26.3a,d,g,i).

Spirit spent the remainder of its mission at and around the Home Plate low mesa (Squyres et al., 2007; Arvidson et al., 2011). The rocks making up Home Plate and many of the surrounding layered deposits were also clastic in nature, with higher 535-nm band depths than basalts, low 904-nm band depths, and flat to negative 754- to 1009-nm slopes. Light-toned nodules, observed in the valley to the east of Home Plate, were discussed by Rice et al. (2010) and Ruff & Farmer (2016). These nodules are highly enriched in Si, as determined by in situ APXS measurements and Mini-TES thermal emissivity measurements (Ruff et al., 2011). The Pancam spectra of these nodules are similar to those of the Si-rich soils described elsewhere in this chapter.

26.3.1.3 Photometry Results in Gusev Crater

Spectrophotometric observations using Pancam multispectral images to investigate the scattering properties of specific geologic units (e.g., Figure 26.4) have been analyzed using a Hapke radiative transfer model (Hapke, 1993) in combination with corrections for diffuse sky illumination and local surface facet orientations derived from stereo models (e.g., Johnson et al., 2006a,b; 2015a,b). At the Spirit site all typical "Soil" class units had similar single scattering albedo (w) values. Variations in w among dusty "Red" class rocks appeared weakly correlated with elevation, suggesting subtle differences in weathering regimes, or the degree/type of coating materials. Less dusty "Gray" class rocks were typically more forward scattering (consistent with irregular to rough particles with few internal scatterers at the wavelength or larger scale) than typical Soil and Red rocks, whose surface particles or coatings contained many internal scatterers. Silica-rich materials exhibited variations in scattering parameters with wavelength, consistent with microporosity inferred from previous high-resolution in situ images and thermal IR spectra (Ruff et al., 2011).

26.3.2 Multispectral Imaging from the Opportunity Rover in Meridiani Planum

The Opportunity rover landed in Meridiani Planum in early 2004 and successfully explored the region over more than 5100 sols of operation, until being starved of power by a major global dust storm in June of 2018. Just like for the Spirit rover mission, multispectral imaging with the Opportunity Pancam instruments complements and and enhances the results from the mission's elemental chemistry (Chapter 28), IR and Mössbauer spectroscopy (Chapters 25 and 27), and microscopic imaging investigations.

26.3.2.1 Fine-Grained Regolith Materials in Meridiani Planum

Results from the first Mars year of Pancam observations of fine-grained regolith components in Meridiani Planum were reviewed and summarized by Bell et al. (2008). Generally,

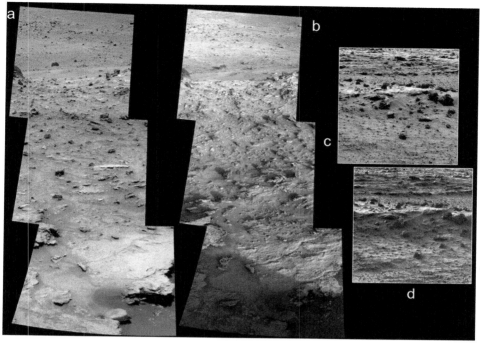

Figure 26.4 Examples of false-color Spirit/Pancam photometric campaign images and mosaics acquired in the Torquas area of the Columbia Hills in Gusev crater, illustrating differences in apparent color and texture due to different sun angles. RGB color composites using 753-nm, 601-nm, and 432-nm filters. Mosaics pointed east: (a) Sol 1148A at 12:24 LTST; (b) Sol 1147A, 17:05 LTST. Single frames pointed west: (c) Sol 1142A, at 13:39 LTST (d) Sol 1147A at 17:15 LTST. (Based on Johnson et al., 2015a.)

the Pancam spectral properties of these same kinds of materials observed since then along the more than 45 km Opportunity traverse (e.g., Squyres et al., 2006b; Arvidson et al., 2011, 2014) have not varied considerably with distance from the landing site. However, additional examples of spectra of rock coatings, veneers, and other fine-grained materials have been observed.

For example, Opportunity's exploration of Cape York, an eroded section of the ancient 22 km diameter Endeavour crater rim, revealed the presence of brighter, finely layered outcrop rocks partially covered by thin, dark veneers. Areas of the most prominent exposures of this outcrop, since named the Matijevic formation, correspond with regions of enhanced ~2200–2400 nm absorption observed from orbit by the Mars Reconnaissance Orbiter CRISM instrument; these features have been interpreted to indicate the presence of ferric-bearing smectite clays (possibly nontronite) in this locale (e.g., Arvidson et al., 2014). In situ multispectral observations from Pancam (Figure 26.5) show that the lighter-toned, fine-grained matrix of the Matijevic formation has a negative 754- to 1009-nm slope

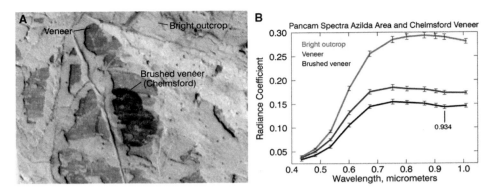

Figure 26.5 Pancam on Opportunity. (a) Opportunity Pancam false-color view showing the Chelmsford veneer after brushing using the RAT. Brushed areas are ~3.8 cm wide. Data acquired on sol 3098. (b) Pancam-based spectra of undisturbed bright layer, together with Chelmsford veneer undisturbed and brushed surfaces. (From Arvidson et al., 2014.)

and a relative reflectance peak at a shorter wavelength than the coatings. When the overlying dust is brushed away, the veneers are darker-toned, have a positive 934- to 1009-nm slope, and have a shallow absorption centered in the 934-nm band (Farrand et al., 2014b). Arvidson et al. (2014) interpret the dark veneers as the carrier of the ferric smectite clay signature detected from orbit, and hypothesize that small amounts of ferric smectites and salts were formed when mildly acidic surface water and/or groundwater was neutralized by interactions with the finely layered basaltic strata.

Also present in the Matijevic formation are a variety of spherules distinct from the so-called "blueberries" encountered earlier in the Burns formation outcrops of the Meridiani plains (Bell et al., 2004b, 2008). These spherules lack the hematite signature of the Burns formation blueberries, but cuttings produced by RAT abrasion of spherule-rich outcrop have a mild increase in 535-nm band depth, which would be consistent with the presence of nanophase ferric oxides (Arvidson et al., 2014).

More evidence of alteration was found in a fracture zone on the Murray Ridge portion of the Endeavour crater rim, where a turn-in-place maneuver by the rover flipped over the rocks Pinnacle and Stuart Islands. These rocks displayed a dark-toned coating with a red, featureless spectral slope and also a light-toned coating displaying a convex NIR shape and a 934- to 1009-nm drop in reflectance (Figure 26.6c). Using a combination of Pancam analysis and APXS-derived elemental chemistry, Arvidson et al. (2016) interpreted the light coating as evidence for the presence of hydrated Mg sulfates and the dark coatings as areas spectrally dominated by the presence of Mn oxides.

26.3.2.2 Rocky Materials in Meridiani Planum

Rocks observed on the Meridiani plains, with the exception of several meteorites, were all sandstones consisting of sulfate-cemented siliciclastic grains that belong to the Burns

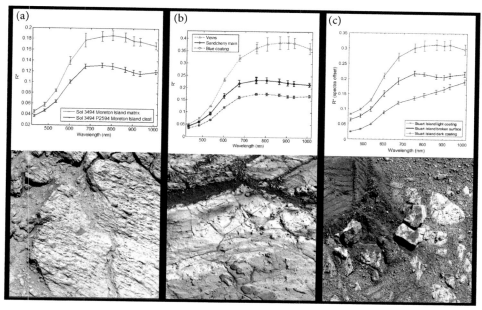

Figure 26.6 Additional examples of Pancam spectra of rock classes observed in the Opportunity rover's mission along the Endeavour crater rim in Meridiani Planum. See text for detailed discussion.

formation (Grotzinger et al., 2005). The Burns formation has been intensely affected by diagenesis (McLennan et al., 2005) resulting in the formation of the gray hematite concretions known colloquially as the "blueberries" and, as observed in crater walls at Endurance and Victoria craters, color changes in layers of the Burns. As outlined by Farrand et al. (2007) and Farrand et al. (2016), the Burns formation outcrop can be divided into two broad spectral groups: lighter-toned rocks (with a relatively elevated 482- to 535-nm slope in Pancam spectra) and darker-toned rocks. The lighter-toned surfaces have a flat to convex NIR shape while the darker-toned surfaces generally display a shallow absorption feature centered in the 904-nm band.

As discussed in Knoll et al. (2007), the lighter-toned Burns formation surfaces are interpreted as a weathered veneer on the outcrop. Grinds into the Burns formation with the rover's Rock Abrasion Tool (RAT) produced cuttings with a red color, and abraded rock surfaces display a spectrum consistent with that of red hematite: a deep 535-nm band depth, positive 754- to 1009-nm and 934- to 1009-nm spectral slopes, and an absorption feature centered in the 864- to 904-nm Pancam bands.

Weathering out of Burns formation outcrop are mm-sized spherical grains informally known as "blueberries." As indicated by both Mössbauer (Klingelhöfer et al., 2004) and Mini-TES observations (Christensen et al., 2004), the blueberries are composed of coarse-grained hematite. In Pancam spectra, the blueberries also have positive 754- to 1009-nm and 934- to 1009-nm slopes, and an absorption feature centered near 864–904 nm. These spectral features are consistent with dark submicron-sized (red) hematite, but not with coarser-grained specular (gray) hematite. These spectral features indicate that the surfaces of the blueberries either have a patina of dark red hematite or that the grain size of the hematite crystals in the blueberries is less than ~90 μm (Lane et al., 1999).

The rocks on the rim of Endeavour crater consist of impact breccias (called the Shoemaker formation); dark-toned basaltic cap rocks; assumed preimpact surfaces (such as the Matijevic formation); and the Grasberg formation, which is exposed along the border of the crater rim and resembles the Burns formation, but is chemically and spectrally distinct (Squyres et al., 2012; Arvidson et al., 2014). The Shoemaker consists of clasts in a matrix, both of which have Pancam reflectance spectra (Figure 26.6a) consistent with basaltic compositions (relatively dark-toned, low 535-nm band depth, negative 754- to 1009-nm slopes, along with in some places a shallow NIR absorption band (Farrand et al., 2013). In various fracture zones along the Endeavour rim (Crumpler et al., 2015), the matrix of the Shoemaker formation has deeper 535-nm band depths and develops an absorption centered in the 904-nm band, consistent with a higher abundance of ferric oxide minerals (Farrand et al., 2014a; 2016).

The Grasberg formation was divided by Crumpler et al. (2015) into upper lighter-toned and lower darker-toned members. The Grasberg is also spectrally distinct from the Burns with stronger Pancam 535- and 904-nm band depths. These spectral features were interpreted by Farrand et al. (2014b) as being consistent with the Grasberg having more disseminated red hematite. The lower Grasberg, the Matijevic formation, and parts of the Shoemaker formation also host light-toned veins that APXS results indicate are composed of $CaSO_4$. In Pancam spectra, the veins (Figure 27.6b) have a steeply negative 934- to 1009-nm spectral slope that is consistent with the presence of a weak ~1 μm H_2O overtone band in some hydrated minerals. Pancam data, together with the APXS $CaSO_4$ detection, suggest that the hydrated calcium sulfate mineral gypsum occurs in these veins (Rice et al., 2010; Squyres et al., 2012; Farrand et al., 2014b).

Fracture fill materials were also observed in the Matijevic formation including, notably, the target Espérance, which was the subject of a RAT grind and multiple APXS measurements. Chemical modeling of the APXS data (Clark et al., 2016) was consistent with the presence of an Al-bearing smectite such as montmorillonite and likely also amorphous silica. Pancam spectra of Espérance had a distinct 934- to 1009-nm downturn, similar to that in the gypsum veins, but potentially caused by a hydrated silica phase instead.

The largest fracture zone examined on the Endeavour rim by Opportunity is the large break in the rim dubbed Marathon Valley. This area was noted from orbital CRISM data as having relatively strong 2.29- and 2.43-μm bands characteristic of the Fe smectite nontronite (Fox et al., 2016). However, unlike in the Cape York Matijevic formation region, the carrier of the Fe smectite appeared in the altered matrix of the Shoemaker formation. The area was criss-crossed by sinuous "red zones," the fragmented rocks of which were enriched in Si and Al, but depleted in Fe. However, they had 535-nm band depths and blue-to-red (482–673 nm) slopes elevated relative to surrounding rocks, indicating a potential enrichment in nanophase ferric oxides (Farrand et al., 2016). The surfaces of those surrounding rocks were not spectrally distinctive, although several sites examined with RAT brushes or grinds showed spectral features consistent with an enrichment of crystalline red hematite.

26.3.2.3 Photometry Results in Meridiani Planum

At the Opportunity site, Johnson et al. (2006b, 2015a) found that outcrop rocks exhibited the highest w values, whereas spherule-rich soil albedos varied substantially, possibly resulting from variable amounts of dust coatings. The spherule-rich soils were typically backscattering and consistent with rough, clear spheres and/or particles containing internal scatterers. Outcrop rocks demonstrated average opposition effect parameters consistent with surfaces of intermediate porosity and/or grain size distribution between those modeled for spherule-rich soils and darker, clast-poor soils. Overall, the respectively similar w values and scattering properties for outcrop rocks, soils, and rover tracks during the rover traverse in Meridiani Planum emphasized the homogeneity of those materials across more than 45 km of rover odometry.

26.3.3 Multispectral Imaging from the Phoenix Lander at High Northern Latitudes

Multispectral imaging on the Phoenix lander, which landed near 68°N, 126°W in 2008 and operated for 157 sols on Mars, provided remote sensing information that significantly enabled and enhanced the in situ exploration of the region around the lander using the robotic arm and its cameras, as well as on board wet chemistry and thermal/evolved gas analysis investigations (e.g., Smith et al., 2008).

26.3.3.1 Soils, Cobbles, and Ice at the Phoenix Landing Site

The Phoenix mission's high latitude landing site was covered with patterned ground shaped by subsurface ice and had few large boulders. The robotic arm was able to operate within an area on the north side of the lander, digging a series of trenches to depths of 3–18 cm, although impenetrable ice was frequently encountered at 3–5 cm (Arvidson et al., 2009; Arvidson, 2016). Samples from the trenches were analyzed by means of analytical instruments and found to be alkaline and to contain $CaCO_3$, aqueous minerals, and salts including perchlorates up to several wt.% (Smith et al., 2009).

The SSI obtained color images of patterned ground on the same sol as landing (Smith et al., 2009). A full-color (Table 26.1) panorama was built up over the duration of the mission (Figure 26.7a). The color was dominated by the dust mantling, consistent with the low rock abundance. SSI spectra of the dust mantling matched nearby orbital CRISM observations as well as HST dust spectra from the 2007 dust storm (Goetz et al., 2010). The bright Dodo-Goldilocks ice had an absorption band through the longest wavelengths SSI observed (Figure 26.7b), deepening from 850 to 1000 nm (Smith et al., 2009). Surveys of the landscape in a limited filter set allowed detection of frost deposits on the shaded side of rocks using the 445- to 967-nm ratio (Cull et al., 2010c).

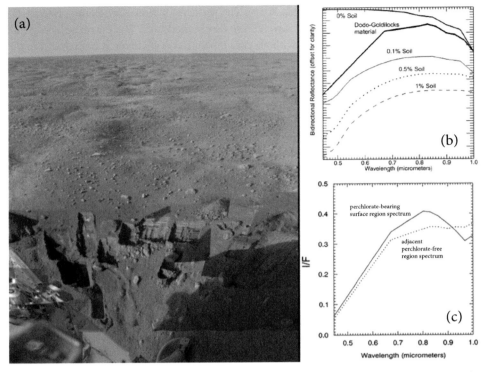

Figure 26.7 Phoenix landing site. (a) SSI Panoramic view of the Phoenix landing site showing several of the trenches dug by the robotic arm, including the ice-bearing Dodo-Goldilocks trench toward the left. (b) SSI spectrum (heavy black line) from Cull et al. (2010b) of bright Dodo-Goldilocks trench deposits show that they are apparently nearly pure water ice, with <0.1% soil contamination. (c) SSI multispectral evidence for small surface regions showing enhanced 967 nm absorption (solid line) interpreted as evidence of perchlorate by Cull et al. (2014). Dashed line is an adjacent region showing no 967-nm band.

Perchlorate salts were identified in multispectral images of the trenching area as isolated clods with a several percent 967-nm absorption (Cull et al., 2010a) (Figure 26.7c). While the feature is not diagnostic of a specific perchlorate composition, hydrated magnesium or calcium perchlorates are consistent with both the spectra and the analytical MECA detection of perchlorates reported by Hecht et al. (2009). Initial observations suggested a subsurface aqueous process concentrating perchlorates (Cull et al., 2010a); a broader survey of multispectral data showed the 967-nm feature on undisturbed material, suggesting an alternative process of atmospheric deposition, aqueous processing into a rock coating, and redistribution by thermally driven mechanical processing (Cull et al., 2014).

The Phoenix RAC and OM provided color images through illumination of a sample by LEDs. RAC images of the scoop showed predominantly fine, reddish soils; brown and black sand-sized (~50 μm) particles were seen in some samples; and rarely whitish fine particles, thought to be salts, were seen (Renno et al., 2009; Goetz et al., 2010). A comprehensive analysis of colors in OM images identified the same set of particle types. There was no definitive detection of fluorescence using the UV illumination with the OM (Goetz et al., 2012).

26.3.3.2 Photometry Results at the Phoenix Landing Site

Multispectral photometric observations of ice and soil with SSI were used to constrain surface scattering and absorption properties (Cull et al., 2010b). The Dodo-Goldilocks type of white ice was determined to be >99% pure ice. Lack of an adequate range of photometric angles left the grain size indeterminate (estimated at 0.1 µm–1 cm), but the spectral slope over 967–1001 nm was inconsistent with the presence of even 1% intermixed soil. The Snow White ice was closer to soil in spectral appearance, but was relatively forward scattering. The resulting estimate of 30 ± 20 wt.% ice (~55 vol.%) was consistent with ice filling the ~50% pore space derived for ice-free soils (Cull et al., 2010b).

26.3.4 Multispectral Imaging from the MSL Curiosity Rover in Gale Crater

Multispectral imaging from the Mastcam instruments on the Curiosity rover (Malin et al., 2017) since its landing in 2012 has helped to provide geologic context for the exploration of Gale crater and its fluvial and sedimentary deposits. In addition, Mastcam observations routinely provide critical contextual and textural information that enhance the interpretation of passive VNIR point spectra (e.g., Johnson et al., 2015e, 2017); active Laser-Induced Breakdown Spectroscopy (LIBS) measurements (Chapter 29) from the rover's ChemCam instrument; active and passive neutron spectroscopy measurements from the rover's Dynamic Albedo of Neutron (DAN) instrument (Chapter 30); and the quantitative chemical, isotopic, and mineralogic results coming from the rover's Sample Analysis on Mars (SAM) and CheMin investigations (e.g., Mahaffy et al., 2012; Blake et al., 2012). An overview of the Curiosity rover's science goals and a description of the rover instrumentation is provided by Grotzinger et al. (2012). Curiosity traversed >14 km during the mission's first 1600 sols, in which time it acquired >400 Mastcam multispectral filter sequences (Table 26.1) on ground targets that have revealed absorption features consistent with iron-bearing silicates, oxides, and sulfates (Figure 26.8).

26.3.4.1 Rocks, Soils, and Other Materials in Gale Crater

Soon after landing, the rover immediately encountered outcrops of cemented pebbles interpreted as fluvial conglomerates (Williams et al., 2013). These were targeted for multispectral observations, aided by the removal of some amount of surficial dust near the landing site by the retrorockets of the descent stage. At least two distinctly different rock spectral classes were found among the conglomerate clasts and unconsolidated pebbles. One class exhibits an NIR absorption feature that is more pronounced and centered at shorter wavelengths. Certain float rocks, including a cluster of spectrally similar rocks

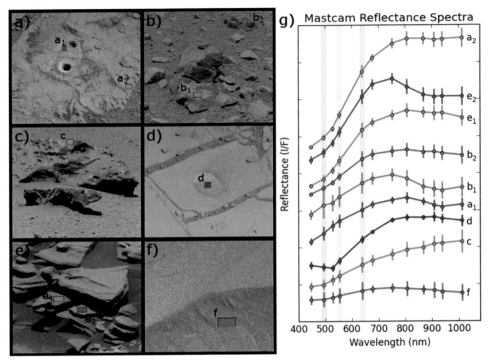

Figure 26.8 MSL/Mastcam M-100 RGB color images of selected multispectral targets presented in sol order. Each image contains one or two rectangular ROIs that are color-coded and labeled to match the plot (g) on the right. The reflectance plot (marked y-axis interval is 0.1) shows combined left and right camera I/F spectra averaged at stereo wavelengths and offset for clarity. (a) John Klein drill tailings (a_1) and dusty surface (a_2), from sol 183, sequence mcam00993. (b) The rock La Reine (b_1) and a nearby float rock (b_2), from sol 346, mcam01405. (c) Lebanon iron meteorite, from sol 641, mcam02729. (d) Confidence Hills Dust Removal Tool spot, from sol 762, mcam03273. (e) Marias pass unnamed rock (e_1) and neighboring Coombs rock target (e_2), from sol 1032, mcam04510. (f) Kubib black sand target, from sol 1183, mcam05362.

observed in the multispectral observation of target "La Reine" on sol 346, have a near-IR downturn that begins at wavelengths as short as ~750 nm (Figure 26.8b). This has been interpreted as possible evidence of an enrichment in orthopyroxene in these rocks (Johnson et al., 2015c,d), which has an absorption band centered at shorter wavelengths (near 900 nm) than clinopyroxenes and olivines (Adams et al., 1974).

To the east of the landing site, the rover encountered lacustrine mudstones enriched in smectite clay (Grotzinger et al., 2014; Vaniman et al., 2014). At Yellowknife Bay the rover made the first uses of its rotary/percussive drill to collect samples, in the process excavating a pile of fine gray tailings adjacent to each of the drill holes. Drill tailings and dump piles of sample material, as well as surfaces brushed by the Dust Removal Tool (DRT), provided Mastcam the opportunity to observe material less spectrally contaminated by reddish, Fe^{3+}-

bearing dust compared to undisturbed surfaces, and as such are routinely targeted for multispectral observations. The first two drill holes ("John Klein" and "Cumberland"), located meters apart at Yellowknife Bay, were spectrally similar and possess an absorption feature near 950 nm (Figure 26.8a). Wellington et al. (2017) compared the Mastcam spectra of drill holes through Pahrump Hills with CheMin XRD results and noted that changes in measured mineral abundances are correlated with changes in the Mastcam spectra, for particular phases. For John Klein and Cumberland, abundant smectite, as well as lesser amounts of primary basaltic phases, likely contribute to the shape and position of the NIR band.

Spectral absorptions near 860 nm observed in a ridge-capping unit in the lower portion of the Gale crater central mound were interpreted as a crystalline hematite absorption band in orbital CRISM data (Fraeman et al., 2013) This feature has also been observed in near-field multispectral imaging in the Pahrump Hills (Figure 26.8d), and in the vicinity of the Bagnold dune field and Murray Buttes. In the latter case, the reappearance of hematite features coincides approximately with weak detections from CRISM orbital observations (Fraeman et al., 2016). Also detected were spectra consistent with a ferric sulfate such as jarosite (e.g., Johnson et al., 2017; Wellington et al., 2017), though only on small surfaces (Figure 26.8e), consistent with its nondetection from orbit.

Throughout the mission, Curiosity has encountered a variety of mm-scale diagenetic features, including nodules, raised ridges, and fracture-filling material. Light-toned veins, identified to be primarily Ca sulfate on the basis of chemical analyses (Nachon et al., 2014), were observed with the longest wavelength Mastcam filters and showed the presence of variable hydration states (Rice et al., 2013a,b).

Unconsolidated fine materials such as dust, soil, and sand have also been extensively imaged in multispectral observations. Bright reddish martian dust thinly coats undisturbed surfaces, and may be incorporated into the interiors of aeolian drifts as well, such as in a thin layer of the Rocknest sand shadow (Wellington et al., 2017). Reddish soils are significantly lower in reflectance compared to the dust, and exhibit a weak NIR band, which may reflect a greater percentage of crystalline basaltic minerals in addition to nanophase iron oxide. The dark sands of the Bagnold Dunes (Figure 26.8f), encountered by the rover at the Namib and High Dunes, have low reflectances and a broad NIR absorption feature that is consistent with an enrichment in olivine. Mastcam and ChemCam passive spectra observations of sampled dune material that was sieved by the rover and redeposited onto the surface showed that the reflectance characteristics vary with different grain sizes, and that spectra of the finer-grained fraction of the dune material are consistent with an enrichment in ferric materials (Johnson et al., 2017).

As was the case with the two MERs, Curiosity has encountered a small number of suspected meteorites. Two large (>1 m) rocks dubbed "Lebanon" and "Littleton" were targeted by the Mastcam and ChemCam Remote Micro-Imager (RMI) instruments after their unusual morphologies were noticed in single-band Navcam imaging. Mastcam spectra on these two rocks (plus a smaller, unnamed fragment situated in front of Lebanon) showed a gradual rise in reflectance toward the NIR (Figure 26.8c), consistent with laboratory spectra of iron meteorites and previous Pancam spectra of less altered portions of meteorites found by the MER rovers (Gaffey, 1976; Schröder et al., 2008; Ashley et al., 2011).

26.3.5 *Multispectral Imaging of Mars Atmospheric and Airfall Dust*

Multispectral imaging on all Mars surface missions has also been used to study materials transported by the martian atmosphere. For example, direct imaging of the Sun has routinely been conducted through multiple color filters in order to estimate the opacity of aerosols (dust and/or water ice) as a function of time, and multispectral sky observations have been used to constrain aerosol scattering properties (e.g., Lemmon et al., 2004, 2015). Spectral ratios from 2-filter Phoenix SSI imaging of the sky were used to infer water-ice content versus dust, showing an increase in ice through the mission (Moores et al., 2009). In addition, the settling rates and properties of airborne dust have been studied with multispectral imaging, conducted as part of dust magnetic properties experiments on several missions (e.g., Kinch et al., 2006, 2015; Madsen et al., 2009; Drube et al., 2010).

Spectra of dust in atmospheric suspension, dust deposited on lander/rover deck elements, and dust observed in bright patches on the ground all exhibit the characteristic strong drop in reflectivity shortward of 600 nm caused by absorption from ferric iron (e.g., Mustard & Bell, 1994; Lemmon et al., 2004, 2015). The ferric iron is generally understood to reside in a component referred to as "nanophase iron oxides" whose detailed composition is still not fully understood (e.g., Morris et al., 1993; Goetz et al., 2005; Morris & Klingelhöfer, 2008).

Dust attracted to permanent magnets on the MER rovers (Madsen et al., 2009) is separable into a dark, larger-grained fraction enriched in magnetite and a brighter, finer-grained fraction dominated by the red nanophase oxide spectral signature (Kinch et al., 2006; Vaughan et al., 2010). It is not clear whether the two populations are truly distinct or two extremes on a continuous distribution. Microscope studies of soils at the Phoenix landing site (Goetz et al., 2010) revealed distinct populations of grains, notably a population of darker, bigger grains and a very fine, reddish population. X-ray diffraction measurements by the Curiosity rover's CheMin instrument of loose surface material at the "Rocknest" site also showed a crystalline component consistent with finely fragmented basaltic rocks, as well as an X-ray amorphous component (Bish et al., 2013).

26.4 Conclusions and Future Work

Multispectral imaging has been an important part of every successful mission that has operated on the surface of Mars. This is partly because of careful selection of filter bandpasses designed to maximize the ability to search for, detect, and characterize a wide variety of materials expected or predicted to occur on Mars, and partly because of the ability to do those search, detection, and characterization activities in the imaging domain.

More specifically, multispectral imaging has been a valuable and unique tool for characterizing martian geologic environments because of (1) the ability to discriminate color units not only based on RGB color but also based on narrowband near-UV to NIR colors, which can provide some diagnostic information on mineralogy; (2) the ability to correlate those color units with local and regional geologic units and contacts; and (3) the ability to use images and derived multispectral parameter maps to directly guide the choice of vehicle traverse paths as well as much more resource-intensive in situ surface activities like spectroscopy, brushing/grinding, or drilling.

Several new multispectral imaging investigations are being developed for future Mars surface missions. For example, NASA's InSight Mars lander, which successfully landed on the flat plains of Elysium Planitia on Mars in late November 2018, is using RGB imaging to provide some color information to help assist with basic geologic context assessment. The joint European Space Agency/Roscosmos ExoMars rover is scheduled to launch in 2020 carrying a pair of 12-filter multispectral imaging cameras called PanCam (e.g., Gunn & Cousins, 2016). NASA will also launch a new rover to Mars in 2020 that will carry RGB-capable engineering cameras and a zoom-capable imaging system called Mastcam-Z (Bell et al., 2016). The filter set for Mastcam-Z will be similar to those used for MSL/Mastcam (Figure 26.1) and MER/Pancam (Table 26.1), but will be modified to be capable of detecting more kinds of materials based on the experience and lessons learned from previous investigations.

Acknowledgments

We thank the scientists, engineers, managers, students, and administrators around the world who have helped to make multispectral observations from Mars surface vehicles such a successful part of so many missions. We also thank B. Horgan and A. Fraeman for very helpful discussions of an earlier draft of this chapter.

References

Adams J.B. (1974) Visible and near-infrared diffuse reflectance spectra of pyroxenes as applied to remote sensing of solid objects in the Solar System. *Journal of Geophysical Research*, **79**, 4829–4836.

Arvidson R.E. (2016) Aqueous history of Mars as inferred from landed mission measurements of rocks, soils, and water ice. *Journal of Geophysical Research*, **121**, 1602–1626.

Arvidson R.E., Squyres S.W., Anderson R.C., et al. (2006) Overview of the spirit Mars exploration rover mission to Gusev crater: Landing site to Backstay Rock in the Columbia Hills. *Journal of Geophysical Research*, **111**, DOI:10.1029/2005JE002499.

Arvidson R.E., Ruff S.W., Morris R.V., et al. (2008) Spirit Mars rover mission to the Columbia Hills, Gusev crater: Mission overview and selected results from the Cumberland Ridge to Home Plate. *Journal of Geophysical Research*, **113**, E12S33, DOI:10.1029/2008JE003183.

Arvidson R.E., Bonitz R.G., Robinson M.L., et al. (2009) Results from the Mars Phoenix Lander robotic arm experiment. *Journal of Geophysical Research*, **114**, DOI:10.1029/2009JE003408.

Arvidson R.E., Bell J.F., Bellutta P., et al. (2010) Spirit Mars rover mission: Overview and selected results from the northern Home Plate Winter Haven to the side of Scamander crater. *Journal of Geophysical Research*, **115**, DOI:10.1029/2010JE003633.

Arvidson R.E., Ashley J.W., Bell J., et al. (2011) Opportunity Mars rover mission: Overview and selected results from Purgatory ripple to traverses to Endeavour crater. *Journal of Geophysical Research*, **116**, DOI:10.1029/2010JE003746.

Arvidson R.E., Squyres S.W., Bell J.F., et al. (2014) Ancient aqueous environments at Endeavour crater, Mars. *Science*, **343**, 1248097.

Arvidson R.E., Squyres S.W., Morris R.V., et al. (2016) High concentrations of manganese and sulfur in deposits on Murray Ridge, Endeavour crater, Mars. *American Mineralogist*, **101**, 1389–1405.

Ashley J.W., Golombek M., Christensen P.R., et al. (2011) Evidence for mechanical and chemical alteration of iron-nickel meteorites on Mars: Process insights for Meridiani Planum. *Journal of Geophysical Research*, **116**, E00F20, DOI:10.1029/2010JE003672.

Bell J.F. III (1996) Iron, sulfate, carbonate, and hydrated minerals on Mars. In: *Mineral spectroscopy: A tribute to Roger G. Burns*. (M.D. Dyar, C. McCammon, & M.W. Schaefer, eds.). Geochemical Society Special Publication 5. Geochemical Society, Houston, TX, 359–380.

Bell J.F. III & Ansty T. (2007) High spectral resolution UV to near-IR observations of Mars during 1999, 2001, and 2003 using HST/STIS. *Icarus*, **191**, 581–602.

Bell J.F. III, McCord T.B., & Owensby P.D. (1990) Observational evidence of crystalline iron oxides on Mars. *Journal of Geophysical Research*, **95**, 14447–14461.

Bell J.F. III, McSween H.Y. Jr., Murchie S.L., et al. (2000) Mineralogic and compositional properties of martian soil and dust: Results from Mars Pathfinder. *Journal of Geophysical Research*, **105**, 1721–1755.

Bell J.F. III, Squyres S., Herkenhoff K., et al. (2003) Mars exploration rover Athena panoramic camera (Pancam) investigation. *Journal of Geophysical Research*, **108**, DOI:10.1029/2003JE002070.

Bell J.F. III, Squyres S.W., Arvidson R.E., et al. (2004a) Pancam multispectral imaging results from the Opportunity rover at Meridiani Planum. *Science*, **306**, 1703–1709.

Bell J.F. III, Squyres S.W., Arvidson R., et al. (2004b) Pancam multispectral imaging results from the Spirit rover at Gusev crater. *Science*, **305**, 800–806.

Bell J.F. III, Savransky D., & Wolff M.J. (2006) Chromaticity of the martian sky as observed by the Mars Exploration Rover Pancam instruments. *Journal of Geophysical Research*, **111**, E12S05, DOI:10.1029/2006JE002687.

Bell J.F. III, Calvin W.M., Farrand W.H., et al. (2008) Mars Exploration Rover Pancam multispectral imaging of rocks, soils, and dust in Gusev crater and Meridiani Planum. In: *The martian surface: Composition, mineralogy, and physical properties* (J.F. Bell III, ed.). Cambridge University Press, Cambridge, 281–314.

Bell J.F. III, Maki J.N., Mehall G.L., Ravine M.A., Caplinger M.A., & Mastcam-Z Team. (2016) Mastcam-Z: Designing a geologic, stereoscopic, and multispectral pair of zoom cameras for the NASA Mars 2020 rover. *3rd International Workshop on Instrumentation for Planetary Missions*, Abstract #4126.

Bell J.F. III, Godber A., McNair S., et al. (2017) The Mars Science Laboratory Curiosity rover Mastcam instruments: Preflight and in-flight calibration, validation, and data archiving. *Earth and Space Science*, **4**, 396–452.

Bibring J.-P., Langevin Y., Gendrin A., et al. (2005) Mars surface diversity as revealed by the OMEGA/Mars Express observations. *Science*, **307**, 1576–1581.

Bish D.L., Blake D., Vaniman D., et al. (2013) X-ray diffraction results from Mars Science Laboratory: Mineralogy of Rocknest at Gale crater. *Science*, **341**, 1238932.

Blake D., Vaniman D., Achilles C., et al. (2012) Characterization and calibration of the CheMin mineralogical instrument on Mars Science Laboratory. *Space Science Reviews*, **170**, 341–399.

Christensen P., Wyatt M., Glotch T., et al. (2004) Mineralogy at Meridiani Planum from the Mini-TES experiment on the Opportunity rover. *Science*, **306**, 1733–1739.

Clark B.C., Arvidson R.E., Gellert R., et al. (2007) Evidence for montmorillonite or its compositional equivalent in Columbia Hills, Mars. *Journal of Geophysical Research*, **112**, E06S01, DOI:10.1029/2006JE002756.

Clark B.C., Morris R.V., Herkenhoff K.E., et al. (2016) Esperance: Multiple episodes of aqueous alteration involving fracture fills and coatings at Matijevic Hill, Mars. *American Mineralogist*, **101**, 1515–1526.

Crumpler L.S., Arvidson R.E., Squyres S.W., et al. (2011) Field reconnaissance geologic mapping of the Columbia Hills, Mars, based on Mars Exploration Rover Spirit and MRO HiRISE observations. *Journal of Geophysical Research*, **116**, E00F24, DOI:10.1029/2010JE003749.

Crumpler L., Arvidson R., Bell J., et al. (2015) Context of ancient aqueous environments on Mars from in situ geologic mapping at Endeavour crater. *Journal of Geophysical Research*, **120**, 538–569.

Cull S.C., Arvidson R.E., Catalano J.G., et al. (2010a) Concentrated perchlorate at the Mars Phoenix landing site: Evidence for thin film liquid water on Mars. *Geophysical Research Letters*, **37**, L22203, DOI:10.1029/2010GL045269.

Cull S., Arvidson R.E., Mellon M.T., Skemer P., Shaw A., & Morris R.V. (2010b) Compositions of subsurface ices at the Mars Phoenix landing site. *Geophysical Research Letters*, **37**, L24203, DOI:10.1029/2010GL045372.

Cull S., Arvidson R.E., Morris R.V., Wolff M., Mellon M.T., & Lemmon M.T. (2010c) Seasonal ice cycle at the Mars Phoenix landing site: 2. Postlanding CRISM and ground observations. *Journal of Geophysical Research*, **115**, E00E19. DOI:10.1029/ 2009JE003410.

Cull S., Kennedy E., & Clark A. (2014) Aqueous and non-aqueous soil processes on the northern plains of Mars: Insights from the distribution of perchlorate salts at the Phoenix lfanding site and in Earth analog environments. *Planetary and Space Science*, **96**, 29–34.

Drube L., Leer K., Goetz W., et al. (2010) Magnetic and optical properties of airborne dust and settling rates of dust at the Phoenix landing site. *Journal of Geophysical Research*, **115**, E00E23. DOI:10.1029/ 2009JE003419.

Edgett K.S., Yingst R.A., Ravine M.A., et al. (2012) Curiosity's Mars hand lens imager (MAHLI) investigation. *Space Science Reviews*, **170**, 259–317.

Ellehoj M.D., Gunnlaugsson H.P., Taylor P.A., et al. (2010) Convective vortices and dust devils at the Phoenix Mars mission landing site. *Journal of Geophysical Research*, **115**, E00E16. DOI:10.1029/ 2009JE003413.

Farrand W.H., Bell J.F. III, Johnson J.R., Squyres S.W., Soderblom J., & Ming D.W. (2006) Spectral variability among rocks in visible and near-infrared multispectral Pancam data collected at Gusev crater: Examinations using spectral mixture analysis and related techniques. *Journal of Geophysical Research*, **111**, E02S15, DOI:10.1029/2005JE002495.

Farrand W.H., Bell J.F., Johnson J.R., et al. (2007) Visible and near-infrared multispectral analysis of rocks at Meridiani Planum, Mars, by the Mars Exploration Rover Opportunity. *Journal of Geophysical Research*, **112**, E06S02, DOI:10.1029/2006JE002773.

Farrand W.H., Bell J.F. III, Johnson J.R., Bishop J.L., & Morris R.V. (2008a) Multispectral imaging from Mars Pathfinder. In: *The martian surface* (J.F. Bell III, ed.). Cambridge University Press, Cambridge, 265–280.

Farrand W.H., Bell J., Johnson J.R., et al. (2008b) Rock spectral classes observed by the Spirit rover's Pancam on the Gusev Crater Plains and in the Columbia Hills. *Journal of Geophysical Research*, **113**, E12S38, DOI:10.1029/2008JE003237.

Farrand W.H., Bell III J.F., Johnson J.R., Rice M.S., & Hurowitz J.A. (2013) VNIR multispectral observations of rocks at Cape York, Endeavour crater, Mars by the Opportunity rover's Pancam. *Icarus*, **225**, 709–725.

Farrand W.H., Bell J.F. III, Johnson J.R., Rice M.S., Jolliff B.L., & Arvidson R.E. (2014a) Observations of rock spectral classes by the Opportunity rover's Pancam on northern Cape York and on Matijevic Hill, Endeavour crater, Mars. *Journal of Geophysical Research*, **119**, 2349–2369.

Farrand W.H., Bell J.F., Johnson J.R., & Mittlefehldt D.W. (2014b) Multispectral VNIR evidence of alteration processes on Solander Point, Endeavour crater, Mars. *8th International Conference on Mars*, Abstract #1354.

Farrand W.H., Johnson J.R., Rice M.S., Wang A., & Bell J.F. III (2016) VNIR multispectral observations of aqueous alteration materials by the Pancams on the Spirit and Opportunity Mars Exploration Rovers. *American Mineralogist*, **101**, 2005–2019.

Fox V., Arvidson R., Guinness E., et al. (2016) Smectite deposits in Marathon Valley, Endeavour crater, Mars, identified using CRISM hyperspectral reflectance data. *Geophysical Research Letters*, **43**, 4885–4892.

Fraeman A.A., Arvidson R.E., Catalano J.G., et al. (2013) A hematite-bearing layer in Gale crater, Mars: Mapping and implications for past aqueous conditions. *Geology*, **41**, 1103–1106.

Fraeman A.A., Johnson J.R., Wellington D.F., et al. (2016) Distribution of iron oxides in lower Mt. Sharp from Curiosity and orbital datasets, and implications for their formation. *AGU Fall Meeting Abstracts*, Abstract #P23B-2173.

Gaffey M.J. (1976) Spectral reflectance characteristics of the meteorite classes. *Journal of Geophysical Research*, **81**, 905–920.

Goetz W., Bertelsen P., Binau C., et al. (2005) Chemistry and minearology of atmospheric dust at Gusev crater: Indication of dryer periods on Mars. *Nature*, **436**, 62–65.

Goetz W., Pike W.T., Hviid S.F., et al. (2010) Microscopy analysis of soils at the Phoenix landing site, Mars: Classification of soil particles and description of their optical and magnetic properties. *Journal of Geophysical Research*, **115**, E00E22, DOI:10.1029/2009JE003437.

Goetz W., Hecht M.H., Hviid S.F., et al. (2012) Search for ultraviolet luminescence of soil particles at the Phoenix landing site, Mars. *Planetary and Space Science*, **70**, 134–147.

Grotzinger J.P., Arvidson R.E., Bell J.F., et al. (2005) Stratigraphy and sedimentology of a dry to wet eolian depositional system, Burns formation, Meridiani Planum, Mars. *Earth and Planetary Science Letters*, **240**, 11–72.

Grotzinger J.P., Crisp J., Vasavada A.R., et al. (2012) Mars Science Laboratory mission and science investigation. *Space Science Reviews*, **170**, 5–56.

Grotzinger J.P., Sumner D.Y., Kah L.C., et al. (2014) A habitable fluvio-lacustrine environment at Yellowknife Bay, Gale crater, Mars. *Science*, **343**, 1242777.

Gunn M.D. & Cousins C.R. (2016) Mars surface context cameras past, present, and future. *Earth and Space Science*, **3**, 144–162.

Hapke B. (1993) *Theory of reflectance and emittance spectroscopy*. Cambridge University Press, Cambridge.

Hecht M.H., Marshall J., Pike W.T., et al. (2008) Microscopy capabilities of the Microscopy, Electrochemistry, and Conductivity Analyzer. *Journal of Geophysical Research*, **113**, E00A22, DOI:10.1029/2008JE003077.

Hecht M.H., Kounaves S.P., Quinn R.C., et al. (2009) Detection of perchlorate and the soluble chemistry of martian soil at the Phoenix lander site. *Science*, **325**, 64–67.

Holstein-Rathlou C., Gunnlaugsson H.P., Merrison J.P., et al. (2010) Winds at the Phoenix landing site. *Journal of Geophysical Research*, **115**, E00E18. DOI:10.1029/2009JE003411.

Horgan B., Fraeman A.A., Rice M.S., Bell J.F., Wellington D., & Johnson J.R. (2017) New constraints from CRISM and Mastcam spectra on the mineralogy and origin of Mt. Sharp geologic units, Gale crater, Mars. *48th Lunar Planet. Sci. Conf.*, Abstract #3021.

Huck F.O., Jobson D.J., Park S.K., et al. (1977) Spectrophotometric and color estimates of the Viking Lander sites. *Journal of Geophysical Research*, **82**, 4401–4411.

Johnson J.R., Grundy W.M., & Lemmon M.T. (2003) Dust deposition at the Mars Pathfinder landing site: Observations and modeling of visible/near-infrared spectra. *Icarus*, **163**, 330–346.

Johnson J.R., Grundy W.M., Lemmon M.T., et al. (2006a) Spectrophotometric properties of materials observed by Pancam on the Mars Exploration Rovers: 1. Spirit. *Journal of Geophysical Research*, **111**, DOI:10.1029/2005JE002494.

Johnson J.R., Grundy W.M., Lemmon M.T., et al. (2006b) Spectrophotometric properties of materials observed by Pancam on the Mars Exploration Rovers: 2. Opportunity. *Journal of Geophysical Research*, **111**, DOI:10.1029/2006JE002762.

Johnson J.R., Bell J.F. III, Cloutis E.A., et al. (2007) Mineralogic constraints on sulfur-rich soils from Pancam spectra at Gusev crater, Mars. *Geophysical Research Letters*, **34**, L13202, DOI:10.1029/2007GL029894.

Johnson J.R., Bell J.F. III, Geissler P., et al. (2008) Physical properties of the martian surface from spectrophotometric observations. In: *The martian surface: Composition, mineralogy, and physical properties* (J.F. Bell III, ed.). Cambridge University Press, Cambridge, 428–450.

Johnson J.R., Bell J.F. III, Hayes A., et al. (2013) Preliminary Mastcam visible/near-infrared spectrophotometric observations at the Curiosity landing site, Mars. *44th Lunar Planet. Sci. Conf.*, Abstract #1374.

Johnson J.R., Bell J.F. III, Gasnault O., et al. (2014a) First iron meteorites observed by the Mars Science Laboratory (MSL) rover Curiosity. *AGU Fall Meeting Abstracts*, Abstract #P51E-3989.

Johnson J.R., Bell J.F. III, Hayes A., et al. (2014b) New Mastcam and Mahli visible/near-infrared spectrophotometric observations at the Curiosity landing site, Mars. *8th International Conference on Mars*, Abstract #1073.

Johnson J.R., Grundy W.M., Lemmon M.T., JBell J.F. III, & Deen R.G. (2015a) Spectrophotometric properties of materials observed by Pancam on the Mars Exploration Rovers: 3. Sols 500–1525. *Icarus*, **248**, 25–71.

Johnson J.R., Bell J.F. III, Guinness E., & Deen R. (2015b) The Mars Exploration Rovers Planetary Data System Archive of Pancam Photometry QUBS. *Geologic Society of America* Annual Meeting, Baltimore, MD, November 1–4, 2015, Abstract #260213.

Johnson J.R., Bell J.F. III, Hayes A., et al. (2015c) Recent Mastcam and MAHLI visible/near-infrared spectrophotometric observations: Kimberley to Hidden Valley. *46th Lunar Planet. Sci. Conf.*, Abstract #1424.

Johnson J.R., Bell J.F. III, Deen R., et al. (2015d) Recent Mastcam and MAHLI visible/near-infrared spectrophotometric observations: Pahrump Hills to Marias Pass. *AGU Fall Meeting*, Abstract #P43B-2125.

Johnson J.R., Bell J.F. III, Bender S., & MSL Science Team. (2015e) ChemCam Passive Reflectance Spectroscopy of surface materials at the Curiosity landing site, Mars. *Icarus*, **249**, 74–92.

Johnson J.R., Bell J.F., Bender S., et al. (2016) Constraints on iron sulfate and iron oxide mineralogy from ChemCam visible/near-infrared reflectance spectroscopy of Mt. Sharp basal units, Gale crater, Mars. *American Mineralogist*, **101**, 1501–1514.

Johnson J.R., Achilles C., Bell J.F., et al. (2017) Visible/near-infrared spectral diversity from in situ observations of the Bagnold Dune Field sands in Gale crater, Mars. *Journal of Geophysical Research*, **122**, 2655–2684.

Keller H.U., Goetz W., Hartwig H., et al. (2008) Phoenix Robotic Arm Camera. *Journal of Geophysical Research*, **113**, E00A17, DOI:10.1029/2007JE003044.

Kinch K.M., Merrison J.P., Gunnlaugsson H.P., Bertelsen P., Madsen M.B., & Nørnberg P. (2006) Preliminary analysis of the MER magnetic properties experiment using a computational fluid dynamics model. *Planetary and Space Science*, **54**, 28–44.

Kinch K.M., Bell J.F., Goetz W., et al. (2015) Dust deposition on the decks of the Mars Exploration Rovers: 10 years of dust dynamics on the Panoramic Camera calibration targets. *Earth and Space Science*, **2**, 144–172.

Klingelhöfer G., Morris R.V., Bernhardt B., et al. (2004) Jarosite and Hematite at Meridiani Planum from Opportunity's Mössbauer Spectrometer. *Science*, **306**, 1740–1745.

Knoll A.H., Jolliff B.L., Farrand W.H., et al. (2008) Veneers, rinds, and fracture fills: Relatively late alteration of sedimentary rocks at Meridiani Planum, Mars. *Journal of Geophysical Research*, **113**, E06S16, DOI:10.1029/2007JE002949.

Lane M., Morris R.V., & Christensen P.R. (1999) Spectral behavior of hematite at visible/near infrared and mid-infrared wavelengths. *5th International Conference on Mars*, Abstract #6085.

Lane M., Bishop J., Dyar M.D., King P., Parente M., & Hyde B. (2008) Mineralogy of the Paso Robles soils on Mars. *American Mineralogist*, **93**, 728–739.

Lemmon M.T., Wolff M.J., Bell J.F. III, Smith M.D., Cantor B.A., & Smith P.H. (2015)Dust aerosol, clouds, and the atmospheric optical depth record over 5 Mars years of the Mars Exploration Rover mission. *Icarus*, **251**, 96111, DOI:10.1016/j.icarus.2014.03.029.

Lemmon M., Smith P.H., Shinohara C., et al. (2008) The Phoenix surface stereo imager (SSI) investigation. *39th Lunar Planet. Sci. Conf.*, Abstract #2156.

Lemmon M.T., Wolff M.J., Smith M.D., et al. (2004) Atmospheric imaging results from the Mars Exploration Rovers: Spirit and Opportunity. *Science*, **306**, 1753–1756.

Madsen M.B., Goetz W., Bertelsen P., et al. (2009) Overview of the magnetic properties experiments on the Mars Exploration Rovers. *Journal of Geophysical Research*, **114**, E06S90, DOI:10.1029/2008je003098.

Mahaffy P.R., Webster C.R., Cabane M., et al. (2012) The sample analysis at Mars Investigation and Instrument Suite. *Space Science Reviews*, **170**, 401–478.

Malin M.C., Ravine M.A., Caplinger M.A., et al. (2017) The Mars Science Laboratory (MSL) Mast cameras and Descent imager: Investigation and instrument descriptions. *Earth and Space Science*, **4**, 506–539.

Matijevic J.R., Crisp J., Bickler D.B., et al. (1997) Characterization of the martian surface deposits by the Mars Pathfinder rover, Sojourner. *Science*, **278**, 1765–1768.

McLennan S.M., Bell J.F., Calvin W.M., et al. (2005) Provenance and diagenesis of the evaporite-bearing Burns formation, Meridiani Planum, Mars. *Earth and Planetary Science Letters*, **240**, 95–121.

McSween H.Y., Wyatt M.B., Gellert R., et al. (2006a) Characterization and petrologic interpretation of olivine-rich basalts at Gusev crater, Mars *Journal of Geophysical Research*, **111**, E02S10, DOI:10.1029/2005JE002477.

McSween H.Y., Ruff S.W., Morris R.V., et al. (2006b) Alkaline volcanic rocks from the Columbia Hills, Gusev crater, Mars. *Journal of Geophysical Research*, **111**, E09S91, DOI:10.1029/2006JE002698.

Mellon M.T., Arvidson R.E., Sizemore H.G., et al. (2009) Ground ice at the Phoenix landing site: Stability state and origin. *Journal of Geophysical Research*, **114**, E00E07. DOI:10.1029/2009JE003417.

Milliken R.E., Grotzinger J.P., & Thomson B.J. (2010) Paleoclimate of Mars as captured by the stratigraphic record in Gale crater. *Geophysical Research Letters*, **37**, L04201. DOI:10.1029/2009GL041870.

Ming D.W., Mittlefehldt D.W., Morris R.V., et al. (2006) Geochemical and mineralogical indicators for aqueous processes in the Columbia Hills of Gusev crater, Mars. *Journal of Geophysical Research*, **111**, E02S12, DOI:10.1029/2005JE002560.

Ming D.W., Gellert R., Morris R.V., et al. (2008) Geochemical properties of rocks and soils in Gusev crater, Mars: Results of the Alpha Particle X-Ray Spectrometer from Cumberland Ridge to Home Plate. *Journal of Geophysical Research*, **113**, E12S39, DOI:10.1029/2008JE003195.

Moores J.E., Lemmon M.T., Smith P.H., Komguem L., & Whiteway J.A. (2010) Atmospheric dynamics at the Phoenix landing site as seen by the Surface Stereo Imager. *Journal of Geophysical Research*, **115**, E00E08. DOI:10.1029/2009JE003409.

Moores J.E., Komguem L., Whiteway J.A., Lemmon M.T., Dickinson C., & Daerden F. (2011) Observations of near-surface fog at the Phoenix Mars landing site. *Geophysical Research Letters*, **38**, L04203, DOI:10.1029/2010GL046315.

Morris R.V. & Klingelhöfer G. (2008) Iron mineralogy and aqueous alteration on Mars from the MER Mössbauer spectrometers. In: *The martian surface: Composition, mineralogy and physical properties* (J.F. Bell III, ed.). Cambridge University Press, Cambridge, 339–365.

Morris R.V., Golden D.C., Bell J.F., Lauer H.V., & Adams J.B. (1993) Pigmenting agents in martian soils: Inferences from spectral, Mössbauer, and magnetic properties of nanophase and other iron oxides in Hawaiian palagonitic soil PN-9. *Geochimica et Cosmochimica Acta*, **57**, 4597–4609.

Morris R.V., Klingelhöfer G., Schröder C., et al. (2006) Mössbauer mineralogy of rock, soil, and dust at Meridiani Planum, Mars: Opportunity's journey across sulfate-rich outcrop, basaltic sand and dust, and hematite lag deposits. *Journal of Geophysical Research*, **111**, DOI:10.1029/2006JE002791.

Morris R.V., Ming D.W., Yen A., et al. (2007) Possible evidence for iron sulfates, iron sulfides, and elemental sulfur at Gusev crater, Mars, from MER, CRISM, and analog data. *7th International Conference on Mars*, Abstract #3933.

Morris R.V., Klingelhofer G., Schroder C., et al. (2008) Iron mineralogy and aqueous alteration from Husband Hill through Home Plate at Gusev crater, Mars: Results from the Mössbauer instrument on the Spirit Mars Exploration Rover. *Journal of Geophysical Research*, **113**, DOI:10.1029/2008JE003201.

Morris R.V., Ruff S.W., Gellert R., et al. (2010) Identification of carbonate-rich outcrops on Mars by the Spirit rover. *Science*, **329**, 421–424.

Murchie S., Kirkland L., Erard S., Mustard J., & Robinson M. (2000) Near-infrared spectral variations of martian surface materials from ISM Imaging Spectrometer Data. *Icarus*, **147**, 444–471.

Mustard J.F. & Bell J.F. III (1994) New composite reflectance spectra of Mars from 0.4 to 3.14 µm. *Geophysical Research Letters*, **21**, 353–356.

Nachon M., Mangold N., Forni O., et al. (2017) Chemistry of diagenetic features analyzed by ChemCam at Pahrump Hills, Gale crater, Mars. *Icarus*, **281**, 121–136.

Parente M., Bishop J.L., & Bell J.F. (2009) Spectral unmixing for mineral identification in pancam images of soils in Gusev crater, Mars. *Icarus*, **203**, 421–436.

Pollack J.B., Ockert-Bell M.E., & Shepard M.K. (1995) Viking Lander image analysis of martian atmospheric dust. *Journal of Geophysical Research*, **100**, 5235–5250.

Renno N.O., Bos B.J., Catling D.C., et al. (2009) Physical and thermodynamical evidence for liquid water on Mars. *Journal of Geophysical Research*, **114**, E00E03. DOI:10.1029/2009JE003362.

Rice M.S. & Bell J.F. III (2011) Mapping hydrated materials with MER Pancam and MSL Mastcam: Results from Gusev crater and Meridiani Planum, and plans for Gale crater. *AGU Fall Meeting*, Abstract #P22A-02.

Rice M.S., Bell J.F. III, Cloutis E.A., et al. (2010) Silica-rich deposits and hydrated minerals at Gusev crater, Mars: Vis-NIR spectral characterization and regional mapping. *Icarus*, **205**, 375–395.

Rice M.S., Bell J.F. III, Cloutis E.A., et al. (2011) Temporal observations of bright soil exposures at Gusev crater, Mars. *Journal of Geophysical Research*, **116**, E00F14, DOI:10.1029/2010JE003683.

Rice M.S., Bell J.F. III, Godber A., et al. (2013a) Mastcam Multispectral Imaging results from the Mars Science Laboratory investigation in Yellowknife Bay. *European Planetary Science Congress*, Abstract #762.

Rice M.S., Cloutis E.A., Bell J.F., et al. (2013b) Reflectance spectra diversity of silica-rich materials: Sensitivity to environment and implications for detections on Mars. *Icarus*, **223**, 499–533.

Rice M.S., Bell J.F. III, Wellington D.F., et al. (2013c) Hydrated minerals at Yellowknife Bay, Gale crater, Mars: Observations from Mastcam's science filters. *AGU Fall Meeting Abstracts*, Abstract #P23C-1795.

Ruff S.W. and J.D. Farmer (2016) Silica deposits on Mars with features resembling hot spring biosignatures at El Tatio in Chile, *Nature Communications*, **7**, 13554, DOI:10.1038/ncomms13554.

Ruff S.W., Christensen P.R., Blaney D.L., et al. (2006) The rocks of Gusev crater as viewed by the Mini-TES instrument. *Journal of Geophysical Research*, **111**, DOI:10.1029/2006JE002747.

Ruff S.W., Farmer J.D., Calvin W.M., et al. (2011) Characteristics, distribution, origin, and significance of opaline silica observed by the Spirit rover in Gusev crater, Mars. *Journal of Geophysical Research*, **116**, E00F23, DOI:10.1029/2010JE003767.

Schröder C., Rodionov D.S., McCoy T.J., et al. (2008) Meteorites on Mars observed with the Mars Exploration Rovers. *Journal of Geophysical Research*, **113**, E06S22. DOI:10.1029/2007JE002990.

Seelos K.D., Seelos F.P., Viviano-Beck C.E., et al. (2014) Mineralogy of the MSL Curiosity landing site in Gale crater as observed by MRO/CRISM. *Geophysical Research Letters*, **41**, 4880–4887.

Shaw A., Arvidson R.E., Bonitz R., et al. (2009) Phoenix soil physical properties investigation. *Journal of Geophysical Research*, **114**, E00E05. DOI:10.1029/2009JE003455.

Smith P.H., Tomasko M., Britt D., et al. (1997) The imager for Mars Pathfinder experiment. *Journal of Geophysical Research*, **102**, 4003–4025.

Smith P.H., Tamppari L., Arvidson R.E., et al. (2008) Introduction to special section on the phoenix mission: Landing site characterization experiments, mission overviews, and expected science. *Journal of Geophysical Research*, **113**, E00A18, DOI:10.1029/2007JE003083.

Smith P.H., Tamppari L., Arvidson R., et al. (2009) Water at the Phoenix landing site. *Science*, **325**, 58–61.

Squyres S.W., Arvidson R.E., Bell J.F., et al. (2004) The Spirit rover's Athena science investigation at Gusev crater, Mars. *Science*, **305**, 794–799.

Squyres S.W., Arvidson R.E., Blaney D.L., et al. (2006a) Rocks of the Columbia Hills. *Journal of Geophysical Research*, **111**, E02S11, DOI:10.1029/2005JE002562.

Squyres S.W., Arvidson R.E., Bollen D., et al. (2006b) Overview of the Opportunity Mars Exploration Rover mission to Meridiani Planum: Eagle crater to Purgatory ripple. *Journal of Geophysical Research*, **111**, DOI:10.1029/2006JE002771.

Squyres S.W., Aharonson O., Clark B.C., et al. (2007) Pyroclastic activity at Home Plate in Gusev crater. *Science*, **316**, 738–742.

Squyres S.W., Arvidson R.E., Ruff S., et al. (2008) Detection of silica-rich deposits on Mars. *Science*, **320**, 1063–1067.

Squyres S.W., Arvidson R.E., Bell J., et al. (2012) Ancient impact and aqueous processes at Endeavour crater, Mars. *Science*, **336**, 570–576.

Stoker C.R., Zent A., Catling D.C., et al. (2010) Habitability of the Phoenix landing site. *Journal of Geophysical Research*, **115**, E00E20. DOI:10.1029/ 2009JE003421.

Thomson B., Bridges N., Milliken R., et al. (2011) Constraints on the origin and evolution of the layered mound in Gale crater, Mars using Mars Reconnaissance Orbiter data. *Icarus*, **214**, 413–432.

Tomasko M.G., Doose L.R., Lemmon M., Smith P.H., & Wegryn E. (1999) Properties of dust in the martian atmosphere from the Imager on Mars Pathfinder. *Journal of Geophysical Research*, **104**, 8987–9007.

Vaniman D.T., Bish D.L., Ming D.W., et al. (2014) Mineralogy of a Mudstone at Yellowknife Bay, Gale crater, Mars. *Science*, **343**, 1243480.

Vaughan A.F., Johnson J.R., Herkenhoff K.E., et al. (2010) Pancam and Microscopic Imager observations of dust on the Spirit rover: Cleaning events, spectral properties, and aggregates. *MARS*, **5**, 129–145.

Wang A. & Ling Z. (2011) Ferric sulfates on Mars: A combined mission data analysis of salty soils at Gusev crater and laboratory experimental investigations. *Journal of Geophysical Research*, **116**, E00F17, DOI:10.1029/2010JE003665.

Wang A., Bell J.F. III, Li R., et al. (2008) Light-toned salty soils and coexisting Si-rich species discovered by the Mars Exploration Rover Spirit in Columbia Hills. *Journal of Geophysical Research*, **113**, E12S40, DOI:10.1029/2008JE003126.

Wellington D.F., Bell J.F., Johnson J.R., et al. (2017) Visible to near-infrared MSL/Mastcam multispectral imaging: Initial results from select high-interest science targets within Gale crater, Mars. *American Mineralogist*, **102**, 1202–1217.

Williams R.M., Grotzinger J.P., Dietrich W., et al. (2013) Martian fluvial conglomerates at Gale crater. *Science*, **340**, 1068–1072.

Wolff M.J., Smith M.D., Clancy R.T., et al. (2009) Wavelength dependence of dust aerosol single scattering albedo as observed by the Compact Reconnaissance Imaging Spectrometer. *Journal of Geophysical Research*, **114**, DOI:10.1029/2009JE003350.

Yen A.S., Morris R.V., Clark B.C., et al. (2008) Hydrothermal processes at Gusev crater: An evaluation of Paso Robles class soils. *Journal of Geophysical Research*, **113**, E06S10, DOI:10.1029/2007JE002978.

27

Mössbauer Spectroscopy at Gusev Crater and Meridiani Planum

Iron Mineralogy, Oxidation State, and Alteration on Mars

RICHARD V. MORRIS, CHRISTIAN SCHRÖDER, GÖSTAR KLINGELHÖFER,
AND DAVID G. AGRESTI

27.1 Introduction and Background

The Mars Exploration Rover (MER) named Spirit landed on the plains of Gusev crater on January 4, 2004, and its twin Opportunity landed on the opposite side of the planet at Meridiani Planum on January 24, 2004 (Squyres et al., 2004a,b). Both MER rovers carried the Athena science instrument package (Squyres et al., 2003) that included a miniature Mössbauer (MB) spectrometer MIMOS II (Klingelhöfer et al., 2003). The instrument detects only ^{57}Fe (~2% of natural Fe), and peak positions in MB spectra constrain Fe speciation according to oxidation (e.g., Fe^0, Fe^{2+}, and Fe^{3+}) and coordination (e.g., tetrahedral and octahedral) states and the structure of Fe-bearing minerals/phases. Peak areas provide quantitative information on the relative distribution of Fe among Fe-bearing phases and oxidation and coordination states. Chapter 7 describes MB spectra in the laboratory and how to use these spectra to identify minerals.

In a remote sensing environment, MB spectra provide geochemical and mineralogic information that is not readily obtained by other methods. Primary igneous rocks are normally dominated by Fe^{2+} (e.g., olivine, pyroxene, ilmenite, magnetite, and chromite) and secondary alteration products are normally dominated by Fe^{3+} (e.g., oxides and oxyhydroxides, sulfates, and amorphous materials), so that the parameter $Fe^{3+}/\Sigma Fe$ is a first-order measure of the alteration state of primary igneous material. The actual situation is more complex, because some minerals can be primary or secondary and can have both Fe^{2+} and Fe^{3+} (e.g., magnetite Fe_3O_4), and some secondary alteration products are predominantly Fe^{2+} (e.g., siderite $FeCO_3$). Thus to determine the extent and style of alteration, the redox information and Fe phase assignments must be considered together. The style of alteration (e.g., hydrolytic vs. sulfatic) is inferred by phase assignment of Fe-bearing alteration products.

In this chapter, we describe the functionality of the MER MB instrument and summarize the salient Mössbauer results obtained at Gusev crater and Meridiani Planum, where 155 and 114 MB spectra were acquired, respectively. Spirit's mission at Gusev crater ended on sol 2210 with the last MB spectrum acquired on sol 2071. A sol is one martian day with landing day sol 0. Opportunity's last communication occurred on sol 5111, but MB spectra were not acquired after sol 2871 because the low ^{57}Co source intensity at that point required unacceptably long integration times. Mission overviews relevant to Mössbauer activity include Squyres et al. (2006, 2009), Arvidson et al. (2006, 2008, 2010, 2011), and Ashley et al. (2011).

27.2 Instrument and Methods

For a full discussion of the Mössbauer effect and Mössbauer spectroscopy, the reader is referred to Chapter 7 in this volume and the literature (e.g., Bancroft, 1973; Hawthorne, 1988; Burns, 1993; Gütlich et al., 2011; Gütlich & Schröder, 2012; Yoshida & Langouche, 2013). Most laboratory MB spectrometers employ transmission geometry where the sample is located between the ^{57}Co source and the detector. For a planetary surface mission involving in situ measurements, the MER MIMOS II instruments (Klingelhöfer et al., 2003) employed instead backscatter geometry (source and detector are on the same side of the sample) because no sample preparation is required (Figure 27.1). To acquire a MIMOS II spectrum (emitted counts vs. source velocity in units of mm/s) for a martian surface target (rock or soil), the instrument's sensor head was placed in physical contact with the target by the rover's robotic arm (Figure 27.2). The ^{57}Co source activity at landing was ~150 mCi and <1 mCi when acquisition of MB spectra ceased. Martian MIMOS II spectra were acquired

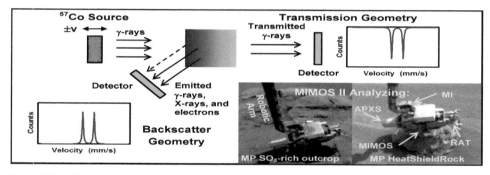

Figure 27.1 Measurement geometry for Mössbauer spectrometers. In transmission geometry, the sample (absorber) is between nuclear source (^{57}Co/Rh; 14.4 keV) and detector(s), peaks are negative features, and the sample should be thin with respect to absorption of gamma rays to minimize nonlinear effects. In backscatter geometry, source and detector are on the same side of the sample, peaks are positive features corresponding to recoilless emission of gamma rays, internal conversion X-rays, and electrons. MIMOS II instruments employ backscatter geometry and simultaneously detect emitted 14.4-keV gamma rays and ~6.4-keV X-rays. MIMOS spectra are counts per velocity channel as a function of velocity in units of mm/s relative to the midpoint of the spectrum of α-Fe^0 which is obtained from an internal standard. (Modified after Klingelhöfer et al., 2003.) Inset: Front Hazcam images show MIMOS II acquiring data on Meridiani Planum (MP) SO$_3$-rich outcrop (left, sol 31B image 1F130936200EDN0454P1131L0M1-BR) and HeatShieldRock (right, sol 350B image 1F159253265EDN40DPP1131L0M1-BR). The MIMOS instrument is hidden from view by other robotic arm instruments. APXS = Alpha Particle X-Ray Spectrometer; MI = Microscopic Imager; RAT = Rock Abrasion Tool.

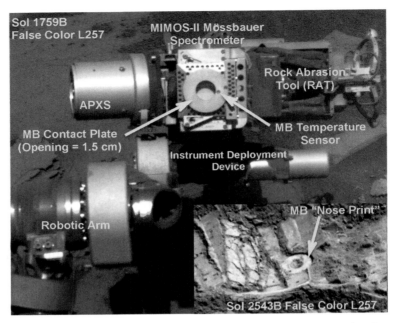

Figure 27.2 False color image of the MIMOS II Mössbauer spectrometer mounted on Opportunity's Instrument Deployment Device. Hole in MB contact plate (1.5 cm diameter) defines the field of view. Contact plate has a sensor for measurement of surface temperatures. Inset shows "nose print" of contact plate made during analysis of soil in a trench made by the churning action of rover wheels.

in temperature intervals 10 K wide during real-time temperature binning using the temperature sensor on the MB contact plate in order to detect temperature-dependent changes in MB spectra. The MER instruments are configured with an internal standard (α-Fe metal foil) and an additional detector in order to simultaneously acquire an α-Fe metal transmission MB spectrum for each surface target (Klingelhöfer et al., 2003).

Peak positions in MB spectra are described by the center shift (δ in mm/s) relative to zero velocity taken as the center point of the spectrum of α-Fe metal foil (standard practice), the quadrupole splitting (ΔE_Q in mm/s), and (for sextets) the magnetic hyperfine field (B_{hf} in T). These parameters for MER targets were, in most cases, calculated from spectra that are the sum of all individual temperature channels for the emitted 14.4-keV gamma rays (to maximize counting statistics). To investigate temperature dependence of parameters, spectra from individual temperature channels are summed over spectra acquired at different locations exhibiting similar mineral composition. MIMOS II integrations were acquired during the nighttime nominally between ~200 K and ~270 K (with 235 ± 15 K taken as the median temperature).

Because the reference α-Fe foil and martian surface targets are at the same temperature in our application, the values obtained for δ may be directly compared with databases compiled from measurements made with source and absorber at ambient laboratory temperatures (e.g., Burns & Solberg, 1990; Burns, 1993; McCammon, 1995; Stevens et al., 1998). However, both ΔE_Q and B_{hf} may exhibit a temperature dependence, which if significant would require either a database obtained on samples measured at the same temperatures as the MER targets or extrapolation of MER results to laboratory temperatures based on the temperature variation observed by the MER instruments.

Mössbauer parameters calculated from MIMOS II spectra reported in Morris et al. (2006a,b), and in all subsequent publications by the MER MB instrument team, are based on a refinement of the velocity calibration employed in the two initial MER MB publications (Morris et al., 2004; Klingelhöfer et al., 2004). This refinement included the temperature dependence of B_{hf} for the α-Fe metal foil standard (Morris et al., 2006b). Therefore, comparisons of laboratory measurements with MER results should be based on data in MER publications from 2006 and beyond. Morris et al. (2006a,b) report for the Fe^{3+} doublets of npOx, jarosite, and Fe3D3 that their values of ΔE_Q are temperature independent over the measurement interval on the martian surface, so that the average values at 235 ± 15 K are the same within uncertainty (±0.02 mm/s) as the values obtained by extrapolation to ambient terrestrial temperatures. The same relationship held for Fe^{2+} doublets assigned to pyroxene, but a weak temperature dependence is indicated for the Fe^{2+} doublet assigned to olivine (martian-surface average and ambient-terrestrial extrapolated values are 2.99 ± 0.03 mm/s and 2.94 ± 0.03 mm/s, respectively).

MB spectra are linear sums of subspectra from all distinct Fe sites (i.e., distinct sites relative to oxidation and coordination states and mineralogic speciation). MB spectra provide quantitative data on the relative distribution of total Fe among Fe-bearing phases but does not provide information on absolute abundances of Fe-bearing phases themselves unless the concentration of Fe in those phases is known or can be estimated. McSween et al. (2008) provide a template for deriving such estimates. Subspectral areas (A), which here are the sum of peak areas associated with each Fe-bearing phase, and their defining MB parameters (δ, ΔE_Q, field B_{hf}) were obtained by least squares fitting procedures (see Morris et al., 2006a,b). Subspectral areas include a correction factor that accounts for recoil-free fraction differences between Fe oxidation states ($f(Fe^{3+})/f(Fe^{2+}) = 1.21$) (De Grave & Van Alboom, 1991; Morris et al., 1995). In general, MER MB spectra are characterized by multiple occurrences of the same Fe-bearing phases present in different proportions. This repetition aided in Fe speciation assignments and provided fitting constraints (e.g., constrain all MB parameters except subspectral areas) late in the mission when low source intensities resulted in MB spectra with poor counting statistics.

MIMOS II can "see through" thin layers of dust and provide mineralogic information about underlying surfaces. From laboratory measurements detecting the 14.4 keV radiation, a layer of air-fall basaltic dust more than ~3 mm thick will fully mask underlying surfaces from detection (Graff et al., 2001; Morris et al., 2001a,b; Klingelhöfer et al., 2003). The dust layer is characterized by particle diameters ≤10 μm, and the palagonitic tephra used as the dust source (<1 mm size fraction) has total Fe concentrations equal to 15.49 wt.% and 22.33 wt.% for the <1 mm and <5 μm size fractions, respectively.

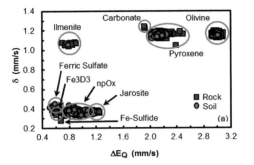
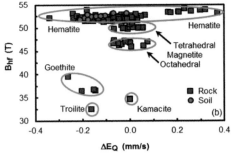

Figure 27.3 MIMOS II phase identification diagrams (235 ±.15 K). (a) Center shift (δ) versus quadrupole splitting ($\Delta E[SUB]_Q$) for doublet subspectra aside from chromite. Ilmenite, carbonate, pyroxene, and olivine are Fe^{2+}- bearing phases. Ferric sulfate, Fe3D3, npOx, and jarosite are Fe^{3+}-bearing phases. Fe sulfide is FeS_2 (pyrite/marcasite). (b) Hyperfine field strength (B_{hf}) versus ΔE_Q for sextet subspectra. Hematite and goethite are Fe^{3+} phases, magnetite is an Fe^{2+}–Fe^{3+} phase with distinct tetrahedral (Fe^{3+}) and octahedral (Fe^{2+}, Fe^{3+}) sites, troilite is an FeS sulfide, and kamacite is an Fe–Ni alloy (~5% Ni). Mössbauer velocity calibration after Morris et al. (2006a,b).

27.3 Mineralogic Assignment of Fe-Bearing Phases

Sixteen Fe-bearing phases, mutually consistent with chemistry measured by the Alpha-Particle X-ray Spectrometer (APXS; see Chapter 28), were identified at Gusev crater and Meridiani Planum (Figure 27.3) (Morris et al., 2006a,b, 2008, 2010). The most frequent and coupled detections at Gusev crater, and also common at Meridiani Planum, are Fe^{2+} in olivine ((Fe^{2+},Mg)$_2$SiO$_4$) and pyroxene ((Fe^{2+},Mg,Ca)SiO$_3$) from primary igneous minerals and Fe^{3+} in npOx (nanophase ferric oxide). NpOx is a generic name for alteration products having doublets assigned to octahedrally coordinated Fe^{3+} and includes any combination of a number of phases such as ferrihydrite, hissingerite, schwertmannite, akaganéite, and superparamagnetic hematite and goethite. The Fe^{2+} doublet assigned to olivine was instead assigned to hydrous iron sulfate by Lane et al. (2004); however, the sulfate assignment is in conflict with the negative correlation of the Fe^{2+} doublet with the concentration of sulfur (S from APXS) and the positive correlation of the npOx Fe^{3+} doublet with S for basaltic soils (Morris et al., 2006a, 2008). Other Fe-bearing igneous minerals found by MB at Gusev crater together with olivine and pyroxene are multiple detections of magnetite (Fe^{3+}($Fe^{2+}Fe^{3+}$)O$_4$) and ilmenite (Fe^{2+}TiO$_3$) and a singular detection of chromite (Fe^{2+}, Fe^{3+}, Mg, Al)Cr$_2$O$_4$). All other Fe-bearing phases at Gusev crater with two exceptions are the Fe^{3+} alteration products hematite (α-Fe$_2$O$_3$), goethite (α-FeOOH), and ferric sulfate. We have not made specific mineralogic assignments for ferric sulfate because its Mössbauer parameters are not mineralogically specific. Instead, we have adopted the generic term "Fe3Sulfate." Possible assignments consistent with other MER data include ferricopiapite, rhomboclase, and amorphous ferric sulfate (Lane et al., 2008; Dyar et al., 2013) but not the hydroxy sulfate jarosite (discussed later). The two exceptions are singular detections of pyrite/marcasite (FeS$_2$) and Fe^{2+}-bearing carbonate (Fe^{2+}, Mg, Ca)CO$_3$.

Lane et al. (2008) and Hausrath et al. (2013) have suggested that certain ferric phosphates could contribute in part to the Fe3Sulfate doublet at the Paso Robles location. While the relative proportions of phosphate versus sulfate is equivocal for MB measurements of targets PasoRobles and PasoLight1, P_2O_5 concentrations for Inner Basin and Home Plate targets (e.g., AradSamra) are insufficient for ferric phosphates to make a significant contribution to Fe3Sulfate (Dyar et al., 2014; Yen et al., 2008).

The most frequent detections at Meridiani Planum are Fe^{3+} assigned to jarosite $((K,Na,H_3O)Fe_3(SO_4)_2(OH)_6)$, hematite, and an unassigned Fe^{3+}-bearing phase termed "Fe3D3," and they occur together in the SO_3-rich bedrock. The uniqueness of the jarosite assignment was questioned by Dyar et al. (2013), but their interpretation was based on the velocity calibration of Klingelhöfer et al. (2004) rather than the revised calibration (Morris et al., 2006b) employed thereafter. The MB parameters for Fe3D3, Fe3Sulfate, and npOx are not mutually exclusive but do not generally overlap (Figure 27.3), implying distinct Fe^{3+} speciation associated with different provenances (sulfate-bearing outcrop, sulfate-bearing soil, and basaltic soil and rock, respectively). Fleischer et al. (2010a) suggest a continuum of hematite particle diameters is present, with coarse represented by the hematite concretions, intermediate represented by outcrop hematite with sextet subspectra, and fine by the Fe3D3 doublet interpreted as superparamagnetic hematite. The concretions ("blueberries") are imbedded in the SO_3-rich outcrop and occur as discrete spherules admixed with, and as lag upon, basaltic soil. The remaining Fe mineralogic assignments, Fe^0 in kamacite (a low-Ni Fe/Ni alloy) and the Fe sulfide troilite (FeS), are all associated with meteorites identified on the surface of the SO_3-rich outcrop at Meridiani Planum. Some combination of Fe-carbide cohenite (Fe_3C) and Fe-phosphide schreibersite $(Fe, Ni)_3P$ may also be present, but are considered tentative identifications because their aggregate concentration is at detection limits.

27.4 Mössbauer Mineralogy and Fe Oxidation State of Rocks and Soils

27.4.1 Scope

Our purpose is to summarize mineralogic composition and redox state of rocks and soils and their distributions across the MER landing sites at Gusev crater and Meridiani Planum. Encompassed are all MB results from Gusev crater and all those through sol 557 and a few thereafter for Meridiani Planum. The MB results alone provide first-order information concerning identification and mineralogic variability of igneous rocks and equivalent information about nonigneous rocks. We take unaltered to weakly altered basaltic rock and soil to have >75% of their total Fe associated with olivine, pyroxene, ilmenite, chromite, magnetite, kamacite, and troilite (i.e., $Fe_{igneous} \equiv Ol + Px + Ilm + Chr + Mt + Kam + Tr$ >75%). The magnitudes of $Fe_{igneous}$ and $Fe^{3+}/\Sigma Fe$ are shown in Figure 27.4 as a function of sol number as a surrogate for physical rover location. Note that even for wholly igneous rocks, $Fe^{3+}/\Sigma Fe$ can vary in accordance with the magnetite content ($Fe^{3+}/\Sigma Fe = 0.67$ for stoichiometric magnetite). Nevertheless, the magnitude of $Fe^{3+}/\Sigma Fe$ provides a quantitative measure on the extent of oxidative alteration, and the associated

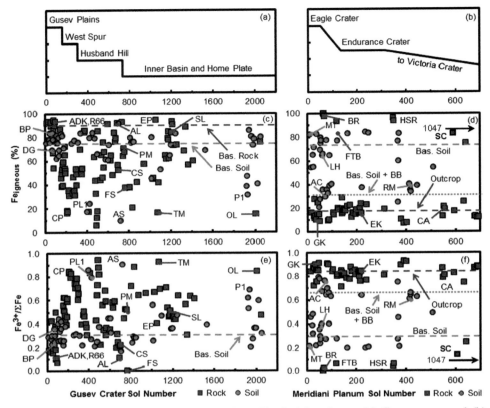

Figure 27.4 Mössbauer measurements over time. Physical location at (a) Gusev crater and (b) Meridiani Planum over time in Sols. Values of igneous $Fe_{igneous}$ for rocks and soils for analysis targets at (c) Gusev crater and (d) Meridiani Planum. Ratio of Fe^{3+} to total Fe for analysis targets at (e) Gusev crater and (f) Meridiani Planum. $Fe_{igneous} \equiv Ol + Px + Ilm + Chr + Mt + Kam + Tr$ (see text); BB = hematite concretions (blueberries). Subspectral areas and key to target names are given in Table 27.1. Locations from Arvidson et al. (2006, 2008).

mineralogic compositions of Fe^{3+}-bearing phases (e.g., hematite, goethite, and Fe^{3+} sulfate) constrain the alteration style.

Subspectral areas and values of $Fe_{igneous}$ and $Fe^{3+}/\Sigma Fe$ are compiled in Table 27.1 for the representative MB targets named in Figure 27.4, and MB spectra for specific targets are plotted in Figures 27.5 and 27.6 for Gusev crater and in Figure 27.7 for Meridiani Planum. The discussion that follows for Gusev crater is based on publications by Morris et al. (2004, 2006a, 2008, 2010), Klingelhöfer et al. (2005), Yen et al. (2005, 2008), Clark et al. (2007), Morris and Klingelhöfer (2008), and Squyres et al. (2008). Relevant publications for Meridiani Planum are Klingelhöfer et al. (2004), Morris et al. (2006b), Clark et al. (2005), Morris and Klingelhöfer (2008), Fleischer et al. (2010a,b, 2011), Schröder et al. (2008, 2010, 2016), and Zipfel et al. (2011).

Table 27.1 Sol number, target type, target surface state, subspectral areas, $Fe^{3+}/\Sigma Fe$, and $Fe_{igneous}$ for representative targets from Gusev crater and Meridiani Planum.

Target	Sol	Type	Surface	Ol (%)	Px (%)	Mt (%)	npOx (%)	Hm (%)	Other phase	(%)	$Fe^{3+}/\Sigma Fe$	$Fe_{igneous}$ (%)	Key
Gusev crater													
Adirondack	34	R	R	48	32	13	6	—	—	—	0.29	94	ADK
DesertGobi	69	S	U	30	33	7	28	2	—	—	0.36	70	DG
BearPawPanda	73	S	D	37	35	12	13	3	—	—	0.25	84	BP
Route66	100	R	B	57	37	—	7	—	—	—	0.07	93	R66
ClovisPlano	218	R	R	1	14	2	25	18	Gt	40	0.84	17	CP
PasoLight1	421	S	D	10	8	6	—	7	Fe3Sulfate	69	0.86	24	PL1
AlgonquinIroquet	690	R	B	71	13	6	8	—	Ilm	2	0.11	92	AL
ComancheSpur	702	R	B	51	—	1	16	5	Carb	27	0.22	52	CS
AradSamra	723	S	D	7	3	—	—	4	Fe3Sulfate	86	0.90	10	AS
PoseyManager	754	R	B	17	23	31	27	3	—	—	0.53	70	PM
FuzzySmith	769	R	U	4	26	—	—	—	Ilm	8	0.00	30	FS
									FeS2	63			
EsperanzaPalma	1056	R	U	—	4	45	4	1	—	—	0.40	95	EP
TrollMontalva	1073	R	U	—	3	14	5	78	—	—	0.93	17	TM
ExamineThisSlide	1177	R	B	4	33	41	22	—	—	—	0.51	78	SL
TroyPenina1	1937	S	D	15	17	—	—	—	Fe3Sulfate	68	0.68	32	P1
TroyOliveLeaf	2000	R	U	6	9	—	—	—	Fe3Sulfate	85	0.85	15	OL
Meridiani Planum													
MerlotTarmack	11	S	U	39	37	6	14	4	—	—	0.22	82	MT
GuadalupeKing3	35	R	R	1	9	—	—	36	Jar	38	0.90	10	GK
									Fe3D3	16			
MtBlancLesHauches	60	S	U	28	32	5	30	5	—	—	0.39	65	LH

Table 27.1 (cont.)

Target	Sol	Type	Surface	Ol (%)	Px (%)	Mt (%)	npOx (%)	Hm (%)	Other phase	(%)	$Fe^{3+}/\Sigma Fe$	$Fe_{igneous}$ (%)	Key
BounceRockCase	69	R	R	—	100	—	—	—	—	—	0.00	100	BR
SeasAegeanCrest	73	S	U	11	12	8	5	65	—	—	0.76	31	AC
FigTreeBarberton2	121	R	U	48	32	—	6	—	Kam Tr	11 3	0.06	94	FTB
EscherKirchner	219	R	R	1	15	—	—	35	Jar Fe3D3	30 20	0.84	16	EK
HeatShieldRock	351	R	B	—	—	—	6	—	Kam	94	0.06	94	HSR
MattsRippleMobarak	415	S	U	17	21	2	11	48	—	—	0.61	40	RM
CobblesArkansas	551	R	U	6	15	—	52	8	Jar	19	0.79	21	CA
SantaCatarina	1047	R	U	52	26	—	14	—	Kam Tr	1 6	0.13	85	SC

Notes: (1) Type: R = rock; S = soil. (2) Surface state for rocks: U = undisturbed surface; B = brushed surface; R = surface removed with Rock Abrasion Tool. (3) Surface state for soils: U = undisturbed surface; D = surface disturbed by rover wheels. (4) Subspectral areas (corrected for recoil-free fractions; see text): Ol = Fe^{2+} associated with olivine; Px = Fe^{2+} associated with pyroxene; Ilm = Fe^{2+} associated with ilmenite; Mt = Fe^{2+} and Fe^{3+} associated with magnetite; Chr = Fe^{2+} and Fe^{3+} associated with chromite; npOx = Fe^{3+} associated with nanophase ferric oxide; Hm = Fe^{3+} associated with hematite; Gt = Fe^{3+} associated with goethite; Carb = Fe^{2+} associated with carbonate; Fe3Sulfate = Fe^{3+} associated with unassigned doublet sulfate phase; FeS2 = Fe sulfide associated with pyrite/marcasite; Kam = Fe metal alloy with ~5 wt.% Ni; Jar = Fe^{3+} associated with jarosite; Fe3D3 = Fe^{3+} associated with unassigned doublet phase at Meridiani Planum; Tr = Fe associated with troilite. (5) $Fe_{igneous}$ = Ol + Px +Ilm + Mt + Chr + Kam + Tr. (6) Formally SpongeBobSquidward but popularly known as HeatShieldRock. Kamacite for HSR includes minor contributions from cohenite and/or schreibersite.

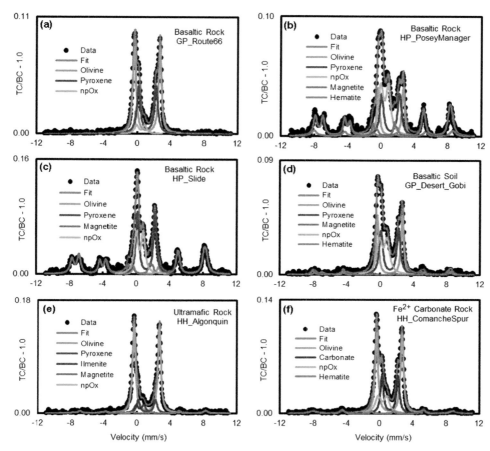

Figure 27.5 Mössbauer spectra for Gusev crater basaltic rocks. (a) Route66, (b) PoseyManager, and (c) Slide, (d) basaltic soil DesertGobi, (e) ultramafic rock Algonquin, and (f) Fe^{2+}-bearing carbonate rock ComancheSpur. TC/BC = (Total Counts)/(Baseline Counts); GP = Gusev plains; HP = Home Plate; HH = Husband Hill.

27.4.2 Gusev Crater

Basaltic rocks ($Fe_{igneous} > 75\%$) are prevalent at the Gusev crater landing site (Figure 27.4). Olivine-rich basaltic rocks are common on the Gusev plains (e.g., Route66; Figure 27.5a), and magnetite-rich basaltic rocks with relatively more Fe^{2+} in pyroxene and less in olivine are common in the Columbia Hills at West Spur, Husband Hill, and Home Plate (e.g., PoseyManager and Slide; Figure 27.5b,c). Basaltic soil (Figure 27.5d) at Gusev crater is most akin to rock Route66 and similar rocks (e.g., Adirondack and Humphrey) on the Gusev plains, implying the abundance of that mineralogic combination across Mars as opposed to magnetite-rich assemblages. Rocks with ultramafic Fe^{2+} mineral assemblages (olivine dominant) were analyzed on Husband Hill (e.g., Algonquin; Figure 27.5e), and just downslope Spirit analyzed outcrop with Fe^{2+}-bearing carbonate (Figure 27.5f). The ComancheSpur outcrop is the only carbonate-bearing target detected by the MER rovers.

The dominant Fe^{3+} alteration product on the Gusev plains is npOx. Its concentration is highly variable in basaltic soil (e.g., 13% to 28% for BearPawPanda and DesertGobi, respectively; Table 27.1), and npOx is also associated with basaltic dust. The dust coats rock surfaces, decreasing $Fe_{igneous}$ when the dust coating is not effectively removed by brushing or abrasion by the Rock Abrasion Tool (RAT). Consistent with the assignment of the npOx ferric doublet to that phase and not, for example, an assignment to an igneous phase, is its correlation with the concentrations of SO_3 and Cl.

Highly altered rocks with goethite (detected only in rocks), hematite, and npOx (not associated with dust) as Fe^{3+}-bearing alteration products (e.g., WS_ClovisPlano; Figure 27.6a) were found at West Spur and Husband Hill. At West Spur and Home Plate, the churning action of Spirit's wheels revealed, just beneath the surface, light-toned soil with Fe^{3+}-bearing sulfate as the dominant Fe-bearing phase (e.g., HH_PasoLight1 and HP_AradSamra; Figure 27.6b). Fe^{3+}-bearing sulfate was detected only in soil with the exception of a small piece of crust (i.e., cemented soil). In close physical association with the Fe^{3+}-bearing sulfate soils at Home Plate are outcrops where the dominant Fe^{3+}-bearing alteration product is hematite

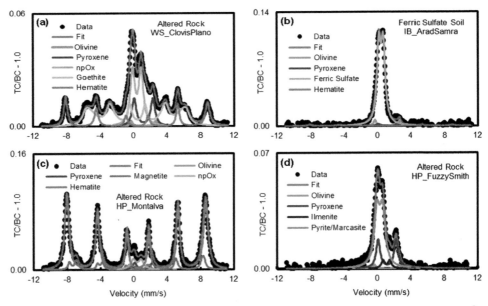

Figure 27.6 Mössbauer spectra for Gusev crater. (a) Highly altered rock WS_ClovisPlano. (b) Fe^{3+}-bearing sulfate soil IB_AradSamra. (c) Hematite-rich outcrop HP_Montalva. (d) Pyrite/marcasite-bearing float rock HP_FuzzySmith. TC/BC = (Total Counts)/(Baseline Counts); WS = West Spur; IB = Inner Basin;. HP = Home Plate.

(e.g., HP_Montalva; Figure 27.6c). A singular detection of pyrite-marcasite (FeS$_2$) was the float rock FuzzySmith (Figure 27.6d) located at Home Plate. The Home Plate structure itself is considered to be a fumarolic, acid-sulfate environment where Fe^{3+}-bearing sulfates are precipitation products of leachate solutions derived from (presumably local) basaltic progenitors. Co-located hematite is interpreted as a primary precipitate, possibly under hydrothermal conditions, or as a diagenetic alteration product of Fe^{3+}-bearing sulfates.

27.4.3 Meridiani Planum

MB measurements at Meridiani Planum are dominated by analyses of the SO$_3$-rich and jarosite-bearing bedrock, with average Fe$_{igneous}$ ~17 % and Fe^{3+}/ΣFe ~0.85 (Figure 27.4d,f). The MB data in the figure refer to outcrop matrix and do not include contributions from imbedded hematite concretions (a.k.a. blueberries). The Fe mineralogy of the outcrop matrix (Figure 27.7a) is approximately subequal proportions of Fe^{3+} associated with jarosite, hematite, and an unassigned Fe^{3+}-bearing phase (Fe3D3), possibly superparamagnetic hematite. Soils are either basaltic (Figure 27.7b) or are mechanical mixtures of basaltic soil and hematite concretions winnowed from the outcrop. Basaltic soil MtBlancLesHauches has high proportions of npOx and is very similar to DesertGobi at Gusev crater with respect to mineralogic composition (Table 27.1). The concretions tend to be concentrated as lag deposits on ripple crests (Figure 27.7c). No soil mineralogically equivalent to the outcrop was found, and no outcrops with basaltic mineralogic compositions were analyzed by MB while the instrument was operational. The MB detection of jarosite was particularly important because the mineral precipitates from aqueous acid sulfate solutions (pH <3–4), implying that the outcrop is the product of acid-sulfate alteration of basaltic progenitors.

Float rocks ranging from pebble to boulder size (referred to as cobbles) were analyzed by MB on the surface of the SO$_3$-rich and jarosite-bearing bedrock and were interpreted as impact ejecta (e.g., BounceRock; Figure 27.7d), impact breccia (e.g., CobblesArkansas), stony meteorites (e.g., FigTreeBarberton; Figure 27.7e), and iron meteorites (e.g., HeatShieldRock; Figure 27.7f). Interestingly, BounceRock is, chemically and mineralogically, a Shergottite-like basalt. The MB detection of kamacite and/or troilite firmly established that cobbles like FigTreeBarberton are in fact stony meteorites on the martian surface and not indigenous basaltic rocks. Furthermore, the chemical and mineralogic similarity of FigTreeBarberton, SantaCatarina, Santorini, and Kasos implies that they are paired meteorites and possibly fragments of the impactor that created Victoria crater.

27.5 Seeing through the Dust on Mars

The Rock Abrasion Tool was designed to brush rock surfaces relatively free of airfall dust and saltating soil particles and to grind into rock surfaces to remove this material and also sufficiently thin "alteration" rinds in order to analyze "pristine" surfaces beneath. Because of the significant mineralogic and chemical contrast between basaltic soil (high Fe$_{igneous}$

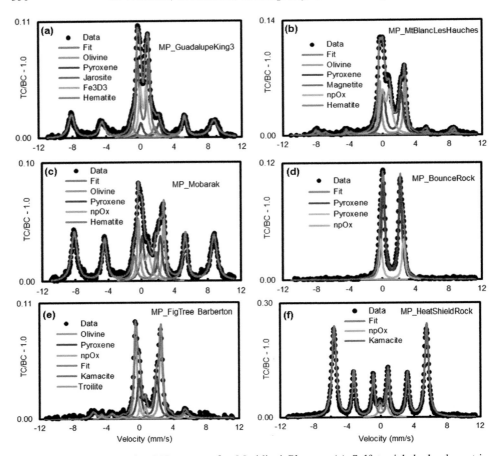

Figure 27.7 Representative MB spectra for Meridiani Planum. (a) Sulfate-rich bedrock matrix. (b) Basaltic soil MontBlancLesHauches. (c) Basaltic soil with hematite concretions (blueberries) as a ripple-crest lag deposit. (d) Impact ejecta BounceRock. (e) Stony meteorite FigTreeBarberton. (f) Iron meteorite HeatShieldRock. TC/BC = (Total Counts)/(Baseline Counts); MP = Meridiani Planum.

and low SO_3 concentration) and the outcrop (low $Fe_{igneous}$ and high SO_3), measurement of RATed and undisturbed outcrop surfaces at Meridiani Planum are ideal for demonstrating "contamination" of the mineralogy and chemistry of rock surfaces by basaltic air-fall dust and saltating particles with respect to measurements by MB and APXS instruments.

Figure 27.8a is a binary plot using data derived from only MB measurements (14.4-keV emitted gamma rays). The y-axis ($Fe_{igneous}$ + npOx) is the percentage of total Fe associated with blueberry-free basaltic soil at Meridiani Planum except for minor hematite, and the x-axis is the percentage of total Fe associated with jarosite. The plot shows that

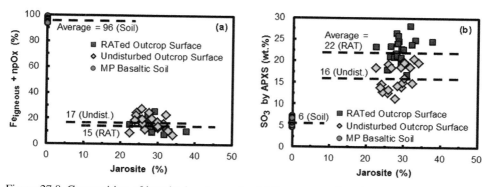

Figure 27.8 Composition of jarosite-bearing rocks. (a) Percentage of total Fe present as Fe$_{igneous}$ + npOx and (b) SO$_3$ concentration from APXS as a function of the percentage of total Fe present as jarosite for Meridiani Planum (MP) basaltic soil, undisturbed surfaces, and RATed outcrop rock surfaces. RAT = Rock Abrasion Tool.

there is no meaningful difference in MB data obtained from undisturbed and RATed surfaces. In Figure 27.8b, where the y-axis is the SO$_3$ concentration by APXS, there is a clear offset in SO$_3$ concentrations, with RATed surfaces on average having elevated SO$_3$ concentrations compared to undisturbed surfaces. The offset is interpreted as a deeper sampling depth for MB compared to APXS-class instruments, and it underscores the science value of RATing (or at least brushing) to obtain reliable rock chemistry in the presence of dust/soil coatings. On the basis of laboratory experiments, a layer of basaltic dust >3 mm thick is required to completely mask MB detection (14.4 keV) of underlying substrate (see Section 27.2).

Therefore, an important feature of MB is that it can "see through" thin layers of dust with no significant difference between undisturbed and brushed surfaces registered in spectra. More substantial layers of dust and weathering rinds or coatings do register in the spectra. Although not extensively investigated to date, the standard 14.4-keV gamma-ray spectra and the 6.4-keV X-ray spectra simultaneously acquired by the MIMOS II instrument can provide information on layer thickness. The difference in energy results in different penetration depths as a result of greater attenuation of lower energy radiation, and the difference allowed the successful identification and characterization of coatings such as on the rock Mazatzal in Gusev crater (e.g., Fleischer et al., 2008).

27.6 Summary

The Mössbauer spectrometers on the MER rovers provided first-order information on the mineralogic composition, diversity, and oxidation state of igneous materials and their mineralogic alteration products. A total of 16 Fe-bearing phases were identified. At Gusev crater, two major basaltic lithologies were identified: one with olivine and pyroxene and the other with pyroxene, variable olivine, and magnetite as the dominant

Fe-bearing phases. Basaltic soil mineralogically similar to the olivine–pyroxene lithology but with variable proportions of npOx (associated with bright martian dust) was present throughout Spirit's traverse. The Columbia Hills of Gusev crater are strongly altered compared to the Gusev plains and are characterized by a variety of Fe^{3+} bearing alteration products including npOx, goethite, hematite, and Fe^{3+}-bearing sulfate. An outcrop containing Fe^{2+}-bearing carbonate was also detected in the Columbia Hills. This diversity in alteration products is a manifestation of a variety of alteration processes ranging from circum-neutral (goethite and carbonate) to acid-sulfate (Fe^{3+} sulfate). The assemblage of Fe-bearing minerals is very different at Meridiani Planum, because the bedrock is dominated by the Fe^{3+}-bearing minerals hematite, jarosite, and an unassigned Fe^{3+}-bearing phase. Basaltic soil similar to the olivine-pyroxene soil at Gusev crater is present on outcrop surfaces, both with and without admixed hematite concretions winnowed from the outcrop. Rocks that litter the Meridiani outcrop surface are interpreted as meteorites by the presence of kamacite and/or troilite or as indigenous basaltic ejecta from impacts into the martian surface by the absence of those phases. MB instruments can obtain mineralogic information on rock surfaces in the presence of dust coatings that would compromise chemical measurements by APXS-class instruments.

Acknowledgments

R. V. M. acknowledges support of NASA's Mars Exploration Program, MER mission, and the NASA Johnson Space Center. C.S. acknowledges support of the UK Space Agency through grant ST/R001278/1 administered by the Science and Technology Facilities Council. Part of the work described in this chapter was conducted at the Jet Propulsion Laboratory, California Institute of Technology, under a contract with NASA. The authors acknowledge steadfast and thorough support of the MER Science and JPL Operations Teams. Our friend and colleague Göstar Klingelhöfer passed unexpectedly on January 13, 2019. He was a leading expert in Mössbauer spectroscopy, developed the highly successful Mössbauer MIMOS II instruments that were an essential part of the MER mission, and continued to actively advocate Mössbauer instrumentation for planetary exploration and geoscience. He is missed.

References

Arvidson R.E., Squyres S.W., Anderson R.C., et al. (2006) Overview of the spirit Mars Exploration Rover mission to Gusev crater: Landing site to Backstay Rock in the Columbia Hills. *Journal of Geophysical Research*, **111**, E02S01, DOI:10.1029/2005JE002499.

Arvidson R.E., Ruff S.W., Morris R.V., et al. (2008) Spirit Mars rover mission to the Columbia Hills, Gusev crater: Mission overview and selected results from the Cumberland Ridge to Home Plate. *Journal of Geophysical Research*, **113**, E12S33, DOI:10.1029/2008JE003183.

Arvidson R.E., Bell J.F. III, Bellutta P., et al. (2010) Spirit Mars rover mission: Overview and selected results from the northern Home Plate Winter Haven to the side of Scamander crater. *Journal of Geophysical Research*, **115**, E00F15, DOI:10.1029/2010JE003746.

Arvidson R.E., Ashley J.W., Bell J.F., et al. (2011) Opportunity Mars rover mission: Overview and selected results from Purgatory ripple to traverses to Endeavour crater. *Journal of Geophysical Research*, **116**, E00F15, DOI:10.1029/2010JE003746.

Ashley J.W., Golombek M., Christensen P.R., et al. (2011) Evidence for mechanical and chemical alteration of iron-nickel meteorites on Mars: Process insights for Meridiani Planum. *Journal of Geophysical Research*, **116**, E00F20, DOI:10.1029/2010JE003672.

Bancroft G.M. (1973) *Mössbauer spectroscopy: An introduction for inorganic chemists and geochemists.* McGraw-Hill, New York.

Burns R.G. (1993) *Mössbauer spectral characterization of iron in planetary surface materials.* Cambridge University Press, Cambridge, 539–556.

Burns R.G. & Solberg T.C. (1990) ^{57}Fe-bearing oxide, silicate, and aluminosilicate minerals, crystal structure trends in Mössbauer spectra. In: *Spectroscopic characterization of minerals and their surfaces* (L.M. Coyne, S. W.S. McKeever, & D.F. Blake, eds.). American Chemical Society, Washington, DC, 262–283.

Clark B.C., Morris R., McLennan S., et al. (2005) Chemistry and mineralogy of outcrops at Meridiani Planum. *Earth and Planetary Science Letters*, **240**, 73–94.

Clark B.C., Arvidson R.E., Gellert R., et al. (2007) Evidence for montmorillonite or its compositional equivalent in Columbia Hills, Mars. *Journal of Geophysical Research*, **112**, E06S01, DOI:10.1029/2006JE002756.

De Grave E. & Van Alboom A. (1991) Evaluation of ferrous and ferric Mössbauer fractions. *Physics and Chemistry of Minerals*, **18**, 337–342.

Dyar M.D., Breves E., Jawin E., et al. (2013) Mössbauer parameters of iron in sulfate minerals. *American Mineralogist*, **98**, 1943–1965.

Dyar M.D., Jawin E.R., Breves E., et al. (2014) Mössbauer parameters of iron in phosphate minerals: Implications for interpretation of martian data. *American Mineralogist*, **99**, 914–942.

Fleischer I., Klingelhoefer G., Schröder C., et al. (2008) Depth selective Mössbauer spectroscopy: Analysis and simulation of 6.4 keV and 14.4 keV spectra obtained from rocks at Gusev crater, Mars, and layered laboratory samples. *Journal of Geophysical Research*, **113**, E06S21, DOI:10.1029/2007JE003022.

Fleischer I., Agresti D., Klingelhöfer G., & Morris R. (2010a) Distinct hematite populations from simultaneous fitting of Mössbauer spectra from Meridiani Planum, Mars. *Journal of Geophysical Research*, **115**, E00F06, DOIg: 10.1029/2010JE003622.

Fleischer I., Brueckner J., Schröder C., et al. (2010b) Mineralogy and chemistry of cobbles at Meridiani Planum, Mars, investigated by the Mars Exploration Rover Opportunity. *Journal of Geophysical Research*, **115**, E00F05, DOI:10.1029/2010JE003621.

Fleischer I., Schroeder C., Klingelhoefer G., et al. (2011) New insights into the mineralogy and weathering of the Meridiani Planum meteorite, Mars. *Meteoritics and Planetary Science*, **46**, 21–34.

Graff T., Morris R., & Christensen P. (2001) Effects of palagonitic dust coatings on thermal emission spectra of rocks and minerals: Implications for mineralogical characterization of the martian surface by MGS-TES. *32nd Lunar Planet. Sci. Conf.*, Abstract #1899.

Gütlich P. & Schröder C. (2012) Mössbauer spectroscopy. In: *Methods in physical chemistry* (R. Schäfer & P. C. Schmidt, eds.). Wiley-VCH, Weinheim, Germany, 351–389.

Gütlich P., Bill E., & Trautwein A.X. (2011) *Mössbauer spectroscopy and transition metal chemistry.* Springer, Berlin and Heidelberg.

Hausrath E., Golden D., Morris R., Agresti D., & Ming D. (2013) Acid sulfate alteration of fluorapatite, basaltic glass and olivine by hydrothermal vapors and fluids: Implications for fumarolic activity and secondary phosphate phases in sulfate-rich Paso Robles soil at Gusev crater, Mars. *Journal of Geophysical Research*, **118**, 1–13.

Hawthorne F.C. (1988) *Mössbauer spectroscopy.* Mineralogical Society of America, 255–340.

Klingelhöfer G., Morris R.V., Bernhardt B., et al. (2003) Athena MIMOS II Mössbauer spectrometer investigation. *Journal of Geophysical Research*, **108**, 8067, DOI:10.1029/2003JE002138.

Klingelhöfer G., Morris R.V., Bernhardt B., et al. (2004) Jarosite and hematite at Meridiani Planum from Opportunity's Mössbauer spectrometer. *Science*, **306**, 1740–1745.

Klingelhöfer G., DeGrave E., Morris R.V., et al. (2005) Mössbauer spectroscopy on Mars: Goethite in the Columbia Hills at Gusev crater. *Hyperfine Interactions*, **166**, 549–554.

Lane M.D., Dyar M.D., & Bishop J.L. (2004) Spectroscopic evidence for hydrous iron sulfate in the martian soil. *Geophysical Research Letters*, **31**, L19702, DOI:10.1029/2004GL021231.

Lane M.D., Bishop J.L., Darby Dyar M., King P.L., Parente M., & Hyde B.C. (2008) Mineralogy of the Paso Robles soils on Mars. *American Mineralogist*, **93**, 728–739.

McCammon C. (1995) Mössbauer spectroscopy of minerals. In: *Mineral physics and crystallography: A handbook of physical constants* (T.J. Ahrens, ed.). American Geophysical Union, Washington, DC, 332–347.

McSween H.Y., Ruff S.W., Morris R.V., et al. (2008) Mineralogy of volcanic rocks in Gusev crater, Mars: Reconciling Mössbauer, Alpha Particle X-Ray Spectrometer, and Miniature Thermal Emission Spectrometer spectra. *Journal of Geophysical Research*, **113**, E06S04, DOI:10.1029/2007JE002970.

Morris R.V. & Klingelhöfer G. (2008) Iron mineralogy and aqueous alteration on Mars from the MER Mössbauer spectrometers. In: *The martian surface* (J.F. Bell III, ed.). Cambridge University Press, Cambridge, 339–365.

Morris R.V., Golden D., Bell J.F. III, & Lauer H. Jr. (1995) Hematite, pyroxene, and phyllosilicates on Mars: Implications from oxidized impact melt rocks from Manicouagan crater, Quebec, Canada. *Journal of Geophysical Research*, **100**, 5319–5328.

Morris R.V., Golden D., Ming D., et al. (2001a) Phyllosilicate-poor palagonitic dust from Mauna Kea Volcano (Hawaii): A mineralogical analogue for magnetic martian dust? *Journal of Geophysical Research*, **106**, 5057–5083.

Morris R.V., Graff T., Shelfer T., & Bell J. III (2001b) Effects of palagonitic dust coatings on visible, near-IR, and Mössbauer spectra of rocks and minerals: Implication for mineralogical remote sensing of Mars. *32nd Lunar Planet. Sci. Conf.*, Abstract #1912.

Morris R.V., Klingelhöfer G., Bernhardt B., et al. (2004) Mineralogy at Gusev crater from the Mössbauer spectrometer on the Spirit rover. *Science*, **305**, 833–836.

Morris R.V., Klingelhöfer G., Schröder C., et al. (2006a) Mössbauer mineralogy of rock, soil, and dust at Gusev crater, Mars: Spirit's journey through weakly altered olivine basalt on the plains and pervasively altered basalt in the Columbia Hills. *Journal of Geophysical Research*, **111**, E02S13, DOI:10.1029/2005JE002584.

Morris R.V., Klingelhöfer G., Schröder C., et al. (2006b) Mössbauer mineralogy of rock, soil, and dust at Meridiani Planum, Mars: Opportunity's journey across sulfate-rich outcrop, basaltic sand and dust, and hematite lag deposits. *Journal of Geophysical Research*, **111**, E12S15, DOI:10.1029/2006JE002791.

Morris R.V., Klingelhöfer G., Schröder C., et al. (2008) Iron mineralogy and aqueous alteration from Husband Hill through Home Plate at Gusev crater, Mars: Results from the Mössbauer instrument on the Spirit Mars Exploration Rover. *Journal of Geophysical Research*, **113**, E12S42, DOI:10.1029/2008JE003201.

Morris R.V., Ruff S.W., Gellert R., et al. (2010) Identification of carbonate-rich outcrops on Mars by the Spirit rover. *Science*, **329**, 1189667.

Schröder C., Rodionov D.S., McCoy T.J., et al. (2008) Meteorites on Mars observed with the Mars Exploration Rovers. *Journal of Geophysical Research*, **113**, E06S22, DOI:10.1029/2007JE002990.

Schröder C., Herkenhoff K.E., Farrand W.H., et al. (2010) Properties and distribution of paired candidate stony meteorites at Meridiani Planum, Mars. *Journal of Geophysical Research*, **115**, E00F09, DOI:10.1029/2010JE003616.

Schröder C., Bland P.A., Golombek M.P., Ashley J.W., Warner N.H., & Grant J.A. (2016) Amazonian chemical weathering rate derived from stony meteorite finds at Meridiani Planum on Mars. *Nature Communications*, **7**, 13459.

Stevens J.G., Khasanov A.M., Miller J.W., Pollak H., & Li Z. (1998) *Mössbauer mineral handbook*. Biltmore Press, Ashville, NC.

Squyres S.W., Arvidson R.E., Baumgartner E.T., et al. (2003) The Athena Mars rover science investigation. *Journal of Geophysical Research*, **108**, 8062, DOI:10.1029/2003JE002121.

Squyres S.W., Arvidson R.E., Bell J.F., et al. (2004a) The Opportunity rover's Athena science investigation at Meridiani Planum, Mars. *Science*, **306**, 1698–1703.

Squyres S.W., Arvidson R.E., Bell J.F., et al. (2004b) The Spirit rover's Athena science investigation at Gusev crater, Mars. *Science*, **305**, 794–799.

Squyres S.W., Arvidson R.E., Bollen D., et al. (2006) Overview of the Opportunity Mars Exploration Rover mission to Meridiani Planum: Eagle crater to Purgatory ripple. *Journal of Geophysical Research*, **111**, DOI:10.1029/2006JE002771.

Squyres S.W., Arvidson R.E., Ruff S., et al. (2008) Detection of silica-rich deposits on Mars. *Science*, **320**, 1063–1067.

Squyres S.W., Knoll A.H., Arvidson R.E., et al. (2009) Exploration of Victoria crater by the Mars rover Opportunity. *Science*, **324**, 1058–1061.

Yen A.S., Gellert R., Schröder C., et al. (2005) An integrated view of the chemistry and mineralogy of martian soils. *Nature*, **436**, 49–54.

Yen A.S., Morris R.V., Clark B.C., et al. (2008) Hydrothermal processes at Gusev crater: An evaluation of Paso Robles class soils. *Journal of Geophysical Research*, **113**, E06S10, DOI:10.1029/2007JE002978.

Yoshida Y. & Langouche G. (2013) *Mössbauer spectroscopy*. Springer, Berlin.

Zipfel J., Schröder C., Jolliff B.L., et al. (2011) Bounce Rock: A shergottite-like basalt encountered at Meridiani Planum, Mars. *Meteoritics and Planetary Science*, **46**, 1–20.

28

Elemental Analyses of Mars from Rovers Using the Alpha-Particle X-Ray Spectrometer

RALF GELLERT AND ALBERT S. YEN

28.1 Introduction

The Alpha-Particle X-ray Spectrometer (APXS) is a contact instrument that has been on the scientific payload of all four Mars rovers as of 2018 (Figure 28.1). It determines the chemical composition of rocks and soils using X-ray spectroscopy. When the sensor head is deployed in close contact to the sample, radioactive ^{244}Cm sources irradiate the rock or soil with high energy alpha particles and X-rays. This irradiation stimulates element specific X-ray emissions that are recorded with an energy-sensitive detector. In principle, all elements heavier than fluorine can be identified and quantified by their characteristic X-ray emission lines. Typically, 16 elements are quantified for each martian sample, and an additional 10 trace elements can be quantified for higher than usual abundances.

The APXS method is theoretically well understood and similar to standard methods applied in terrestrial labs. Sample preparation is not needed; however, brushing dust off rock surfaces has been highly beneficial for obtaining the composition of rocks. Subsurface soils have been exposed by the rover wheels. The interior of rocks can be accessed through removal of surface coatings and alteration rinds by the Rock Abrasion Tool (RAT) on the Mars Exploration Rovers (MERs) or by the drill on the Mars Science Laboratory (MSL).

The APXS instruments have provided bulk compositional data at four landing sites analyzing >1000 samples along a combined total traverse of ~70 km across Mars. The diverse composition of soils and rocks has provided important clues about martian geology and the environmental conditions that produced the observed elemental trends. The typical soils at all landing sites are very similar to each other (O'Connell-Cooper et al., 2017) and are basaltic in composition, but enriched in S, Cl, and Zn, likely from volcanic exhalations (Yen et al., 2005). The composition of the wind mixed soils has been used as a proxy for the average martian crust (Taylor & McLennan, 2008) and was used for calibration of orbital instruments (Boynton et al., 2007). Unusual soils were also measured at each landing site that exhibit elemental trends different from the globally similar soil as discussed later in this chapter.

A variety of igneous rocks have been classified using the bulk compositional data. These range from primitive basalts to felsic mugearites. High sulfur concentrations were found at

Figure 28.1 Images of the three generations of APXS instruments on Mars. (a) The 1997 Pathfinder rover Sojourner had an APXS mounted at the front of the rover. (b) The identical MER rovers each had an Instrument Deployment Device (IDD) with four contact instruments, including the APXS. (c) View of the APXS instrument on Opportunity (d) View of the APXS instrument on Spirit. (e) The much larger MSL arm with the APXS in the center and its rectangular contact plate.

multiple sites including Ca sulfate veins, ferric sulfate deposits, and the extensive Burns formation sandstone with ~ 30% sulfate, indicative of past interactions with acidic fluids.

The APXS bulk geochemistry complements mineralogy data and images from all missions and often delivers crucial constraints for the interpretation of other investigations, such as ground truth for orbital remote sensing instruments. It also enables direct comparison of the chemistry on Mars with martian meteorites.

28.2 The APXS Method

28.2.1 Physics Principles of the APXS Method

In terrestrial laboratories X-ray spectroscopy is one of the standard methods to determine the chemical composition of bulk samples. In the laboratory the excitation can be performed either by X-rays using X-ray tubes (XRF, X-ray Fluorescence) or high-energy charged particle beams (PIXE, particle induced X-ray emission) requiring an accelerator. The APXS instrument on the Mars rovers uses a combination of PIXE and XRF (Figure 28.2). This is achieved by using radioactive curium sources. ^{244}Cm undergoes alpha decay, emitting a ~5.8 MeV alpha particle. In ~10% of all decays, the alpha particle is followed immediately by plutonium X-rays, as the decay proceeds into an excited Pu daughter state. The curium sources are covered by thin titanium foils to prevent self-sputtering, which lower the energy to about 5.2 MeV (Rieder et al., 2003).

The APXS instrument consists of a main electronics box housed in the rover's body, and the sensor head, mounted on an arm, which can deploy the sensor head to the intended

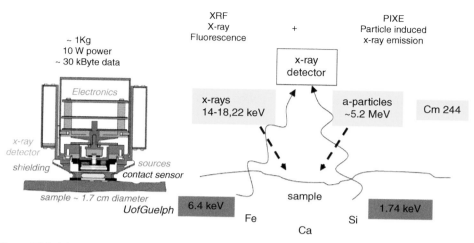

Figure 28.2 Diagram illustrating how APXS works. (Left) Schematic view across the MSL APXS sensor head. The centered silicon drift detector (SDD) X-ray detector is shielded from direct radiation to reduce the background signal. The right side illustrates the APXS method as a combination of XRF and PIXE. The energies of the stimulated X-rays identify the elements, and abundance is extracted from the number of detected element-specific X-rays.

sample. Once deployed, the radioactive sources irradiate the sample with energetic alpha particles and X-rays. This irradiation excites atoms of the sample and induces X-ray emission that is measured with a high energy resolution X-ray detector inside the sensor head.

The excitation of X-rays in the sample is a multistep process. To stimulate the emission of X-rays from an atom in the sample, one of the innermost bound electrons of this atom needs to be ejected by the irradiation. This can be achieved in two different ways: either by simple Coulomb interaction of the energetic alpha particles with the bound electron, or by photoeffect, in which an energetic X-ray is absorbed. It transfers all of its energy to the innermost bound electron and ejects it with excess kinetic energy. Notably, the inner shell ionization of light elements is most efficient with charged particles resulting from PIXE, while the photoeffect interaction probability is dominant for heavier elements.

After generation of an inner-shell vacancy, there are two competing processes to return the excited atom back to the ground state. One is X-ray emission, in which an outer electron jumps into the vacant inner shell, emitting the difference in the binding energies as an X-ray photon. Because electron binding energies are specific for each atom, the X-ray energies are characteristic for each element. Depending on the original outer shell electron energy, several different, but still element-specific energies can be emitted. The competing process to X-ray emission is Auger electron emission, in which an electron is emitted, carrying away the excess energy as kinetic energy. The ratio of emitted X-rays per inner shell vacancy is called fluorescence yield. It is strongly dependent on the atomic number (Z) and the yield is only a few percent for low-Z elements such as oxygen or silicon. The fluorescence yield increases to ~30% for the high-Z elements such as Fe (Campbell et al., 2012).

To fully understand the APXS method, one also needs to discuss the depth range of the incoming X-rays and alpha particles, as well as the probability that the X-rays generated within the sample will actually reach the surface and be detected.

Alpha particles of ~5 MeV penetrate to a depth of ~20 μm in a typical sample. They lose energy by Coulomb interaction with the weakly bound electrons (~1–10 eV) and the innermost bound electrons (~1–80 keV). Due to their high energy the alpha particles typically generate several inner-shell vacancies in the topmost 20 μm of the sample along their path until they lose all their kinetic energy. The X-rays emitted by the ^{244}Cm source have several specific energies from 14 to 21 keV. These X-rays are typically absorbed primarily by interacting once through photoeffect and penetrate to a depth of ~200 μm into the sample.

The most important factor governing the depth of the X-rays detected in an APXS measurement comes from the range of the characteristic X-rays emitted by each element within the sample. The probability (or cross section) for photoeffect on any atom in the sample is always the highest if the X-ray energy is just above the binding energy of the electrons. It decreases drastically when the X-ray energy is much higher, which means that photoeffect usually targets only the innermost bound electrons. Therefore X-rays from light elements such as Na or Si, with energies of 1.04 and 1.74 keV respectively, have a range of only 1–5 μm, because these energies are already quite effectively absorbed by the electrons bound in inner shells of oxygen atoms, which typically comprise ~50% of the mass of rocks and soils. Fe X-rays of 6.4 keV on the other hand can travel to a depth of ~50 μm in typical samples, because with their higher energy they are absorbed effectively by elements such as Ca or Ti, which typically make up only a few percent of the sample.

To summarize, X-rays from low-Z elements are mainly excited by alpha particles, i.e., PIXE. They typically originate from the upper 1–5 μm on the surface of the sample. For elements with higher Z, the excitation through incoming X-rays (XRF) increases. Titanium, Z = 22, is approximately the element where PIXE and XRF have equal contributions. Higher Z elements such as iron are nearly exclusively excited by XRF. The strength of the signal is also impacted by the fluorescence yield and the thickness of the layer that contributes to the overall signal from the sample. Low-Z elements below Ti have small fluorescence yields and are emitted from thinner layers because excited X-rays in larger depths are absorbed before they reach the surface to be detected. However, with the strong alpha emitting sources in the APXS, X-rays from low-Z elements are excited very efficiently through PIXE, which can also generate many X-rays per alpha particle. High-Z-element signals benefit from the higher fluorescence yield and from the fact that X-rays with higher energies can escape the sample from greater depths. Overall, the ^{244}Cm sources deliver a nicely balanced signal for all elements, as can be seen in Figure 28.4.

^{244}Cm has been the first choice of radioisotope for all APXS instruments (Radchenko et al., 2000), as its long half-life of 18 years makes it suitable for long-duration missions such as MER, MSL, or Rosetta, which can last more than a decade. A key property of ^{244}Cm is the extremely low intensity of emitted higher energy gamma rays from internal nucleus transitions. These high-energy photons are not easily shielded inside the instrument and produce a mostly flat background through Compton scattering inside the sample and the detector.

28.2.2 The APXS Design – Changes and Improvements over Time

The APXS has undergone significant design changes and improvements from Pathfinder to MER and subsequently to MSL. The goal of these changes was to improve the sensitivity of the instrument, i.e., to measure samples in less time by higher counting rates and to improve the detection limits for trace elements.

The original Pathfinder instrument (Rieder et al., 1997) was mainly optimized for Rutherford Backscattering (RBS) with a thin alpha detector in the center of a ring of ^{244}Cm sources. The Si-PIN detector for X-rays was attached to the side. Because RBS requires near 180° backscattering angles, the design required a larger distance to the sample.

The MER instruments, designed and developed at the Max-Planck-Institute for Chemistry (Rieder et al., 2003), introduced the then-new silicon drift detector (SDD), which was placed in the center of the ring of sources. Alpha detectors, which were still required for MER by NASA for detecting carbon and oxygen, were arranged as a ring outside the sources. However, the RBS mode was found to be very complicated to analyze because of the ~6 mbar CO_2 atmosphere on Mars that contributes a signal overlapping the much smaller one from the samples.

The design for the MSL APXS was based on the experience gained from the first years of MER operations. The alpha detectors were abandoned, which allowed the distance from the sources and detector to the sample to be shortened by 1 cm, resulting in a gain of about three times higher count rate compared to the MER APXS.

The next design change on MSL concerned the ^{244}Cm sources. The low X-ray energy spectrum containing the lower Z elements sodium to calcium has the highest count rates from the PIXE excitation, enabling short integration times of some 10 minutes for good quality data. This high sensitivity for low-Z elements with the alpha particles gives the APXS a significant advantage over pure XRF instruments like the XRF mode of CheMin, or the future PIXL instrument on Mars 2020. However, as important high-Z trace elements above nickel typically have low concentrations (~10 ppm), a more intense X-ray source was desired to increase the XRF part of the spectrum. Therefore, for MSL, the six 5 mCi ^{244}Cm sources, each emitting alphas and X-rays, were replaced by three sources of 10 mCi each. This freed up space directly around the X-ray detector to add three sealed 30 mCi ^{244}Cm sources, which only emit the X-rays. This resulted in an increase in sensitivity from MER to MSL by a factor of 3 for low-Z elements through the close-up geometry and a factor of ~6 for high-Z elements by essentially doubling the XRF excitation sources.

The MSL instrument, built by MDA, Brampton and supported by the Canadian Space agency, has several other improvements over MER. While the MER instrument relied on a mechanical contact plate, the MSL APXS introduced the so-called proximity mode, where short measurements are taken during deployment to determine the approximate distance to the sample. This provides for an optimized deployment with defined small standoff for uneven surfaces. Also, the X-ray detector on MSL employs a Peltier cooler that allows for cooling of the detector by ~35°C. This enables high-quality measurement during a much larger portion of the martian day and made rasters possible with multiple spots measured within a few hours. For MER, typically one sample is measured per night where good full width at half–maximum (FWHM) energy resolution is achieved only late at night when temperatures drop below –50°C.

28.3 Analysis of APXS Spectra

APXS spectra are plotted as counts as a function of energy (Figure 28.3). The first step in the data reduction is to determine the areas for all characteristic elemental emission peaks. This is typically performed with a nonlinear least square fit algorithm that models all theoretical known energies as Gaussian peaks with variable peak area and width. The minimization routine then varies peak areas and widths until model and experimental data are in optimal agreement within the given uncertainties. The FWHM, typically determined for the Mn K_α line at 5.9 keV for a given spectrum, is an exact function of energy, so all peak widths are correlated. In addition, detector artifacts such as exponential tailing, background, and other known effects can be modeled as well. This is important for spectra that were not measured under optimal conditions, e.g., at elevated temperatures, where thermal noise broadens the peaks.

Once peak areas for all elements are determined there are two different ways to transfer peak areas into concentrations. One way is to calculate from scratch the emission of X-rays for the given excitation by X-rays and alpha particles. In theory, all cross sections and fluorescence yields are known and for any given hypothetical sample composition the emitted X-ray peak areas can be predicted (Campbell et al., 2012). Varying the composition until the predicted peak areas agree with the measured ones, leads to the unknown sample composition.

Another way of extracting the composition of unknown samples is by empirical calibration of the spectrometer with samples of well-known composition. This approach was performed for the published APXS results from MER and MSL (Gellert et al., 2006).

The calibration determines the response or yield for all detected elements, which is defined as peak area per second per weight percent abundance of the element in the sample. It is given for a standard distance of the sample in the lab. All detectable elements are assumed to be present as their standard oxides, e.g., two oxygen atoms are added to the detected silicon (SiO_2), and all oxides are normalized to 100%, assuming no undetectable light compounds such as water or carbonates. This is done to account for varying sample distances on Mars.

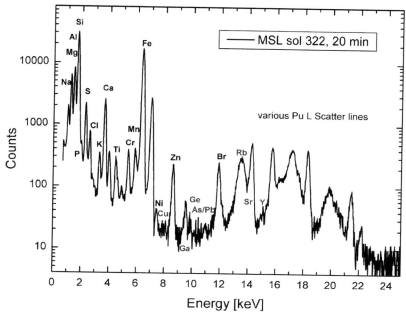

Figure 28.3 An APXS spectrum measured on MSL on sol 322 of the rock sample Aillik. This is a histogram of the number of detected X-rays for each energy channel. The position of the identified elements is indicated in the graph. Black labels indicate the standard 16 elements; blue labels show the trace elements that can be quantified for elevated abundances. The spectrum is plotted on a logarithmic scale to accommodate the large range of counts from thousands per channel for low-Z elements to about 10 in the energy range around 10 keV for high-Z trace elements. The uncertainty of each channel is given as the square root of the number of counts, which is the standard procedure in nuclear decay experiments.

In both methods, self-absorption of X-rays within the sample can be taken into account when the sample is assumed to be homogeneous. This means that depending on the composition of the sample, X-rays emitted within the sample are absorbed in a predictable way. As an extreme example, nickel X-rays, whose energy is just above the binding energy of iron, are well absorbed by iron. Therefore in iron meteorites, typically consisting of 90% iron, nickel X-rays are about three times more strongly absorbed than in usual rocks, with only 10%–20% iron. This effect can be accounted for by iteratively calculating the incoming and outgoing absorption cross sections for all elements and correcting for the different absorption probabilities. This effect typically requires a correction of ~10% for usual rocks for all elements.

For the self-absorption process, it is necessary to take into account the distribution of elements within the sample volume that is contributing to the spectra. As discussed earlier, the signals are collected from an area of about 2 cm diameter and a depth of a few µm for low-Z elements to 100 µm for high-Z elements. The assumption of a homogeneous sample

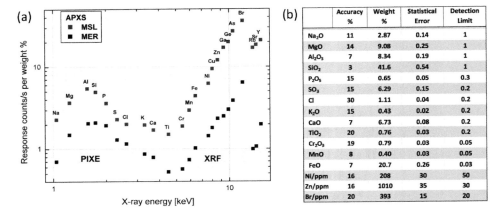

Figure 28.4 Sensitivity for the MER and MSL APXS. (Left) The peak area is plotted in counts per second per weight percent abundance of the element in the sample. High values indicate a high sensitivity, i.e., higher peaks in the spectrum. PIXE and XRF combined provide a balanced high signal for all elements. The lowest elements, especially Na and Mg, are impacted by absorption in the 6 mbar CO_2 atmosphere and the Be window covering the X-ray detector. (Right) An example of the results extracted from the spectrum in Figure 28.3. The table includes the detection limits and accuracy found during calibration and the typical precision, i.e., reproducibility errors.

is the key reason for imperfections in the APXS analysis method, as the sample is made up of minerals with distinct compositions that often have grain sizes much larger than the ~few μm escape depth of the low-Z elements. Therefore, the absorption for all distinct minerals differs from that of a glass-like homogeneous sample. This is the reason that in terrestrial XRF analyses the powdered sample is melted into borate pellets, breaking up the minerals and producing a compositionally homogeneous glass. However, preparing such pellets is not feasible on Mars and would also result in the loss of many geochemically important volatile elements.

The imperfections in the analysis method are represented by the accuracy given in Figure 28.4 for all elements. It is the average percentage deviation of APXS result to real composition found in the ~100 geologic samples used during the calibration. For comparison of rocks along the traverse, the precision (listed as statistical error in Figure 28.4) can be used, because errors in accuracy generally cancel out in similar samples.

With all possible caveats discussed earlier, the APXS method was tested with the available sample preparation tools on the different rovers. Dust-covered rocks typically have compositions similar to those of the soil, especially for the low-Z elements. Brushing removes most of the dust and reveals a more representative composition for the rock. The drill on MSL has a diameter of 1.6 cm and penetrates ~5 cm, mixing up and exposing subsurface material. Results from the homogenized powder are usually in good agreement with the brushed surface. Laterally heterogeneous samples such as small veins are routinely measured with multiple well-defined offset measurements called rasters, revealing the different areal coverage of the different minerals present. Here the strength of the precision of the APXS method comes into play, since, for example, the calcium and sulfur abundances are essentially independently determined. If the extracted abundances point to a 1:1 molar ratio, there is a strong indication for a $CaSO_4$ mineral.

28.4 APXS Results from Mars

APXS geochemical data have been utilized in a variety of studies at the four landing sites including petrology (McSween et al., 1999, 2008; Zipfel et al., 2011; Stolper et al., 2013), alteration analyses (Clark et al., 2005; Ming et al., 2008; Yen et al., 2017), categorization of geologic units (Squyres et al., 2006a; Thompson et al., 2016), mapping of trace elements (Yen et al., 2006; Schmidt et al., 2008; Berger et al., 2017), meteorite studies (Schröder et al., 2008; Ashley et al., 2011; Fleischer et al., 2011), and atmospheric analyses (VanBommel et al., 2018), among many others. Several planet-wide inferences and trends seen in the APXS data are discussed in the following sections. All APXS data discussed in this chapter are archived in the Planetary Data System.

28.4.1 Soils

Fine-grained basaltic soils and dust are remarkably uniform in chemical composition across the MER and MSL landing sites. The range of concentrations measured in soils by the three rovers covers comparable values for major, minor, and trace elements (Figure 28.5a), indicative of planet-wide similarities. Fe–Mn correlations (see Section 28.4.3; Figure 28.5b) reflect the igneous protolith for the soils, and the Si–Al trends (Figure 28.5c) suggest variable amounts of plagioclase feldspar with a molar Si/Al ratio of ~1.8:1. Not all trends, however, are identical across the landing sites: For example, Fe–Mg associations at Gale crater (Figure 28.5d) reflect varying concentrations of olivine (~Fo50), but similar correlations are not observed in the data from Meridiani or Gusev. Meridiani soils exhibit larger concentrations of Fe, likely due to local contributions of hematite. Large deviations from a Mars-average soil composition are observed in analyses containing coarse sand grains (>500 μm; these samples have been excluded from Figure 28.5), which generally have clear chemical signatures of local rocks.

The volatile elements S, Cl, Zn, and P are well correlated and vary by roughly a factor of two in the soils at a given landing site (Figure 28.5e–g). Even with SO_3 concentration in excess of 8 wt.% in the finest-grained samples, however, there are no clear correlations between S and Ca, Mg, Fe, or any other abundant cation. These volatile elements likely originated from volcanic outgassing and subsequently condensed on dust and sand grain surfaces without forming distinct sulfates or chlorides. The finest grains, those with the highest specific surface area, contain the largest quantities of these volcanic volatiles (Berger et al., 2016). Given the expected mobility of S and Cl under aqueous conditions, soil samples maintaining a molar S/Cl ratio of ~3.7:1 experienced minimal alteration post deposition. In contrast, several analyses of fine-grained soils (not included in Figure 28.5; e.g., Boroughs

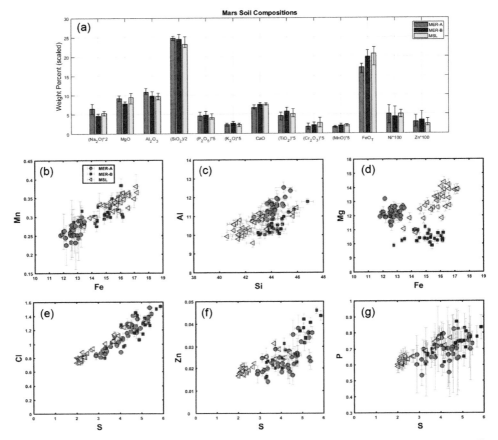

Figure 28.5 APXS data for martian soils. (a) Average chemical composition for basaltic soils measured by APXS at Gusev crater ($n = 35$), Meridiani Planum ($n = 30$), and Gale crater ($n = 28$). Data are shown on an S–Cl–Br-free basis, and error bars represent the range of the measured values in the dataset and do not account for precision or accuracy of the measurements. All oxide values except MgO, Al_2O_3, CaO, and FeO_T have been scaled to facilitate comparison. (b–g) Molar scatterplots for fine-grained basaltic soils at the MER and MSL landing sites.

trench at Gusev crater) have higher S/Cl ratios and are inferred to have experienced a history of aqueous alteration and differential mobilization of S versus Cl (Gellert et al., 2006).

28.4.2 Surface Coatings

Rocks at the surface of Mars, especially those that appear to have been recently exhumed, exhibit surface coatings rich in S and Cl, and occasionally Zn and Ni. These coatings are apparent in images (Figure 28.6a) as well as in comparisons between rock interiors and brushed surfaces. These enhancements are not simply due to insufficient removal of dust accumulations as either the S/Cl ratios depart from the dust trend (Figure 28.5e) or other chemical characteristics are inconsistent with a surface layer of dust. The brushed surface of Mojave2 in Gale crater (Figure 28.6b), for example, has a molar S/Cl ratio of ~3.6, which is comparable to that of martian dust (see Section 28.4.1). However, Mg concentrations are <60% of the value for dust while S and Cl on the brushed surface exceed values for dust. These rock coatings are not an artifact of incomplete dust removal and are found planet-wide.

The formation of these rock coatings likely involves the action of small amounts of liquid water to mobilize the S and Cl. The process may involve thin films of water seeking near-subsurface cold traps, such as buried rocks, during diurnal or seasonal cycling continuing over long time scales. The weathering process that resulted in the formation of S- and Cl-rich coatings also produced alteration rinds. The surface of Mazatzal, for example, exhibits a ferric oxide, likely hematite, not found in dust or in unaltered Gusev plains basalts (Morris et al., 2004). In addition, alteration of the surface of Wishstone is also evident by the significant depletion of P, likely due to dissolution of phosphate, at the surface of the rock (Figure 28.6b).

28.4.3 Fe/Mn Ratios

Fe and Mn are among the elements that are well established by the APXS method. Fe^{2+} and Mn^{2+} have nearly the same ionic radii and distribute similarly in primary igneous rocks, maintaining a consistent Fe/Mn ratio. During exposure to weathering environments, higher oxidation states of Fe and Mn are commonly formed. Differing mobility of these species results in elemental fractionation. Thus, altered samples will typically exhibit Fe/Mn ratios different from those of precursor materials.

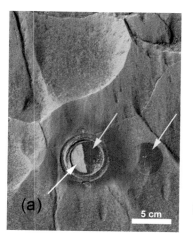
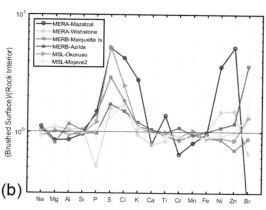

Figure 28.6 APXS data of rock coatings. (a) Pancam image of target "Mazatzal" (p2590, sol 082), a Gusev plains basalt showing a darker surface coating (yellow arrows) after brushing to remove the reddish dust. Brighter, unaltered rock interior is exposed after grinding (white arrow). (b) Examples from the MER and MSL landing sites of ratios of brushed rock surfaces to their interiors showing coatings enhanced in S, Cl, and occasionally Zn and Ni.

Basaltic rocks and unaltered soils measured by the APXS define an igneous Fe/Mn ratio for Mars. The mean and standard deviation of the Fe/Mn value for ~100 basaltic data points (fit with the dashed line in Figure 28.7a) is 48 ± 3. Families of altered rocks plot away from the igneous trendline (Figure 28.7a). The samples with high Fe/Mn ratios include (1) alteration halos resulting from acidic leaching of sandstones at Gale crater (Yen et al., 2017), (2) hematitic concretions at Meridiani formed in ground water (Squyres et al., 2006b), (3) Cl-rich bench deposits skirting the rim of Endeavour crater (Mittlefehldt et al., 2018), and (4) sulfate-rich soils at Gusev crater that are likely products of fumarolic activity (Yen et al., 2008). Samples with Fe/Mn ratios lower than those of the unaltered material include (1) fracture fill material with montmorillonite-like composition at the rim of Endeavour crater (Clark et al., 2016), (2) Mn-rich rock coatings associated with sulfur and other trace elements resulting from subsurface fluid flow at the rim of Endeavour crater (Arvidson et al., 2016), and (3) Mn-oxides precipitating from late-stage fluid flow (Thompson et al.,

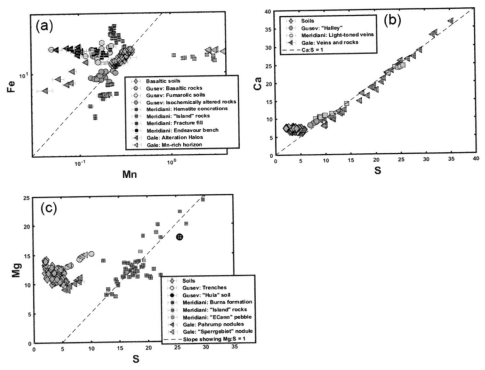

Figure 28.7 Elemental trends observed in APXS data. (a) Molar Fe versus Mn for selected targets at Gusev crater, Meridiani Planum, and Gale crater. The unaltered trend is shown by the fit through the basaltic rocks and soils (dashed line through gray points). Other plotted targets represent altered rocks at the three landing sites. (b) Molar Ca versus S and (c) molar Mg versus S showing that Mg and Ca sulfates are found at Gusev, Meridiani, and Gale. Dashed lines represent trend lines for pure $MgSO_4$ and $CaSO_4$ (various hydration states are possible).

2016). All of these targets that have Fe/Mn ratios departing from the basaltic trend line have clearly experienced aqueous alteration. In certain cases, however, oxidative weathering occurs nearly isochemically and without significant changes to the Fe/Mn ratio. The "Watchtower" class of rocks at Gusev crater, for example (green circles in Figure 28.7a), is extensively altered with samples having $Fe^{3+}/Fe_T > 0.9$ (Morris et al., 2006), but the process did not involve sufficient liquid water to mobilize cations and fractionate Fe and Mn.

28.4.4 Sulfates

Sulfur is ubiquitous on the surface of Mars. Of the ~1400 distinct APXS measurements made by the MER and MSL rovers, only ~25 analyses have less than 2 wt.% SO_3. Elemental correlations with sulfur indicate the presence of Ca and Mg sulfates at all three landing sites (Figure 28.7b,c). In situ detections of Fe sulfates and hydroxysulfates (Klingelhöfer et al., 2004; Yen et al., 2008) have been made in conjunction with data from other rover instruments (Rampe et al., 2017). Mg, Ca, and Fe sulfates detected by APXS or other means at the surface of Mars are clear indicators of past alteration processes. These are secondary phases that form in the presence of liquid water.

Ca sulfates are pervasive at Gale crater as visually distinct light-toned veins or disseminated in the matrix of the sandstones and mudstones. X-ray diffraction analyses have confirmed the presence of gypsum, bassanite, and anhydrite (Ca/S ratio = 1) in drill samples from Gale crater (Vaniman et al., 2018). Isolated occurrences of Ca sulfate veins are found at the rim of Endeavour crater (Squyres et al., 2012), and there is one clear example of Ca sulfate at Gusev (Figure 28.7b). Mg sulfates have generally been encountered less frequently than Ca sulfates and are found in nodules and other diagenetic features at Gale crater, minor deposits at the rim of Endeavour crater, and in subsurface soils and likely fumarolic deposits at Gusev (Figure 28.8c). The Burns

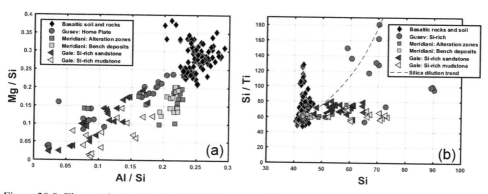

Figure 28.8 Elemental ratios for selected APXS targets. (a) Molar Mg/Si versus Al/Si for selected APXS samples. Basaltic samples are indicated in black; high-Si samples trend toward the origin. (b) Molar Si/Ti versus Si for selected samples. The dashed line represents the expected trend for silica addition (dilution of Ti). Consistent Si/Ti ratios are signatures of acidic alteration.

formation rocks at Meridiani appear to exhibit a bimodal distribution of Mg sulfates with higher abundances in certain samples, possibly correlated with elevation (Clark et al., 2005). Oddly, even with nearly 30 wt.% SO_3 in several analyses of Burns formation rocks, no other distinct sulfates are apparent from the chemical data. An Fe hydroxysulfate, namely jarosite, was identified on the basis of Mössbauer spectroscopy at Meridiani (Klingelhöfer et al., 2004), but there are no clear elemental correlations with K or Na, common cations in jarosite. Hydronium jarosite (H_3O^+ cation, not detectable by APXS) may be the most consistent with the chemical analyses. Jarosite was also detected at Gale crater by X-ray diffraction, but at an abundance of ~3 wt.% (Rampe et al., 2017), no obvious elemental trends are expected (or seen) in APXS data. Similarly, ferric iron sulfates were detected by Mössbauer spectroscopy in subsurface soil samples at Gusev crater that were inferred to have a fumarolic origin. Many of these targets were highly variable in chemistry over short distance scales, and an insufficient number of APXS analyses were collected to clearly establish the presence of Fe sulfates on the basis of chemistry. The exception is the Penina soil near Home Plate that shows a clear Fe–S correlation in APXS data.

28.4.5 Silica

The average SiO_2 content of primary rocks on the Gusev plains and of typical soils at the three rover landing sites is ~46 wt.%, yet numerous samples analyzed by the APXS exhibit >50 wt.% SiO_2 and substantially lower Al/Si and Mg/Si ratios (Figure 28.8). Pathways for achieving elevated SiO_2 concentrations in martian samples include silicic volcanism, leaching of other rock-forming elements resulting in Si-rich phases, and precipitation of silica from the influx of Si-rich fluids.

At Gale crater, detrital tridymite, a high-temperature (>870°C) polymorph of SiO_2, originating from silicic volcanics was presented as an explanation for the nearly 75 wt.% SiO_2 in "Buckskin" samples (Morris et al., 2016). Also at Gale crater, chemical and mineralogic analyses within and outside alteration halos indicate fluids crosscutting mudstone and sandstone strata resulting in >60 wt.% SiO_2, largely present as amorphous silica (Yen et al., 2017). In these alteration zones, the Si/Ti ratios are largely unaffected (Figure 28.8b), characteristic of acidic leaching.

The highest SiO_2 abundance (62 wt.%) measured by the Opportunity rover was in analyses of the "Esperance" veins at the rim of Endurance crater, which had a montmorillonite-like composition, possibly with an additional siliceous component (Clark et al., 2016). Multiple episodes of aqueous interaction are suggested, including alteration of igneous rocks to produce a clay-like composition, possibly followed by acidic diagenesis to further enhance the SiO_2 concentrations (Clark et al., 2016). Separate from the Esperance analyses, amorphous silica of unconstrained origin is also modeled to be present in Burns formation rocks at Meridiani Planum (Clark et al., 2005). Alteration zones and

Table 28.1 *Examples of sample compositions from the MER and MSL landing sites*

Sol	Name	Na$_2$O	MgO	Al$_2$O$_3$	SiO$_2$	P$_2$O$_5$	SO$_3$	Cl	K$_2$O	CaO	TiO$_2$	Cr$_2$O$_3$	MnO	FeO	Ni	Zn	Br
C0673	Sourdough, soil	2.8	8.3	9.4	43.1	0.8	5.1	0.64	0.45	7.2	1.1	0.5	0.44	20.1	514	354	36
A0401	Paso Robles, soil	1.6	5.5	4.1	21.8	5.6	31.7	0.55	0.19	6.8	0.6	0.0	0.25	21.0	109	98	494
A1190	Kenosha Comets, soil	0.3	2.3	1.7	90.5	0.3	1.0	0.14	0.00	0.7	1.2	0.3	0.03	1.4	151	278	17
A0034	Adirondack_RAT	2.4	10.8	10.9	45.7	0.5	1.2	0.20	0.07	7.8	0.5	0.6	0.41	18.8	165	81	14
A0511	Backstay_Brush	4.1	8.3	13.3	49.5	1.4	1.5	0.35	1.07	6.0	0.9	0.1	0.24	13.0	191	269	26
B0068	BounceRockCase_RAT	1.7	6.8	10.5	51.6	0.9	0.6	0.10	0.11	12.1	0.7	0.1	0.40	14.4	81	38	39
C0048	JakeM	6.6	4.6	14.6	48.9	0.8	2.8	0.95	1.89	6.8	0.7	0.0	0.23	10.9	59	318	107
A0335	WshstoneChisel_RAT	5.0	4.5	15.0	43.8	5.2	2.2	0.35	0.57	8.9	2.6	0.0	0.22	11.6	67	64	22
B3935	JeanBaptisteCharbonneau	3.3	4.0	16.2	52.7	0.9	3.2	0.51	0.44	7.3	0.9	0.0	0.18	10.3	0	50	86
C0809	Mojave2_DRT	2.8	4.5	12.4	51.8	1.4	5.8	0.52	0.65	4.3	1.1	0.4	0.34	13.5	839	1739	84
B0036	Guadalupe_RAT	1.7	8.4	5.8	36.2	1.0	24.9	0.50	0.53	4.9	0.6	0.2	0.30	14.8	589	324	30
C0627	Stephen	1.0	18.3	4.1	33.4	0.7	4.4	3.36	2.47	4.2	0.9	0.3	4.05	21.6	1082	8119	1886

The sol includes the landing site (C = Curiosity, Gale crater; B = Opportunity, Meridiani Planum; A = Spirit, Gusev crater). Accuracy and precision errors are similar to the ones given in Figure 28.4b for each element.

bench deposits at Meridiani show elevated silica and flat Si/Ti ratios, consistent with interactions with acidic fluids.

Concentrations of SiO_2 in excess of 90 wt.% were measured in the environs of Home Plate at Gusev crater, a likely pyroclastic construct (Squyres et al., 2008). Associations with TiO_2 for a few of these high silica samples (Figure 28.8) indicate that they are likely products of acidic alteration (Yen et al., 2017). Several of the high-silica samples, however, do not exhibit corresponding increases in Ti and may have been precipitated from Si-rich hydrothermal fluids (Ruff et al., 2011).

28.5 Summary

The APXS instrument encountered a remarkable chemical diversity at the four landing sites where it was used. Together with data from other instruments, the results shed light on martian surface processes, including indications of extensive aqueous activity in the planet's past. Table 28.1 gives a brief excerpt of some of the most remarkable compositions.

The soil Sourdough represents the well-mixed global soil found at all landing sites with similar composition. In Gusev crater two distinct light-toned subsurface soil types were exposed by the wheels. PasoRobles type soils contain ferric sulfates, excess SiO_2, and likely other sulfates, while KenoshaComets is nearly pure SiO_2, likely present as amorphous opaline silica (Gellert et al., 2006; Squyres et al., 2008). Various igneous rock types were identified, often with several near identical members with the same composition. Adirondack type rocks are more primitive, olivine-bearing rocks, generally similar to the global soil (McSween et al., 2004). Felsic rocks were identified on all four landing sites. Pathfinder discovered rocks interpreted as andesitic (Rieder et al., 1997 or McSween et al., 1999). JakeM (Stolper et al., 2013), Wishstone (Ming et al., 2008), and JeanBaptisteCharbonneau (Gellert et al., 2016), each from a different landing site, are unique in their low abundance of Ni and Zn. The alkaline basalt Backstay (Ming et al., 2008) and BounceRock, similar in composition and mineralogy to a martian meteorite (Zipfel et al., 2011), are additional distinct examples of igneous rocks. Widespread and heavily altered sediments were documented: The Murray Formation at Gale represented by Mojave2 (Grotzinger et al., 2015) and the Burns Formation at Meridiani Planum (Squyres et al., 2006a) with the sample Guadalupe. The precision of the APXS instrument enabled monitoring of compositions along kilometers of traverse and determining possible changes in the water chemistry during sedimentation (Gellert and Clark, 2015; Hurowitz et al., 2017). The trace elements measured by the APXS instrument such as zinc and germanium have been used as tracers for aqueous alteration (Berger et al., 2017). The sample Stephen, a fracture filling at Kimberly in Gale crater, is an example with exceptional trace element concentrations (including Co and Cu), indicative of mobilization through aqueous processes (Lanza et al., 2016).

Acknowledgments

The MSL APXS was financed and managed by the Canadian Space Agency with MDA, Brampton, as prime subcontractor to build the instrument. Support for operations of the MSL APXS is provided by CSA and NASA. The MER APXS was developed and built in the Cosmochemistry Department of the Max-Planck Institute for Chemistry, Mainz, Germany, partly supported by the German Space Agency, DLR. A portion of the research was carried out at the Jet Propulsion Laboratory, California Institute of Technology, under a contract with NASA.

References

Arvidson R.E., Squyres S.W., Morris R.V., et al. (2016) High concentrations of manganese and sulfur in deposits on Murray Ridge, Endeavour crater, Mars. *American Mineralogist*, **101**, 1389–1405.

Ashley J.W., Golombek M., Christensen P.R., et al. (2011) Evidence for mechanical and chemical alteration of iron-nickel meteorites on Mars: Process insights for Meridiani Planum. *Journal of Geophysical Research*, **116**, E00F20, DOI:10.1029/2010JE003672.

Berger J.A., Schmidt M.E., Gellert R., et al. (2016) A global Mars dust composition refined by the Alpha-Particle X-ray Spectrometer in Gale crater. *Geophysical Research Letters*, **43**, 67–75.

Berger J.A., Schmidt M.E., Gellert R., et al. (2017) Zinc and germanium in the sedimentary rocks of Gale crater on Mars indicate hydrothermal enrichment followed by diagenetic fractionation. *Journal of Geophysical Research*, **122**, 1747–1772.

Boynton W.V., Taylor G.J., Evans L.G., et al. (2007) Concentration of H, Si, Cl, K, Fe, and Th in the low- and mid-latitude regions of Mars. *Journal of Geophysical Research*, **112**, DOI:10.1029/2007JE002887.

Campbell J.L., Perrett G.M., Gellert R., et al. (2012) Calibration of the Mars Science Laboratory Alpha Particle X-ray Spectrometer. *Space Science Reviews*, **170**, 319–340.

Clark B.C., Morris R.V., McLennan S.M., et al. (2005) Chemistry and mineralogy of outcrops at Meridiani Planum. *Earth and Planetary Science Letters*, **240**, 73–94.

Clark B.C., Morris R.V., Herkenhoff K.E., et al. (2016) Esperance: Multiple episodes of aqueous alteration involving fracture fills and coatings at Matijevic Hill, Mars. *American Mineralogist*, **101**, 1515–1526.

Fleischer I., Schröder C., Klingelhöfer G., et al. (2011) New insights into the mineralogy and weathering of the Meridiani Planum meteorite, Mars. *Meteoritics and Planetary Science*, **46**, 21–34.

Gellert R., Rieder R., Brückner J., et al. (2006) Alpha Particle X-ray Spectrometer (APXS): Results from Gusev crater and calibration report. *Journal of Geophysical Research*, **111**, E02S05, DOI:10.1029/2005JE002555.

Gellert R., Clark B.C. & MSL and MER Science Teams. (2015) In situ compositional measurements of rocks and soils with the Alpha Particle X-ray Spectrometer on NASA's Mars rovers. *Elements*, **11**, 39–44.

Gellert R., Arvidson R.E., Clark B.C., et al. (2016) Igneous and sedimentary compositions from four landing sites on Mars from the Alpha Particle X-ray Spectrometer (APXS). *Meteoritics and Planetary Science*, **51**, A280.

Grotzinger J.P., Gupta S., Malin M.C., et al. (2015) Deposition, exhumation, and paleoclimate of an ancient lake deposit, Gale crater, Mars. *Science*, **350**, aac7575.

Hurowitz J.A., Grotzinger J.P., Fischer W.W., et al. (2017) Redox stratification of an ancient lake in Gale crater, Mars. *Science*, **356**, eaah6849.

Klingelhöfer G., Morris R.V., Bernhardt B., et al. (2004) Jarosite and Hematite at Meridiani Planum from Opportunity's Mössbauer spectrometer. *Science*, **306**, 1740–1745.

Lanza N.L., Wiens R.C., Arvidson R.E., et al. (2016) Oxidation of manganese in an ancient aquifer, Kimberley formation, Gale crater, Mars. *Geophysical Research Letters*, **43**, 7398–7407.

McSween H.Y., Murchie S.L., Crisp J., et al. (1999) Chemical, multispectral, and textural constraints on the composition and origin of rocks at the Mars Pathfinder landing site. *Journal of Geophysical Research*, **104**, 8679–8715.

McSween H.Y., Arvidson R.E., Bell J.F. III, et al. (2004) Basaltic rocks analyzed by the Spirit rover in Gusev crater. *Science*, **305**, 842–845.

McSween H.Y., Ruff S.W., Morris R.V., et al. (2008) Mineralogy of volcanic rocks in Gusev crater, Mars: Reconciling Mössbauer, Alpha Particle X-Ray Spectrometer, and Miniature Thermal Emission Spectrometer spectra. *Journal of Geophysical Research*, **113**, E06S04, DOI:10.1029/2007JE002970.

Ming D.W., Gellert R., Morris R.V., et al. (2008) Geochemical properties of rocks and soils in Gusev crater, Mars: Results of the Alpha Particle X-Ray Spectrometer from Cumberland Ridge to Home Plate. *Journal of Geophysical Research*, **113**, E12S39, DOI:10.1029/2008JE003195.

Mittlefehldt D.W., Gellert R., vanBommel S., et al. (2018) Diverse lithologies and alteration events on the rim of Noachian-aged Endeavour crater, Meridiani Planum, Mars: In situ compositional evidence. *Journal of Geophysical Research*, **123**, 1255–1306.

Morris R.V., Klingelhöfer G., Bernhardt B., et al. (2004) Mineralogy at Gusev crater from the Mössbauer spectrometer on the Spirit rover. *Science*, **305**, 833–836.

Morris R.V., Klingelhöfer G., Schröder C., et al. (2006) Mössbauer mineralogy of rock, soil, and dust at Meridiani Planum, Mars: Opportunity's journey across sulfate-rich outcrop, basaltic sand and dust, and hematite lag deposits. *Journal of Geophysical Research*, **111**, DOI:10.1029/2006JE002791.

Morris R.V., Vaniman D.T., Blake D.F., et al. (2016) Silicic volcanism on Mars evidenced by tridymite in high-SiO_2 sedimentary rock at Gale crater. *Proceedings of the National Academy of Sciences of the USA*, **113**, 7071–7076.

O'Connell-Cooper C.D., Spray J.G., Thompson L.M., et al. (2017) APXS-derived chemistry of the Bagnold dune sands: Comparisons with Gale crater soils and the global martian average. *Journal of Geophysical Research*, **122**, 2623–2643.

Radchenko V., Andreichikov B., Wänke H., et al. (2000) Curium-244 alpha-sources for space research. *Applied Radiation and Isotopes*, **53**, 821–824.

Rampe E.B., Ming D., Blake D., et al. (2017) Mineralogy of an ancient lacustrine mudstone succession from the Murray formation, Gale crater, Mars. *Earth and Planetary Science Letters*, **471**, 172–185.

Rieder R., Economou T., Wänke H., et al. (1997) The chemical composition of martian soil and rocks returned by the Mobile Alpha Proton X-ray Spectrometer: Preliminary results from the X-ray mode. *Science*, **278**, 1771–1774.

Rieder R., Gellert R., Brückner J., et al. (2003) The new Athena alpha particle X-ray spectrometer for the Mars Exploration Rovers. *Journal of Geophysical Research*, **108**, DOI:10.1029/2003JE002150.

Ruff S.W., Farmer J.D., Calvin W.M., et al. (2011) Characteristics, distribution, origin, and significance of opaline silica observed by the Spirit rover in Gusev crater, Mars. *Journal of Geophysical Research*, **116**, DOI: E00F23 10.1029/2010JE003767.

Schmidt M.E., Ruff S.W., McCoy T.J., et al. (2008) Hydrothermal origin of halogens at Home Plate, Gusev crater. *Journal of Geophysical Research*, **113**, E06S12. DOI:10.1029/2007JE003027.

Schröder C., Rodionov D.S., McCoy T.J., et al. (2008) Meteorites on Mars observed with the Mars Exploration Rovers. *Journal of Geophysical Research*, **113**, E06S22. DOI:10.1029/2007JE002990.

Squyres S.W., Arvidson R.E., Blaney D.L., et al. (2006a) Rocks of the Columbia Hills. *Journal of Geophysical Research*, **111**, E02S11, DOI:10.1029/2005JE002562.

Squyres S.W., Arvidson R.E., Bollen D., et al. (2006b) Overview of the Opportunity Mars Exploration Rover mission to Meridiani Planum: Eagle crater to Purgatory ripple. *Journal of Geophysical Research*, **111**, DOI:10.1029/2006JE002771.

Squyres S.W., Arvidson R.E., Ruff S., et al. (2008) Detection of silica-rich deposits on Mars. *Science*, **320**, 1063–1067.

Squyres S.W., Arvidson R.E., Bell J.F., et al. (2012) Ancient impact and aqueous processes at Endeavour crater, Mars. *Science*, **336**, 570–576.

Stolper E.M., Baker M.B., Newcombe M.E., et al. (2013) The petrochemistry of Jake_M: A martian mugearite. *Science*, **341**, 1239463.

Taylor R. & McLennan S.M. (2008) *Planetary crusts: Their composition, origin, and evolution*. Cambridge University Press, Cambridge.

Thompson L.M., Schmidt M.E., Spray J.G., et al. (2016) Potassium-rich sandstones within the Gale impact crater, Mars: The APXS perspective. *Journal of Geophysical Research*, **121**, 1981–2003.

VanBommel S.J., Gellert R., Clark B.C., & Ming D.W. (2018) Seasonal atmospheric argon variability measured in the equatorial region of Mars by the Mars Exploration Rover Alpha Particle X-Ray Spectrometers: Evidence for an annual argon-enriched front. *Journal of Geophysical Research*, **123**, 544–558.

Vaniman D., Martínez G.M., Rampe E.B., et al. (2018) Gypsum, bassanite, and anhydrite at Gale crater, Mars. *American Mineralogist*, **103**, 1011–1020.

Yen A.S., Gellert R., Schröder C., et al. (2005) An integrated view of the chemistry and mineralogy of martian soils. *Nature*, **436**, 49–54.

Yen A.S., Mittlefehldt D.W., McLennan S.M., et al. (2006) Nickel on Mars: Constraints on meteoritic material at the surface. *Journal of Geophysical Research*, **111**, DOI:10.1029/2006JE002797.

Yen A.S., Morris R.V., Clark B.C., et al. (2008) Hydrothermal processes at Gusev crater: An evaluation of Paso Robles class soils. *Journal of Geophysical Research*, **113**, E06S10, DOI:10.1029/2007JE002978.

Yen A.S., Ming D.W., Vaniman D.T., et al. (2017) Multiple stages of aqueous alteration along fractures in mudstone and sandstone strata in Gale crater, Mars. *Earth and Planetary Science Letters*, **471**, 186–198.

Zipfel J., Schroder C., Jolliff B., et al. (2011) Bounce Rock: A shergottite-like basalt encountered at Meridiani Planum, Mars. *Meteoritics and Planetary Science*, **46**, 1–20.

29

Elemental Analyses of Mars from Rovers with Laser-Induced Breakdown Spectroscopy by ChemCam and SuperCam

NINA L. LANZA, ROGER C. WIENS, SYLVESTRE MAURICE, AND JEFFREY R. JOHNSON

29.1 The ChemCam Instrument on Mars

29.1.1 Why Laser-Induced Breakdown Spectroscopy?

The first Laser-Induced Breakdown Spectroscopy (LIBS) experiment for extraterrestrial applications is part of the ChemCam instrument suite on board the Mars Science Laboratory (MSL) rover called Curiosity (Figure 29.1). ChemCam consists of a LIBS instrument and a Remote Micro-Imager (RMI). LIBS provides microbeam (350–500 μm diameter) chemical analyses of targets using a pulsed laser at target standoff distances between 1.6 and 7 m from the rover, while the RMI provides a 20 mrad field of view context image for the LIBS sampling location (Maurice et al., 2012; Wiens et al., 2012). The LIBS technique (see Chapter 8 for details) is well suited to the requirements of planetary surface missions for a number of reasons. First, LIBS analyses require no sample preparation, and samples can be probed at a standoff distance from the rover. Analyses are also rapid; each laser pulse produces an entire spectrum containing 6144 channels, and with typical repetition rates of 3–10 Hz LIBS can thus acquire hundreds of spectra within the space of several minutes. LIBS is also sensitive to every element, both low and high Z (atomic number), including light elements such as H, Li, F, and C that are challenging to detect with other rover instrument analysis techniques such as X-ray fluorescence. Prior to ChemCam, elements lower in atomic number than sodium (11) were not routinely assessed on Mars. LIBS also has relatively low detection limits for many elements (~2–1000 ppm for many species) (Cremers & Radziemski, 2013). Because the laser ablates small amounts of material (micrograms) with each pulse, multiple LIBS pulses can be used at a single location to obtain information about chemical trends with depth, including the presence of rock coatings or weathering rinds. Along the same lines, dust is removed from rock surfaces by the shockwave generated by the LIBS plasma (Lasue et al., 2018). The small LIBS analysis footprint allows for measurements of individual mineral grains in coarse-grained rocks within a target even at a standoff distance. With all of these capabilities, LIBS is extremely well suited to the requirements of a surface mission to Mars.

Because of versatility of the LIBS technique, ChemCam has made many discoveries on Mars that may not have been possible without the inclusion of a LIBS instrument on the

Figure 29.1 The ChemCam instrument on board the Curiosity rover on sol 1463 of the mission at the "Murray Buttes" location in Gale crater, Mars. ChemCam's mast unit (white box near center-top of image) sits atop the rover's mast and is composed of the LIBS laser and telescope. The mast unit is connected by an optical fiber and electrical cables to the body unit in the rover body, which houses the instrument's three spectrometers. The Mastcam cameras are visible directly below the ChemCam mast unit, identified by their two different-sized, square-shaped apertures. The ChemCam calibration targets are located on the back of the rover as a small plate containing round targets underneath the ultra-high frequency (UHF) antenna (gray cylinder). This image is part of a mosaic obtained by the Mars Hand Lens Imager (MAHLI) on the rover arm. Image credit: MSSS/JPL/NASA.

Curiosity rover payload (Section 29.2). Results such as the detection of hydrogen in martian dust (Section 29.2.1), the first-time observation of many minor and trace elements (Section 29.2.2), the overall broad range of chemical compositions observed (Section 29.2.3), and observations of fine-scale depth trends in rock surfaces (Section 29.2.4) are all due in part to the unique capabilities of LIBS. The goal of this chapter is to provide an overview of how LIBS has been used on the martian surface as part of the ChemCam instrument, including science planning, operational modes, coordination with other rover instruments, some examples of scientific results, and the future of LIBS on Mars as part of the Mars 2020 SuperCam instrument suite.

29.1.2 ChemCam Operational Modes on Mars

During normal rover operations, ChemCam may be used in a number of operational modes depending on the science questions of interest to be answered. A typical analysis on a martian target involves obtaining both spectral and image data in one of several standard

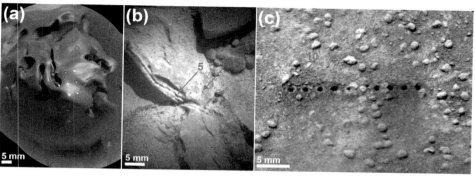

Figure 29.2 Examples of typical ChemCam raster types. Each image shows a post-LIBS RMI mosaic with first and last LIBS analysis locations marked in red. All targets were analyzed with 30 shots/location. LIBS analysis pits are visible in each analysis location as dark spots. (a) The rock target Egg Rock (likely a metallic meteorite) was analyzed with a 3 × 3 grid on sol 1505 (ccam05504). An initial RMI was obtained centered at location 1 prior to LIBS analysis, and a final RMI was obtained at location 9 after all LIBS analyses were complete. (b) The rock target McGrath 5 was analyzed with a 1 × 5 horizontal raster on sol 234 (ccam02234); as for Egg Rock, pre- and post-LIBS RMIs were obtained. (c) The soil and pebble target CC BT 745a was analyzed with a 1 × 10 horizontal raster on sol 1399 (ccam15025); because of the relatively large area covered, four RMIs were obtained to ensure coverage of the LIBS analysis area. Image credit: LANL-CNES/JPL/NASA.

raster patterns (Figure 29.2). ChemCam can also be used in a passive mode, which is discussed in Section 29.1.3.

For a typical rock or soil analysis, ChemCam performs a raster of LIBS shots. A raster is typically comprised of 1–10 LIBS sampling locations, each of which receives 30–150 laser pulses ("shots"), returning the same number of spectra. The default setting is 30 shots per location, while 150-shot analyses are used for depth profiling, discussed in Section 29.2.4. The most commonly used rasters are lines of 5 or 10 sampling locations and a 3 × 3 square, with 30 LIBS shots per location. Prior to LIBS analysis, an RMI image of the starting sampling location is obtained. The LIBS raster is then obtained, moving whenever possible from right to left and down to up to avoid pointing inconsistencies due to backlash in the mast movement mechanism. After each LIBS analysis of 30 shots, a dark spectrum is obtained in which the spectrometer collects ambient light without the laser firing; this dark spectrum is later subtracted from the active spectra (for more details on LIBS data processing for ChemCam see Chapter 8). After the last LIBS analysis in the raster, another RMI image is obtained at the final location. Raster spacing is optimized to ensure that the entire raster is contained within the two RMI images, and is typically 2 mrad; if larger spacing is desired, another RMI image is obtained in the middle of the raster sequence to ensure that the entire analysis region is documented. These RMI images can be stitched into a mosaic with either the "before" or "after" image on top to compare the target's appearance before and after LIBS analysis (Figure 29.2). Each LIBS laser pulse will move unconsolidated materials such as dust away from the analysis location due to the shock wave produced by the plasma, thus revealing the target below by dusting off its surface (Lasue et al., 2018). Because dust is ubiquitous on the martian surface, the first ~5 shots of a ChemCam LIBS analysis are typically considered to be a mixture of dust on the rock surface and the rock itself and are not included in reported bulk compositions (Maurice et al., 2016a). The raw data products produced by a typical ChemCam analysis include 2–3 RMI images and 150–300 LIBS spectra.

During rover planning, ChemCam analysis targets are selected using image and range data for targets close to the rover. However, up-to-date downlink data are not always available. In these cases a "blind" target activity may be scheduled, in which the rover points to a location to the lower right of the rover and samples whatever materials happen to be there (Cousin et al., 2014). This kind of blind targeting is straightforward to plan and allows the science team to collect a random sampling of materials along the traverse, but does not enable analysis of targets of potentially greater scientific interest. To optimize ChemCam data collection on blind targeting sols, a software tool for automatic target selection called the Automated Exploration for Gathering Increased Science (AEGIS) was developed (Francis et al., 2016) based on similar software designed for the Mars Exploration Rover mission (Estlin et al., 2012). AEGIS uses computer vision techniques to identify geologic features of interest in image scenes, including rocks, veins, and concretions. Using on board image data, AEGIS identifies a feature of interest in the scene and then performs a standard ChemCam 5×1 horizontal raster on the selected target. In this way, targeted data may be obtained even on sols that do not allow for targeted planning, thereby increasing the science return from ChemCam (Francis et al., 2017).

In addition to analyzing martian rock and soil targets, ChemCam also performs regular analyses of 10 calibration targets located on the rover body (Figure 29.1). These calibration targets are composed of well-characterized materials with known compositions that were analyzed with ChemCam on Earth pre-launch and are reanalyzed repeatedly on Mars to help interpret LIBS observations of unknown martian materials (Wiens et al., 2013). Four targets are composed of sintered, granular ceramic materials that are compositionally and texturally similar to sediments (Vaniman et al., 2012). Four targets are composed of igneous glasses (Fabre et al., 2011). The calibration target suite also includes graphite to calibrate for carbon and a titanium plate for LIBS wavelength calibration (Wiens et al., 2013; also see Chapter 8 for LIBS element quantification). During normal operations, the calibration targets are analyzed on a regular basis, typically during untargeted sols when science targets cannot be selected by the science team.

29.1.3 Using ChemCam's Passive Spectroscopy Mode

Along with the active LIBS mode, ChemCam may also be used in a passive mode whereby the spectrometers collect ambient light without the laser firing to determine mineralogy using reflectance spectroscopy. The ChemCam spectrometers exhibit sufficient radiometric sensitivity in the 400–840 nm region to enable collection of surface spectral reflectance in passive mode (Johnson et al., 2015; Wiens et al., 2015; Maurice et al., 2016a). The unprecedented high spatial (0.65 mrad field of view) and spectral (<1 nm) resolution of these measurements allows unique investigations of the spectral band positions, shapes, depths, and slopes of spectra collected of targets both near and far from the rover. Johnson et al. (2015, 2016a,b) presented ChemCam relative reflectance spectra (400–840 nm) of rocks, dust, and soils, where spatial resolutions of the point locations ranged from 1.3 to 4.5 mm for targets near the rover (~2–7 m) and 1–3 m for distant targets (~2–5 km). Analyses of the band depths, spectral slopes, and ratios among spectra collected during the first year of rover operations revealed spectral endmembers characterized by variations in ferrous and ferric components (Johnson et al., 2015).

Identification of these species through ChemCam's passive mode helps constrain mineralogy and crystallinity of Fe-bearing minerals relevant for understanding the aqueous history of Mars. For example, exposed materials (brushed or freshly broken surfaces) exhibit low 535-nm band depths and low 670/440-nm ratios, indicative of the less oxidized nature of these more ferrous materials relative to, e.g., martian dust. Typical reddish, oxidized, dusty surfaces exhibit similarities to altered volcanic material. Dark float rocks exhibit low relative reflectance and significant downturns longward of 700 nm that are consistent with the presence of orthopyroxene. Mg-rich raised ridges tend to exhibit distinct, negative near-infrared slopes (Léveillé et al., 2014). Ca sulfate veins found throughout Curiosity's traverse through Gale crater exhibit the highest relative reflectance values, but are still relatively red owing to the effects of dust and/or minor structural Fe^{3+} contamination (Nachon et al., 2014). Such dust was less prominent on rocks within the "blast zone" surrounding the Bradbury landing site. These samples were likely affected by the landing thrusters, which partially removed the ubiquitous dust layer. The first evidence for the presence of ferric sulfates at Gale was found in in the passive spectrum of the freshly broken rock target Perdido2 (Figure 29.3) (Johnson et al., 2016a). An absorption band near 433 nm, paired with a spectral downturn longward of 700 nm, is consistent with detection of

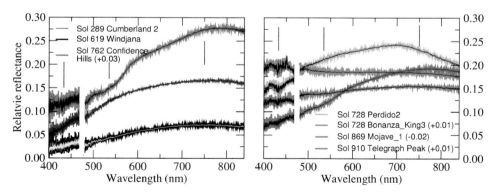

Figure 29.3 ChemCam relative reflectance (passive) spectra of representative locations for drill tailings and fresh rock surfaces. Central yellow or black lines represent 5-channel average of spectra shown in colors. Some spectra are offset for clarity, as shown in legend. Vertical bars correspond to wavelength locations 433, 535, and 750 nm (after Johnson et al., 2016a).

a ferric sulfate such as jarosite (e.g., Bishop & Murad, 2005; Cloutis et al., 2006; McCollom et al., 2014, Sobron et al.; 2014; Sklute et al., 2015). In addition, spectral features consistent with enrichments in ferric oxides such as hematite and magnetite were found in select drill tailings (Figure 29.3) and in landscape-scale features higher on Mt. Sharp (Johnson et al., 2016a). Recent investigations of dune sands have shown spectral signatures consistent with olivine interspersed with more ferric-rich phases (Johnson et al., 2016b,c).

In addition to its utility for ground targets, the ChemCam passive mode may also be used to examine water and aerosols in the martian atmosphere. By aiming at two different sky elevation angles, scattered sunlight is collected from two different path lengths through the martian atmosphere and then used to model gas abundance (McConnochie et al., 2017, 2018). Passive measurements of the atmosphere have found seasonal variations in water vapor and aerosol content, with evidence for diurnal interactions between atmospheric water vapor and surface materials (McConnochie et al., 2017, 2018). As the mission continues through multiple martian seasons and years, additional passive sky data will fill in the current gaps in our understanding of how water vapor, aerosols, and other atmospheric components vary over time.

29.1.4 Synergy with Other Rover Instruments

ChemCam observations are greatly enhanced by the acquisition of Mastcam images at the same targets. Mastcam consists of two cameras mounted on the mast below the ChemCam mast unit (Figure 29.1), each with different focal lengths and science color filters (Malin et al., 2010). While ChemCam obtains small-scale images of the LIBS analysis location with its RMI instrument, Mastcam images provide larger scale context images. Mastcam images of ChemCam targets are typically acquired after ChemCam analysis and thus capture the appearance of the target surface post LIBS, when some or all surface dust has been removed.

ChemCam analysis locations may also be targeted by the Mars Hand Lens Imager (MAHLI) instrument for more detailed imaging. Located on the arm, MAHLI provides high resolution color images of targets (Edgett et al., 2012). MAHLI images are taken of ChemCam targets after LIBS analysis to provide detailed images of LIBS pits, dust movement and particle size, and overall surface characteristics. Images of LIBS pits in particular can provide insight into target hardness from pit size and shape, and may also show whether a rock coating is present if there are significant color differences between the surface and the pit interior (e.g., Lanza et al., 2015, 2016).

ChemCam data are also complemented by mineralogy data from the Chemistry and Mineralogy (CheMin) X-ray diffraction (XRD) instrument (Blake et al., 2013). While LIBS data provide information about chemistry, they do not directly assess mineral structures. Although some molecular lines may be observed in the plasma, mineralogy of LIBS targets is typically inferred by comparison with laboratory standards (e.g., Wiens et al., 2013). In contrast, CheMin provides direct measurements of mineralogy. Bedrock drilling for CheMin allows ChemCam to sample materials in which mineralogy has been directly measured by CheMin, allowing for a more robust interpretation of rock type than from chemistry alone. Such interpretations are often challenging in Gale crater, where most materials are sedimentary in nature but igneous in composition (Grotzinger et al., 2015a), and where many rocks are composed of mineral grains that did not necessarily form together (e.g., Williams et al., 2013).

In addition to CheMin data, ChemCam data are also complemented by chemistry measurements from the Alpha-Particle X-ray Spectrometer (APXS) instrument (see Chapter 28 for additional details). Located on the rover arm, APXS interrogates targets with alpha particles and X-rays and measures the generated X-rays that are returned to obtain the elemental composition of a target (Gellert et al., 2015). The APXS detector is 1.7 cm in diameter, while the ChemCam analysis spot size is ~350–500 µm in diameter; this difference in footprint makes the chemistry data from these two instruments highly complementary, with ChemCam analyzing individual grains or otherwise small regions and APXS providing a bulk composition.

29.2 Examples of ChemCam Science Discoveries

29.2.1 Hydrogen in Rocks and Soils

A strength of the LIBS technique is its ability to detect almost every element, including light elements and particularly hydrogen. H is an important marker of hydrated mineral species containing either water (H_2O) or hydroxyl (OH^-), as well as being a constituent of many organic compounds. As a result, the detection and quantification of hydrogen on the martian surface are critical to answering questions regarding habitability and potential biosignatures. Given overwhelming geomorphologic evidence for abundant liquid water on the martian surface, the remaining inventory of water in the present day is likely to be significant (Lasue et al., 2013). Prior to the Curiosity mission, hydrogen was observed on the martian surface as water ice polar caps (Boynton et al., 2002), near surface water ice (Smith et al., 2009), hydrated mineral species (e.g., Murchie et al., 2009; Carter et al., 2013), and atmospheric vapor (Whiteway et al., 2009). More puzzling were observations of broad swaths of hydrogen-bearing materials by the Neutron and Gamma-Ray Spectrometer on board the Mars Odyssey orbiter (Feldman et al., 2004). These materials are not clearly affiliated with hydrated mineral species or water- and ice-formed morphologies.

Initial results from ChemCam LIBS data provided clarification for the presence of previously unknown hydrated materials on the martian surface. ChemCam results show that dust and soils in Gale crater are always hydrated to some extent (Meslin et al., 2013). Soils in Gale crater have ~1.5–3 wt.% H_2O in fines <150 µm in size (Leshin et al., 2013). In both soils and dust, the hydration appears to be carried by an amorphous component (Leshin et al., 2013; Meslin et al., 2013). Dust is ubiquitous on the martian surface, making it a likely candidate for the host of the hydration signature observed from orbit. Because of this ubiquity, the first few LIBS shots of each rock target (and some soil targets) analysis likely contains some dust (Lasue et al., 2018). This phenomenon was observed on the first

ChemCam target, the rock Coronation analyzed on sol 14 of the mission. Coronation is a cobble-sized, fine-grained float rock that was analyzed in a single location with 50 shots. A large hydrogen peak at 656.5 nm was observed in the first shot but was significantly attenuated in the subsequent three shots, becoming essentially unobservable by shot 5 and above (Figure 29.4a). Subsequent study of martian targets showed this pattern to be typical of most martian rock surfaces (Meslin et al., 2013; Melikechi et al., 2014; Lanza et al., 2015; Lasue et al., 2018).

In addition to its presence in an amorphous component in dust and soils, hydrogen has also been observed in rock targets, most notably in Ca sulfate veins (Nachon et al., 2014). H peak areas in these vein target observations are consistent with bassanite in almost all cases (Nachon et al., 2014; Rapin et al., 2016), although this hydration state may be a surface-only effect; CheMin observations from drilled samples have found three different hydration states of Ca sulfate (Vaniman et al., 2018). ChemCam has also identified the hydration state of opals in a silica-rich region of Gale (Rapin et al., 2018). Work is ongoing to improve H quantification in geologic materials (Schröder et al., 2015; Rapin et al., 2016;

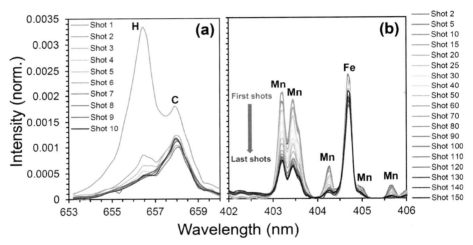

Figure 29.4 Example shot-to-shot LIBS spectral data obtained on Mars. (a) Hydrogen peaks in the first 10 shots on the rock target Coronation, analyzed by ChemCam on sol 13 (ccam02013). The rock surface was visibly dusty at the start of the analysis. As the laser shock wave moves dust away from the analysis area, the amount of H present in the spectra decreases. Subsequent work by Meslin et al. (2013) and Lasue et al. (2018) found that martian dust is hydrated and can be removed within the first ~5 shots of an analysis. The carbon present in the spectra is due to carbon species in the atmosphere. (b) Depth trends in manganese in the rock target Stephen (ccam03619). Initial shots show high Mn abundance, while subsequent shots show a significant attenuation of the Mn peaks. The sampling depth increases by shot number as additional material is ablated from the LIBS pit. The nearby Fe peak shows less shot-to-shot variation than the Mn peaks, suggesting that Fe does not have a significant variation in abundance with depth. This systematic decrease in peak height with shot/depth is characteristic of a thin layer atop a compositionally different substrate, in this case an exposed fracture fill. (After Lanza et al., 2016.)

Thomas et al., 2018; Ytsma et al., 2017). Nonetheless, ChemCam observations of hydrogen on the martian surface have provided new insights into the nature of hydrated materials on Mars.

29.2.2 First-Time Observation and Quantification of Minor and Trace Elements

ChemCam's ability to analyze all elements has enabled the first in situ observation of several elements on Mars, including Li (Ollila et al., 2014), B (Gasda et al., 2017), and F (Forni et al., 2015). Such observations of trace or unusual elements can provide key information about mineralogy and geochemical formation context that might otherwise be difficult to discern from major element chemistry alone. Observations of B in Ca sulfate veins provide constraints on the nature of groundwater circulating within Gale bedrock, and is itself an important building block of prebiotic chemistry (Gasda et al., 2017). The association of F with Al and H in some targets has revealed the presence of micas, a family of phyllosilicate minerals not previously observed by Mars surface missions (Treiman & Medard, 2016). The relatively frequent occurrences of F in associated with Ca and P suggest the presence of fluorapatites (Forni et al., 2015). More broadly, the elements Li, Rb, Sr, and Ba are routinely quantified in ChemCam targets (Ollila et al., 2014; Payré et al., 2017), while Zn and Cu have been identified in several locations (Lasue et al., 2016; Goetz et al., 2017; Payré et al., 2019). By providing a fuller view of elemental compositions beyond major elements, ChemCam LIBS can help to constrain the types and evolution of materials present on the martian surface.

29.2.3 A Wide Range of Chemical Compositions

ChemCam has obtained an unprecedented >600,000 individual spectra from the martian surface. To date ChemCam has fired ≥600,000 LIBS shots, each of which returns a spectrum with 6144 channels. With this wealth of data, ChemCam measurements have revealed a surprisingly wide range of compositions at Gale crater. While the vast majority of rocks in Gale appear to be sedimentary in nature, there are many igneous grains and clasts present within these sedimentary rocks that have provided intriguing new insights into martian magmatism. At the Bradbury Landing site, ChemCam discovered felsic compositions within alluvial pebbles, conglomerate clasts, and igneous float rocks (Williams et al., 2013; Meslin et al., 2013; Sautter et al., 2014; Mangold et al., 2016; Cousin et al., 2017). Both chemical compositions and rock textures suggest the presence of coarse-grained feldspars (Sautter et al., 2014). Feldspar-rich clasts in Bradbury Rise provide the highest Al compositions along the plagioclase–feldspar join; the majority of these observations are consistent with the Ca-rich feldspar bytownite (Gasda et al., 2016) (Figure 29.5). In addition to their high alkali content, these felsic igneous materials are also relatively high in silica, with compositions similar to Earth's continental crust. The presence of these compositions on Mars suggests that the geologic evolution of early Mars may be more similar to that of Earth than previously recognized (Sautter et al.,

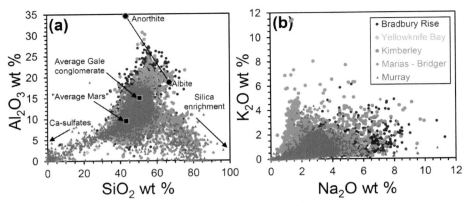

Figure 29.5 ChemCam LIBS data for Si, Al, Na, and K for >10 000 observations along the traverse in Gale crater. Data are separated by location along the traverse, with Bradbury at the landing site and Murray–Stimson at the rover's most recent location. All data for targets 2.0–4.5 m standoff distance from ChemCam were used. (a) In the Si versus Al plot, two loci of points can be seen, one near the average Mars point (Taylor & McLennan, 2009) and the other near the average Gale conglomerate point (Mangold et al., 2016). The data extend to the albite–anorthite join, notably for targets at Bradbury Rise, suggesting the presence of feldspars. Significant silica enrichment is observed in targets at Marias Pass and Bridger Basin. Ca sulfates were observed in all regions along the traverse except Bradbury Rise. (b) The alkali plot of Na versus K shows the strong K trend at the Kimberley, which also appears at lower levels in targets from Marias Pass and Bridger Basin.

2015). At Marias Pass, ChemCam also discovered an outcrop composed almost entirely of silica (Figure 29.5) (Frydenvang et al., 2017). Subsequent analysis by CheMin X-ray diffraction found both opal-A and tridymite there (Morris et al., 2016), which on Earth is a telltale sign of high-temperature silicic volcanism. This interpretation is strengthened by the observation of high-K sandstones containing a significant fraction of sanidine, a high-temperature potassium feldspar (Figure 29.5) (Le Deit et al., 2016; Treiman et al., 2016). Overall, the abundance of alkali- and silica-rich rocks at Gale crater shows that Mars clearly had evolved magmatism and is not the simple basaltic planet it was once thought to be.

In addition to providing insights about igneous materials, an examination of the overall relationship between alkalis, Si, and Al can also provide information about the amount of aqueous alteration experienced by the rocks in Gale crater. Gale once hosted a large lake, which left behind a >300 m thick section of fine-grained, finely laminated mudstone called the Murray formation (e.g., Grotzinger et al., 2015b). An aqueous origin for the Murray is strengthened by the observation of a marked increase in Al (e.g., above average Mars) in these materials (Figure 29.5) and the attendant change in the observed chemical index of alteration, which indicates an increase of chemical weathering in the Murray formation as compared to materials down section (Mangold et al., 2019). This interpretation of subtle chemical shifts within large-scale stratigraphic sections would not be possible without the thousands of elemental observations provided by ChemCam's LIBS instrument.

29.2.4 Probing the Near-Surface of Rocks with Depth Profiles

Because of the nature of the LIBS sampling technique, the ChemCam LIBS instrument can obtain compositional information from the near subsurface (~100s of μm) of a target along a depth profile. Multiple laser pulses are used in a single location, so each pulse samples material at a greater depth than the previous pulse to provide a LIBS spectrum at each depth. The amount and depth of material that is ablated is variable depending on the target's hardness and composition. The ablation depth averages ~0.3–2 μm per pulse (Wiens et al., 2012; Lanza et al., 2015), although the ablation rate is typically higher in the first shots and appears to be nonlinear with depth (Maurice et al., 2016a).

ChemCam's depth profiling capability may be used to study a number of surface features and target characteristics. Coatings and rinds on rock surfaces can provide constraints on the amount and type of aqueous alteration experienced by a rock, which gives valuable information about the overall weathering environment (Coleman & Pierce, 1981; Dixon et al., 2002; Salvatore et al., 2013; Lanza et al., 2015). Fracture fills exposed at the surface may also produce thin compositional layers that can be analyzed with LIBS depth profiles. Such fills are produced when fluids flowing along bedrock fractures precipitate minerals, which then subsequently become exposed at the surface as the bedrock is eroded. Fracture fills of a range of compositions have been observed in Gale crater, suggesting that a wide range of fluid compositions have circulated within bedrock units in this area (e.g., Léveillé et al., 2014; Nachon et al., 2014; Lanza et al., 2016).

An example of how ChemCam may be used to identify and analyze fracture fills can be found in the Kimberley region of Gale crater, where Curiosity spent ~60 sols performing analyses on outcrops in several geologic units, including the Dillinger member (Rice et al., 2017). This is characterized by fin-like, erosion-resistant features that are parallel to the bedding (Lanza et al., 2016). An initial ChemCam analysis of one such feature named Stephen on sol 611 found high concentrations of Mn (>30 wt.%) on the surface that decreased in abundance through 30 shots (Lanza et al., 2016). Subsequent analyses of the same target consisting of 150-shot depth profiles showed a clear trend in Mn abundance with depth, suggesting the presence of a thin high-Mn layer (Figure 29.4b). This depth trend is interpreted as indicating a thin (likely ≤10 μm) layer of fracture-filling Mn oxides that became exposed after erosion of the surrounding bedrock. By contrast, analysis by APXS found only ~4 wt.% MnO in the target Stephen because it integrates composition over a larger volume than does ChemCam. The example of the target Stephen demonstrates the utility of the depth profiling capabilities of ChemCam LIBS for discerning thin layers of discrete composition when they are present within martian rock targets.

29.3 Future Applications of LIBS on Planetary Surfaces

A new LIBS system for planetary exploration has been selected as part of the SuperCam instrument suite on board the NASA Mars 2020 rover mission. As with ChemCam, SuperCam combines several analytical techniques into one payload instrument, which include LIBS, remote Raman spectroscopy, time-resolved luminescence (TRL), visible and near-infrared reflectance spectroscopy (VISIR), a color Remote Micro-Imager (RMI), and a microphone (Maurice et al., 2015, 2016b; Murdoch et al., 2019; Wiens et al., 2016; Wiens et al., 2017). In addition to acquiring elemental data from LIBS, SuperCam can also acquire information about molecular structures (mineralogy) with Raman, TRL, and VISIR. SuperCam will use a Nd:YAG laser emitting at 1064 nm for LIBS; the beam is doubled to 532 nm for Raman measurements. The LIBS and Raman are co-boresighted to allow for direct measurement of fine-scale chemistry and mineralogy at a single location. In addition to the same VISIR range as ChemCam (~400–840 nm), the SuperCam spectrometer range has been extended to include 1.3–2.6 μm to enhance passive reflectance spectroscopy capabilities (Clegg et al., 2015; Wiens et al., 2017). With these new capabilities, SuperCam will be able to directly measure the chemistry, mineralogy, and organic content of rocks and soils at a single analysis location. For example, SuperCam can assess surface features such as rock coatings and rinds (as described for ChemCam in Section 29.2.4) and also provide mineralogy for the same locations. Raman and TRL can observe and identify organic molecules that may be present in targets up to 12 m from the rover (Wiens et al., 2016). The addition of a microphone to the SuperCam payload allows for the recording of the sound of LIBS plasma shock waves on different material types, which can provide needed information about the nature of the target and the depth of laser penetration into the sample (Chide et al., 2019a; 2019b). Overall, SuperCam provides measurements of fine-scale chemistry, mineralogy, and organics with color context imaging within a single instrument, which will provide an unprecedented and diverse data set for martian surface materials.

Although a relative newcomer to the suite of planetary exploration instruments, LIBS has many potential applications beyond the martian surface. The challenges facing landed spacecraft will only increase as additional Solar System bodies become targets for exploration. These future missions may require the analysis of solid, liquid, or even vapor samples, and with a wide range of extreme environmental conditions possible on target bodies, many of these missions will be time limited. ChemCam has demonstrated the versatility and scientific capabilities of LIBS on Mars. The many advantages of the LIBS technique could be applied to a diverse range of environments on other Solar System bodies in the future.

Acknowledgments

This work was supported in the U.S. by NASA's Mars Exploration Program. Thanks to R. Williams for assistance with image processing.

References

Bishop J.L. & Murad E. (2005) The visible and infrared spectral properties of jarosite and alunite. *American Mineralogist*, **90**, 1100–1107.
Blake D.F., Morris R.V., Kocurek G., et al. (2013) Curiosity at Gale crater, Mars: Characterization and analysis of the Rocknest sand shadow. *Science*, **341**, DOI:10.1126/science.1239505.

Boynton W.V., Feldman W.C., Squyres S.W., et al. (2002) Distribution of hydrogen in the near surface of Mars: Evidence for subsurface ice deposits. *Science*, **297**, 81–85.

Carter J., Poulet F., Bibring J.-P., Mangold N., & Murchie S. (2013) Hydrous minerals on Mars as seen by the CRISM and OMEGA imaging spectrometers: Updated global view. *Journal of Geophysical Research*, **118**, 831–858.

Chide B., Maurice S., Murdoch N., et al. (2019a) Listening to laser sparks: A link between Laser-Induced Breakdown Spectroscopy, acoustic measurements and crater morphology. *Spectrochimica Acta B*, **153**, 50–60.

Chide B., Maurice S., Bousquet B., et al. (2019b) The Mars 2020 SuperCam Microphone to constrain rock hardness and LIBS crater volume. *50th Lunar Planet. Sci. Conf.*, Abstract #1411.

Clegg S.M., Wiens R.C., Maurice S., et al. (2015) Remote geochemical and mineralogical analysis with SuperCam for the Mars 2020 rover. *46th Lunar Planet. Sci. Conf.*, Abstract #2781.

Cloutis E.A., Hawthorne F.C., Mertzman S.A., et al. (2006) Detection and discrimination of sulfate minerals using reflectance spectroscopy. *Icarus*, **184**, 121–157.

Coleman S.M. & Pierce K.L. (1981) *Weathering rinds on andesitic and basaltic stones as a Quaternary age indicator, western United States*. US Geological Survey Professional Paper 1210.

Cousin A., Clegg S., Dehouck E., et al. (2014) ChemCam blind targets: A helpful way of analyzing soils and rocks along the traverse. *45th Lunar Planet. Sci. Conf.*, Abstract #1278.

Cousin A., Sautter V., Payré V., et al. (2017) Classification of igneous rocks analyzed by ChemCam at Gale crater, Mars. *Icarus*, **288**, 265–283.

Cremers D. & Radziemski L.L. (2013) *Handbook of Laser-Induced Breakdown Spectroscopy*. John Wiley & Sons, Hoboken, NJ.

Dixon J.C., Thorn C.E., Darmody R.G., & Campbell S.W. (2002) Weathering rinds and rock coatings from an Arctic alpine environment, northern Scandinavia. *GSA Bulletin*, **114**, 226–238.

Edgett K.S., Yingst R.A., Ravine M.A., et al. (2012) Curiosity's Mars Hand Lens Imager (MAHLI) investigation. *Space Science Reviews*, **170**, 259–317.

Estlin T.A., Bornstein B.J., Gaines D.M., et al. (2012) AEGIS automated science targeting for the MER Opportunity rover. *ACM Transactions on Intelligent Systems and Technology*, **3**, 1–19.

Fabre C., Maurice S., Cousin A., et al. (2011) Onboard calibration igneous targets for the Mars Science Laboratory Curiosity rover and the Chemistry Camera laser induced breakdown spectroscopy instrument. *Spectrochimica Acta B: Atomic Spectroscopy*, **66**, 280–289.

Feldman W.C., Prettyman T.H., Maurice S., et al. (2004) Global distribution of near-surface hydrogen on Mars. *Journal of Geophysical Research*, **109**, E09006, DOI:10.1029/2003JE002160.

Forni O., Gaft M., Toplis M.J., et al. (2015) First detection of fluorine on Mars: Implications for Gale crater's geochemistry. *Geophysical Research Letters*, **42**, 1020–1028.

Francis R., Estlin T., Gaines D., et al. (2016) AEGIS intelligent targeting deployed for the Curiosity rover's ChemCam instrument. *47th Lunar Planet. Sci. Conf.*, Abstract #2487.

Francis R., Estlin T., Doran G., et al. (2017) AEGIS autonomous targeting for ChemCam on Mars Science Laboratory: Deployment and results of initial science team use. *Science Robotics*, **2**, eaan4582.

Frydenvang J., Gasda P.J., Hurowitz J.A., et al. (2017) Discovery of silica-rich lacustrine and eolian sedimentary rocks in Gale crater, Mars. *Geophysical Research Letters*, **4**, DOI:10.1002/2017GL073323.

Gasda P.J., DeLapp D.M., McInroy R.E., et al. (2016) Identification of fresh feldspars in Gale crater using ChemCam. *47th Lunar Planet. Sci. Conf.*, Abstract #1604.

Gasda P.J., Haldeman E.B., Wiens R.C., et al. (2017) In situ detection of boron by ChemCam on Mars. *Geophysical Research Letters*, **44**, 8739–8748.

Gellert R., Clark B., & MSL and MER Science Teams (2015) In situ compositional Measurements of Rocks and Soils with the Alpha Particle X-ray Spectrometer on NASA's Mars rovers. *Elements*, **11**, 39–44.

Goetz W., Payre V., Wiens R.C., et al. (2017) Detection of copper by the ChemCam instrument along the traverse of the Curiosity rover, Gale crater, Mars. *48th Lunar Planet. Sci. Conf.*, Abstract #2894.

Grotzinger J., Crisp J., Vasavada A.R., & MSL Science Team. (2015a) Curiosity's Mission of Exploration at Gale crater, Mars. *Elements*, **11**, 19–26.

Grotzinger J.P., Gupta S., Malin M.C., et al. (2015b) Deposition, exhumation, and paleoclimate of an ancient lake deposit, Gale crater, Mars. *Science*, **350**, aac7575.

Johnson J.R., Bell J.F., Bender S., et al. (2015) ChemCam passive reflectance spectroscopy of surface materials at the Curiosity landing site, Mars. *Icarus*, **249**, 74–92.

Johnson J.R., Bell J.F., Bender S., et al. (2016a) Constraints on iron sulfate and iron oxide mineralogy from ChemCam visible/near-infrared reflectance spectroscopy of Mt. Sharp basal units, Gale crater, Mars. *American Mineralogist*, **101**, 1501–1514.

Johnson J.R., Cloutis E., Fraeman A.A., et al. (2016b) ChemCam passive reflectance spectroscopy of recent drill tailings, hematite-bearing rocks, and dune sands. *47th Lunar Planet. Sci. Conf.*, Abstract #1155.

Johnson J.R., Achilles C., Bell J.F., et al. (2017) Visible/near-infrared spectral diversity from in situ observations of the Bagnold Dune Field sands in Gale crater, Mars. *Journal of Geophysical Research*, **122**, 2655–2684.

Lanza N.L., Fischer W.W., Wiens R.C., et al. (2014) High manganese concentrations in rocks at Gale crater, Mars. *Geophysical Research Letters*, **41**, 5755–5763.

Lanza N.L., Ollila A.M., Cousin A., et al. (2015) Understanding the signature of rock coatings in Laser-Induced Breakdown Spectroscopy Data. *Icarus*, **249**, 62–73.

Lanza N.L., Wiens R.C., Arvidson R.E., et al. (2016) Oxidation of manganese in an ancient aquifer, Kimberley formation, Gale crater, Mars. *Geophysical Research Letters*, **43**, 7398–7407.

Lasue J., Mangold N., Hauber E., et al. (2013) Quantitative assessments of the martian hydrosphere. *Space Science Reviews*, **174**, 155–212.

Lasue J., Clegg S.M., Forni O., et al. (2016) Observation of >5 wt % zinc at the Kimberley outcrop, Gale crater, Mars. *Journal of Geophysical Research*, **121**, 338–352.

Lasue J., Maurice S., Cousin A., et al. (2018) Martian eolian dust probed by ChemCam. *Geophysical Research Letters*, **45**(20), 10,968–10,977.

Le Deit L., Mangold N., Forni O., et al. (2016) The potassic sedimentary rocks in Gale crater, Mars, as seen by ChemCam on board Curiosity. *Journal of Geophysical Research*, **121**, 784–804.

Leshin L.A., Mahaffy P.R., Webster C.R., et al. (2013) Volatile, isotope, and organic analysis of martian fines with the Mars Curiosity rover. *Science*, **341**, 1238937.

Léveillé R.J., Bridges J., Wiens R.C., et al. (2014) Chemistry of fracture-filling raised ridges in Yellowknife Bay, Gale crater: Window into past aqueous activity and habitability on Mars. *Journal of Geophysical Research*, **119**, 2398–2415.

Malin M.C., Caplinger M.A., Edgett K.S., et al. (2010) The Mars Science Laboratory (MSL) Mast-Mounted Cameras (Mastcams) flight instruments. *41st Lunar Planet. Sci. Conf.*, Abstract #1123.

Mangold N., Thompson L.M., Forni O., et al. (2016) Composition of conglomerates analyzed by the Curiosity rover: Implications for Gale crater crust and sediment sources. *Journal of Geophysical Research*, **121**, 353–387.

Mangold N., Dehouck E., Fedo C., et al. (2019) Chemical alteration of fine-grained sedimentary rocks at Gale crater. *Icarus*, **321**, 619–631.

Maurice S., Wiens R.C., Saccoccio M., et al. (2012) The ChemCam Instrument Suite on the Mars Science Laboratory (MSL) rover: Science objectives and mast unit description. *Space Science Reviews*, **170**, 95–166.

Maurice S., Wiens R.C., Le Mouélic S., et al. (2015) The SuperCam instrument for the Mars 2020 rover. *European Planetary Science Congress*, Abstract #EPSC2015-185.

Maurice S., Clegg S.M., Wiens R.C., et al. (2016a) ChemCam activities and discoveries during the nominal mission of the Mars Science Laboratory in Gale crater, Mars. *Journal of Analytical Atomic Spectrometry*, **31**, 863–889.

Maurice S., Wiens R.C., Rapin W., et al. (2016b) A microphone supporting LIBS investigation on Mars. *47th Lunar Planet. Sci. Conf.*, Abstract #3044.

McCollom T.M., Ehlmann B.L., Wang A., Hynek B., Moskowitz B., & Berquó T.S. (2014) Detection of iron substitution in natroalunite-natrojarosite solid solutions and potential implications for Mars. *American Mineralogist*, **99**, 948–964.

McConnochie T.H., Smith M.D., Bender S., et al. (2017) Water vapor and aerosols from ChemCam passive sky observations. *6th International Workshop on the Mars Atmosphere: Modeling and Observations*. Abstract #3201.

McConnochie T.H., Smith M.D., Wolff M.J., et al. (2018) Retrieval of water vapor column abundance and aerosol properties from ChemCam passive sky spectroscopy. *Icarus*, **307**, 294–326.

Melikechi N., Mezzacappa A., Cousin A., et al. (2014) Correcting for variable laser-target distances of Laser-Induced Breakdown Spectroscopy measurements with ChemCam using emission lines of martian dust spectra. *Spectrochimica Acta B: Atomic Spectroscopy*, **96**, 51–60.

Meslin P.-Y., Gasnault O., Forni O., et al. (2013) Soil diversity and hydration as observed by ChemCam at Gale crater, Mars. *Science*, **341**, 1238670.

Morris R.V., Vaniman D.T., Blake D.F., et al. (2016) Silicic volcanism on Mars evidenced by tridymite in high-SiO_2 sedimentary rock at Gale crater. *Proceedings of the National Academy of Sciences of the USA*, **113**, 7071–7076.

Murchie S.L., Mustard J.F., Ehlmann B.L., et al. (2009) A synthesis of martian aqueous mineralogy after 1 Mars year of observations from the Mars Reconnaissance Orbiter. *Journal of Geophysical Research*, **114**, E00D06, DOI:10.1029/2009JE003342.

Murdoch N., Chide B., Lasue J., et al. (2019) Laser-induced breakdown spectroscopy acoustic testing of the Mars 2020 microphone. *Planetary and Space Science*, **165**, 260–271.

Mustard J.F., Adler M., Allwood A., et al. (2013) Report of the Mars 2020 Science Definition Team.

Nachon M., Clegg S.M., Mangold N., et al. (2014) Calcium sulfate veins characterized by ChemCam/Curiosity at Gale crater, Mars. *Journal of Geophysical Research*, **119**, 1991–2016.

Ollila A.M., Newsom H.E., Clark B., et al. (2014) Trace element geochemistry (Li, Ba, Sr, and Rb) using Curiosity's ChemCam: Early results for Gale crater from Bradbury landing site to Rocknest. *Journal of Geophysical Research*, **119**, 255–285.

Payré V., Fabre C., Sautter V., et al. (2019) Copper enrichments in Kimberley formation, Gale crater, Mars, Evidence for a Cu deposit at the source. *Icarus*, **321**, 736–751.

Payré V., Fabre C., Cousin A., et al. (2017) Alkali trace elements in Gale crater, Mars, with ChemCam: Calibration update and geological implications. *Journal of Geophysical Research*, **122**, 650–679.

Rapin W., Meslin P.Y., Maurice S., et al. (2016) Hydration state of calcium sulfates in Gale crater, Mars: Identification of bassanite veins. *Earth and Planetary Science Letters*, **452**, 197–205.

Rapin W., Chauviré B., Gabriel T., et al. (2018) In situ analysis of opal in Gale crater, Mars. *Journal of Geophysical Research*, **123**, 1955–1972.

Rice M.S., Gupta S., Treiman A.H., et al. (2017) Geologic overview of the Mars Science Laboratory rover mission at the Kimberley, Gale crater, Mars. *Journal of Geophysical Research*, **122**, 2–20.

Salvatore M.R., Mustard J.F., Head J.W., Cooper R.F., Marchant D.R., & Wyatt M.B. (2013) Development of alteration rinds by oxidative weathering processes in Beacon Valley, Antarctica, and implications for Mars. *Geochimica et Cosmochimica Acta*, **115**, 137–161.

Sautter V., Fabre C., Forni O., et al. (2014) Igneous mineralogy at Bradbury Rise: The first ChemCam campaign at Gale crater. *Journal of Geophysical Research*, **119**, 30–46.

Sautter V., Toplis M.J., Wiens R.C., et al. (2015) *In situ* evidence for early continental crust on Mars. *Nature Geoscience*, **8**, 605–609.

Schröder S., Meslin P.Y., Gasnault O., et al. (2015) Hydrogen detection with ChemCam at Gale crater. *Icarus*, **249**, 43–61.

Sklute E.C., Jensen H.B., Rogers A.D., & Reeder R.J. (2015) Morphological, structural, and spectral characteristics of amorphous iron sulfates. *Journal of Geophysical Research*, **120**, 809–830.

Smith P.H., Tamppari L.K., Arvidson R.E., et al. (2009) H_2O at the Phoenix landing site. *Science*, **325**, 58–61.

Sobron P., Bishop J.L., Blake D.F., Chen B., & Rull F. (2014) Natural Fe-bearing oxides and sulfates from the Rio Tinto Mars analog site: Critical assessment of VNIR reflectance spectroscopy, laser Raman spectroscopy, and XRD as mineral identification tools. *American Mineralogist*, **99**, 1199–1205.

Taylor S.R. & McLennan S. (2009) *Planetary crusts: Their composition, origin and evolution.* Cambridge University Press, New York.

Thomas N.H., Ehlmann B.L., Anderson D.E., et al. (2018) Characterization of hydrogen in basaltic materials with *Laser-Induced Breakdown Spectroscopy* (LIBS) for application to MSL ChemCam data. *Journal of Geophysical Research*, **123**, 1996–2021.

Treiman A.H. & Medard E. (2016) Mantle metasomatism in Mars: Potassic basaltic sandstone in Gale crater derived from partial melt of phlogopite-peridotite. *GSA Annual Meeting*, Abstract #49–12.

Treiman A.H. Bish D.L., Vaniman D.T., et al. (2016) Mineralogy, provenance, and diagenesis of a potassic basaltic sandstone on Mars: CheMin X-ray diffraction of the Windjana sample (Kimberley area, Gale crater). *Journal of Geophysical Research*, **121**, 75–106.

Vaniman D., Dyar M.D., Wiens R., et al. (2012) Ceramic ChemCam calibration targets on Mars Science Laboratory. *Space Science Reviews*, **170**, 229–255.

Vaniman D., Martinez G.M., Rampe E., et al. (2018) Gypsum, bassanite, and anhydrite at Gale crater, Mars. *American Mineralogist*, **103**(7), 1011–1020.

Whiteway J.A., Komguem L., Dickinson C., et al. (2009) Mars water-ice clouds and precipitation. *Science*, **325**, 68–70.

Wiens R.C., Maurice S., Barraclough B., et al. (2012) The ChemCam Instrument Suite on the Mars Science Laboratory (MSL) rover: Body unit and combined system tests. *Space Science Reviews*, **170**, 167–227.

Wiens R.C., Maurice S., Lasue J., et al. (2013) Pre-flight calibration and initial data processing for the ChemCam *Laser-Induced Breakdown Spectroscopy* instrument on the Mars Science Laboratory rover. *Spectrochimica Acta B: Atomic Spectroscopy*, **82**, 1–27.

Wiens R.C., Maurice S., & the MSL Team. (2015) ChemCam: Chemostratigraphy by the first Mars microprobe. *Elements*, **11**, 33–38.

Wiens R.C., Maurice S., McCabe K., et al. (2016) The SuperCam Remote Sensing Instrument Suite for Mars 2020. *47th Lunar Planet. Sci. Conf.*, Abstract #1322.

Wiens R.C., Maurice S., & Rull Perez F. (2017) The SuperCam remote sensing instrument suite for the Mars 2020 rover mission: A preview. *Spectroscopy* **32**(5), 50–55.

Williams R.M.E., Grotzinger J.P., Dietrich W.E., et al. (2013) Martian fluvial conglomerates at Gale crater. *Science*, **340**, 1068–1072.

Ytsma C.R., Dyar M.D., Lepore K.H., Wagoner C.M., & Hanlon A.E. (2017) Normalization and baseline removal effects on univariate and multivariate hydrogen prediction accuracy using laser-induced breakdown spectroscopy. *48th Lunar Planet. Sci. Conf.*, Abstract #2979.

30

Neutron, Gamma-Ray, and X-Ray Spectroscopy of Planetary Bodies

THOMAS H. PRETTYMAN, PETER A. J. ENGLERT, NAOYUKI YAMASHITA, AND MARGARET E. LANDIS

30.1 Introduction

The elemental composition of the shallow subsurface of airless planetary bodies and planets with thin atmospheres can be determined from measurements of secondary particles produced by the interaction of space radiation with surface materials. Sources include galactic cosmic rays, solar energetic particles, and solar X-rays. In addition, gamma rays are generated by the decay of radioisotopes (K, Th, and U) present in minerals that make up the regolith. These measurements can be conducted from a spacecraft orbiting in close proximity to a planetary body or in situ with instrumentation deployed on a rover or lander. Artificial radiation sources can also be deployed for active interrogation of the subsurface in the vicinity of the lander/rover. For planets with thick atmospheres, such as Venus, active interrogation is required to measure rock-forming elements, such as Si and Fe. The elemental data provide insights into geological processes underlying the formation and evolution of the regolith and interior. The techniques outlined here can also be used to characterize planetary atmospheres.

Chapter 9 describes methods and instrumentation used to determine regolith elemental composition from measurements of neutrons, gamma rays, and X-rays. Here, we provide a brief review of planetary missions that deployed radiation detection instrumentation and summarize their main discoveries. Passive measurements on orbital missions are emphasized. Active, in situ X-ray measurements are described in Chapter 28. A supplementary table summarizes missions deploying radiation instrumentation and, when available, includes links to the archived data (please see the supplementary table online, at www.cambridge.org/9781107186200). The aim of this chapter is to provide a historical background of planetary radiation measurements for those entering the field. Upcoming missions are described along with future directions for instrument technology.

The development and implementation of nuclear spectroscopy (with neutrons and gamma rays) is intertwined with X-ray spectroscopy. Both methods utilize similar radiation measurement technology, which has undergone many cycles of innovation and maturation since the dawn of the Space Age. Nevertheless, because they are optimal for different settings, missions utilizing X-rays and nuclear radiation are described in separate sections.

Solar X-rays are most useful for studies of airless bodies in the inner Solar System and are sensitive to depths of a few hundred μm. Nuclear spectrometers can be deployed at any heliocentric distance and probe deeper into the surface (depths up to about a meter). In addition, they have different sensitivities to elements present within planetary surfaces. For example, under quiet-Sun conditions, X-ray spectroscopy is well-suited to orbital measurements of intermediate-Z elements, such as Mg, Al, and Si. Under the same conditions, nuclear spectroscopy is sensitive to additional elements, for example H, Ca, Ti, Fe, K, U, and Th (see Chapter 9). Depending on mission science requirements, these modalities have been used separately and in combination on numerous missions spanning roughly 60 years. Although neutron, gamma-ray, and X-ray spectroscopy are now well established for planetary remote sensing, the state-of-the-art continues to advance as sensor technology improves.

30.2 Nuclear Spectroscopy: Gamma Rays and Neutrons

30.2.1 First Missions – Moon, Mars, and Venus

In the early 1960s, the NASA Ranger program sent several probes to acquire images of the Moon. Rangers 3 through 5 each included a boom-mounted, gamma-ray spectrometer (cesium iodide scintillator with plastic anticoincidence shield), with the goal of measuring lunar gamma rays while in close proximity to the Moon. This objective was not achieved; however, gamma-ray spectra were acquired by two of the probes in deep space (Rangers 3 and 5, both in 1962), providing an early measurement of interstellar gamma rays (Arnold et al., 1962; Metzger et al., 1964) and setting the stage for missions that followed.

The earliest orbital measurements of lunar gamma rays were made in 1966 by sodium iodide scintillators on Soviet space probes Luna 10 and 12 (e.g., Surkov, 1984). The Soviet Venera and Vega missions (1961–1987) included the first probes to acquire data on the surface of Venus. Venera 8, 9, and 10 and Vega 1 and 2 made passive measurements of radioelements (K, Th, and U) using sodium iodide (Venera) and cesium iodide (Vega) scintillators (Surkov, 1984; Surkov et al., 1987). Venus' thick atmosphere prevented passive measurements using cosmogenic gamma rays. Concentrations of radioelements from Venera/Vega, compiled by Surkov et al. (1987), are consistent with venusian rocks having similar composition to the tholeiitic basalts and gabbros that make up Earth's crust. Similar instrumentation was flown on Mars 5 (1974). While Mars 5 failed to reach its intended target (the martian moon Phobos), the probe did acquire gamma-ray spectra from Mars at an altitude of about 2000 km, providing the first measurements of the concentration of radioelements within the martian regolith (Surkov et al., 1980). Additional elemental data were acquired by the Phobos 2 probe in 1989 (Surkov et al., 1989).

The Apollo 15 and 16 missions (1971–1972) deployed sodium iodide gamma-ray spectrometers with plastic anticoincidence shields. The spectrometers were mounted on extendable booms to minimize background contributions from the Lunar Command Module (Harrington et al., 1974). The Apollo missions mapped the elemental composition of about 20% of the lunar surface in a near-equatorial band (e.g., Adler et al., 1973b). Concentrations of radioelements and major rock-forming elements, including Ti, K, Fe, and Th, were measured (Bielefeld et al., 1976; Metzger et al., 1977, 1979; Haines et al., 1978; Metzger & Parker, 1979; Davis, 1980). The elemental data provided constraints on early crustal evolution and processes underlying mare volcanism (e.g., Haines & Metzger, 1980).

There were no organized efforts to preserve the nuclear spectroscopy data from these early missions. Although some attempts have been made to recover the Apollo data, gamma-ray spectra from this period are not available in digital form. Starting in 1989, NASA required scientific data to be preserved in a digital archive and the NASA Planetary Data System (PDS) was formed. Other space agencies have begun to follow suit. When available, links to digital archives are provided in the supplementary table.

30.2.2 Near-Earth Asteroid Eros

The Near-Earth Asteroid Rendezvous (NEAR) mission (1996–2001) was the first mission selected in the NASA Discovery program (low-cost, principal-investigator–led missions for Solar System exploration). The NEAR-Shoemaker space probe traveled to the stony, Amor group asteroid 433 Eros ($34.4 \times 11.2 \times 11.2$ km). The probe deployed a deck-mounted, gamma-ray spectrometer consisting of a cesium iodide scintillator with an anticoincidence shield (Goldsten et al., 1997). The anticoincidence shield, a bismuth germanate scintillator, was designed to reduce background contributions from the spacecraft while allowing gamma rays from Eros to enter the cesium iodide scintillator when pointed at the asteroid. The orbital altitudes sampled by NEAR were greater than 2 body radii and the gamma-ray signal was too weak for use in elemental analyses. The spectrum of gamma rays from Eros was acquired after the probe landed successfully on the surface. The ratio of Mg/Si, Si/O and concentration of K at the landing was consistent with ordinary chondrites, likely meteorite analogs for Eros (Evans et al., 2001). Deviations of Fe/Si and Fe/O from chondritic values and ratios determined by X-ray spectrometry were initially interpreted as resulting from regolith grain segregation processes; however, these ratios were revised in a follow on study that found a homogeneous regolith with composition similar to L or LL chondrites (Peplowski et al., 2015).

30.2.3 The Moon

Lunar Prospector (1998–1999), a NASA Discovery mission, mapped the elemental composition of the lunar surface from low-altitude, circular polar mapping orbits (100- and 30-km above the surface). While planetary neutron spectroscopy was envisioned during the Apollo era (Lingenfelter et al., 1961, 1972), Lunar Prospector was the first mission to successfully deploy both neutron and gamma-ray spectrometers (Feldman et al., 2004c). The gamma-ray spectrometer was a high-efficiency, bismuth germanate scintillator with a boron-loaded plastic anticoincidence shield. The use of boron-loaded plastic enabled the measurement of fast neutrons. Thermal and epithermal neutrons were measured using Sn- and Cd-covered ^{3}He gas proportional counters.

Global maps of the average atomic mass and thermal neutron macroscopic absorption cross section were determined by neutron spectroscopy (Elphic et al., 1998, 2000; Feldman et al., 2000; Gasnault et al., 2001; Maurice et al., 2004). Suppression of epithermal neutron counts at the poles provided evidence for water ice cold-trapped in permanently shadowed craters (Feldman et al., 2001) (Figure 30.1a), motivating future observations and missions. Gamma-ray measurements of radioelements and neutron absorption constraints on rare earth element concentrations (Gd + Sm) revealed extensive regions rich in incompatible and heat producing elements surrounding the near side mare (Lawrence et al., 1998; Elphic et al., 2000). High concentrations of long-lived radioisotopes may have served as a heat source for mare volcanism. The K/Th ratio was found to be consistent with lunar samples and meteorites, supporting formation of the Moon from a volatile depleted source (Prettyman et al., 2006). Orbital measurements of the neutron leakage flux enabled accurate determination of the concentration of elements such as Fe from neutron capture gamma rays (Lawrence et al., 2002). Gamma-ray spectral unmixing was applied to determine the concentration of radioelements, K and Th, and major oxides, including MgO, Al_2O_3, CaO, TiO_2, and FeO (Prettyman et al., 2006). The elemental data constrain lunar thermal evolution, crustal processes, and volcanism, and provide global context for lunar samples and meteorites.

Lunar Prospector was followed by other lunar missions deploying nuclear spectrometers, including the NASA Lunar Reconnaissance Orbiter (LRO), the Chinese missions Chang'E

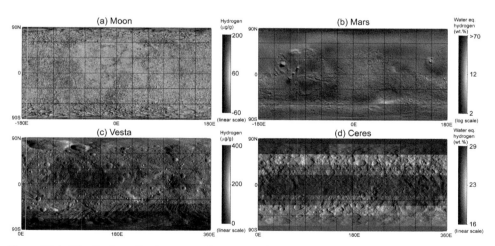

Figure 30.1 Distribution of hydrogen on planetary surfaces. (a) the Moon, (b) Mars, (c) Vesta, and (d) Ceres determined from neutron spectroscopy data acquired by Lunar Prospector, 2001 Mars Odyssey, and Dawn, respectively. Lunar, vestan, and cerean maps are available from the PDS (see supplementary table online). The map of martian hydrogen was determined from epithermal neutron counting data using the procedure described by Feldman et al. (2004d). Scale bars display hydrogen concentration in μg/g for the Moon and Vesta and equivalent (eq.) H_2O for Mars and Ceres. (Examples of elemental maps determined by gamma-ray and X-ray spectroscopy appear in Chapter 9.)

1 and 2, and the Japanese Kaguya mission. The LRO Lunar Exploration Neutron Detector (LEND) implemented a collimator with the goal of increasing the spatial resolution of the instrument to the scale of permanently shadowed craters. While the effectiveness of the collimator is debated, LEND maps of uncollimated neutron counts are similar to maps from Lunar Prospector (Litvak et al., 2012). The Chang'E missions deployed gamma-ray spectrometers, including cesium iodide and lanthanum bromide scintillators on Chang'E 1 and 2, respectively. Chang'E confirmed the distribution of K and Th measured by Lunar Prospector (Zhu et al., 2010, 2013).

The SELENE mission (whose main orbiter is known as Kaguya) by the Japan Aerospace Exploration Agency (JAXA) carried a gamma-ray spectrometer with a high-energy-resolution, high-purity germanium (HPGe) detector to the Moon to constrain the elemental composition of the lunar surface (Hasebe et al., 2008). Kaguya orbited the Moon in a polar, circular orbit at 100 km altitude for about 15 months beginning in October 2007. Then, in February 2009, the altitude was lowered to an approximately 30 × 50 km elliptical orbit (with perilune near the south pole) in which data were acquired for about four months. Global maps of U, Th, K, Ca, and Fe were reported by the Kaguya gamma-ray spectrometer (GRS) team (e.g., Naito et al., 2018) and Ti, Si, Al, Mg, H and more are in preparation. The high-energy resolution of SELENE enabled improved mapping of Ca (Yamashita et al., 2012) and the direct measurement of U (Yamashita et al., 2010).

30.2.4 Mars

The 2001 Mars Odyssey spacecraft (2001–present) included a suite of nuclear spectrometers, consisting of a large-volume HPGe GRS, a neutron spectrometer (NS), and a high-energy neutron detector (HEND) (Boynton et al., 2004). The GRS subsystem ceased operation in 2009; however, the neutron spectrometers continue to collect data. Odyssey acquires global mapping data from a low altitude (400 km), circular polar mapping orbit. Mars' atmosphere is thin (10- to 20-g/cm^2 column), such that cosmic rays can reach the surface, enabling the determination of regolith elemental composition. The spectrometers are also sensitive to seasonal cycling of the atmosphere, including changes in atmospheric composition and deposition of CO_2 frost on the surface. Seasonal variations are prominent during three Mars years of data collected by the GRS subsystem and more than nine Mars years of data acquired by the neutron spectrometers.

The gamma-ray and neutron spectrometers detected an enhancement in subsurface hydrogen at high latitudes (poleward of about 60°) in both hemispheres during late spring and summer, when seasonal CO_2 frost was not present. The observations were consistent with thermophysical models that predicted the presence of water ice within the cryosphere, stable at shallow depths near the poles (e.g., Mellon et al., 2004). Early results were reported for the southern hemisphere (Boynton et al., 2002; Feldman et al., 2002; Mitrofanov et al., 2002). Additional accumulation enabled global mapping of the frost-free hydrogen concentration via neutron spectroscopy (Feldman et al., 2004d) (Figure 30.1b). Studies of hydrogen layering using neutron spectroscopy indicated ice emplacement and modification at southern high latitudes was not exclusively due to diffusive exchange of vapor with the atmosphere (e.g., Prettyman et al., 2004). The presence of extensive subsurface ice deposits at high latitudes was confirmed by the detection of ice-exposing impacts (Byrne et al., 2009; Dundas et al., 2014) and in situ characterization by the Phoenix lander (e.g., Arvidson et al., 2009).

The concentrations of H, Si, Cl, K, Fe, and Th within the martian regolith were determined by gamma-ray spectroscopy in equatorial to mid-latitude regions unaffected by seasonal frost accumulation and for which thermal models predicted ice at depths greater than sensed by nuclear spectroscopy (Boynton et al., 2007). Analyses of gamma-ray and neutron measurements gave consistent results for the distribution of hydrogen in equatorial to mid-latitude regions. The data indicate variability in the abundance and hydration state of hydrogen-bearing minerals (Feldman et al., 2004a,b). The elemental data indicate Mars regolith and crust consist primarily of tholeiitic basalts (McSween et al., 2009).

Seasonal cycling of the martian atmosphere was characterized using gamma rays and neutrons. The data were used to determine the spatiotemporal distribution of seasonal frost in both hemispheres (Kelly et al., 2006; Litvak et al., 2006; Prettyman et al., 2009), enabling interannual comparisons of frost deposition. In the southern hemisphere, the condensation flow is associated with the formation of a strong polar vortex, resulting in the enrichment of noncondensable gases in the polar atmosphere. Concentrations of Ar and N_2 + Ar, measured respectively by gamma-ray and neutron spectroscopy, provide constraints on polar atmospheric processes (Sprague et al., 2007; Prettyman et al., 2009).

The Mars Science Laboratory (MSL) rover deployed the Dynamic Albedo of Neutrons (DAN) instrument, a neutron spectrometer with a pulsed, 14-MeV neutron source for active interrogation of subsurface elemental composition (Litvak et al., 2008). The instrument provided constraints on hydrogen and chlorine concentrations in context with geologic units traversed by MSL (e.g., Litvak et al., 2016).

The European Space Agency's (ESA) ExoMars Trace Gas Orbiter began orbital operations at Mars in 2016. The payload includes the Fine Resolution Epithermal Neutron Detector (FREND), which is similar in design to LEND. The stated objective of the investigation is to determine the spatial distribution of water ice and seasonal frost at high spatial resolution. Data are available online (see supplementary table online).

30.2.5 Mercury

The MESSENGER spacecraft spent several years at Mercury, acquiring data from an elliptical, polar orbit with perihermion near the north pole at roughly 200 km altitude. This enabled elemental mapping of the northern hemisphere using X-rays, gamma rays, and neutrons. The gamma-ray spectrometer was an HPGe detector with active cooling, mounted to the body of the spacecraft (Goldsten et al., 2007). The neutron spectrometer was a separate unit consisting of a block of boron-loaded plastic with two lithium-loaded glass scintillators covering opposite sides of the block. The scintillators were read out separately with photomultiplier tubes. The arrangement of scintillators and design of readout electronics enabled acquisition of thermal, epithermal, and fast neutrons (e.g., Lawrence et al., 2010). Analyses treated wide variations in spacecraft pointing geometry and kinematical ram effects for low-energy neutrons (e.g., Lawrence et al., 2010; Peplowski et al., 2012). The range of sampled altitudes (with an apohermion of 15,000 km) facilitated separation of background and planetary contributions.

Gamma-ray spectra were analyzed to determine abundances of rock-forming elements Al, Ca, Cl, Na, S, Fe (Evans et al., 2012; Peplowski et al., 2012; Evans et al., 2015) and radioelements K and Th (Peplowski et al., 2011). Neutron and gamma-ray analyses support carbon as the primary regolith darkening agent and graphite as a significant component of Mercury's crust (Peplowski et al., 2016). The epithermal and fast neutron data are consistent with high concentrations of cold-trapped water ice in permanently shadowed craters in the northern hemisphere, with possible delivery of water by comets and water-rich asteroids (Lawrence et al., 2013b). Synthesis of elemental data from X-ray and nuclear spectroscopy shows Mercury's silicate-rich and Fe-poor crust contains a diversity of rock types, with mineralogy consistent with a highly reducing environment during formation of the crust/mantle (Vander Kaaden et al., 2017).

30.2.6 Main Asteroid Belt

The NASA Dawn mission explored the two most massive bodies in the main asteroid belt: 4 Vesta and 1 Ceres (see Chapter 20). Instruments on board the spacecraft acquired data during a close flyby of Mars and in circular polar mapping orbits around Vesta and Ceres. The payload included the Gamma Ray and Neutron Detector (GRaND) instrument (Prettyman et al., 2011) that acquired global mapping data at ~0.8 body radius altitude at both Vesta and Ceres during Dawn's primary mission. GRaND's gamma-ray spectrometer was a high-efficiency bismuth germanate scintillator; however, the instrument also included a 4 × 4 array of room temperature semiconductors (cadmium zinc telluride) as a demonstration technology. The gamma-ray spectrometers were surrounded by plastic scintillators that served as neutron spectrometers and anticoincidence shields. The primary neutron spectrometer consisted of a lithium-loaded glass plate, optically coupled to a gadolinium-wrapped block of boron-loaded plastic and read out by a single photomultiplier tube. The configuration enabled separate measurements of neutrons in the thermal, epithermal, and fast energy ranges.

At Vesta, globally averaged mass ratios of Fe/O and Fe/Si and the concentration of K and Th were determined by gamma-ray spectroscopy (Prettyman et al., 2012). The elemental data strengthen the link between Vesta and the howardite, eucrite, and diogenite (HED) meteorites, implying Vesta underwent igneous differentiation to form an iron-rich core, ultramafic mantle, and basaltic crust. The measured K/Th ratio is consistent with that of howardite, indicating Vesta formed from a volatile depleted source (Prettyman et al., 2015b). The distribution of H measured by neutron spectroscopy reveals large portions of Vesta's basaltic surface are contaminated by debris from carbonaceous chondrite impactors, providing clues to the transport of water in the early Solar System (Prettyman et al.,

2012) (Figure 30.1c). The GRaND data were analyzed to determine the global distribution of iron, thermal neutron absorption cross section, average atomic mass, and a compositional parameter derived from high-energy gamma-ray measurements (Lawrence et al., 2013a; Peplowski et al., 2013; Prettyman et al., 2013; Yamashita et al., 2013). Petrologic variations derived from elemental mapping data indicate the oldest portions of Vesta's surface are rich in basaltic eucrite and that the floor of the Rheasilvia basin is rich in diogenite, which constrains Vesta's interior structure and evolution (Toplis et al., 2013).

At Ceres, the GRaND data were analyzed to determine the global distribution of H and Fe, the equatorial-average concentration of K, and bounds on regolith carbon content (Prettyman et al., 2017; Prettyman et al., 2019a). High concentrations of hydrogen near the poles indicate the presence of widespread, subsurface ice, consistent with thermal models that predict ice is stable within the shallow depths sensed by GRaND at high latitudes (Figure 30.1d). The measurements constrain ice-table depth and regolith thermophysical properties. Ceres' ice free regolith contains abundant endogenic hydrogen in the form of aqueously altered minerals and possibly organic matter. The elemental data indicate that unlike the smaller parent bodies of the carbonaceous chondrite meteorites, Ceres underwent ice rock fractionation, consistent with geophysical observations that indicate a partially differentiated interior.

In 2018, the Dawn spacecraft was placed into a highly eccentric orbit around Ceres with periapsis at 35 km altitude in order to acquire elemental data with high spatial resolution and sensitivity. The orbits sampled a narrow band of longitudes around Occator crater, with periapsis gradually drifting from the northern to the southern hemisphere, crossing Urvara crater. With 50-km spatial resolution at periapsis, the new GRaND data offer a view of Ceres at scales compatible with geology, mineralogy, topography, and gravity data sets, yielding fresh insights into crustal processes and interior evolution (Prettyman et al., 2019b).

30.3 X-Ray Spectroscopy

30.3.1 The Moon

The history of missions to measure fluorescent X-rays from planetary surfaces began with the lunar orbiting spacecraft Luna 10 and Luna 12 (Mandel'shtam et al., 1966, 1968). Several small Geiger counters with aluminum and organic windows were sensitive to X-rays between 3 and 14 angstroms (Å). The Luna 10 measurement results could not unambiguously discern lunar X-rays from background. Luna 12 Geiger counters, having a band pass between 8 and 14 Å, were looking alternatively at the lunar surface and into deep space. The Luna 12 data show for the first time that solar X-rays produced measurable fluxes of fluorescent X-rays from the lunar surface.

The first measurements of the chemical composition of the lunar surface using X-rays were performed during the Apollo 15 and 16 missions (Adler et al., 1972a,b, 1973a,b, 1974). The success of X-ray measurements was based on the selection and design of the measurement equipment: large area (25 cm^2) gas proportional counters with thin beryllium windows were chosen, with signal outputs that were proportional to the energy of absorbed X-rays. The counters function in principle as an energy dispersive spectrometer with low resolution. To improve discrimination between X-rays of rock-forming elements, three detectors were used, of which two had thin elemental Mg and Al filters (Yin et al., 1993).

The combined X-ray fluorescence (XRF) measurements of Apollo 15 and 16 covered about 10% of the lunar surface and occurred during solar minimum or quiet Sun conditions. These measurements resulted in maps of Al/Si and Mg/Si weight concentration ratios (Andre et al., 1977, 1978). XRF data of both missions overlap within some areas and are reproducible within about 10% (Yin et al., 1993). In addition to proving conclusively the feasibility of remote sensing X-ray spectrometry, these results provided support for new scientific knowledge about the Moon. Al/Si and Mg/Si ratios are inversely correlated with topography: Al/Si ratios are high in the highlands and low in the mare regions (Adler et al., 1972b), while Mg/Si ratios are low in the highlands and high in the mare regions (Andre et al., 1978). The results indicate that the highlands are anorthositic and the maria are basaltic. Other important correlations include the Al/Si ratio and the optical albedo (Adler et al., 1973b), suggesting that the optical albedo is associated with high Al_2O_3 surface abundances.

After the Apollo missions, lunar X-ray studies tested or utilized different types of semi-conductor detectors. Mission outcomes were influenced by the sensitivity of these detectors to space radiation, the timing of the missions in the framework of the solar cycle, and the duration of the missions.

The SMART-1 mission tested swept charge devices (SCD), an innovative semiconductor device for use in remote X-ray spectrometry and an advanced version of CCDs. Orbital parameters of SMART-1 did not support systematic lunar exploration with fluorescent X-rays, but was able to detect a number of rock-forming elements from the Moon during solar flare events including the first unambiguous detection of Ca by X-rays (Grande et al., 2007).

An SCD-based X-ray spectrometer was also flown during the Chandrayaan-1 mission, which investigated the lunar surface during 293 days (3400 orbits, see supplementary table online) at 100- and 200-km altitudes and with a 90° inclination. Reports on a cluster of selected flare observations (Narendranath et al., 2011; Weider et al., 2012; Athiray et al., 2014) provide estimates of Mg/Si and Al/Si ratios for two separate regions on the near side of the Moon, Ca and Fe, as well as the first and unambiguous direct measurement of Na on the lunar surface. Where spatial overlap exists, results agree with existing lunar sample compositions and the Apollo 14 average soil composition.

SELENE and Chang'E 1 XRF arrays were operated at solar minimum, and quantitative elemental analyses were not reported (Okada et al., 2009; Peng et al., 2009). Chang'E 2 results for a selected solar flare were reported by Ban et al. (2014). Low-resolution, global maps of Mg/Si and Al/Si were also determined (Dong et al., 2016). Further characterization of the lunar surface with X-ray spectroscopy would be a desirable goal for future missions.

30.3.2 Asteroids

The robustness of the Apollo 15 and 16 remote sensing experiments encouraged the use of unfiltered and Mg- and Al-filtered gas proportional counters for X-ray spectrometry on missions to asteroids. Advanced gas proportional counter systems were utilized on the Near Earth Asteroid Rendezvous (NEAR) Mission to asteroid 433 Eros. NEAR collected the first remote X-ray data from an asteroid during eight major solar flares and measured abundance ratios for Mg, Al, S, Ca, and Fe to Si on 433 Eros' surface (Nittler et al., 2001; Lim & Nittler, 2009). Significant effort was placed into precise characterization of the solar monitor spectra obtained simultaneously with the fluorescent X-ray spectra from the asteroid (Lim & Nittler, 2009), supporting the conclusion that 433 Eros had a major-element composition of ordinary chondrites, except for a strong depletion of sulfur due to space weathering (Nittler et al., 2001).

The first asteroid mission to fly semiconductor detectors was the Hayabusa mission to asteroid 25143 Itokawa, which carried a total of four charged-coupled devices (CCDs). The purpose of the mission was to collect and return asteroid sample material. In orbits of 20 km to 55 m altitude the spacecraft collected X-ray data during a 38 day period. Arai et al. (2008) studied Mg/Si, Al/Si, and S/Si elemental abundance ratios and determined that these can be used to classify 25143 Itokawa as an ordinary chondrite or a primitive achondrite.

30.3.3 Mercury

The MESSENGER mission to Mercury took advantage of the robustness of gas proportional counters proven by Apollo and NEAR. The primary 1-year Mercury science mission, complemented by extended missions (3.5 years), resulted in the first global X-ray derived elemental abundance ratio and abundance maps of a Solar System body. The major elemental composition of Mercury's surface was characterized by Nittler et al. (2011) and Weider et al. (2014, 2015, 2016). Measurements obtained during solar flare and quiet-Sun periods produced spatially complete maps of Mg/Si and Al/Si elemental weight ratios, and less complete maps of S/Si, Ca/Si, and Fe/Si that were derived from measurements from flare periods only (Weider et al., 2015). These maps can be used, together with other MESSENGER datasets, to study the geochemical characteristics of Mercury's surface. For example, the degree of melting within Mercury's mantle can be inferred from measurements of Mg/Si, S/Si, Ca/Si, and Al/Si. Results from MESSENGER's X-Ray Spectrometer, Gamma-Ray Spectrometer, and Neutron Spectrometer have been used to identify nine distinct geochemical regions on Mercury, spanning a wider range in SiO_2 abundance than terrains on other terrestrial objects except Earth (Vander Kaaden et al., 2017). X-ray spectrometer measurements of Mercury's largest identified pyroclastic deposit, combined with neutron and reflectance spectroscopy data, constrain the composition of volatiles involved in the eruption that emplaced the pyroclastic material (Weider et al., 2016). The rich X-ray datasets will facilitate additional studies of Mercury's global-scale compositional variations, as well as smaller-scale studies focused on specific features.

30.4 Future Missions and Instrument Technology

Nuclear spectrometers are planned for several upcoming missions, including the ESA/JAXA BepiColombo mission to Mercury, the JAXA Martian Moons Exploration mission, the NASA Lunar Polar Hydrogen Mapper (LunaH-Map) CubeSat mission (Hardgrove et al., 2018), and the NASA Psyche mission to main belt asteroid 16 Psyche (Lawrence et al., 2016). BepiColombo, because of its proximity to the Sun, carries an innovative X-ray spectrometer array.

Missions in development will use gamma-ray sensor technology with flight heritage, including HPGe (Psyche) and new scintillators, such as the cerium-doped Cs_2LiYCl_6 (CLYC) used by LunaH-Map and $CeBr_3$ flown on BepiColombo (Kozyrev et al., 2016). Both CLYC and $CeBr_3$ have highly proportional light yields, resulting in improved gamma-ray energy resolution compared to most previously flown scintillators. Lanthanum bromide scintillators produce excellent energy resolution; however, backgrounds from radiolanthanum and actinium may limit the ability to measure planetary gamma rays below 3 MeV (Zhu et al., 2013). Passive gamma-ray spectroscopy requires scintillators with low self-activity. Europium-doped strontium iodide is a potential alternative that is being explored for future missions (Prettyman et al., 2015a). The use of enriched Li in CLYC, along with its ability to discriminate particles by pulse shape, enables dual gamma–neutron spectroscopy using a single crystal (Glodo et al., 2008).

Scintillators can be operated without cooling and are relatively insensitive to radiation damage compared to HPGe. Consequently, the development of alternative room-temperature sensors with high resolution and efficiency is an active area of research. Other areas of interest include gamma-ray and neutron sensors with imaging and/or directional capability to improve spatial resolution of orbital measurements.

The Mercury Imaging X-ray Spectrometer (MIXS) on the BepiColombo Mercury Planetary Orbiter (MPO) is a two-component instrument based on semiconductor technology (Fraser et al., 2010). Both components have identical focal plane assemblies of monolithic 19.2×19.2 mm^2 active pixel sensors (Zhang et al., 2006), but differ in their X-ray optics. A state-of-the-art collimator provides measurements on scales of 70- to 270–km resolution for the MIXS-C component. The second component, MIXS-T, is the first imaging X-ray telescope for planetary remote sensing with an anticipated resolution of 10 km for major elements during solar flares. The spatial resolution achieved by MIXS-T is made possible by a novel, low-mass microchannel plate X-ray optics, in a Wolter type-I optical geometry (Fraser et al., 2010). The instrument is expected to enhance the compositional data set acquired by MESSENGER.

Finally, student involvement in planetary missions is on the rise. This is in part due to increased opportunities for student participation in low cost, small satellite missions, such as LunaH-Map, and larger missions led by principle investigators at universities, such as Psyche. For example, the OSIRIS-REX mission X-ray spectrometer (REXIS) investigation has involved over 60 students at undergraduate and graduate levels (Masterson et al., 2018). Projects such as these are necessary to train the next generation of planetary nuclear and X-ray spectroscopists, enabling continued use of these techniques on future missions.

Acknowledgments

This work was supported by NASA's SSERVI Toolbox for Research and Exploration (TREX) project and by NASA's PICASSO program.

References

Adler I., Trombka J., Gerard J., et al. (1972a) The Apollo 15 X-ray fluorescence experiment. *Science*, **175**, 436–440.
Adler I., Trombka J., Gerard J., et al. (1972b) Apollo 15 Geochemical X-ray Fluorescence experiment: Preliminary report. *Science*, **175**, 436–440.
Adler I., Trombka J.I., Yin L.I., Gorenstein P., Bjorkholm P., & Gerard J. (1973a) Lunar composition from Apollo orbital measurements. *Naturwissenschaften*, **60**, 231–242.
Adler I., Trombka J.I., Lowman P., et al. (1973b) Apollo 15 and 16 results of the integrated geochemical experiment. *The Moon*, **7**, 487–504.
Adler I., Podwysocki M., Andre C.G., et al. (1974) The role of horizontal transport as evaluated from the Apollo 15 and 16 orbital experiments. *Proceedings of the 5th Lunar Sci. Conf.*, 975–979.
Andre C.G., Bielefeld M.J., Elaison E., Soderblom L.A., Adler I., & Philpotts J.A. (1977) Lunar surface chemistry: A new imaging technique. *Science*, **197**, 986–989.
Andre C.G., Wolfe R., & Adler I. (1978) Evidence for a high-magnesium subsurface basalt in Mare Crisium from orbital X-ray fluorescence data. *Mare Crisium: The View from Luna24*, 1–12.
Arai T., Okada T., Yamamoto Y., Ogawa K., Shirai K., & Kato M. (2008) Sulfur abundance of asteroid 25143 Itokawa observed by X-ray fluorescence spectrometer onboard Hayabusa. *Earth, Planets and Space*, **60**, 21–31.
Arnold J.R., Metzger A.E., Anderson E.C., & Van Dilla M.A. (1962) Gamma rays in space, Ranger 3. *Journal of Geophysical Research*, **67**, 4878–4880.
Arvidson R.E., Bonitz R.G., Robinson M.L., et al. (2009) Results from the Mars Phoenix Lander Robotic Arm experiment. *Journal of Geophysical Research*, **114**, DOI:10.1029/2009je003408.
Athiray P.S., Narendranath S., Sreekumar P., & Grande M. (2014) C1XS results: First measurement of enhanced sodium on the lunar surface. *Planetary and Space Science*, **104**, 279–287.
Ban C., Zheng Y., Zhu Y., Zhang F., Xu L., & Zou Y. (2014) Research on the inversion of elemental abundances from Chang'E-2 X-ray spectrometry data. *Chinese Journal of Geochemistry*, **33**, 289–299.
Bielefeld M.J., Reedy R.C., Metzger A.E., Trombka J., & Arnold J.R. (1976) Surface chemistry of selected lunar regions. *Proceedings of the 7th Lunar Sci. Conf.*, 2661–2676.
Boynton W., Feldman W., Squyres S., et al. (2002) Distribution of hydrogen in the near surface of Mars: Evidence for subsurface ice deposits. *Science*, **297**, 81–85.
Boynton W., Feldman W., Mitrofanov I., et al. (2004) The Mars Odyssey gamma-ray spectrometer instrument suite. *Space Science Reviews*, **110**, 37–83.
Boynton W.V., Taylor G.J., Evans L.G., et al. (2007) Concentration of H, Si, Cl, K, Fe, and Th in the low- and mid-latitude regions of Mars. *Journal of Geophysical Research*, **112**, DOI:10.1029/2007je002887.
Byrne S., Dundas C.M., Kennedy M.R., et al. (2009) Distribution of mid-latitude ground ice on Mars from new impact craters. *Science*, **325**, 1674–1676.
Davis P.A. (1980) Iron and titanium distribution on the moon from orbital gamma ray spectrometry with implications for crustal evolutionary models. *Journal of Geophysical Research*, **85**, 3209–3224.
Dong W.-D., Zhang X.-P., Zhu M.-H., Xu A.-A., & Tang Z.-S. (2016) Global Mg/Si and Al/Si distributions on the lunar surface derived from Chang'E-2 X-ray Spectrometer. *Research in Astronomy and Astrophysics*, **16**(4), DOI:10.1088/1674–4527/16/1/004.
Dundas C.M., Byrne S., McEwen A.S., et al. (2014) HiRISE observations of new impact craters exposing martian ground ice. *Journal of Geophysical Research*, **119**, 109–127.
Elphic R.C., Lawrence D.J., Feldman W.C., et al. (1998) Lunar Fe and Ti abundances: Comparison of Lunar Prospector and Clementine data. *Science*, **281**, 1493–1496.

Elphic R.C., Lawrence D.J., Feldman W.C., et al. (2000) Lunar rare earth element distribution and ramifications for FeO and TiO_2: Lunar Prospector neutron spectrometer observations. *Journal of Geophysical Research*, **105**, 20,333–20,345.

Evans L.G., Starr R.D., Brückner J., et al. (2001) Elemental composition from gamma-ray spectroscopy of the NEAR-Shoemaker landing site on 433 Eros. *Meteoritics and Planetary Science*, **36**, 1639–1660.

Evans L.G., Peplowski P.N., Rhodes E.A., et al. (2012) Major-element abundances on the surface of Mercury: Results from the MESSENGER Gamma-Ray Spectrometer. *Journal of Geophysical Research*, **117**, DOI:10.1029/2012JE004178.

Evans L.G., Peplowski P.N., McCubbin F.M., et al. (2015) Chlorine on the surface of Mercury: MESSENGER gamma-ray measurements and implications for the planet's formation and evolution. *Icarus*, **257**, 417–427.

Feldman W.C., Lawrence D.J., Elphic R.C., Vaniman D.T., Thomsen D.R., & Barraclough B.L. (2000) Chemical information content of lunar thermal and epithermal neutrons. *Journal of Geophysical Research*, **105**, 20,347–20,363.

Feldman W.C., Maurice S., Lawrence D.J., et al. (2001) Evidence for water ice near the lunar poles. *Journal of Geophysical Research*, **106**, 23,231–23,251.

Feldman W.C., Boynton W.V., Tokar R.L., et al. (2002) Global distribution of neutrons from Mars: Results from Mars Odyssey. *Science*, **297**, 75–78.

Feldman W., Head J., Maurice S., et al. (2004a) Recharge mechanism of near-equatorial hydrogen on Mars: Atmospheric redistribution or sub-surface aquifer. *Geophysical Research Letters*, **31**, L18701, DOI:10.1029/2004GL020661.

Feldman W., Mellon M., Maurice S., et al. (2004b) Hydrated states of $MgSO_4$ at equatorial latitudes on Mars. *Geophysical Research Letters*, **31**, L16702, DOI:10.1029/2004GL020181.

Feldman W.C., Ahola K., Barraclough B.L., et al. (2004c) Gamma-ray, neutron, and alpha-particle spectrometers for the Lunar Prospector mission. *Journal of Geophysical Research*, **109**, E07S06, DOI:10.1029/2003JE002207.

Feldman W.C., Prettyman T.H., Maurice S., et al. (2004d) Global distribution of near-surface hydrogen on Mars. *Journal of Geophysical Research*, **109**, DOI:10.1029/2003JE002160.

Fraser G.W., Carpenter J.D., Rothery D.A., et al. (2010) The Mercury Imaging X-ray Spectrometer (MIXS) on bepicolombo. *Planetary and Space Science*, **58**, 79–95.

Gasnault O., Feldman W.C., Maurice S., et al. (2001) Composition from fast neutrons: Application to the Moon. *Geophysical Research Letters*, **28**, 3797–3800.

Glodo J., Higgins W.M., van Loef E.V.D., & Shah K.S. (2008) Scintillation properties of 1 Inch Cs_2LiYCl_6:CeCrystals. *IEEE Transactions on Nuclear Science*, **55**, 1206–1209.

Goldsten J.O., Mcnutt R.L., Gold R.E., et al. (1997) The X-ray/gamma-ray spectrometer on the Near Earth Asteroid Rendezvous Mission. In: *The near Earth asteroid rendezvous mission* (C.T. Russell, ed.). Springer, Dordrecht, 169–216.

Goldsten J.O., Rhodes E.A., Boynton W.V., et al. (2007) The MESSENGER gamma-ray and neutron spectrometer. *Space Science Reviews*, **131**, 339–391.

Grande M., Kellett B.J., Howe C., et al. (2007) The D-CIXS X-ray spectrometer on the SMART-1 mission to the Moon: First results. *Planetary and Space Science*, **55**, 494–502.

Haines E.L. & Metzger A.E. (1980) Lunar highland crustal models based on iron concentrations: Isostasy and center-of-mass displacement. *Proceedings of the 11th Lunar Planet. Sci. Conf.*, 689–718.

Haines E.L., Etchegaray-Ramirez M.I., & Metzger A.E. (1978) Thorium concentrations in the lunar surface. II: Deconvolution modeling and its application to the regions of Aristarchus and Mare Smythii. *Proceedings of the 9th Lunar Planet. Sci. Conf.*, 2985–3013.

Hardgrove C., West S.T., Heffern L.E., et al. (2018) Development of the Miniature Neutron Spectrometer for the Lunar Polar Hydrogen Mapper mission. *49th Lunar Planet. Sci. Conf.*, Abstract #2341.

Harrington T.M., Marshall J.H., Arnold J.R., Peterson L.E., Trombka J.I., & Metzger A.E. (1974) The Apollo gamma-ray spectrometer. *Nuclear Instruments and Methods*, **118**, 401–411.

Hasebe N., Shibamura E., Miyachi T., et al. (2008) Gamma-ray spectrometer (GRS) for lunar polar orbiter SELENE. *Earth, Planets and Space*, **60**, 299–312.

Kelly N.J., Boynton W.V., Kerry K., et al. (2006) Seasonal polar carbon dioxide frost on Mars: CO_2 mass and columnar thickness distribution. *Journal of Geophysical Research*, **112**, DOI:10.1029/2006je002678.

Kozyrev A., Mitrofanov I., Owens A., et al. (2016) A comparative study of LaBr3(Ce(3+)) and CeBr3 based gamma-ray spectrometers for planetary remote sensing applications. *Review of Scientific Instruments*, **87**(8), 085112. DOI:10.1063/1.4958897.

Lawrence D.J., Feldman W.C., Barraclough B.L., et al. (1998) Global elemental maps of the Moon: The Lunar Prospector gamma-ray spectrometer. *Science*, **281**, 1484–1489.

Lawrence D.J., Feldman W., Elphic R., et al. (2002) Iron abundances on the lunar surface as measured by the Lunar Prospector gamma-ray and neutron spectrometers. *Journal of Geophysical Research*, **107**, 5130.

Lawrence D.J., Feldman W.C., Goldsten J.O., et al. (2010) Identification and measurement of neutron-absorbing elements on Mercury's surface. *Icarus*, **209**, 195–209.

Lawrence D.J., Peplowski P.N., Prettyman T.H., et al. (2013a) Constraints on Vesta's elemental composition: Fast neutron measurements by Dawn's Gamma Ray and Neutron Detector. *Meteoritics and Planetary Science*, **48**, 2271–2288.

Lawrence D.J., Feldman W.C., Goldsten W.C., et al. (2013b) Evidence for water ice near Mercury's north pole from MESSENGER neutron spectrometer measurements. *Science*, **339**, 292–296.

Lawrence D.J., Peplowski P.N., Goldsten J.O., et al. (2016) The Psyche gamma-ray and neutron spectrometer: Characterizing the composition of a metal-rich body using nuclear spectroscopy. *47th Lunar Planet. Sci. Conf.*, Abstract #1622.

Lim L.F. & Nittler L.R.J.I. (2009) Elemental composition of 433 Eros: New calibration of the NEAR-Shoemaker XRS data. *Icarus*, **200**, 129–146.

Lingenfelter R.E., Canfield E.H., & Hess W.N. (1961) The lunar neutron flux. *Journal of Geophysical Research*, **66**, 2665–2671.

Lingenfelter R.E., Canfield E.H., & Hampel V.E. (1972) The lunar neutron flux revisited. *Earth and Planetary Science Letters*, **16**, 355–369.

Litvak M., Mitrofanov I., Kozyrev A., et al. (2006) Comparison between polar regions of Mars from HEND/Odyssey data. *Icarus*, **180**, 23–37.

Litvak M.L., Mitrofanov I.G., Barmakov Y.N., et al. (2008) The Dynamic Albedo of Neutrons (DAN) experiment for NASA's 2009 Mars Science Laboratory. *Astrobiology*, **8**, 605–612.

Litvak M.L., Mitrofanov I.G., Sanin A., et al. (2012) Global maps of lunar neutron fluxes from the LEND instrument. *Journal of Geophysical Research*, **117**, DOI:10.1029/2011JE003949.

Litvak M.L., Mitrofanov I.G., Hardgrove C., et al. (2016) Hydrogen and chlorine abundances in the Kimberley formation of Gale crater measured by the DAN instrument on board the Mars Science Laboratory Curiosity rover. *Journal of Geophysical Research*, **121**, 836–845.

Mandel'shtam S.L., Tindo I.T., & Karev V.I. (1966) Investigation of lunar X-Ray emission with the help of the lunar satellite Luna-10. *Kosmicheskie Issledovaniia*, **4**, 827–837.

Mandel'shtam S.L., Tindo I.P., Cheremukhin G.S., Sorokin L.S., & Dmitriev A.B. (1968) X radiation of the Moon and X-ray cosmic background in the lunar Sputnik Luna-12. *Kosmicheskie Issledovaniia*, **6**, 119–127.

Masterson R.A., Chodas M., Bayley L., et al. (2018) Regolith X-Ray Imaging Spectrometer (REXIS) aboard the - OSIRIS-REx asteroid sample return mission. *Space Science Reviews*, **214**, 48. DOI:10.1007/s11214-018-0483-8.

Maurice S., Lawrence D.J., Feldman W.C., Elphic R.C., & Gasnault O. (2004) Reduction of neutron data from Lunar Prospector. *Journal of Geophysical Research*, **109**, E07S04, DOI:10.1029/2003JE002208.

McSween H.Y., Jr., Taylor G.J., & Wyatt M.B. (2009) Elemental composition of the martian crust. *Science*, **324**, 736–739.

Mellon M.T., Feldman W.C., & Prettyman T.H. (2004) The presence and stability of ground ice in the southern hemisphere of Mars. *Icarus*, **169**, 324–340.

Metzger A.E. & Parker R.E. (1979) The distribution of titanium on the lunar surface. *Earth and Planetary Science Letters*, **45**, 155–171.

Metzger A.E., Anderson E.C., Van Dilla M.A., & Arnold J.R. (1964) Detection of an interstellar flux of gamma rays. *Nature*, **204**, 766–767.

Metzger A.E., Haines E., Parker R., & Radocinski R. (1977) Thorium concentrations in the lunar surface. I-Regional values and crustal content. *Proceedings of the 10th Lunar Sci. Conf.*, 10, 949–999.

Metzger A.E., Haines E., Etchegaray-Ramirez M., & Hawke B. (1979) Thorium concentrations in the lunar surface. III-Deconvolution of the Apenninus region. *Proceedings of the 10th Lunar Planet. Sci. Conf.*, 1701–1718.

Mitrofanov I., Anfimov D., Kozyrev A., et al. (2002) Maps of subsurface hydrogen from the high energy neutron detector, Mars Odyssey. *Science*, **297**, 78–81.

Naito M., Hasebe N., Nagaoka H., et al. (2018) Iron distribution of the Moon observed by the Kaguya gamma-ray spectrometer: Geological implications for the south pole-Aitken basin, the Orientale basin, and the Tycho crater. *Icarus*, **310**, 21–31.

Narendranath S., Athiray P.S., Sreekumar P., et al. (2011) Lunar X-ray fluorescence observations by the Chandrayaan-1 X-ray Spectrometer (C1XS): Results from the nearside southern highlands. *Icarus*, **214**, 53–66.

Nittler L.R., Starr R.D., Lev L., et al. (2001) X-ray fluorescence measurements of the surface elemental composition of asteroid 433 Eros. *Meteoritics and Planetary Science*, **36**, 1673–1695.

Nittler L.R., Starr R.D., Weider S.Z., et al. (2011) The major-element composition of Mercury's surface from MESSENGER X-ray spectrometry. *Science*, **333**, 1847–1850.

Okada T., Shirai K., Yamamoto Y., et al. (2009) X-ray fluorescence spectrometry of Lunar Surface by XRS onboard SELENE (Kaguya). *Transactions of the Japan Society for Aeronautical and Space Sciences, Space Technology Japan*, **7**, Tk_39–Tk_42.

Peng W.-X., Wang H.-Y., Zhang C.-M., et al. (2009) Prospective results of CHANG'E-2 X-ray spectrometer. *Chinese Physics C*, **33**(10), 819–825.

Peplowski P.N., Evans L.G., Hauck S.A., et al. (2011) Radioactive elements on Mercury's surface from MESSENGER: Implications for the planet's formation and evolution. *Science*, **333**, 1850–1852.

Peplowski P.N., Rhodes E.A., Hamara D.K., et al. (2012) Aluminum abundance on the surface of Mercury: Application of a new background-reduction technique for the analysis of gamma-ray spectroscopy data. *Journal of Geophysical Research*, **117**, DOI:10.1029/2012JE004181.

Peplowski P.N., Lawrence D.J., Prettyman T.H., et al. (2013) Compositional variability on the surface of 4 Vesta revealed through GRaND measurements of high-energy gamma rays. *Meteoritics and Planetary Science*, **48**, 2252–2270.

Peplowski P.N., Bazell D., Evans L.G., Goldsten J.O., Lawrence D.J., & Nittler L.R. (2015) Hydrogen and major element concentrations on 433 Eros: Evidence for an L- or LL-chondrite-like surface composition. *Meteoritics and Planetary Science*, **50**, 353–367.

Peplowski P.N., Klima R.L., Lawrence D.J., et al. (2016) Remote sensing evidence for an ancient carbon-bearing crust on Mercury. *Nature Geoscience*, **9**, 273–276.

Prettyman T.H., Feldman W., Mellon M., et al. (2004) Composition and structure of the martian surface at high southern latitudes from neutron spectroscopy. *Journal of Geophysical Research*, **109**, DOI:10.1029/2003je002139.

Prettyman T.H., Hagerty J.J., Elphic R.C., et al. (2006) Elemental composition of the lunar surface: Analysis of gamma ray spectroscopy data from Lunar Prospector. *Journal of Geophysical Research*, **111**, E12007, DOI:10.1029/2005JE002656.

Prettyman T.H., Feldman W.C., & Titus T.N. (2009) Characterization of Mars' seasonal caps using neutron spectroscopy. *Journal of Geophysical Research*, **114**, 10.1029/2008je003275.

Prettyman T.H., Feldman W.C., McSween H.Y., Jr., et al. (2011) Dawn's gamma ray and neutron detector. *Space Science Reviews*, **163**, 371–459.

Prettyman T.H., Mittlefehldt D.W., Yamashita N., et al. (2012) Elemental mapping by Dawn reveals exogenic H in Vesta's regolith. *Science*, **338**, 242–246.

Prettyman T.H., Mittlefehldt D.W., Yamashita N., et al. (2013) Neutron absorption constraints on the composition of 4 Vesta. *Meteoritics and Planetary Science*, **48**, 2211–2236.

Prettyman T.H., Yamashita N., Lambert J.L., Stassun K.G., & Raymond C.A. (2015a) Ultra-bright scintillators for planetary gamma-ray spectroscopy. *SPIE Newsroom*, DOI:10.1117/2.1201510.006162.

Prettyman T.H., Yamashita N., Reedy R.C., et al. (2015b) Concentrations of potassium and thorium within Vesta's regolith. *Icarus*, **259**, 39–52.

Prettyman T.H., Yamashita N., Toplis M.J., et al. (2017) Extensive water ice within Ceres' aqueously altered regolith: Evidence from nuclear spectroscopy. *Science*, **355**, 55–59.

Prettyman T.H., Yamashita N., Ammannito E., et al. (2019a) Elemental composition and mineralogy of Vesta and Ceres: Distribution and origins of hydrogen-bearing species. *Icarus*, **318**, 42–55.

Prettyman T.H., Yamashita N., Landis M.E., et al. (2019b) Dawn's GRaND finale: High spatial-resolution elemental measurements reveal an anomaly at Occator crater. *50th Lunar Planet. Sci. Conf.*, Abstract #1356.

Sprague A.L., Boynton W.V., Kerry K.E., et al. (2007) Mars' atmospheric argon: Tracer for understanding martian atmospheric circulation and dynamics. *Journal of Geophysical Research*, **112**, DOI:10.1029/2005je002597.

Surkov Y.A., Moskalyova L.P., Manvelyan O.S., Basilevsky A.T., & Kharyukova V.P. (1980), Geochemical interpretation of the results of measuring gamma-radiation of Mars. *Proceedings of the 11th Lunar Planet. Sci. Conf.*, 669–676.

Surkov Y.A. (1984) Nuclear-physical methods of analysis in lunar and planetary investigations. *Isotopenpraxis Isotopes in Environmental and Health Studies*, **20**, 321–329.

Surkov Y.A., Kirnozov F.F., Glazov V.N., Dunchenko A.G., Tatsy L.P., & Sobornov O.P. (1987) Uranium, thorium, and potassium in the Venusian rocks at the landing sites of Vega 1 and 2. *Journal of Geophysical Research*, **92**, E537–E540.

Surkov Y.A., Barsukov V.L., Moskaleva L.P., et al. (1989) Determination of the elemental composition of martian rocks from Phobos 2. *Nature*, **341**, 595–598.

Toplis M.J., Mizzon H., Monnereau M., et al. (2013) Chondritic models of 4 Vesta: Implications for geochemical and geophysical properties. *Meteoritics and Planetary Science*, **48**, 2300–2315.

Vander Kaaden K.E., McCubbin F.M., Nittler L.R., et al. (2017) Geochemistry, mineralogy, and petrology of boninitic and komatiitic rocks on the mercurian surface: Insights into the mercurian mantle. *Icarus*, **285**, 155–168.

Weider S.Z., Kellett B.J., Swinyard B., et al. (2012) The Chandrayaan-1 X-ray Spectrometer: First results. *Planetary and Space Science*, **60**, 217–228.

Weider S.Z., Nittler L.R., Starr R.D., McCoy T.J., & Solomon S.C. (2014) Variations in the abundance of iron on Mercury's surface from MESSENGER X-Ray Spectrometer observations. *Icarus*, **235**, 170–186.

Weider S.Z., Nittler L.R., Starr R.D., et al. (2015) Evidence for geochemical terranes on Mercury: Global mapping of major elements with MESSENGER's X-Ray Spectrometer. *Earth and Planetary Science Letters*, **416**, 109–120.

Weider S.Z., Nittler L.R., Murchie S.L., et al. (2016) Evidence from MESSENGER for sulfur- and carbon-driven explosive volcanism on Mercury. *Geophysical Research Letters*, **43**, 3653–3661.

Yamashita N., Hasebe N., Reedy R.C., et al. (2010) Uranium on the Moon: Global distribution and U/Th ratio. *Geophysical Research Letters*, **37**, L10201, DOI:10.1029/2010GL043061.

Yamashita N., Gasnault O., Forni O., et al. (2012) The global distribution of calcium on the Moon: Implications for high-Ca pyroxene in the eastern mare region. *Earth and Planetary Science Letters*, **353–354**, 93–98.

Yamashita N., Prettyman T.H., Mittlefehldt D.W., et al. (2013) Distribution of iron on Vesta. *Meteoritics and Planetary Science*, **48**, 2237–2251.

Yin L.I., Trombka J.I., Adler I., & Bielefeld M. (1993) X-ray remote sensing techniques for geochemical analysis of planetary surfaces. In: *Remote geochemical analysis: Elemental and mineralogical composition* (C. M. Pieters & P.A.J. Englert, eds.). Cambridge University Press, Cambridge, 199–212.

Zhang C., Lechner P., Lutz G., et al. (2006) Development of DEPFET Macropixel detectors. *Nuclear Instruments and Methods in Physics Research Section A: Accelerators, Spectrometers, Detectors and Associated Equipment*, **568**, 207–216.

Zhu M.H., Ma T., & Chang J. (2010) Chang'E-1 gamma ray spectrometer and preliminary radioactive results on the lunar surface. *Planetary and Space Science*, **58**, 1547–1554.

Zhu M.H., Chang J., Ma T., et al. (2013) Potassium map from Chang'E-2 constraints the impact of Crisium and Orientale basin on the Moon. *Scientific Reports*, **3**, 1611, DOI:10.1038/srep01611.

31

Radar Remote Sensing of Planetary Bodies

JEFFREY J. PLAUT

31.1 Introduction

Radar (RAdio Detection And Ranging) is an active remote sensing technique in which signals are transmitted, interact with a target, and echoes are received and recorded for analysis. Radar has been used in planetary studies for decades. Platforms for radar systems can be stationary, mobile along the surface, airborne, or space-borne. Radar data are used to infer a variety of properties of a target: its position, motion, and geometric and electrical properties. Of particular interest in planetary studies, radar can be used to map terrain and to probe beneath a planet's surface. The theory behind radar is given in Chapter 10. This chapter will briefly review planetary radar studies, including those conducted from Earth-based radio telescopes and from orbiting spacecraft. Detailed accounts of these observations are available elsewhere (see, for example, the thorough review of Campbell, 2002). Much of the chapter will focus on results obtained in the twenty-first century to date, including radar sounding of the Moon, Mars, and small bodies by space-borne systems.

31.2 Earth-Based Radar Observations of Planetary Bodies

31.2.1 Overview

Radar observations of planetary bodies conducted from the surface of Earth most commonly utilize a single stationary antenna for both transmit and receive (monostatic configuration). An alternative method utilizes one station to transmit, and one or more additional stations to receive the echoes. This latter configuration (known as bistatic, meaning separate positions of transmitter and receiver), allows for interferometric analysis of combinations of received echo streams to isolate the positions of the echoes and to measure the topography of the target, and for longer integration times in some instances. Radar observations conducted from Earth's surface are limited to a "window" in frequency, owing to Earth's atmosphere at shorter wavelengths and the ionosphere at longer wavelengths. In practice, useful scientific Earth-based radar observations have been limited to a range between X-band (~3 cm wavelength, ~10.0 GHz frequency) and P-band (~1 m wavelength, ~300 MHz frequency). Following World War II, radar techniques had advanced to a point where radar echoes were successfully obtained from Earth of the surface of the Moon in 1946 (Dewitt & Stodola, 1949). The first radar echoes of Venus were obtained using the Goldstone tracking and communications station in 1961 (Victor & Stevens, 1961). Analysis of the Doppler character of the echoes led to the discovery that Venus rotates very slowly, and allowed refinements of the values of Venus' radius and of the Astronomical Unit.

31.2.2 The Moon

Radar mapping of the Moon for most of the last several decades has been performed from radio telescopes on Earth, thus limiting the coverage to the lunar nearside. Recent radar observations from lunar orbit have now extended the coverage to the entire surface (see Section 31.3.4). Earth-based observations have been conducted at a range of radar wavelengths, primarily 3–4 cm, 12.6 cm, 70 cm, and 7.5 m (Thompson et al., 1970; Zisk et al., 1974; Thompson, 1978, 1987; Stacy et al., 1997). While the geomorphology of the Moon's nearside is well known from telescopic and lunar orbital imaging, the Earth-based radar observations provided insights into the character of the lunar regolith that are not obtained in visible imaging. Regolith surfaces surrounding lunar impact craters show highly variable characteristics in imaging radar data (Figure 31.1a). At a given wavelength, the amount of radar energy scattered back to the receiver from a flat surface is controlled primarily by

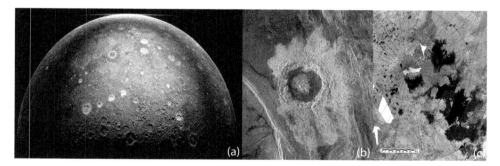

Figure 31.1 Radar images of the Moon, Venus, and Titan. (a) An image of part of the northern hemisphere of the Moon acquired using the Arecibo and Green Bank radio telescopes in combination. The Moon's north pole is at bottom center, with the near side and equator toward the top. Note the higher backscatter associated with certain impact craters, indicative of less weathered, youthful surfaces. (Image courtesy of Bruce Campbell, Smithsonian Institution.) (b) Magellan image of impact crater Dickinson on Venus. Rough surfaces associated with ejecta and flow materials show high backscatter, while the interior is smooth, possibly resulting from later volcanic infilling. Bands on the lower left are a small "ridge belt" that appears to have deflected some ejecta flows. Crater diameter is 69 km. (NASA/JPL image PIA00479.) (c) This mosaic of *Cassini* images of the north polar region of Titan is colorized according to radar backscatter to emphasize the contrast between the low-backscatter lake surfaces (blue and black) and the surrounding terrain (yellow). The lakes are believed to consist primarily of liquid methane and ethane. (Image: NASA/JPL-Caltech/ASI/USGS.)

surface roughness near the scale of the wavelength. A lunar crater ejecta surface will generally become smoother with time due to micrometeroid flux and gravitational stabilization. Thus, a younger crater will appear "rougher" at a given wavelength, and older craters that may have originally retained rough ejecta will now appear smooth to the radar. For these reasons radar observations provide a tool for relative dating of individual craters (Ghent et al., 2014). Further details of the scales of the roughness differences are obtained when features are observed at multiple wavelengths. Radar images of lunar craters have also enabled detection of impact melt deposits that are often invisible to other imaging techniques due to burial by ballistic ejecta (Carter et al., 2012).

An additional dimension of information about the surface of the Moon and other targets is provided by radar systems using multiple polarizations. Typically, the radar signals are transmitted in one polarization (either linearly or circularly polarized), and the echoes are received in two polarizations (orthogonal linear, i.e., horizontal and vertical, or in both circular polarizations). A smooth planar interface will produce an "expected" polarization, which is the same in linear polarizations (H transmit, H receive; or V–V), and opposite in circular polarizations. Deviations from the expected polarizations provide information on the scattering mechanisms responsible for the echo. This "depolarization" effect can be caused by roughness near the wavelength scale, even-number reflections ("corner reflectors"), and buried scatterers (Campbell, 2012). Most commonly, Earth-based systems use circular polarizations to generate images of both the unexpected and expected echo polarizations, and to measure their ratio, known as the Circular Polarization Ratio (CPR).

High CPRs are observed on icy surfaces of Mercury, Mars, and the Galilean satellites. This can be explained by the coherent backscatter effect (Black et al., 2001). This effect results from constructive interference of waves that follow the same path through a heterogeneous volume, but in opposite directions, thereby emerging with the same phase shift. The keys to enhancement of this effect in planetary ices are the low loss experienced by radio waves in cold ice and the presence of sufficient scatterers in the volume (cracks, voids, objects) to multiple-scatter the waves so they have a high probability of reflecting in the backscatter direction. The effect has been shown observationally and theoretically to produce high polarization ratios and overall high radar reflections, even in the "expected" polarization (Black et al., 2001).

The successful use of polarization ratios to identify candidate icy regions on planetary surfaces has inspired a search for such signatures in permanently shadowed craters near the lunar poles. This effort has included both Earth-based and orbital radar observations (see Section 31.3 for orbital examples). The Earth-based observations, at least, have not provided unequivocal evidence for ice deposits in these craters. Stacy et al. (1997) used the Arecibo radio telescope to construct 12.6-cm polarimetric radar maps of the lunar poles, and found no evidence of enhanced CPR indicative of ice. Campbell et al. (2006) used the Arecibo system to transmit and the Green Bank Telescope in West Virginia to receive the echoes to study the permanently shadowed craters at high resolution. They reported the lack of correlation of CPR and solar illumination conditions, and ascribed the enhancement of polarization ratios in some craters to surface roughness effects on crater walls and ejecta.

31.2.3 Mars

Forming radar images of Mars from Earth is a challenge because of the relatively rapid rotation rate of Mars. A variety of approaches have been used to overcome this problem, as summarized by Simpson et al. (1992) and Campbell (2002). One early solution was to use the Goldstone station in combination with the Very Large Array (VLA) to synthesize a large aperture (Muhleman et al., 1991). This study identified a zone of no detected reflections (called "Stealth") that was noted to be associated with the Medusae Fossae Formation. The authors also noted the unusual polarization signature of the residual south polar ice cap, which they ascribed to coherent backscatter within a depth of ≥ 2 m.

Harmon et al. (2012) used data from the Arecibo Observatory over multiple Mars apparitions to create images of the depolarized (same-sense circular) echo of the major volcanic provinces of Tharsis and Elysium at ~3 km resolution (Figure 31.2a). They found high CPRs on many lava flows indicating extremely rough and/or blocky surfaces. The images provide important complementary information to visible and infrared orbital imagery for distinguishing and mapping lava flows and other volcanic surfaces, and constraining eruptive processes that generate unique roughness signatures detected by radar. Harmon and Nolan (2017) further analyzed Arecibo data for other regions, including the variable radar response of polar capsurfaces.

31.2.4 Outer Planet Moons

The first robust observations of the Galilean satellites of Jupiter were conducted with the Arecibo system at 12.6-cm wavelength (Campbell et al., 1977). The moons' surfaces were observed to be highly reflective, especially Europa, and to have strong depolarization that would be expected for volume scattering in an icy medium. Reviews of subsequent observations can be found in Ostro (1982) and Ostro et al. (1992). Observations at multiple wavelengths (3.5 cm, 12.6 cm, 70 cm) all showed radar reflectivities higher than for any other known Solar System targets. The depolarization signature, expressed in the circular polarization ratio, was consistent with the coherent backscatter effect with penetration into the icy regolith to depths of many wavelengths. Both the mean surface reflectivities and the polarization ratios increase from Callisto to Ganymede to Europa, consistent with their optical albedos and the inferred geologic youthfulness of their surfaces (Europa being the optically brightest and youngest). The Saturnian satellites Rhea, Dione, Tethys, and Enceladus were observed with Arecibo and displayed reflectivities and polarization ratios similar to the Galilean satellites, though none were as extreme in their values as Europa (Black et al., 2007). Saturn's large moon Titan, which is the main target of the *Cassini* space-borne radar (see Section 31.3.5), was observed by Earth-based systems (Campbell et al., 2003; Black et al., 2011). Evidence was found for very smooth surfaces, consistent with liquid lakes, and the polarization behavior was seen to be intermediate between rocky surfaces and clean ice such as on the Galilean satellites, suggesting a less efficient coherent backscatter mechanism.

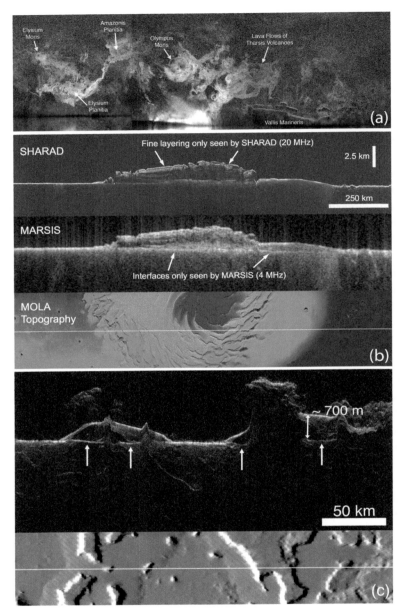

Figure 31.2 Radar images of Mars. (a) An Arecibo image of mid-latitudes of Mars, where high-backscatter surfaces are primarily younger rough lava flows. (Image courtesy of Smithsonian Institution.) Data appear in Harmon et al. (2012). (b) SHARAD (top) and MARSIS (middle) radargrams and ground track are shown on shaded relief topography (bottom) of the north polar plateau of Mars. The radargrams have been corrected for the speed of radio waves in ice for all points below the nadir ground track surface echo. MARSIS detects the basal interface across the entire width

31.3 Space-Borne Radar Systems

31.3.1 Overview

For planetary studies, space-borne radar systems fall into two distinct categories: imaging radar (also known as Synthetic Aperture Radar [SAR]) and sounding radar. One or both have been utilized to observe planetary surfaces (and subsurfaces) with great success, at the Moon, Venus, Mars, and Titan. The two techniques use the same basic principle of monostatic (co-located transmitter and receiver) radar, but differ in geometry of observation, choice of frequency and bandwidth, and data processing algorithms.

Imaging radar is primarily for mapping terrain, to generate images that superficially at least, resemble optical images. SAR images are normally projected to approximately square pixels, but in acquisition they are defined by different resolutions in azimuth (along-track) and range (across-track). The primary controls on the brightness of a SAR image pixel are slope with respect to the look direction, surface roughness at scales near the wavelength, and the inherent reflectivity of the surface (determined by the complex dielectric permittivity). Since SAR data are acquired with a side-looking geometry, surfaces tilted toward the radar will generally scatter more energy back toward the radar, and those tilted away, less. For flat terrain, smooth surfaces will scatter energy forward, away from the radar, while rough surfaces will scatter more isotropically, including a fraction of energy scattered back to the radar. Since the roughness sensitivity is related to the scale of the wavelength, surfaces observed at different wavelengths can show different contrast, depending on whether the surface is smooth or rough relative to the wavelength. For planetary applications, SAR system wavelengths are in the range of ~1 cm to 10s of cm. SAR systems may operate at a single polarization, dual-polarization, or full 4-polarization ("quad-pol") that captures all possible combinations of transmit and receive polarizations. Interferometric SAR uses multiple receive antennas or repeat passes to measure phase difference on terrain elements when viewed from slightly different positions or times. This technique can be used to obtain topographic maps, and to measure small changes (on the order of the wavelength) in surface height due, to for example, tectonic deformation, subsidence, or uplift.

The primary objective of radar sounding is to capture echoes from the subsurface. The resulting data product is a two-dimensional image known as a "radargram," with the dimensions of azimuth (along-track) and range (time-delay, a function of depth). Unlike

Caption for Figure 31.2 (cont.)

of the plateau, while SHARAD provides fine details of internal layering due to its higher vertical resolution. (Image: NASA/JPL/ASI/University of Rome.) (c) SHARAD radargram (top) and ground track shown on shaded relief topography (bottom) of a region of lobate debris aprons in Deuteronilus Mensae. The radargram has been corrected for the speed of radio waves in ice. Arrows indicate subsurface echoes that appear to lie at the same elevation as the surrounding plains. The maximum echo depth is at right, at 700 m. (Image: NASA/JPL/ASI/University of Rome.)

imaging radar, which is normally side-looking, sounding radar obtains its echoes from the nadir direction, which includes the surface of the terrain directly beneath the sensor. The first echoes are from the surface, and in this sense the sounding radar obtains an altimetric profile regardless of whether further penetration occurs or not. Later echoes may result from the signal encountering interfaces between subsurface layers of contrasting dielectric constant. Another difference between sounding and imaging radar is the "pixel" aspect ratio. Rather than the approximately square pixels in imaging radar, sounding radar azimuth pixels are typically 10s of times the scale of the range resolution. To maximize the penetration capability, much longer wavelengths are used in sounding radar than in imaging radar. Sounding radar wavelengths are typically in the range of several meters (frequency of 50–100 MHz) to >100 m (frequency of <3 MHz). Antennas operating at such frequencies are quite large (e.g., a half-wavelength) and must be of a simple design, such as a dipole, to have a reasonably low mass to be accommodated on a spacecraft. As a result, space-borne sounding radar antennas usually have little to no directivity, and methods must be applied to handle echoes coming from areas outside of the nadir position of interest. This is the so-called "clutter problem." Along-track clutter can be suppressed by Doppler filtering to limit the source of echoes to a narrow along-track footprint. Cross-track clutter is more problematic. Echoes arising from cross-track surface topography may overprint echoes from the nadir subsurface that arrive with the same time delay, hampering interpretation. The most common solution is to use digital elevation data (when available) to generate a synthetic radargram that portrays the apparent position and expected intensity of cross-track clutter. In some cases, the cross-track clutter obscures subsurface echoes and separating them is impossible; the scientist may be forced to discard such data if no other techniques for clutter suppression are available.

31.3.2 Venus

The Pioneer Venus orbiter began mapping the surface with radar in 1978. The radar was primarily an altimeter, which obtained a global map of surface heights at ~25 km resolution. The major continental-scale landforms of Venus were revealed in this map, and the unimodal distribution of elevation (hypsometry) was discovered, which contrasts with Earth's bimodal distribution caused by the seafloor and continents (Pettengill et al., 1980). The Soviet Union sent a pair of radar-mapping orbiters, Venera 15 and 16 into Venus orbit in 1983. The SAR systems operated at 8 cm wavelength, and mapped the northern ~1/3 of Venus at a resolution of 1–2 km (Barsukov et al., 1986). The major terrain types of the northern hemisphere were revealed, including extensive plains, mountain belts, volcanic centers, and circular deformation features called coronae. The population of craters of apparent impact origin was relatively sparse, and the corresponding surface age was estimated as 0.5–1.0 billion years (Barsukov et al., 1986).

The Magellan spacecraft's radar mapped Venus from orbit from 1990 to 1992. The radar system included a SAR and altimeter, both operating at 12.6 cm wavelength. The resulting SAR image map covered 98% of the surface, at a resolution of 100–250 m (Saunders et al.,

1992; Figure 31.1b). The altimetry resolution was 10–30 km. From the altimetry echoes, estimates of the inherent reflectivity and root mean square slope were also derived. Finally, the radar receiver was used in a passive mode ("radiometry") to estimate the emissivity of the surface at the 12.6-cm wavelength. This provided an independent measure of the electrical properties of the surface. The multiple datasets provided researchers with a wealth of information on the geology, geomorphology, and surface properties of Venus (see articles in special sections of *Journal of Geophysical Research*, Saunders, 1992; Saunders et al., 1992; and *Guide to Magellan Image Interpretation*, Ford et al., 1993). While the bulk of Magellan results were published in the 1990s, the data have continued to yield findings, as no follow-on radar mapping mission has been implemented. Published studies have included syntheses of geologic history (Basilevsky & Head, 2002; Basilevsky & McGill, 2007), studies of impact features (Basilevsky et al., 2003; Cook et al., 2003; Bondarenko & Head, 2004; Whitten & Campbell, 2016) and studies of volcanic features (e.g., Byrnes & Crown, 2002; Zimbelman, 2003; Romeo and Turcotte, 2009; Mouginis-Mark, 2016). Geologic maps of much of the surface of Venus have been compiled by researchers in the scientific community and published in a program sponsored by NASA and administered by the U.S. Geological Survey.

31.3.3 Mars

Radar sounding of the subsurface of Mars had great promise, as evidence for the presence of ice and potentially liquid water near the surface was widespread, and sounding techniques seemed well suited to probing and detecting such deposits. The European Space Agency (ESA) selected MARSIS, the Mars Advanced Radar for Subsurface and Ionospheric Sounding, to fly on the Mars Express mission in 2003. MARSIS was a complex multifrequency sounder, with modes for subsurface and ionospheric sounding (Jordan et al., 2009). In the subsurface mode, the radar could simultaneously acquire data in one or two channels with center frequencies of 1.8, 3.0, 4.0, and 5.0 MHz, each operated with a 1-MHz bandwidth that provides a range (vertical, depth) resolution of 150 m in free space. The ionospheric sounding mode covered these and lower frequencies, down to 0.1 MHz. The antenna system consisted of a main two-element dipole that when deployed was 40 m in length, and a secondary receive-only monopole antenna for identifying off-nadir clutter in the subsurface modes. The primary objective of the MARSIS subsurface experiment was to detect, map, and characterize interfaces related to water, both liquid and solid, in the subsurface of the martian crust. Secondary objectives for subsurface sounding were to detect geologic interfaces and to characterize the surface (for roughness and radar properties). The ionospheric mode was intended to measure ionospheric properties that might affect the subsurface signals, and to study the characteristics of the ionosphere itself.

The polar layered deposits (PLDs) have proven to be ideal targets for radar sounding (Figure 31.2b). The full thickness of the polar plateaus (Plana Boreum and Australe) are typically penetrated by MARSIS, including the lower "basal unit" in the north (Picardi et al., 2005; Plaut et al., 2007a; Selvans et al., 2010). This allows mapping of topography on the basal contact; the contact has substantial relief in the south and is relatively flat in the north. MARSIS mapping has allowed new estimates of the volume of the PLD and hence the size of the H_2O reservoirs. In terms of global equivalent water layer, the MARSIS-based values are 10–12 m for the south (Plaut et al., 2007a) and 8–10 m for the north (Selvans et al., 2010). The bulk composition of the polar plateaus is constrained by the interaction with the radar signals. The position of basal reflections in time delay is consistent with the continuation of the surrounding terrain under the PLD in both the north and south, and this observation leads to the determination of the dielectric constant, which within errors is indistinguishable from that of pure ice. Further constraints on composition are obtained from the strength of deep reflections, which again show indications of only minor impurities in the ice. Estimates of the mass fraction of impurities are generally <10%, with values <5% considered likely. The unit surrounding the SPLD known as the Hesperian Dorsa Argentea Formation contains a reflective horizon in MARSIS data over much of its mapped occurrence (Plaut et al., 2007b). The reflectors are observed at time delays consistent with a maximum depth between 500 and 1000 m. The relatively strong returns and the morphology of surface features both suggest an ice-rich layer overlying a lithic substrate.

The equatorial Medusae Fossae Formation showed deep reflectors in MARSIS data (Watters et al., 2007). Geometric constraints allowed determination of the dielectric constant to be near 3, which would be consistent with a low-density or high-porosity lithic deposit such as ash or dust, or with ice (or a combination of these). MARSIS reflectors were seen as deep as 2 km.

A goal for MARSIS was to detect liquid water in the subsurface. While recent analyses of MARSIS data suggest that a 20-km-wide liquid water zone may be present below the south polar layered ice deposits (Orosei et al., 2018), no unequivocal detections of groundwater have been made in bedrock or regolith sites. The reasons for this lack of detection are not entirely clear, but are likely the result of one or a combination of factors (Clifford et al., 2010). Among these is the possibility that there may be no subsurface liquid reservoirs in the upper several km of the subsurface. The lower heat flow values inferred from minimal deflection of the PLD mean that the melting temperature of ice may not be reached in the upper several km. Alternatively, liquid water may be hosted at shallow depths, but beneath crustal rocks that are impervious to the current radar sounder signals. Small aquifers may exist in the shallow subsurface but their presence may not be obvious in the radargrams due to their limited lateral extent. Finally, if the upper boundaries of aquifers are transitional in filled porosity, the dielectric contrast may not be sufficiently sharp to produce a detectable echo.

The ionospheric sounding part of the MARSIS experiment was highly successful, with the first results reported by Gurnett et al. (2005). By sweeping through the range of frequencies both above and below the natural plasma frequency of the ionosphere, MARSIS was able to characterize variations in plasma density as a function of altitude, solar illumination, solar activity, and geographic position. Structures were detected in the ionosphere that are linked to the presence of remnant magnetic fields in the crust of Mars (Gurnett et al., 2005, 2008). When comet C/2013 ("Siding Spring") passed within 135,000 km of Mars in 2014, it triggered the formation of a transient ionized layer of particles that was observed with the MARSIS ionospheric sounder (Gurnett et al., 2015).

The SHAllow RADar (SHARAD) was carried on NASA's Mars Reconnaissance Orbiter, which arrived at Mars in 2006. SHARAD operates at a central frequency of 20 MHz, with a 10-MHz bandwidth that provides a range resolution of 15 m in free space (Seu et al., 2007). While MARSIS was designed to maximize penetration into the subsurface by using the lowest frequencies that could be accommodated on an orbiter, SHARAD was intended to obtain higher resolution sounding data, using a higher frequency, at the expense of deep penetration capability. Thus MARSIS and SHARAD provide a complementary view into the subsurface of Mars (Figure 31.2b). The higher vertical resolution of SHARAD allowed the delineation of the fine structure of the layers of the PLD, and provided evidence for the sequence of events responsible for the evolution of the PLD. Phillips et al. (2008) identified groups of laterally continuous packets of reflectors, which they associated with the million-year time scale periodicities in Mars' orbital parameters. They also confirmed the earlier MARSIS observation that the base of the NPLD is not measurably deflected by the mass load of the deposits, implying a cold and mechanically thick lithosphere. Putzig et al. (2009) mapped the detailed 3D stratigraphy of the interior PLD using SHARAD data. Holt et al. (2010) used such stratigraphic evidence to demonstrate that the dominant canyon feature of the NPLD, Chasma Boreale, was not erosional but rather the result of nonuniform deposition. Similarly, Smith and Holt (2010) showed that the spiral trough structures characterizing the NPLD topography developed in place as migrating features resulting from wind transport and depositional patterns.

Away from the polar regions, SHARAD data provided a remarkable glimpse into the structure and composition of a class of features found in the mid-latitudes called Lobate Debris Aprons (LDA; Figure 31.2c). These features, along with the related lineated valley fill and concentric crater fill, were observed in image and topographic data from earlier missions to occur adjacent to steep topography and to display banding features and other morphology suggestive of flow. SHARAD data confirmed that many of these features are cored with thick deposits of relatively pure ice, and it was proposed that they represent the remnants of glacial ice from an earlier climate regime (Holt et al., 2008; Plaut et al., 2009a). The position of the apparent basal reflectors in time delay allowed estimation of the dielectric constant of the penetrated material, which was indistinguishable from that of pure water ice. In addition, the intensity of the basal reflections in many cases indicated minimal loss of radar energy in the medium, providing corroboration of the relative purity of the ice. The observations were consistent with a model of LDA formation that begins with a widespread regional ice sheet, which eventually sublimates in most areas except near steep topography, where material shed from exposed outcrops provides a regolith cover ~10 m in thickness that protects the ice cores from further sublimation (e.g., Head et al., 2005). The ice-rich mid-latitude remnant glaciers are promising targets for further in situ exploration, for climate history, habitability and resource utilization. Several plains regions in the northern lowlands showed SHARAD reflectors in the upper ~100 m of the subsurface that suggest the presence of substantial ice-charged regolith (Plaut et al., 2009b; Bramson et al., 2015; Stuurman et al., 2016). Other shallow reflectors apparently unrelated to ice were found in younger volcanic terrains, including Amazonis Planitia (Campbell et al., 2008) and the Cereberus Fossae region (Morgan et al., 2013). Stillman and Grimm (2011) suggested that penetration and detection of subsurface interfaces in rocky units on Mars is limited by scattering or absorption, which is more likely to occur in ancient rock units that have undergone fracturing, or that may contain radar-absorbing hydrous clay minerals from earlier, wetter climate epochs. However, the Medusae Fossae Formation, which is likely Hesperian in age, is easily penetrated by SHARAD signals (Carter et al., 2009; Morgan et al., 2015).

31.3.4 The Moon

The pioneer of planetary radar sounding was the Apollo Lunar Sounder Experiment (ALSE), which flow on the Apollo 17 Service Module during its orbits of the Moon in 1973 (Phillips et al., 1973; Porcello et al., 1974). ALSE was operated at three frequency bands, centered at 5.3 MHz, 15.8 MHz, and 160 MHz. Unlike modern digitally recorded radars, ALSE echoes were displayed on a cathode ray tube which was recorded on optical film that was retrieved by the astronauts and returned to Earth. Optical processing techniques previously developed for airborne SAR systems were then applied to generate radargram images (Porcello et al., 1974). In addition, a hybrid optical/digital system was developed to allow computer processing of the data. While unambiguous subsurface interfaces were difficult to identify in ALSE data, Peeples et al. (1978) used multiple parallel tracks to isolate off-nadir clutter from candidate subsurface reflectors. They identified features at depths of ~0.9 and ~1.6 km that they hypothesized were stratigraphic boundaries in mare basin flow materials, which developed a low-porosity regolith during hiatuses in deposition, providing the needed contrast in electrical properties.

Radar sounding returned to the Moon on the Japanese lunar orbiter, Kaguya, which collected data with the Lunar Radar Sounder (LRS) from 2007 to 2009 (Ono et al., 2009). LRS operated at 5 MHz central frequency, with a bandwidth of 2 MHz, which provided a free space range resolution of 75 m. While LRS mapping of the Moon was conducted on a global basis, unambiguous subsurface reflectors were found only in limited areas, specifically the floors of a subset of the nearside mare basins (Pommerol et al., 2010). The maximum depth of the detected interfaces was about 500 m. The mare basin interfaces are generally interpreted to be paleo-surfaces with developed regoliths, as proposed in the ALSE studies. The lack of detection of any interfaces as deep as the ALSE detections (1–1.5 km) is puzzling, as the two systems operated at similar frequencies. Pommerol et al. (2010) found a negative correlation between mare regions with detected interfaces and maps of the inferred presence of ilmenite, which is known to be a radar absorber. Oshigami et al. (2014) used inferred depths to mare unit interfaces to estimate volumes of mare basalt flows, which were comparable to volumes of continental flood basalts on Earth.

A rover carrying a ground-penetrating radar was successfully deployed on the lunar surface by China in 2013. The rover, called "Yutu," was part of the Chang'e-3 lunar lander mission. The rover completed traverses totaling about 100 m. The Lunar Penetrating Radar (LPR) had channels at 60-MHz and 500-MHz frequencies. Lai et al. (2016) analyzed data from the 500-MHz channel and identified a three-layer regolith structure in the first ~7 m of depth. Su et al. (2014) reported a detection of reflectors at an estimated depth of 330 m using the 60 MHz channel, which they attribute to a buried paleoregolith surface.

Two imaging radar systems mapped the Moon from lunar orbit, starting in 2008. The radars, called "Mini-SAR" and "Mini-RF," flew on the Indian Space Research Organization's Chandrayaan-1 spacecraft and on NASA's Lunar Reconnaissance Orbiter, respectively. Both were designed to demonstrate lightweight radio-frequency technology, with scientific data collection as secondary objectives (Nozette et al., 2010). Both operated at S-band (12.6-cm wavelength) and Mini-RF also carried an X-band (4.2-cm) system. The radars had polarization diversity, with transmission at circular polarization, and receive at both horizontal and vertical polarizations. The main scientific objective of the radar experiments was to locate and verify the presence of ice-rich regolith in permanently shadowed lunar craters. Mini-SAR operated continuously for several months and successfully completed its planned mapping of the polar regions. Mini-RF operated in its nominal modes from 2007 to 2010. It completed polar maps in both X-band and S-band, and 67% global coverage in S-band at a resolution of 30 m (Cahill et al., 2014). Cahill et al. (2014) found global patterns in radar response and mapped the lunar surface into three distinct "terranes": Nearside Radar Dark, Highlands Radar Bright, and Basins Radar Bright. Carter et al. (2012) and Neish et al. (2014) examined lunar impact melt features and found that many of them have radar properties that are common in volcanic flows, such as relatively high polarization ratios.

The prime targets of Mini-SAR and Mini-RF were the permanently shadowed interiors of polar craters, as it was expected that the polarization signatures could be used to definitively detect the presence of near-surface ice. Both radars acquired the planned data of the polar regions, but their interpretation remains controversial. The ambiguity arises because elevated polarization ratios are observed in some polar craters that are consistent with the presence of ice, but they are not so high as to rule out "dry" explanations, due to a rough surface or shallow regolith scattering. Spudis et al. (2010) identified "anomalous" craters in Mini-SAR data of the north polar region that displayed elevated CPRs only in their interiors. Spudis et al. (2010) attributed the anomalous craters' elevated interior CPR to the presence of ice in the regolith, which was supported by the occurrence in craters with permanently shadowed interiors. Using Mini-RF data, Spudis et al. (2013) identified more "anomalous" polar craters, totaling 43 in the north and 28 in the south, ranging from 4 to 20 km in diameter. They estimated the total mass of water ice responsible for the scattering anomalies at 600 million metric tons. Neish et al. (2011) examined Mini-RF data for the floor of the south polar Cabeus crater, which was impacted by the LCROSS spacecraft and displayed near-infrared spectral evidence of ejected water vapor and ice particles. No elevated CPR values were observed at the Cabeus impact site, leading Neish et al. (2011) to conclude that the spectrally detected water ice was not in the form of decimeter-to-meter thick ice deposits, but rather in the form of small grains or coatings distributed within the regolith. Patterson et al. (2017) used the Mini-RF receiver in a bistatic geometry with The Arecibo Observatory and found an "opposition surge" in a nonshadowed area of Cabeus crater that may be caused by the presence of ice. Thomson et al. (2012) examined Mini-RF data for the south polar Shackleton crater and found that the elevated CPR values could be explained by surface roughness effects, and that the upper limit on the presence of ice was 5–10%. Campbell (2012) provided theoretical and observational bases for interpretation of elevated circular polarization ratios, noting that ice-free rough surfaces can be responsible for elevated ratios, and that scattering in thick ice has distinctive polarization signatures that are not observed on the Moon. Fa and Cai (2013) challenged the conclusions of Spudis et al. (2010, 2013) that the Mini-RF CPR signature of anomalous polar craters was due to the presence of ice. They pointed out the effects of crater interior slopes on CPR, and concluded that there was no statistically significant difference in CPR characteristics among anomalous and fresh craters, either in polar or nonpolar regions. They suggested that the CPR values observed in anomalous crater interiors (polar and nonpolar) are due to rocks at the surface and in the regolith. Similarly, Eke et al. (2014) noted additional geometric effects that may affect interpretation of crater interior CPR. They found that anomalous craters have steeper interior slopes than other craters, which may explain the persistence of rocky debris and cause a local incidence angle enhancement of CPR. In summary, while radar held promise for definitively detecting and mapping lunar polar ice, the Moon is not sufficiently "cooperative" with the configuration of its potential ice deposits to allow the current radar datasets to be used for an unambiguous detection.

31.3.5 Titan

The *Cassini* radar was used to study Saturn's large moon Titan beginning in 2004 (Elachi et al., 2005). The radar operated in Ku-band, at a wavelength of 2.17 cm. It had a number of modes, including SAR, altimetry, radiometry, and scatterometry. Its primary goal was to determine the physical state, topography, and composition of Titan's surface (Elachi et al., 2004). During its prime and extended missions, *Cassini* executed many dozens of flybys of Titan, allowing near-global sampling of Titan's terrains by the radar. The SAR resolution is ~500 m from altitudes below 1500 km, and several km at altitudes up to 4000 km. Radar data from the first few flybys revealed a complex surface, with large low-relief radar-dark areas, sinuous channels and relatively few impact craters (Elachi et al., 2006 Stofan et al., 2006). These early data suggested a role for liquids in shaping Titan's surface. One of the more remarkable findings based on the radar observations was confirmation of the presence of lakes of liquid hydrocarbons (likely methane) in the north polar region (Stofan et al., 2007; Figure 31.1c). The lake surfaces were extremely radar-dark, and associated radar-dark sinuous features were interpreted to be drainage channels. An inventory of lake features was provided by Hayes et al. (2008). Enhanced processing of *Cassini* radar altimetry data over the large lake/sea Ligeia Mare allowed generation of a bathymetric profile that traced the sea floor for ~300 km, finding a maximum depth of 160 m (Mastrogiuseppe et al., 2014). The remarkable transparency of the liquid to the radar signals implied a composition of nearly pure methane–ethane. A low-resolution topographic map of Titan was produced using a combination of *Cassini* altimetry and multiple-beam SAR data (Lorenz et al., 2013). Scatterometry results were reported by Wye et al. (2007) and a review of the radiometry results was provided by Janssen et al. (2016).

31.3.6 Comet 67P/Churyumov–Gerasimenko

The Rosetta spacecraft carried a bistatic radar system known as CONSERT (Comet Nucleus Sounding Experiment by Radiowave Transmission) to probe the interior of the comet 67P/Churyumov–Gerasimenko (Kofman et al., 1998). The radar system included a 90-MHz transmitter and receiver on both the Rosetta main spacecraft and the lander known as Philae. The goal was to capture signals that propagate through the body of the comet nucleus to constrain the composition and structure of the interior. The CONSERT radars functioned properly, but unfortunately the landing of the Philae craft was not nominal, and its instruments functioned only briefly. Nevertheless, during its operation, CONSERT recorded nearly two hours of data, including signals that clearly traversed paths through the nucleus (Kofman et al., 2015; Figure 31.3). The travel time of the signals through well-constrained distances through the nucleus allowed the estimation of the refractive index, and hence the real permittivity (dielectric constant) of the material. The estimated permittivity from these calculations was 1.27 ± 0.1. This is significantly lower than that of pure water ice of density ~1, and implies that a large amount of porosity is present within the nucleus. Kofman et al. (2015) estimated the porosity at 75–85%, concluding that most of the volume of the comet nucleus was in fact empty space.

31.4 Future Directions

The success of radar techniques in past planetary studies has led space agencies and researchers to continue to plan and implement radar systems on new missions. Key targets for upcoming missions are the icy satellites of Jupiter. Both ESA and NASA are developing missions for studying these icy moons, and both spacecraft will carry sounding radars. ESA's Jupiter Icy Moon Explorer (JUICE) will first orbit Jupiter and conduct flybys of the Galilean satellites Europa, Ganymede, and Callisto. The spacecraft will then be placed into orbit around Ganymede to conduct detailed mapping surveys. The radar sounder selected for JUICE is the Radar for Icy Moon Exploration (RIME), a joint development of the Italian Space Agency and NASA (Bruzzone et al., 2013). RIME operates at a central frequency of 9 MHZ, with a selectable bandwidth of 1 or 3 MHz. While the primary target is Ganymede from a local orbit, at least two flybys are planned of Europa to search for evidence of shallow liquid water over targets with surface indications of recent resurfacing activity. While direct detection of liquid water at Ganymede is unlikely due to its expected depth (~100 km), RIME will probe the icy crust to search for evidence of past liquid/ice interactions and to reveal the subsurface expression of tectonic and impact features observed at the surface. NASA's Europa Clipper mission is planned to orbit Jupiter and

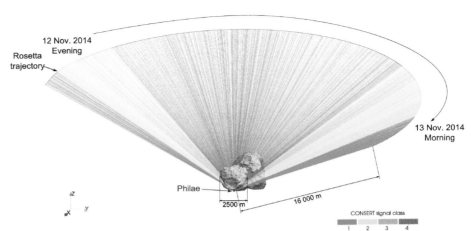

Figure 31.3 Result from CONSERT bistatic radar probing of Comet 67P/Churyumov-Gerasimenko on the Rosetta mission. The CONSERT unit on the Philae lander was in contact with the unit on the Rosetta spacecraft during the periods shown in green, yellow, and orange. Red indicates no signal. All signal paths shown passed through part of the body of the comet nucleus. (Image: ESA/Rosetta/Philae/CONSERT.)

perform several dozen flybys of Europa. It carries the Radar for Europa Assessment and Sounding, Ocean to Near-Surface (REASON), provided by NASA (Patterson et al., 2015). REASON is a dual-frequency system, operating at 9 MHz and 60 MHz. This allows near-simultaneous acquisition of deep-penetrating long wavelength and high-resolution shorter wavelength sounding data. The natural radio emissions from Jupiter are strong in frequencies up ~42 MHz, so for both RIME and REASON the 9-MHz data are planned to be acquired primarily on the side of the target moons shielded from the Jupiter noise. REASON's 60-MHz channel is outside of the range of the expected noise and will be operated in both sub- and anti-Jovian geometries. Both RIME and REASON are targeting the upper 10 km of the icy crust for detailed study, but the radars are designed to probe to greater depths if the crust "cooperates." It is possible that echoes may be obtained from the base of the ice shell of Europa, where it is in contact with the global subsurface ocean.

Detailed concepts have been proposed for imaging radar missions to both Mars and Venus (e.g., Campbell, 2006; Hensley et al., 2016). While these missions have not been approved as of this writing, there is widespread support in the planetary science community for them. For Mars, a SAR system would probe the upper 10s of meters of the crust for ice deposits, and provide a new picture of the geomorphology currently hidden beneath layers of dust and regolith. For Venus, an interferometric SAR system could map the global topography at sub-km resolution, and search for active surface deformation associated with volcanic or seismic events. The success of CONSERT in probing a comet and the high interest in ice-rich asteroidal bodies has led to proposals to probe the interiors of small bodies with sounding radar (e.g., Asphaug et al., 2014). Further developments in miniaturization of electronics and on-board processing capabilities will increase the feasibility of placing advanced radar systems on smaller spacecraft and/or at lower costs to mission and spacecraft resources.

References

Asphaug E., Belton M., Bockelee-Morvan D., et al. (2014) The Comet Radar Explorer mission. *DPS Annual Meeting*, Abstract # 209.07.

Barsukov V.L., Basilevsky A.T., Burba G.A., et al. (1986) The geology and geomorphology of the Venus surface as revealed by the radar images obtained by Veneras 15 and 16. *Journal of Geophysical Research*, **91**, 378–398.

Basilevsky A. & Head J.W. (2002) Venus: Timing and rates of geologic activity. *Geology*, **30**, 1015–1018.

Basilevsky A.T. & McGill G.E. (2007) Surface evolution of Venus. In: *Exploring Venus as a terrestrial planet* (L.W. Esposito, E.R. Stofan, & T.E. Cravens, eds.). American Geophysical Union, Washington, DC, 23–43.

Basilevsky A.T., Head J.W., & Setyaeva I.V. (2003) Venus: Estimation of age of impact craters on the basis of degree of preservation of associated radar-dark deposits. *Geophysical Research Letters*, **30**, DOI:10.1029/2003GL017504.

Benner L.A.M., Nolan M.C., Ostro S.J., et al. (2006) Near-Earth Asteroid 2005 CR37: Radar images and photometry of a candidate contact binary. *Icarus*, **182**, 474–481.

Benner L.A.M., Busch M.W., Giorgini J.D., Taylor P.A., & Margot J.-L. (2015) Radar observations of near-Earth and main-belt asteroids. In: *Asteroids IV* (P. Michel, F. DeMeo, & W. Bottke, eds.). University of Arizona Press, Tucson, 165–182.

Black G.J., Campbell D.B., & Nicholson P.D. (2001) Icy Galilean satellites: Modeling radar reflectivities as a coherent backscatter effect. *Icarus*, **151**, 167–180.

Black G.J., Campbell D.B., & Carter L.M. (2007) Arecibo radar observations of Rhea, Dione, Tethys, and Enceladus. *Icarus*, **191**, 702–711.

Black G.J., Campbell D.B., & Carter L.M. (2011) Ground-based radar observations of Titan: 2000–2008. *Icarus*, **212**, 300–320.

Bondarenko N.V. & Head J.W. (2009) Crater-associated dark diffuse features on Venus: Properties of surficial deposits and their evolution. *Journal of Geophysical Research*, **114**, DOI:10.1029/2008JE003163.

Bramson A.M., Byrne S., Putzig N.E., et al. (2015) Widespread excess ice in Arcadia Planitia, Mars. *Geophysical Research Letters*, **42**, 6566–6574.

Bruzzone L., Plaut J.J., Alberti G., et al. (2013) RIME: Radar for Icy Moon Exploration. *IEEE IGARSS*, 3907–3910.

Butler B.J., Muhleman D.O., & Slade M.A. (1993) Mercury: Full-disk radar images and the detection and stability of ice at the north pole. *Journal of Geophysical Research*, **98**, 15,003–15,023.

Butrica A.J. (1996) *To see the unseen: A history of planetary radar astronomy.* NASA History Office, Washington, DC.

Byrnes J.M. & Crown D.A. (2002) Morphology, stratigraphy, and surface roughness properties of Venusian lava flow fields. *Journal of Geophysical Research*, **107**, DOI:10.1029/2001JE001828.

Cahill J.T.S., Thomson B.J., Patterson G.W., et al. (2014) The Miniature Radio Frequency instrument's (Mini-RF) global observations of Earth's Moon. *Icarus*, **243**, 173–190.

Campbell B.A. (2002) *Radar remote sensing of planetary surfaces.* Cambridge University Press, Cambridge.

Campbell B.A. (2006) Eagle: A synthetic aperture radar mapper for the Mars Scout Program. *37th Lunar Planet. Sci. Conf.*, Abstract #2188.

Campbell B.A. (2012) High circular polarization ratios in radar scattering from geologic targets. *Journal of Geophysical Research*, **117**, DOI:10.1029/2012JE004061.

Campbell B., Carter L., Phillips R., et al. (2008) SHARAD radar sounding of the Vastitas Borealis Formation in Amazonis Planitia. *Journal of Geophysical Research*, **113**, DOI:10.1029/2008JE003177.

Campbell B.A., Campbell D.B., Morgan G.A., Carter L.M., Nolan M.C., & Chandler J.F. (2015) Evidence for crater ejecta on Venus tessera terrain from Earth-based radar images. *Icarus*, **250**, 123–130.

Campbell D.B. & Burns B.A. (1980) Earth-based radar imagery of Venus. *Journal of Geophysical Research*, **85**, 8271–8281.

Campbell D.B., Chandler J.F., Pettengill G.H., & Shapiro I.I. (1977) Galilean satellites of Jupiter: 12.6-Centimeter radar observations. *Science*, **196**, 650–653.

Campbell D.B., Black G.J., Carter L.M., & Ostro S.J. (2003) Radar evidence for liquid surfaces on Titan. *Science*, **302**, 431–434.

Campbell D.B., Campbell B.A., Carter L.M., Margot J.-L., & Stacy N.J.S. (2006) No evidence for thick deposits of ice at the lunar south pole. *Nature*, **443**, 835–837.

Carter L.M., Campbell D.B., & Campbell B.A. (2004) Impact crater related surficial deposits on Venus: Multipolarization radar observations with Arecibo. *Journal of Geophysical Research*, **109**, DOI:10.1029/2003JE002227.

Carter L.M., Campbell D.B., & Campbell B.A. (2006) Volcanic deposits in shield fields and highland regions on Venus: Surface properties from radar polarimetry. *Journal of Geophysical Research*, **111**, DOI:10.1029/2005JE002519.

Carter L.M., Campbell B.A., Watters T.R., et al. (2009) Shallow Radar (SHARAD) sounding observations of the Medusae Fossae Formation, Mars. *Icarus*, **199**, 295–302.

Carter L.M., Neish C.D., Bussey D.B.J., et al. (2012) Initial observations of lunar impact melts and ejecta flows with the Mini-RF radar. *Journal of Geophysical Research*, **117**, DOI:10.1029/2011JE003911.

Chabot N.L., Ernst C., Harmon J.K., et al. (2012) Craters hosting radar-bright deposits in Mercury's north polar region: Areas of persistent shadow determined from MESSENGER images. *Journal of Geophysical Research*, **118**, 26–36.

Clifford S.M., Lasue J., Heggy E., Boisson J., McGovern P., & Max M.D. (2010) Depth of the martian cryosphere: Revised estimates and implications for the existence and detection of subpermafrost groundwater. *Journal of Geophysical Research*, **115**, DOI:10.1029/2009JE003462.

Cook C.M., Melosh H.J., & Bottke W.F. (2003) Doublet craters on Venus. *Icarus*, **165**, 90–100.

Dewitt J.H. & Stodola E.K. (1949) Detection of radio signals reflected from the Moon. *Proceedings of the IRE*, **37**(3), 229–242.

Eke V.R., Bartram S.A., Lane D.A., Smith D., & Teodoro L.F.A. (2014) Lunar polar craters – Icy, rough or just sloping? *Icarus*, **241**, 66–78.

Elachi C., Wall S., Allison M., et al. (2005) *Cassini* radar views the surface of Titan. *Science*, **308**, 970–974.

Elachi C., Wall S., Janssen M., et al. (2006) Titan Radar Mapper observations from Cassini's TA and T3 flybys. *Nature*, **441**, 709–713.

Fa W. & Cai Y. (2013) Circular polarization ratio characteristics of impact craters from Mini-RF observations and implications for ice detection at the polar regions of the Moon. *Journal of Geophysical Research*, **118**, 1582–1608.

Ford J.P., Plaut J.J., Weitz C.M., et al. (1993) *Guide to Magellan image interpretation*. JPL Publication #93-24.

Ghent R.R., Hayne P.O., Bandfield J.L., et al. (2014) Constraints on the recent rate of lunar ejecta breakdown and implications for crater ages. *Geology*, **42**, 1059–1062.

Gurnett D.A., Kirchner D.L., Huff R.L., et al. (2005) Radar soundings of the ionosphere of Mars. *Science*, **310**, 1929–1933.

Gurnett D.A., Huff R.L., Morgan D.D., et al. (2008) An overview of radar soundings of the martian ionosphere from the Mars Express spacecraft. *Advances in Space Research*, **41**, 1335–1346.

Gurnett D.A., Morgan D.D., Persoon A.M., et al. (2015) An ionized layer in the upper atmosphere of Mars caused by dust impacts from comet Siding Spring. *Geophysical Research Letters*, **42**, 4745–4751.

Harmon J.K. & Nolan M.C. (2017) Arecibo radar imagery of Mars: II. Chryse–Xanthe, polar caps, and other regions. *Icarus*, **281**, 162–199.

Harmon J.K., Slade M.A., Vélez R.A., Crespo A., Dryer M.J., & Johnson J.M. (1994) Radar mapping of Mercury's polar anomalies. *Nature*, **369**, 213–215.

Harmon J.K., Slade M.A., & Rice M.S. (2011) Radar imagery of Mercury's putative polar ice: 1999–2005 Arecibo results. *Icarus*, **211**, 37–50.

Harmon J.K., Nolan M.C., Husmann D.I., & Campbell B.A. (2012) Arecibo radar imagery of Mars: The major volcanic provinces. *Icarus*, **220**, 990–1030.

Hayes A., Aharonson O., Callahan P., et al. (2008) Hydrocarbon lakes on Titan: Distribution and interaction with a porous regolith. *Geophysical Research Letters*, **35**, DOI:10.1029/2008GL033409.

Head J.W., Neukum G., Jaumann R., et al. (2005) Tropical to mid-latitude snow and ice accumulation, flow and glaciation on Mars. *Nature*, **434**, 346–351.

Hensley S., Smrekar S.E., Nunes D.C., & The_VERITAS_Science_Team. (2016) VERITAS: Towards the next generation of cartography for the planet Venus. *47th Lunar Planet. Sci. Conf.*, Abstract #1965.

Holt J.W., Safaeinili A., Plaut J.J., et al. (2008) Radar sounding evidence for buried glaciers in the southern mid-latitudes of Mars. *Science*, **322**, 1235–1238.

Holt J.W., Fishbaugh K.E., Byrne S., et al. (2010) The construction of Chasma Boreale on Mars. *Nature*, **465**, 446–449.

Janssen M., Le Gall A., Lopes R.M., et al. (2016) Titan's surface at 2.2-cm wavelength imaged by the *Cassini* RADAR radiometer: Results and interpretations through the first ten years of observation. *Icarus*, **270**, 443–459.

Jordan R., Picardi G., Plaut J., et al. (2009) The Mars express MARSIS sounder instrument. *Planetary and Space Science*, **57**, 1975–1986.

Jurgens R.F., Slade M.A., & Saunders R.S. (1988a) Evidence for highly reflecting materials on the surface and subsurface of Venus. *Science*, **240**, 1021–1023.

Jurgens R.F., Slade M.A., Robinett L., et al. (1988b) High resolution images of Venus from ground-based radar. *Geophysical Research Letters*, **15**, 577–580.

Kofman W., Barbin Y., Klinger J., et al. (1998) Comet nucleus sounding experiment by radiowave transmission. *Advances in Space Research*, **21**, 1589–1598.

Kofman W., Herique A., Barbin Y., et al. (2015) Properties of the 67P/Churyumov–Gerasimenko interior revealed by CONSERT radar. *Science*, **349**, aab0639.

Lai J., Xu Y., Zhang X., & Tang Z. (2016) Structural analysis of lunar subsurface with Chang'E-3 lunar penetrating radar. *Planetary and Space Science*, **120**, 96–102.

Lawrence D.J., Feldman W.C., Goldsten J.O., et al. (2012) Evidence for water ice near Mercury's north pole from MESSENGER neutron spectrometer measurements. *Science*, **339**, 292–296.

Lorenz R.D., Stiles B.W., Aharonson O., et al. (2013) A global topographic map of Titan. *Icarus*, **225**, 367–377.

Magri C., Nolan M.C., Ostro S.J., & Giorgini J.D. (2007) A radar survey of main-belt asteroids: Arecibo observations of 55 objects during 1999–2003. *Icarus*, **186**, 126–151.

Mastrogiuseppe M., Poggiali V., Hayes A., et al. (2014) The bathymetry of a Titan sea. *Geophysical Research Letters*, **41**, 1432–1437.

Morgan G.A., Campbell B.A., Carter L.M., Plaut J.J., & Phillips R.J. (2013) 3D Reconstruction of the source and scale of buried young flood channels on Mars. *Science*, **340**, 607–610.

Morgan G.A., Campbell B.A., Carter L.M., & Plaut J.J. (2015) Evidence for the episodic erosion of the Medusae Fossae Formation preserved within the youngest volcanic province on Mars. *Geophysical Research Letters*, **42**, 7336–7342.

Mouginis-Mark P.J. (2016) Geomorphology and volcanology of Maat Mons, Venus. *Icarus*, **277**, 433–441.

Muhleman D.O., Butler B.J., Grossman A.W., & Slade M.A. (1991) Radar images of Mars. *Science*, **253**, 1508–1513.

Neish C.D., Bussey D.B.J., Spudis P., et al. (2011) The nature of lunar volatiles as revealed by Mini-RF observations of the LCROSS impact site. *Journal of Geophysical Research*, **116**, DOI:10.1029/2010JE003647.

Neish C.D., Madden J., Carter L.M., et al. (2014) Global distribution of lunar impact melt flows. *Icarus*, **239**, 105–117.

Nozette S., Spudis P., Bussey B., et al. (2010) The lunar Reconnaissance Orbiter Miniature Radio Frequency (Mini-RF) technology demonstration. *Space Science Reviews*, **150**, 285–302.

Ono T., Kumamoto A., Nakagawa H., et al. (2009) Lunar Radar Sounder observations of subsurface layers under the nearside Maria of the Moon. *Science*, **323**, 909–912.

Orosei R., Lauro S.E., Pettinelli E., et al. (2018) Radar evidence of subglacial liquid water on Mars. *Science*, **361**, 490–493.

Oshigami S., Watanabe S., Yamaguchi Y., et al. (2014) Mare volcanism: Reinterpretation based on Kaguya Lunar Radar Sounder data. *Journal of Geophysical Research*, **119**, 1037–1045.

Ostro S.J. (1982) Radar properties of Europa, Ganymede, and Callisto. In: *Satellites of Jupiter* (D. Morrison, ed.). University of Arizona Press, Tucson, 213–236.

Ostro S.J. (1989) Radar observations of asteroids. In: *Asteroids II* (R.P. Binzel, T. Gehrels, & M.S. Matthews, eds.). University of Arizona Press, Tucson, 192–212.

Ostro S.J., Campbell D.B., Simpson R.A., et al. (1992) Europa, Ganymede, and Callisto: New radar results from Arecibo and Goldstone. *Journal of Geophysical Research*, **97**, 18227–18244.

Ostro S.J., Hudson R.S., Benner L.A.M., et al. (2002) Asteroid radar astronomy. In: *Asteroids III* (W. Bottke, A. Cellino, P. Paolicchi, & R.P. Binzel, eds.). University of Arizona Press, Tucson, 151–168.

Patterson G.W., Blankenship D., Moussessian A., et al. (2015) REASON for Europa. *DPS Annual Meeting*, Abstract #312.09.

Patterson G.W., Stickle A.M., Turner F.S., et al. (2017) Bistatic radar observations of the Moon using Mini-RF on LRO and the Arecibo Observatory. *Icarus*, **283**, 2–19.

Peeples W.J., Sill W.R., May T.W., et al. (1978) Orbital radar evidence for lunar subsurface layering in Maria Serenitatis and Crisium. *Journal of Geophysical Research*, **83**, 3459–3470.

Pettengill G.H., Eliason E., Ford P.G., Loriot G.B., Masursky H., & McGill G.E. (1980) Pioneer Venus radar results altimetry and surface properties. *Journal of Geophysical Research*, **85**, 8261–8270.

Phillips R.J., Adams G.F., Brown W.E. Jr., et al. (1973) Apollo Lunar Sounder experiment. In: *Apollo 17 Preliminary Science Report*. NASA, Washington, DC.

Phillips R.J., Zuber M.T., Smrekar S.E., et al. (2008) Mars north polar deposits: Stratigraphy, age, and geodynamical response. *Science*, **320**, 1182–1185.

Picardi G., Plaut J.J., Biccari D., et al. (2005) Radar soundings of the subsurface of Mars. *Science*, **310**, 1925–1928.

Plaut J.J., Ivanov A., Safaeinili A., et al. (2007a) Radar sounding of subsurface layers in the south polar plains of Mars: Correlation with the Dorsa Argentea formation. *39th Lunar Planet. Sci. Conf.*, Abstract #2144.

Plaut J.J., Picardi G., Safaeinili A., et al. (2007b) Subsurface radar sounding of the south polar layered deposits of Mars. *Science*, **316**, 92–95.

Plaut J.J., Safaeinili A., Holt J.W., et al. (2009a) Radar evidence for ice in lobate debris aprons in the mid-northern latitudes of Mars. *Geophysical Research Letters*, **36**, DOI:10.1029/ 2008GL036379.

Plaut J.J., Safaeinili A., Campbell B.A., et al. (2009b) A widespread radar-transparent layer detected by SHARAD in Arcadia Planitia, Mars. *40th Lunar Planet. Sci. Conf.*, Abstract #2312.

Pommerol A., Kofman W., Audouard J., et al. (2010) Detectability of subsurface interfaces in lunar maria by the LRS/SELENE sounding radar: Influence of mineralogical composition. *Geophysical Research Letters*, **37**, DOI:10.1029/2009GL041681.

Porcello L.J., Jordan R.L., Zelenka J.S., et al. (1974) The Apollo lunar sounder radar system. *Proceedings of the IEEE*, **62**, 769–783.

Putzig N.E., Phillips R.J., Campbell B.A., et al. (2009) Subsurface structure of Planum Boreum from Mars Reconnaissance Orbiter Shallow Radar soundings. *Icarus*, **204**, 443–457.

Romeo I. & Turcotte D.L. (2009) The frequency-area distribution of volcanic units on Venus: Implications for planetary resurfacing. *Icarus*, **203**, 13–19.

Saunders R.S. (1992) Foreword to special section on *Magellan* at Venus. *Journal of Geophysical Research*, **97**, 15921, DOI:10.1029/92JE02288.

Saunders R.S., Spear A.J., Allin P.C., et al. (1992) *Magellan* mission summary. *Journal of Geophysical Research*, **97**, 13,067–13,090.

Selvans M.M., Plaut J.J., Aharonson O., & Safaeinili A. (2010) Internal structure of Planum Boreum, from Mars advanced radar for subsurface and ionospheric sounding data. *Journal of Geophysical Research*, **115**, DOI:10.1029/2009JE003537.

Seu R., Phillips R.J., Biccari D., et al. (2007) SHARAD sounding radar on the Mars Reconnaissance Orbiter. *Journal of Geophysical Research*, **112**, DOI:10.1029/2006JE002745.

Shepard M.K., Taylor P.A., Nolan M.C., et al. (2015) A radar survey of M- and X-class asteroids. III. Insights into their composition, hydration state, & structure. *Icarus*, **245**, 38–55.

Simpson R.A., Harmon J.K., Zisk S.H., Thompson T., & Muhleman D.O. (1992) Radar determination of Mars surface properties. In: *Mars* (H.H. Kieffer, B. Jakosky, C.W. Snyder, & M.S. Matthews, eds.). University of Arizona Press, Tucson, 652–685.

Slade M.A., Butler B.J., & Muhleman D.O. (1992) Mercury radar imaging: Evidence for polar ice. *Science*, **258**, 635–640.

Smith I.B. & Holt J.W. (2010) Onset and migration of spiral troughs on Mars revealed by orbital radar. *Nature*, **465**, 450–453.

Spudis P.D., Bussey D.B.J., Baloga S.M., et al. (2010) Initial results for the north pole of the Moon from Mini-SAR, Chandrayaan-1 mission. *Geophysical Research Letters*, **37**, DOI:10.1029/2009GL042259.

Spudis P.D., Bussey D.B.J., Baloga S.M., et al. (2013) Evidence for water ice on the Moon: Results for anomalous polar craters from the LRO Mini-RF imaging radar. *Journal of Geophysical Research*, **118**, 2016–2029.

Stacy N.J.S., Campbell D.B., & Ford P.G. (1997) Arecibo radar mapping of the lunar poles: A search for ice deposits. *Science*, **276**, 1527–1530.

Stillman D.E. & Grimm R.E. (2011) Radar penetrates only the youngest geological units on Mars. *Journal of Geophysical Research*, **116**, DOI:10.1029/2010JE003661.

Stofan E.R., Lunine J.I., Lopes R., et al. (2006) Mapping of Titan: Results from the first Titan radar passes. *Icarus*, **185**, 443–456.

Stofan E.R., Elachi C., Lunine J.I., et al. (2007) The lakes of Titan. *Nature*, **445**, 61–64.

Stuurman C.M., Osinski G.R., Holt J.W., et al. (2016) SHARAD detection and characterization of subsurface water ice deposits in Utopia Planitia, Mars. *Geophysical Research Letters*, **43**, 9484–9491.

Su Y., Fang G.-Y., Feng J.-Q., et al. (2014) Data processing and initial results of Chang'e-3 lunar penetrating radar. *Research in Astronomy and Astrophysics*, **14**, 1623–1632.

Talpe M.J., Zuber M.T., Yang D., et al. (2012) Characterization of the morphometry of impact craters hosting polar deposits in Mercury's north polar region. *Journal of Geophysical Research*, **117**, DOI:10.1029/2012JE004155.

Thompson T.W. (1978) High resolution lunar radar map at 7.5 meter wavelength. *Icarus*, **36**, 174–188.

Thompson T.W. (1987) High-resolution lunar radar map at 70-cm wavelength. *Earth, Moon, and Planets*, **37**, 59–70.

Thompson T.W., Pollack J.B., Campbell M.J., & O'Leary B.T. (1970) Radar maps of the moon at 70-cm wavelength and their interpretation. *Radio Science*, **5**, 253–262.

Thomson B.J., Bussey D.B.J., Neish C.D., et al. (2012) An upper limit for ice in Shackleton crater as revealed by LRO Mini-RF orbital radar. *Geophysical Research Letters*, **39**, DOI:10.1029/2012GL052119.

Victor W.K. & Stevens R. (1961) Exploration of Venus by radar. *Science*, **134**, 46–48.

Watters T.R., Campbell B., Carter L., et al. (2007) Radar sounding of the Medusae Fossae Formation Mars: Equatorial ice or dry, low-density deposits? *Science*, **318**, 1125–1128.

Whitten J.L. & Campbell B.A. (2016) Recent volcanic resurfacing of Venusian craters. *Geology*, **44**, 519–522.

Wye L.C., Zebker H.A., Ostro S.J., & the *Cassini* Research Team. (2007) Electrical properties of Titan's surface from *Cassini* RADAR scatterometer measurements. *Icarus*, **188**, 367–385.

Yan S., Guang-You F., Jian-Qing F., et al. (2014) Data processing and initial results of Chang'e-3 lunar penetrating radar. *Research in Astronomy and Astrophysics*, **14**, 1623.

Zimbelman J.R. (2003) Flow field stratigraphy surrounding Sekmet Mons Volcano, Kawelu Planitia, Venus. *Journal of Geophysical Research*, **108**, DOI:10.1029/2002JE001965.

Zisk S.H., Pettengill G.H., & Catuna G.W. (1974) High-resolution radar maps of the lunar surface at 3.8-cm wavelength. *The Moon*, **10**, 17–50.

Index

absorption, 3–17, 21, 24–26, 29, 42–43, 68–71, 73–74, 80, 85–89, 92, 102–107, 110, 112–114, 123–124, 126, 130, 139, 147–150, 153–157, 159, 171, 174, 192, 197–199, 208–210, 218, 224–225, 261–271, 291–294, 296–298, 307, 309, 315, 317, 327, 351–354, 357, 360–361, 363, 368–372, 374–385, 396–398, 401–406, 419, 424–425, 430–436, 438, 442–444, 447, 449–450, 453–456, 458–460, 462, 466–468, 490, 493, 514, 517, 521–530, 539, 561–562, 577, 591, 595, 614

abundance, 16, 21, 26–29, 38, 69, 74–75, 78–80, 87–89, 104, 109, 149, 176, 181, 192, 194, 200, 207–208, 218, 220, 231–233, 261, 263, 265–268, 271, 295, 297–300, 310, 313, 315, 319, 324–329, 332, 338, 343–344, 352, 356–357, 361–364, 374, 379–380, 382–383, 395, 416, 419–420, 422, 424, 434, 442, 446–447, 449, 466, 473, 484–487, 489–492, 499, 502–503, 505, 509, 524, 529, 541, 547, 555, 557, 560–563, 567–568, 570, 578, 580, 582–583, 593–594, 596–597

albedo, 25–27, 33–37, 108, 113, 115, 311, 313, 318, 331, 334, 353, 355–356, 364, 374–376, 381–382, 396, 398, 400–401, 404–406, 416–418, 420–422, 425, 428–429, 432–435, 437–438, 447, 449, 458, 484–485, 489–490, 493, 516–517, 520, 525, 593, 596, 607

alteration, 55, 75, 91, 113, 192, 297, 300–301, 331, 337, 341–344, 375, 384, 393, 395–396, 404, 417, 425, 453, 455–457, 462–466, 469–472, 487–488, 491–493, 499, 503–504, 510, 522, 538, 542–544, 548–549, 551–552, 555, 563–568, 570, 582–583

ammonia, 130, 143

amorphous, 30, 37, 68, 90, 104–106, 111, 138, 282, 293, 297, 331, 351, 369, 378, 383, 395–396, 431, 444, 466–467, 469, 484–486, 492–493, 504–507, 509–510, 517, 524, 530, 538, 542, 568, 570, 579–580

analog, 92, 112, 138, 141, 268, 274–275, 279–280, 290, 294–296, 352, 398, 402–403, 406–407, 459, 464, 506, 590

anisothermality, 327, 333–334

asteroid, 10, 14, 17, 30, 53, 61–62, 69, 90–91, 111, 114, 169, 191–193, 215, 225, 290–294, 297–298, 300, 316, 333, 351, 355, 368, 393–407, 413, 416–417, 425, 590, 594, 597–598, 618

asteroids/minor planets
 1 Ceres, 191, 231, 293–294, 397, 404, 405, 413, 414, 417, 418, 420, 421, 422, 424, 425, 591, 594, 595
 4 Vesta, 10, 91, 191, 194, 230–231, 300, 355, 397, 406–407, 413–414, 417–418, 421–422, 424, 591, 594–595
 16 Psyche, 407, 598
 21 Lutetia, 405, 413, 416, 418, 422–423
 22 Kalliope, 401, 405
 130 Electra, 401
 243 Ida, 404
 433 Eros, 191–192, 355, 394, 397, 401, 404, 590, 597
 624 Hektor, 401, 406
 951 Gaspra, 404
 2867 Šteins, 394, 416
 5535 Annefrank, 394
 9969 Braile, 394
 25143 Itokawa, 192, 215, 394, 396, 404, 597

atmosphere, 17, 30, 32, 34, 37, 52, 109, 111, 151, 169–170, 172, 184–186, 191–193, 231, 253–254, 257, 261–262, 264, 268, 289–292, 294, 296–297, 300–301, 331, 353, 399, 418, 428, 437–439, 443–444, 447–448, 450, 453, 458, 465, 490, 499, 501, 509, 513, 530, 559, 562, 578, 580, 589, 592–593, 604

atmospheric correction, 263, 331–332

backscatter cross section, 244, 255

backscattering, 27–28, 136, 149, 525, 559

basalt, 90–92, 182, 209, 326, 331, 338–343, 352, 369–370, 372, 374–375, 377, 379–381, 384–385, 397, 406, 417, 424, 454–455, 457, 460, 462, 464–466, 469, 471, 485–489, 491, 493, 499, 501, 503–504, 506–508, 510, 517–520, 522, 524, 529–530, 541–543, 547–552, 555, 563–567, 570, 582, 589, 593–596, 614

Index

beam splitter, 45, 47, 50, 59
bedrock, 380, 439, 445, 459–460, 463, 469, 487, 499, 510, 543, 549–550, 552, 578, 581, 583
biconical, 47–48, 73–74, 113
biosignature, 278, 506, 579
blackbody, 22, 43, 46–47, 325, 384, 500
breccia, 92, 370, 375, 380–381, 383, 425, 464, 487, 524, 549
bromide, 59, 224, 592, 598

calibration, 21, 44, 46–47, 49, 152, 173, 175–176, 180–182, 186, 222, 227, 232, 262–263, 267–269, 271, 278, 360, 382–383, 386–387, 399–400, 402, 404, 407, 418, 431, 435, 484, 500, 541–543, 555, 560, 562, 574, 576
carbon, 87, 108, 110–114, 129, 198, 293, 351, 356, 362–363, 404, 425, 435, 448, 559, 576, 580, 594–595
center shift, 154, 157, 160–161, 540, 542
chemical index of alteration (CIA), 337, 341–342, 346, 582
chemistry, 3–4, 9–10, 16–17, 25, 42, 44, 62, 69, 106, 109, 174, 231, 245–246, 257, 268, 270–271, 297, 361, 396, 404, 428, 444, 446, 450, 469, 472, 487–488, 517, 520, 522, 525, 542, 550–551, 556, 567–568, 570, 578–579, 581, 584
coating, 49–50, 52, 90–92, 264, 266–268, 297, 300, 327, 454, 458, 462, 503–504, 520–522, 525–526, 548, 551–552, 555, 564–566, 573, 578, 583–584, 615
comets, 69, 106, 111–112, 114–115, 191, 296–297, 383, 394–395, 413–419, 425, 435, 594, 612, 617–618
  comet 9P/Tempel 1, 419
  comet 67P/Churyumov-Gerasimenko (67P/CG), 106, 413–415, 417–423, 617–618
  comet 81P/Wild 2, 111–112
  comet C/2013 A1 Siding Spring, 612
cosmic ray, 111, 191–194, 197, 199, 212–217, 220, 229, 278, 297, 300, 588, 592
crust, 17, 342, 361–363, 369, 375–376, 378–380, 417, 424, 445, 453, 455, 457, 462–464, 466, 487, 490–491, 548, 555, 581, 588–589, 593–594, 611–612, 617–618
crustal evolution, 375–376, 380, 487, 493, 590
crystal axis, 50–51, 62
curium, 556

data acquisition, 191, 398–399
dehydration, dehydrated, 292, 425, 469
detector, 17, 21, 26, 31, 45–50, 69, 73–74, 132, 134–137, 141, 150, 152, 168, 184–185, 193, 195, 211, 213, 215, 218–228, 232, 240, 261–262, 275–276, 279, 353, 370, 372, 381, 386–387, 414–415, 419, 445, 499–500, 515, 539–540, 555, 557, 559–560, 562, 579, 592–594, 596–597

diagenesis, 82, 456, 492, 523, 568
dispersion, 23–24, 43, 134, 278, 401, 415
dust, 50, 111–112, 115, 172, 176, 262, 266, 278, 298, 312, 331–345, 414, 418, 433, 435, 438, 454–455, 458–461, 469, 472, 486–487, 490, 493, 499–504, 506–509, 515, 517, 520, 522, 525, 528–530, 541, 548–552, 555, 562–565, 573–580, 612, 618

electronic processes, 16–17
  charge transfer, 7–8, 12–16, 69, 76, 454
  crystal field splitting, 10, 12
  crystal field theory (CFT), 4, 69, 368, 403
  oxygen-metal charge transfer (OMCT), 12, 357–360, 363–364
elemental analysis, 168, 170–171, 179, 186, 590
elemental composition, 171, 173, 184, 191–192, 194, 199, 231–232, 361–362, 579, 581, 588–590, 592–593, 596
emission, 31–33, 36–38, 46, 50, 60, 109, 124, 132–133, 148–150, 168, 170, 173–177, 180, 183–184, 186, 192, 197–198, 200–201, 207–208, 214–215, 217–220, 222, 225, 262, 289–290, 300, 330–332, 380–381, 384, 398, 400, 418–420, 509, 539, 555, 557, 560, 618
emissivity, 22, 30–37, 43, 46–48, 51–61, 290, 295, 325–326, 330–334, 344, 396, 400–401, 403, 405, 419, 466, 484, 490, 492–493, 500–502, 509, 520, 611

fluid inclusions, 138
fluorescence, 123–124, 133, 136, 141, 143–145
formation and evolution, 21, 341, 393, 402, 406, 588

Galilean satellites, 296, 606–607, 617
gamma rays, 148–151, 191–195, 197–198, 200–202, 204, 206–208, 211–214, 216–217, 220–227, 229–231, 357, 361, 370, 414, 551, 588–595, 597–598
garnet, 8, 54, 74–75, 168
geochemical, 78, 169, 174, 176, 180, 182, 184–185, 211, 231, 329, 339, 343, 345–346, 361–362, 364, 393, 396, 414, 502, 505, 538, 562–563, 588, 597
glass, 12, 16, 37, 47, 90–92, 120, 123, 125, 139, 141, 147, 227, 297–298, 300, 341, 351, 354, 368–370, 376–377, 381, 384–385, 396, 453–454, 460–464, 466, 485, 491, 504, 516, 562, 576, 593–594
grain size, 25, 29, 37–38, 52, 54, 68, 72, 77, 83, 92, 102–104, 150, 159, 173, 220, 261, 265–268, 270, 295–299, 313, 316, 326–327, 344, 357, 419–421, 431, 446, 524–525, 527, 529, 562
graphite, 113, 151, 356–357, 362–363, 576, 594

halide, 31, 44, 59, 186, 469
Halon, 73–74

harmonic oscillator, 23, 43
hemispherical, 33, 47–48, 73, 113, 400
hydrated silica, 457, 464–467, 469, 472, 491–492, 517, 524
hydrocarbon, 103, 111–113, 433–435, 437–439, 448, 616
hydrothermal, 55, 82, 90, 280, 395, 425, 457, 463–464, 469–470, 472, 485, 499, 505–506, 510, 518, 549, 568
hyperspectral, 307, 309–311, 313–314, 316–320, 415–416, 419, 455–456

ice
  carbon dioxide (CO_2), 138, 198, 231, 431, 433–435, 438, 448, 453, 592
  carbon monoxide (CO), 110–111, 138, 443, 445
  methane (CH_4), 103
  nitrogen (N_2), 442
  oxygen (O_2), 138
  planetary, 107–108, 114–115, 138, 606
  water (H_2O), 103–105, 107, 114, 138, 192, 198, 230–231, 266, 277, 294, 331, 381, 419, 421, 425, 428, 430–439, 444–446, 515, 526, 530, 579, 591–595, 613, 615, 617
icy moons, 433, 617
igneous, 9, 33, 192, 200, 342, 425, 455, 460, 462, 471, 487–488, 491, 493, 503, 510, 538, 542–544, 548, 551, 555, 563, 565–566, 568, 570, 576, 579, 581–582, 594
imaging spectrometers, 263–264, 267–268, 270–271, 372, 378, 386–387, 421, 424, 455–456, 513
in situ resource utilization (ISRU), 270–271
index of refraction, 4, 21–24, 29–30, 52, 267
infrared, see spectroscopy, mid-infrared and near-infrared
instruments on planetary missions
  Alpha Particle X-ray Spectrometer (APXS), 170, 192, 343, 502, 505, 507–508, 517, 520, 522, 524, 539, 542, 550–552, 555–568, 570, 579, 583
  ChemCam, 168–170, 174–177, 180, 184–186, 275, 527, 529, 573–584
  Compact Reconnaissance Imaging Spectrometer for Mars (CRISM), 25, 263–264, 269, 271, 313, 453, 456–461, 463–464, 467–468, 472–473, 521, 524–525, 529
  Comet Nucleus Sounding Experiment by Radiowave Transmission (CONSERT), 617–618
  Diviner Lunar Radiometer Experiment, 31, 333–334, 382, 384
  Gamma Ray and Neutron Detector (GRaND), 195, 227, 229, 594–595
  Gamma Ray Spectrometer (GRS), 221, 223, 225, 344, 487–488, 579, 589–590, 592–594, 597
  Imaging Spectrometer for Mars (ISM), 454–455, 459, 472, 514
  Linear Etalon Imaging Spectral Array (LEISA), 444–446, 449–450
  Long Range Reconnaissance Imager (LORRI), 447, 449
  Mars Advanced Radar for Subsurface and Ionosphere Sounding (MARSIS), 608, 611–613
  Mast Camera (Mastcam), 514–516, 527–529, 574, 578
  Mastcam-Z, 531
  Mercury Atmospheric and Surface Composition Spectrometer (MASCS), 351, 353–360
  Mercury Dual Imaging System (MDIS), 351, 353–354, 356–358, 360, 362
  Mercury Visible and Infrared Spectrograph (VIRS), 351, 353–355, 357–360
  Messenger Wide-Angle Camera (WAC), 351, 353–354, 358
  Miniature Thermal Emission Spectrometer (Mini-TES), 170, 326–327, 330–331, 485, 499–510, 517, 520, 523
  Miniaturized Mössbauer Spectrometer (MIMOS II), 538–542, 551
  Moon Mineralogy Mapper (M^3), 25, 316, 334, 369, 371–374, 377–379, 381–384, 387
  Multispectral Visible Imaging Camera (MVIC), 444–445, 448, 450
  Neutron Spectrometer (NS), 590, 592–594, 597
  Observatoire pour la Minéralogie, l'Eau, les Glaces et l'Activité (OMEGA), 25, 28, 453, 455–461, 463, 467, 472–473, 481, 514
  Optical, Spectroscopic, and Infrared Remote Imaging System (OSIRIS), 414, 423
  Panoramic Camera (Pancam), 312, 316–317, 499, 501, 504–509, 513, 516–524, 530–531, 565
  Particle-Induced X-ray Emission (PIXE), 556–559, 562
  Radar for Europa Assessment and Sounding, Ocean to Near-Surface (REASON), 618
  Radar for Icy Moon Exploration (RIME), 617–618
  Ralph Instrument, 444–445
  Raman Laser Spectrometer (RLS), 274
  Rock Abrasion Tool (RAT), 504, 508, 522–525, 539, 546, 548–551, 555
  SHAllow RADar (SHARAD), 608–609, 613–614
  SHERLOC, 133, 170, 274
  SuperCam, 146, 168, 170, 184, 274–276, 283, 574, 583–584
  Synthetic Aperture Radar (SAR), 239–244, 249–250, 252, 609–610, 614, 616–618
  Thermal Emission Imaging System (THEMIS), 36, 327–328, 332–333, 461, 463, 467, 472, 484–493
  Thermal Emission Spectrometer (TES), 57, 316, 327–328, 330–333, 462–463, 484–493, 502–503, 506–507, 510
  Visual and Infrared Mapping Spectrometer (VIMS) on Cassini, 266, 271, 428–433, 435–439

Visible and Infrared Mapping Spectrometer (VIR-MS) on Dawn, 416, 418, 422, 424
Visible and Infrared Thermal Imaging Spectrometer (VIRTIS), 405, 415–416, 418, 421, 425
X-Ray Spectrometer (XRS), 219, 232, 596–598
interferometers, 46, 132, 134–135, 249, 250
intrinsic dimensionality, 308, 310–311
iodide, 59, 224–225, 589–590, 592, 598
isomer shift, 147, 153, 155, 159–160, 162–163
isotopic, 125, 184–185, 200, 337, 339–341, 345, 527

Kirchhoff's Law, 22, 35, 37, 47–48, 61–62, 400
Kuiper Belt, 296, 395–396, 428, 435–436, 450–451

Lambert/Lambertian, 26–27, 47–48, 197
linear spectral mixture analysis (LSMA), 309–310, 316, 319, 325, 502
lunar
  rock, 91, 232, 368–369, 372, 383, 397
  soil, 297–299, 353, 368, 372, 374, 384, 397

machine learning, 174, 268, 307, 314, 316, 318, 320
Mars
  Gale crater, 343, 457, 462, 467–468, 527, 529, 563–570, 574, 577, 579, 581–583
  Gusev crater, 342–344, 346, 499, 503–504, 509–510, 516–521, 538, 542–545, 547–549, 551–552, 564, 566, 569–570
  Meridiani Planum, 330–331, 343, 467–468, 472, 499, 503, 506–507, 509–510, 520, 522–523, 525, 538–539, 542–546, 549–552, 564, 566, 568–570
matched filter, 175, 243, 314
melts, 139, 141, 147, 370, 464
Mercury, 7, 17, 32, 53, 69, 191–193, 219, 224, 231, 233, 290, 294–296, 298, 300, 351–359, 361–364, 368, 593–594, 597–598, 606
metamorphic, 33, 90, 457, 464–465
meteorites
  Abee, 61–62
  ALH 84001, 92, 487
  ALH A77005, 487
  ALH A81005, 92
  Allende, 61–62
  Almahata Sitta, 91–92
  Chassigny, 92, 487
  chondrite, 57, 62, 91–92, 111, 139, 196, 199, 209, 292, 356, 402, 404–407, 416–417, 424, 590, 594–595, 597
  Dar el Gani 862, 61–62
  Dhofar 007, 61–62
  EET 87521, 92
  EET A79001, 92
  Goalpara, 92
  HED, 594
  Juvinas, 92
  LEW 87009, 92
  Los Angeles, 487
  lunar, 91, 377, 591
  martian, 92, 342, 485, 487–488, 556, 570
  Nakhla, 487
  Northwest Africa (NWA) 2737, 61
  Northwest Africa (NWA) 7034, 487
  Northwest Africa (NWA) 7325, 61
  Yamato 984028, 61
  Zagami, 487
Michelson interferometer, 45–46, 135, 484, 499
Mie theory, 33–37
Milanković, 446
mineralogic, 30, 168, 170, 185, 191, 274–275, 342, 345, 352–353, 368, 402, 453, 460, 486–488, 502–503, 507, 527, 538, 541–544, 547, 549, 551–552, 568
mineralogy, 21, 25, 28, 32–33, 44, 62, 107, 140, 170–171, 185, 245, 268, 270–271, 297, 316, 324, 329, 334, 343, 352, 376, 380, 396, 401–403, 407, 417, 419, 422, 453, 455, 459, 469, 471, 484, 487–488, 492–493, 499, 502, 507–508, 510, 513–514, 518, 531, 538, 543, 549–550, 556, 570, 576–578, 581, 584, 594–595
minerals
  actinolite, 76, 78
  akaganéite, 76–77, 159, 542
  allophane, 81–82, 466–467
  almandine, 54, 55
  alunite, 15, 56–57, 76, 78, 85, 140, 264–265, 466–467
  alunogen, 85
  amblygonite, 58, 60
  amphibole, 8, 54, 78, 163, 485
  analcime, 80, 82
  andalusite, 54–55
  andesine, 54–55
  andradite, 54–55
  anhydrite, 85–86, 277, 567
  anorthosite, 17, 376, 378–379, 382, 462, 471, 596
  apatite, 58, 60, 85, 87, 581
  apophyllite, 54–55
  aragonite, 56–57, 82–84, 131–132
  baricite, 58, 60, 85, 87
  barrerite, 55–56
  bassanite, 85, 580
  beryl, 54–55
  biotite, 54–55, 80–81
  brucite, 76, 78, 294, 417
  calcite, 51–52, 56–57, 82–84, 123, 131–132, 140
  carbonate, 9, 14, 17, 53, 56–57, 69–70, 82–84, 88–90, 92, 130–131, 278, 331, 395–396, 404–405, 407, 424, 453–454, 457, 463–465, 471–473, 485, 490–491, 493, 499, 504–505, 510, 514, 542, 546–548, 552, 560
  cerussite, 56–57, 84
  chalcopyrite, 59–60

minerals (cont.)
  chamosite, 80–81
  childrenite, 58, 60, 85–86
  chloride, 36, 53, 59–61, 87, 93, 327, 466, 472, 490–491, 493, 563
  chlorite, 14–15, 54–55, 78–81, 463–464
  chromite, 56, 59, 369, 538, 542–543, 546
  clinochlore, 54–55, 80–81
  coesite, 140
  copiapite, 15, 85, 280, 517, 542
  coquimbite, 56–57, 85
  crysotile, 80
  dioptase, 54–55
  dolomite, 56–57, 82–84, 89, 170
  enstatite, 10, 24, 54–55, 62, 75, 88–89, 362, 405, 416, 425
  eosphorite, 58, 60, 86
  feldspar, 9, 11–12, 16, 54, 56, 74–75, 90, 92, 138, 140, 170, 341–342, 369, 376, 396, 454, 485, 491, 499, 507–508, 563, 581–582
  feldspathoid, 54
  ferrihydrite, 76–77, 90, 542
  ferrosilite, 10, 54–55, 404
  Fe-smectite, 54–55, 466, 524
  fluoride, 59, 415
  forsterite, 10–11, 54–55, 75, 88–89, 299, 362, 485
  franklinite, 56, 59
  galena, 59–60
  gibbsite, 76, 78
  glauberite, 56–57
  glauconite, 54–55, 80–81
  goethite, 12–13, 56, 59, 76–77, 140, 160, 499, 503, 507, 510, 542, 544, 546, 548, 552
  gormanite, 58, 60, 85–86
  gypsum, 51–52, 85–86, 88–90, 140, 170, 280, 468–469, 472, 524, 567
  halite, 36, 57–61, 506
  halloysite, 79–80
  hematite, 12–15, 17, 56, 58, 59, 76–77, 108, 126, 140, 330–331, 433, 435, 454, 456, 468, 472, 485, 499, 503, 506–508, 510, 522–525, 529, 542–544, 546, 548–550, 552, 563, 565, 578
  heulandite, 55–56
  hexahydrite, 85
  hornblende, 54–55
  hydroxide, 49, 53, 56, 58, 59, 75–76, 78, 92
  illite, 80–81, 342, 463
  ilmenite, 8, 56, 59, 340–341, 351–352, 357, 369, 380, 538, 542–543, 546, 614
  imogolite, 80, 82, 466–467
  iron oxide/hydroxide (FeOx), 9, 12–13, 17, 69, 75–77, 90, 92, 108, 162, 343, 428, 433, 435, 438, 454, 460, 469, 503, 516–517, 520, 522, 524–525, 529–530, 542, 546, 565, 578
  jarosite, 13, 15, 56–57, 76, 78, 85, 140, 264–265, 457, 466–467, 469–470, 508, 529, 541–543, 546, 549–552, 567, 578
  kamacite, 542–543, 546, 549, 552
  kaolinite, 54, 79–80, 342, 457, 465–466
  kieserite, 85–86, 468–469
  kulanite, 58, 60, 85–86
  kutnahorite, 56–57
  laumontite, 55–56, 82
  lazulite, 58, 60
  lepidocrocite, 12, 76–77
  maghemite, 56, 59, 76–77, 126
  magnesite, 56–57, 82–84, 88–89, 140
  magnetite, 8, 12, 56, 59, 76–77, 112, 126, 264, 356, 404, 530, 538, 542–543, 546–547, 551, 578
  mesolite, 81–82
  metavariscite, 58, 60
  minrecordite, 56–57
  montmorillonite, 54–55, 79–80, 140, 264, 466, 524, 566, 568
  nitrate, 83–84, 92, 123, 140
  nontronite, 71–72, 78, 80, 89, 170, 466–467, 499, 508, 521, 524
  olivine, 6, 9–12, 16–17, 29, 32, 35, 52, 54, 69, 74–75, 88–90, 92, 138–139, 159, 170, 295, 299, 301, 316, 326, 328, 331, 341–344, 346, 351, 354, 362, 369, 377–380, 384–385, 395–396, 402, 404, 406, 419, 424, 453–455, 457, 460–462, 464–465, 471–473, 485–490, 493, 499, 503, 505, 507, 510, 516–517, 528–529, 538, 541–543, 546–548, 551–552, 563, 570, 578
  opal, 54–55, 80–82, 457, 466–467, 469–470, 499, 504–505, 508, 510, 570, 580, 582
  oxide, 5, 12–13, 17, 49, 53, 56, 57, 59, 76–77, 126, 159, 163, 330, 342–343, 351, 396, 413, 448, 453, 460, 471, 522, 527, 530, 538, 560, 564, 566, 583, 591
  oxyhydroxide, 69, 75, 159, 163, 453, 469, 538
  perchlorate, 53, 59–61, 70, 87, 93, 130, 470, 525–526
  phlogopite, 54–55
  phosphate, 53, 58–60, 85–86, 92, 131, 162, 517, 543, 565
  phyllosilicate, 14–15, 17, 49, 53–55, 72, 78–81, 86, 88–90, 114, 152, 269, 293, 396–397, 404–405, 407, 417, 422, 425, 453, 455–457, 461, 464, 466–467, 470, 492, 504, 581
  plagioclase, 11–12, 16, 29, 138, 326, 331, 353, 362, 369, 376–378, 380, 384, 454, 462, 471, 485, 487, 491–492, 499, 503–504, 507–508, 563, 581
  prehnite, 78, 140, 463–464
  pyrite, 59–60, 542, 546, 548–549
  pyrolusite, 56, 59
  pyromorphite, 58, 60
  pyroxene, 6–10, 16–17, 24, 29, 54, 74–75, 78, 88–90, 92, 139, 161, 163, 265, 267, 270, 295, 316, 326–328, 351–354, 362, 369, 374–380, 384–385, 395–396, 402, 404–406, 417, 422,

425, 453–455, 460–462, 471, 485–487, 489, 491, 507, 510, 516–517, 520, 528, 538, 541–543, 546–547, 551–552, 577
quartz, 35–36, 51–52, 88–89, 140, 264, 280–281, 283, 485–486, 491–493
rhodochrosite, 56–57
rozenite, 84–85
rutile, 56, 58–59
saponite, 79–80, 88–89, 295
schwertmannite, 76–77, 542
scolecite, 81–82
sepiolite, 78, 80–81
serpentine, 14–15, 54–55, 78–81, 396, 463–464
siderite, 56–57, 82–83, 538
sinjarite, 59–61, 87
smectite, 14, 70, 78–82, 170, 292, 313, 455, 463–466, 470, 486, 521–522, 524, 528–529
smithsonite, 56–57, 82
sodalite, 54–55
sphalerite, 59–60
sphene, 54
spinel, 11–13, 17, 74–75, 369, 376–379, 384–385
spodumene, 54–55
starkeyite, 84–85, 468–469
stibnite, 59–60
stilbite, 55–56
strengite, 58, 60
strontianite, 56–57, 84
strunzite, 58, 60
sulfate, 9, 14–15, 17, 45, 48–49, 53, 56–59, 70, 73, 78, 82, 84–86, 90, 92, 130–131, 182, 277–278, 280–281, 291, 293–294, 331, 342, 344–345, 407, 453–454, 456–457, 460, 463, 466–472, 485, 490–491, 499, 503, 507–508, 510, 514, 516–518, 522, 524, 527, 529, 538, 542–544, 546, 548–550, 552, 556, 563, 566–567, 570, 577–578, 580–582
sulfide, 16, 53, 59–60, 112, 296–297, 361–363, 396, 425, 542–543, 546
sylvite, 59–61
szomolnokite, 15, 85–86
talc, 78, 80–81
thenardite, 56–57
thomsonite, 55–56
titanite, 54–55
topaz, 54–55
tremolite, 76, 78, 265, 267, 270
triphylite, 58, 60
triplite, 58, 60
troilite, 425, 542–543, 546, 549, 552
turquoise, 58, 60
vivianite, 58, 60
voltaite, 85
wavellite, 58, 60
whitlockite, 58, 60
witherite, 56–57, 84
yavapaiite, 85–86

zeolite, 54–56, 80, 82, 342–343, 463–464, 469, 486, 492–493
mixtures
areal, 264–265, 267–268, 270
intimate, 25–27, 264–266, 420–421
linear, 87, 264, 311
mineral, 87, 89–90, 92, 120
molecular, 264, 266–267
Moon, 17, 30, 53, 69, 91, 169, 186, 191–193, 224–225, 229–230, 232, 268, 290, 294, 298–300, 333, 337, 339, 341, 345, 351–353, 355–357, 368–372, 374–376, 378–381, 383–387, 395–397, 399, 424, 589–592, 595–596, 604–606, 609, 614–616
multispectral, 31, 299, 307, 309, 327, 333, 354, 357, 362, 373, 386, 445, 456, 461, 484, 513–517, 520–521, 525–531
multivariate analysis, 173, 180

nanophase, 72, 76–77, 82, 158–160, 162, 267, 297–299, 362, 374, 385, 428, 433, 435, 454, 460, 517, 522, 525, 529–530, 542, 546
nuclear process, 193

obliquity, 397, 446
optical constants, 21, 23–24, 28–29, 33, 35–38, 54–55, 57, 104–105, 266, 297, 310, 313, 320, 403, 420
organic, 16, 68, 74, 102–103, 109, 111–115, 124, 134, 136, 138, 182, 255, 262, 267, 274, 278, 283, 292–293, 356, 395–396, 405–407, 413, 421–422, 425, 431–435, 437–439, 445–448, 579, 584, 595
oxidation state, 3, 5–6, 8–9, 155, 162, 291, 351, 541, 543, 551, 565
oxygen fugacity, 16, 147, 396

partial least squares, 175, 179, 181, 328
particle size, 12, 27–32, 35, 38, 47, 51–52, 62, 72, 87, 102, 109, 158, 220, 231–232, 290, 313, 329, 375, 403, 432, 486, 493, 499, 578
photometry, 520, 525, 527
Planck
  constant, 22, 125, 148, 210
  function, 22
  Planck's Law, 46
  radiance, 500
planetary missions
  2001 Mars Odyssey, 227, 344, 463, 484, 579, 591–592
  Apollo, 147, 192, 225, 227, 232, 339–340, 346, 368–370, 374, 376–377, 379, 381, 589–590, 595–597, 614
  Bepi-Colombo, 32, 294
  Cassini, 115, 254, 266, 271, 381, 428–431, 436–439, 605, 607, 616–617
  Chandrayaan-1, 232, 371, 372, 615
  Chang'E, 591

planetary missions (cont.)
- Clementine, 337, 338, 340–341, 370–372
- Dawn, 195, 225, 227, 231, 397, 404–406, 413–416, 420–422, 424, 591, 594–595
- Deep Impact, 381, 419
- ExoMars, 274, 531, 593
- Hayabusa, 228, 396, 404, 407, 597
- Hubble Space Telescope (HST), 450, 514, 525
- Kaguya, 223, 371, 592, 614
- Lucy, 397, 406–407
- Luna, 11, 370, 589, 595
- Lunar Prospector, 198, 216, 222–223, 225–227, 229–230, 370–371, 381, 590–592
- Lunar Reconnaissance Orbiter (LRO), 31, 333, 591–592, 615
- Mars Exploration Rover (MER), 151, 170, 316–317, 326–327, 330, 333, 344, 460, 464, 467, 469, 485, 499, 513, 516, 529–531, 538–543, 548, 551–552, 555–556, 559–560, 562–566, 569–570, 576
- Mars Express, 455, 514, 611
- Mars Global Surveyor (MGS), 57, 316, 463, 484
- Mars Pathfinder (MPF), 316, 513
- Mars Reconnaissance Orbiter (MRO), 263, 271, 456, 521, 613
- Mars Science Laboratory (MSL) Curiosity rover, 168, 170, 186–187, 275, 457, 468, 485, 514–516, 527–531, 555–557, 559–566, 569–570, 573–574, 577, 579, 583, 593
- Mars 2020, 133, 168, 170, 184, 186–187, 274–276, 282–283, 559, 574, 583
- MErcury Surface, Space ENvironment, GEochemistry, and Ranging (MESSENGER), 192, 219, 227, 231–233, 294, 351–355, 357–358, 361–363, 593, 597–598
- Near Earth Asteroid Renderzvous (NEAR) - Shoemaker, 225, 227, 590, 597
- New Horizons, 106, 109, 115, 443–444, 446–451
- Opportunity, MER, 464, 467, 485, 499–501, 503, 506–510, 513–514, 516, 520–525, 538, 540, 556, 568–569
- OSIRIS-REx, 31, 399, 407, 598
- Phobos 2, 454, 514, 589
- Phoenix, 514–516, 525–527, 530, 593
- Psyche, 405, 407, 598
- Rosetta, 106, 405, 413–417, 420, 425, 559, 617–618
- SELENE (Kaguya), 223, 228, 592, 596
- Spirit, MER, 312, 316–317, 342–343, 460, 469, 499–501, 503–505, 509–510, 513–514, 516–521, 538, 548, 552, 556, 569
- Vega, 185, 225, 589
- Venera, 185, 225, 589, 610
- Viking, 333

planetary surface, 16, 21, 24–25, 28, 32, 34, 51, 68, 73, 77, 84, 87, 105–106, 108–109, 114–115, 147, 164, 192, 194–195, 198–199, 207–208, 211, 214, 217–218, 220, 257, 261, 291, 301–302, 307, 324, 333–334, 539, 573, 583, 589, 591, 595, 606, 609

Pluto
- atmosphere, 443–444, 447–448, 450
- Charon, 106–107, 443–446, 448–450
- climate, 446–447
- composition, 446, 448, 450
- Cthulhu Macula, 447, 449
- haze, 443–444, 448
- Hydra, 450
- Kerberos, 450
- Nix, 450
- seasons, 446, 448, 450
- Sputnik Planitia, 446–447

polarization, 9–10, 13, 15, 50, 122–123, 140, 156–157, 247–249, 255, 401, 606–607, 609, 615–616

polarizing power, 70–71

principal component analysis (PCA), 176–179, 181, 337, 344, 346, 402

quadrupole splitting, 147, 152, 154–155, 159, 162, 540, 542

Radar
- cross section, 248, 255
- radargram, 608–610, 612, 614
- sounding, 604, 609, 611, 614

radiation, 17, 21–22, 24, 27, 32–35, 42, 45–46, 50, 52, 68–70, 73, 92, 107, 121–124, 133–136, 150, 154, 175, 191, 193, 200, 207–209, 211, 213, 216, 218, 220, 224–225, 227, 240, 296–297, 351, 372, 375, 399, 434, 448, 541, 551, 557, 588, 596, 598

Raman scattering, 120–122, 124–126, 140
- normal Raman scattering, 121, 125–126
  - depolarization ratio, 123
  - energy diagram, 124
  - Stokes and anti-stokes Raman lines, 120, 122, 124–125
- resonance Raman scattering, 126
- stimulated Raman scattering, 125

Raman spectra
- CO_2, 128–129
- crystals, 123, 131
- effect of polarization, 123
- glasses and melts, 139, 141
- high pressure and temperature, 140, 142–143, 145
- H_2O, 127–128
- molecules, 120, 122–125
- MX_3, 129
- MX_4, 129–130

Rayleigh scattering, 34, 122, 124, 208, 211, 267, 278, 433, 435

recoil-free fraction, 149, 160–162, 541, 546

reflection, 25–26, 50–52, 104, 127, 134, 246–247, 265, 351, 606–607, 612–613

reflectivity, 43, 47, 50–52, 60–61, 108, 173, 278, 339, 403, 432, 530, 609, 611
regolith, 22, 30–34, 36, 191–199, 201, 207, 209–211, 213–218, 223, 229–230, 297–299, 324, 333, 337, 342–344, 351–352, 368, 378–380, 382, 396–397, 403, 417, 420, 425, 454, 458–459, 464, 487, 489, 499, 517, 520, 588–590, 592–595, 605, 607, 613–616, 618
remote sensing, 3, 9, 14, 16–17, 22, 30, 32, 36, 44, 48, 51, 53, 62, 68–69, 90, 102, 107, 109, 111, 113–115, 148, 174, 183, 191–192, 208, 211–212, 227, 239–240, 243, 253, 262–263, 266, 271, 282, 289–291, 301, 312, 319, 337–341, 343, 345, 368, 370, 420, 429, 431, 444, 484, 487, 499, 513, 525, 538, 556, 589, 596–598, 604
Reststrahlen, 30–32, 35–36, 51
rock abundance, 333, 525
rock types, 376, 379, 383, 503, 570, 579, 594

Saturn
  Dione, 428–432, 435–437, 607
  Enceladus, 428–429, 431–432, 437–438, 607
  Hyperion, 429–431, 433–435, 437–438
  Iapetus, 108, 267, 429–431, 433–435, 437–438, 447
  Mimas, 429–432, 437
  Phoebe, 267, 429–431, 433–435, 437–438
  Rhea, 430–432, 437, 607
  Tethys, 428–433, 436–437, 607
  Titan, 106, 108–109, 111, 192, 254–255, 257, 262–264, 290, 428–429, 437–439, 605, 607, 609, 616–617
scattering, 4, 24–30, 32–38, 51–52, 54, 102, 104, 114–115, 120–126, 134–137, 140, 151, 196–199, 201–202, 208–211, 218, 221–223, 232, 244–249, 255, 261–262, 264–267, 271, 283, 310, 313, 351–352, 378, 385, 399–400, 403, 420, 454, 458, 493, 516, 520, 525, 527, 530, 559, 606, 614–616
sedimentary, 14, 111, 455, 485, 499, 503, 506–508, 527, 579, 581
spatial heterodyne Raman spectrometer(SHRS), 135, 138, 142, 144
silicate, 9, 14, 16, 30, 36, 53–56, 74–75, 78, 80, 82, 87, 90, 125, 130, 139, 147, 162, 198, 267, 295, 298, 300, 330–331, 333, 340–341, 344, 351–357, 361, 363, 369, 382, 385, 395–396, 404–407, 413, 424–425, 438, 454, 457, 460, 463–464, 466–467, 471, 485–486, 490, 493, 504–505, 514, 516–517, 520, 527, 594
solar radiation, 232, 261
solid solution, 16–17, 54–55, 57, 60, 105, 107–108, 114, 442–443, 485–486
space weathering, 16, 90, 267, 289–290, 294, 297–300, 351–352, 356–357, 368, 372, 374–375, 378, 383, 396–397, 407, 425, 597
spectral end members, 44, 502

spectral library, 44, 51, 53–54, 57, 113, 271, 313, 320, 325–327, 332, 485, 504
spectral modeling
  Hapke, 23, 25–26, 28–29, 33–37, 88, 102, 104, 114, 266, 313, 401, 403, 420, 434, 446, 520
  mixing/mixture, 25–26, 32–33, 312–313, 319, 341, 363, 462
  Modified Gaussian Model (MGM), 16, 75, 88–89, 403
  radiative transfer, 21–23, 25–26, 28, 32–38, 88, 102, 104, 263, 266–268, 271, 310, 313, 325, 331–332, 368, 403, 420, 438, 443, 446, 458, 520
  Shkuratov, 23, 25–26, 28–29, 88, 313, 403
  unmixing, 89–90, 229, 307, 312, 325, 591
spectrometer, 42, 45–49, 73–74, 132–138, 141, 150, 168, 170, 175–176, 184–185, 191–192, 211, 215–216, 219–222, 224–229, 231, 233, 274–278, 283, 346, 353, 355, 357, 359, 361, 363, 370, 372–373, 379, 381, 386–387, 399, 404–405, 414–416, 460, 499–500, 502, 513, 517, 538–540, 551, 560, 574–576, 584, 589–594, 596–598
spectroscopy
  atomic emission, 176, 178, 183, 186
  attenuated total reflection (ATR), 44, 49–50
  emission, 21, 30–31, 38, 44–47, 54, 62, 102, 111, 172, 174, 176, 292, 300, 310, 334, 344, 351, 353, 397, 399–400, 462
  Fourier Transform Infrared (FTIR), 37, 45–51, 73–74
  Fourier Transform Raman (FT-Raman), 133, 135, 141, 143, 146
  gamma-ray, 191, 221, 223–225, 227, 588–589, 591, 593–594, 598
  imaging, 91, 261, 267, 271, 418
  Laser-Induced Breakdown Spectroscopy (LIBS), 168–187, 275–276, 329, 527, 573–576, 578–584
  micro-Raman, 132, 134, 136–141
  middle infrared (mid-IR), 21–24, 26, 29, 33, 36–38, 42–45, 47–48, 51–62, 68, 83–84, 115, 309, 385, 396, 399–401, 403–406, 517
  Mössbauer, 147, 150, 152–153, 157, 161, 163–164, 517, 520, 539, 567
  near-infrared (NIR), 71, 78, 80, 82–84, 112–113, 378, 381, 384, 399, 401–406, 429, 492, 520, 522–524, 527
  neutron, 191–192, 226, 363, 527, 589–594, 598
  nuclear, 191–192, 193, 195, 212, 216, 218, 228–229, 231, 588–590, 593
  Raman, 68, 70, 120, 126, 132–135, 138–141, 168, 274–276, 279–282, 584
  reflectance, 13, 21, 42, 44, 47, 62, 69–70, 92, 231, 301, 361, 364, 397–398, 470, 576, 584, 597
  remote Raman, 132, 136–138, 184, 277, 281, 584

spectroscopy (cont.)
   short-wave infrared (SWIR), 68, 307, 310, 312, 319, 351–353, 359, 453–454, 459–460, 462, 470, 472–473
   thermal infrared (TIR), 29–32, 294–296, 324, 326–334, 346, 386–387, 472–473, 484, 487, 493, 499, 502, 509
   time-resolved Raman, 138
   transmittance, 44, 49, 55, 72, 92, 262–263, 265
   vibrational, 44, 53, 70, 89
   visible/near-infrared (VNIR), 3, 12–13, 15–17, 22, 26, 30, 38, 68–70, 72–84, 86–88, 90–92, 191, 307, 309–310, 312, 319, 428, 513–514, 527
stand-off, 135, 171, 176, 183, 274–276, 278–283
sublimation, 432, 446–447, 613
sulfur, 9, 16–17, 111, 277, 344, 361, 404, 466, 517, 542, 555, 563, 566, 597
surface roughness, 22, 28, 31, 36, 246–247, 255, 257, 333–334, 420, 606, 609, 616

target transformation, 326, 329–332, 502
Tetracorder, 268–270
tholin, 108–111, 115, 428, 435, 438, 445, 448–450
T-matrix, 33–34, 36

Venus, 7, 17, 75, 170, 183, 185–186, 191–192, 254, 257, 262, 290, 588–589, 605, 609–611, 618
vibrations
   asymmetric bending, 70
   asymmetric stretching, 70
   bending, 42, 69, 72, 78
   combination band, 44, 71–72, 77–80, 82–86, 102–104, 113, 292–293, 454
   factor group, 131, 140–141
   group theory, 127, 129–130, 140
   normal modes, 126–130, 140
   selection rules, 126–127, 143
   overtone band, 44, 68, 70–72, 77–80, 82–86, 102–104, 113, 129, 293, 454, 524
   stretching, 72, 77–78, 86–87, 114, 280
   symmetric bending, 69, 127, 130
   symmetric stretching, 69, 128, 130, 139
viewing geometry, 4, 25, 27, 73, 113, 232, 244, 247, 261, 313, 402, 456, 516
volatile, 68, 102–103, 106–107, 109, 113, 115, 316, 344–345, 363, 413, 425, 428, 435, 443, 445–447, 450, 484, 562–563, 591, 594, 597
volcanic, 10, 16, 55, 59, 90–91, 252, 256, 280, 309, 315–316, 318, 331, 341, 354, 361, 363, 370, 445, 450, 454–455, 462–463, 465–467, 469, 471, 489–490, 499, 503, 505, 510, 555, 563, 568, 577, 605, 607, 610–611, 614–615, 618
volcanism, 17, 333, 363, 375, 380–382, 425, 455, 469, 471, 489, 568, 582, 590–591
volume scattering, 30–31, 44, 52–53, 68, 493, 607

weathering, 16–17, 90, 211, 267, 289–290, 294, 297–300, 341–344, 351–352, 356–357, 368, 372, 374–375, 378, 383, 385, 396–397, 407, 419, 425, 456, 465, 467–469, 472, 488, 520, 523, 551, 565–566, 573, 582–583, 597

X-ray diffraction (XRD), 86, 170, 185, 295, 329, 485, 529–530, 567, 578, 582
X-ray fluorescence (XRF), 192, 208, 214, 219–220, 222, 227, 232–233, 556–559, 562, 573, 596
X-rays, 151, 191–193, 207–209, 211–215, 217–220, 227, 539, 555–561, 579, 588–589, 593, 595–596